Handbook of Refinery Desulfurization

CHEMICAL INDUSTRIES
A Series of Reference Books and Textbooks

Founding Editor

HEINZ HEINEMANN
Berkeley, California

Series Editor

JAMES G. SPEIGHT
CD & W, Inc.
Laramie, Wyoming

MOST RECENTLY PUBLISHED

Handbook of Refinery Desulfurization, Nour Shafik El-Gendy and James G. Speight

Refining Used Lubricating Oils, James Speight and Douglas I. Exall

The Chemistry and Technology of Petroleum, Fifth Edition, James G. Speight

Educating Scientists and Engineers for Academic and Non-Academic Career Success, James Speight

Transport Phenomena Fundamentals, Third Edition, Joel Plawsky

Synthetics, Mineral Oils, and Bio-Based Lubricants: Chemistry and Technology, Second Edition, Leslie R. Rudnick

Modeling of Processes and Reactors for Upgrading of Heavy Petroleum, Jorge Ancheyta

Synthetics, Mineral Oils, and Bio-Based Lubricants: Chemistry and Technology, Second Edition, Leslie R. Rudnick

Fundamentals of Automatic Process Control, Uttam Ray Chaudhuri and Utpal Ray Chaudhuri

The Chemistry and Technology of Coal, Third Edition, James G. Speight

Practical Handbook on Biodiesel Production and Properties, Mushtaq Ahmad, Mir Ajab Khan, Muhammad Zafar, and Shazia Sultana

Introduction to Process Control, Second Edition, Jose A. Romagnoli and Ahmet Palazoglu

Fundamentals of Petroleum and Petrochemical Engineering, Uttam Ray Chaudhuri

Advances in Fluid Catalytic Cracking: Testing, Characterization, and Environmental Regulations, edited by Mario L. Occelli

Advances in Fischer-Tropsch Synthesis, Catalysts, and Catalysis, edited by Burton H. Davis and Mario L. Occelli

Transport Phenomena Fundamentals, Second Edition, Joel Plawsky

Asphaltenes: Chemical Transformation during Hydroprocessing of Heavy Oils, Jorge Ancheyta, Fernando Trejo, and Mohan Singh Rana

Handbook of Refinery Desulfurization

Nour Shafik El-Gendy
Egyptian Petroleum Research Institute, Cairo, Egypt

James G. Speight
CD&W Inc., Laramie, Wyoming, USA

CRC Press
Taylor & Francis Group
Boca Raton London New York

CRC Press is an imprint of the
Taylor & Francis Group, an **informa** business

CRC Press
Taylor & Francis Group
6000 Broken Sound Parkway NW, Suite 300
Boca Raton, FL 33487-2742

First issued in paperback 2020

© 2016 by Taylor & Francis Group, LLC
CRC Press is an imprint of Taylor & Francis Group, an Informa business

No claim to original U.S. Government works

ISBN 13: 978-0-367-57549-6 (pbk)
ISBN 13: 978-1-4665-9671-9 (hbk)

Library of Congress Cataloging-in-Publication Data

El-Gendy, Nour Shafik.
  Handbook of refinery desulfurization / Nour Shafik El-Gendy and James G. Speight.
    pages cm. -- (Chemical industries)
  Includes bibliographical references and index.
    ISBN 978-1-4665-9671-9 (hardcover : alk. paper) 1. Petroleum--Refining--Desulfurization. I.
  Speight, James G. II. Title.

  TP690.45.E43 2016
  665.5'3--dc23
                                                                          2015021151

Visit the Taylor & Francis Web site at
http://www.taylorandfrancis.com

and the CRC Press Web site at
http://www.crcpress.com

Contents

Preface

Sulfur, nitrogen, and metals in petroleum cause expensive processing problems in the refinery. As conventional technology does not exist to economically remove these contaminants from crude oil, the problem is left for the refiners to handle downstream at a high cost. In addition, regulations in various countries restricting the allowable levels of sulfur in end products continue to become increasingly stringent. This creates an ever more challenging technical and economic situation for refiners, as the sulfur levels in available crude oils continue to increase, conferring a market disadvantage for producers of high-sulfur crudes. Lower-sulfur crudes continue to command a premium price in the market, while higher sulfur crude oils sell at a discount. Desulfurization would offer producers the opportunity to economically upgrade their resources.

The sulfur content of petroleum varies from <0.05% to >14% wt. but generally falls in the range of 1–4% wt. Petroleum having <1% wt. sulfur is referred to as *low-sulfur*, and that having >1% wt. sulfur is referred to as *high-sulfur*. Heavy oils, residua, and bitumen are generally considered to be high-sulfur feedstocks by the refining industry. In addition, petroleum refining has entered a significant transition period as the industry moves into the 21st century. Refinery operations have evolved to include a range of next-generation processes as the demand for transportation fuels and fuel oil has shown a steady growth. These processes are different from one another in terms of the method and product slates, and they will find employment in refineries according to their respective features. The primary goal of these processes is to convert the heavy feedstocks to lower-boiling products, and during the conversion there is a reduction in the sulfur content.

With the inception of hydrogenation as a process by which both coal and petroleum could be converted into lighter products, it was also recognized that hydrogenation would be effective for the simultaneous removal of nitrogen, oxygen, and sulfur compounds from the feedstock. However, with respect to the prevailing context of fuel industries, hydrogenation seemed to be uneconomical for application to petroleum fractions. At least two factors dampened interest: (1) the high cost of hydrogen and (2) the adequacy of current practices for meeting the demand for low-sulfur products by refining low-sulfur crude oils, or even by alternative desulfurization techniques.

Nevertheless, it became evident that reforming processes instituted in many refineries were providing substantial quantities of by-product hydrogen, enough to tip the economic balance in favor of hydrodesulfurization processes. In fact, the need for such commercial operations has become more acute because of a shift in supply trends that has increased the amount of high-sulfur crude oils employed as refinery feedstocks.

Overall, there has, of necessity, been a growing dependence on high-sulfur heavier oils and residua as a result of continuing increases in the prices of the conventional crude oils coupled with the decreasing availability of these crude oils through the depletion of reserves in various parts of the world. Furthermore, the ever growing tendency to convert as much as possible lower-grade feedstocks to liquid products is causing an increase in the total sulfur content in refined products. Refiners must, therefore, continue to remove substantial portions of sulfur from the lighter products; however, residua and the heavier crude oils pose a particularly difficult problem. Indeed, it is now clear that there are other problems involved in the processing of the heavier feedstocks and that these heavier feedstocks, which are gradually emerging as the liquid fuel supply of the future, need special attention.

The hydrodesulfurization of petroleum fractions has long been an integral part of refining operations, and in one form or another, hydrodesulfurization is practiced in every modern refinery. The process is accomplished by the catalytic reaction of hydrogen with the organic sulfur compounds in the feedstock to produce hydrogen sulfide, which can be separated readily from the liquid (or gaseous) hydrocarbon products. The technology of the hydrodesulfurization process is well established, and petroleum feedstocks of every conceivable molecular weight range can be treated to remove sulfur. Thus, it is not surprising that an extensive knowledge of hydrodesulfurization has been acquired along with the development of the process during the last few decades. However, most of the available information pertaining to the hydrodesulfurization process has been obtained with the lighter and more easily desulfurized petroleum fractions, but it is, to some degree, applicable to the hydrodesulfurization of the heavier feedstocks such as the heavy oils and residua. On the other hand, the processing of the heavy oils and residua present several problems that are not found with distillate processing and that require process modifications to meet the special requirements that are necessary for heavy feedstock desulfurization.

In the last three decades, there have been many changes in the refining industry. The overall character of the feedstocks entering refineries has changed to such an extent that the difference can be measured by a decrease of several points on the API gravity scale. It is, therefore, the object of the present text to discuss the processes by which various feedstocks may, in the light of current technology, be treated to remove sulfur and, at the same time, afford maximum yields of low-sulfur liquid products. Thus, this text is designed for those scientists and engineers who wish to be introduced to desulfurization concepts and technology, as well as those scientists and engineers who wish to make more detailed studies of how desulfurization might be accomplished. Chapters relating to the composition and evaluation of heavy oils and residua are considered necessary for a basic understanding of the types of feedstock that will necessarily need desulfurization treatment. For those readers requiring an in-depth theoretical treatment, a discussion of the chemistry and physics of the desulfurization process has been included. Attention is also given to the concept of desulfurization during the more conventional refinery operations.

The effects of reactor type, process variables and feedstock type, catalysts, and feedstock composition on the desulfurization process provide a significant cluster of topics through which to convey the many complexities of the process. In the concluding chapters, examples and brief descriptions of commercial processes are presented and, of necessity, some indications of methods of hydrogen production are also included. In addition, environmental issues have become of such importance that a chapter on the cleanup of refinery gases is included. Moreover, the environmental effects of sulfur-containing gases are also addressed.

Finally, as refineries and feedstocks evolve, biocatalytic processes for reducing sulfur offers the petroleum industry potentially great rewards by use of such processes (biocatalytic desulfurization). Generally, biological processing of petroleum feedstocks offers an attractive alternative to conventional thermochemical treatment due to the mild operating conditions and greater reaction specificity afforded by the nature of biocatalysis. Efforts in microbial screening and development have identified microorganisms capable of petroleum desulfurization, denitrogenation, and demetallization.

Biological desulfurization of petroleum may occur either oxidatively or reductively. In the oxidative approach, organic sulfur is converted to sulfate and may be removed in process water. This route is attractive because it would not require further processing of the sulfur and may be amenable for use at the wellhead, where process water may then be reinjected. In the reductive desulfurization scheme, organic sulfur is converted into hydrogen sulfide, which may then be catalytically converted into elemental sulfur, an approach of utility at the refinery. Regardless of the mode of biodesulfurization, key factors affecting the economic viability of such processes are biocatalyst activity and cost, differential in product selling price, sale or disposal of coproducts or wastes from the treatment process, and the capital and operating costs of unit operations in the treatment scheme. Furthermore, biocatalytic approaches to viscosity reduction and the removal of metals and nitrogen are additional approaches to fuel upgrading.

Nour Shafik El-Gendy
Egyptian Petroleum Research Institute

James G. Speight
CD&W Inc.

Authors

Nour Shafik El-Gendy is an associate professor in the field of environmental biotechnology and head manager of Petroleum Biotechnology Lab, Egyptian Petroleum Research Institute, Cairo, Egypt. She is the author of two books in the fields of biofuels and petroleum biotechnology and more than 100 research papers in the fields of oil pollution, bioremediation, biosorption, biofuels, macro- and micro-corrosion, green chemistry, wastewater treatment, and nano-biotechnology and its applications in the petroleum industry and biofuels. Dr. El-Gendy is also an editor in 12 international journals in the field of environmental biotechnology and microbiology, and she has supervised 20 MSc and PhD theses in the fields of biofuels, micro–macro fouling, bioremediation, wastewater treatment, biodenitrogenation, and biodesulfurization. Dr. El-Gendy is a member of many international associations concerned with environmental health and sciences. She is also a lecturer and supervisor for undergraduate research projects at the British University in Egypt BUE and Faculty of Chemical Engineering, Cairo University, Egypt, and also teaches an environmental biotechnology course for postgraduates at the Faculty of Science, Monufia University, Egypt. Her biography is recorded in *Who's Who in Science and Engineering*, 9th edition, 2006–2007.

James G. Speight, who has doctoral degrees in chemistry, geological sciences, and petroleum engineering, is the author of more than 60 books on petroleum science, petroleum engineering, and environmental sciences. He has served as adjunct professor in the Department of Chemical and Fuels Engineering at the University of Utah and in the Departments of Chemistry and Chemical and Petroleum Engineering at the University of Wyoming. In addition, he has been a visiting professor in chemical engineering at the following universities: the University of Missouri, Columbia, the Technical University of Denmark, and the University of Trinidad and Tobago.

As a result of his work, Dr. Speight has been honored as the recipient of the following awards:

- Diploma of Honor, United States National Petroleum Engineering Society. For Outstanding Contributions to the Petroleum Industry. 1995.
- Gold Medal of the Russian Academy of Sciences. For Outstanding Work in the Area of Petroleum Science. 1996.
- Einstein Medal of the Russian Academy of Sciences. In recognition of Outstanding Contributions and Service in the field of Geologic Sciences. 2001.
- Gold Medal—Scientists without Frontiers, Russian Academy of Sciences. In recognition of His Continuous Encouragement of Scientists to Work Together across International Borders. 2005.
- Methanex Distinguished Professor, University of Trinidad and Tobago. In Recognition of Excellence in Research. 2006.
- Gold Medal—Giants of Science and Engineering, Russian Academy of Sciences. In Recognition of Continued Excellence in Science and Engineering. 2006.

1 Desulfurization

1.1 INTRODUCTION

Desulfurization as practiced in petroleum refineries is a catalytic process that is widely used to remove sulfur from petroleum feedstocks, refined petroleum products, and natural gas (Jones, 1995; Speight and Ozum, 2002; Parkash, 2003; Hsu and Robinson, 2006; Mokhatab et al., 2006; Gary et al., 2007; Speight, 2014). The purpose of removing the sulfur is to reduce the emissions of sulfur dioxide (SO_2), which also converts to sulfur trioxide (SO_3) in the presence of oxygen, when fuels or petroleum products are used in automotive fuels and fuel oils, as well as fuels for railroad locomotives, ships, gas or oil burning power plants, residential and industrial furnaces, and any other forms of fuel combustion. In addition, the increase in the use of high-sulfur feedstock in refineries (and the regulations restricting the amount of sulfur in fuels and other products) is reflected in the increase in capacity during the last 32 years of refinery desulfurization processes from 6,781,060 barrels/day in 1982 to 17,094,540 barrels/day at the end of 2014 (Energy Information Administration, 2015).

Another example of removing sulfur (especially within the refinery) is the desulfurization of the product streams (such as naphtha). Sulfur, even in extremely low concentrations, has detrimental effects on process catalysts in the catalytic reforming units that are subsequently used to produce high-octane naphtha as a blend stock for gasoline manufacture. Typically, the sulfur is removed by a hydrodesulfurization process (Jones, 1995; Speight and Ozum, 2002; Parkash, 2003; Hsu and Robinson, 2006; Gary et al., 2007; Speight, 2014), and the refinery includes facilities for the capture and removal of the resulting hydrogen sulfide (H_2S) gas. In another part of the petroleum refinery, the H_2S gas is then subsequently removed from the gas stream (Chapter 11) and converted into the by-product elemental sulfur or sulfuric acid (H_2SO_4).

The chemical reactions that take place during petroleum upgrading can be very simply represented as reactions involving hydrogen transfer (Speight, 2000, 2014; Ancheyta and Speight, 2007). In the case of hydrotreating, much of the hydrogen is supplied from an external source, and hydrogenation and various hydrogenolysis reactions consume the hydrogen with a resulting reduction in the molecular weight of the starting material (Stanislaus and Cooper, 1994). In the case of coking, hydrogen is supplied by the feedstock material from aromatization and condensation reactions, with the ultimate result that some carbon is rejected in the process. Aromatization reactions produce more aromatic carbon in the products than in the starting material, whereas condensation reactions generally redistribute the form of aromatic carbon in terms of protonated (tertiary) and nonprotonated (quaternary) aromatic carbons. In both cases, nuclear magnetic resonance methods can be used to determine these forms of aromatic carbon. By combining solid- and liquid-state nuclear magnetic resonance techniques with elemental and mass balances, the modes of extrinsic and intrinsic hydrogen consumption during residua upgrading can be ascertained.

A refinery is composed of various thermal and catalytic processes to convert molecules in the heavier fractions to smaller molecules in fractions distilling at these lower temperatures (Figures 1.1 through 1.3) (Jones, 1995; Speight and Ozum, 2002; Parkash, 2003; Hsu and Robinson, 2006; Gary et al., 2007; Speight, 2014). This efficiency translates into a strong economic advantage, leading to the widespread use of conversion processes in refineries. However, understanding the principles of catalytic cracking and of adsorption and reaction on solid surfaces is valuable (Samorjai, 1994; Masel, 1995).

Understanding refining chemistry not only allows an explanation of the means by which these products can be formed from crude oil but also offers a chance of predictability (Speight and Ozum,

1

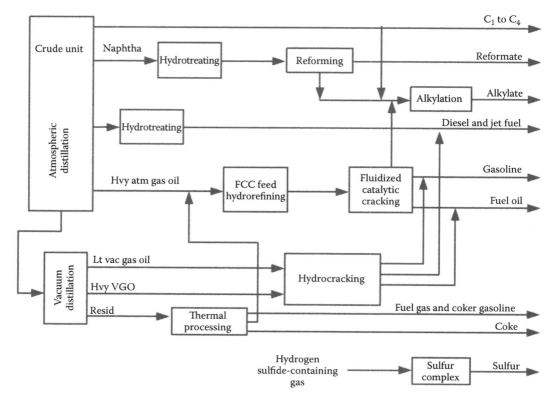

FIGURE 1.1   Schematic overview of a refinery. (From Speight, J.G. 2014. *The Chemistry and Technology of Petroleum.* 5th Edition. CRC Press, Taylor & Francis Publishers. Boca Raton, FL, Figure 15.1, p. 392.)

2002; Parkash, 2003; Hsu and Robinson, 2006; Gary et al., 2007; Speight, 2014). This is necessary when the different types of crude oil accepted by refineries are considered (Speight and Ozum, 2002; Parkash, 2003; Hsu and Robinson, 2006; Gary et al., 2007; Speight, 2014). Furthermore, the major processes by which these products are produced from crude oil constituents involve thermal decomposition. There have been many simplified efforts to represent refining chemistry that, under

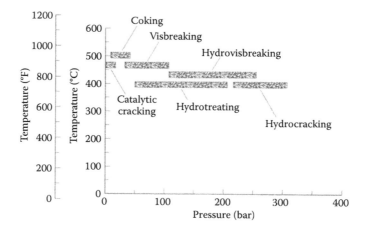

FIGURE 1.2   Temperature and pressure ranges for refinery processes. (From Speight, J.G. 2007. *The Desulfurization of Heavy Oils and Residua.* 2nd Edition. Marcel Dekker Inc., New York. Figure 4.2, p. 129.)

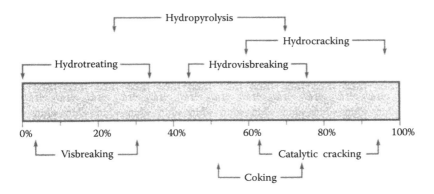

FIGURE 1.3 Feedstock conversion in refinery processes. (From Speight, J.G. 2007. *The Desulfurization of Heavy Oils and Residua*. 2nd Edition. Marcel Dekker Inc., New York. Figure 4.3, p. 129.)

certain circumstances, are adequate to the task. However, refining is much more complicated than such simplified representations would indicate (Speight, 2014).

The different reactivities of molecules that contain sulfur, nitrogen, and oxygen can be explained by the relative strength of the carbon–sulfur, carbon–nitrogen, and carbon–oxygen bonds, in aromatic and saturated systems. This allows some explanation of the differences in reactivity toward hydrodesulfurization, hydrodenitrogenation, and hydrodemetallization; however, it does not account for the reactivity differences due to stereochemistry and the interactions between different molecular species in the feedstock. In fact, the chemistry of the conversion process may be complex (Speight and Ozum, 2002; Parkash, 2003; Hsu and Robinson, 2006; Gary et al., 2007; Speight, 2014), and an understanding of the chemistry involved in the conversion of a crude oil to a variety of products is essential to the understanding of refinery operations.

Bond energies offer some guidance about the preferential reactions that occur at high temperature and, for the most part, can be an adequate guide to the thermal reactions of the constituents of petroleum. However, it is not that simple. Often, the bond energy data fail to include the various steric effects that are a consequence of complex molecules containing three-dimensional structures, especially structures in the resin and asphaltene constituents (Speight, 1994; Schabron and Speight, 1998). Furthermore, the complexity of the individual reactions occurring in an extremely complex mixture and the interference of the products with those from other components of the mixture is unpredictable. Moreover, the interference of secondary and tertiary products with the course of a reaction, and hence with the formation of primary products, may also be a cause for concern. Hence, caution is advised when applying the data from model compound studies to the behavior of petroleum, especially the molecularly complex heavy oils. These have few, if any, parallels in organic chemistry.

Petroleum refining is based on two premises that relate to the fundamental properties and structure of the constituents of petroleum. The premises are as follows: (1) hydrocarbons are less stable than the elements (carbon and hydrogen) from which they are formed at temperatures >25°C (77°F), and (2) if reaction conditions ensure a rapid reaction, any system of hydrocarbons tends to decompose into carbon and hydrogen as a consequence of its thermodynamic instability. Furthermore, hydrodesulfurization chemistry is based on careful (or more efficient) hydrogen management (Speight and Ozum, 2002; Parkash, 2003; Hsu and Robinson, 2006; Gary et al., 2007; Speight, 2014). This involves not only the addition of hydrogen but also the removal of (molecular) hydrogen sinks (in which hydrogen is used with little benefit to the process) and the addition of a pretreatment step to remove or control any constituents that will detract from the reaction. This latter detraction can be partially resolved by the application of deasphalting, coking, or hydrotreating steps (Chapters 8 and 10) as pretreatment options.

This information is useful for relating process conditions to the overall process chemistry, and for optimizing process conditions for a desired product slate. With this type of information, it will be possible to develop a correlation between analytical measurements and processability. Furthermore, the method is relatively simple and will be adaptable for refinery use on a regular basis (Speight, 2015). In addition, the method will be useful for predicting specific or general processability requirements. It is therefore the purpose of this chapter to serve as an introduction to desulfurization so that the subsequent chapters dealing with refinery desulfurization are easier to visualize and understand.

1.2   HYDRODESULFURIZATION

Hydrodesulfurization is a chemical method to remove sulfur from the hydrocarbon compounds that comprise petroleum products. This is usually accomplished through a catalyzed reaction with hydrogen at moderate to high temperature and pressure—as, e.g., in the hydrotreating and hydrocracking processes. Variations of both processes are used in refineries.

1.2.1   REACTION MECHANISM

In the desulfurization of benzothiophene, two different parallel reactions with hydrogen are catalyzed: (1) the hydrogenation pathway and (2) the hydrogenolysis pathway. In the hydrogenation pathway, the thiophene ring is hydrogenated before desulfurization, while in the hydrogenolysis pathway the thiophene ring is split owing to the attack of surface-adsorbed hydrogen at the sulfur atom. For benzothiophene desulfurization, the hydrocarbon products are styrene and ethylbenzene. The hydrogen sulfide formed inhibits the hydrogenolysis but not the hydrogenation reactions.

The proposed hydrodesulfurization reaction network for dibenzothiophene (Houalla et al., 1980) proceeds via the path of minimal hydrogen consumption, and the hydrogenation of the initial products (biphenyl and cyclohexylbenzene) proceeds at a slow rate. The rate of dibenzothiophene increases (at the expense of the hydrogenolysis reaction) at higher concentrations of hydrogen sulfide. Furthermore, the concentration of cyclohexylbenzene in the products depends on the catalyst type applied.

The presence of alkyl substituents on the starting materials (benzothiophene and dibenzothiophene) might favor one of the possible hydrodesulfurizations; however, this depends on the position(s) of the alkyl substituents(s) and, thus, on the extent of the alteration of the electron density by the electron-donating effect of alkyl groups. In addition, substituents in the vicinity of the sulfur atom cause steric hindrance and influence the hydrodesulfurization reaction (Kabe et al., 1992). For example, dibenzothiophene and alkyl derivatives substituted adjacent to the sulfur atom are refractory (difficult to desulfurize) (Table 1.1) to the hydrodesulfurization reaction using conventional catalysts. The key sulfur compounds present in diesel oil fractions after conventional hydrodesulfurization are 4-methyldibenzothiophene and 4,6-dimethyldibenzothiophene.

Finally, the application of microwave technology to desulfurization has also received some attention. The theory of the use of microwaves in desulfurization of petroleum streams at relatively low temperatures is supported by the concept that hydrocarbon molecules are more transparent to or less affected by microwaves than organosulfur or organosulfur–metallic compounds (Shang et al., 2013). Thus, microwave energy would preferentially activate the sulfur compounds.

The application of microwave irradiation in petroleum refinery processes (catalytic reforming, catalytic cracking, catalytic hydrocracking, hydrodealkylation, and catalytic polymerization) was developed in the 1990s (Shang et al., 2013). It was found to be important as it achieves sustainable savings in capital and operating costs because catalytic reactions can be accelerated by the microwave energy, performed under less severe conditions, i.e., lower temperature and pressure and shorter catalyst contact time. These make it possible to use smaller reaction vessels with reduced catalyst consumption (Loupy, 2006). In addition, pulsed mode microwave input is better than continuous microwave input for hydrodesulfurization. Cause by continuous microwave input tends to

TABLE 1.1

Increase in Molecular Size and Sulfur Sheltering Increase the Difficulty of Hydrodesulfurization

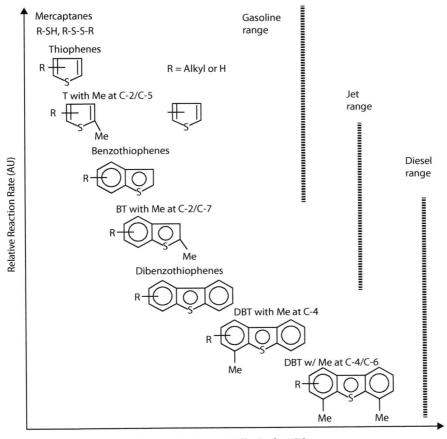

lead to induced hotspots, especially for the material whose dielectric properties increase with temperature. Temperature runaway is caused by continuous microwave input. This can be controlled by the application of pulsed microwave input (Purta et al., 2004).

Several catalysts are reported in microwave hydrodesulfurization—powdered iron, charcoal on iron, palladium oxide–silica-based material, calcium oxide, alkali metal oxide catalysts, and traditional hydrotreating catalysts, e.g., molybdenum sulfide supported on porous γ-alumina promoted by cobalt or nickel ($Co-Mo/Al_2O_3$, $Ni-Mo/Al_2O_3$). In addition, additives like boron or phosphorous or silica can also be used. Microwave sensitizers such as diethanolamine, silicon carbide, activated charcoal, and serpentine are commonly used to improve microwave absorption (Miadonye et al., 2009).

Unsupported catalysts such as metal hydrides and metal powder have proved to be effective in microwave hydrodesulfurization, acting as hydrogen donors. Metal powders have been reported to be effective for the desulfurization of coal tar pitch and of hydrocracked residuum from Athabasca bitumen (Wan and Kriz, 1985; Mutyala et al., 2010).

As promising as microwave technology may seem, the challenge is to apply microwave-assisted hydrodesulfurization in the petroleum industry on a commercial scale since the reactor materials are either polytetrafluoroethylene or glass/quartz, which limit the maximum operating temperature and pressure required for the hydrodesulfurization process to achieve ultralow sulfur levels. In the design for an improved microwave hydrodesulfurization vessel for a continuous scale process (Gomez, 2005), an impeller shelf drives multiple impellers, and the microwave magnetrons and waveguides run through along the vessel.

1.2.2 Catalysts

γ-Alumina (γ-Al$_2$O$_3$)-supported molybdenum oxide catalysts promoted with cobalt or nickel have been widely used in conventional hydrodesulfurization processes. Active sites are formed when molybdenum oxide (MoO$_3$) is converted to molybdenum sulfide (MoS$_2$) by sulfurization (Arnoldy et al., 1985). The hydrogenation route is the most important pathway in the hydrodesulfurization of dibenzothiophene with substituents in the 4- and 6-positions (Kabe et al., 1993). The direct hydrogenolysis route is less favorable because of the steric hindrance (Robinson et al., 1999a). The molecule becomes more flexible upon hydrogenation of (one of) the aromatic rings, and the steric hindrance is relieved (Landau et al., 1996). Consequently, catalysts with a relatively high hydrogenation activity must be considered.

Nickel-promoted mixed sulfide catalysts are known for their high hydrogenation activity. Furthermore, noble catalysts (containing Pt or Pd) are attractive to use because of their high hydrogenation activity (Robinson et al., 1999b). It has also been observed (Kabe et al., 2001) that, under deep desulfurization conditions, the partial pressure of hydrogen sulfide has a strong inhibitory effect on the catalytic activity and product selectivity of the hydrodesulfurization reactions of dibenzothiophene and 4,6-dimethyldibenzothiophene. The inhibiting effect is the result of the stronger adsorption of hydrogen sulfide compared with dibenzothiophene and 4,6-dimethyldibenzothiophene on the catalyst, and thus the process is dependent on the catalyst type. Noble catalysts are characterized by sensitivity for elevated levels of hydrogen sulfide (Stanislaus and Cooper, 1994). If deep desulfurization is performed in a separate process stage, i.e., after the initial removal of the bulk of organic sulfur, alternative catalyst types can be applied because high hydrogen sulfide concentrations are minimized and catalyst supports can also play a role in the progress of the deep hydrodesulfurization reaction (Robinson et al., 1999a).

A new concept of bifunctional catalysts has been proposed to increase the sulfur resistance of noble metal hydrotreating catalysts (Song, 1999, 2003; Song and Ma, 2003). It combines catalyst supports with bimodal pore size distribution (e.g., zeolites) and two types of active sites. The first type of sites, placed in large pores, is accessible for organosulfur compounds and is sensitive to sulfur inhibition. The second type of active sites, placed in small pores, is not accessible for large S-containing molecules and is resistant to poisoning by hydrogen sulfide. Since hydrogen can easily access the sites located in small pores, it can be adsorbed and transported within the pore system to regenerate the poisoned metal sites in the large pores. The practical applications of this concept need to be further demonstrated.

Various supports have been used to enhance the catalytic activity in the hydrodesulfurization reaction: carbon (Farag et al., 1999), silica (Cattaneo et al., 1999), zeolites (Breysse et al., 2002), titania and zirconia (Afanasiev et al., 2002), and silica–alumina (Qu et al., 2003). Combining new types of catalytic species with advanced catalyst supports such as amorphous silica–alumina can result in an extremely high desulfurization performance (Babich and Moulijn, 2003). The application of amorphous silica–alumina-supported noble metal–based catalysts for the second-stage deep desulfurization of gas oil is an example (Reinhoudt et al., 1999).

The platinum (Pt) and platinum–palladium (Pt–Pd) catalysts are very active in the deep hydrodesulfurization of prehydrotreated straight run gas oil under industrial conditions. These catalysts are able to reduce the sulfur content to 6 ppm while simultaneously reducing the aromatics to 75% of their initial amount (Reinhoudt et al., 1999). At high sulfur levels, the amorphous silica–alumina-supported noble metal catalysts are poisoned by sulfur, and nickel–tungsten/amorphous silica–alumina catalysts become preferable for deep hydrodesulfurization and dearomatization. Amorphous silica–alumina-supported nickel–tungsten and platinum catalysts showed a much better performance in 4-ethyl, 6-methyl-dibenzothiophene desulfurization because of their high hydrogenation activity, especially at low hydrogen sulfide levels (Robinson et al., 1999b).

In addition, it has been reported (Reinhoudt et al., 1999) that amorphous silica–alumina-supported platinum–palladium catalysts are very promising to apply in deep desulfurization, provided that any hydrogen sulfide in the products is removed efficiently. A major drawback is the price of the noble metals. During hydrodesulfurization reactions, the catalysts age and deactivate as a result of coke and metal deposition on the catalyst (Seki and Yoshimoto, 2001). The deposition severity is greatly dictated by the feedstock properties. Since asphaltene constituents are precursors of coke formation, higher-boiling fractions and nonvolatile fractions increase catalyst deactivation. Furthermore, other hydrode-sulfurization reaction conditions (such as temperature and pressure) enhance the rate of deactivation.

1.2.3 Reactor Configuration

Apart from the catalyst type involved, optimal process configurations to minimize the suppression of hydrogen sulfide on the catalyst activity are important. The hydrogen sulfide produced from sulfur compounds with higher reactivity in the early stage of desulfurization negatively influences the hydrodesulfurization of less reactive sulfur compounds. To circumvent this problem, a two-stage principle carried out in conventional concurrent trickle-flow reactor can be applied. After the removal of the bulk of easily convertible sulfur compounds in the first step, the more refractory compounds are removed in the second step with pure hydrogen. This approach also enables the use of the most appropriate catalyst types in different stages (Reinhoudt et al., 1999).

The conventional type of a hydrotreating reactor is a fixed bed with a cocurrent supply of oil stream and hydrogen. These systems have an unfavorable hydrogen sulfide profile concentration over the reactor due to a high hydrogen sulfide concentration at the reactor outlet. The removal of the last vestiges (ppm levels) of sulfur is inhibited. Countercurrent operation can provide a more preferable concentration profile, since in this operation mode the oil feed is introduced at the top and hydrogen at the bottom of the reactor. Hydrogen sulfide is removed from the reactor at the top, avoiding a possible recombination with olefins at the reactor outlet.

One commercial example of this approach is the hydrotreating process based on cocurrent/countercurrent technology (Figure 1.4). In the first stage, the feed and hydrogen cocurrently contact the catalyst bed and organosulfur compounds are converted. The hydrogen sulfide is then removed from the reactant flow. In the second stage, the reactor system operates in the countercurrent mode, providing more favorable concentration profiles of hydrogen sulfide and hydrogen over the length of the reactor. This configuration allows application of catalysts sensitive to sulfur poisoning, i.e., noble metal–containing catalysts, in the second stage of the process. Moreover, nitrogen and aromatics can also be removed (Babich and Moulijn, 2003).

Reactors using monolithic catalyst supports may be an attractive alternative to conventional multi-phase reactors (Kapteijn et al., 2001). Monolithic catalysts can be prepared in different ways. They can be produced by direct extrusion of support material (often cordierite is used; however, different types of clays or typical catalyst carrier materials such as alumina are also used), or of a paste also containing catalyst particles (e.g., zeolites, V-based catalysts), or a precursor of catalyst active species (e.g., polymers for carbon monoliths); in this case, the catalyst loading of the reactor can be high (Kapteijn et al., 2001). Alternatively, catalysts, supports, or their precursors can be coated into a monolith structure by wash coating (Beers et al., 2000) and different types of monoliths can be used (Figure 1.5) (Egorova and Prins, 2004). Instead of a catalyst, trickle-bed, monolithic channels are present where bubble-train flow occurs. Gas bubbles and liquid slugs move with constant velocity through the monolith channels approaching plug flow behavior. Gas is separated from the catalyst by a very thin liquid film, and during their travel through the channels the liquid slugs show internal recirculation. These two properties result

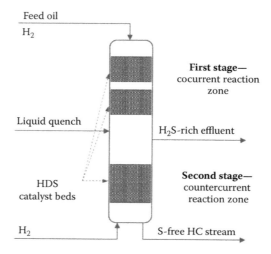

FIGURE 1.4   Cocurrent/countercurrent hydrodesulfurization technology.

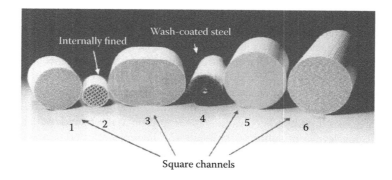

FIGURE 1.5   Monolith structures of various shapes. Square channel cordierite structures (1, 3, 5, 6), internally finned channels (2), wash-coated steel monolith (4). (From Egorova, M. 2003. Study of the Aspects of Deep Hydrodesulfurization by Means of Model Reactions. PhD Thesis, Swiss Federal Institute of Technology, Zurich, Switzerland.)

in optimal mass transfer properties (Kapteijn et al., 2001). Furthermore, very sharp residence time distributions for gas and liquid compared with trickle flow can be achieved (Nijhuis et al., 2001). Currently, the application of monoliths in various forms and applications is an object of research (Kapteijn et al., 2001). Larger channel geometries (internally finned monolith channels) might allow countercurrent flow at a relevant industrial scale, and the scale-up properties are promising.

Although the concentrations of benzothiophene and dibenzothiophene are considerably decreased by hydrodesulfurization (Monticello, 1998) and it has been commercially used for a long time, it has several disadvantages: (1) hydrodesulfurization of diesel feedstock for a low-sulfur product requires a larger reactor volume, longer processing times, and substantial hydrogen and energy inputs; (2) for refractory sulfur compounds, hydrodesulfurization requires higher temperature and pressure and longer residence time, which adds to the cost of the process due to the requirement of stronger reaction vessels and facilities (McHale, 1981); (3) the application of extreme conditions to desulfurize refractory compounds results in the deposition of carbonaceous coke on the catalysts; (4) exposure of crude oil fractions to severe conditions including temperatures above about 360°C (680°F) decreases the fuel value of the treated product; (5) deep hydrodesulfurization processes need large new capital investments and/or have higher operating costs; (6) the hydrogen sulfide that is generated poisons the catalysts and shortens their useful life; (7) deep hydrodesulfurization is affected by components in the reaction mixture such as organic heterocompounds and polynuclear aromatic hydrocarbons (Egorova, 2003); (8) for older units that are not competent to meet the new sulfur removal levels, erection of new hydrodesulfurization facilities and heavy load of capital cost is inevitable; (9) hydrodesulfurization removes sulfur compounds such as thiol derivatives, sulfide derivatives, and disulfide derivatives effectively—some aromatic sulfur-containing compounds such as 4- and 4,6-substituted dibenzothiophene, and polynuclear aromatic sulfur heterocycles are resistant to hydrodesulfurization and form the most abundant organosulfur compounds after hydrodesulfurization (Monticello, 1998); (10) hydrogen atmosphere in hydrodesulfurization results in the hydrogenation of olefin compounds and reduces the calorific value of fuel—in order to increase the calorific value, the hydrodesulfurization-related stream is sent to the fluid catalytic cracking unit, which adds to the cost (Hernández-Maldonado and Yang, 2004); (11) although hydrodesulfurization is considered a cost-effective method for fossil fuel desulfurization, the cost of sulfur removal from refractory compounds by hydrodesulfurization is high. To reduce the sulfur content from 200 to 50 ppm, the desulfurization cost would be four times higher. However, further reduction of sulfur concentration by hydrodesulfurization, to <1 ppm, is still a challenging research target.

Transportation fuels, such as gasoline, jet fuel, and diesel, are ideal fuels due to their high energy density, ease of storage and transportation, and established distribution network. However, their sulfur concentration must be <10 ppm to protect the deactivation of catalysts in reforming process and electrodes in a fuel cell system (Wild et al., 2006). Accordingly, it is necessary to develop and establish alternative technologies and suitable catalyst/support systems to meet current specifications and reduce the energy requirements and capital cost of the hydrotreating process.

1.3 THERMODYNAMIC ASPECTS

The thermodynamics of the hydrodesulfurization reaction has been evaluated from the equilibrium constants of typical desulfurization or partial desulfurization reactions such as (1) hydrogenation of model compounds to yield saturated hydrocarbons (R–H) and hydrogen sulfide (H_2S), (2) decomposition of model compounds to yield unsaturated hydrocarbons $R–CH = CHR^1$) and hydrogen sulfide (H_2S), (3) decomposition of alkyl sulfides to yield thiols (RSH) and olefins ($RCH = CHR^1$), (4) condensation of thiols (RSH) to yield alkyl sulfides (RSR^1) and hydrogen sulfide (H_2S), and (5) hydrogenation of disulfides ($R-SS-R^1$) to yield thiols ($R-SH$, R^1-SH). The logarithms of the equilibrium constants for the reduction of sulfur compounds to saturated hydrocarbons over a wide temperature range are almost all positive, indicating that the reaction can proceed to completion if hydrogen is present in the stoichiometric quantity (Speight, 2000).

The equilibrium constant does, however, decrease with increasing temperature for each particular reaction but still retains a substantially positive value at 425°C (795°F), which is approaching the maximum temperature at which many of the hydrodesulfurization (especially nondestructive) reactions would be attempted. The data also indicate that the decomposition of sulfur compounds to yield unsaturated hydrocarbons and hydrogen sulfide is not thermodynamically favored at temperatures <325°C (615°F), and such a reaction has no guarantee of completion until temperatures of about 625°C (1155°F) are reached. However, substantial decomposition of thiols can occur at temperatures <300°C (570°F); in fact (with only few exceptions), the decomposition of all saturated sulfur compounds is thermodynamically favored at temperatures <425°C (795°F).

The data indicate the types of reactions that can occur during the hydrodesulfurization reaction and include those reactions that will occur at the upper end of the temperature range of the hydrodesulfurization process, whether it is a true hydrodesulfurization reaction or a cracking reaction. Even though some of the reactions given here may only be incidental, they must nevertheless be taken into account because of the complex nature of the feedstock. Several process variations (in addition to the fact that the overall hydrodesulfurization process is exothermic) also contribute to the complexity of the product mix (Speight, 2000).

1.4 KINETICS OF HYDRODESULFURIZATION

Kinetic studies using individual sulfur compounds have usually indicated that simple first-order kinetics with respect to sulfur is the predominant mechanism by which sulfur is removed from the organic material as hydrogen sulfide. However, there is still much to be learned about the relative rates of reaction exhibited by the various compounds present in petroleum. The reactions involving the hydrogenolysis of sulfur compounds encountered in hydroprocessing are exothermic and thermodynamically complete under ordinary operating conditions. The various molecules have very different reactivities, with mercaptan sulfur being much easier to eliminate than thiophene sulfur or dibenzothiophene sulfur.

The structural differences between the various sulfur-containing molecules make it impractical to have a single rate expression applicable to all reactions in hydrodesulfurization. Each sulfur-containing molecule has its own hydrogenolysis kinetics that is usually complex because several successive equilibrium stages are involved, and these are often controlled by internal diffusion limitations during refining (Ancheyta and Speight, 2007).

Thiophene compounds are the most refractory of the sulfur compounds. Consequently, thiophene is frequently chosen as representative of the sulfur compounds in light feedstocks. However, the decomposition of thiophene is extremely complex and has not been fully resolved. Nevertheless, as complex as the desulfurization of thiophene might appear, projection of the reaction kinetics to benzothiophene and dibenzothiophene, and to their derivatives, is even more complex. As has already been noted for bond energy data, kinetic data derived from model compounds cannot be expected to include contributions from the various steric effects than are a consequence of complex molecules containing three-dimensional structures. Furthermore, steric effects can lead to the requirement of additional catalyst properties and process parameters for sulfur removal (Isoda et al., 1996a,b).

Furthermore, the complexity of the individual reactions occurring in an extremely complex mixture and the interference of the products with those from other components of the mixture is unpredictable. Or the interference of secondary and tertiary products with the course of a reaction and, hence, with the formation of primary products may also be a cause for concern. Hence, caution is advised when applying the data from model compound studies to the behavior of petroleum, especially the molecularly complex heavy oils. These have few, if any, parallels in organic chemistry. All such contributions may be missing from the kinetic data that must be treated with some degree of caution.

However, there are several generalizations that come from the available thermodynamic data and investigations of pure compounds, as well as work carried out on petroleum fractions. Thus, at room temperature, hydrogenation of sulfur compounds to hydrogen sulfide is thermodynamically favorable and the reaction will essentially proceed to completion in the presence of a stoichiometric amount of hydrogen. Sulfides, simple thiophene derivatives, and benzothiophene derivatives are generally easier to desulfurize than the dibenzothiophene derivatives and the higher molecular weight condensed thiophene derivatives.

Nevertheless, the development of general kinetic data for the hydrodesulfurization of different feedstocks is complicated by the presence of a large number of sulfur compounds, each of which may react at a different rate because of structural differences and differences in molecular weight. This may be reflected in the appearance of a complicated kinetic picture for hydrodesulfurization in which the kinetics is not, apparently, first order (Scott and Bridge, 1971). The overall desulfurization reaction may be satisfied by a second-order kinetic expression when it can, in fact, also be considered as two competing first-order reactions. These reactions are (1) the removal of nonasphaltene sulfur and (2) the removal of asphaltene sulfur (Figure 1.6). It is the sum of these reactions that gives the second-order kinetic relationship.

In addition, the sulfur compounds in a feedstock may cause changes in the catalyst upon contact and, therefore, every effort should be made to ensure that the kinetic data from such investigations

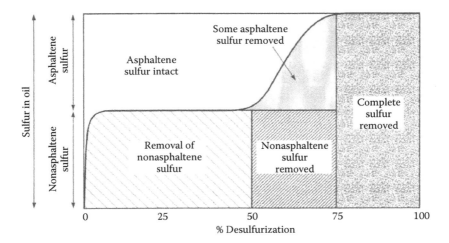

FIGURE 1.6 Representation of the removal of the various sulfur types during desulfurization. Nonasphaltene sulfur: sulfur that is partly protected in simple ring systems. Asphaltene sulfur: sulfur that is protected in complex ring systems.

are derived under standard conditions. In this sense, several attempts have been made to accomplish standardization of the reaction conditions by presulfiding the catalyst, passing the feedstock over the catalyst until the catalyst is stabilized, obtaining the data at various conditions, and then rechecking the initial data by repetition.

Thus, it has become possible to define certain general trends that occur in the hydrodesulfurization of petroleum feedstocks. One of the more noticeable facets of the hydrodesulfurization process is that the rate of reaction declines markedly with the molecular weight of the feedstock (Scott and Bridge, 1971), which is indicative of the increased amount of resin constituents and asphaltene constituents in the feedstock (Speight, 2014). For example, examination of the thiophene portion of a (narrow-boiling) feedstock and the resulting desulfurized product provides excellent evidence that benzothiophene derivatives are removed in preference to the dibenzothiophene derivatives and other condensed thiophene derivatives. The sulfur compounds in heavy oils and residua are presumed to react (preferentially) in a similar manner.

It is also generally accepted that the simpler sulfur compounds (e.g., thiol derivatives, R-SH, and thioether derivatives, R-S-R[1]) are (unless steric influences offer resistance to the hydrodesulfurization) easier to remove from petroleum feedstocks than the more complex cyclic sulfur compounds such as the benzothiophene derivatives. It should be noted here that, because of the nature of the reaction, steric influences would be anticipated to play a lesser role in the hydrocracking process.

Residua hydrodesulfurization is considerably more complex than the hydrodesulfurization of model organic sulfur compounds or, for that matter, narrow-boiling petroleum fractions. In published studies of the kinetics of residua hydrodesulfurization, one of three approaches has generally been taken: (1) the reactions can be described in terms of simple first-order expressions; (2) the reactions can be described by use of two simultaneous first-order expressions—one expression for easy-to-remove sulfur and a separate expression for difficult-to-remove sulfur; and (3) the reactions can be described using a pseudo-second-order treatment. Each of the three approaches has been used to describe hydrodesulfurization of residua under a variety of conditions with varying degrees of success; however, it does appear that pseudo-second-order kinetics are favored. In this particular treatment, the rate of hydrodesulfurization is expressed by a simple second-order equation:

$$C/1 - C = k \, (1/\text{LHSV})$$

where C is the wt.% sulfur in product/wt.% sulfur in the charge, k is the reaction rate constant, and LHSV is the liquid hourly space velocity (volume of liquid feed per hour per volume of catalyst).

The application of this model to a residuum desulfurization gave a linear relationship (Beuther and Schmid, 1963). However, it is difficult to accept the desulfurization reaction as a reaction that requires the interaction of two sulfur-containing molecules (as dictated by the second-order kinetics). To accommodate this anomaly, it has been suggested that, as there are many different types of sulfur compounds in residua and each may react at a different rate, the differences in reaction rates offered a reasonable explanation for the apparent second-order behavior. For example, an investigation of the hydrodesulfurization of an Arabian Light–atmospheric residuum showed that the overall reaction could not be adequately represented by a first-order relationship (Scott and Bridge, 1971). However, the reaction could be represented as the sum of two competing first-order reactions and the rates of desulfurization of the two fractions (the oil fraction and the asphaltene fraction) could be well represented as an overall second-order reaction.

If each type of sulfur compound is removed by a reaction that was first order with respect to sulfur concentration, the first-order reaction rate would gradually, and continually, decrease as the more reactive sulfur compounds in the mix became depleted. The more stable sulfur species would remain, and the residuum would contain the more difficult-to-remove sulfur compounds. This sequence of events will presumably lead to an apparent second-order rate equation, which is, in fact, a compilation of many consecutive first-order reactions of continually decreasing rate constant. Indeed, the desulfurization of model sulfur-containing compounds exhibits first-order kinetics, and the concept that the residuum consists of a series of first-order reactions of decreasing rate constant leading to an overall second-order effect has been found to be acceptable.

Application of the second-order rate equation to the hydrodesulfurization process has been advocated because of its simplicity and use for extrapolating and interpolating hydrodesulfurization data over a wide variety of conditions. However, while the hydrodesulfurization process may appear to exhibit second-order kinetics at temperatures near 395°C (745°F), at other temperatures the data (assuming second-order kinetics) does not give a linear relationship (Ozaki et al., 1963).

On this basis, the use of two simultaneous first-order equations may be more appropriate. The complexity of the sulfur compounds tends to increase with an increase in boiling point and the reactivity tends to decrease with complexity of the sulfur compounds, and residua (and, for that matter, the majority of heavy oils) may be expected to show substantial proportions of difficult to desulfurize sulfur compounds. It is anticipated that such an approach would be more consistent with the relative reactivity of various sulfur compound types observed for model compounds and for the various petroleum fractions that have been investigated.

Other kinetic work has shown that, for a fixed level of sulfur removal, the order of a reaction at constant temperature can be defined with respect to pressure:

$$k = 1/\text{LHSV} \ (P_h)^n$$

where P_h is the hydrogen partial pressure, LHSV is the liquid volume hourly space velocity, k is a constant, and n is the order of the reaction. It has been concluded, on the basis of this equation, that the hydrodesulfurization of residuum is first order with respect to pressure over the range 800–2300 psi, although it does appear that the response to pressure diminishes markedly (and may even be minimal) above 1000 psi.

One marked effect of a hydrodesulfurization process is the buildup of hydrogen sulfide, and the continued presence of this reaction product in the reactor reduces the rate of hydrodesulfurization. Thus, using the two first-order models, the effect of hydrogen sulfide on the process can be represented as

$$k/k_0 = 1/1 + k_1 P_{H_2S}$$

In this equation, k is the rate constant with hydrogen sulfide present, k_0 is the rate constant in the absence of hydrogen sulfide, and k_1 is a constant.

Data obtained using this equation showed that a change in hydrogen sulfide concentration from 1% to 12% (by volume) could reduce by 50% the rate constants for the easy-to-desulfurize and the difficult-to-desulfurize reactions. On the basis of the data available from kinetic investigations, the kinetics of residuum hydrodesulfurization may be represented by the following general equation:

$$-ds/dt = \left[P_H^n /(1 + k_a A = k_s P_{H_2S})^m \right] \Sigma k_i S^i$$

where S is the weight fraction of sulfur in the liquid phase, t is the residence time, P_H is the partial pressure of hydrogen, A is the weight fraction of asphaltene constituents in the liquid phase, P_{H_2S} is the partial pressure of hydrogen sulfide, S is the weight fraction of sulfur associated with component i in the range of i to j, k_a is the adsorption constant for asphaltene constituents, k is the adsorption constant for hydrogen sulfide, and K_i is the specific reaction rate constant for component i.

The latter constant k_i in the above relationship is a function of the chemistry of the component, the catalyst activity, and the reaction temperature. Therefore

$$k_i = k_0 \, A/A_o \, e^{-\Delta E/RT}$$

where k_0 is the reaction rate constant at standard catalyst activity, A_o is the standard catalyst activity, A is the catalyst activity, ΔE is the activation energy, R is the gas constant, and T is the absolute temperature.

This relationship gives activation energies for the hydrodesulfurization of various residua in the range 27–35 kcal g-mol^{-1}. In this context, it was interesting to note that the deasphalting of Khafji residuum had no effect on the activation energy of 30 kcal g-mol^{-1}, and it was suggested that the activation energies of the various components in a particular residuum might be approximately the same.

In spite of all the work, the kinetics and mechanism of alkyl-substituted dibenzothiophene, where the sulfur atom may be sterically hindered, are not well understood and these compounds are in general very refractory to hydrodesulfurization. Other factors that influence the desulfurization process such as catalyst inhibition or deactivation by hydrogen sulfide, the effect of nitrogen compounds, and the effect of various solvents need to be studied to obtain a comprehensive model that is independent of the type of model compound or feedstock used.

The role of various elements in the supported catalyst requires additional work to fully understand their function. There is also a dearth of information dealing with the nature of the surface of the newer catalysts subjected to various pretreatment and exposed to model compounds such as alkyl-substituted dibenzothiophene or other feedstocks. There is a need to correlate that data from such characterization with kinetic and mechanistic studies.

Throughout this section, the focus has been on the kinetic behavior of various organic molecules during refinery operations, specifically desulfurization. However, it must be remembered that the kinetic properties of the catalyst also deteriorate because of the deposits on its surface. Such deposits typically consist of coke and metals that are products of the various chemical reactions.

1.5 SULFUR REMOVAL DURING REFINING

Sulfur removal, as currently practiced in the petroleum industry, can be achieved using three options. The first option involves the use of thermal methods such as the various cracking techniques that concentrate the majority of the sulfur into the nonvolatile products, i.e., coke. Such processes are located in the *conversion* section of a refinery (Figure 1.1). The remainder of the sulfur may occur in the gases and as low-boiling organic sulfur compounds. The second option involves the use of chemical methods such as alkali treating (sweetening), which may be located in the product *finishing* section of a refinery (Figure 1.1). The third option is hydrodesulfurization that occurs in the conversion (hydrocracking) section or in the finishing (hydrotreating) section of the refinery (Figure 1.1). It is this latter process that will receive the major emphasis throughout our text in the manner in which it relates to the processing of heavy oils and residua.

1.5.1 THERMAL CRACKING

The term *cracking* applies to the decomposition of petroleum constituents that is induced by elevated temperatures (>350°C, >660°F) whereby the higher molecular weight constituents of petroleum are converted to lower molecular weight products. Cracking reactions involve carbon–carbon bond rupture and are thermodynamically favored at high temperature.

Thus, cracking is a phenomenon by which higher-boiling (higher molecular weight) constituents in petroleum are converted into lower-boiling (lower molecular weight) products. However, certain products may interact with one another to yield products having higher molecular weights than the constituents of the original feedstock. Some of the products are expelled from the system as, say, gases, naphtha-range materials, kerosene-range materials, and the various intermediates that produce other products such as coke. Materials that have boiling ranges higher than naphtha and kerosene may (depending on the refining options) be referred to as *recycle* stock, which is recycled in the cracking equipment until conversion is complete. However, the recycle stock, having been through the reactor at least once, will have a lower reactivity than the original feedstock, thereby reducing the reactivity of the feedstock in each recycle event. However, the overall conversion is increased.

Two general types of reaction occur during cracking:

1. The decomposition of large molecules into small molecules (primary reactions)

$$CH_3CH_2CH_2CH_3 \rightarrow CH_4 + CH_3CH{=}CH_2$$

    Butane   methane   propene

and

$$CH_3CH_2CH_2CH_3 \rightarrow CH_3CH_3 + CH_2{=}CH_2$$

    Butane   ethane   ethylene

2. Reactions by which some of the primary products interact to form higher molecular weight materials (secondary reactions)

$$CH_2{=}CH_2 + CH_2{=}CH_2 \rightarrow CH_3{\cdot}CH_2{\cdot}CH{=}CH_2$$

or

$$RCH{=}CH_2 + R^1{\cdot}CH = CH_2 \rightarrow \text{cracked residuum} + \text{coke} + \text{other products}$$

Thermal cracking is a free radical chain reaction; a free radical is an atom or group of atoms possessing an unpaired electron. Free radicals are very reactive, and it is their mode of reaction that actually determines the product distribution during thermal cracking. Free radical reacts with a hydrocarbon by abstracting a hydrogen atom to produce a stable end product and a new free radical. Free radical reactions are extremely complex, and it is hoped that these few reaction schemes illustrate potential reaction pathways. Any of the preceding reaction types are possible; however, it is generally recognized that the prevailing conditions and those reaction sequences that are thermo-dynamically favored determine the product distribution.

One of the significant features of hydrocarbon free radicals is their resistance to isomerization, e.g., migration of an alkyl group; as a result, thermal cracking does not produce any degree of branching in the products other than that already present in the feedstock.

Data obtained from the thermal decomposition of pure compounds indicate certain decomposition characteristics that permit predictions to be made of the product types that arise from the thermal cracking of various feedstocks. For example, normal paraffins are believed to form, initially, higher molecular weight material, which subsequently decomposes as the reaction progresses. Other paraffin derivative materials and terminal olefins ($RCH_2CH=CH_2$) are produced. An increase in pressure inhibits the formation of low molecular weight gaseous products and therefore promotes the formation of higher molecular weight materials.

Branched paraffins react somewhat differently to the normal paraffins during cracking processes and produce substantially higher yields of olefins having one fewer carbon atom than the parent hydrocarbon. Cycloparaffins (naphthenes) react differently to their noncyclic counterparts and are somewhat more stable. For example, cyclohexane produces hydrogen, ethylene, butadiene, and benzene: alkyl-substituted cycloparaffins decompose by means of scission of the alkyl chain to produce an olefin and a methyl or ethyl cyclohexane.

The aromatic ring is considered fairly stable at moderate cracking temperatures (350–500°C, 660–930°F). Alkyl aromatics, like the alkyl naphthenes, are more prone to dealkylation than to ring destruction. However, ring destruction of the benzene derivatives occurs above 500°C (930°F); however, condensed aromatics may undergo ring destruction at somewhat lower temperatures (450°C, 840°F).

1.5.2 Catalytic Cracking

Catalytic cracking is the thermal decomposition of petroleum constituents in the presence of a catalyst (Pines, 1981; Decroocq, 1984). The catalyst causes the activation of specific bonds within the organic substrate by the catalyst, and the desired reaction is achieved. However, in petroleum refining, the reactions that yield products that are lower boiling than those in the feedstock are accompanied by secondary processes that lead to the formation of higher-boiling products and coke. Thus, coke deposited on the surface of the catalyst modifies, by its own chemical action, the properties of the active sites and gradually renders the sites inaccessible to the reactants. However, residual coke on the catalyst autocatalyzes the formation of more carbonaceous deposits either in the catalyst or in the fluid. In short, the kinetic activity of the catalyst changes (Chapter 6).

Catalytic cracking processes evolved in the 1930s from research on petroleum and coal liquids. The petroleum work came to fruition with the invention of acid cracking. The work to produce liquid fuels from coal, most notably in Germany, resulted in metal sulfide hydrogenation catalysts. In the 1930, a catalytic cracking catalyst for petroleum that used solid acids as catalysts was developed using acid-treated clay minerals. The first catalysts used for catalytic cracking were acid-treated clay minerals, formed into beads. In fact, clay minerals are still employed as catalysts in some cracking processes (Speight, 2014).

Clay minerals are a family of crystalline aluminosilicate solids that interact with a variety of organic compounds (Theng, 1974). Acid treatment develops acidic sites by removing aluminum from the structure and often enhances the reactivity of the clay with specific families of organic compounds. The acid sites also catalyze the formation of coke, and Houdry developed a moving bed process that continuously removes the cooked beads from the reactor for regeneration by oxidation with air.

Clay minerals are natural compounds of silica and alumina, containing major amounts of the oxides of sodium, potassium, magnesium, calcium, and other alkali and alkaline earth metals. Iron and other transition metals are often found in natural clay minerals, substituted for the aluminum cations. Oxides of virtually every metal are found as impurity deposits in clay minerals.

Clay minerals are layered crystalline materials. They contain large amounts of water within and between the layers (Keller, 1985). Heating the clay minerals above 100°C can drive out some or all of this water; at higher temperatures, the clay structures themselves can undergo complex solid-state reactions. Such behavior makes the chemistry of clay minerals a fascinating field of study in its own right. Typical clay minerals include kaolinite, montmorillonite, and illite (Keller, 1985). They are found in most natural soils and in large, relatively pure deposits from which they are mined for applications ranging from adsorbents to paper making.

Strong mineral acids react with clay minerals to hydrolyze aluminum–oxygen bonds. The aluminum dissolves, leaving an altered structure. Protons in the acid also exchange with sodium and other alkali cations, leaving an acidic material. Houdry took advantage of such reactions in making the active acidic catalysts for his catalytic cracking process. The action of the silica–alumina catalysts is a function of the Lewis acid and Brønsted acid sites on the catalyst. Both acid sites appear to be functional, with the ratio dependent on the degree of hydration.

Zeolites are crystalline aluminosilicates that are used as catalysts and that have a particular chemical and physical structure (Decroocq, 1984; Goursot et al., 1997). They are highly porous crystals veined with submicroscopic channels. The channels contain water (hence, the bubbling at high temperatures), which can be eliminated by heating (combined with other treatments) without altering the crystal structure (Occelli and Robson, 1989).

Typical naturally occurring zeolites include *analcite* (also called *analcime*), $Na(AlSi_2O_6)$, and *faujasite*, $Na_2Ca(AlO_2)_2(SiO_2)_4.H_2O$, which are the structural analogues of the synthetic *zeolite X* and *zeolite Y*, respectively. *Sodalite* ($Na_8[(Al_2O_2)_6(SiO_2)_6]Cl_2$) contains the truncated octahedral structural unit known as the *sodalite cage* that is found in several zeolites. The corners of the faces of the cage are defined by either four or six Al/Si atoms, which are joined together through oxygen atoms. The zeolite structure is generated by joining sodalite cages through the four silicon–aluminum rings, thus enclosing a cavity or *supercage* bounded by a cube of eight sodalite cages and readily accessible through the faces of that cube (channels or pores). The structural frameworks of faujasite, zeolite X, and zeolite Y are generated by joining sodalite cages together through the six-Si/Al faces. In zeolites, the effective width of the pores is usually controlled by the nature of the cation (M^+ or M^{2+}).

Natural zeolites form hydrothermally (e.g., by the action of hot water on volcanic ash or lava), and synthetic zeolites can be made by mixing solutions of aluminates and silicates and maintaining the resulting gel at temperatures of 100°C (212°F) or higher for appropriate periods (Swaddle, 1997). Zeolite-A can form at temperatures <100°C (212°F); however, most zeolite syntheses require hydrothermal conditions (typically 150/300°F at the appropriate pressure). The reaction mechanism appears to involve redissolution of the gel and reprecipitation as the crystalline zeolite, and the identity of the zeolite produced depends on the composition of the solution. Aqueous alkali metal hydroxide solutions favor zeolites with relatively high aluminum contents, while the presence of organic molecules such as amines or alcohols favors highly siliceous zeolites such as silicalite or ZSM-5. Various tetra-alkyl ammonium cations favor the formation of certain specific zeolite structures and are known as *template* ions, although it should not be supposed that the channels and cages form simply by the wrapping of aluminosilicate fragments around suitably shaped cations. Although zeolite catalysts offer benefits for many processes, it is not yet fully agreed that zeolites offer any benefits over conventional catalysts when used under mild cracking process conditions.

The acid catalysts first used in catalytic cracking were amorphous solids composed of approximately 87% silica (SiO_2) and 13% alumina (Al_2O_3), and were designated low-alumina and 75% silica. However, this type of catalyst is now being replaced by crystalline aluminosilicates (zeolites) that are more common for the hydrodesulfurization of residua (O'Connor et al., 1998; Speight and Ozum, 2002; Parkash, 2003; Hsu and Robinson, 2006; Gary et al., 2007; Speight, 2014).

Although thermal cracking is a free radical (neutral) process, catalytic cracking is an ionic process involving carbonium ions, which are hydrocarbon ions having a positive charge on a carbon atom. The formation of carbonium ions during catalytic cracking can occur by the (1) addition of a proton from an acid catalyst to an olefin or (2) abstraction of a hydride ion (H^-) from a hydrocarbon by the acid catalyst or by another carbonium ion—carbonium ions are not formed by cleavage of a carbon–carbon bond.

Once the carbonium ions are formed, the modes of interaction constitute an important means by which product formation occurs during catalytic cracking, e.g., isomerization either by hydride ion shift or by methyl group shift, both of which occur readily. The trend is for stabilization of the carbonium ion by movement of the charged carbon atom toward the center of the molecule, which accounts for the isomerization of α-olefins to internal olefins when carbonium ions are produced. Cyclization can occur by the internal addition of a carbonium ion to a double bond, which, by continuation of the sequence, can result in aromatization of the cyclic carbonium ion.

Like the paraffins, naphthenes do not appear to isomerize before cracking. However, the naphthenic hydrocarbons (from C_9 upward) produce considerable amounts of aromatic hydrocarbons during catalytic cracking. Reaction schemes similar to that outlined here provide possible routes for the conversion of naphthenes to aromatics.

Alkyl benzenes undergo nearly quantitative dealkylation to benzene without apparent ring degradation at <500°C (930°F). However, polymethylbenzene derivatives undergo disproportionation and isomerization with very little benzene formation.

Catalytic cracking can be represented by simple reaction schemes (see, e.g., Speight, 2014). However, questions have arisen about how the cracking of paraffins is initiated. Several hypotheses for the initiation step in catalytic cracking of paraffins have been proposed (Cumming and Wojciechowski, 1996). The Lewis site mechanism is the most obvious, as it proposes that a carbenium ion is formed by the abstraction of a hydride ion from a saturated hydrocarbon by a strong Lewis acid site: a tricoordinated aluminum species. On Brønsted sites, a carbenium ion may be readily formed from an olefin by the addition of a proton to the double bond or, more rarely, via the abstraction of a hydride ion from a paraffin by a strong Brønsted proton. This latter process requires the formation of hydrogen as an initial product. This concept, for various reasons, was often neglected.

It is therefore not surprising that the earliest cracking mechanisms postulated that the initial carbenium ions are formed only by the protonation of olefins generated either by thermal cracking or present in the feed as an impurity. For a number of reasons, this proposal was not convincing, and in the continuing search for initiating reactions it was even proposed that electrical fields associated with the cations in the zeolite are responsible for the polarization of reactant paraffins, thereby activating them for cracking. More recently, however, it has been convincingly shown that a pentacoordinated carbonium ion can be formed on the alkane itself by protonation, if a sufficiently strong Brønsted proton is available (Cumming and Wojciechowski, 1996).

Coke formation is considered, with just cause to a malignant side reaction of normal carbenium ions. However, while chain reactions dominate events occurring on the surface and produce the majority of products, certain less desirable bimolecular events have a finite chance of involving the same carbenium ions in a bimolecular interaction with one another. Of these reactions, most will produce a paraffin and leave carbene/carboid-type species on the surface. This carbene/carboidtype species can produce other products; however, the most damaging product will be the one that remains on the catalyst surface and cannot be desorbed and results in the formation of coke, or remains in a non-coke form but effectively blocks the active sites of the catalyst.

A general reaction sequence for coke formation from paraffins involves oligomerization, cyclization, and dehydrogenation of small molecules at active sites within zeolite pores:

$$Alkanes \rightarrow alkenes$$

$$Alkenes \rightarrow oligomers$$

$$Oligomers \rightarrow naphthenes$$

$$Naphthenes \rightarrow aromatics$$

$$Aromatics \rightarrow coke$$

Whether these are the true steps to coke formation can only be surmised. The problem with this reaction sequence is that it ignores sequential reactions in favor of consecutive reactions. Moreover, it must be accepted that the chemistry leading up to coke formation is a complex process, consisting of many sequential and parallel reactions.

There is a complex and little understood relationship between coke content, catalyst activity, and the chemical nature of the coke. For instance, the C-to-H ratio of coke depends on how the coke was formed; its exact value will vary from system to system (Cumming and Wojciechowski, 1996). Furthermore, it seems that catalyst decay is not related in any simple way to the hydrogen-to-carbon atomic ratio of the coke, or to the total coke content of the catalyst, or any simple measure of coke properties. Moreover, despite many and varied attempts, there is currently no consensus about the detailed chemistry of coke formation. There is, however, much evidence and good reason to believe that catalytic coke is formed from carbenium ions that undergo addition, dehydrogenation, and cyclization, and elimination side reactions in addition to the main-line chain propagation processes (Cumming and Wojciechowski, 1996).

1.5.3 HYDROGENATION

The purpose of hydrogenating petroleum constituents is to (1) improve existing petroleum products or develop new products or even new uses, (2) convert inferior or low-grade materials into valuable products, and (3) transform higher molecular weight constituents into liquid fuels. The distinguishing feature of the hydrogenating processes is that, although the composition of the feedstock is relatively unknown and a variety of reactions may occur simultaneously, the final product may actually meet all the required specifications for its particular use (Speight, 2014).

Hydrogenation processes for the conversion of petroleum and petroleum products may be classified as *destructive* and *nondestructive*. The former (hydrogenolysis or hydrocracking) is characterized by the rupture of carbon–carbon bonds and is accompanied by hydrogen saturation of the fragments to produce lower-boiling products. Such treatment requires rather high temperatures and high hydrogen pressures, the latter to minimize coke formation. Many other reactions, such as isomerization, dehydrogenation, and cyclization, can occur under these conditions (Dolbear et al., 1987).

On the other hand, nondestructive, or simple, hydrogenation is generally used for the purpose of improving product (or even feedstock) quality without appreciable alteration of the boiling range. Treatment under such mild conditions is often referred to as *hydrotreating* or *hydrofining* and is essentially a means of eliminating nitrogen, oxygen, and sulfur as ammonia, water, and hydrogen sulfide, respectively.

The hydrodesulfurization process can fall into either the destructive or nondestructive category. However, for heavy feedstocks, some hydrocracking is preferred, if not necessary, to remove the sulfur. Thus, hydrodesulfurization in this context falls into the hydrocracking or destructive hydrogenation category. The basic chemical concept of the process remains the same—to convert the organic sulfur in the feedstock to hydrogen sulfide:

$$Feedstock_{sulfur} + H_2 \rightarrow H_2S + products_{sulfur\text{-}deficient}$$

Although the definition of the two processes is purely arbitrary, it is generally assumed that destructive hydrogenation (which is characterized by the cleavage of carbon-to-carbon linkage and is accompanied by hydrogen saturation of the fragments to produce lower-boiling products) requires temperatures in excess of 350°C (660°F). Nondestructive hydrogenation processes are more generally used for the purpose of improving product quality without any appreciable alteration of the boiling range. Mild processing conditions (temperatures <350°C or <660°F) are employed so that only the more unstable materials are attached. Thus, sulfur, nitrogen, and oxygen compounds eliminate hydrogen sulfide (H_2S), ammonia (NH_3), and water (H_2O), respectively. Unsaturated thermal products such as olefins ($RCH=CHR^1$) are hydrogenated to produce the more stable hydrocarbons ($RCH_2CH_2R^1$).

While the definitions of the various hydroprocesses are (as has been noted above) quite arbitrary, it may be difficult, if not impossible, to limit the process to any one particular reaction in a commercial operation. The prevailing conditions may, to a certain extent, minimize, say, cracking reactions during a hydrotreating operation. However, with respect to the heavier feedstocks, the ultimate aim of the operation is to produce as much low-sulfur liquid products as possible from the feedstock. Any hydrodesulfurization process that has been designed for application to the heavier oils and residua may require that hydrocracking and hydrodesulfurization occur simultaneously.

1.5.3.1 Hydrocracking

Hydrocracking (Figure 1.7) is a thermal process (>350°C, >660°F) in which hydrogenation accompanies cracking. Relatively high pressure (100–2000 psi) is employed, and the overall result is usually a change in the character or quality of the products.

The wide range of products possible from hydrocracking is the result of combining catalytic cracking reactions with hydrogenation and the multiplicity of reactions that can occur (Speight,

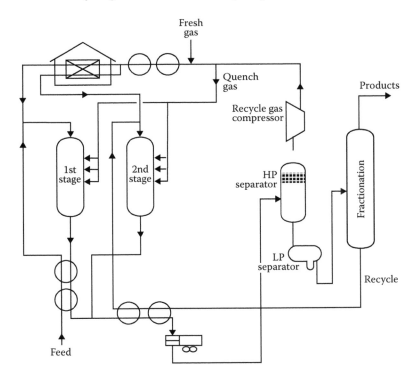

FIGURE 1.7 Single-stage or two-stage (optional) hydrocracking unit. (From OSHA Technical Manual, Section IV, Chapter 2: Petroleum Refining Processes. http://www.osha.gov/dts/osta/otm/otm_iv/otm_iv_2.html; Speight, J.G. 2014. *The Chemistry and Technology of Petroleum*. 4th Edition. CRC Press, Taylor & Francis Publishers. Figure 15.11, p. 411.)

2000; Ancheyta and Speight, 2007). Dual-function catalysts in which the cracking is provided by silica–alumina (or zeolite) catalysts, platinum, tungsten oxide, or nickel.

Essentially all the initial reactions of catalytic cracking occur; however, some of the secondary reactions are inhibited or stopped by the presence of hydrogen. For example, the yields of olefins and the secondary reactions that result from the presence of these materials are substantially diminished and branched-chain paraffins undergo demethylation. The methyl groups attached to secondary carbons are more easily removed than those attached to tertiary carbon atoms, whereas methyl groups attached to quaternary carbons are the most resistant to hydrocracking.

The effect of hydrogen on naphthenic hydrocarbons is mainly that of ring scission followed by immediate saturation of each end of the fragment produced. The ring is preferentially broken at favored positions, although generally all the carbon–carbon bond positions are attacked to some extent. For example, methyl-cyclopentane is converted (over a platinum–carbon catalyst) to 2-methylpentane, 3-methylpentane, and *n*-hexane.

Aromatic hydrocarbons are resistant to hydrogenation under mild conditions; however, under more severe conditions, the main reactions are conversion of the aromatic to naphthenic rings and scissions within the alkyl side chains. The naphthenes may also be converted to paraffins. Polynuclear aromatic hydrocarbons are more readily attacked than the single-ring compounds, the reaction proceeding by a stepwise process in which one ring at a time is saturated and then opened (Speight, 2000, 2014; Ancheyta and Speight, 2007).

The presence of hydrogen changes the nature of the products (especially the coke yield). The yield of coke is decreased by preventing the buildup of precursors that are incompatible in the liquid medium and eventually form coke (Magaril and Aksenova, 1967, 1968, 1972; Magaril and Ramazaeva, 1969; Magaril et al., 1970, 1971; Speight and Moschopedis, 1979). In fact, the chemistry involved in the reduction of asphaltene constituents to liquids using models in which the polynuclear aromatic system borders on graphitic is difficult to visualize. However, the *paper chemistry* derived from the use of a molecularly designed model composed of smaller polynuclear aromatic systems is much easier to visualize (Speight, 2014). However, precisely how asphaltene constituents react with the catalysts is open to much more speculation because of the complexity of the catalyst as well as the complexity of the asphaltene three-dimensional structure and molecular size (Speight et al., 1985; Schabron and Speight, 1998).

In contrast to the visbreaking process, in which the general principle is the production of products for use as fuel oil, hydroprocessing is employed to produce a slate of products for use as liquid fuels. Nevertheless, the decomposition of asphaltene constituents is, again, an issue. Moreover, just as models consisting of large polynuclear aromatic systems are inadequate to explain the chemistry of visbreaking, they are also of little value for explaining the chemistry of hydrocracking.

Deposition of solids or incompatibility is still possible when asphaltene constituents interact with catalysts, especially acidic support catalysts, through the functional groups, e.g., the basic nitrogen species just as they interact with adsorbents. Furthermore, there is a possibility for interaction of the asphaltene with the catalyst through the agency of a single functional group in which the remainder of the asphaltene molecule remains in the liquid phase. There is also a less desirable option in which the asphaltene reacts with the catalyst at several points of contact, causing immediate incompatibility on the catalyst surface.

1.5.3.2 Hydrotreating

It is generally recognized that the higher the hydrogen content of a petroleum product, especially the fuel products, the better is the quality of the product. This knowledge has stimulated the use of hydrogen-adding processes in the refinery.

Thus, hydrogenation without simultaneous cracking is used for saturating olefins or for converting aromatics to naphthenes. Under atmospheric pressure, olefins can be hydrogenated up to about 500°C (930°F); however, beyond this temperature, dehydrogenation commences. Application of pressure and the presence of catalysts make it possible to effect complete hydrogenation at room or even cooler temperatures; the same influences are helpful in minimizing dehydrogenation at higher temperatures.

A wide variety of metals are active hydrogenation catalysts; those of most interest are nickel, palladium, platinum, cobalt, iron, nickel-promoted copper, and copper chromite. Special preparations of the first three are active at room temperature and atmospheric pressure. The metallic catalysts are easily poisoned by sulfur- and arsenic-containing compounds, and even by other metals. To avoid such poisoning, less effective but more resistant metal oxides or sulfides are frequently employed, generally those of tungsten, cobalt, chromium, or molybdenum. Alternatively, catalyst poisoning can be minimized by mild hydrogenation to remove nitrogen, oxygen, and sulfur from feedstocks in the presence of more resistant catalysts, such as cobalt–molybdenum–alumina ($Co–Mo–Al_2O_3$).

1.6 MACROMOLECULAR CONCEPTS

In a mixture as complex as petroleum, the reaction processes can only be generalized because of difficulties in analyzing not only the products but also the feedstock as well as the intricate and complex nature of the molecules that make up the feedstock. The concentration of the heteroatoms in the higher molecular weight and polar constituents (Figure 1.8) is detrimental to process efficiency and to catalyst performance (Speight, 1987, 2014; Dolbear, 1998).

There are a variety of sulfur-containing molecules in a residuum or heavy crude oil that produce different products as a result of a hydrodesulfurization reaction. Although the deficiencies of current analytical techniques dictate that the actual mechanism of desulfurization remains largely speculative, some attempt has been made to determine the macromolecular chemical concepts that are involved in the hydrodesulfurization of heavy oils and residua.

Under the usual commercial hydrodesulfurization conditions (elevated temperatures and pressures, high hydrogen-to-feedstock ratios, and the presence of a catalyst), the various reactions that result in the removal of sulfur from the organic feedstock occur (Speight, 2000; Ancheyta and

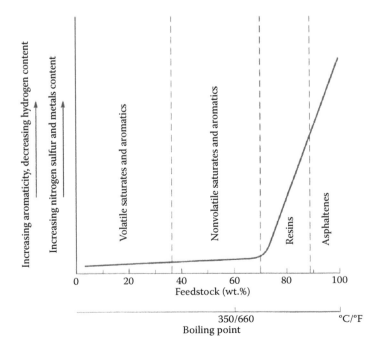

FIGURE 1.8 Relative occurrence of heteroatoms in feedstock fractions. (From Speight, J.G. 2007. *The Desulfurization of Heavy Oils and Residua*. 2nd Edition. Marcel Dekker Inc., New York. Figure 4.13, p. 153.)

Speight, 2007). Thus, thiols as well as open-chain and cyclic sulfides are converted to saturated and/or aromatic compounds depending, of course, on the nature of the particular sulfur compound involved. Benzothiophene derivatives are converted to alkyl aromatics, while dibenzothiophene derivatives are usually converted to biphenyl derivatives. In fact, the major reactions that occur as part of the hydrodesulfurization process involve carbon–sulfur bond rupture and saturation of the reactive fragments (as well as saturation of olefins).

During the course of the reaction, aromatic rings are not usually saturated, even though hydrogenation of the aromatic rings may be thermodynamically favored. The saturation of some of the aromatic rings may appear to have occurred because of the tendency of partial ring saturation to occur before carbon–sulfur bond rupture, as is believed to be true in the case of dibenzothiophene derivatives. It is generally recognized that the ease of desulfurization is dependent on the type of compound, and the lower-boiling fractions are desulfurized more easily that the higher-boiling fractions. The difficulty of sulfur removal increases in the order

Paraffins < naphthenes < aromatics

The wide ranges of temperature and pressure employed for the hydrodesulfurization process virtually dictate that many other reactions will proceed concurrently with the desulfurization reaction. Thus, the isomerization of paraffins and naphthenes may occur, and hydrocracking will increase as the temperature and pressure increase. Furthermore, at the higher temperatures (but low pressures), naphthenes may dehydrogenate to aromatics and paraffins dehydrocyclize to naphthenes, while at lower temperature (high pressures) some of the aromatics may be hydrogenated.

These reactions do not all occur equally, which is due, to some extent, to the nature of the catalyst. The judicious choice of a catalyst will lead to the elimination of sulfur (and the other heteroatoms nitrogen and oxygen) and, although some hydrogenation and hydrocracking may occur, the extent of such reactions may be relatively minor (Speight, 2000, 2014; Ancheyta and Speight, 2007). In the present context, the hydrodesulfurization of the heavier feedstocks (heavy oils and residua) may require that part or almost all of the feedstock be converted to lower-boiling products. If this be the case (as is now usual), hydrocracking will, of course, compete on an almost equal footing with the hydrodesulfurization reaction for the production of the low-sulfur, low-boiling products.

Thus, the hydrodesulfurization process is a very complex sequence of reactions due, no doubt, to the complexity of the feedstock. Furthermore, the fact that feedstocks usually contain nitrogen and oxygen compounds (in addition to metal compounds) increases the complexity of the reactions that occur as part of the hydrodesulfurization process. The nitrogen compounds that may be present are typified by pyridine, quinoline, carbazole, indole, and pyrrole derivatives. Oxygen may be present as phenols (Ar–OH, where Ar is an aromatic moiety) and carboxylic acids ($-CO_2H$). The most common metals to occur in petroleum are nickel (Ni) and vanadium (V) (Reynolds, 1998).

In conventional crude oil, these other atoms (nitrogen, oxygen, and metals) may be of little consequence. Heavy oil contains substantial amounts of these atoms, and the nature of the distillation process dictates that virtually all of the metals and substantial amounts of the nitrogen and oxygen originally present in the petroleum will be concentrated in the residua.

The simultaneous removal of nitrogen during the processing is a very important aspect of hydrodesulfurization. Compounds containing nitrogen have pronounced deleterious effects on the storage stability of petroleum products, and nitrogen compounds in charge stocks to catalytic processes can severely limit (and even poison) the activity of the catalyst. Oxygen compounds are corrosive (especially the naphthenic acids) and can promote gum formation as part of the deterioration of the hydrocarbons in the product. Metals in feedstocks that are destined for catalytic processes can poison the selectivity of the catalyst and, like the nitrogen and oxygen compounds, should be removed.

Fortunately, the continued developments of the hydrodesulfurization process over the last two decades has resulted in the production of catalysts that can tolerate substantial amounts of nitrogen compounds, oxygen compounds, and metals without serious losses in catalyst activity or in catalyst life (Chapter 6). Thus, it is possible to use the hydrodesulfurization process not only as a means of producing low-sulfur liquid products but also as a means of producing low-sulfur, low-nitrogen, low-oxygen, and low-metal streams that can be employed as feedstocks for processes where catalyst sensitivity is one of the process features.

Refining the constituents of heavy oil and bitumen has become a major issue in modern refinery practice (Speight, 2013, 2014). The limitations of processing heavy oils and residua depend to a large extent on the amount of higher molecular weight constituents (i.e., asphaltene constituents) present in the feedstock that are responsible for high yields of thermal and catalytic coke (Figure 1.9) (Speight, 1984, 2000, 2014; LePage et al., 1992).

In many attempts to determine the macromolecular changes that occur as a result of hydrode-sulfurization, the focus has been on the changes that occur in the bulk fractions, i.e., the asphaltene constituents, resin constituents, and oil fraction. Furthermore, most of the attention has been focused on the asphaltene fraction because of the undesirable effects (e.g., coke deposition on the catalyst) that this particular fraction (which is found in substantial proportions in residua and heavy oils) has on refining processes.

Indeed, the thermal decomposition of asphaltene constituents is a complex phenomenon insofar as the asphaltene constituents are not uniform in composition (Speight, 1994, 2014). Each asphaltene subfraction will decompose to give a different yield of thermal coke (Speight, 2014), and the thermal decomposition of asphaltene constituents also produces thermal coke with different solubility profiles that are dependent on the reaction or process conditions (Speight, 1987, 1989, 2014).

Because of their high molecular weight and complexity, the asphaltene constituents remain an unknown entity in the hydrodesulfurization process. There are indications that, with respect to some residua and heavy oils, removal of the asphaltene constituents before the hydrodesulfurization step brings out a several-fold increase in the rate of hydrodesulfurization and that, with these particular residua (or heavy oils), the asphaltene constituents must actually inhibit hydrodesulfurization.

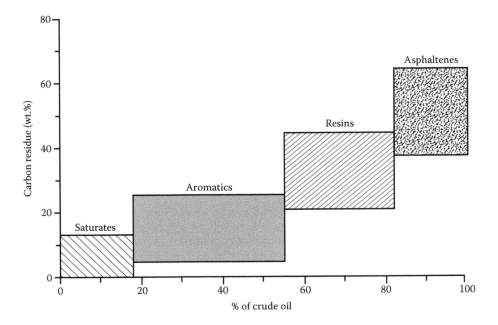

FIGURE 1.9   Yield of carbon residue (thermal coke) for various feedstock fractions. (From Speight, J.G. 2007. *The Desulfurization of Heavy Oils and Residua*. 2nd Edition. Marcel Dekker Inc., New York. Figure 4.14, p. 157.)

As a result of their behavior, there have been several attempts to focus attention on the asphaltene constituents during hydrodesulfurization studies. The other fractions of a residuum or heavy oil (i.e., the resins and the oils) have, on the other hand, been considered as higher molecular weight extensions of heavy gas oils about which desulfurization is fairly well understood.

Nevertheless, studies of the thermal decomposition of asphaltene constituents can provide relevant information about the mechanism by which asphaltene constituents are desulfurized. In fact, the thermal decomposition of petroleum asphaltene constituents has received some attention and has been well defined (Magaril and Aksenova, 1967, 1968, 1972; Magaril and Ramazaeva, 1969; Magaril et al., 1970, 1971). Special attention has been given to the nature of the volatile products of asphaltene decomposition mainly because of the difficulty of characterizing the non-volatile coke.

It has been generally assumed that the chemistry of coke formation involves immediate condensation reactions to produce higher molecular weight, condensed aromatic species. However, the initial reactions in the coking of petroleum feedstocks that contain asphaltene constituents involve the thermolysis of asphaltene aromatic-alkyl systems to produce volatile species (paraffins and olefins) and nonvolatile species (aromatics) (Speight, 1987, 1992, 1998a,b, 2014; Wiehe, 1992, 1993, 1994; Mushrush and Speight, 1995, 1998; Speight and Long, 1996; Schabron and Speight, 1997; Storm et al., 1997). In addition, the rate parameters for thermal decomposition vary from feedstock to feedstock and differ even for asphaltene constituents derived from a similar origin (Neurock et al., 1990, 1994). Each feedstock must possess a unique set of rate parameters that suggests that the molecular composition of the feedstocks (Chapter 7) influence the thermal reactions, and general relationships such as those derived from elemental analytical data must be treated with caution and only deduced to be general trends.

An additional corollary to this work is that conventional models of petroleum asphaltene constituents (which, despite evidence to the contrary, invoked the concept of a large polynuclear aromatic system) offer little, if any, explanation of the intimate events involved in the chemistry of coking. Models that invoke the concept of asphaltene constituents as a complex solubility class with molecular entities composed of smaller polynuclear aromatic systems (Speight, 2014) are more in keeping with the present data. Indeed, such models are more in keeping with the natural product origins of petroleum and with the finding that the hydrocarbon backbone of the polar fraction of petroleum is consistent with the hydrocarbon types in the nonpolar fractions (Speight, 1994, 2014).

If the hypothetical models, used as examples elsewhere (Speight, 1992, 1994, 1998b, 2000), can be used as a guide to the process chemistry, it should be anticipated that there might be some initial fragmentation and aromatization—indeed, data available suggest that this does occur (Speight, 2000, 2014). The isolatable products from the early stages of asphaltene decomposition are more aromatic than the original asphaltene constituents and have a slightly reduced molecular weight. By definition and character (Speight, 1987, 2014, 2015), these are carbenes. The next step will be the more complete fragmentation of the asphaltene constituents to produce the carboids that are, by definition through solubility, the true precursors to coke.

Molecular breakdown commences at the extremities of the molecules to leave a polar core that eventually separates from the liquid reaction medium. In the case of the decomposition of the amphoteric molecule, thermal degradation could just as easily commence at the aliphatic carbon sulfur bonds followed by thermal scission of the alkyl moieties from the aromatic systems. If the removal of the sulfur-containing moiety is not sterically hindered, this could well be the case, based on the relative strengths of aliphatic carbon–carbon and carbon–sulfur bonds. The end result would be the same. In the case of the neutral polar asphaltene molecule, the nonthiophene sulfur occurs between two aromatic rings and therefore is somewhat stronger than the aliphatic carbon–sulfur bond, and the molecule at this point receives a degree of steric protection against thermal scission.

There is also the distinct possibility that there will be some reactions that occur almost immediately with the onset of heating. Such reactions will most certainly include the elimination of carbon dioxide from carboxylic fragments and, perhaps, even intermolecular coupling through phenolic moieties.

There may even be prompt reactions that occur almost immediately. In other words, these are reactions that are an inevitable consequence of the nature of asphaltene. An example of such reactions is the rapid aromatization of selected hydroaromatic rings to create a more aromatic asphaltene.

Whereas such reactions will, obviously, play a role in coking, the precise role is difficult to define. Decarboxylation and the reactions of phenol moieties are not believed to be a major force in the coking chemistry. Perhaps, it is the immediate aromatization of selected, or all of the, hydroaromatic systems that renders inter- and intramolecular hydrogen management difficult and coke formation a relatively simple operation.

As these reactions are progressing, sulfur in aliphatic locations is released as, in the presence of hydrogen, hydrogen sulfide. Alternatively, the formation of hydrogen sulfide in the absence of added hydrogen suggests hydrogen abstraction by the sulfur atoms. The occurrence of sulfur or sulfur atoms in the reaction mixture would complicate the situation by participating in (or catalyzing) the intermolecular condensation of the polynuclear aromatic systems.

In keeping with the known data, the sulfur in polynuclear aromatic systems is difficult to remove in the absence of hydrogen. The heterocyclic rings tend to remain intact and are incorporated into the coke. The presence of hydrogen facilitates the removal of heterocyclic sulfur. Indeed, there is also the possibility that the hydrogen before cracking or hydrodesulfurization may render sulfur removal and hydrocarbon production even more favorable.

1.7 SEDIMENT FORMATION AND FOULING

One possible disadvantage of refining the heavier feedstocks is the occurrence of a solid sediment (phase separation of incompatible by-products) or the formation of coke during processing of heavy feedstocks, which can be a major limitation on process (reactor) performance and product yield (Mushrush and Speight, 1995, 1998). Because of the nature of the process chemistry (Speight, 2014), sediment formation can be considered in terms of ternary phase behavior (Speight, 1992, 2014). Solid formation occurs as a result of changes in the relative amounts of the lower-boiling hydrocarbons and higher molecular weight polar species.

Thus, as the feedstock reacts to the high temperatures, the increasing hydrocarbon nature of the liquid and the increasing polarity of the reacted asphaltene constituents may be likened to asphaltene precipitation. However, at the high temperatures in the reactor, a more reasonable model is that the liquids in the reactor split into a light aliphatic phase and a heavy, more aromatic phase. The formation of a heavy, aromatic liquid will be the initial step before phase separation of the aromatic material leading to solid deposition and ultimately to coke formation. It might be surmised that the solubility of the hydrogen gas in the aromatic phase may be anticipated to be low, leading to irreversible aromatization of the constituents.

A variety of processing approaches has been tested to control solid formation, typically by modifying the composition of the feed mixture or the recycle. The addition of hydrogenated middle distillates increases conversion and suppresses solid formation (Carlson et al., 1958), and the success of this approach is normally attributed to the donation of hydrogen from the solvent to the bitumen.

Other more specific examples include the Eureka process (Chapter 5), which controls solid formation by stripping the cracked products from the reactor with superheated steam. This leaves a higher-boiling product in the reactor that has a reduced coking propensity because of a decreased tendency for the high molecular weight aromatic polar compounds to separate in the reactor. Catalytic processes, such as the H–oil process (Chapter 13), control coke formation by the hydrogenation activity of the catalyst, which mainly serves to transfer hydrogen to the higher molecular weight species. The ebullating bed configuration also tends to remove the lower molecular weight products and leave a more homogeneous material in the liquid phase. A different approach is to use catalytic reactions to modify the least soluble components and thereby suppress solids. To achieve significant hydrogenation of aromatics, a low-temperature hydrogenation is required before high-severity cracking (Speight, 2013, 2014).

REFERENCES

Afanasiev, P., Cattenot, M., Geantet, C., Matsubayashi, N., Sato, K., and Shimada, S. 2002. (Ni)W/ZrO₂ Hydrotreating Catalysts Prepared in Molten Salts. *Appl. Catal. A* 237(1/2): 227–237.

Ancheyta, J., and Speight, J.G. (Editors) 2007. *Hydroprocessing Heavy Oils and Residua.* CRC Press, Taylor & Francis Group, Boca Raton, FL.

Arnoldy, P., Van den Heikant, J.A.M., De Bok, G.D., and Moulijn, J.A. 1985. Temperature-Programmed Sulfiding of MoO₃Al₂O₃ Catalysts. *J. Catal.* 92: 35–55.

Babich, I.V., and Moulijn, J.A. 2003. Science and Technology of Novel Processes for Deep Desulfurization of Oil Refinery Streams: A Review. *Fuel* 82(6): 607–631.

Beers, A.E.W., Hoek, I., Nijhuis, T.A., Downing, R.S., Kapteijn, F., and Moulijn, J.A. 2000. Structured Catalysts for the Acylation of Aromatics. *Top. Catal.* 13(3): 275–280.

Beuther, H., and Schmid, B.K. 1963. Reaction Mechanisms and Rates in Residua Hydrodesulfurization. Paper No, 20, Section III. In: *Proceedings 6th World Petroleum Congress*, Vol. 3, p. 297.

Breysse, M., Furimsky, E., Kasztelan, S., Lacroix, M., and Perot, G. 2002. Hydrogen Activation by Transition Metal Sulfides. *Catal. Rev. Sci. Eng.* 44: 651–735.

Carlson, C.S., Langer, A.W., Stewart, J., and Hill, R.M. 1958. Thermal Hydrogenation. Transfer of Hydrogen from Tetralin to Cracked Residua. *Ind. Eng. Chem.* 50: 1067–1070.

Cattaneo, R., Shido, T., and Prins, R. 1999. *Hydrotreatment and Hydrocracking of Oil Fractions.* Elsevier, Amsterdam.

Cumming, K.A., and Wojciechowski, B.W. 1996. Hydrogen Transfer, Coke Formation, and Catalyst Decay and Their Role in the Chain Mechanism of Catalytic Cracking. *Catal. Rev. Sci. Eng.* 38: 101–157.

Decroocq, D. 1984. *Catalytic Cracking of Heavy Petroleum Fractions.* Editions Technip, Paris, France.

Dolbear, G.E. 1998. Hydrocracking: Reactions, Catalysts, and Processes. In: *Petroleum Chemistry and Refining*, J.G. Speight (Editor). Taylor & Francis Publishers, Washington, DC, Chapter 7, pp. 175–198.

Dolbear, G.E., Tang, A., and Moorehead, E.L. 1987. Upgrading Studies with California, Mexican, and Middle Eastern Heavy Oils. *Fuel* 66: 267.

Egorova, M. 2003. Study of the Aspects of Deep Hydrodesulfurization by Means of Model Reactions. PhD Thesis, Swiss Federal Institute of Technology, Zurich, Switzerland.

Egorova, M., and Prins, R. 2004. Mutual Influence of the HDS of Dibenzothiophene and HDN of 2-Methyl-pyridine. *J. Catal.* 221: 11–19.

Energy Information Administration. 2015. Energy Information Administration, US Department of Energy, Washington, DC. http://www.eia.gov/dnav/pet/hist/LeafHandler.ashx?n=PET&s=8_NA_8CD0_NUS_5 &f=A; accessed January 31, 2015.

Farag, H., Whitehurst, D.D., Sakanishi, K., and Mochida, I. 1999. Carbon versus Alumina as a Support for Co–Mo Catalysts Reactivity towards HDS of Dibenzothiophenes and Diesel Fuel. *Catal. Today* 50: 9–17.

Gary, J.H., Handwerk, G.E., and Kaiser, M.J. 2007. *Petroleum Refining: Technology and Economics*, 5th Edition. CRC Press, Taylor & Francis Group, Boca Raton, FL.

Gomez, R.A.M. 2005. Treatment of Crude Oils. United States Patent 6,955,753, October 18.

Goursot, A., Vasilyev, V., and Arbuznikov, A. 1997. Modeling of Adsorption Properties of Zeolites: Correlation with the Structure. *J. Phys. Chem. B* 101(33): 6420–6428.

Hernández-Maldonado, A.J., and Yang, R.T. 2004. Desulfurization of Transportation Fuels by Adsorption. *Cat. Rev. Sci. Eng.* 46: 111–150.

Houalla, M., Broderick, D.H., Sapre, A.V., Nag, N.K., De Beer, V.H.J., Gates, B.C., and Kwart, H. 1980. Hydrodesulfurization of Methyl-Substituted Dibenzothiophenes Catalyzed by Sulfided Co-Mo/γ-Al₂O₃. *J. Catal.* 61: 523–527.

Hsu, C.S., and Robinson, P.R. (Editors) 2006. *Practical Advances in Petroleum Processing*, Vols. 1–2. Springer Science, New York.

Isoda, T., Nagao, S., Ma, X., Korai, Y., and Mochida, I. 1996a. Hydrodesulfurization of Refractory Sulfur Species. 1. Selective Hydrodesulfurization of 4,6-Dimethyldibenzothiophene in the Major Presence of Naphthalene over Como/Al₂O₃ and Ru/Al₂O₃ Blend Catalysts. *Energy Fuels* 10: 482–486.

Isoda, T., Nagao, S., Ma, X., Korai, Y., and Mochida, I. 1996b. Hydrodesulfurization of Refractory Sulfur Species. 2. Selective Hydrodesulfurization of 4,6-Dimethyldibenzothiophene in the Dominant Presence of Naphthalene over Ternary Sulfides Catalyst. *Energy Fuels* 10: 487–492.

Jones, D.S.J. 1995. *Elements of Petroleum Processing.* John Wiley & Sons Inc., Hoboken, NJ.

Kabe, T., Ishihara, A., and Tajima, H. 1992. Hydrodesulfurization of Sulfur-Containing Polyaromatic Compounds in Light Oil. *Ind. Eng. Chem. Res.* 31: 1577–1580.

Kabe, T., Ishihara, A., and Zhang, Q. 1993. Deep Desulfurization of Light Oil. Part 2: Hydrodesulfurization of Dibenzothiophene, 4-Methyldibenzothiophene and 4,6-Dimethyldibenzothiophene. *Appl. Catal. A-Gen.* 97: L1–L9.

Kabe, T., Aoyama, Y., Wang, D., Ishihara, A., Qian, W., Hosoya, M., and Zhang, Q. 2001. Effects of H_2S on Hydrodesulfurization of Dibenzothiophene and 4,6-Dimethyldibenzothiophene on Alumina-Supported NiMo and NiW Catalysts. *Appl. Catal. A-Gen.* 209: 237–247.

Kapteijn, F., Nijhuis, T.A., Heiszwolf, J.J., and Moulijn, J.A. 2001. New Non-Traditional Multiphase Catalytic Reactors Based on Monolithic Structures. *Catal. Today* 66: 133–144.

Keller, W.D. 1985. Clays. In: *Kirk Othmer Concise Encyclopedia of Chemical Technology*, M. Grayson (Editor). Wiley Interscience, New York, p. 283.

Landau, M.V., Berger, D., and Herskowitz, M. 1996. Hydrodesulfurization of Methyl-Substituted Dibenzothiophenes: Fundamental Study of Routes to Deep Desulfurization. *J. Catal.* 159: 236–245.

LePage, J.F., Chatila, S.G., and Davidson, M. 1992. *Resid and Heavy Oil Processing*. Editions Technip, Paris, France.

Loupy, A. 2006. *Microwaves in Organic Synthesis*, 2nd Edition. Wiley-VCH Verlag GmbH & Co kGaA, Weinheim, Germany.

Magaril, R.Z., and Aksenova, E.L. 1967. Mechanism of Coke Formation during the Cracking of Petroleum Tars. *Izv. Vyssh. Zaved., Neft Gaz.* 10(11): 134–136.

Magaril, R.A., and Aksenova, E.L. 1968. Study of the Mechanism of Coke Formation in the Cracking of Petroleum Resins. *Int. Chem. Eng.* 8: 727–729 [first published in *Vysshikh Uchebn. Zavendenii, Neft Gaz.* 11: 134–136, 1967].

Magaril, R.Z., and Ramazaeva, L.F. 1969. Study of Coke Formation in the Thermal Decomposition of Asphaltene Constituents in Solution. *Izv. Vyssh. Ucheb. Zaved., Neft Gaz.* 12(1): 61–64.

Magaril, R.Z., and Aksenova, E.I. 1972. Coking Kinetics and Mechanism of the Thermal Decomposition of Asphaltene Constituents. *Khim. Tekhnol. Tr. Tyumen. Ind. Inst.* 169–172.

Magaril, R.Z., Ramazaeva, L.F., and Askenova, E.I. 1970. Kinetics of Coke Formation in the Thermal Processing of Petroleum. *Khim. Tekhnol. Topliv Masel.* 15(3): 15–16.

Magaril, R.Z., Ramazeava, L.F., and Aksenova, E.I. 1971. Kinetics of Coke Formation in the Thermal Processing of Crude Oil. *Int. Chem. Eng.* 11: 250.

Masel, R.I. 1995. *Principles of Adsorption and Reaction on Solid Surfaces*. John Wiley & Sons Inc., New York.

McHale, W.D. 1981. Process for Removing Sulfur from Petroleum Oils. United States Patent 4,283,270.

Miadonye, A., Snow, S., Irwin, D.J.G., and Khan, M.R. 2009. Desulfurization of Heavy Crude Oil by Microwave Irradiation. *WIT Trans. Eng. Sci.* 63: 455–465.

Mokhatab, S., Poe, W.A., and Speight, J.G. 2006. *Handbook of Natural Gas Transmission and Processing*. Elsevier, Amsterdam.

Monticello, D.J. 1998. Biodesulfurization of Diesel Fuels. *Chem. Tech.* 28(7): 38–45.

Mushrush, G.W., and Speight, J.G. 1995. *Petroleum Products: Instability and Incompatibility*. Taylor & Francis Publishers, Philadelphia, PA.

Mushrush, G.W., and Speight, J.G. 1998. Instability and Incompatibility of Petroleum Products. In: *Petroleum Chemistry and Refining*, J.G. Speight (Editor). Taylor & Francis, Washington, DC, Chapter 8.

Mutyala, S., Fairbridge, C., Pare, J.R.J., Belanger, J.M.R., Hawkins, S., and Ng, R. 2010. Microwave Applications to Oil Sands and Petroleum: A Review. *Fuel Process. Technol.* 91: 127–135.

Neurock, M., Libanati, C., Nigam, A., and Klein, M.T. 1990. Monte Carlo Simulation of Complex Reaction Systems: Molecular Structure and Reactivity in Modelling Heavy Oils. *Chem. Eng. Sci.* 45: 2083–2088.

Neurock, M., Nigam, A., Trauth, D., and Klein, M.T. 1994. Molecular Representation of Complex Hydrocarbon Feedstocks through Efficient Characterization and Stochastic Algorithms. *Chem. Eng. Sci.* 49: 4153–4177.

Nijhuis, T.A., Kreutzer, M.T., Romijn, A.C.J., Kapteijn, F., and Moulijn, J.A. 2001. Monolithic Catalysts as More Efficient Three-Phase Reactors. *Catal. Today* 66: 157–165.

Occelli, M.L., and Robson, H.E. (Editors) 1989. *Zeolite Synthesis*. Symposium Series No. 398. American Chemical Society, Washington, DC.

O'Connor, P., Verlaan, J.P.J., and Yanik, S.J. 1998. Challenges, Catalyst Technology and Catalytic Solutions in Resid FCC. *Catal. Today* 43(3–4): 305–313.

Ozaki, H., Satomi, Y., and Hisamitsu, T. 1963. Studies on Coking of Residual Oils. In: *Proceedings 6th World Petroleum Congress*, Vol. 6, p. 97.

Parkash, S. 2003. *Refining Processes Handbook*. Gulf Professional Publishing, Elsevier, Amsterdam.

Pines, H. 1981. *The Chemistry of Catalytic Hydrocarbon Conversions*. Academic Press, New York.

Purta, D.A., Portnoff, M.A., Pourarian, F., Nasta, M.A., and Zhang, J. 2004. Catalyst for the Treatment of Organic Compounds. United States Patent Application 2004077485.

Qu, L., Zhang, W., Kooyman, P.J., and Prins, R. 2003. MAS NMR, TPR, and TEM Studies of the Interaction of NiMo with Alumina and Silica-Alumina Supports. *J. Catal.* 215: 7–13.

Reinhoudt, H.R., Troost, R., Van Langeveld, A.D., Sie, S.T., Van Veen, J.A.R., and Moulijn, J.A. 1999. Catalysts for Second-Stage Deep Hydrodesulfurization of Gas Oils. *Fuel Process. Technol.* 61: 133–147.

Reynolds, J.G. 1998. Metals and Heteroatoms in Heavy Crude Oils. In: *Petroleum Chemistry and Refining*, J.G. Speight (Editor). Taylor & Francis Publishers, Washington, DC, Chapter 3.

Robinson, W.R.A.M., Van Veen, J.A.R., De Beer, V.H.J., and Van Santen, R.A. 1999a. Development of Deep Hydrodesulfurization Catalysts: I. CoMo and NiMo Catalysts Tested with (Substituted) Dibenzothiophene. *Fuel Process. Technol.* 61: 89–101.

Robinson, W.R.A.M., Van Veen, J.A.R., De Beer, V.H.J., and Van Santen, R.A. 1999b. Development of Deep Hydrodesulfurization Catalysts: II. NiW, Pt and Pd Catalysts Tested with (Substituted) Dibenzothiophene. *Fuel Process. Technol.* 61: 103–116.

Samorjai, G.A. 1994. *Introduction to Surface Chemistry and Catalysis*. John Wiley & Sons Inc., Hoboken, NJ.

Schabron, J.F., and Speight, J.G. 1997. An Evaluation of the Delayed Coking Product Yield of Heavy Feedstocks Using Asphaltene Content and Carbon Residue. *Rev. Inst. Fr. Petrole* 52: 73.

Schabron, J.F., and Speight, J.G. 1998. The Solubility and Three-Dimensional Structure of Asphaltene Constituents. *Petrol. Sci. Technol.* 16: 361.

Scott, W., and Bridge, A.G. 1971. The Continuing Development of Hydrocracking. In: *Origin and Refining of Petroleum*, H.G. McGrath and M.E. Charles (Editors). Advances in Chemistry Series No. 103. American Chemical Society, Washington, DC, Chapter 6, pp. 113–129.

Seki, H., and Yoshimoto, M. 2001. Deactivation of HDS Catalyst in Two-Stage RDS Process: II. Effect of Crude Oil and Deactivation Mechanism. *Fuel Process. Technol.* 69: 229–238.

Shang, H., Du, W., Liu, Z., and Zhang, H. 2013. Development of Microwave Induced Hydrodesulfurization of Petroleum Streams: A Review. *J. Ind. Eng. Chem.* 19: 1061–1068.

Song, C. 1999. Designing Sulfur-Resistant, Noble-Metal Hydrotreating Catalysts. *Chemtech* 3: 26–30.

Song, C., and Ma, X. 2003. New Design Approaches to Ultra-Clean Diesel Fuels by Deep Desulfurization and Deep Dearomatization. *Appl. Cat. B: Environ.* 41(1–2): 207–238.

Song, C.S. 2003. An Overview of New Approaches to Deep Desulfurization for Ultra-Clean Gasoline, Diesel Fuel and Jet Fuel. *Catal. Today* 86: 211–263.

Speight, J.G. 1984. Upgrading Heavy Oils and Residua: The Nature of the Problem. In: *Catalysis on the Energy Scene*, S. Kaliaguine and A. Mahay (Editors). Elsevier, Amsterdam, p. 515.

Speight, J.G. 1987. Initial Reactions in the Coking of Residua. *Prepr. Div. Petrol. Chem. Am. Chem. Soc.* 32(2): 413.

Speight, J.G. 1989. Thermal Decomposition of Asphaltene Constituents. *Neftekhimiya* 29: 732.

Speight, J.G. 1992. A Chemical and Physical Explanation of Incompatibility during Refining Operations. In: *Proceedings 4th International Conference on the Stability and Handling of Liquid Fuels*. US Department of Energy (DOE/CONF-911102), p. 169.

Speight, J.G. 1994. Chemical and Physical Studies of Petroleum Asphaltene Constituents. In: *Asphalts and Asphaltene Constituents*, Vol. 1, T.F. Yen and G.V. Chilingarian (Editors). Elsevier, Amsterdam, Chapter 2.

Speight, J.G. 1998a. *Petroleum Chemistry and Refining*, J.G. Speight (Editor). Taylor & Francis Publishers, Washington, DC, Chapter 5.

Speight, J.G. 1998b. The Chemistry and Physical of Coking. *Korean J. Chem. Eng.* 15(1): 1–8.

Speight, J.G. 2000. *The Desulfurization of Heavy Oils and Residua*. Marcel Dekker Inc., New York.

Speight, J.G. 2013. *Heavy and Extra Heavy Oil Upgrading Technologies*. Gulf Professional Publishing, Elsevier, Oxford, UK.

Speight, J.G. 2014. *The Chemistry and Technology of Petroleum*, 5th Edition. CRC Press, Taylor & Francis Group, Boca Raton, FL.

Speight, J.G. 2015. *Handbook of Petroleum Product Analysis*, 2nd Edition. John Wiley & Sons Inc., Hoboken, New Jersey.

Speight, J.G., and Long, R.B. 1996. The Concept of Asphaltene Constituents Revisited. *Fuel Sci. Technol. Int.* 14: 1.

Speight, J.G., and Moschopedis, S.E. 1979. The Production of Low-Sulphur Liquids and Coke from Athabasca Bitumen. *Fuel Process. Technol.* 2: 295.

Speight, J.G., and Ozum, B. 2002. *Petroleum Refining Processes*. Marcel Dekker Inc., New York.

Speight, J.G., Wernick, D.L., Gould, K.A., Overfield, R.E., Rao, B.M.L., and Savage, D.W. 1985. Molecular Weights and Association of Asphaltene constituents: A Critical Review. *Rev. Inst. Fr. Pétrol.* 40: 51.

Stanislaus, A., and Cooper, B.H. 1994. Aromatic Hydrogenation Catalysis: A Review. *Catal. Rev. Sci. Eng.* 36: 75–123.

Storm, D.A., Decanio, S.J., Edwards, J.C., and Sheu, E.Y. 1997. Sediment Formation during Heavy Oil Upgrading. *Petrol. Sci. Technol.* 15: 77.

Swaddle, T.W. 1997. *Inorganic Chemistry.* Academic Press Inc., New York.

Theng, B.K.G. 1974. *The Chemistry of Clay-Organic Reactions.* John Wiley & Sons Inc., New York.

Wan, J.K.S., and Kriz, J.F. 1985. Hydrodesulfurization of Hydrocracked Pitch. United States Patent 4,545,879.

Wiehe, I.A. 1992. A Solvent-Resid Phase Diagram for Tracking Resid Conversion. *Ind. Eng. Chem. Res.* 31: 530–536.

Wiehe, I.A. 1993. A Phase-Separation Kinetic Model for Coke Formation. *Ind. Eng. Chem. Res.* 32: 2447–2454.

Wiehe, I.A. 1994. The Pendant-Core Building Block Model of Petroleum Residua. *Energy Fuels* 8: 536–544.

Wild, P.J., Nyqvist, R.G., Bruijn, F.A., and Stobbe, E.R. 2006. Removal of Sulphur-Containing Odorants from Fuel Gases for Fuel Cell-Based Combined Heat and Power Applications. *J. Power Sources* 159: 995–1002.

2 Feedstocks

2.1 INTRODUCTION

Petroleum is scattered throughout the earth's crust, which is divided into natural groups or strata, categorized in order of their antiquity (Table 2.1) (Speight, 2014a). These divisions are recognized by the distinctive systems of organic debris (as well as fossils, minerals, and other characteristics) that form a chronological time chart that indicates the relative ages of the earth's strata. It is generally acknowledged that carbonaceous materials such as petroleum occur in all these geological strata from the Precambrian to the recent, and the origin of petroleum within these formations is a question that remains open to conjecture and the basis for much research. The answer cannot be given in this text, nor for that matter can it be presented in any advanced treatise.

Petroleum is by far the most commonly used source of energy, especially as the source of liquid fuels (Speight, 2008, 2014a). Indeed, because of the wide use of petroleum, the past 100 years could very easily been variously called the "Oil Century," the "Petroleum Era," or the "New Rock Oil Age" (Speight, 2011a,c). Currently, the majority of the energy consumed by humans is produced from the fossil fuels with smaller amounts of energy coming from nuclear and hydroelectric sources. The continuing and expanding use of petroleum has led to various projections about its longevity. Some projections are disastrous and predict that petroleum availability is very short lived, while other projections are not quite so disastrous (Speight, 2011a). In addition, energy from alternate sources is projected to be stable while energy from other sources (including renewable sources) is projected to show steady growth but not sufficient to influence the consumption of petroleum as an energy source. As a result, fossil fuels are projected to be the major sources of energy for the next 50 years (Speight, 2011a,b, 2014a). In this respect, petroleum and its associates (heavy oil and bitumen) are extremely important in any energy scenario, especially those scenarios that relate to the production of liquid fuels.

The growing interest in *renewable energy sources* has been prompted, in part, by increasing concern over the pollution, resource depletion, and possible climate change (Speight, 1996a, 2011b). However, it will more than likely be the mid-21st century before substantial amounts of fuels are available from nonfossil fuel sources. The precise fraction of energy that will be provided by renewable energy sources beyond the year 2050 is not known and is difficult to estimate. Until that time, the world must be prepared to use available energy sources in the most efficient and environmentally acceptable manner.

It is a fact that in recent years, the average quality of crude oil has become worse. This is reflected in a progressive decrease in American Petroleum Institute (API) gravity (i.e., increase in density) and an increase in sulfur content and the tendency for the quality of crude oil feedstocks to deteriorate (Speight, 2011a). Nevertheless, the nature of crude oil refining has been changed considerably (Speight and Ozum, 2002; Hsu and Robinson, 2006; Gary et al., 2007; Speight, 2008, 2011b, 2014a). This, of course, has led to the need to manage crude quality more effectively through evaluation and desired product slate (Speight, 2001, 2014a, 2015). Indeed, the declining reserves of lighter crude oil have resulted in an increasing need to develop options (Speight, 2014a) to desulfurize and upgrade the heavy feedstocks, specifically heavy oil and bitumen. This has resulted in a variety of process options that specialize in sulfur removal during refining (Speight and Ozum, 2002; Hsu and Robinson, 2006; Gary et al., 2007; Speight, 2008, 2014a). It is worthy of note at this point that microbial desulfurization is becoming a recognized technology for desulfurization (Monticello, 1995; Armstrong et al., 1997).

TABLE 2.1

General Description and Approximate Age of Geological Strata

Era	Period	Epoch	Age (Years × 10^6)
Cenozoic	Quaternary	Recent	0.01
		Pleistocene	3
	Tertiary	Pliocene	12
		Miocene	25
		Oligocene	38
		Eocene	55
		Paleocene	65
Mesozoic	Cretaceous		135
	Jurassic		180
	Triassic		225
Paleozoic	Permian		275
	Carboniferous		
	Pennsylvanian		350
	Mississippian		
	Devonian		413
	Silurian		430
	Ordovician		500
	Cambrian		600

Source: Speight, J.G. 2000. *The Desulfurization of Heavy Oils and Residua.* Marcel Dekker Inc., New York. Table 2.1, p. 2.

The increasing supply of heavy crude oils is a matter of serious concern for the petroleum industry. To satisfy the changing pattern of product demand, significant investments in refining conversion processes will be necessary to profitably utilize these heavy crude oils (Speight and Ozum, 2002; Hsu and Robinson, 2006; Gary et al., 2007; Speight, 2008, 2014a). Although the most efficient and economical solution to this problem will depend, to a large extent, on individual country and company situations, the most promising technologies will likely involve the conversion of vacuum residua and extra heavy crude oil (bitumen) into light- and middle-distillate products.

Petroleum is not a uniform material. In fact, its chemical and physical (fractional) composition can vary not only with the location and age of the oil field but also with the depth of the individual well. Indeed, two adjacent wells may produce petroleum with markedly different characteristics. On a molecular basis, petroleum is a complex mixture of hydrocarbons with small amounts of organic compounds containing sulfur, oxygen, and nitrogen, as well as compounds containing metallic constituents, particularly vanadium nickel, iron, and copper (Table 2.2), all of which must be given consideration when refining options are planned. The hydrocarbon content may be as high as 97% w/w, e.g., in a light paraffinic crude oil, or as low as 50% w/w in heavy crude oil and bitumen (Speight, 1999, 2001, 2014a, 2015).

The fuels that are derived from petroleum supply more than half of the world's total supply of energy. Naphtha, kerosene, and diesel oil provide fuel for automobiles, tractors, trucks, aircraft, and ships. Fuel oil and natural gas are used to heat homes and commercial buildings, as well as to generate electricity. Petroleum products are the basic materials used for the manufacture of synthetic fibers for clothing and in plastics, paints, fertilizers, insecticides, soaps, and synthetic rubber. The uses of petroleum as a source of raw material in manufacturing are central to the functioning of modern industry.

Petroleum is a carbon-based resource. Therefore, the geochemical carbon cycle is also of interest to fossil fuel usage in terms of petroleum formation, use, and the buildup of atmospheric carbon

TABLE 2.2
Compound Types in Petroleum

Class	Compound Types
Saturated hydrocarbons	*n*-Paraffins
	Isoparaffins and other branched paraffins
	Cycloparaffins (naphthenes)
	Condensed cycloparaffins (including steranes and hopanes)
	Alkyl side chains on ring systems
Unsaturated hydrocarbons	Olefins not indigenous to petroleum; present in products of thermal reactions
Aromatic hydrocarbons	Benzene systems
	Condensed aromatic systems
	Condensed aromatic-cycloalkyl systems
	Alkyl side chains on ring systems
Saturated heteroatomic systems	Alkyl sulfides
	Cycloalkyl sulfides
	Alkyl side chains on ring systems
Aromatic heteroatomic systems	Furans (single-ring and multi-ring systems)
	Thiophenes (single-ring and multi-ring systems)
	Pyrroles (single-ring and multi-ring systems)
	Pyridines (single-ring and multi-ring systems)
	Mixed heteroatomic systems
	Amphoteric (acid–base) systems
	Alkyl side chains on ring systems
Metals	Nickel, vanadium
	Iron, copper[a]

[a] The presence of these and other metals may be dependent on the history of the crude oil after production.

dioxide. Petroleum technology, in one form or another, is with us until suitable alternative forms of energy are readily available (Speight, 2008, 2011b). It is for this reason that the more efficient use of petroleum and, indeed, the conversion of petroleum to sulfur-free products is of paramount importance.

Therefore, a thorough understanding of the benefits and limitations of the conversion of petroleum to liquid products and desulfurization during conversion is necessary and will be introduced within the pages of this book. However, the efficient conversion and desulfurization of petroleum, heavy oils, and residua is not merely a matter of passing the feedstock through a series of reactors. A thorough knowledge of the behavior of the feedstock under refinery conditions is not only beneficial but also necessary.

With the necessity of processing heavy oil, bitumen, and residua to obtain more gasoline and other liquid fuels, it has been recognized that knowledge of the constituents of these higher boiling feedstocks is also of some importance. Indeed, the problems encountered in processing the heavier feedstocks can be equated to the chemical character and the amount of complex, higher-boiling constituents in the feedstock. Refining these materials is not just a matter of applying know-how derived from refining conventional crude oils but requires knowledge of the chemical structure and chemical behavior of these more complex constituents (Speight and Ozum, 2002; Hsu and Robinson, 2006; Gary et al., 2007; Speight, 2008, 2014a).

However, heavy crude oil and bitumen are extremely complex, and very little direct information can be obtained by distillation. It is not possible to isolate and identify the constituents of the heavier feedstocks (using analytical techniques that rely on volatility). Other methods of identifying the chemical constituents must be employed. Such techniques include a myriad of fractionation procedures (Speight, 2001, 2014a, 2015) as well as methods designed to draw inferences about the hydrocarbon skeletal structures and the nature of the heteroatomic functions.

Petroleum (and the related materials heavy oil and bitumen) originates from living matter and exists in the solid, liquid, or gaseous state. Petroleum is formed in the ground by chemical and physical changes in plant (mainly) and animal residues under various conditions of temperature and pressure over long geological time periods (Speight, 1999). Carbon and hydrogen are the major elemental constituents with minor amounts of nitrogen, oxygen, sulfur, and metals. The types and relative amounts of hydrocarbons and other chemical compounds (compounds of carbon and hydrogen that also contain nitrogen, oxygen, sulfur, and/or metals) present in petroleum from different sources vary widely, but except for gasoline, the ultimate compositions and heating values fall within fairly narrow limits.

Of the data that are available (Speight, 1999), the ultimate composition of petroleum and heavy oil vary over fairly narrow limits of elemental distribution:

Carbon: 83.0–87.0% w/w
Hydrogen: 10.0–14.0% w/w
Nitrogen: 0.1–2.0% w/w
Oxygen: 0.05–1.5% w/w
Sulfur: 0.05–6.0% w/w
Metals: Ni plus V, <1000 ppm w/w; Fe plus Cu, <200 ppm w/w

The origin of iron (Fe) and copper (Cu) is debatable; one theory suggests that they arose from the original precursors, while another theory proposes that these two metals may originate from contact with steel piping and other containers. Both theories remain unproven and both may be correct. However, it is necessary to acknowledge the presence of these two metals as both can influence catalyst behavior during refining, regardless of their origin (Speight and Ozum, 2002; Hsu and Robinson, 2006; Gary et al., 2007; Speight, 2008, 2014a).

The elemental composition is fairly constant despite the wide variation in physical properties from the lighter, more mobile crude oils at one extreme to the heavy asphaltic crude oils at the other extreme. Indeed, when the many localized or regional variations in maturation conditions are assessed, it is perhaps surprising that the ultimate compositions are very similar. Perhaps this observation, more than any other observation, is indicative of the similarity in nature of the precursors from one site to another.

Because of the narrow range of carbon and hydrogen content, it is not possible to classify petroleum or heavy oil on the basis of carbon content, as coal is classified; carbon contents of coal can vary from as low as 75% w/w in lignite to 95% w/w in anthracite (Speight, 2013a). Of course, other subdivisions are possible within the various carbon ranges of the coals; however, petroleum is restricted to a much narrower range of elemental composition. However, it is possible to compare the atomic hydrogen-to-carbon (H/C) ratio of the heavy feedstocks with petroleum (Figure 2.1). This gives a general indication of the molar volumes of hydrogen that are required for upgrading but does not give any indication of the chemistry of upgrading.

The elemental analysis of oil sand bitumen (extra heavy oil) has also been widely reported (Speight, 1990); however, the data suffer from the disadvantage that identification of the source is too general (i.e., Athabasca bitumen, which covers several deposits) and is often not site specific. In addition, the analysis is quoted for separated bitumen, which may have been obtained by any one of several procedures and may therefore not be representative of the total bitumen on the sand. However, recent efforts have focused on a program to produce sound, reproducible data from samples for which the origin is carefully identified (Wallace et al., 1988; Speight, 2015). It is to be hoped that this program continues, as it will provide a valuable database for tar sand and bitumen characterization.

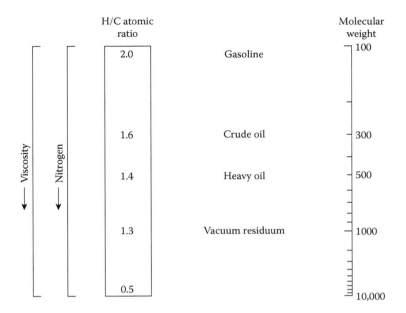

FIGURE 2.1 Representation of the atomic carbon/hydrogen ratio of various feedstocks. (From Speight, J.G. 2000. *The Desulfurization of Heavy Oils and Residua*. Marcel Dekker Inc., New York. Figure 2.2, p. 7.)

Like conventional petroleum, of the data that are available, the elemental composition of oil sand bitumen is generally constant and, like the data for petroleum, falls into a narrow range (Speight, 2014a):

Carbon: 83.4 ± 0.5% w/w
Hydrogen: 10.4 ± 0.2% w/w
Nitrogen: 0.4 ± 0.2% w/w
Oxygen: 1.0 ± 0.2% w/w
Sulfur: 5.0 ± 0.5% w/w
Metals, Ni plus V: >1000 ppm w/w; Fe plus Cu: 200 ± 50 ppm w/w

The major exception to these narrow limits is the oxygen content of bitumen, which can vary from as little as 0.2% to as high as 4.5%. This is not surprising, since when oxygen is estimated by difference, the analysis is subjected to the accumulation of all of the errors in the other elemental data. In addition, bitumen is susceptible to aerial oxygen and the oxygen content is very dependent on the sample history. In addition, the ultimate composition of the Alberta bitumen does not appear to be influenced by the proportion of bitumen in the oil sand or by the particle size of the tar sand minerals.

The *heating values* of petroleum, heavy oil, and bitumen are measured as the *gross heats of combustion* that are given with reasonable accuracy by the equation

$$Q = 12,400 - 2100d^2$$

In this equation, d is the 60/60°F specific gravity. Deviation is generally <1%, although many highly aromatic crude oils show considerably higher values. The ranges for petroleum, heavy oil, and bitumen are 10,000–11,600 cal/g (Speight, 1999). For comparison, the heat of combustion of gasoline is 11,000–11,500 cal/g and for kerosene (and diesel fuel) the heat of combustion falls in the range 10,500–11,200 cal/g. The heat of combustion for fuel oil is on the order of 9500–11,200 cal/g.

With all of the scenarios in place, there is no doubt that petroleum and its relatives—residua, heavy oil, and extra heavy oil (bitumen)—will be required to produce a considerable proportion of liquid fuels into the foreseeable future. Desulfurization processes will be necessary to remove sulfur in an environmentally acceptable manner to produce environmentally acceptable products. Refining strategies will focus on upgrading the heavy oils and residua, and will emphasize the differences between the properties of the heavy crude feedstocks. This will dictate the choice of methods or combinations thereof for conversion of these materials to products currently and in the future (Speight and Ozum, 2002; Hsu and Robinson, 2006; Gary et al., 2007; Speight, 2008, 2011b, 2014a).

Refinery processes (Figure 2.2) can be categorized as hydrogen addition processes (e.g., hydroprocesses such as hydrotreating and hydrocracking, hydrovisbreaking, and donor–solvent processes) and as carbon rejection processes (e.g., catalytic cracking, coking, visbreaking, and other processes such as solvent deasphalting). All have serious disadvantages when applied singly to upgrading of heavy oils or residua. Removal of heteroatoms and metals by exhaustive hydrodenitrogenation (HDN), hydrodesulfurization (HDS), and hydrodemetallization (HDM) is very expensive. The catalytic processes suffer from the disadvantage of excessive catalyst use due to metal and carbon deposition. Development of more durable and cheaper catalysts and additives to prevent coke laydown has not yet solved the problem. The noncatalytic processes alone yield uneconomically large amounts of coke.

Both hydrogen addition and carbon rejection processes will be necessary in any realistic scheme of heavy oil upgrading (Speight, 2013c, 2014a). Most coker products require hydrogenation, and most hydrotreated products require some degree of fractionation. For example, to maximize yields of transport fuels from Maya crude, efficient carbon rejection followed by hydrogenation may be

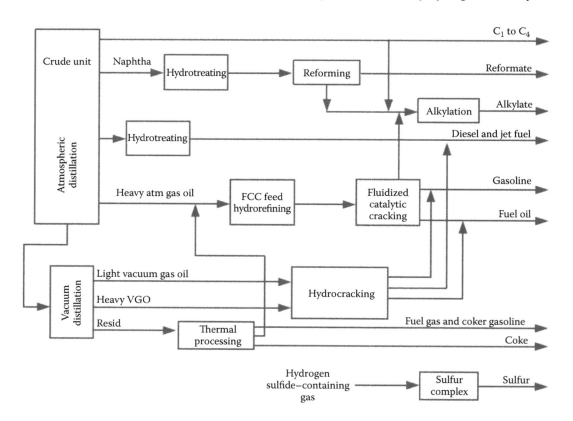

FIGURE 2.2 General flow diagram for a typical refinery showing the distillation section (atmospheric distillation and vacuum distillation) and the hydrotreating (desulfurization) sections.

necessary. There are various other approaches to the processing of other heavy oil residua (Bakshi and Lutz, 1987). As of now, it is not known which combination of processes best converts a heavy feedstock into salable products.

HDS and HDM activities cannot be predicted by such conventional measurements as total sulfur, metals, or asphaltene contents, or Conradson carbon values (Dolbear et al., 1987). To choose effective processing strategies, it is necessary to determine properties from which critical reactivity indices can be developed. Indeed, properties of heavy oil vacuum residua determined by conventional methods are not good predictors of behavior of substrates in upgrading processes (Speight and Ozum, 2002; Hsu and Robinson, 2006; Gary et al., 2007; Speight, 2008, 2011b, 2014a). The properties of residua vary widely, and the existence of relatively large numbers of polyfunctional molecules results in molecular associations that can affect reactivity. Therefore, it is evident that more knowledge is needed about the components of residua that cause specific problems in processing, and how important properties change during processing (Speight, 2014a).

In summary, upgrading heavy oils and residua must, at some stage of the refinery operations, utilize HDS. Indeed, HDS processes are used at several places in virtually every refinery (1) to protect catalysts, (2) to meet product specifications related to refinery processes, and (3) to conform to environmental regulations (Table 2.3) (Speight, 1996b, 1999). Thus, several types of chemistry might be anticipated as occurring during HDS (Ancheyta and Speight, 2007; Speight, 2014a). Similarly, HDN is commonly used only in conjunction with hydrocracking, to protect catalysts. Other hydrotreating processes are used to saturate olefins and aromatics to meet product specifications or to remove metals from residual oils.

TABLE 2.3
Outcome of Hydroprocesses during Refining

Reaction	Feedstock	Purpose
HDS	Catalytic reformer feedstocks	Reduce catalyst poisoning
	Diesel fuel	Environmental specifications
	Distillate fuel oil	Environmental specifications
	Hydrocracker feedstocks	Reduce catalyst poisoning
	Coker feedstocks	Reduce sulfur content of coke
HDN	Lubricating oil	Improve stability
	Catalytic cracking feedstocks	Reduce catalyst poisoning
	Hydrocracker feedstocks	Reduce catalyst poisoning
HDM	Catalytic cracking feedstocks	Avoid metals deposition
		Avoid coke buildup
		Avoid catalyst destruction
	Hydrocracker feedstocks	Avoid metals deposition
		Avoid coke buildup
		Avoid catalyst destruction
CRR	Catalytic cracker feedstocks	Reduce coke buildup on catalyst
	Residua	Reduce coke yield
	Heavy oils	Reduce coke yield

Source: Speight, J.G. 2000. *The Desulfurization of Heavy Oils and Residua.* Marcel Dekker Inc., New York. Table 2.3, p. 10.

Note: CRR, Conradson carbon residue; HDM, hydrodemetallization; HDN, hydrodenitrogenation; HDS, hydrodesulfurization.

2.2 NATURAL FEEDSTOCKS

Definitions are the means by which scientists and engineers communicate the nature of a material to each other and to the world, through either the spoken or the written word. Thus, the definition of a material can be extremely important and have a profound influence on how the technical community and the public perceive that material.

Historically, there may be the impression that the industry is coming full circle. The early uses of petroleum focused on the use of the higher-boiling (heavier) constituents that remained after the more volatile fractions had evaporated or been removed under the prevailing conditions. This leads to a series of definitions by which petroleum and a variety of other materials are defined.

2.2.1 PETROLEUM

Petroleum and the equivalent term *crude oil* cover a wide assortment of materials consisting of mixtures of hydrocarbons and other compounds containing variable amounts of sulfur, nitrogen, and oxygen, which may vary widely in volatility, specific gravity, and viscosity. Metal-containing constituents, notably those compounds that contain vanadium and nickel, usually occur in the more viscous crude oils in amounts up to several thousand parts per million (ppm) and can have serious consequences during processing of these feedstocks (Speight, 2014a and references cited therein). Because petroleum is a mixture of widely varying constituents and proportions, its physical properties also vary widely, and the color from colorless to black (Speight and Ozum, 2002; Hsu and Robinson, 2006; Gary et al., 2007; Speight, 2008, 2011b, 2014a).

Petroleum is by far the most commonly used source of energy, especially as the source of liquid fuels. In recent years, the average quality of crude oil has become worse. This is reflected in a progressive decrease in API gravity (i.e., increase in density) and an increase in sulfur content. Thus, the nature of crude oil refining has been changed considerably and will continue to evolve (Speight and Ozum, 2002; Hsu and Robinson, 2006; Gary et al., 2007; Speight, 2008, 2011a,b, 2014a).

The declining reserves of lighter crude oil has resulted in an increasing need to develop options to upgrade the abundant supply of known heavy oil reserves (Speight and Ozum, 2002; Hsu and Robinson, 2006; Gary et al., 2007; Speight, 2008, 2011a, 2014a). In addition, there is considerable focus and renewed efforts on adapting recovery techniques to the production of heavy oil (Speight, 2011a, 2014a).

In addition, the crude oils available today to the refinery are quite different in composition and properties to those available some 30 years ago (Speight, 2011a, 2014a). The current crude oils are somewhat heavier insofar as they have higher proportions of nonvolatile (asphaltic) constituents. In fact, by the standards of yesteryear, many of the crude oils currently in use would have been classified as heavy feedstocks, bearing in mind that they may not approach the definitions used today for heavy crude oils. Changes in feedstock character, such as this tendency to heavier materials, require adjustments to refinery operations to handle these heavier crude oils, in order to reduce the amount of coke formed during processing and to balance the overall product slate. In fact, the outlook for liquid fuels into the next century and for the next several decades is believed to hinge on the availability and conversion of heavy oils, residua, extra heavy oil, and tar sand bitumen (Speight, 2008, 2011a, 2014a).

2.2.2 NATURAL GAS AND GAS CONDENSATE

The generic term *natural gas* applies to gases commonly associated with petroliferous (petroleum-producing, petroleum-containing) geologic formations. Natural gas generally contains high proportions of methane (a single carbon hydrocarbon compound, CH_4), and some of the higher molecular weight higher paraffins (C_nH_{2n+2}) generally containing up to six carbon atoms may also be present in small quantities (Table 2.2). The hydrocarbon constituents of natural gas are combustible;

however, nonflammable nonhydrocarbon components such as carbon dioxide, nitrogen, and helium are often present in the minority and are regarded as contaminants.

In addition to the natural gas fund in petroleum reservoirs, there are also those reservoirs in which natural gas may be the sole occupant. The principal constituent of natural gas is methane; however, other hydrocarbons, such as ethane, propane, and butane, may also be present. Carbon dioxide is also a common constituent of natural gas. Trace amounts of rare gases, such as helium, may also occur, and certain natural gas reservoirs are a source of these rare gases. Just as petroleum can vary in composition, so can natural gas. Differences in natural gas composition occur between different reservoirs, and two wells in the same field may also yield gaseous products that are different in composition (Mokhatab et al., 2006; Speight, 2007, 2014a).

There are several general definitions that have been applied to natural gas. Thus, *lean gas* is gas in which methane is the major constituent. *Wet gas* contains considerable amounts of the higher molecular weight hydrocarbons. *Sour gas* contains hydrogen sulfide, whereas *sweet gas* contains very little, if any, hydrogen sulfide. *Residue gas* is natural gas from which the higher molecular weight hydrocarbons have been extracted, and *casing head gas* is derived from petroleum but is separated at the separation facility at the wellhead.

To further define the terms *dry* and *wet* in quantitative measures, the term *dry natural gas* indicates that there is <0.1 gallon (1 gallon US = 264.2 m³) of gasoline vapor (higher molecular weight paraffins) per 1000 ft³ (1 ft³ = 0.028 m³). The term *wet natural gas* indicates that there are such paraffins present in the gas, in fact >0.1 gal/1000 ft³.

Associated or *dissolved natural gas* occurs either as free gas or as gas in solution in the petroleum. Gas that occurs as a solution in the petroleum is dissolved gas, whereas the gas that exists in contact with the petroleum (gas cap) is associated gas.

Other components such as carbon dioxide, hydrogen sulfide, mercaptan derivatives (thiol derivatives, RSH), and trace amounts of other constituents may also be present. Thus, there is no single composition of components that might be termed *typical* natural gas. Methane and ethane constitute the bulk of the combustible components; carbon dioxide and nitrogen are the major noncombustible (inert) components.

By definition, natural gas condensate (sometimes also referred to as *natural gasoline*) is a mixture of low-boiling liquid hydrocarbons isolated from petroleum and natural gas wells suitable for blending with light naphtha to produce gasoline (Mokhatab et al., 2006; Speight, 2007, 2014a). Because of the presence of low-boiling hydrocarbons, light naphtha can become extremely explosive, even at relatively low ambient temperatures. Some of these gases may be burned off (flared) at the field wellhead; however, others remain in the liquid products extracted from the well.

2.2.3 Opportunity Crudes

There is also the need for a refinery to be configured to accommodate *opportunity crude oils* and/or *high-acid crude oils*, which, for many purposes, are often included with heavy feedstocks (Speight, 2014a,b).

Opportunity crude oils are either new crude oils with unknown or poorly understood processing issues or are existing crude oils with well-known processing concerns. Opportunity crude oils are often, but not always, heavy crude oils but in either case are more difficult to desalt, most commonly due to high solids content, high levels of acidity, viscosity, electrical conductivity, or contaminants. They may also be oils that are incompatible, causing excessive equipment fouling when processed either in blends or separately.

Typically, opportunity crude oils are often dirty and need cleaning before refining by removal of undesirable constituents such as high-sulfur, high-nitrogen, and high-aromatic (such as polynuclear aromatic) components (Speight, 2014a,b). A controlled visbreaking treatment would clean up such crude oils by removing these undesirable constituents (which, if not removed, would cause problems further down the refinery sequence) as coke or sediment. There is also the need for a refinery to be configured to accommodate opportunity crude oils and/or high-acid crude oils, which, for many purposes, are often included with heavy feedstocks.

In addition to taking preventative measure for the refinery to process these feedstocks without serious deleterious effects on the equipment, refiners will need to develop programs for detailed and immediate feedstock evaluation so that they can understand the qualities of a crude oil very quickly and it can be valued appropriately, and management of the crude processing can be planned meticulously.

Compatibility of opportunity crudes with other opportunity crudes and with conventional crude oil and heavy oil is a very important property to consider when making decisions regarding which crude to purchase. Blending crudes that are incompatible can lead to extensive fouling and processing difficulties due to unstable asphaltene constituents (Speight, 2014a). These problems can quickly reduce the benefits of purchasing the opportunity crude in the first place. For example, extensive fouling in the crude preheat train may occur, resulting in decreased energy efficiency, increased emissions of carbon dioxide, and increased frequency at which heat exchangers need to be cleaned. In a worst-case scenario, crude throughput may be reduced leading to significant financial losses.

Opportunity crude oils, while offering initial pricing advantages, may have composition problems that can cause severe problems at the refinery, harming infrastructure, yield, and profitability. Before refining, there is the need for comprehensive evaluations of opportunity crudes, giving the potential buyer and seller the needed data to make informed decisions regarding fair pricing and the suitability of a particular opportunity crude oil for a refinery. This will assist the refiner to manage the ever-changing crude oil quality input to a refinery—including quality and quantity requirements and situations, crude oil variations, contractual specifications, and risks associated with such opportunity crudes.

2.2.4 HIGH-ACID CRUDES

One of the newest terms in the petroleum lexicon is arbitrarily (even erroneously) named *shale oil*, which is crude oil that is produced from tight shale formation and should not be confused with the shale oil that is produced by the thermal treatment of oil shale and the decomposition of kerogen contained therein (Speight, 2012). The tight shale formations are those same formations that produce gas (*tight gas*) (Speight, 2013b). The introduction of the term *shale oil* to define crude oil from tight shale formations is the latest to add confusion to the system of nomenclature of petroleum–heavy oil–bitumen materials. The term has been used without any consideration of the original-term shale oil produced by the thermal decomposition of kerogen in oil shale. It is not quite analogous, but is certainly similarly confusing, to the term *black oil* that has been used to define petroleum by color rather than by any meaningful properties.

Typical of the oil from tight shale formations is the Bakken crude oil, which is a light crude oil. Briefly, Bakken crude oil is a light sweet (low-sulfur) crude oil that has a relatively high proportion of volatile constituents. The production of the oil yields not only petroleum but also a significant amount of volatile gases (including propane and butane) and low-boiling liquids (such as pentane and natural gasoline), which are often referred to collectively as (low-boiling or light) naphtha (Speight, 2014a).

The liquid stream produced from the Bakken formation will include the crude oil, low-boiling liquids, and gases that were not flared, along with the materials and by-products of the fracking process. These products are then mechanically separated into three streams: (1) produced salt water, often referred to as brine; (2) gases; and (3) petroleum liquids, which include condensates, natural gas liquids, and light oil. Depending on the effectiveness and appropriate calibration of the separation equipment that is controlled by the oil producers, varying quantities of gases remain dissolved and/or mixed in the liquids, and the whole is then transported from the separation equipment to the well-pad storage tanks, where emissions of volatile hydrocarbons have been detected as emanating from the oil.

Bakken crude oil is considered to be a low-sulfur (sweet) crude oil, and there have been increasing observations of elevated levels of hydrogen sulfide in the oil. Hydrogen sulfide is a toxic, highly flammable, corrosive, explosive gas, and there have been increasing observations of elevated levels of hydrogen sulfide in Bakken oil.

High-acid crude oil is crude oil that contains considerable proportions of naphthenic acids, which, as commonly used in the petroleum industry, refers collectively to all of the organic acids present in the crude oil (Shalaby, 2005; Rikka, 2007). By the original definition, a naphthenic acid is a monobasic carboxyl group attached to a saturated cycloaliphatic structure. However, it has been a convention accepted in the oil industry that all organic acids in crude oil are called naphthenic acids. Naphthenic acids in crude oils are now known to be mixtures of low to high molecular weight acids, and the naphthenic acid fraction also contains other acidic species.

Naphthenic acids can be very water soluble to oil soluble depending on their molecular weight, process temperatures, salinity of waters, and fluid pressures. In the water phase, naphthenic acids can cause stable reverse emulsions (oil droplets in a continuous water phase). In the oil phase with residual water, these acids have the potential to react with a host of minerals, which are capable of neutralizing the acids. The main reaction product found in practice is the calcium naphthenate soap (the calcium salt of naphthenic acids). The total acid matrix is therefore complex, and it is unlikely that a simple titration, such as the traditional methods for measurement of the total acid number, can give meaningful results to use in predictions of problems. An alternative way of defining the relative organic acid fraction of crude oils is therefore a real need in the oil industry, both upstream and downstream.

In addition to taking preventative measure for the refinery to process these feedstocks without serious deleterious effects on the equipment, refiners will need to develop programs for detailed and immediate feedstock evaluation so that they can understand the qualities of a crude oil very quickly and it can be valued appropriately, and management of the crude processing can be planned meticulously (Mayes, 2015).

2.2.5 Oil from Tight Shale

As well as gas (Speight, 2013b), oil is also produced from tight shale formations, and there has been a tendency to refer to this oil as shale oil. However, this is not only incorrect terminology (according to the accepted technology) but also confusing and illogical, and the use of such terminology should be discouraged—shale oil is produced from oil shale by thermal decomposition of the kerogen contained therein (Speight, 1990, 2012, 2014a).

The challenges associated with the production of shale oils are a function of their compositional complexities and the varied geological formations where they are found. These oils are light but they are very waxy and reside in oil-wet formations. These properties create some of the main difficulties associated with oil extraction from the shale. Such problems include scale formation, salt deposition, paraffin wax deposits, destabilized asphaltene constituents, corrosion, and bacterial growth. Multicomponent chemical additives are added to the stimulation fluid to control these problems.

Oil from tight shale formation is characterized by low-asphaltene content, low-sulfur content, and a significant molecular weight distribution of the paraffinic wax content. Paraffin carbon chains of C_{10} to C_{60} have been found, with some shale oils containing carbon chains up to C_{72}. To control deposition and plugging in formations due to paraffins, dispersants are commonly used. In upstream applications, these paraffin dispersants are applied as part of multifunctional additive packages where asphaltene stability and corrosion control are also addressed simultaneously.

Scale deposits of calcite, carbonates, and silicates must be controlled during production, or plugging problems arise. A wide range of scale additives is available. These additives can be highly effective when selected appropriately. Depending on the nature of the well and the operational conditions, a specific chemistry is recommended or blends of products are used to address scale deposition.

Another challenge encountered with oil from tight shale formations is the transportation infrastructure. Rapid distribution of shale oils to the refineries is necessary to maintain consistent plant throughput. Some pipelines are in use, and additional pipelines are being constructed to provide consistent supply. During the interim, barges and railcars are being used, along with a significant expansion in trucking to bring these oils to the refineries. Eagle Ford production is estimated to increase by a factor of 6 from 350,000 bpd to approximately 2,000,000 bpd by 2017. Thus, a more reliable infrastructure is needed to distribute this oil to multiple locations. Similar expansion in oil production is estimated for Bakken and other identified (and perhaps as yet unidentified) tight shale formations.

2.2.6 HEAVY OIL

When petroleum occurs in a reservoir that allows the crude material to be recovered by pumping operations as a free-flowing dark- to light-colored liquid, it is often referred to as *conventional petroleum*. Heavy oils are the other types of petroleum that are different from conventional petroleum insofar as they are much more difficult to recover from the subsurface reservoir (Speight, 2009). The definition of heavy oils is usually based on the API gravity or viscosity, and the definition is quite arbitrary although there have been attempts to rationalize the definition based on viscosity, API gravity, and density. Heavy oil has a much higher viscosity (and lower API gravity) than conventional petroleum, and primary recovery of heavy oil usually requires thermal stimulation of the reservoir (Speight, 2001; Speight and Ozum, 2002; Hsu and Robinson, 2006; Gary et al., 2007; Speight, 2008, 2011a, 2014a).

In addition to the attempt to define petroleum, heavy oil, bitumen, and residua, there have been several attempts to classify these materials by the use of properties such as API gravity, sulfur content, or viscosity (Speight, 1999). However, any attempt to classify petroleum, heavy oil, and bitumen on the basis of a single property is no longer sufficient to define the nature and properties of petroleum and petroleum-related materials, perhaps even being an exercise in futility.

Classification will not be discussed in this text, having been described in more detail elsewhere (Speight, 2014a). Let it suffice to use general definitions for the purposes of this text.

For many years, petroleum and heavy oil were very generally defined in terms of physical properties. For example, heavy oils were considered to be those crude oils that had gravity of somewhat <20° API, with the heavy oils falling into the API gravity range 10–15°. For example, Cold Lake heavy crude oil has an API gravity equal to 12° and extra heavy oils, such as tar sand bitumen, usually have an API gravity in the range 5–10° (Athabasca bitumen = 8° API). Residua would vary depending on the temperature at which distillation was terminated; however, usually, vacuum residua are in the range of 2–8° API.

In a more general sense, the generic term *heavy oil* is often applied to a petroleum that has an API gravity of <20° and usually, but not always, a sulfur content >2% by weight. In contrast to conventional crude oils, heavy oils are darker in color and may even be black. The term *heavy oil* has also been arbitrarily used to describe both the heavy oils that require thermal stimulation of recovery from the reservoir and the bitumen in bituminous sand (tar sand) formations from which the heavy bituminous material is recovered by a mining operation.

However, the term *extra heavy oil* is used to define the subcategory of petroleum that occurs in the near-solid state and is incapable of free flow under ambient conditions. Bitumen from tar sand deposits is often termed as extra heavy oil.

2.2.7 EXTRA HEAVY OIL

Extra heavy oil is a term that has arisen recently as a means of differentiation between material that occurs in the solid or near-solid state and generally has mobility under reservoir conditions and tar sand bitumen, which is immobile under deposit conditions (Speight, 2009, 2014a). In terms of scientific meaning, the term is questionable and is subjected to much verbal and written variation. While this material may resemble tar sand bitumen and does not flow easily, extra heavy oil can generally recognized (for the purposes of this text) as being more viscous than heavy oil but having mobility in the reservoir—tar sand bitumen is typically incapable of mobility (free flow) under the conditions in the tar sand deposit.

For example, the tar sand bitumen located in Alberta, Canada, is not mobile in the deposit and requires extreme methods to recover the bitumen. On the other hand, much of the extra heavy oil located in the Orinoco belt of Venezuela requires recovery methods that are less extreme because of the mobility of the material in the reservoir. Whether the mobility of extra heavy oil is due to a high reservoir temperature (that is higher than the pour point of the extra heavy oil) or due to other factors is variable and subject to local conditions in the reservoir. In addition, test methods for conventional crude oil are not always applicable to extra heavy oil, and modification of the test method(s) may be necessary—as illustrated for tar sand bitumen (Section 2.2.8) (Wallace, 1988; Wallace et al., 1988; Speight, 2014a).

2.2.8 TAR SAND BITUMEN

The term *bitumen* (also, on occasion, referred to as *native asphalt* and *extra heavy oil*) includes a wide variety of naturally occurring reddish brown to black materials of semisolid, viscous to brittle character that can exist in nature with no mineral impurity or with mineral matter contents that exceed 50% by weight. Bitumen is frequently found filling pores and crevices of sandstone, limestone, or argillaceous sediments, in which case the organic and associated mineral matrix is known as *rock asphalt* (Abraham, 1945; Hoiberg, 1964). Tar sand bitumen is a high-boiling material with little, if any, material boiling below 350°C (660°F), and the boiling range is close to the boiling range of an atmospheric residuum.

Bitumen in tar sand deposits represents a potentially large supply of energy. However, many of these reserves are only available with some difficulty, and optional refinery scenarios will be necessary for conversion of these materials to low-sulfur liquid products because of the substantial differences in character between conventional petroleum and tar sand bitumen (Speight, 2014a). Bitumen recovery requires the prior application of reservoir fracturing procedures before the introduction of thermal recovery methods. Currently, commercial operations in Canada use mining techniques for bitumen recovery.

Because of the diversity of available information and the continuing attempts to delineate the various tar sand deposits in the world, it is virtually impossible to present accurate numbers that reflect the extent of the reserves in terms of the barrel unit. Indeed, investigations into the extent of many of the world's deposits are continuing at such a rate that the numbers vary from one year to the next. Accordingly, the data quoted here must be recognized as approximate, with the potential of being quite different at the time of publication.

Throughout this text, frequent reference is made to tar sand bitumen; however, because commercial operations have been in place for >30 years (Speight, 1990, 2009, 2014a), it is not surprising that more is known about the Alberta (Canada) tar sand reserves than any other reserves in the world. Therefore, when discussion is made of tar sand deposits, reference is made to the relevant deposit; however, when information is not available, the Alberta material is used for the purposes of the discussion.

Tar sand deposits are widely distributed worldwide (Speight, 1990, 1999, and references cited therein). The potential reserves of bitumen that occur in tar sand deposits have been variously estimated to exceed the reserves of petroleum (Speight, 2014a). That commercialization has taken place in Canada does not mean that commercialization is imminent for other tar sands deposits. There are considerable differences between the Canadian and the US deposits that could preclude across-the-board application of the Canadian principles to the US sands (Speight, 1990, 2014a). The key is accessibility and recoverability. Various definitions have been applied to energy reserves but the crux of the matter is the amount of a resource that is recoverable using current technology. Although tar sands are not a principal energy reserve, they certainly are significant with regard to projected energy consumption over the next several generations.

Thus, in spite of the high estimations of the reserves of bitumen, the two conditions of vital concern for the economic development of tar sand deposits are the concentration of the resource, or the percent bitumen saturation, and its accessibility, usually measured by the overburden thickness. Recovery methods are based either on mining combined with some further processing or operation on the oil sands *in situ*. The mining methods are applicable to shallow deposits, characterized by an overburden ratio (i.e., overburden depth to thickness of tar–sand deposit). For example, indications are that for the Athabasca deposit, no more than 10% of the in-place deposit is mineable within current concepts of the economics and technology of open-pit mining; this 10% portion may be considered as the *proven reserves* of bitumen in the deposit.

It is incorrect to refer to bitumen as *tar* or *pitch*. Although the word tar is somewhat descriptive of the black bituminous material, it is best to avoid its use with respect to natural materials. More correctly, the name *tar* is usually applied to the heavy product remaining after the destructive distillation of coal or other organic matter. *Pitch* is the distillation residue of the various types of tar.

Thus, alternative names, such as *bituminous sand* or *oil sand*, are gradually finding usage, with the former name (bituminous sands) more technically correct. The term *oil sand* is also used in the same way as the term *tar sand*, and these terms are used interchangeably throughout this text.

However, to define conventional petroleum, heavy oil, and bitumen, the use of a single physical parameter such as viscosity is not sufficient. Other properties such as API gravity, elemental analysis, composition, and, most of all, the properties of the bulk deposit must also be included in any definition of these materials. Only then will it be possible to classify petroleum and its derivatives (Speight, 2014a).

2.3 REFINERY-PRODUCED FEEDSTOCKS

For the purposes of this book, the different types of feedstock that may be fed to a desulfurization unit may be natural feedstocks as well as the more typical refinery-produced feedstocks. For convenience, such feedstocks are described here, and are (1) naphtha, (2) middle distillates, and (3) residua.

2.3.1 Naphtha

The naphtha fraction if present in the crude oil (but is usually not present to any great extent in heavy oil, extra heavy oil, and tar sand bitumen) is typically composed of saturated constituents with lesser amounts of monoaromatic constituents and diaromatic constituents. Within the saturated constituents in petroleum gases and naphtha, every possible paraffin from methane (CH_4) to *n*-decane (*n*-$C_{10}H_{22}$, normal decane) is present. The isoparaffins begin at C_4 with isobutane as the only isomer of *n*-butane. The number of isomers grows rapidly with carbon number, and there may be increased difficulty in dealing with multiple isomers during analysis.

In addition to aliphatic molecules, the saturated constituents consist of cycloalkanes (also called *naphthenes*) with predominantly five- or six-carbon rings. Methyl derivatives of cyclopentane and cyclohexane, which are commonly found at higher levels than the parent unsubstituted structures, may be present (Tissot and Welte, 1978). Fused-ring dicycloalkane derivatives such as *cis*-decahydronaphthalene (*cis*-decalin) and *trans*-decahydronaphthalene (*trans*-decalin) and hexahydroindan are also common; however, bicyclic naphthenes separated by a single bond, such as cyclohexyl cyclohexane, are not.

The numerous aromatic constituents in petroleum naphtha begin with benzene, and the C_1 to C_3 alkylated derivatives are also present (Tissot and Welte, 1978). Each of the alkyl benzene homologues through the 20 isomeric C_4 alkyl benzenes has been isolated from crude oil along with various C_5 derivatives. Benzene derivatives having fused cycloparaffin rings (naphthene-aromatics), such as indane and tetralin, have been isolated along with a number of their methyl derivatives. Naphthalene is included in this fraction, while 1-methyl naphthalene and 2-methyl naphthalene and higher homologues of fused two-ring aromatics appear in the middle-distillate fraction.

Sulfur-containing compounds are the common heteroatom compounds that occur in the naphtha fraction (Rall et al., 1972)—most derivatives of the other heteroatom derivative (nitrogen, oxygen, and metals) boil at higher temperatures than the naphtha boiling range. Generally, the total amount of sulfur in the gases and naphtha fraction is <1% w/w of the total sulfur in the crude oil. In naphtha from high-sulfur (sour) petroleum, 50–70% of the sulfur may be in the form of mercaptans (thiol derivatives, RSH) (Rall et al., 1972). In naphtha from low-sulfur (sweet) crude oil, the sulfur is distributed between sulfide derivatives (thioether derivatives) and thiophene derivatives—the sulfide derivatives may be in the form of linear structures (alkyl sulfide derivatives) as well as five- or six-ring cyclic structures (thiacyclane derivatives). The sulfur structure distribution tends to follow the distribution hydrocarbons; that is, naphthenic oils with high content of cycloalkane derivatives tend to have a high thiacyclane content. Methyl disulfide and ethyl disulfide have been confirmed to be present in some crude oils in analyses that minimized their possible formation by oxidative coupling of thiol derivatives (Rall et al., 1972).

2.3.2 MIDDLE DISTILLATES

For the purposes of this text, the term *middle distillates* includes all of the feedstock constituents that boil between the end point of naphtha and the initial boiling point of the vacuum residuum. Thus, kerosene, jet fuel, diesel fuel, gas oil, and vacuum gas oil are all derived from raw middle distillate, which can also be obtained from cracked and hydroprocessed refinery streams. The composition of the distillates must be carefully monitored to ensure that fouling (phase separation) does not occur during blending of the various streams into saleable products.

Saturated species and aromatics species occur in the lower-boiling middle-distillate fraction of petroleum, and the aromatic derivatives include simple compounds with up to three aromatic rings as well as heterocyclic compounds. Within the saturated constituents, the concentration of n-paraffins decreases regularly from C_{11} to C_{20}. Mono- and dicycloparaffins with five or six carbons per ring constitute the bulk of the naphthenes in the middle-distillate boiling range, decreasing in concentration as the carbon number increases (Tissot and Welte, 1978), and the alkylated naphthenes may have a single long side chain as well as one or more methyl or ethyl groups.

The most abundant aromatics in the mid-distillate boiling fractions are di- and tri-methyl naphthalene derivatives. Other one- and two-ring aromatics are undoubtedly present in small quantities as either naphthene homologs or alkyl homologs in the C_{11}–C_{20} range. In addition to these homologues of alkylbenzenes, tetralin derivatives, and naphthalene derivatives, the mid-distillate contains some fluorene derivatives and phenanthrene derivatives. The phenanthrene structure appears to be favored over the anthracene structure (Tissot and Welte, 1978), and this preference appears to continue into the higher-boiling fractions of petroleum (Speight, 2014a).

The five-membered heterocyclic constituents in the mid-distillate range are primarily the thiacyclane derivatives, benzothiophene derivatives, and dibenzothiophene derivatives with lesser amounts of dialkyl-, diaryl-, and aryl-alkyl sulfide derivatives; alkylthiophene derivatives are also present. As with the naphtha fractions, these sulfur species account for a minimal fraction of the total sulfur in the crude.

Although only trace amounts (usually ppm levels) of nitrogen are found in the middle-distillate fractions, both neutral and basic nitrogen compounds have been isolated and identified in fractions boiling below 343°C (650°F) (Speight, 2014a). Pyrrole derivatives and indole derivatives account for the about two-thirds of the nitrogen, while the remainder is found in the basic alkylated pyridine and alkylated quinoline compounds.

The saturate constituents contribute less to the gas oil fractions than the aromatic constituents but more than the polar constituents that is present at percentage levels rather than trace levels. Within the gas oil fractions, the composition of the saturates (paraffins, isoparaffins, and naphthenes) is highly dependent on the petroleum source. Generally, the naphthene constituents account for approximately two-thirds of the saturate constituents; however, the overall range of variation is from <20% to >80% v/v. In most samples, the n-paraffins from C_{20} to C_{44} are present in sufficient quantity to be detected as distinct peaks in gas chromatographic analysis.

The bulk of the saturated constituents in vacuum gas oil consist of isoparaffins and especially naphthene derivatives. The aromatic derivatives in vacuum gas oil may contain one to six fused aromatic rings that may bear additional naphthene rings and alkyl substituents in keeping with their boiling range. Mono- and di-aromatic derivatives account for about 50% of the aromatics in petroleum vacuum gas oil samples. Analytical data show the presence of up to four fused naphthenic rings on some aromatic derivatives. This is consistent with the suggestion that these species originate from the aromatization of steroid precursors. Although present at lower concentration, alkyl benzene derivatives and naphthalene derivatives show one long side chain and multiple short side chains. In addition, phenanthrene derivatives tend to outnumber anthracene derivatives by as much as 100:1, and chrysene derivatives appear to be favored over pyrene derivatives.

Heterocyclic constituents are significant contributors to the vacuum gas oil fraction. In terms of sulfur compounds, thiophene-type and thiacyclane-type sulfur predominate over sulfide-type sulfur, and some molecular constituents may even contain more than one sulfur atom. The benzothiophene derivatives and dibenzothiophene derivatives are the prevalent thiophene forms of sulfur, which confers stability on the sulfur atom (Speight, 2000, 2014a) and can lead to fouling through the deposition of sediment. In the vacuum gas oil range, the nitrogen-containing compounds include higher molecular weight pyridine derivatives, quinoline derivatives, benzoquinoline derivatives, amides, indoles, and carbazole, and molecules with two nitrogen atoms (diaza compounds) with three and four aromatic rings are especially prevalent. Typically, about one-third of the compounds are basic, i.e., pyridine and its benzo-derivatives, while the remainder is present as neutral species (amide derivatives and carbazole derivatives). Although benzo- and dibenzoquinoline derivatives found in petroleum are rich in structurally hindered molecular species, hindered and unhindered structures have been found to be present at equivalent concentrations in source rocks. This has been rationalized as "geo-chromatography" in which the less polar (hindered) structures move more readily to the reservoir. Oxygen levels in the vacuum gas oil parallel the nitrogen content. Thus, the most commonly identified oxygen compounds are the carboxylic acids and phenols, collectively called naphthenic acids.

2.3.3 RESIDUUM

Of all of the feedstocks described in this section, *residua* are the only manufactured products. The constituents of residua do occur naturally as part of the native material, but residua are specifically produced during petroleum refining and the properties of the various residua depend on the cut point or boiling point at which the distillation is terminated (Speight and Ozum, 2002; Hsu and Robinson, 2006; Gary et al., 2007; Speight, 2008, 2014a).

A *residuum* (pl. residua, also shortened to *resid*, pl. resids) is the residue obtained from petroleum after nondestructive distillation has removed all the volatile materials. The temperature of the distillation is usually maintained below 350°C (660°F) since the rate of thermal decomposition of petroleum constituents is minimal below this temperature but the rate of thermal decomposition of petroleum constituents is substantial above 350°C (660°F) (Figure 2.3). Residua are black, viscous materials and are obtained by distillation of a crude oil under atmospheric pressure (atmospheric residuum) or under reduced pressure (vacuum residuum) (Figure 2.4). They may be liquid at room

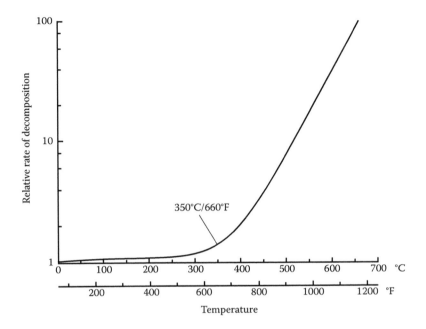

FIGURE 2.3 Simplified representation of the rate of thermal decomposition of petroleum constituents with temperature. (From Speight, J.G. 2000. *The Desulfurization of Heavy Oils and Residua.* Marcel Dekker Inc., New York. Figure 2.3, p. 16.)

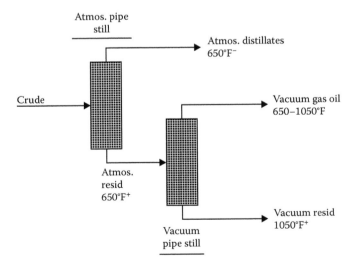

FIGURE 2.4 General representation of the production of residua by distillation. (From Speight, J.G. 2000. *The Desulfurization of Heavy Oils and Residua.* Marcel Dekker Inc., New York. Figure 2.4, p. 16.)

temperature (generally atmospheric residua) or almost solid (generally vacuum residua) depending on the cut point of the distillation (Tables 2.4 and 2.5).

As refineries have evolved, distillation has remained the prime means by which petroleum is refined (Speight, 2014a). Indeed, the distillation section of a modern refinery is the most flexible unit in the refinery since conditions can be adjusted to process a wide range of refinery feedstocks from the lighter crude oils to the heavier, more viscous crude oils. However, the maximum permissible temperature (in the vaporizing furnace or heater) to which the feedstock can be subjected is 350°C (660°F). The rate of thermal decomposition increases markedly above this temperature: if decomposition occurs within a distillation unit, it can lead to coke deposition in the heater pipes or in the tower itself, resulting in the failure of the unit.

The atmospheric distillation tower is divided into a number of horizontal sections by metal trays or plates, and each is the equivalent of a still. The more trays, the more redistillation, and hence the better is the fractionation or separation of the mixture fed into the tower. A tower for fractionating crude petroleum may be 13 ft in diameter and 85 ft in height; however, a tower stripping unwanted volatile material from gas oil may be only 3 or 4 ft in diameter and 10 ft in height. Towers concerned with the distillation of liquefied gases are only a few feet in diameter but may be up to 200 ft in height. A tower used in the fractionation of crude petroleum may have from 16 to 28 trays, but one used in the fractionation of liquefied gases may have 30–100 trays. The feed to a typical tower enters the vaporizing or flash zone, an area without trays. The majority of the trays are usually located above this area. The feed to a bubble tower, however, may be at any point from top to bottom with trays above and below the entry point, depending on the kind of feedstock and the characteristics desired in the products.

Liquid collects on each tray to a depth of, say, several inches, and the depth is controlled by a dam or weir. As the liquid level rises, excess liquid spills over the weir into a channel (downspout) that carries the liquid to the tray below.

TABLE 2.4
Properties of Tia Juana Crude Oil and the 650°F, 950°F, and 1050°F Residua

	Whole Crude	Residua		
		650°F+	950°F+	1050°F+
Yield, vol.%	100.0	48.9	23.8	17.9
Sulfur, wt.%	1.08	1.78	2.35	2.59
Nitrogen, wt.%		0.33	0.52	0.60
API gravity	31.6	17.3	9.9	7.1
Carbon residue, wt.%				
Conradson		9.3	17.2	21.6
Metals				
Vanadium, ppm		185		450
Nickel, ppm		25		64
Viscosity				
Kinematic				
100°F	10.2	890		
210°F		35	1010	7959
Furol				
122°F		172		
210°F		484	3760	
Pour point, °F	−5	45	95	120

Source: Speight, J.G. 2000. *The Desulfurization of Heavy Oils and Residua.* Marcel Dekker Inc., New York. Table 2.4, p. 17.

TABLE 2.5

Miscellaneous Properties of Heavy Oils and Residua

Crude Oil Origin	Kuwait[a]	Kuwait		Venezuela (East)	Buzurgan	Boscan	Khafji		California	Cabimas
Residuum Type	Vacuum Gas Oil	Atmospheric	Vacuum	Atmospheric			Atmospheric	Vacuum	Vacuum	Vacuum
Fraction of crude, vol.%	–	42	21	74	52	78	–	–	23	34
Gravity, °API	22.4	13.9	5.5	9.6	3.1	5.0	14.4	6.5	4.3	6.8
Viscosity SUS, 210°F	–	–	–	–	–	–	–	–	–	–
SFS, 122°F	–	553	500,000	27,400	–	–	–	–	–	–
SFS, 210°F	–	–	–	–	–	–	429	–	–	–
cSt, 100°F	–	–	–	–	–	–	–	–	–	–
cSt, 210°F	–	55	1900	–	3355	5250	–	–	15,000	7800
Pour point, °F	–	65	–	95	–	–	–	–	–	162
Sulfur, wt.%	2.97	4.4	5.45	2.6	6.2	5.9	4.1	5.3	2.3	3.26
Nitrogen, wt.%	0.12	0.26	0.39	0.61	0.45	0.79	–	–	0.98	0.62
Metals, ppm										
Nickel	0.2	14	32	94	76	133	37	53	120	76
Vanadium	0.04	50	102	218	233	1264	89	178	180	614
Asphaltenes, wt.%										
Pentane insolubles	0	–	11.1	–	–	–	–	12.0	19.0	12.9
Hexane insolubles	0	–	–	–	–	–	–	–	–	–
Heptane insolubles	0	2.4	7.1	9	18.4	15.3	–	–	–	10.5
Resins, wt.%	0	–	39.4	–	–	–	–	–	–	–
Carbon residue, wt.%										
Ramsbottom	<0.1	9.8	–	14.5	–	–	–	–	–	–
Conradson	0.09	12.2	23.1	–	22.5	18.0	–	21.4	24.0	18.7

(Continued)

TABLE 2.5 (CONTINUED)

Miscellaneous Properties of Heavy Oils and Residua

Crude Oil Origin	Arabian Light	Louisiana	Saudi Arabia	Alaska (North Slope)		Boscan	Tar Sand Triangle	P.R. Spring	N.W. Asphalt Ridge
Residuum Type	Atmospheric	Vacuum	Vacuum	Atmospheric	Vacuum				
Fraction of crude, vol.%	–	13.1	20	58	22	100	100	100	100
Gravity, °API	16.7	11.3	5.0	15.2	8.2	10.3	11.1	10.3	14.4
Viscosity SUS, 210°F	–	–	–	1281	–	–	–	–	–
SFS, 122°F	–	–	–	–	–	–	–	–	–
SFS, 210°F	–	–	–	–	–	–	–	–	–
cSt, 100°F	–	–	–	–	–	20,000	7000[b]	200,000[b]	15,000[b]
cSt, 210°F	27	700	2700	42	1950	–	–	–	–
Pour point, °F	–	–	–	75	–	37	–	–	–
Sulfur, wt.%	3.00	0.93	5.2	1.6	2.2	5.6	4.38	0.75	0.59
Nitrogen, wt.%	–	0.38	0.30	0.36	0.63	–	0.46	1.00	1.02
Metals, ppm									
Nickel	11	20	28	18	47	117	53	98	120
Vanadium	28	–	75	30	82	1220	108	25	25
Asphaltenes, wt.%									
Pentane insolubles	–	6.5	15.0	4.3	8.0	12.6	26.0	16.0	6.3
Hexane insolubles	–	–	–	–	–	11.4	–	–	–
Heptane insolubles	2.0	–	–	31.5	–	–	–	–	–
Resins, wt.%	–	–	–	–	–	24.1	–	–	–
Carbon residue, wt.%									
Ramsbotton	8	12.0	20.0	8.4	17.3	14.0	–	–	–
Conradson	–	–	–	–	–	–	21.6	12.5	3.5

(Continued)

TABLE 2.5 (CONTINUED)
Miscellaneous Properties of Heavy Oils and Residua

Crude Oil Origin	Athabasca	Lloydminster	Cold Lake	Qayarah	Jobo	Bachaquero		Heavy Arabian	West Texas	
Residuum Type						Atmospheric	Vacuum	Vacuum	Vacuum	Atmospheric
Fraction of crude, vol.%	–	100	100	100	100	34	–	27	–	–
Gravity, °API	5.9	14.5	10.0	15.3	8.6	17	2.8	4	9.4	18.4
Viscosity SUS, 210°F	513	260	–	–	–	–	–	–	313	–
SFS, 122°F	–	294	–	–	247	–	–	–	–	86
SFS, 210°F	820	–	–	–	–	–	–	–	–	–
cSt, 100°F	–	–	–	–	–	–	–	–	–	–
cSt, 210°F	–	–	79	–	–	–	–	–	–	–
Pour point, °F	50	38	–	–	–	–	–	–	–	–
Sulfur, wt.%	4.9	4.3	4.4	8.4	3.9	2.4	3.7	5.3	3.3	2.5
Nitrogen, wt.%	0.41	–	0.39	0.7	0.7	0.3	0.6	0.4	0.5	0.6
Metals, ppm										
Nickel	86	40	62	60	103	450	100	230	27	11
Vanadium	167	100	164	130	460	–	900	–	57	20
Asphaltenes, wt.%										
Pentane insolubles	17.0	12.9	15.9	20.4	18	10	–	25	–	–
Hexane insolubles	13.5	–	–	14.3	–	–	–	–	–	–
Heptane insolubles	11.4	–	10.8	13.5	–	–	–	–	–	–
Resin, wt.%	34.0	38.4	31.2	36.1	–	–	–	–	–	–
Carbon residue, wt.%										
Ramsbottom	14.9	–	–	–	–	–	–	–	–	–
Conradson	18.5	9.1	13.6	15.6	14	12	27.5	–	16.9	6.6

(Continued)

TABLE 2.5 (CONTINUED)
Miscellaneous Properties of Heavy Oils and Residua

Crude Oil Origin	Tia Juana (Light)		Safaniya		PR Spring	Asphalt Ridge	Tar Sand Triangle	Sunnyside
Residuum Type	Atmospheric	Vacuum	Atmospheric	Vacuum	Bitumen	Bitumen	Bitumen	Bitumen
Fraction of crude, vol.%	49	18	40	22	100	100	100	100
Gravity, °API	17.3	7.1	11.1	2.6	10.3	14.4	11.1	–
Viscosity SUS, 210°F	165	–	–	–	–	–	–	–
SFS, 122°F	172	–	–	–	–	–	–	–
SFS, 210°F	–	–	–	–	–	–	–	–
cSt, 100°F	890	–	–	–	–	–	–	–
cSt, 210°F	35	7959	79	–	–	–	–	–
Pour point, °F	–	–	–	–	–	–	–	–
Sulfur, wt.%	1.8	2.6	4.3	5.3	0.8	0.6	4.4	0.5
Nitrogen, wt.%	0.3	0.6	0.4	0.4	1.0	1.0	0.5	0.9
Metals, ppm								
Nickel	25	64	26	46	98	120	53	–
Vanadium	185	450	109	177	25	25	108	–
Asphaltenes, wt.%								
Pentane insolubles	–	–	17.0	30.9	–	–	–	–
Hexane insolubles	–	–	–	–	–	–	–	–
Heptane insolubles	–	–	–	–	–	–	–	–
Resin, wt.%	–	–	–	–	–	–	–	–
Carbon residue, wt.%								
Ramsbottom	–	–	–	–	12.5	3.5	21.6	–
Conradson	9.3	21.6	14.0	25.9	–	–	–	–

Source: Speight, J.G. 2000. *The Desulfurization of Heavy Oils and Residua.* Marcel Dekker Inc., New York. Table 2.5, pp. 18–21.
a Included for comparison.
b Estimated.

The temperature of the trays is progressively cooler from bottom to top (Figure 2.5). The bottom tray is heated by the incoming heated feedstock, although in some instances a steam coil (reboiler) is used to supply additional heat. As the hot vapors pass upward in the tower, condensation occurs onto the trays until refluxing (simultaneous boiling of a liquid and condensing of the vapor) occurs on the trays. Vapors continue to pass upward through the tower, whereas the liquid on any particular tray spills onto the tray below, and so on until the heat at a particular point is too intense for the material to remain liquid. It then becomes vapor and joins the other vapors passing upward through the tower. The whole tower thus simulates a collection of several (or many) stills, with the composition of the liquid at any one point or on any one tray remaining fairly consistent. This allows part of the refluxing liquid to be tapped off at various points as *sidestream* products. Thus, in the distillation of crude petroleum, light naphtha and gases are removed as vapor from the top of the tower; heavy naphtha, kerosene, and gas oil are removed as sidestream products; and reduced crude (atmospheric residuum) is taken from the bottom of the tower.

The *topping* operation differs from normal distillation procedures insofar as the majority of the heat is directed to the feed stream rather than by reboiling the material in the base of the tower. In addition, products of volatility intermediate between that of the overhead fractions and bottoms (residua) are withdrawn as sidestream products. Furthermore, steam is injected into the base of the column and the sidestream strippers to adjust and control the initial boiling range (or point) of the fractions.

Topped crude oil must always be stripped with steam to elevate the flash point or to recover the final portions of gas oil. The composition of the topped crude oil is a function of the temperature of the vaporizer (or flasher).

Vacuum distillation as applied to the petroleum refining industry is truly a technique of the 20th century and has since been widely use in petroleum refining (Speight and Ozum, 2002; Hsu and Robinson, 2006; Gary et al., 2007; Speight, 2008, 2014a). Vacuum distillation evolved because of the need to separate the less volatile products, such as lubricating oils, from the petroleum without subjecting these high-boiling products to cracking conditions (Table 2.6). The boiling point of the heaviest cut obtainable at atmospheric pressure is limited by the temperature (ca. 350°C; ca. 660°F) at which the residue starts to decompose or crack, unless cracking distillation is preferred (Speight, 2014a). When the feedstock is required for the manufacture of lubricating oils, further fractionation without cracking is desirable and this can be achieved by distillation under vacuum conditions.

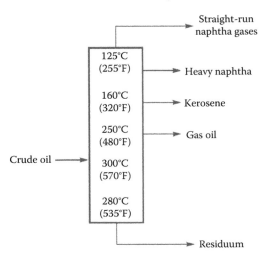

FIGURE 2.5 General representation of the temperature profile in a noncracking atmospheric distillation tower. (From Speight, J.G. 2000. *The Desulfurization of Heavy Oils and Residua*. Marcel Dekker Inc., New York. Figure 2.10, p. 32.)

TABLE 2.6
Products Obtained from Atmospheric Distillation and Vacuum Distillation Units and Further Processing Required to Produce Low-Sulfur Liquid Fuels

Product	Further Treatment	Typical Use
Atmospheric Distillation		
Gases	Separation	LPG
	Sulfur removal (gas processing)	Petrochemical feed
Light naphtha	Hydrotreating	Gasoline blend stock
Medium naphtha	Hydrotreating	Gasoline blend stock
Heavy naphtha	Hydrotreating	Gasoline blend stock
FCC naphtha	Hydrotreating	Gasoline blend stock
Coker naphtha	Hydrotreating	Gasoline blend stock
Kerosene	Hydrotreating	Diesel blend stock
Gas oil		Lube oil base stocks
	FCC	FCC naphtha
		Gasoline blend stock
	Hydrocracking	Gasoline blend stock
Atmospheric residuum	Coking unit	Distillates
		Coke
	FCC	FCC naphtha
		Kerosene
	Hydrocracking	Gasoline blend stock
Lube oil base stocks	Hydrotreating	Lube base oil
		Cosmetic oil
		Medicinal oil
Vacuum Distillation		
Vacuum gas oil (light)	Hydrotreating	Lube base oil
		Cosmetic oil
		Medicinal oil
	FCC	FCC naphtha
	Hydrocracking	Gasoline blend stock
Vacuum gas oil (heavy)	Hydrotreating	Lube base oil
		Cosmetic oil
		Medicinal oil
	FCC	FCC naphtha
Vacuum residuum	Asphalt plant	Road asphalt
		Asphalt derivatives
	Coking unit	Distillates
		Coke
	Hydrocracking	Gasoline blend stock
	Resid conversion unit	Distillates
		Coke

Note: Not all options are presented, but only sufficient options to give an indication of the processing future of the selected fractions as well as the need for hydrotreating and sulfur removal.

The distillation of high-boiling lubricating oil stocks may require pressures as low as 15–30 mm Hg, but operating conditions are more usually 50–100 mm Hg. Volumes of vapor at these pressures are large and pressure drops must be small to maintain control, so vacuum columns are necessarily of a large diameter. Differences in vapor pressure of different fractions are relatively larger than for lower boiling fractions, and relatively few plates are required. Under these conditions, heavy gas oil may be obtained as an overhead product at temperatures of about 150°C (300°F). Lubricating oil fractions may be obtained as sidestream products at temperatures of 250–350°C (480–660°F). The feedstock and residue temperatures must be kept below the temperature of 350°C (660°F), above which (as has already been noted) the rate of thermal decomposition increases substantially and cracking occurs (Figure 2.3). The partial pressure of the hydrocarbons is effectively reduced yet further by the injection of steam. The steam added to the column, principally for the stripping of bitumen in the base of the column, is superheated in the convection section of the heater.

The fractions obtained by vacuum distillation of reduced crude depend on whether the run is designed to produce lubricating or vacuum gas oils (Speight and Ozum, 2002; Hsu and Robinson, 2006; Gary et al., 2007; Speight, 2014). In the former case, the fractions include (1) heavy gas oil, an overhead product and is used as catalytic cracking stock or, after suitable treatment, a light lubricating oil; (2) lubricating oil (usually three fractions: light, intermediate, and heavy), obtained as a sidestream product; and (3) residuum, the nonvolatile product that may be used directly as asphalt or to produce asphalt—the residuum may also be used as a feedstock for a coking operation or blended with gas oils to produce a heavy fuel oil.

The continued use of atmospheric and vacuum distillation has been a major part of refinery operations during this century and no doubt will continue to be employed, at least into the first several decades of the 21st century, as the primary refining operation (Speight, 2011a).

When a residuum is obtained from a crude oil and thermal decomposition has commenced, it is more usual to refer to this product as *pitch* (Speight, 2014a). The differences between the parent petroleum and the residua are due to the relative amounts of various constituents present, which are removed or remain by virtue of their relative volatility (Figure 2.6).

The chemical composition of a residuum from an asphaltic crude oil is complex. Physical methods of fractionation usually indicate high proportions of asphaltene constituents and resin constituents, even in amounts up to 50% (or higher) of the residuum. In addition, the presence of ash-forming metallic constituents, including such organometallic compounds as those of vanadium and nickel, is also a distinguishing feature of residua and heavy oils. Furthermore, the deeper the cut into the

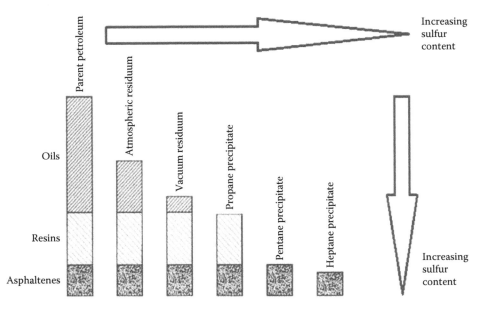

FIGURE 2.6 Simplified relationship between petroleum, two residua, propane asphalt, and two asphaltene fractions. (From Speight, J.G. 2000. *The Desulfurization of Heavy Oils and Residua*. Marcel Dekker Inc., New York. Figure 2.5, p. 22.)

crude oil, the greater is the concentration of sulfur and metals in the residuum and the greater the deterioration in physical properties (Tables 2.4 and 2.5).

The definitions that are used to describe petroleum reserves are often misunderstood because they are not adequately defined at the time of use. Therefore, as a means of alleviating this problem, it is pertinent at this point to consider the definitions used to describe the amount of petroleum that remains in subterranean reservoirs (Speight, 2014a).

2.4 SULFUR IN PETROLEUM

Organic sulfur compounds vary in polarity and chromatographic behavior such that some elute during liquid chromatography with the aromatic hydrocarbon fraction, while others elute with the more polar nitrogen, oxygen, and sulfur fractions that include the resin constituents and asphaltene constituents. Until recently, precise organic molecular structures have been established only for relatively low molecular weight sulfur compounds. These compounds contain lower molecular weight constituents, generally with <15 carbon atoms and have boiling points below 250–300°C (480–570°F). This unsatisfactory state of knowledge is aggravated by the fact that generally 60–80% of the sulfur is in the asphaltene and resin fractions that contain innumerable individual constituents of speculative chemical structure (Speight, 2001, 2014a).

Sulfur is preferentially associated with the higher molecular weight constituents of crude oil (Figure 2.6), and when crude oil is refined the sulfur concentrates into the higher molecular weight fractions. Although many of the lower-boiling sulfur compounds in petroleum have been identified (Speight, 2014a), it can only be surmised (with some degree of confidence) that the same types of organic sulfur extend into uncharacterized fractions. However, it is also believed that the distributions of sulfur functional types differ in the high-boiling fractions. In fact, there is a distinct likelihood that many of the more polar and higher molecular weight compounds contain more than one heteroatom (Rall et al., 1972; Orr and White, 1990).

To summarize all of this work in very general terms, the distribution of the sulfur-containing constituents of petroleum has been determined to include the following compound types: (1) non-thiophene sulfur, i.e., sulfides R-S-R[1]; (2) thiophene derivatives; (3) benzothiophene derivatives; (4) dibenzothiophene derivatives; (5) benzo-naphthothiophene derivatives; and (6) di-naphthothiophene derivatives. It is the removal of sulfur from such systems in the various feedstocks to produce hydrocarbon products that this book addresses.

After carbon and hydrogen, sulfur is considered as the most abundant element in petroleum. Crude oils with higher densities contain more sulfur compounds. Distillation fractions with higher boiling points contain higher concentrations of sulfur compounds (Schulz et al., 1999). Organic and inorganic sulfur-containing compounds naturally occur in crude oil. Inorganic sulfur-containing compounds include elemental sulfur, hydrogen sulfide, and pyrite. These compounds may be in dissolved or suspended form. About 200 organic sulfur compounds have been identified in crude oil (Figure 2.7). Organic sulfur compounds in crude oil are generally aromatic or saturated forms of thiol derivatives, sulfide derivatives, and heterocycle derivatives. Light fractions boiling below 200°C contain mainly sulfides and thiol derivatives. These can easily be removed by chemical methods. Thiophene derivatives remaining in the heavier fractions are resistant to chemical processes. Among these, aromatic compounds such as dibenzothiophene or its derivatives are significant because they have higher boiling points (>200°C), and it is difficult to remove them from atmospheric tower outlet streams (e.g., middle distillates) (Kawatra and Eisele, 2001). Derivatives of benzothiophene and dibenzothiophene are the major sulfur compounds in certain types of crude oil. Thiophene derivatives comprise up to 70% of the sulfur compounds present in Texas gas oil and up to 60% of the sulfur compounds present in Middle East gas oil. These sulfur compounds end up in heavy petroleum products such as diesel and Bunker-C oil. Thiophene derivatives are also the major sulfur compounds found in naphtha produced by cracking heavy oil (Kim et al., 1996). Benzothiophene, non-β-, single β-, and di-β-substituted benzothiophene derivatives (boiling point, >219°C) are the typical thiophene derivatives that are found in diesel. Among organic sulfur compounds, some are considered recalcitrant. These compounds are chiefly stable aromatic sulfur-containing compounds that need more invasive desulfurization procedure to remove their sulfur atom. The refractory portion of distillate/diesel fuels is attributed to thiophene derivative compounds such as dibenzothiophene derivatives with 4- and/or 6-alkyl substituting groups (Soleimani et al., 2007).

The sulfur content of petroleum varies from <0.05% to >14% wt., but generally falls in the range 1–4% wt. Petroleum having <1% wt. sulfur are referred to as *low-sulfur petroleum* and those with >1% wt. as *high-sulfur petroleum*. Most of the sulfur present in petroleum is organically bound, and any dissolved hydrogen sulfide and/or elemental sulfur are usually representing only a minor part of the total sulfur. One exception that springs to mind is the heavy crude oil that used to be recovered by steam injection from the formation at Qayarah in Northeast Iraq. The total sulfur content of this oil is approximately 8% wt., of which 6% wt. is organically bound sulfur and the remaining 2% wt. is elemental sulfur. Sulfur is spread throughout most fractions of petroleum but, generally, the largest (ca. 60% or more of the sulfur) fraction is in the high molecular weight components.

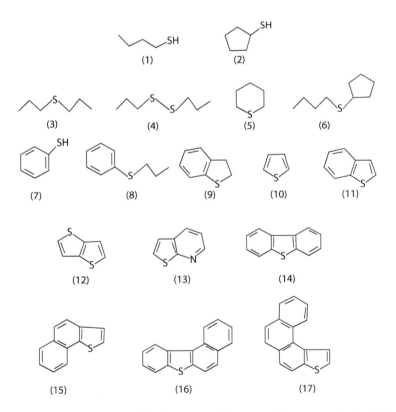

FIGURE 2.7 Representative structures of various types of organosulfur compounds in petroleum. 1, alkane thiol derivatives; 2, cycloalkane thiol derivatives; 3, dialkylsulfide derivatives; 4, polysulfides; 5, cyclic sulfide derivatives; 6, alkylcycloalkylsulfide derivatives; 7, arene thiol derivatives; 8, alkyl aryl sulfide derivatives; 9, thiaindane derivatives; 10, thiophene derivatives; 11, benzothiophene derivatives; 12, thienothiophene derivatives; 13, thienopyridine derivatives; 14, dibenzothiophene derivatives; 15, naphtha-thiophene derivatives; 16, benzonaphthothiophene derivatives; 17, phenanthrothiophene derivatives.

2.5   SULFUR LEVELS AND LEGISLATIVE REGULATIONS

The sulfur content of crude oil can vary from 0.01% to 8% (w/w) depending on the geographical source and its origin (Tables 2.7 and 2.8) (De Krom, 2002; Kilbane and Le Borgne, 2004; Speight, 2014a). Depletion of continental crude oil deposits has forced the exploitation of deeper reservoirs containing petroleum rich in polynuclear aromatic sulfur heterocyclic compounds, and other unconventional oil reserves, including heavy oil, extra heavy oil, and oil sands and bitumen, which comprise 70% of the world's total oil resources (US EIA, 2010). Also, with the rising of oil prices, the production of synthetic crude oil from unconventional resources such as oil sands is becoming increasingly economically viable (US EIA, 2010). Unconventional oil reserves are characterized by large quantities of sulfur, nitrogen, nickel, and vanadium; for example, Athabasca oil sand bitumen has a sulfur content of 4.86% (Bunger et al., 1979). The sulfur content in heavy crude oil varies from 0.1% to 15% (w/w) (Shang et al., 2013). Bunker oil is derived from crude oil as a bottom product of petroleum refining and is the most commonly used form of marine fuel for shipping. Banker oil composition is complex with high asphaltene content and characterized by high viscosity, that it increases the difficulty of its refinery process. Bunker oil has high level of sulfur varying from 1.5 to 4 wt.% that generates high levels of SO_x upon combustion (Mohan et al., 2007). As land-based sources of sulfur dioxide (SO_2) are decreasing, those from shipping are increasing; annual SO_2 emissions from ships were estimated to be 16.2 million tonnes in 2006 and are increasing to 22.7 million tonnes in 2020. There is an urgent need for removing sulfur from bunker oil (Tang et al., 2013).

A typical flue gas from the combustion of fossil fuels contains quantities of NO_x, SO_x, and particulate matter. SO_2 gas at elevated levels can cause bronchial irritation and trigger asthma attacks in susceptible individuals. Potential health risks expand to a broader section of the public when the gas turns to particulate matter (see http://www.epa.gov/acidrain/). Long-term exposure to combustion-related fine particulate air pollution is an important environmental risk factor for cardiopulmonary and lung cancer mortality (Pope et al., 2002). SO_2 can react with moisture in the air and cause acid

TABLE 2.7
Crude Oil Characteristics Based on Density and Sulfur Content

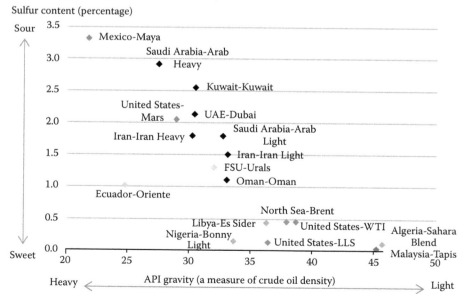

Source: EIA, 2012. http://www.eia.gov/todayinenergy/detail.cfm?id=7110.

TABLE 2.8
Organic Sulfur Content in Crude Oil from Various Countries

Source	Sulfur Content, % w/w
Argentina	0.06–0.42
Australia	0–0.1
Canada	0.12–4.29
Cuba	7.03
Denmark	0.2–0.25
Egypt	0.04–4.19
Indonesia	0.01–0.66
Iran	0.25–3.23
Iraq	2.26–3.3
Italy	1.98–6.36
Kuwait	0.01–3.48
Libya	0.01–1.79
Mexico	0.9–3.48
Nigeria	0.04–0.26
Norway	0.03–0.67
Russia	0.08–1.93
Saudi Arabia	0.04–2.92
United Kingdom	0.05–1.24
USA	0.05–5
Venezuela	0.44–4.99

Source: De Krom H. 2002. Shell Global Solutions International B.V. (Amsterdam). Personal Communication.

rain or low pH fogs. The acid formed in this way can accelerate the erosion of historical buildings. It can be transferred to soil, damage the foliage, depress the pH of the lakes with low buffer capacity, and endanger marine life (US EPA, 2006). As SO_2 is transported by air streams, it can be produced in one area and show its adverse impact in another remote place thousands of kilometers away (Soleimani et al., 2007). Moreover, sulfur is considered as poison for catalytic process in petroleum processing (Leflaive et al., 2000).

Nations worldwide have recognized the problem of SO_2 release into the atmosphere due to the combustion of petroleum-derived fuels, and have decided to reduce sulfur emissions through legislation. Therefore, to control SO_2 emissions, international cooperation is required. Since 1979, Canada, Japan, the United States, and the European nations in particular have signed several agreements to reduce and monitor emissions of sulfur dioxide. Most of these agreements targeted transport fuel because it was one of the important sources of sulfur dioxide. The maximum allowable sulfur content in diesel is targeted at 10 ppm by 2010 in the United States (Kilbane, 2006) and 10–15 ppm by 2011 in the European Union (Stanislaus et al., 2010). In China, the sulfur content in fuel oil is targeted to be no more than 0.01 mg/mL by 2012 (Yang et al., 2014). The acceptable sulfur level in gasoline has been restricted to be <10 ppm in most of the developed countries (Shang et al., 2013).

The simplest way to decrease the amount of sulfur dioxide emitted into the air is to limit the amount of sulfur in fuel. Sulfur removal is also important for the new generation of engines, which are equipped with nitrogen oxide (NO_x) storage catalyst, since sulfur in the fuel has poisoning effects on the catalyst (König et al., 2001). The significant reduction of sulfur-induced corrosion and the slower acidification of engine lubricating oil would lead to lower maintenance costs and longer maintenance intervals. These are also additional benefits of using ultralow sulfur diesel in diesel-powered vehicles (Stanislaus et al., 2010).

It is believed that refining industries will spend about $37 billion on new desulfurization equipment and an additional $10 billion on annual operating expenses during the next 10 years to meet the new sulfur regulations. In addition to this opportunity in the refinery, there is also a large potential in the desulfurization of crude oil itself (Mohebali and Ball, 2008).

REFERENCES

Abraham, H. 1945. *Asphalts and Allied Substances*. Van Nostrand Reinhold Company, New York.

Ancheyta, J., and Speight, J.G. 2007. *Hydroprocessing of Heavy Oils and Residua*. CRC Press, Taylor & Francis Group, Boca Raton, FL.

Armstrong, S.M., Sankey, B.M., and Voordouw, G. 1997. Evaluation of Sulfate Reducing Bacteria for Desulfurizing Bitumen or Its Fractions. *Fuel* 76: 223–227.

Bakshi, A., and Lutz, I. 1987. Adding Hydrogen Donor to Visbreaking Improves Distillate Yields. *Oil Gas J.* 85(28): 84–87.

Bunger, J.W., Thomas, K.P., and Dorrence, S.M. 1979. Compound Types and Properties of Utah and Athabasca Tar Sand Bitumens. *Fuel* 58: 183–195.

De Krom, H. 2002. Shell Global Solutions International B.V. (Amsterdam). Personal Communication.

Dolbear, G., Tang, A., and Moorhead, E. 1987. Upgrading Studies with Californian, Mexican and Middle Eastern Heavy Oils. In: *Metal Complexes in Fossil Fuels*, pp. 221–232. Edited by R. Filby and J. Branthaver. American Chemical Society, Washington, DC.

Gary, J.G., Handwerk, G.E., and Kaiser, M.J. 2007. *Petroleum Refining: Technology and Economics*, 5th Edition. CRC Press, Taylor & Francis Group, Boca Raton, FL.

Hoiberg, A.J. 1964. *Bituminous Materials: Asphalts, Tars, and Pitches*. John Wiley & Sons, New York.

Hsu, C.S., and Robinson, P.R. (Editors) 2006. *Practical Advances in Petroleum Processing*. Volumes 1 and 2. Springer Science, New York.

Kawatra, S.K., and Eisele, T.C. 2001. *Coal Desulfurization, High-Efficiency Preparation Methods*. Taylor & Francis, New York.

Kilbane, J.J. 2006. Microbial Biocatalyst Developments to Upgrade Fossil Fuels. *Curr. Opin. Biotechnol.* 17: 305–314.

Kilbane, J.J., and Le Borgne, S. 2004. Petroleum Biorefining: The Selective Removal of Sulfur, Nitrogen, and Metals. In: *Petroleum Biotechnology, Developments and Perspectives*, pp. 29–65. Edited by R. Vazquez-Duhalt and R. Quintero-Ramirez. Elsevier, Amsterdam.

Kim, B.H., Shin, P.K., Na, J.U., Park, D.H., and Bang, S.H. 1996. Microbial Petroleum Desulfurization. *J. Microbiol. Biotechnol.* 6(5): 299–308.

König, A., Herding, G., Hupfeld B., Richter, T., and Weidmann, K. 2001. Current Tasks and Challenges for Exhaust after Treatment Research, a Viewpoint from the Automotive Industry. *Top. Catal.* 16/17: 23–31.

Leflaive, P., Lemberton, J.L., Perot, G., Mirgain, C., Carriat, J.Y., and Colin, J.M. 2000. Formation of Sulfur-Containing Compounds under Fluid Catalytic Cracking Reaction Conditions. *Stud. Surf. Sci. Catal.* 130: 2465–2470.

Mayes, J.M. 2015. What Are the Possible Impacts on US Refineries Processing Shale Oils? *Hydrocarbon Process.* 94(2): 67–70.

Mohan, S.R., Vicente, S., Ancheyta, J., and Diaz, J.A.I. 2007. A Review of Recent Advances on Process Technologies for Upgrading Heavy Oils and Residua. *Fuel* 86: 1216–1231.

Mohebali, G., and Ball, A.S. 2008. Biocatalytic Desulfurization (BDS) of Petrodiesel Fuels. *Microbiology* 154: 2169–2183.

Mokhatab, S., Poe, W.A., and Speight, J.G. 2006. *Handbook of Natural Gas Transmission and Processing*. Elsevier, Amsterdam.

Monticello, D.J. 1995. Multistage Process for Deep Desulfurization of Fossil Fuels. United States Patent 5,387, 523. February 7.

Orr, W.L., and White, C.L. (Editors). 1990. *Geochemistry of Sulfur in Fossil Fuels*. Symposium Series No. 429. American Chemical Society, Washington, DC.

Pope, C.A., Burnett, R.T., Thun, M.J., Calle, E.E., Krewski, D., Ito, K., and Thurston, G.D. 2002. Lung Cancer, Cardiopulmonary Mortality, and Long-Term Exposure to Fine Particulate Air Pollution. *JAMA* 287: 1132–1141.

Rall, H.T., Thompson, C.J., Coleman, H.J., and Hopkins, R.L. 1972. Bulletin No. 659. United States Bureau of Mines. United States Government Printing Office, Washington, DC.

Rikka, P. 2007. MSc Thesis. Spectrometric Identification of Naphthenic Acids Isolated from Crude Oil. Department of Chemistry and Biochemistry, Texas State University-San Marcos.

Schulz, H., Böhringer, W., Ousmanov, F., and Waller, P. 1999. Refractory Sulfur Compounds in Gas Oils. *Fuel Process. Technol.* 61: 5–41.

Shalaby, H.M. 2005. Refining of Kuwait's Heavy Crude Oil: Materials Challenges. Proceedings, Workshop on Corrosion and Protection of Metals. Arab School for Science and Technology, December 3–7, Kuwait.

Shang, H., Du, W., Liu, Z., and Zhang, H. 2013. Development of Microwave Induced Hydrodesulfurization of Petroleum Streams: A Review. *J. Ind. Eng. Chem.* 19: 1061–1068.

Soleimani, M., Bassi, A., and Margaritis, A. 2007. Biodesulfurization of Refractory Organic Sulfur Compounds in Fossil Fuels. *Biotechnol. Adv.* 25: 570–596.

Speight, J.G. 1990. Chapters 12–16. In: *Fuel Science and Technology Handbook.* J.G. Speight (Editor). Marcel Dekker, New York.

Speight, J.G. 1996a. *Environmental Technology Handbook.* Taylor & Francis Publishers, Philadelphia.

Speight, J.G. 1996b. Petroleum Refinery Processes. In: *Kirk-Othmer Encyclopedia of Chemical Technology.* 4th Edition. 18: 433.

Speight, J.G. 1999. *The Chemistry and Technology of Petroleum.* 3rd ed. Marcel Dekker, New York.

Speight, J.G. 2001. *Handbook of Petroleum Analysis.* John Wiley & Sons Inc., Hoboken, NJ.

Speight, J.G. 2007. *Natural Gas: A Basic Handbook.* GPC Books, Gulf Publishing Company, Houston, TX.

Speight, J.G. 2008. *Handbook of Synthetic Fuels.* McGraw-Hill, New York.

Speight, J.G. 2011a. *The Refinery of the Future.* Gulf Professional Publishing. Elsevier, Oxford, UK.

Speight, J.G. 2011b. (Editor). *The Biofuels Handbook.* The Royal Society of Chemistry, London.

Speight, J.G. 2011c. *An Introduction to Petroleum Technology, Economics, and Politics.* Scrivener Publishing LLC, Beverly, MA.

Speight, J.G. 2012. *Shale Oil Production Processes.* Gulf Professional Publishing, Elsevier, Oxford, UK.

Speight, J.G. 2013a. *The Chemistry and Technology of Coal,* 3rd Edition. CRC Press, Taylor & Francis Group, Boca Raton, FL.

Speight, J.G. 2013b. *Shale Gas Production Processes.* Gulf Professional Publishing, Elsevier, Oxford, UK.

Speight, J.G. 2013c. *Heavy and Extra Heavy Oil Upgrading Technologies.* Gulf Professional Publishing, Elsevier, Oxford, UK.

Speight, J.G. 2014a. *The Chemistry and Technology of Petroleum,* 5th Edition. CRC Press, Taylor & Francis Group, Boca Raton, FL.

Speight, J.G. 2014b. *High Acid Crudes.* Gulf Professional Publishing, Elsevier, Oxford, UK.

Speight, J.G. 2015. *Handbook of Petroleum Product Analysis,* 2nd Edition. John Wiley & Sons Inc., Hoboken, NJ.

Speight, J.G., and Ozum, B. 2002. *Petroleum Refining Processes.* Marcel Dekker Inc., New York.

Stanislaus, A., Marafi, A., and Rana, M.S. 2010. Recent Advances in the Science and Technology of Ultra-Low Sulfur Diesel (ULSD) Production. *Catal. Today* 153: 1–68.

Tang, Q., Lin, S., Cheng, Y., Liu, S.-J., and Xiong, J.-R. 2013. Enhanced Biodesulfurization of Bunker Oil by Ultrasound Pre-Treatment with Native Microbial Seeds. *Biochem. Eng. J.* 77: 58–65.

Tissot, B.P., and Welte, D.H. 1978. Petroleum Formation and Occurrence. Springer-Verlag, New York.

US Energy Information Administration, Petroleum, 2010.

US EPA. 2006. Environmental Protection Agency, 2006; http://www.epa. gov/air/urbanair/so$_2$.

Wallace, D. (Editor). 1988. *A Review of Analytical Methods for Bitumens and Heavy Oils.* Alberta Oil Sands Technology and Research Authority, Edmonton, Alberta, Canada.

Wallace, D., Starr, J., Thomas, K.P., and Dorrence, S.M. 1988. *Characterization of Oil Sands Resources.* Alberta Oil Sands Technology and Research Authority, Edmonton, Alberta, Canada.

Yang, Y.-Z., Liu, X.-G., and Xu, B.-S. 2014. Recent Advances in Molecular Imprinting Technology for the Deep Desulfurization of Fuel Oils. *New Carbon Mater.* 29(1): 1–14.

3 Feedstock Evaluation

3.1 INTRODUCTION

The chemical composition of petroleum and the related heavy oil, as well as extra heavy oil and tar sand bitumen petroleum products, is complex (Speight, 2001, 2014, 2015), which makes it essential that the most appropriate analytical methods are selected from a comprehensive list of methods and techniques for the analysis and evaluation of refinery feedstocks (Dean, 1998; Miller, 2000; Budde, 2001; Speight, 2001, 2005, 2014; Speight and Arjoon, 2012).

In the early days of petroleum processing, because of the character and easy-to-refine nature of petroleum, there was no need to understand the character and behavior of refinery feedstocks in the detail that is currently required. Refining was relatively simple and involved distillation of the valuable kerosene fraction that was then sold as an illuminant. After the commercialization of the internal combustion engine, the desired product became gasoline, and it was also obtained by distillation. Even when crude oil that contained little natural gasoline was used, cracking (i.e., thermal decomposition with simultaneous removal of distillate) became the modus operandi. However, with the demands for fuels placed on the petroleum industry during and after World War II, as well as the emergence of the age of petrochemicals and plastics, there was a need to produce materials not even considered as products in the decade before World War II. Thus, petroleum refining took on the role of technological innovator as newer and more innovative processes came on-stream with the accompanying advances in the use of catalysts and in reactor structure. In addition, there became a necessity to learn more about the character of refinery feedstocks so that refiners could engage in process predictability and plan a product slate that was based on market demand. It was a difficult task when the character of the crude oil was unknown, and the idea that petroleum refining should be a hit-and-miss affair was not acceptable.

In the modern refinery, strategies for upgrading petroleum emphasize differences in properties that, in turn, influence the choice of methods or combinations thereof for conversion of petroleum to various products (Speight and Ozum, 2002; Parkash, 2003; Hsu and Robinson, 2006; Gary et al., 2007; Speight, 2014). Naturally, similar principles are applied to heavy feedstocks (such as heavy oil, extra heavy oil, tar sand bitumen, and residua), and the availability of processes that can be employed to convert these feedstocks to usable products has increased significantly in recent years (Speight and Ozum, 2002; Parkash, 2003; Hsu and Robinson, 2006; Gary et al., 2007; Speight, 2014). However, before dealing with the methods used to evaluate petroleum feedstocks, it is necessary to briefly recap the nature of refinery processes. This will assist in putting evaluation methods into the correct perspective.

In general terms, refinery processes (Figure 3.1) can be conveniently divided into three different types (Speight and Ozum, 2002; Parkash, 2003; Hsu and Robinson, 2006; Gary et al., 2007; Speight, 2014): (1) *separation*—division of the feedstock into various streams (or fractions) depending on the nature of the crude material; (2) *conversion*—production of salable materials from the feedstock by skeletal alteration, or even by alteration of the chemical type, of the feedstock constituents; and (3) *finishing*—purification of the various product streams by a variety of processes that essentially remove impurities from the product.

In some cases, a fourth category can be added and includes processes such as the *reforming (molecular rearrangement) processes*. For the purposes of this text, reforming (and other) processes are included in the finishing processes because that is precisely what they are—processes designed to *finish* various refinery streams and render them ready for sale.

The separation and finishing processes may involve distillation or treatment with a wash solution. The conversion processes are usually regarded as those processes that change the number of

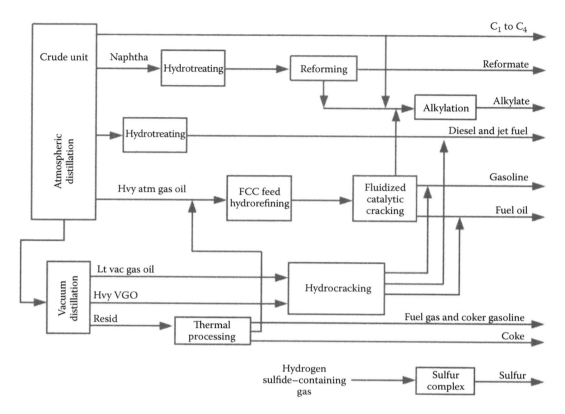

FIGURE 3.1   Representation of a petroleum refinery. (From Speight, J.G. 2014. *The Chemistry and Technology of Petroleum*. 5th Edition. CRC Press, Taylor & Francis Group, Boca Raton, FL, Figure 15.1, p. 392.)

carbon atoms per molecule (*thermal decomposition*), alter the molecular hydrogen–carbon ratio (*aromatization, hydrogenation*), or even change the molecular structure of the material without affecting the number of carbon atoms per molecule (*isomerization*).

Although it is possible to classify refinery operations using the three general terms just outlined, the behavior of various feedstocks in these refinery operations is not simple. The atomic ratios from ultimate analysis (Table 3.1) can give an indication of the nature of a feedstock and the generic hydrogen requirements to satisfy the refining chemistry (Speight, 2014); however, it is not possible to predict with any degree of certainty how the feedstock will behave during refining. Any deductions made from such data are pure speculation and are open to much doubt.

Thus, to determine the processability of petroleum, a series of consistent and standardized characterization procedures are required (Speight, 2015). These procedures can be used with a wide variety of feedstocks to develop a general approach to predict processability. The ability to predict the outcome of feedstock processing offers (1) the choice of processing sequences, (2) the potential for coke lay-down on the catalyst, (3) determining the catalyst tolerance to different feedstocks, (4) predictability of product distribution and quality, and (5) incompatibility during processing and incompatibility of the products on storage.

Upgrading of heavy feedstocks (such as heavy oil, extra heavy oil, tar sand bitumen, and residua) can be designed in an optimal manner by performing selected evaluations of chemical and structural features of these feedstocks (Schabron and Speight, 1996). However, molecular-level characterization of heavy feedstocks is difficult (if not impossible) because of low volatility, high polarity, and compositional polydispersity (Speight, 2001, 2014, 2015); however, the interest in the characterization of heavy feedstocks is the increased significance of bottom-of-the-barrel upgrading processes

TABLE 3.1

Elemental Composition and Ratios for Petroleum, Heavy Oil, Bitumen, and Petroleum Products

	%C	%H	%O	%N	%S	H/C (atomic)	N/C	O/C	S/C
Athabasca bitumen	83.1	10.3	1.3	0.4	4.9	1.487	0.004	0.012	0.022
Cold Lake heavy oil	83.7	10.4	1.1	0.4	4.4	1.491	0.004	0.010	0.020
Qayarah heavy oil	80.7	10.2	–	0.7	8.4	1.519	0.007	–	0.039
Crude oil	85.8	13.0	–	0.2	1.0	1.818	0.002	–	0.004
Gasoline[a]	86.0	14.0	–	–	–	1.903	–	–	–
Natural gas[a]	75.8	24.2	–	–	–	3.831	–	–	–

Source: Speight, J.G. 2000. *The Desulfurization of Heavy Oils and Residua.* Marcel Dekker Inc., New York, Table 3.1, p. 41.

[a] Included for comparison.

in modern refineries (Speight, 2000; Speight and Ozum, 2002; Parkash, 2003; Hsu and Robinson, 2006; Ancheyta and Speight, 2007; Gary et al., 2007; Speight, 2011, 2014). Therefore, proper characterization in terms of the refinability of heavy feedstocks is of utmost importance for individual unit performance and overall refinery optimization. The characterization schemes do not need to be complex but must focus on key parameters that affect processability. For example, the identification of the important features can be made with a saturates–aromatics–resin constituents–asphaltene separation (Speight, 2014). Subsequent analysis of the fractions also provides further information about the processability of the feedstock. In this follow-on analysis, there is often emphasis on the asphaltene fraction since the solubility of asphaltene constituents and the solubility of the thermal products have a dramatic affect on solids deposition and coke formation during upgrading. Further evaluation of the asphaltene constituents can lead to correlation of asphaltene properties with process behavior. In addition, useful information can be gained from knowledge of elemental composition and molecular weight, as well as size exclusion chromatography and high-performance liquid chromatography profiles.

Heavy feedstocks exhibit a wide range of physical properties, and several relationships can be made between various physical properties (Speight, 2014). Whereas properties such as the viscosity, density, boiling point, and color of petroleum may vary widely, the ultimate or elemental analysis varies, as already noted, over a narrow range for a large number of samples. The carbon content is relatively constant, while the hydrogen and heteroatom contents are responsible for the major differences between petroleums. Nitrogen, oxygen, and sulfur can be present in only trace amounts in some petroleums, which as a result consists primarily of hydrocarbons.

On the other hand, petroleum containing 9.5% heteroatoms may contain essentially hydrocarbon constituents insofar as the constituents contain at least one or more nitrogen, oxygen, and/or sulfur atoms within the molecular structures. Furthermore, it is the heteroelements that can have substantial effects on the distribution of refinery products. Coupled with the changes brought to the feedstock constituents by refinery operations, it is not surprising that refining the heavy feedstocks is a monumental task.

In the present context, heavy feedstocks (such as heavy oil, extra heavy oil, tar sand bitumen, and residua) can also be assessed in terms of sulfur content, carbon residue, nitrogen content, and metal content. Properties such as the API gravity and viscosity also help the refinery operator to gain an understanding of the nature of the material that is to be processed. The products from high-sulfur feedstocks often require extensive treatment to remove (or change) the corrosive sulfur compounds. Nitrogen compounds and the various metals that occur in crude oils will cause serious loss of catalyst life. The carbon residue presents an indication of the amount of thermal coke that may be formed to the detriment of the liquid products.

Thus, initial inspection of the feedstock (conventional examination of the physical properties) is necessary. From this, it is possible to make deductions about the most logical means of refining. In fact, evaluation of crude oils from physical property data about which refining sequences should be employed for any particular crude oil is a predominant part of the initial examination of any material that is destined for use as a refinery feedstock.

From a chemical standpoint, petroleum is an extremely complex mixture of hydrocarbon compounds, usually with minor amounts of nitrogen-, oxygen-, and sulfur-containing compounds, as well as trace amounts of metal-containing compounds (Speight, 2001, 2014, 2015). However, the composition of petroleum from different sources is not constant and, therefore, the chemical composition of a feedstock is a much truer indicator of refining behavior. Whether the composition is represented in terms of compound types or in terms of generic compound classes, it can enable the refiner to determine the nature of the reactions. Hence, chemical composition can play a large part in determining the nature of the products that arise from the refining operations. It can also play a role in determining the means by which a particular feedstock should be processed (Speight and Ozum, 2002; Parkash, 2003; Hsu and Robinson, 2006; Gary et al., 2007; Speight, 2014). Therefore, the judicious choice of a feedstock to produce any given product is just as important as the selection of the product for any given purpose. Alternatively, and more in keeping with the current context, the judicious choice of a processing sequence to convert a heavy feedstock to liquid products is even more important.

Heavy feedstocks are exceedingly complex and structured mixtures consisting predominantly of hydrocarbons, and containing sulfur, nitrogen, oxygen, and metals as minor constituents. Although sulfur has been reported in elemental form in some crude oils, most of the minor constituents occur in combination with carbon and hydrogen.

The physical and chemical characteristics of crude oils, and the yields and properties of products or factions prepared from them, vary considerably and are dependent on the concentration of the various types of hydrocarbons and minor constituents present. Some types of petroleum have economic advantages as sources of fuels and lubricants with highly restrictive characteristics because they require less specialized processing than that needed for the production of the same products from many types of crude oil. Others may contain unusually low concentrations of components that are desirable fuel or lubricant constituents, and the production of these products from such crude oils may not be economically feasible.

To satisfy specific needs with regard to the type of petroleum to be processed, as well as to the nature of the product, most refiners have, through time, developed their own methods of petroleum analysis and evaluation. However, such methods are considered proprietary and are not normally available. Consequently, various standards organizations, such as the American Society for Testing and Materials (ASTM) (Speight, 2015), have devoted considerable time and effort to the correlation and standardization of methods for the inspection and evaluation of petroleum and petroleum products. A complete discussion of the large number of routine tests available for petroleum fills an entire book (Speight, 2015). However, it seems appropriate that in any discussion of the physical properties of petroleum and petroleum products, reference should be made to the corresponding test, and accordingly, the various test numbers have been included in the text.

Thus, initial inspection of the nature of the feedstock will provide deductions about the most logical means of refining or correlation of various properties to structural types present, and hence attempted classification of the petroleum (Speight, 2014). Indeed, careful evaluation from physical property data is a major part of the initial study of any refinery feedstock. Proper interpretation of the data resulting from the inspection of crude oil requires an understanding of their significance.

Finally, not all tests that are applied to the evaluation of conventional petroleum are noted here. It is the purpose of this chapter to present the tests that are regularly applied to the evaluation of the more difficult feedstocks in a refinery, such as heavy oil, extra heavy oil, tar sand bitumen, and residua. However, tests such as those used to determine volatility—a characteristic that is not obvious in the heavy feedstocks—are omitted. The reader is referred to a more detailed discussion of petroleum evaluation for these tests (Speight, 2014).

3.2   FEEDSTOCK EVALUATION

3.2.1   ELEMENTAL (ULTIMATE) ANALYSIS

The analysis of petroleum feedstocks for the percentages of carbon, hydrogen, nitrogen, oxygen, and sulfur is perhaps the first method used to examine the general nature, and perform an evaluation, of a feedstock. The atomic ratios of the various elements to carbon (i.e., H/C, N/C, O/C, and S/C) are frequently used for indications of the overall character of the feedstock. It is also of value to determine the amounts of trace elements, such as vanadium and nickel, in a feedstock since these materials can have serious deleterious effects on catalyst performance during refining by catalytic processes.

However, it has become apparent, with the introduction of the heavier feedstocks into refinery operations, that these ratios are not the only requirement for predicting feedstock character before refining. The use of more complex feedstocks (in terms of chemical composition) has added a new dimension to refining operations. Thus, although atomic ratios, as determined by elemental analyses, may be used on a comparative basis between feedstocks, there is now no guarantee that a particular feedstock will behave as predicted from these data. Product slates cannot be predicted accurately, if at all, from these ratios.

The ultimate analysis (elemental composition) of petroleum is not reported to the same extent as it is for coal (Speight, 1994). Nevertheless, there are ASTM procedures for the ultimate analysis of petroleum and petroleum products; however, many such methods may have been designed for other materials. For example, carbon content can be determined by the method designated for coal and coke (ASTM D3178) or by the method designated for municipal solid waste (ASTM E777). There are also methods designated for the following:

1. Hydrogen content (ASTM D1018, ASTM D3178, ASTM D3343, ASTM D3701, and ASTM E777)
2. Nitrogen content (ASTM D3179, ASTM D3228, ASTM E148, ASTM E258, and ASTM E778)
3. Oxygen content (ASTM E385)
4. Sulfur content (ASTM D1266, ASTM D1552, ASTM D4045 and ASTM D4294)

However, because of the importance of sulfur in the current context, it is worth making several comments on the analyses of sulfur compounds, which is especially important to analysts in the petroleum industry. In the petroleum chemical industry, analysts must analyze various hydrocarbon matrices for sulfur compounds, to monitor trace odor problems, determine the recovery of sulfur from crude oil, and prevent catalyst poisoning. Environmental analysts detect sulfur compounds to monitor pollution and determine the origin and subsequent fate of various sulfur compounds. Although several methods can be used to monitor sulfur compounds in petroleum and petroleum products, there are several important advantages to using gas chromatography for these analyses (Speight, 2001, 2015). In contrast to methods that simply indicate total sulfur levels, gas chromatography allows individual compounds to be identified and quantified in a wide variety of samples, often at sensitivities of parts per billion or less. Samples can be gaseous, liquid, or solid. The method is particularly well suited to analyses of volatile sulfur compounds, which often are the compounds most important to the analyst.

More generally, of the data that are available, the proportions of the elements in petroleum vary only slightly over narrow limits (Speight, 2014, 2015). Yet, there is a wide variation in physical properties from the lighter, more mobile crude oils at one extreme to the heavier, asphaltic crude oils at the other extreme. Most of the more aromatic species and the heteroatoms occur in the higher-boiling fractions of feedstocks. The heavier feedstocks are relatively rich in these higher-boiling fractions.

Heteroatoms affect every aspect of refining. Sulfur is usually the most concentrated and is fairly easy to remove; many commercial catalysts are available that routinely remove 90% of the sulfur.

Nitrogen is more difficult to remove than sulfur, and there are fewer catalysts that are specific for nitrogen. If the nitrogen and sulfur are not removed, the potential for the production of nitrogen oxides (NO_x) and sulfur oxides (SO_x) during processing and use becomes real.

Perhaps the more pertinent property in the present context is the sulfur content, which along with the API gravity represents the two properties that have the greatest influence on the value of heavy feedstocks. The sulfur content varies from about 0.1% to about 3% by weight for the more conventional crude oils, to as much as 5% to 6% for heavy feedstocks (Table 3.2) (Speight, 2014). Depending on the sulfur content of the crude oil feedstock, the high-boiling fractions of petroleum, including the nonvolatile residua, typically have substantially higher sulfur content

TABLE 3.2
Variation in the Sulfur Content of Crude Oils

Petroleum	Specific Gravity	API Gravity	Sulfur (% w/w)
Agbami (Africa)	0.790	48.1	0.04
Alba (North Sea)	0.936	19.5	1.25
Athabasca bitumen (Alberta, Canada)[a]	1.014	8.0	4.80
Azeri BTC (Asia)	0.843	36.4	0.15
Bonny light (Nigeria)	0.850	35.3	0.15
Bozhong (Bohai Bay, China)	0.950	16.9	0.29
Brent (North Sea)	0.830	38.2	0.38
Cabinda (Angola)	0.860	32.2	0.15
Calypso (Trinidad and Tobago)	0.971	30.8	0.59
Captain North Sea	0.940	19.8	0.64
Clair (North Sea)	0.910	23.4	0.45
Coco (Democratic Republic of the Congo)	0.870	30.6	0.15
Cold Lake heavy oil (Alberta, Canada)[a]	0.933	20.4	3.0
Djeno (Republic of Congo)	0.890	27.3	0.39
Doba (Chad)	0.930	21.1	0.10
Draugen (Norwegian Sea)	0.826	39.9	0.16
Duri (Sumatra)	0.930	20.3	0.21
Eocene (partitioned zone—Kuwait/Saudi Arabia)	0.940	18.3	4.57
Escravos (Nigeria)	0.860	33.5	0.17
Hibernia blend (Canada)	0.850	35.0	0.53
Kuito (Angola)	0.920	22.1	0.74
Nemba (Angola)	0.830	39.8	0.21
Nigerian light	0.845	36.0	0.20
N'Kossa (Republic of Congo)	0.830	39.9	0.06
Palanca (Angola)	0.840	37.3	0.18
Pennington (Nigeria)	0.850	35.4	0.10
Qinghuangdao (China)	0.960	16.5	0.28
Ratawi (partitioned zone—Kuwait/Saudi Arabia)	0.910	24.2	4.10
Saudi heavy	0.887	28.0	2.10
Saudi light	0.855	34.0	2.00
USA (midcontinent, sweet)	0.825	40.0	0.40
USA (West Texas, sour)	0.865	32.0	1.90
Venezuela heavy	0.876	30.0	2.30
Venezuela light	0.855	24.0	1.50

[a] Included for comparison.

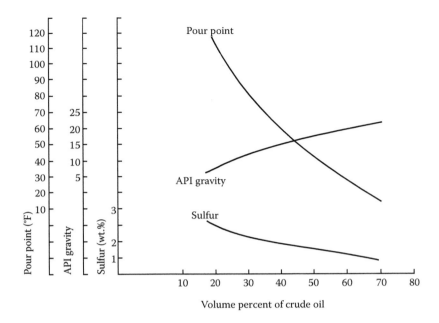

FIGURE 3.2 Variation of sulfur content, API gravity, and pour point with the depth of distillation cut. (From Speight, J.G. 2014. *The Desulfurization of Heavy Oils and Residua.* 2nd Edition. CRC Press, Taylor & Francis Group, Boca Raton, FL, Figure 1.6, p. 22.)

(Figure 3.2). Indeed, the very nature of the distillation process by which residua are produced, i.e., removal of distillate without thermal decomposition, dictates that the majority of the sulfur, which is predominantly in the higher molecular weight fractions, be concentrated in the residuum (Speight and Ozum, 2002; Parkash, 2003; Hsu and Robinson, 2006; Gary et al., 2007; Speight, 2014).

3.2.2 Metal Content

Metals (particularly vanadium and nickel) are found in most crude oils. Heavy feedstocks contain relatively high proportions of metals either in the form of salts or as organometallic constituents (such as the metallo-porphyrins), which are extremely difficult to remove from the feedstock. Indeed, the nature of the process by which residua are produced virtually dictates that all the metals in the original crude oil are concentrated in the residuum (Speight, 2014). Those metallic constituents that may actually volatilize under the distillation conditions and appear in the higher-boiling distillates are the exceptions here.

Metals cause particular problems because they poison catalysts used for sulfur and nitrogen removal as well as other processes such as catalytic cracking (Speight and Ozum, 2002; Parkash, 2003; Hsu and Robinson, 2006; Gary et al., 2007; Speight, 2014). Thus, serious attempts are being made to develop catalysts that can tolerate a high concentration of metals without serious loss of catalyst activity or catalyst life.

A variety of tests (ASTM D1026, ASTM D1262, ASTM D1318, ASTM D3341, and ASTM D3605) have been designated for the determination of metals in petroleum products. Determination of metals in whole feeds can be accomplished by combustion of the sample so that only inorganic ash remains. The ash can then be digested with an acid, and the solution examined for metal species by atomic absorption spectroscopy or by inductively coupled argon plasma spectrometry.

3.2.3 DENSITY AND SPECIFIC GRAVITY

Density (or specific gravity) of all of the physicochemical properties of heavy feedstocks can give valuable information about heavy feedstock behavior; however, it is preferable that density is used along with other physical property data since one property alone does not give adequate information (Speight, 2001, 2014, 2015).

Density is the mass of a unit volume of material at a specified temperature and has the dimensions of grams per cubic centimeter (a close approximation to grams per milliliter). *Specific gravity* is the ratio of the mass of a volume of the substance to the mass of the same volume of water and is dependent on two temperatures, those at which the masses of the sample and the water are measured. When the water temperature is 4°C (39°F), the specific gravity is equal to the density in the centimeter–gram–second (cgs) system, since the volume of 1 g of water at that temperature is, by definition, 1 mL. Thus, the density of water, for example, varies with temperature, and its specific gravity at equal temperatures is always unity. The standard temperatures for a specific gravity in the petroleum industry in North America are 60/60°F (15.6/15.6°C).

The density and specific gravity of crude oil (ASTM D287, ASTM D1298, ASTM D1217, and ASTM D1555) are two properties that have found wide use in the industry for preliminary assessment of the character of the crude oil. In particular, density was used to give an estimate of the most desirable product, i.e., kerosene, in crude oil. Thus, a conventional crude oil with a high content of paraffins (high kerosene) and excellent mobility at ambient temperature and pressure may have a specific gravity (density) of about 0.8. Heavy feedstocks with a high content of asphaltene and resin constituents (low kerosene) and poor mobility at ambient temperature and pressure, thereby requiring vastly different processing sequences, may have a specific gravity (density) on the order of 0.95 (Table 3.2) (Speight, 2014).

Specific gravity is influenced by the chemical composition of petroleum; however, quantitative correlation is difficult to establish. Nevertheless, it is generally recognized that increased amounts of aromatic compounds result in an increase in density, whereas an increase in saturated compounds results in a decrease in density. It is also possible to recognize certain preferred trends between the API gravity of crude oils and residua, and one or more of the other physical parameters. For example, a correlation exists between the API gravity and sulfur content, Conradson carbon residue, viscosity, asphaltene plus resin content, and nitrogen content (Speight, 2014). However, the derived relationships between the density of petroleum and its fractional composition were valid only if they were applied to a certain type of petroleum, and lost some of their significance when applied to different types of petroleum. Nevertheless, density is still used to give a rough estimation of the nature of petroleum and petroleum products.

The values for density (and specific gravity) cover an extremely narrow range (Table 3.2) considering the differences in the feedstock appearance and behavior. In an attempt to inject a more meaningful relationship between the physical properties and processability of the various crude oils, the American Petroleum Institute devised a measurement of gravity based on the Baumé scale for industrial liquids. The Baumé scale for liquids lighter than water was used initially:

$$\text{Degree Baumé} = 140/\text{sp gr at } 60/60°F - 130$$

However, a considerable number of hydrometers calibrated according to the Baumé scale were found at an early period to be in error by a consistent amount, and this led to the adoption of the equation

$$\text{Degree API} = 141.5/\text{sp gr at } 60/60°F - 131.5$$

The specific gravity of petroleum usually ranges from about 0.8 (45.3° API) for the lighter crude oils to over 1.0 (10° API) for heavy crude oil and tar sand bitumen (Tables 3.2 and 3.3). Distillation

TABLE 3.3

Variation in the Properties of Various Residua

	API Gravity	Sulfur % w/w	Nitrogen % w/w	Nickel % w/w	Vanadium % w/w	Asphaltene Constituents (C_7)% w/w	Carbon Residue % w/w
Atmospheric Residua							
Alaska, North Slope crude oil, >650°F	15.2	1.6	0.4	18.0	30.0	2.0	8.5
Arabian heavy crude oil, >650°F	11.9	4.4	0.3	27.0	103.0	8.0	14
Arabian light crude oil, >650°F	17.7	3	0.2	10.0	26.0	1.8	7.5
Kuwait crude oil, >650°F	13.9	4.4	0.3	14.0	50.0	2.4	12.2
Lloydminster heavy oil (Canada), >650°F	10.3	4.1	0.3	65.0	141.0	14.0	12.1
Taching crude oil, >650°F	27.3	0.2	0.2	5.0	1.0	4.4	3.8
Vacuum Residua							
Alaska, North Slope crude oil, >1050°F	8.2	2.2	0.6	47.0	82.0	4.0	18
Arabian heavy crude oil, >1050°F	7.3	5.1	0.3	40.0	174.0	10.0	19
Arabian light crude oil, >1050°F	8.5	4.4	0.5	24.0	66.0	4.3	14.2
Kuwait crude oil, >1050°F	5.5	5.5	0.4	32.0	102.0	7.1	23.1
Lloydminster heavy oil (Canada) >1050°F	8.S	4.4	0.6	115.0	252.0	18.0	21.4
Taching crude oil, >650°F	21.5	0.3	0.4	9.0	2.0	7.6	7.9

residua are expected to have API gravity on the order of 5° to 10° API. This is in keeping with the general trend that increased aromaticity leads to a decrease in API gravity (or, more correctly, an increase in specific gravity) (Speight, 2014).

Density or specific gravity or API gravity may be measured by using a hydrometer (ASTM D287 and D1298) or a pycnometer (D1217). The variation of density with temperature, effectively the coefficient of expansion, is a property of great technical importance since most petroleum products are sold by volume, and specific gravity is usually determined at the prevailing temperature (21°C, 70°F) rather than at the standard temperature (60°F, 15.6°C). The tables of gravity corrections (ASTM D1555) are based on an assumption that the coefficient of expansion of all petroleum products is a function (at fixed temperatures) of density only.

The specific gravity of bitumen shows a fairly wide range of variation. The largest degree of variation is usually due to local conditions that affect material close to the faces, or exposures, occurring in surface oil sand beds. There are also variations in the specific gravity of the bitumen found in beds that have not been exposed to weathering or other external factors. The range of specific gravity variation is usually of the order of 0.9–1.04 (Table 3.2).

A very important property of the Athabasca bitumen (which also accounts for the success of the hot water separation process) is the variation in density (specific gravity) of the bitumen with temperature (Speight, 2009, 2011, 2013a,b, 2014). Over the temperature range 30–130°C (85–265°F), there is a density–water inversion in which the bitumen is lighter than water. Flotation of the bitumen (with aeration) on the water is facilitated, hence the logic of the hot water separation process (Speight, 2009, 2011, 2013a,b, 2014).

3.2.4 Viscosity

Viscosity is an important single fluid characteristic governing the motion of petroleum and petroleum products, and is actually a measure of the internal resistance to motion of a fluid by reason of the forces of cohesion between molecules or molecular groupings.

By definition, viscosity is the force in dynes required to move a plane of 1 cm² area at a distance of 1 cm from another plane of 1 cm² area through a distance of 1 cm in 1 s. In the cgs system, the unit of viscosity is the poise (P) or centipoise (cP) (1 cP = 0.01 P). Two other terms in common use are *kinematic viscosity* and *fluidity*. Kinematic viscosity is the viscosity in centipoise divided by the specific gravity. The unit of kinematic viscosity is the stoke (cm²/s), although centistokes (0.01 cSt) is in more common use; fluidity is simply the reciprocal of viscosity.

In the early days of the petroleum industry, viscosity was regarded as the *body* of an oil—a significant number for lubricants or for any liquid pumped or handled in quantity. The changes in viscosity with temperature, pressure, and rate of shear are pertinent not only in lubrication but also for such engineering concepts as heat transfer. The viscosity and relative viscosity of different phases, such as gas, liquid oil, and water, are determining influences in producing the flow of reservoir fluids through porous oil-bearing formations. The rate and amount of oil production from a reservoir are often governed by these properties.

The viscosity (ASTM D445, ASTM D88, ASTM D2161, ASTM D341, and ASTM D2270) of crude oils varies markedly over a very wide range. Values vary from <10 cP at room temperature to many thousands of centipoise at the same temperature. In the present context, oil sand bitumen occurs at the higher end of this scale, where a relationship between viscosity and density between various crude oils has been noted (Speight, 2014).

Many types of instruments have been proposed for the determination of viscosity. The simplest and most widely used are capillary types (ASTM D445), and the viscosity is derived from the equation

$$n = Br^4P/8nl$$

In the equation, r is the tube radius, l the tube length, P the pressure difference between the ends of a capillary, n the coefficient of viscosity, and B the quantity discharged in unit time. Not only are such capillary instruments the most simple, but when designed in accordance with known principle and used with known necessary correction factors, they are probably the most accurate viscometers available. It is usually more convenient, however, to use relative measurements, and for this purpose the instrument is calibrated with an appropriate standard liquid of known viscosity.

Batch flow times are generally used; in other words, the time required for a fixed amount of sample to flow from a reservoir through a capillary is the datum that is actually observed. Any features of a technique that contribute to longer flow times are usually desirable. Some of the principal capillary viscometers in use are those of Cannon-Fenske, Ubbelohde, Fitzsimmons, and Zeitfuchs.

The Saybolt universal viscosity (SUS) (ASTM D88) is the time in seconds required for the flow of 60 mL of petroleum from a container, at constant temperature, through a calibrated orifice. The Saybolt furol viscosity (SFS) (ASTM D88) is determined in a similar manner except that a larger orifice is employed.

As a result of the various methods for viscosity determination, it is not surprising that much effort has been spent on interconversion of the several scales, especially converting Saybolt to kinematic viscosity (ASTM D2161),

$$\text{Kinematic viscosity} = a \times \text{Saybolt } s + b/\text{Saybolt } s$$

In this equation, a and b are constants.

The SUS equivalent to a given kinematic viscosity varies slightly with the temperature at which the determination is made because the temperature of the calibrated receiving flask used in the Saybolt method is not the same as that of the oil. Conversion factors are used to convert kinematic viscosity from 2 to 70 cSt at 38°C (100°F) and 99°C (210°F) to equivalent Saybolt universal viscosity in seconds. Appropriate multipliers are listed to convert kinematic viscosities >70 cSt. For a kinematic viscosity determined at any other temperature, the equivalent Saybolt universal value is calculated by use of the Saybolt equivalent at 38°C (100°F) and a multiplier that varies with the temperature:

$$\text{Saybolt } s \text{ at } 100°F \ (38°C) = cSt \times 4.635$$

$$\text{Saybolt } s \text{ at } 210°F \ (99°C) = cSt \times 4.667$$

Various studies have also been made on the effect of temperature on viscosity since the viscosity of petroleum, or a petroleum product, decreases as the temperature increases. The rate of change appears to depend primarily on the nature or composition of the petroleum; however, other factors, such as volatility, may also have a minor effect. The effect of temperature on viscosity is generally represented by the equation

$$\log \log (n + c) = A + B \log T$$

where n is absolute viscosity, T is temperature, and A and B are constants. This equation has been sufficient for most purposes and has come into very general use. The constants A and B vary widely with different oils, but c remains fixed at 0.6 for all oils having a viscosity over 1.5 cSt; it increases only slightly at lower viscosity (0.75 at 0.5 cSt). The viscosity–temperature characteristics of any oil, so plotted, thus create a straight line, and the parameters A and B are equivalent to the intercept and slope of the line. To express the viscosity and viscosity–temperature characteristics of an oil, the slope and the viscosity at one temperature must be known; the usual practice is to select 38°C (100°F) and 99°C (210°F) as the observation temperatures.

Suitable conversion tables are available (ASTM D341), and each table or chart is constructed in such a way that for any given petroleum or petroleum product, the viscosity–temperature points result in a straight line over the applicable temperature range. Thus, only two viscosity measurements need be made at temperatures far enough apart to determine a line on the appropriate chart from which the approximate viscosity at any other temperature can be read.

The charts can be applicable only to measurements made in the temperature range in which petroleum is assumed to be a Newtonian liquid. The oil may cease to be a simple liquid near the cloud point because of the formation of wax particles or, near the boiling point, because of vaporization. Thus, the charts do not give accurate results when either the cloud point or boiling point is approached. However, they are useful over the Newtonian range for estimating the temperature at which an oil attains a desired viscosity. The charts are also convenient for estimating the viscosity of a blend of petroleum liquids at a given temperature when the viscosity of the component liquids at the given temperature is known (Speight, 2014).

Since the viscosity–temperature coefficient of lubricating oil is an important expression of its suitability, a convenient number to express this property is very useful, and hence, a viscosity index (ASTM D2270) was derived. It is established that naphthenic oils have higher viscosity–temperature coefficients than do paraffinic oils at equal viscosity and temperatures. The Dean and Davis scale was based on the assignment of a zero value to a typical naphthenic crude oil and that of 100 to a typical paraffinic crude oil; intermediate oils were rated by the formula

$$\text{Viscosity index} = L - U/L - H \times 100$$

where L and H are the viscosities of the 0 and 100 index reference oils, both having the same viscosity at 99°C (210°F), and U is that of the unknown, all at 38°C (100°F). Originally, the viscosity index was calculated from Saybolt viscosity data; however, subsequently, figures were provided for kinematic viscosity.

3.2.5 CARBON RESIDUE

The carbon residue (ASTM D189 and ASTM D524) of a crude oil is a property that can be correlated with several other properties of petroleum (Speight, 2014). The carbon residue presents indications of the volatility or gasoline-forming propensity of the feedstock and, for the most part in this text, the coke-forming propensity of a feedstock. Tests for carbon residue are sometimes used to evaluate the carbonaceous depositing characteristics of fuels used in certain types of oil-burning equipment and internal combustion engines.

There are two older well-used methods for determining the carbon residue: the Conradson method (ASTM D189) and the Ramsbottom method (ASTM D524). Both are equally applicable to the high-boiling fractions of crude oils that decompose to volatile material and coke when distilled at a pressure of 1 atm. Heavy feedstocks that contain metallic constituents (and distillation of crude oils concentrates these constituents in the residua) will have erroneously high carbon residues. Thus, the metallic constituents must first be removed from the oil, or they can be estimated as ash by complete burning of the coke after carbon residue determination.

Although there is no exact correlation between the two methods, it is possible to interconnect the data; however, caution is advised when using that portion of the curve below 0.1% w/w Conradson carbon residue.

Recently, a newer method (ASTM D4530) has also been accepted that requires smaller sample amounts and was originally developed as a thermogravimetric method. The carbon residue produced by this method is often referred to as the microcarbon residue (MCR). Agreements between the data from the three methods are good, making it possible to interrelate all of the data from carbon residue tests (Long and Speight, 1989).

Although the three methods have their relative merits, there is a tendency to advocate use of the more expedient microcarbon method to the exclusion of the Conradson method and Ramsbottom method because of the lesser amounts required in the microcarbon method.

The mechanical design and operating conditions of such equipment have such a profound influence on carbon deposition during service that comparison of carbon residues between oils should be considered as giving only a rough approximation of relative deposit-forming tendencies. Recent work has focused on the carbon residue of the different fractions of crude oils, especially the asphaltene constituents (Speight, 2014). A more precise relationship between carbon residue and hydrogen content, H/C atomic ratio, nitrogen content, and sulfur content has been shown to exist (Speight, 2014). These data can provide more precise information about the anticipated behavior of a variety of feedstocks in thermal processes.

Because of the extremely small values of carbon residue obtained by the Conradson and Ramsbottom methods when applied to the lighter distillate fuel oils, it is customary to distill such products to 10% residual oil and determine the carbon residue thereof. Such values may be used directly in comparing fuel oils, as long as it is kept in mind that the values are carbon residues on 10% residual oil and are not to be compared with straight carbon residues.

3.2.6 SPECIFIC HEAT

Specific heat is defined as the quantity of heat required to raise a unit mass of material through one degree of temperature (ASTM D2766).

Specific heats are extremely important engineering quantities in refinery practice because they are used in all calculations on heating and cooling petroleum products. Many measurements have been made on various hydrocarbon materials; however, the data for most purposes may be summarized by the general equation

$$C = 1/d \ (0.388 + 0.00045t)$$

where C is the specific heat at $t°F$ of an oil whose specific gravity 60/60°F is d; thus, specific heat increases with temperature and decreases with specific gravity.

3.2.7 HEAT OF COMBUSTION

The gross heat of combustion of crude oil and its products are given with fair accuracy by the equation

$$Q = 12,400 - 2100d^2$$

where d is the 60/60°F specific gravity. Deviation is generally <1%, although many highly aromatic crude oils show considerably higher values; the ranges for crude oil is 10,000–11,600 cal/g (Speight, 2014). For gasoline, the heat of combustion is 11,000–11,500 cal/g, and for kerosene (and diesel fuel) it falls in the range 10,500–11,200 cal/g. Finally, the heat of combustion for fuel oil is on the order of 9500–11,200 cal/gm. Heats of combustion of petroleum gases may be calculated from the analysis and data for the pure compounds. Experimental values for gaseous fuels may be obtained by measurement in a water flow calorimeter, and heats of combustion of liquids are usually measured in a bomb calorimeter.

For thermodynamic calculation of equilibria useful in petroleum science, combustion data of extreme accuracy are required because the heats of formation of water and carbon dioxide are large in comparison with those in the hydrocarbons. Great accuracy is also required of the specific heat data for the calculation of free energy or entropy. Much care must be exercised in selecting values from the literature for these purposes, since many of those available were determined before the development of modern calorimetric techniques.

3.3 CHROMATOGRAPHIC METHODS

Feedstock evaluation by separation into various fractions has been used quite successfully for several decades. Knowledge of the components of the feedstock on a before-refining and after-refining basis is a valuable aid to process development and monitoring (Speight and Ozum, 2002; Parkash, 2003; Hsu and Robinson, 2006; Gary et al., 2007; Speight, 2014). In addition, such information has also been a valuable aid to process development, and there are several standard test methods for feedstock/product evaluation. These are (1) separation of aromatic and nonaromatic fractions from high-boiling oils (ASTM D2549), (2) determination of hydrocarbon groups in rubber extender oils by clay–gel adsorption (ASTM D2007), and (3) determination of hydrocarbon types in liquid petroleum products by a fluorescent indicator adsorption test (ASTM D1319).

Gel permeation chromatography is an attractive technique for the determination of the number average molecular weight (M_n) distribution of petroleum fractions, especially the heavier constituents, and petroleum products (Reynolds and Biggs, 1988). Ion exchange chromatography is also widely used in the characterization of petroleum constituents and products. For example, cation exchange chromatography can be used primarily to isolate the nitrogen constituents in petroleum (McKay et al., 1981), thereby giving an indication of how the feedstock might behave during refining and also an indication of any potential deleterious effects on catalysts.

Liquid chromatography (also called adsorption chromatography) has helped characterize the group composition of crude oils and hydrocarbon products since the beginning of this century. The type and relative amount of certain hydrocarbon classes in the matrix can have a profound effect on the quality and performance of the hydrocarbon product. The fluorescent indicator adsorption method (ASTM D1319) has been used to measure the paraffinic, olefinic, and aromatic content of gasoline, jet fuel, and liquid products in general.

High-performance liquid chromatography (HPLC) has found great utility in separating different hydrocarbon group types and identifying specific constituent types. Of particular interest is the application of the HPLC technique to the identification of the molecular types in the heavier feedstocks, especially the molecular types in the asphaltene fraction. This technique is especially useful for studying such materials on a before-processing and after-processing basis (Speight, 1986, 2014; Long and Speight, 1998). The general advantages of HPLC are as follows: (1) each sample may be analyzed as received even though the boiling range may vary over a considerable range; (2) the total time per analysis is usually on the order of minutes; and, perhaps most important, (3) the method can be adapted for on-stream analysis in a refinery.

In recent years, supercritical fluid chromatography has found use in the characterization and identification of petroleum constituents and products. A supercritical fluid is defined as a substance above its critical temperature. A primary advantage of chromatography using supercritical mobile phases results from the mass transfer characteristics of the solute. The increased diffusion coefficients of supercritical fluids compared with liquids can lead to greater speed in separations or greater resolution in complex mixture analyses.

3.4 MOLECULAR WEIGHT

Even though refining produces, in general, lower molecular weight species than those originally in the feedstock, there is still the need to determine the molecular weight of the original constituents as well as the molecular weights of the products as a means of understanding the process. For those original constituents and products, e.g., resin constituents and asphaltene constituents, that have little or no volatility, vapor pressure osmometry has been proven to be of considerable value.

A particularly appropriate method involves the use of different solvents (at least two), and the data are then extrapolated to infinite dilution. There has also been the use of different temperatures for a particular solvent after which the data are extrapolated to room temperature (Speight et al., 1985; Speight, 1987). In this manner, different solvents are employed and the molecular weight of a petroleum fraction (particularly the asphaltene constituents) can be determined for which it can be assumed that there is little or no influence from any intermolecular forces. In summary, the molecular weight may be as close to the real value as possible.

In fact, it is strongly recommended that to negate concentration effects and temperature effects, the molecular weight determination be carried out at three different concentrations at three different temperatures. The data for each temperature are then extrapolated to zero concentration, and the zero concentration data at each temperature are then extrapolated to room temperature (Speight, 1987).

A correlation relating molecular weight, asphaltene content, and heteroatom content with the carbon residues of whole residua has been developed and has been extended to molecular weight and carbon residue (Schabron and Speight, 1997a,b). The linear correlation holds for the whole residua and residue-forming materials such as asphaltene constituents, saturates, aromatics, and polar constituents. At molecular weights greater than about 3000, the result suggests that the correlation can be used as a tool to quantitatively gauge association for asphaltene constituents. Subsequent results suggest that the inclusion of heteroatoms is not necessary in the relationship, since heteroatom associative affects are already taken into account by measuring the apparent molecular weights of the asphaltene constituents.

3.5 OTHER PROPERTIES

Petroleum and the majority of petroleum products are liquids at ambient temperature, and problems that may arise from solidification during normal use are not common. Nevertheless, the melting point is a test (ASTM D87 and ASTM D127) that is widely used by suppliers of wax and by the wax consumers; it is particularly applied to the highly paraffinic or crystalline waxes. In the present context, tar sand bitumen and many residua may be solid at ambient temperature, and it may be necessary to know the melting point to prevent solidification in pipes. Two other properties that may be of value (but only to a lesser extent) to assist in the evaluation of heavy feedstocks are the pour point (ASTM D97) and the aniline point (ASTM D611).

The *pour point* of a crude oil product is the lowest temperature at which the oil will pour or flow under prescribed conditions. It is an approximate indicator of the relative paraffinic character and aromatic character of the material. For heavy feedstocks, the pour point is typically high (above 0°C; 32°F) and is more an indication of the temperatures (or conditions) required to move the material from one point in the refinery to another.

The *aniline point* is the temperature at which equal parts of aniline and the oil are completely miscible. For oils of a given type, the aniline point increases slightly with molecular weight but increases markedly with paraffinic character, and may therefore be used to obtain an approximate estimation of aromatics content. Aniline point determinations are only infrequently applied to heavy feedstocks since their very character, and the other evaluation methods outlined here, indicates such feedstocks to be extremely complex with high proportions of ring systems (aromatic constituents and naphthene constituents).

Spectroscopic methods have played an important role in the evaluation of petroleum and of petroleum products for the last three decades, and many of the methods are now used as standard methods of analysis for refinery feedstocks and products. Application of these methods to feedstocks and products is a natural consequence for the refiner.

Nowhere is the contribution of spectroscopic studies more emphatic than in application to the delineation of structural types in the heavier feedstocks. This has been necessary because of the unknown nature of these feedstocks by refiners. One particular example is the n.d.M. method (refractive index–density–molecular weight method) (ASTM D3238), which is designed for the carbon distribution and structural group analysis of petroleum oils. Later investigators have taken structural group analysis several steps further than the n.d.M. method (Speight, 2014, 2015).

Conventional infrared spectroscopy yields information about the functional features of various petroleum constituents. For example, infrared spectroscopy will aid in the identification of N–H and O–H functions, the nature of polymethylene chains, the C–H out-of-place bending frequencies, and the nature of any polynuclear aromatic systems (Speight, 1994, 2014).

With the recent progress of Fourier transform infrared spectroscopy, quantitative estimates of the various functional groups can also be made. This is particularly important for application to the higher molecular weight solid constituents of petroleum (i.e., the asphaltene fraction). It is also possible to derive structural parameters from infrared spectroscopic data, and these are (1) saturated hydrogen to saturated carbon ratio, (2) paraffinic character, (3) naphthenic character, (4) methyl group content, and (5) paraffin chain length.

Nuclear magnetic resonance has frequently been employed for general studies and for the structural studies of petroleum constituents (Speight, 2014). In fact, proton magnetic resonance studies (along with infrared spectroscopic studies) were, perhaps, the first studies of the modern era that allowed structural inferences to be made about the polynuclear aromatic systems that occur in the high molecular weight constituents of petroleum. In general, the proton (hydrogen) types in petroleum fractions can be subdivided into different types, such as (1) aromatic ring hydrogen, (2) aliphatic hydrogen adjacent to an aromatic ring, (3) aliphatic hydrogen remote from an aromatic ring, (4) naphthenic hydrogen, (5) methylene hydrogen, and (6) terminal methyl hydrogen remote from an aromatic ring. Other ratios are also derived from which a series of structural parameters can be calculated. However, it must be remembered that the structural details of the carbon backbone obtained from proton spectra are derived by inference, but it must be recognized that protons at peripheral positions can be obscured by intermolecular interactions. This, of course, can cause errors in the ratios that can have a substantial influence on the outcome of the calculations.

It is in this regard that carbon-13 magnetic resonance can play a useful role. Since carbon magnetic resonance deals with analyzing the carbon distribution types, the obvious structural parameter to be determined is the aromaticity, and a direct determination from the various carbon type environments is one of the better methods for the determination of aromaticity. Thus, through a combination of proton and carbon magnetic resonance techniques, refinements can be made on the structural parameters and, for solid-state high-resolution carbon magnetic resonance, additional structural parameters can be obtained.

Mass spectrometry can play a key role in the identification of the constituents of feedstocks and products. The principal advantages of mass spectrometric methods are (1) the high reproducibility of quantitative analyses, (2) the potential for obtaining detailed data on the individual components and/or carbon number homologues in complex mixtures, and (3) the minimal sample size required for analysis. The ability of mass spectrometry to identify individual components in complex mixtures is unmatched by any modern analytical technique. Perhaps the exception is gas chromatography.

However, there are disadvantages arising from the use of mass spectrometry: (1) the limitation of the method to organic materials that are volatile and stable at temperatures up to 300°C (570°F), and (2) the difficulty of separating isomers for absolute identification. The sample is usually destroyed; however, this is seldom a disadvantage.

Nevertheless, in spite of these limitations, mass spectrometry does furnish useful information about the composition of feedstocks and products even if this information is not as exhaustive as might be required. There are structural similarities that might hinder identification of individual components. Consequently, identification by type or by homologue will be more meaningful since similar structural types may be presumed to behave similarly in processing situations. Knowledge of the individual isomeric distribution may add only a little to the understanding of the relationships between composition and processing parameters.

Mass spectrometry should be used discriminately where a maximum amount of information can be expected. The heavier nonvolatile feedstocks are for practical purposes, beyond the useful range of routine mass spectrometry. At the elevated temperatures necessary to encourage volatility, thermal decomposition will occur in the inlet, and any subsequent analysis would be biased to the low molecular weight end and to the lower molecular products produced by the thermal decomposition.

3.6   USE OF THE DATA

It has been asserted that more needs to be done in correlating analytical data obtained for heavy feedstocks with processability (Speight, 1984, 2014; Reynolds 1991). In particular, measurements of properties are needed that reflect the uniqueness of individual heavy feedstocks and processes such as hydrodesulfurization and hydrodemetallization (Reynolds, 1991). Currently, used coke yield predictors are simplistic and feedstock specific (Beret and Reynolds, 1990; Hsu and Robinson, 2006; Gary et al., 2007; Speight, 2014).

Standard analyses on whole heavy feedstocks—such as determinations of elemental compositions and molecular weight—have not served to be reliable predictors of processability. Many techniques have been demonstrated as being applicable to the characterization of the constituents of heavy feedstocks; however, few have been developed to a satisfactory stage of systematic use. Inverse gas–liquid chromatography can be used to indicate the increase in processing difficulties with increasing polarity of the constituents of heavy feedstocks; however, it does not suggest how such difficulties may be solved. Determining average structural features is of limited value and does not appear to be greatly helpful (Speight, 2014).

The data derived from any one or more of the evaluation techniques described here give an indication of the feedstock behavior. The data can also be employed to give the refiner a view of the differences between different residua (Speight and Ozum, 2002; Parkash, 2003; Hsu and Robinson, 2006; Gary et al., 2007; Speight, 2014) (and other feedstocks), thereby presenting an indication of the means by which the crude feedstock should be processed and for the prediction of product properties (Dolbear et al., 1987).

It must be emphasized that to proceed from the raw evaluation data to full-scale production is not the preferred step. Further evaluation of the processability of the feedstock is usually necessary through the use of a pilot-scale operation from which it will be possible to correlate the data obtained from the actual plant operations (as well as the pilot plant data) with one or more of the physical properties determined as part of the feedstock evaluation.

Predicting processability is a matter of understanding the chemical and physical composition of the feedstock as it relates to properties, refining behavior, and product yields (Speight, 1992). Measurements can be made to give the refiner an indication of the feedstock behavior during upgrading and, also, the prediction of process options that may lead to improvements in product character and/or yield. Because of feedstock complexity, there are disadvantages on relying on the use of bulk properties as the sole means of predicting behavior (Dolbear et al., 1987). Although the early studies primarily focused on the composition and behavior of asphalt, the techniques developed for those investigations have provided an excellent means of studying heavy feedstocks (Speight, 2014).

Fractionation of heavy feedstocks into components of interest and then studying the components appears to be a better approach than obtaining data on whole residua. By careful selection of a characterization scheme, it may be possible to obtain a detailed overview of feedstock composition that can be used for process predictions. Thus, fractionation methods also play a role, along with the physical testing methods, of evaluating heavy oil, extra heavy oil, and tar sand bitumen—residua are already present in the refinery as products of the distillation process when applied to lighter crude oils. For example, by careful selection of an appropriate technique, it is possible to obtain a detailed "map" of feedstock or product composition that can be used for process predictions (Speight, 2001, 2014, 2015).

REFERENCES

Ancheyta, J., and Speight, J.G. 2007. *Hydroprocessing of Heavy Oils and Residua*. CRC-Taylor & Francis Group, Boca Raton, FL.

ASTM D87. 2014. Standard Test Method for Melting Point of Petroleum Wax (Cooling Curve). In: *Annual Book of Standards*. ASTM International, West Conshohocken, PA.

ASTM D88. 2014. Standard Test Method for Saybolt Viscosity. In: *Annual Book of Standards*. ASTM International, West Conshohocken, PA.

ASTM D97. 2014. Standard Test Method for Pour Point of Petroleum Products. In: *Annual Book of Standards*. ASTM International, West Conshohocken, PA.

ASTM D127. 2014. Standard Test Method for Drop Melting Point of Petroleum Wax, Including Petrolatum. In: *Annual Book of Standards*. ASTM International, West Conshohocken, PA.

ASTM D189. 2014. Standard Test Method for Conradson Carbon Residue of Petroleum Products. In: *Annual Book of Standards*. ASTM International, West Conshohocken, PA.

ASTM D287. 2014. Standard Test Method for API Gravity of Crude Petroleum and Petroleum Products (Hydrometer Method). In: *Annual Book of Standards*. ASTM International, West Conshohocken, PA.

ASTM D341. 2014. Standard Practice for Viscosity–Temperature Charts for Liquid Petroleum Products. In: *Annual Book of Standards*. ASTM International, West Conshohocken, PA.

ASTM D445. 2014. Standard Test Method for Kinematic Viscosity of Transparent and Opaque Liquids (and Calculation of Dynamic Viscosity). In: *Annual Book of Standards*. ASTM International, West Conshohocken, PA.

ASTM D524. 2014. Standard Test Method for Ramsbottom Carbon Residue of Petroleum Products. In: *Annual Book of Standards*. ASTM International, West Conshohocken, PA.

ASTM D611. 2014. Standard Test Methods for Aniline Point and Mixed Aniline Point of Petroleum Products and Hydrocarbon Solvents Active Standard (Latest Version). In: *Annual Book of Standards*. ASTM International, West Conshohocken, PA.

ASTM D1018. 2014. Standard Test Method for Hydrogen in Petroleum Fractions. In: *Annual Book of Standards*. ASTM International, West Conshohocken, PA.

ASTM D1026-82. Method of Test for Sodium in Lubricating Oils and Additives (Gravimetric Method).

ASTM D1262-81. Method of Test for Lead in New and Used Greases.

ASTM D1217. 2014. Standard Test Method for Density and Relative Density (Specific Gravity) of Liquids by Bingham Pycnometer. In: *Annual Book of Standards*. ASTM International, West Conshohocken, PA.

ASTM D1266. 2014. Standard Test Method for Sulfur in Petroleum Products (Lamp Method). In: *Annual Book of Standards*. ASTM International, West Conshohocken, PA.

ASTM D1298. 2014. Standard Test Method for Density, Relative Density, or API Gravity of Crude Petroleum and Liquid Petroleum Products by Hydrometer Method. In: *Annual Book of Standards*. ASTM International, West Conshohocken, PA.

ASTM D1318. 2014. Standard Test Method for Sodium in Residual Fuel Oil (Flame Photometric Method). In: *Annual Book of Standards*. ASTM International, West Conshohocken, PA.

ASTM D1319. 2014. Standard Test Method for Hydrocarbon Types in Liquid Petroleum Products by Fluorescent Indicator Adsorption. In: *Annual Book of Standards*. ASTM International, West Conshohocken, PA.

ASTM D1552. 2014. Standard Test Method for Sulfur in Petroleum Products (High-Temperature Method). In: *Annual Book of Standards*. ASTM International, West Conshohocken, PA.

ASTM D1555. 2014. Standard Test Method for Calculation of Volume and Weight of Industrial Aromatic Hydrocarbons and Cyclohexane. In: *Annual Book of Standards*. ASTM International, West Conshohocken, PA.

ASTM D2007. 2014. Standard Test Method for Characteristic Groups in Rubber Extender and Processing Oils and Other Petroleum-Derived Oils by the Clay–Gel Absorption Chromatographic Method. In: *Annual Book of Standards*. ASTM International, West Conshohocken, PA.

ASTM D2161. 2014. Standard Practice for Conversion of Kinematic Viscosity to Saybolt Universal Viscosity or to Saybolt Furol Viscosity. In: *Annual Book of Standards*. ASTM International, West Conshohocken, PA.

ASTM D2270. 2014. Standard Practice for Calculating Viscosity Index from Kinematic Viscosity at 40 and 100°C. In: *Annual Book of Standards*. ASTM International, West Conshohocken, PA.

ASTM D2549. 2014. Standard Test Method for Separation of Representative Aromatics and Nonaromatics Fractions of High-Boiling Oils by Elution Chromatography. In: *Annual Book of Standards*. ASTM International, West Conshohocken, PA.

ASTM D2766. 2014. Standard Test Method for Specific Heat of Liquids and Solids. In: *Annual Book of Standards*. ASTM International, West Conshohocken, PA.

ASTM D3178. 2014. Standard Guide for Characterization of Coal Fly Ash and Clean Coal Combustion Fly Ash for Potential Uses. In: *Annual Book of Standards*. ASTM International, West Conshohocken, PA.

ASTM D3179. 2014. Standard Practice for Ultimate Analysis of Coal and Coke. In: *Annual Book of Standards*. ASTM International, West Conshohocken, PA.

ASTM D3228. 2014. Standard Test Method for Total Nitrogen in Lubricating Oils and Fuel Oils by Modified Kjeldahl Method. In: *Annual Book of Standards*. ASTM International, West Conshohocken, PA.

ASTM D3238. 2014. Standard Test Method for Calculation of Carbon Distribution and Structural Group Analysis of Petroleum Oils by the n–d–M Method. In: *Annual Book of Standards*. ASTM International, West Conshohocken, PA.

ASTM D3341. 2014. Standard Test Method for Lead in Gasoline–Iodine Monochloride Method. In: *Annual Book of Standards*. ASTM International, West Conshohocken, PA.

ASTM D3343. 2014. Standard Test Method for Estimation of Hydrogen Content of Aviation Fuels. In: *Annual Book of Standards*. ASTM International, West Conshohocken, PA.

ASTM D3605. 2014. Standard Test Method for Trace Metals in Gas Turbine Fuels by Atomic Absorption and Flame Emission Spectroscopy. In: *Annual Book of Standards*. ASTM International, West Conshohocken, PA.

ASTM D3701. 2014. Standard Test Method for Hydrogen Content of Aviation Turbine Fuels by Low Resolution Nuclear Magnetic Resonance Spectrometry. In: *Annual Book of Standards*. ASTM International, West Conshohocken, PA.

ASTM D4045. 2014. Standard Test Method for Sulfur in Petroleum Products by Hydrogenolysis and Rateometric Colorimetry. In: *Annual Book of Standards*. ASTM International, West Conshohocken, PA.

ASTM D4294. 2014. Standard Test Method for Sulfur in Petroleum and Petroleum Products by Energy Dispersive X-ray Fluorescence Spectrometry. In: *Annual Book of Standards*. ASTM International, West Conshohocken, PA.

ASTM D4530. 2014. Standard Test Method for Determination of Carbon Residue (Micro Method). In: *Annual Book of Standards*. ASTM International, West Conshohocken, PA.

ASTM E148. 2014. Standard Specification for Apparatus For Microdetermination of Carbon and Hydrogen in Organic and Organo-Metallic Compounds. In: *Annual Book of Standards*. ASTM International, West Conshohocken, PA.

ASTM E258. 2014. Standard Test Method for Total Nitrogen in Organic Materials by Modified Kjeldahl Method. In: *Annual Book of Standards*. ASTM International, West Conshohocken, PA.

ASTM E385. 2014. Standard Test Method for Oxygen Content Using a 14-MeV Neutron Activation and Direct-Counting Technique. In: *Annual Book of Standards*. ASTM International, West Conshohocken, PA.

ASTM E777. 2014. Standard Test Method for Carbon and Hydrogen in the Analysis Sample of Refuse-Derived Fuel. In: *Annual Book of Standards*. ASTM International, West Conshohocken, PA.

ASTM E778. 2014. Standard Test Methods for Nitrogen in the Analysis Sample of Refuse-Derived Fuel. In: *Annual Book of Standards*. ASTM International, West Conshohocken, PA.

Beret, S., and Reynolds, J.G. 1990. Effect of Prehydrogenation on Hydroconversion of Maya Residuum. II. Hydrogen Incorporation. *Fuel Sci. Technol. Int.* 8: 191–220.

Budde, W.L. 2001. *The Manual of Manuals*. Office of Research and Development, Environmental Protection Agency, Washington, DC.

Dean, J.R. 1998. *Extraction Methods for Environmental Analysis*. John Wiley & Sons Inc., New York.

Dolbear, G.E., Tang, A., and Moorehead, E.L. 1987. Upgrading Studies with California, Mexican, and Middle Eastern Heavy Oils. In: *Metal Complexes in Fossil Fuels*. R.H. Filby and J.F. Branthaver (Editors). Symposium Series No. 344. American Chemical Society, Washington, DC. pp. 220–232.

Gary, J.G., Handwerk, G.E., and Kaiser, M.J. 2007. *Petroleum Refining: Technology and Economics*, 5th Edition. CRC Press, Taylor & Francis Group, Boca Raton, FL.

Hsu, C.S., and Robinson, P.R. (Editors). 2006. *Practical Advances in Petroleum Processing*, Volume 1 and Volume 2. Springer Science, New York.

Long, R.B., and Speight, J.G. 1989. Studies in Petroleum Composition. I: Development of a Compositional Map for Various Feedstocks. *Rev. Inst. Fr. Petrol.* 44: 205.

McKay, J.F., Amend, P.J., Harnsberger, P.M., Cogswell, T.E., and Latham, D.R. 1981. Composition of Petroleum Heavy Ends 1. Separation of Petroleum >675°C Residues. *Fuel* 60: 14–16.

Miller, M. (Editor). 2000. *Encyclopedia of Analytical Chemistry*. John Wiley & Sons Inc., Hoboken, NJ.

Parkash, S. 2003. *Refining Processes Handbook*. Gulf Professional Publishing, Elsevier, Amsterdam.

Reynolds, J.G. 1991. Can Size Exclusion Chromatography with Element Specific Detection, the D 2007-80 with Asphaltene Precipitation (SARA) Separation, and Hydrogen Distribution by NMR Help at All in Predicting Residuum Processability? *Fuel Sci. Technol. Int.* 9: 613–614.

Reynolds, J.G., and Biggs, W.R. 1988. Analysis of Residuum Desulfurization by Size Exclusion Chromatography with Element Specific Detection. *Fuel Sci. Technol. Int.* 6: 329–354.

Schabron, J.F., and Speight, J.G. 1996. Advances in Predicting Heavy Feedstock Processability. *Arab J. Sci. Eng.* 21: 663–678.

Schabron, J.F., and Speight, J.G. 1997a. An Evaluation of the Delayed Coking Product Yield of Heavy Feedstocks Using Asphaltene Content and Carbon Residue. *Rev. Inst. Fr. Petrol.* 52: 73.

Schabron, J.F., and Speight, J.G. 1997b. Correlation between Carbon Residue and Molecular Weight. *Prepr. Div. Fuel Chem. Am. Chem. Soc.* 42(2): 386.

Speight, J.G. 1984. Upgrading Heavy Oils and Residua: The Nature of the Problem. In: *Catalysis on the Energy Scene.* S. Kaliaguine and A. Mahay (Editors). Elsevier, Amsterdam. p. 515.

Speight, J.G. 1986. Polynuclear Aromatic Systems in Petroleum. *Prepr. Am. Chem. Soc. Div. Petrol. Chem.* 31(4): 818.

Speight, J.G. 1987. Initial Reactions in the Coking of Residua. *Prepr Am. Chem. Soc. Div. Petrol. Chem.* 32(2): 413.

Speight, J.G. 1992. Molecular Modes for Petroleum Asphaltene Constituents and Implications for Processing. In: *Proceedings. Eastern Oil Shale Symposium.* IMMR, Lexington, Kentucky. p. 17.

Speight, J.G. 1994. Chemical and Physical Studies of Petroleum Asphaltene Constituents. In: *Asphaltene Constituents and Asphalts, I. Developments in Petroleum Science.* T.F. Yen and G.V. Chilingarian (Editors). Elsevier, Amsterdam. Chapter 2, p. 40.

Speight, J.G. 2001. *Handbook of Petroleum Analysis.* John Wiley & Sons Inc., Hoboken, NJ.

Speight, J.G. 2005. *Environmental Analysis and Technology for the Refining Industry.* John Wiley & Sons Inc., Hoboken, NJ.

Speight, J.G. 2009. *Enhanced Recovery Methods for Heavy Oil and Tar Sands.* Gulf Publishing Company, Houston, TX.

Speight, J.G. 2013a. *Heavy Oil Production Processes.* Gulf Professional Publishing, Elsevier, Oxford.

Speight, J.G. 2013b. *Heavy and Extra Heavy Oil Upgrading Technologies.* Gulf Professional Publishing, Elsevier, Oxford.

Speight, J.G. 2014. *The Chemistry and Technology of Petroleum.* 5th Edition. CRC Press, Taylor & Francis Group, Boca Raton, FL.

Speight, J.G. 2015. *Handbook of Petroleum Product Analysis.* 2nd Edition. John Wiley & Sons Inc., Hoboken, NJ.

Speight, J.G., and Arjoon, K.K. 2012. *Bioremediation of Petroleum and Petroleum Products.* Scrivener Publishing, Salem, MA.

Speight, J.G., and Ozum, B. 2002. *Petroleum Refining Processes.* Marcel Dekker Inc., New York.

Speight, J.G., Wernick, D.L., Gould, K.A., Overfield, R.E., Rao, B.M.L., and Savage, D.W. 1985. Molecular Weights and Association of Asphaltene Constituents: A Critical Review. *Rev. Inst. Fr. Petrol.* 40: 27.

4 Desulfurization during Refining

4.1 INTRODUCTION

In a very general sense, the use of petroleum or petroleum derivatives—isolated from areas where natural seepage occurred (Abraham, 1945; Forbes, 1958; Hoiberg, 1960; Speight, 2014a)—can be traced back over 5000 years to the times when the nonvolatile constituents (collectively called *asphalt*—also known by the Biblical name *slime*) were used as a building mastic and the lower-boiling constituents (collectively called *naphtha*) were used in the weapon of war known as *Greek Fire*, which also contained small amounts of asphalt as its sticky long-burning component. Isolation of the naphtha constituents and any treatment of the asphalt (such as hardening in the air before use) or of the oil (such as allowing for more volatile components to escape before use in lamps) may be considered to fall under the very general definition of refining—the treatment and modification of petroleum and its constituents for use. However, petroleum refining as currently practiced is a very recent science, and many innovations evolved during the 20th century.

Briefly, petroleum refining is the separation of petroleum into fractions and the subsequent treating of these fractions to yield marketable products (Speight and Ozum, 2002; Parkash, 2003; Hsu and Robinson, 2006; Gary et al., 2007; Chaudhuri, 2011; Speight, 2014a). In fact, a refinery is a collection of integrated manufacturing plants that vary in number with the variety of products, produced (Figure 4.1) according to the market demand for each product. In the early decades of the 20th century, refining processes were developed to extract kerosene for lamps. Any other products were considered to be unusable and were usually discarded. Thus, the first refining processes were developed to purify, stabilize, and improve the quality of kerosene. However, the invention of the internal combustion engine led (at about the time of World War I, 1914–1918) to a demand for gasoline for use in increasing quantities as a motor fuel for cars and trucks. This demand on the lower-boiling products increased, particularly when the market for aviation fuel developed. Thereafter, refining methods had to be constantly adapted and improved to meet the quality requirements of the various products.

The need to remove sulfur compounds from petroleum arises not only from the objectionable odor that the thiols impart to petroleum products but also from the instability that sulfur compounds appear to promote in these products. For example, free sulfur can be formed in a product by the oxidation of hydrogen sulfide:

$$2H_2S + O_2 \rightarrow 2H_2O + 2S$$

In addition, free sulfur is noted for the complex series of reactions that it will undergo with a variety of hydrocarbons of the type that might be found in fuel oil products. On the other hand, thiophene derivatives, aliphatic mercaptans, and general organic sulfides appear to have little effect, whereas disulfides and polysulfides (R-Sn-R, where $n = 2$ or more) actively promote the formation of sludge. A particular class of sulfur compounds may appear to have very little deleterious effect on the product in which it is found. However, it is the reactions that this material can promote or the products that it can form, or even the ultimate use of the petroleum fraction, that dictates whether (and to what extent) the sulfur compounds should be removed. From the general aspects of petroleum refining, it must be presumed that sulfur compounds, regardless of their chemical constitution, should be removed to preserve product quality and to protect refinery equipment.

Sulfur removal, as practiced in various refineries, can take several forms, such as concentration in refinery products such as coke, hydrodesulfurization, or chemical removal (acid treating and caustic treating, i.e., sweetening or finishing processes). Nevertheless, the desulfurization of petroleum is almost universally accomplished by the catalytic reaction of hydrogen with the sulfur-containing

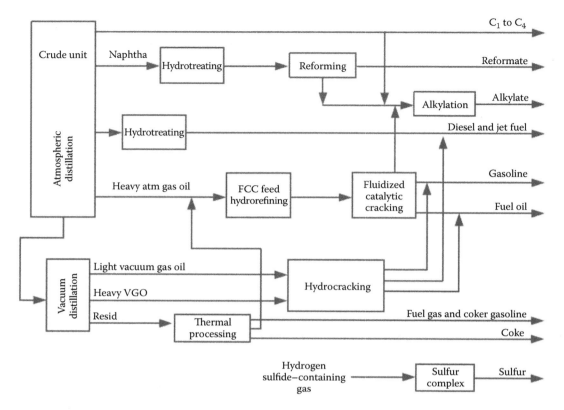

FIGURE 4.1   Schematic overview of a refinery. (From Speight, J.G. 2014. *The Chemistry and Technology of Petroleum*. 5th Edition. CRC Press, Taylor & Francis Publishers, Boca Raton, FL. Figure 15.1, p. 392.)

constituents of petroleum to produce hydrogen sulfide that can be readily separated from the liquid products (Speight and Ozum, 2002; Parkash, 2003; Hsu and Robinson, 2006; Gary et al., 2007; Chaudhuri, 2011; Speight, 2014a).

However, there are certain other refinery processes (Speight, 1992, 1996, 2014a; Pellegrino, 1998; Speight and Ozum, 2002; Parkash, 2003; Hsu and Robinson, 2006; Gary et al., 2007; Chaudhuri, 2011) that have been in service for many years in which the sulfur content of the product(s) is reduced relative to the sulfur content of the feedstock. These processes may not be recognized specifically as sulfur reduction processes, but they are adaptable to residua and heavy oils and may be effective for reducing sulfur content. In a text of this nature, these processes should be given some consideration, and they are, therefore, described briefly in this chapter.

Thus, this chapter presents an introduction to the petroleum refining processes that bring about some degree of desulfurization. Other processes—such as reforming, isomerization, alkylation, and polymerization—that are used for product improvement (Speight, 2014a) are not included here. This will allow the reader to place each process in the correct context of the refinery and to understand the various aspects of refinery operations and the fate of sulfur during the major processes.

4.2   REFINERY CONFIGURATION

As the need for lower-boiling products developed, the desired quantities of the lower-boiling low-sulfur or sulfur-free products (Table 4.1) became less available; thus, refineries had to introduce conversion processes to produce greater quantities of lighter products from the higher-boiling

TABLE 4.1

Petroleum Fractions and Their Uses

Fraction	Boiling Range		Uses
	°C	°F	
Fuel gas	−160 to −40	−260 to −40	Refinery fuel
Propane	−40	−40	Liquefied petroleum gas
Butane(s)	−12 to −1	11–30	Increases volatility of gasoline, advantageous in cold climates
Light naphtha	−1 to 150	30–300	Gasoline components, may be (with heavy naphtha) reformer feedstock
Heavy naphtha	150–205	300–400	Reformer feedstock: with light gas oil, jet fuels
Gasoline	−1 to 180	30–355	Motor fuel
Kerosene	205–260	400–500	Fuel oil
Stove oil	205–290	400–550	Fuel oil
Light gas oil	260–315	500–600	Furnace and diesel fuel components
Heavy gas oil	315–425	600–800	Feedstock for catalytic cracker
Lubricating oil	>400	>750	Lubrication
Vacuum gas oil	425–650	800–1100	Feedstock for catalytic cracker
Residuum	>650	>1100	Heavy fuel oil, asphalt

Source: Speight, J.G. 2000. *The Desulfurization of Heavy Oils and Residua.* 2nd Edition. Marcel Dekker Inc., New York. Table 7.2, p. 257.

fractions, and they took on different configurations (Speight and Ozum, 2002; Parkash, 2003; Hsu and Robinson, 2006; Gary et al., 2007; Chaudhuri, 2011; Speight, 2011, 2014a). However, not all refineries have the same basic configuration nor are they equipped for sulfur removal.

Typically, the configuration of each refinery and operating characteristics are unique and are determined primarily by the refinery location, age, crude oil slate, market requirements for refined products, and quality specifications such as, in the current context, sulfur content of the refined products. In the general refinery context, the term *configuration* denotes the specific set of refining process units in a given refinery, the throughput capacity of the various units, the technical characteristics of the units (sometimes called *process chemistry*), and the interrelationships of the units. Although no two refineries have identical configurations, they can be classified into groups of comparable refineries according to refinery complexity, which can be defined in a general manner varying from low complexity to very high complexity using the terminology (1) topping refinery, (2) hydroskimming refinery, (3) conversion refinery, and (4) deep conversion refinery, which are defined by process configuration and refinery complexity (Table 4.2). Some refineries may be characterized on an arbitrary numerical scale (from 1 to 10, with 10 being the most complex) that denotes (for a given refinery), the extent, capability, and capital intensity of the refining processes downstream of the crude distillation unit. The higher the complexity, the greater the ability of the refinery to produce added value products and the greater the capital investment in the refinery. Thus, a 10.0 refinery will have the ability to (1) convert significant amounts of the heavy feedstocks/crude fractions into lighter, high-value products and (2) produce light products to more stringent quality specifications such as low-sulfur and ultra low-sulfur fuels.

The simplest refinery configuration (Table 4.2) is the *topping refinery*, which is designed to prepare feedstocks for petrochemical manufacture or for the production of industrial fuels in remote oil-production areas. This type of refinery consists of tankage, a distillation unit, recovery facilities for gases and light hydrocarbons, and the necessary utility systems (steam, power, and water-treatment plants). Topping refineries produce large quantities of unfinished oils and are highly

TABLE 4.2
Different Refinery Configurations and Complexity

Refinery Type	Processes[a]	Relative Complexity	Numerical Complexity
Topping	Distillation	Low	1–2
Hydroskimming	Distillation	Moderate	2–4
	Hydrotreating		
Conversion	Distillation	High	5–8
	Deasphalting		
	Coking		
	Catalytic cracking		
	Hydrotreating		
	Hydrocracking		
Deep conversion	Distillation	Very high	9–10
	Deasphalting		
	Coking		
	Catalytic cracking		
	Hydrotreating		
	Hydrocracking		
	Deep desulfurization		

[a] Product improvement processes such as reforming, alkylation, and other product upgrading processes as well as blending facilities are not included here.

dependent on local markets—other than the separation of high-sulfur-containing constituents by distillation, topping refinery is not equipped for desulfurization. The addition of hydrotreating and reforming units to this basic configuration results in a more flexible *hydroskimming refinery* (Table 4.2), which can also produce desulfurized distillate fuels and high-octane naphtha (gasoline blend stock). These refineries may produce residual fuel oil (on the order of 50% v/v of the refinery output) but have to contend with a market in which there is an increasing demand for low-sulfur (even no-sulfur) fuel oil.

The most versatile configuration is the *conversion refinery* (Table 4.2), which incorporates all the basic units found in both the topping and hydroskimming refineries, but also features gas oil conversion plants such as catalytic cracking and hydrocracking units, olefin conversion plants such as alkylation or polymerization units, and, frequently, coking units for sharply reducing or eliminating the production of residual fuels. The conversion refinery may also incorporate solvent extraction processes for manufacturing lubricants and petrochemical units with which to recover propylene, benzene, toluene, and xylenes for further processing into polymers. The deep conversion refinery has the ability to treat heavy feedstocks and produce ultra-low-sulfur products. Furthermore, this type of refinery has sufficient process capacity to convert heavy feedstocks (components and blends) into low-boiling low-sulfur products. Currently, most refineries in the United States are either conversion or deep conversion refineries (MathPro, 2003, 2011). However, the overall configuration of a refinery is complex and will continue to grow in complexity (Speight, 2011, 2014a), much more so than a general verbal or numerical classification can truly represent.

In addition, some refineries may be more oriented toward the production of the precursor to gasoline (naphtha), whereas the configuration of other refineries may be more oriented toward the production of middle distillates such as jet fuel and gas oil. In addition, the yields and quality of refined petroleum products produced by any given oil refinery depends on the mixture of crude oil used as feedstock and the configuration of the refinery facilities. Light/sweet (low-sulfur) crude oil is generally more expensive and produces high yields of high-value low-boiling products such as naphtha, kerosene, and diesel fuel. Heavy sour (high-sulfur) crude oil is generally less expensive and produces higher yields of low-value higher-boiling products that require further treatment before sales.

Once a feedstock has arrived at the refinery, it is sent to the most appropriate process units. This may commence with a blending operation or a dewatering and desalting operation—the dewatering/desalting operation conventionally takes place after the blending operation since attempts at dewatering and desalting each components of the blended feedstocks would require several dewatering/desalting units or create a bottleneck at the sole dewatering/desalting unit.

Typically, the blending operation does not remove sulfur from any of the blended feedstock components; however, the operation can numerically reduce the sulfur content of a feedstock blend (by blending calculated amounts of high-sulfur crude oils with low-sulfur crude oils) to a level that falls within the operational parameters of the refinery processing units. It is at this stage, however, that caution is advised because of the potential for phase separation and fouling when incompatible feedstocks are blended together without subjecting the feedstocks to the appropriate standardized testing protocols.

4.3 DEWATERING AND DESALTING

Before separation of petroleum into its various constituents can proceed, there is the need to clean the petroleum—this is particularly true for the opportunity crudes and high-acid crudes where an alkali wash may remove some (but not all) of the undesirable impurities.

Desalting is a water-washing operation performed at the production field and at the refinery site for additional crude oil cleanup (Figure 4.2). This process is often referred to as *desalting/dewatering* in which the goal is to remove water, the constituents of the brine, and any other impurities that accompany the crude oil from the reservoir to the wellhead during recovery operations. At this point (from the sulfur perspective), any hydrogen sulfide (H_2S) and thiols (RSH) may be removed as the respective (HS^-, RS^-) water-soluble salts.

Petroleum is recovered from the reservoir mixed with a variety of substances—gases, water, and dirt (minerals)—and the desalting/dewatering operation may commence at the wellhead or a field site near the recovery operation and is the first attempt to remove the gases, water, and dirt that accompany crude oil coming from the ground. The separator may be no more than a large vessel that gives a quieting zone for gravity separation into three phases: gases, crude oil, and water containing entrained dirt. Without this treatment (which is crude oil dependent), any soluble ions

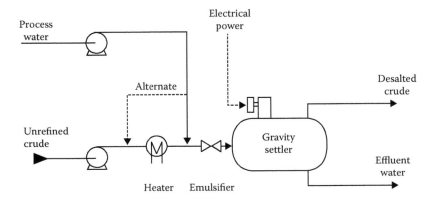

FIGURE 4.2 Electrostatic desalting unit. (From OSHA Technical Manual, Section IV, Chapter 2: Petroleum Refining Processes. http://www.osha.gov/dts/osta/otm/otm_iv/otm_iv_2.html; From Speight, J.G. 2014. *The Chemistry and Technology of Petroleum*. 5th Edition. CRC Press, Taylor & Francis Publishers, Boca Raton, FL. Figure 15.2, p. 394.)

in the brine (such as HS⁻, RS⁻) have the potential to cause pitting corrosion during transportation (Speight, 2014b). Pipeline operators, for instance, are insistent on the quality of the fluids put into the pipelines. Therefore, any crude oil to be shipped by pipeline or, for that matter, by any other form of transportation must meet rigid specifications in regard to water and salt content, especially salts with particularly active anions and cations (sulfur salts fall into either category). Thus, in many instances, sulfur content, nitrogen content, and viscosity of the crude oil may also be specified.

The usual practice is to blend crude oils of similar characteristics, although fluctuations in the properties of the individual crude oils may cause significant variations in the properties of the blend over a period of time. Before transportation, blending several crude oils can eliminate the frequent need for additional pipelines as well as the need to change the processing conditions that may be required to process the individual crude oils as separate feedstocks individually. However, simplification of the refining procedure is not always the end result. Incompatibility of different crude oils can occur when, for example, a paraffinic crude oil (low-sulfur crude oil) is blended with heavy asphaltic oil (high-sulfur crude oil), and this can result in sediment formation in the unrefined feedstock or in the products, thereby complicating the transportation of the oil and the refinery process (Mushrush and Speight, 1995; Speight, 2014a).

4.4 DISTILLATION

Distillation was the first method by which petroleum was refined. The original technique involved a batch operation in which the still was a cast-iron vessel mounted on brickwork over a fire and the volatile materials were passed through a pipe or gooseneck that led from the top of the still to a condenser. The latter was a coil of pipe (*worm*) immersed in a tank of running water (hence the use of the term *worm-end liquid* for products such as naphtha and kerosene). Modern refineries now use a variety of distillation processes, the most common being (1) atmospheric distillation and (2) vacuum distillation.

4.4.1 ATMOSPHERIC DISTILLATION

The present-day petroleum distillation unit is a collection of distillation units (trays) contained in a tower (*atmospheric tower, pipe still*) (Figure 4.3), which brings about a fairly efficient degree of fractionation (separation).

The feed to a distillation tower is heated by flow-through pipes arranged within a large furnace (*pipe still heater, pipe still furnace*), which heats the feed to a predetermined temperature—usually a temperature at which a predetermined portion of the feed will vaporize. The vapor is held under pressure in the furnace until it discharges as a foaming stream into the distillation tower where the feedstock is separated into volatile fractions (primary products—naphtha, kerosene, and gas oil) and nonvolatile material. All of the primary products are in fact equilibrium mixtures that contain some proportion of lower-boiling constituents. The primary fractions are stripped of these constituents (stabilized) before storage or further processing. The unvaporized (nonvolatile) portion of the feedstock descends to the bottom of the tower to be pumped away as a nonvolatile product (bottom product, reduced crude).

The reduced crude may then be processed by vacuum or steam distillation to separate the high-boiling lubricating oil fractions without the danger of decomposition, which occurs at high (>350°C, 660°F) temperatures (Speight and Ozum, 2002; Parkash, 2003; Hsu and Robinson, 2006; Gary et al., 2007; Chaudhuri, 2011; Speight, 2014a). Indeed, atmospheric distillation may be terminated with a lower-boiling fraction (*boiling cut*) if it is thought that vacuum or steam distillation will yield a better-quality product or if the process appears to be economically more favorable.

4.4.2 VACUUM DISTILLATION

Vacuum distillation as applied to the petroleum refining industry is truly a technique of the 20th century and has since found wide use in petroleum refining. Vacuum distillation evolved because of the need

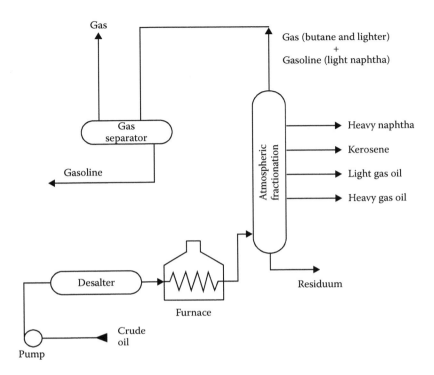

FIGURE 4.3 Atmospheric distillation unit. (From OSHA Technical Manual, Section IV, Chapter 2: Petroleum Refining Processes. http://www.osha.gov/dts/osta/otm/otm_iv/otm_iv_2.html; From Speight, J.G. 2014. *The Chemistry and Technology of Petroleum.* 5th Edition. CRC Press, Taylor & Francis Publishers, Boca Raton, FL. Figure 15.3, p. 397.)

to separate the less volatile products, such as lubricating oils, from the petroleum without subjecting these high-boiling products to cracking conditions. The boiling point of the heaviest cut obtainable at atmospheric pressure is limited by the temperature (ca. 350°C, ca. 660°F) at which the residue starts to decompose (crack). When the feedstock is required for the manufacture of lubricating oils, further fractionation without cracking is desirable, and this can be achieved by distillation under vacuum conditions.

Operating conditions for vacuum distillation (Figure 4.4) are usually 50–100 mm of mercury (atmospheric pressure = 760 mm of mercury). To minimize large fluctuations in pressure in the vacuum tower, the units are necessarily of a larger diameter than the atmospheric units. Some vacuum distillation units have diameters on the order of 45 ft (14 m). By this means, a heavy gas oil may be obtained as an overhead product at temperatures of about 150°C (300°F), and lubricating oil cuts may be obtained at temperatures of 250–350°C (480–660°F), with feed and residue temperatures being kept below 350°C (660°F), above which cracking will occur. The partial pressure of the hydrocarbons is effectively reduced still further by the injection of steam. The steam added to the column, principally for the stripping of asphalt in the base of the column, is superheated in the convection section of the heater.

The fractions obtained by vacuum distillation of the reduced crude (atmospheric residuum) from an atmospheric distillation unit depend on whether or not the unit is designed to produce lubricating or vacuum gas oils. In the former case, the fractions include (1) heavy gas oil, which is an overhead product and is used as catalytic cracking stock or, after suitable treatment, a light lubricating oil; (2) lubricating oil (usually three fractions—light, intermediate, and heavy), which is obtained as a side-stream product; and (3) asphalt (or residuum), which is the bottom product and may be used directly as, or to produce, asphalt and which may also be blended with gas oils to produce a heavy fuel oil.

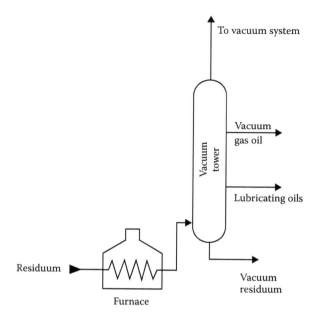

FIGURE 4.4 Vacuum distillation unit. (From OSHA Technical Manual, Section IV, Chapter 2: Petroleum Refining Processes. http://www.osha.gov/dts/osta/otm/otm_iv/otm_iv_2.html; From Speight, J.G. 2014. *The Chemistry and Technology of Petroleum.* 5th Edition. CRC Press, Taylor & Francis Publishers, Boca Raton, FL. Figure 15.4, p. 398.)

In the early refineries, distillation was the prime means by which products were separated from crude petroleum. As the technologies for refining evolved into the 21st century, refineries became much more complex (Figure 4.1); however, distillation remained the prime means by which petroleum is refined. Indeed, the distillation section of a modern refinery (Figures 4.3 and 4.4) is the most flexible section in the refinery since conditions can be adjusted to process a wide range of refinery feedstocks from the lighter crude oils to the heavier more viscous crude oils. However, the maximum permissible temperature (in the vaporizing furnace or heater) to which the feedstock can be subjected is 350°C (660°F). Thermal decomposition occurs above this temperature, which, if it occurs within a distillation unit, can lead to coke deposition in the heater pipes or in the tower itself with the resulting failure of the unit.

The contained use of atmospheric and vacuum distillation has been a major part of refinery operations during this century, and no doubt will continue to be employed throughout the remainder of the century as the primary refining operation.

4.4.3 CRACKING DISTILLATION

Of all the units in a refinery, the distillation unit is required to have the greatest flexibility in terms of variable quality of feedstock and range of product yields. The maximum permissible temperature of the feedstock in the vaporizing furnace (if cracking is to be minimized) is the factor limiting the range of products in a single-stage (atmospheric) column. If cracking is required at this stage, the temperature is raised and the residence time of the feedstock in the hot zone is increased.

Thermal decomposition or *cracking* of the constituents begins as the temperature of the oil approaches 350°C (660°F) and the rate increases markedly above this temperature—the feedstock may be injected into the tower at temperatures on the order of 395°C (740°F); however, the residence time can be adjusted so that minimal (no) cracking, or a measured amount of cracking, occurs. Under the usual tower operations, thermal decomposition is generally regarded as being undesirable because the coke-like material produced tends to be deposited on the tubes with consequent forma-tion of hotspots and eventual failure of the affected tubes. In the processing of lubricating oil stocks, an equally important consideration in the avoidance of these high temperatures is the deleterious effect on the lubricating properties. However, there are occasions when cracking distillation might be regarded as beneficial and the still temperature will be adjusted accordingly. In such a case, the products will be named accordingly using the prefix *cracked*, e.g., cracked residuum, in which case the nondescript term *pitch* (Chapter 1) is applied.

4.4.4 Desulfurization during Distillation

Distillation is not a desulfurization process insofar as the sulfur in the feedstock is not completely (if at all) eliminated. The maximum permissible temperatures to which the feedstock is heated are on the order of 350°C (660°F), which minimizes the tendency for thermal decomposition of the compo-nents of the feedstock to occur. Cracking distillation, in which the feedstock is deliberately heated to temperatures in excess of 350°C (660°F), is the exception. Sulfur will (in most cases) be eliminated as hydrogen sulfide (H_2S) through the agency of intermolecular and intramolecular hydrogen transfer.

$$Molecule^1–S + H–molecule^2 \rightarrow product^1 + product^2 + H_2S$$

$$H_2–molecule^1–S \rightarrow product^1 + H_2S$$

Cracking distillation (thermal decomposition with simultaneous removal of distillate) was rec-ognized as a means of producing the valuable lighter product (kerosene) from heavier nonvolatile materials. In the early days of the process (1870–1900), the technique was very simple—a batch of crude oil was heated until most of the kerosene had been distilled from it and the overhead material had become dark in color. At this point, distillation was discontinued and the heavy oils were held in the hot zone, during which time some of the high molecular weight components were decomposed to produce lower molecular weight products. After a suitable time, distillation was continued to yield light oil (kerosene) instead of the heavy oil that would otherwise have been produced.

Hence, distillation, insofar as the sulfur-containing constituents of crude oils are concerned, is primarily a concentration process in which the majority of the sulfur-containing compounds (which are usually of high molecular weight) are concentrated into the higher-boiling fractions, such as coke (Tables 4.3 and 4.4) (Speight, 2000). Thus, distillation is also a means by which the character, especially the sulfur content, of a residuum may be adjusted. For example, inspections of various

TABLE 4.3
Sulfur Concentration in Petroleum Coke

Feedstock	API Gravity	Sulfur (wt.%)	Sulfur in Coke (wt.%)	S% (Coke) S% (Feedstock)
Elk Basin, Wyoming, residuum	2.5	3.5	6.5	1.83
Hawkins, Texas, residuum	4.5	4.5	7.0	1.55
Kuwait, residuum	6.0	5.37	10.8	2.01
Athabasca (Canada), bitumen	7.3	5.0	7.5	1.5
West Texas, residuum	–	3.5	3.06	0.88
Boscan (Venezuela), crude oil	10.0	5.0	5.0	1.0
East Texas, residuum	10.5	1.26	2.57	2.04
Texas Panhandle, residuum	18.9	0.6	0.6	1.00

Source: Speight, J.G. 2000. *The Desulfurization of Heavy Oils and Residua.* 2nd Edition. Marcel Dekker Inc., New York. Table 7.1, p. 255.

TABLE 4.4

Distillation Products from Arabian Crude Oil

	Gasoline	Kerosene	Gas Oil	Residuum
Yield, wt.%	15	12	17	55
Range, °C	<150	150–230	230–340	340+
Specific gravity	0.70	0.78	0.85	0.97
API gravity	70.6	49.9	35.0	14.3
Sulfur, wt.%	0.02	0.2	1.4	3.9

Source: Speight, J.G. 2000. *The Desulfurization of Heavy Oils and Residua.* 2nd Edition. Marcel Dekker Inc., New York. Table 7.3, p. 259.

TABLE 4.5

Increased Sulfur Content of Vacuum Residua Compared with Atmospheric Residua

	Arabian Light	Arabian Heavy	Alaska (Sag River)
Atmospheric Residuum (650°F⁺)			
Yield on crude, vol.%	43.0	52.8	53.0
Gravity, °API	16.2	11.1	15.5
Sulfur, wt.%	3.1	4.5	1.6
Metals (Ni + V), ppm	33	124	46
Vacuum Residuum (1050°F⁺)			
Yield on crude, vol.%	13.2	26.0	18.3
Gravity, °API	6.5	4.0	7.6
Sulfur, wt.%	4.1	5.7	2.3
Metals (Ni + V), ppm	100	239	126
Carbon residue (Ramsbottom)	22	24	18
Asphaltenes (pentane insolubles), wt.%	12	25	10

Source: Speight, J.G. 2000. *The Desulfurization of Heavy Oils and Residua.* 2nd Edition. Marcel Dekker Inc., New York. Table 7.4, p. 264.

crude oil residua (Table 4.5) show that, for any particular crude oil, the vacuum residuum is virtually always higher in sulfur than an atmospheric residuum from the same crude oil. Thus, although distillation is the usual primary means by which a crude oil is processed, it may be completely bypassed in the case of an extremely heavy crude oil in favor of whole-crude processing by any of the more suitable thermal methods.

With respect to heavy feedstocks, distillation as such may never be applied. Removal of volatile constituents may be achieved using a flash technique in which the whole crude is introduced into a vessel and the temperature is such that all of the volatile material boiling below a predetermined temperature is removed *in toto* as an overhead oil without any attempt at fractionation. This total fraction from the heavy feedstock may be subdivided at a later stage of the refinery operation or serve as a feedstock (without fractionation) for a cracking unit.

In short, distillation is, at best, looked upon as a means by which the lower-boiling fractions can be separated from a feedstock before being subjected to a suitable conversion (or refining) method. It is, in fact, the means by which the undesirable higher molecular weight materials are removed from the feedstock as atmospheric or vacuum residua. It would, indeed, be a very rare occasion if the distillation process actually served as an efficient means of desulfurization rather than a concentration process.

The fractions obtained by vacuum distillation of reduced crude depend on whether the run is designed to produce lubricating or vacuum gas oils. In the former case, the fractions include (1) heavy gas oil, an overhead product and is used as catalytic cracking stock or, after suitable treatment, a light lubricating oil; (2) lubricating oil (usually three fractions: light, intermediate, and heavy), obtained as a sidestream product; and (3) residuum, the nonvolatile product that may be used directly as asphalt or to produce asphalt. The residuum may also be used as a feedstock for a coking operation or blended with gas oils to produce a heavy fuel oil. If the reduced crude is not required as a source of lubricating oils, the lubricating and heavy gas oil fractions are combined or, more likely, removed from the residuum as one fraction and used as a catalytic cracking feedstock.

In summary, sulfur and other heteroatom-containing compounds have, because of their relatively high boiling points, a propensity to remain in either the atmospheric residuum or in the vacuum residuum. Hence, distillation is a concentration process in which the predominantly hydrocarbon (low-heteroatom) material is volatile and the majority of the high-heteroatom (sulfur-containing) constituents remain in the residua. Depending on the properties of the crude oil, >60% w/w of the heteroatom constituents in the original crude oil can remain in the (atmospheric or vacuum) residuum, while some thermally labile heteroatoms may be eliminated (Long and Speight, 1990; Speight and Francisco, 1990; Speight and Ozum, 2002; Parkash, 2003; Hsu and Robinson, 2006; Gary et al., 2007; Chaudhuri, 2011; Speight, 2014a). This concentration effect can concentrate the majority of the heteroatom constituents in a minority of the feedstock. Some heteroatom constituents (especially the sulfur-containing compounds) can occur in the lower-boiling fractions, thereby making sulfur the ubiquitous heteroatom.

4.5 THERMAL PROCESSES

One of the earliest conversion processes used in the petroleum industry is the thermal decomposition of higher-boiling materials into lower-boiling products. This process is known as thermal cracking, and the exact origins of the process are unknown. The process was developed in the early 1900s to produce distillates from the unwanted higher-boiling products of the distillation process. However, it was soon learned that the thermal cracking process also produced a wide slate of products varying from highly volatile gases to nonvolatile high-sulfur coke.

By way of explanation and to alleviate any confusion, steam cracking is a petrochemical process sometimes used in refineries to produce olefins (such as ethylene) from various feedstocks for petrochemicals manufacture. The feedstocks for steam cracking vary most commonly from ethane to butane to naphtha to vacuum gas oil, with higher-boiling feedstocks giving higher yields of by-products such as naphtha. Steam cracking is carried out at temperatures on the order of 870°C (1500–1600°F), and at pressures slightly above atmospheric. Naphtha produced from steam cracking contains benzene, which is extracted before hydrotreating, and the residuals from steam cracking are often used as blend stocks for heavy fuel oil.

Although new thermal cracking units are under development (Speight and Ozum, 2002; Parkash, 2003; Speight, 2003, 2011, 2014a; Hsu and Robinson, 2006; Gary et al., 2007; Chaudhuri, 2011), processes that can be regarded as having evolved from the original concept of thermal cracking are *visbreaking* and the various coking processes (Table 4.6) (Speight and Ozum, 2002; Parkash, 2003; Hsu and Robinson, 2006; Gary et al., 2007; Chaudhuri, 2011; Speight, 2014a). Such processes have been on-stream in refineries for several decades and are well established. Although these processes are not generally recognized as specific desulfurizing processes, they do, nevertheless, cause some desulfurization to occur during feedstock refining.

TABLE 4.6

Parameters for Various Cracking Processes

Visbreaking	Delayed Coking	Fluid Coking
Mild (880–920°F) heating at 50–200 psig	Moderate (900–960°F) heating at 90 psig	Severe (900–1050°F) heating at 10 psig
Reduce viscosity of fuel oil	Soak drums (845–900°F) coke walls	Oil contacts refractory coke
Low conversion (10%) to 430°F	Coked until drum solid	Bed fluidized with steam-even heating
Heated coil or drum	Coke (removed hydraulically) 20–40% on feed	Higher yields of light ends (<C_5)
	Yield 430°F, 30%	Less coke make

Source: Speight, J.G. 2000. *The Desulfurization of Heavy Oils and Residua.* 2nd Edition. Marcel Dekker Inc., New York. Table 7.5, p. 271.

4.5.1 THERMAL CRACKING

The products of cracking processes (temperature: 455–540°C, 850–1005°F; pressure: 100–1000 psi) are typically light gas oil and heavy gas oil (both contain sulfur), as well as a residual oil that could also be used as heavy fuel oil but which contains the majority of the refractory sulfur compounds that were originally in the raw crude oil. Gas oils from catalytic cracking, although containing some sulfur, were deemed suitable as domestic and industrial fuel oils, or as diesel fuels when blended with low-sulfur gas oil.

The gas oil produced by cracking is also a further important source of naphtha. In a once-through cracking operation, all of the cracked material is separated into products and may be used as such. However, the gas oil produced by cracking (cracked gas oil) is more resistant to cracking (more refractory in terms of the hydrocarbon constituents and the sulfur-containing constituents) than gas oil produced by distillation (straight-run gas oils, which also contains sulfur) but could still be cracked to produce more. This was achieved using a later innovation (post-1940) involving a recycle operation in which the cracked gas oil was combined with fresh feed for another trip through the cracking unit. The extent to which recycling was carried out affected the yield of naphtha from the process.

Mild cracking conditions, with a low conversion per cycle, favor a high yield of naphtha, with low gas and coke production (but the naphtha quality is not high; tending to contain low-boiling sulfur compounds), whereas more severe conditions give increased gas and coke production and reduced naphtha yield (but of higher quality). With limited conversion per cycle (to prevent the formation of excessive amounts of high-sulfur coke), the heavier feedstocks must be recycled; however, these recycle oils become increasingly refractory upon repeated cracking and, if they are not required as a fuel oil stock, may be coked to increase naphtha or refined by means of a hydrotreating process.

4.5.2 VISBREAKING

Visbreaking is a relatively mild thermal (noncatalytic) cracking process that is used to reduce the viscosity of residua (Speight and Ozum, 2002; Parkash, 2003; Hsu and Robinson, 2006; Gary et al., 2007; Chaudhuri, 2011; Speight, 2014a). The process uses the approach of mild thermal cracking to improve the viscosity characteristics of a residuum without attempting significant conversion to distillates. Low residence times are required to avoid gas and coke production, and the hot liquid is quenched as it exits the reactor.

Visbreaking conditions range from 450°C (840°F) to about 510°C (950°F) at the heating coil outlet with pressures varying from 50 to 300 psi. There are a number of different configurations for visbreaking units that depend on the product slate and refinery requirements (Pellegrino, 1998).

In a typical visbreaking operation (Figure 4.5), a crude oil residuum is passed through a furnace where it is heated to a temperature of 480°C (895°F) under an outlet pressure of about 100 psi. The

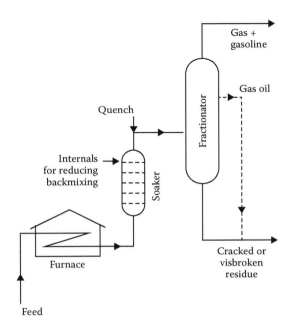

FIGURE 4.5 Soaker visbreaker. (From OSHA Technical Manual, Section IV, Chapter 2: Petroleum Refining Processes. http://www.osha.gov/dts/osta/otm/otm_iv/otm_iv_2.html; From Speight, J.G. 2014. *The Chemistry and Technology of Petroleum.* 5th Edition. CRC Press, Taylor & Francis Publishers, Boca Raton, FL. Figure 15.5, p. 403.)

heating coils in the furnace are arranged to provide a soaking section of low heat density, where the charge remains until the visbreaking reactions are completed and the cracked products are then passed into a flash-distillation chamber. The overhead material from this chamber is then fractionated to produce a low-quality naphtha as an overhead product and light gas oil as bottom. The liquid products from the flash chamber are cooled with a gas oil flux and then sent to a vacuum fractionator. This yields a heavy gas oil distillate and a residual tar of reduced viscosity.

The obvious benefit of the visbreaking process is that the reactions are not allowed to proceed to completion; the hot liquid reaction mix is quenched by the addition of light gas oil and then sent to a vacuum fractionator. This yields a heavy gas oil distillate and a residuum of reduced viscosity. Quench oil may also be used to terminate the reactions. An alternative process design uses lower furnace temperatures and longer reaction (contact) time that is achieved by installing a soaking drum between the furnace and the fractionator. The disadvantage of this approach is high potential for coke deposition in the soaking drum and the subsequent need to remove the coke.

Conversion of residua in visbreaking follows first-order reaction kinetics (Henderson and Weber, 1965). The minimum viscosity of the unconverted residue can lie outside the range of allowable conversion if sediment begins to form (Rhoe and de Blignieres, 1979). When pipelining of the visbreaker product is the process objective, addition of a diluent such as gas condensate can be used to achieve a further reduction in viscosity. However, the main limitation of the visbreaking process, and for that matter all thermal processes, is that the products can be unstable. Thermal cracking at low pressure gives olefins, particularly in the naphtha fraction. These olefins give a very unstable product, which tends to undergo secondary reactions to form gum and intractable residua. Modification of visbreaking (thermal) chemistry is possible, and more stable products are produced by application of the hydrogen-donor solvents as in the hydrogen donor visbreaking process (Carlson et al., 1958; Langer et al., 1962; Bland and Davidson, 1967; Vernon et al., 1984; McConaghy, 1987; Kubo et al., 1996). There also remains the (potential) issue of sediment disposal, especially when the sediment is too contaminated for further use.

TABLE 4.7

Feedstock and Product Data for Visbreaking a Residuum

Feedstock			
Specific gravity		0.987	
API gravity		11.9	
Conradson carbon, wt.%		10.6	
Viscosity SFS (210°F)		495	
Sulfur, wt.%		0.61	
Product Yields, vol.%			
≤C_4, vol.%		1.6	
C_5 to 220°C naphtha, vol.%		6.2	
220–340°C gas oil, vol.%		6.3	
>340°C (residuum), vol.%		88.4	
Product Quality	**Naphtha**	**Gas Oil**	**Residuum**
Specific gravity	0.748	0.851	0.990
API gravity	57.7	34.8	11.4
Conradson carbon, wt.%	–	0.01	15.0
Viscosity SFS (210°F)	–	–	160
Sulfur, wt.%	0.26	0.33	0.58

Source: Speight, J.G. 2000. *The Desulfurization of Heavy Oils and Residua.* 2nd Edition. Marcel Dekker Inc., New York. Table 7.6, p. 273.

Visbreaking is not usually claimed as a process for the reduction of the sulfur content of the feedstock since the sole purpose of the process is to reduce the viscosity of the feedstock. Although the visbreaking process offers a form of desulfurization of heavy feedstocks with low conversion, the process is often not considered to be sufficiently efficient for desulfurization purposes. However, it is capable of producing product streams that have cumulatively reduced sulfur content relative to the sulfur content of the feedstock (Table 4.7) and that are more amenable to catalytic process as well as specific catalytic desulfurization processes. These desulfurization stocks can (depending on their properties) be blended with the original feedstock to produce an overall low-sulfur stream. Visbreaking has been seriously considered as a potential primary process (or pretreating process) for upgrading Athabasca bitumen (Table 4.8).

Important variables in the visbreaking process include temperature, pressure, and residence time. Anyone can be changed (within predetermined limits) to alter the product slate. For example, raising the heater outlet temperature increases the yield of distillates and gaseous products. Because of its relative simplicity, visbreaking offers a means of converting heavy feedstocks under controllable conditions to products that might be suitable for immediate desulfurization in high yields with (relatively) prolonged catalyst life. In addition, additives such a calcium oxide (CaO) may be used to act as centers for sediment deposition and sulfur removal, represented simply as

$$\text{Feedstock}_{\text{sulfur}} + CO \rightarrow \text{products} + CaS$$

Hydrovisbreaking, in which the process is carried out under an atmosphere of hydrogen, is also used on occasion (Speight, 2014a) and *catalytic visbreaking* has been proposed in which the severity of thermal decomposition is increased by the addition of a selenium-, tellurium-, or sulfur-containing catalyst supported on a porous substrate. In both cases, sulfur removal from the products is enhanced (Yan, 1991).

TABLE 4.8

Feedstock and Product Data for Visbreaking Athabasca Tar Sand Bitumen

Feedstock

Gravity, °API	8.6	
Distillation	Vol.%	°F
	IPB	–
	5	430
	10	560
	30	820
	50	1010
Sulfur	4.8	
Nitrogen	0.4	
Conradson carbon, wt.%	13.5	

Product Yields (wt.% on Whole Bitumen)

≤C$_4$	3
C$_5$/400°F	7
400/650°F	21
650/1000°F	35
Residue	34
Total	100

Product Qualities	**C$_5$/380 (195°C)**	**380/650 (195/345°C)**	**650/1000 (345/540°C)**
Gravity, °API	54	25.8	13.2
Sulfur, wt.%	2.02	2.11	3.91
Nitrogen, ppm	110	450	2800
Bromine number	115	38	20
Conradson carbon, wt.%	–	–	0.23

Source: Speight, J.G. 2000. *The Desulfurization of Heavy Oils and Residua.* 2nd Edition. Marcel Dekker Inc., New York. Table 7.7, p. 274.

4.5.3 COKING

Coking is a thermal process for the continuous conversion of heavy, low-grade oils into lighter products. Unlike visbreaking, coking involves compete thermal conversion of the feedstock into volatile products and coke (Table 4.9). The feedstock is typically a residuum, and the products are gases, naphtha, fuel oil, gas oil, and coke. The gas oil may be the major product of a coking operation and serves primarily as a feedstock for catalytic cracking units. The coke obtained is usually used as fuel; however, specialty uses, such as in electrode manufacture and production of chemicals and metallurgical coke, are also possible and increase the value of the coke. For these uses, the coke may require treatment to remove sulfur and metal impurities.

4.5.3.1 Delayed Coking

Delayed coking is a semicontinuous process (semibatch process) (Figure 4.6) that uses two coke drums and a single fractionator tower (distillation column) and coking furnace. The process objective is to convert low-value (high-boiling, high-sulfur) resid to valuable products (naphtha and diesel) and coker gas oil.

Typically, a feedstock stream is introduced to the fractionating tower—the bottom products from the fractionator are heated to approximately 480–540°C (900–1000°F) in the coking furnace, and

TABLE 4.9

Comparison of Visbreaking with Delayed Coking and Fluid Coking

Visbreaking	Delayed Coking	Fluid Coking
Purpose: to reduce viscosity of fuel oil to acceptable levels	Purpose: to produce maximum yields of distillate products	Purpose: to produce maximum yields of distillate products
Conversion is not a prime purpose	Moderate (480–515°C, 900–960°F) heating at pressures of 90 psi	Severe (480–565°C; 900–1050°F) heating at pressures of 10 psi
Mild (470–495°C, 880–920°F) heating at pressures of 50–200 psi	Reactions allowed to proceed to completion	Reactions allowed to proceed to completion
Reactions quenched before going to completion	Complete conversion of the feedstock	Complete conversion of the feedstock
Low conversion (10%) to products boiling below 220°C (430°F)	Soak drums (845–900°F) used in pairs (one on-stream and one off-stream being decoked)	Oil contacts refractory coke
Heated coil or drum (soaker)	Coked until drum solid	Bed fluidized with steam; heat dissipated throughout the fluid bed
	Coke removed hydraulically from off-stream drum	Higher yields of light ends (<C_5) than delayed coking
	Coke yield: 20–40% by weight (dependent upon feedstock)	Less coke make than delayed coking (for one particular feedstock)
	Yield of distillate boiling below 220°C (430°F); ca. 30% (but feedstock dependent)	

Source: Speight, J.G. 2014. *The Chemistry and Technology of Petroleum.* 5th Edition. CRC Press, Taylor & Francis Publishers, Boca Raton, FL, Table 15.2, p. 404.

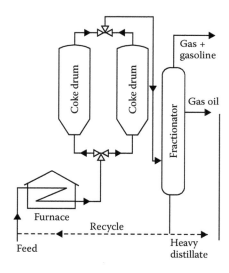

FIGURE 4.6   Delayed coker. (From OSHA Technical Manual, Section IV, Chapter 2: Petroleum Refining Processes. http://www.osha.gov/dts/osta/otm/otm_iv/otm_iv_2.html; From Speight, J.G. 2014. *The Chemistry and Technology of Petroleum.* 5th Edition. CRC Press, Taylor & Francis Publishers, Boca Raton, FL. Figure 15.6, p. 404.)

then fed to an insulated coke drum where thermal cracking produces lower-boiling (cracked) reaction products and coke. The reaction products produced in the coke drum are fed back to the fractionator for product separation. After the coke drum becomes filled with coke, the feed is alternated to the parallel (empty) coke drum, and the filled coke drum is purged and cooled, first by steam injection, and then by water addition.

Thus, in the process, the feedstock is introduced into the product fractionator where it is heated and lighter fractions are removed as a sidestreams. The fractionator bottoms, including a recycle stream of heavy product, are then heated in a furnace whose outlet temperature varies from 480°C to 515°C (895–960°F). The heated feedstock enters one of a pair of coking drums where the cracking reactions continue. The cracked products leave as overheads, and coke deposits form on the inner surface of the drum. To give continuous operation, two drums are used: while one is onstream, the other is being cleaned. The temperature in the coke drum ranges from 415°C to 450°C (780–840°F) with pressures from 15 to 90 psi.

A coke drum blowdown system recovers hydrocarbon and steam vapors generated during the quenching and steaming process. Once cooled, the coke drum is vented to the atmosphere, opened, and then high-pressure water jets are used to cut the coke from the drum. After the coke-cutting cycle, the drum is closed and preheated to prepare the vessel for going back on-line (i.e., receiving heated feed). A typical coking cycle will last for 16–24 h on-line and 16–24 h cooling and decoking.

Overhead products go to the fractionator, where naphtha and heating oil fractions are recovered. The nonvolatile material is combined with preheated fresh feed and returned to the reactor. The coke drum is usually on stream for about 24 h before becoming filled with porous coke, after which the coke is removed hydraulically. Normally, 24 h is required to complete the cleaning operation and to prepare the coke drum for subsequent use on stream.

An alternate continuous coking process is also available (Sullivan, 2011). The process uses a kneading and mixing action to continuously expose new resid surface to the vapor space and causes a more complete removal (and recovery) of volatiles from the produced petroleum coke. The process takes resid from the coker heater directly into a reactor/devolatilizer and, as a result of kneading/mixing action by the reactor/devolatilizer, new surfaces of the residuum mass are continuously exposed to the gas phase, enhancing the rapid mass transfer of volatiles into the gas phase. The volatiles are then rapidly cooled to retard degradation. With the rapid reduction of volatiles content in the resid mass, the carbonization reaction rates are accelerated, enabling continuous and rapid production of solid petroleum coke particles. The shortening of the contact time of the volatiles with the hot residuum minimizes the degradation of volatiles. In addition to the recovery of additional and more valuable volatile products, there are other benefits of the new process compared with delayed coking: (1) the consumption of utilities is less because no steam or water is required; (2) since there is no quenching, energy from the hot coke is recovered; (3) the process is continuous and thus is never opened to the atmosphere; (4) there is no cutting procedure as in the delayed coking process where high-pressure water is used to cut the coke out of the drums and volatiles, and coke particles not are released into the atmosphere.

4.5.3.2 Fluid Coking and Flexicoking

Fluid coking is a continuous process (Figure 4.7) that uses the fluidized solids technique to convert atmospheric and vacuum residua to more valuable products. The process objective is to convert low-value (high-boiling, high sulfur) resid to valuable products (naphtha and diesel) and coker gas oil. The feedstocks charged to a fluid coker (and, hence, to a flexicoking operation) may be any type of heavy feedstocks or residuum where the carbon residue falls in the range 5 to 50 wt.% or those materials having an API gravity <20°.

The preheated feedstock (260–370°C, 500–700°F) is injected directly onto the hot coke (480–565°C, 895–1050°F), at approximately atmospheric pressure, to crack to additional coke and liquid products that leave the reactor as overhead oil. Fluid coking uses two vessels, a reactor and a burner;

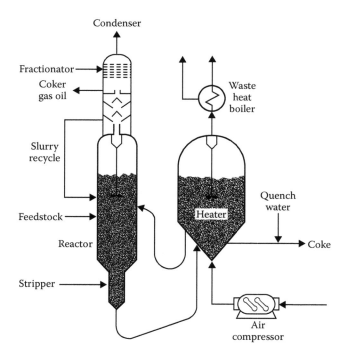

FIGURE 4.7 Fluid coker. (From Speight, J.G. 2014. *The Chemistry and Technology of Petroleum.* 5th Edition. CRC Press, Taylor & Francis Publishers, Boca Raton, FL. Figure 15.7, p. 405.)

coke particles are circulated between these to transfer heat (generated by burning a portion of the coke) to the reactor. The reactor holds a bed of fluidized coke particles, and steam is introduced at the bottom of the reactor to fluidize the bed. These conditions permit the coking reactions to be conducted at higher temperatures and shorter contact times than can be employed in delayed coking. Moreover, decreased yields of coke are realized and greater quantities of more valuable liquid product are recovered in the fluid coking process.

Flexicoking (Figure 4.8) is also a continuous process that is a direct descendant of fluid coking. The unit uses the same configuration as the fluid coker but has a gasification section in which excess coke can be gasified to produce refinery fuel gas. The flexicoking process was designed during the late 1960s and the 1970s as a means by which excess coke yield could be reduced in view of the gradual incursion of the heavier feedstocks in refinery operations. Such feedstocks are notorious for producing high yields of coke (>15% by weight) in thermal and catalytic operations.

Thus, the flexicoking process is essentially the fluid coking process except that a coke gasifier is added that burns nearly all of the produced coke at 925–985°C (1700–1800°F) with steam to produce low-heating-value synthesis gas (syngas). The produced synthesis gas, along with entrained fines, is routed through the heater vessel for fluidization of the hot coke bed and for heat transfer to the solids. The synthesis gas is then treated to remove entrained particles and reduced sulfur compounds, and the syngas can then be used in specially designed boilers or other combustion sources that can accommodate the low heat content of the syngas. Most of the carbon dioxide emissions produced in the flexicoking unit will not be released at the unit, but rather it will be part of the synthesis gas. Some of the carbon dioxide produced in the flexicoking unit is expected to be removed as part of the sulfur removal process and subsequently released in the sulfur recovery plant; the carbon dioxide that remains in the scrubbed syngas will be released from the stationary combustion unit that uses the syngas as fuel (usually a boiler specifically designed to use the low-heating-value-content synthesis gas).

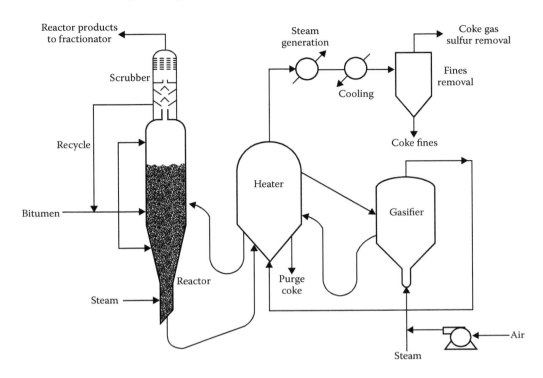

FIGURE 4.8 Flexicoking process. (From Speight, J.G. 2014. *The Chemistry and Technology of Petroleum.* 5th Edition. CRC Press, Taylor & Francis Publishers, Boca Raton, FL. Figure 15.8, p. 406.)

4.5.4 DESULFURIZATION DURING COKING

Coking processes are different from the other thermal processes found in a refinery insofar as the reaction times may be longer and the reactions are usually allowed to proceed to completion (in contrast to, say, visbreaking where the reactions are terminated by quenching with a gas oil fraction). The coke obtained from the coking processes is usually used as a fuel for the process, although marketing for specialty uses, such as electrode manufacture, increases the value of the coke. However, because of the tendency of the process to concentrate the feedstock sulfur in the coke, the coker feedstock may have to be chosen carefully to produce a coke of sufficiently low sulfur content for a specialty use.

Coking processes have the virtue of eliminating the residue fraction of the feed, at the cost of forming a solid carbonaceous product. The yield of coke in a given coking process tends to be proportional to the carbon residue content of the feed (measured as the Conradson carbon residue (Speight, 2001, 2014a, 2015). The data (Tables 4.10 and 4.11) illustrate how the yield of coke from delayed and fluid coking varies with the Conradson carbon residue of the feed. As with delayed coking, the fluid coking process is capable of producing liquid products with substantially lower sulfur contents than the feedstock (Table 4.12); however, a substantial portion of the original feedstock sulfur is concentrated in the coke (Table 4.13). There is elimination of sulfur into the gaseous products; however, uses for the coke depend very much on the amount of feedstock sulfur. The yields of liquid products from flexicoking are the same as from fluid coking, because the coking reactor is unaltered. As with fluid coking, the extent of desulfurization depends on the sulfur content of the feedstock as well as on the chemical nature of the sulfur in the feedstock.

Although coking processes offer a form of desulfurization of heavy feedstocks and high conversion, the processes are considered as extreme insofar as the analogy of using a sledgehammer to put a picture nail in the wall has been used. However, the application of, say, delayed coking as a

TABLE 4.10
Coke Yields from Delayed Coking and Fluid Coking

Carbon Residue (wt.%)	°API	Coke Yield (wt.%)	
		Delayed Coker	Fluid Coker
1		0	
5	26	8.5	3
10	16	18	11.5
15	10	27.5	17
20	6	35.5	23
25	3.5	42	29
30	2		34.5
40	−2.5		46

Source: Speight, J.G. 2000. *The Desulfurization of Heavy Oils and Residua.* 2nd Edition. Marcel Dekker Inc., New York. Table 7.2, p. 257.

first treatment step might remove potentially low-value material that would have an adverse effect on a catalyst, and such a potential use should not be overlooked. However, the increasing attention paid to reducing atmospheric pollution has also served to direct some attention to coking, since the process not only concentrates pollutants such as feedstock sulfur in the coke, but also can usually yield volatile products that can be conveniently desulfurized.

4.6 CATALYTIC CRACKING

Catalytic cracking is a conversion process (Figure 4.9) with the objective of converting low-value gas oils to valuable products (naphtha and diesel) and slurry oil. The process can be applied to a variety of feedstocks ranging from gas oil to heavy oils and residua (Table 4.14) (Evans and Quinn, 1993; Sadeghbeigi, 1995; Speight and Ozum, 2002; Parkash, 2003; Hsu and Robinson, 2006; Gary et al., 2007; Chaudhuri, 2011; Speight, 2014a). It is one of several practical applications used in a refinery that employ a catalyst to improve process efficiency (Table 4.15). The original incentive to develop cracking processes arose from the need to increase supplies of gasoline blend stock (naphtha). Since cracking could virtually double the volume of naphtha from a barrel of crude oil, the purpose of cracking was wholly justified. The process employs a variety of reactors with bed types varying from fixed beds to moving beds to fluidized beds.

Catalytic cracking has a number of advantages over thermal cracking—(1) the naphtha produced has a higher octane number; (2) the catalytically cracked naphtha consists largely of isoparaffins and aromatics, which have high octane numbers and greater chemical stability than mono-olefins and di-olefins, which are present in much greater quantities in thermally cracked naphtha. Substantial quantities of gaseous olefins (suitable for polymer gasoline manufacture) and smaller quantities of methane, ethane, and ethylene are produced by catalytic cracking. Sulfur compounds are changed in such a way that the sulfur content of catalytically cracked naphtha is lower than in thermally cracked naphtha. Catalytic cracking produces less heavy residues or tar and more of the useful gas oils than thermal cracking. The process has considerable flexibility, permitting the manufacture of both motor and aviation gasoline and a variation in the gas oil yield to meet changes in the fuel oil market. However, during the cracking reaction, carbonaceous material is deposited on the catalyst, which markedly reduces its activity, and removal of the deposit is very necessary. This is usually accomplished by burning the catalyst in the presence of air until catalyst activity is reestablished.

The several processes currently employed in catalytic cracking differ mainly in the method of catalyst handling, although there is overlap with regard to catalyst type and the nature of the products.

TABLE 4.11
Feedstock and Product Data from the Delayed Coking of Residua

Source	South Louisiana	Kuwait Virgin	Kuwait Hydrodesulfurized	West Texas Virgin	West Texas Hydrodesulfurized	Oklahoma	California	Gilsonite	Athabasca Bitumen
Feedstock Properties									
Gravity, °API	12.3	6.7	16.1	8.9	15.2	13.0	12.0	2.0	7.3
Sulfur, wt.%	0.68	5.22	0.66	2.96	0.64	1.15	1.6	0.3	5.3
Carbon residue, wt.%	13.0	19.8	9.1	17.8	9.3	14.1	9.4	26.0	17.9
Product Yields									
C_4, wt.%	7.8	7.6	9.0	10.9	9.5	9.8	14.9	17.6	8.2
C_5–400°F, vol.%	22.8	26.7	22.0	28.9	20.1	}20.4	22.5	45.2	20.3
400–650°F, vol.%	18.4	28.0	41.9	16.5	31.7		36.5	}8.5	58.8
Gas oil, vol.%	37.6	18.4	19.1	26.4	27.5	57.2	16.7		
Coke, wt.%	23.7	30.2	18.5	28.4	20.7	23.6	19.1	45.0	21.0
Product Properties									
Naphtha									
Gravity, °API	56.5	54.1	57.5	53.7	57.1	56.4	57.9	57.2	51.9
Sulfur, wt.%	0.2	0.84	0.04	0.6	0.09	–	0.9	–	1.9
Furnace Oil									
Gravity, °API	34	28.5	37.0	33.9	35.3	–	27.5	–	–
Sulfur, wt.%	0.3	3.15	0.18	1.08	019	–	–	–	2.7
Nitrogen, wt.%	0.08	0.04	0.05	0.06	0.08	–	–	–	–
Gas Oil									
Gravity, °API	23.0	18.3	23.3	21.6	23.8	28.0	14.9	15.9	–
Sulfur, wt.%	0.6	4.40	0.66	1.93	0.45	–	1.1	–	3.8
Nitrogen, wt.%	0.13	0.15	0.19	0.25	0.26	–	–	–	–
Coke									
Sulfur, wt.%	1.3	7.5	1.7	4.5	1.6	–	–	–	5–10

Source: Speight, J.G. 2000. *The Desulfurization of Heavy Oils and Residua.* 2nd Edition. Marcel Dekker Inc., New York. Table 7.12, p. 285.

TABLE 4.12

Feedstock and Product Data from the Fluid Coking of Various Feedstocks

	Los Angeles Basin			Texas Panhandle	Kuwait	South Louisiana		Hawkins	Zaca
	Vacuum Residuum	Deasphalter Residuum	Visbreaker Residuum	Vacuum Residuum	Vacuum Residuum	Vacuum Residuum	Atmospheric Residuum	Vacuum Residuum	Atmospheric Residuum
Feedstock Properties									
Specific gravity	1.024	1.078	1.106	0.951	1.032	0.989	0.948	1.043	1.039
API gravity	6.7	—	—	17.3	5.6	11.6	17.8	4.2	4.7
Conradson carbon, wt.%	17	33	41	11	21.8	13	5	24.5	19
Sulfur, wt.%	2.1	2.3	2.1	0.7	5.5	0.6	0.5	4.3	7.8
Product Yields									
C_4, vol.%	11	13	15	9	14	11	8	13	10.5
C_5 to 221°C	17	18	14	21	21	21	17	19.5	20.5
Naphtha, vol. %									
221–545°C gas oil, vol.%	62	45	32	69	48	61	74	52.0	61
Coke, wt.%	21	36	48	12	28	17	8	27.5	17.5
Product Quality									
Naphtha (C_5 to 221°C)									
Specific gravity	0.759	0.763	0.759	0.743	0.755	0.755	0.755	0.755	0.784
API gravity	54.9	63.9	54.9	58.9	55.9	55.9	55.9	55.9	49
Sulfur, wt.%	—	1.7	0.8	0.2	0.5	0.2	0.1	0.9	4.6
Gas Oil (221–545°C)									
Specific gravity	0.953	0.973	0.966	0.882	0.963	0.922	0.910	0.953	0.973
API gravity	17.0	13.9	15.0	28.9	15.4	22.0	24.0	17.0	14
Sulfur, wt.%	1.7	2.2	1.7	0.4	4.7	0.5	0.4	3.7	6.2

Source: Speight, J.G. 2000. *The Desulfurization of Heavy Oils and Residua.* 2nd Edition. Marcel Dekker Inc., New York. Table 7.14, p. 288.

TABLE 4.13

Sulfur and Nitrogen Distribution in the Products from the Delayed Coking of Residua

Product	%	
	Sulfur	Nitrogen
Gas	30	–
Naphtha	5	1
Kerosene	}35	2
Gas oil		22
Coke	30	75
Totals	100	100

Source: Speight, J.G. 2000. *The Desulfurization of Heavy Oils and Residua.* 2nd Edition. Marcel Dekker Inc., New York. Table 7.13, p. 286.

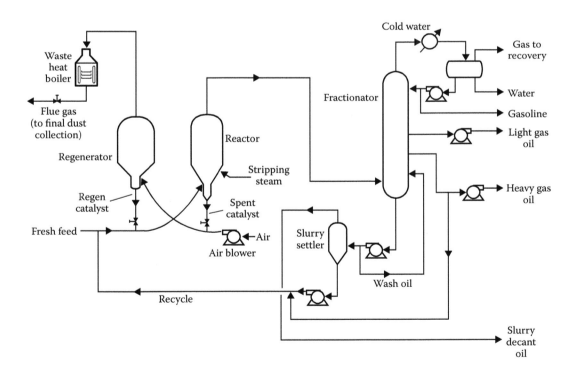

FIGURE 4.9 FCC unit. (From Speight, J.G. 2014. *The Chemistry and Technology of Petroleum.* 5th Edition. CRC Press, Taylor & Francis Publishers, Boca Raton, FL. Figure 15.9, p. 408.)

TABLE 4.14

Summary of Catalytic Cracking Processes

Conditions	Feedstocks	Products	Variations
Solid acidic catalyst (such as silica–alumina, zeolite)	Gas oils and residua	Lower molecular weight than feedstock	Fixed bed
Temperature: 480–540°C (900–1000°F; solid/vapor contact)	Residua pretreated to remove salts (metals)	Some gases (feedstock and process parameters dependent)	Moving bed
Pressure: 10–20 psi	Residua pretreated to remove high molecular weight (asphaltic constituents)	Isoparaffins in product	Fluidized bed
Provisions needed for continuous catalyst replacement with heavier feedstocks (residua)		Coke deposited on catalyst	
Catalyst may be regenerated or replaced			

Source: Speight, J.G. 2014. *The Chemistry and Technology of Petroleum.* 5th Edition. CRC Press, Taylor & Francis Publishers, Boca Raton, FL. Table 15.3, p. 407.

TABLE 4.15

Refinery Processes That Employ Catalysts

Process	Materials Charged	Products Recovered	Temperature of Reaction	Type of Reaction
Cracking	Gas oil, fuel oil, heavy feedstocks	Gasoline, gas, and fuel oil	875–975°F, 470–525°C	Dissociation or splitting of molecules
Hydrogenation	Gasoline to heavy feedstocks	Low-boiling products	400–850°F, 205–455°C	Mild hydrogenation; cracking; removal of sulfur, nitrogen, oxygen, and metallic compounds
Reforming	Gasolines, naphthas	High-octane gasolines, aromatics	850–1000°F, 455–535°C	Dehydrogenation, dehydroisomerization, isomerization, hydrocracking, dehydrocyclization
Isomerization	Butane C_4H_{10}	Isobutane C_4H_{10}		Rearrangement
Alkylation	Butylene and isobutane, C_4H_8 and C_4H_{10}	Alkylate, C_8H_{18}	32–50°F, 0–10°C	Combination
Polymerization	Butylene, C_4H_8	Octene, C_8H_{16}	300–350°F, 150–175°C	Combination

Source: Speight, J.G. 2000. *The Desulfurization of Heavy Oils and Residua.* 2nd Edition. Marcel Dekker Inc., New York. Table 7.8, p. 275.

4.6.1 PROCESS OPTIONS

The *fluid-bed catalytic cracking process* (Figure 4.9) differs from the fixed-bed and moving-bed processes, insofar as the powdered catalyst is circulated essentially as a fluid with the feedstock (Speight and Ozum, 2002; Parkash, 2003; Hsu and Robinson, 2006; Gary et al., 2007; Chaudhuri, 2011; Speight, 2014a). The several fluid catalytic cracking (FCC) processes in use differ primarily in mechanical design. Side-by-side reactor–regenerator construction along with unitary vessel construction (with the reactor either above or below the regenerator) are the two main mechanical variations.

The *fixed-bed process* was the first to be used commercially and uses a static bed of catalyst in several reactors, which allows a continuous flow of feedstock to be maintained. Thus, the cycle of operations consists of (1) flow of feedstock through the catalyst bed, (2) discontinuance of feedstock flow and removal of coke from the catalyst by burning, and (3) insertion of the reactor on-stream. The *moving-bed process* (Sadeghbeigi, 1995; Speight and Ozum, 2002; Parkash, 2003; Hsu and Robinson, 2006; Gary et al., 2007; Chaudhuri, 2011; Speight, 2014a) uses a reaction vessel in which cracking takes place and a kiln in which the spent catalyst is regenerated, catalyst movement between the vessels is provided by various means. The *fluid-bed process* (Sadeghbeigi, 1995; Speight and Ozum, 2002; Parkash, 2003; Hsu and Robinson, 2006; Gary et al., 2007; Chaudhuri, 2011; Speight, 2014a) differs from the fixed-bed and moving-bed processes insofar as the powdered catalyst is circulated essentially as a fluid with the feedstock. The several FCC processes in use differ primarily in mechanical design (Speight and Ozum, 2002; Parkash, 2003; Hsu and Robinson, 2006; Gary et al., 2007; Chaudhuri, 2011; Speight, 2014a). Side-by-side, reactor–regenerator configuration, and the reactor either above or below the regenerator are the main mechanical variations. From a flow standpoint, all FCC processes contact the feedstock and recycle streams with the finely divided catalyst in the reactor.

Catalytic cracking using a fluidized bed is the most popular form of cracking and is the emphasis of this section. In the process, the cracking reactions take place in the riser to form products, including coke. In the riser, the catalyst and the feedstock and products rise up the reactor pipe and, since the reactions are predominantly endothermic, the reaction temperature declines from bottom to top. At the top of the riser, the mixture enters a solid–gas separator, and the product vapors are led away. The coked catalyst passes to the stripper where steam is added and unreacted/reacted feedstocks adsorbed on the catalyst are released. The stripped catalyst is then directed into the regenerator where air is added and the combustion of coke on the catalyst (and any feedstocks/products still adsorbed that were not stripped) occurs with the liberation of heat. Regenerator temperatures are typically 705–760°C (1300–1400°F). Heat exchangers and the circulating catalyst are used to retain/capture the regeneration heat for use in preheating the feedstock.

4.6.2 FEEDSTOCK

The feedstock to a FCC unit is the most important process variable and has the greatest impact on operating conditions, yield, and product quality. A typical feedstock consists of hydrocarbon families, including paraffins and cycloparaffins, which are saturates (aromatic hydrocarbons with a different number of aromatic rings), resin constituents, and asphaltene constituents. Feedstock behavior in the riser also depends on the mechanical and operational conditions of the unit, and the concentration and distribution of the feedstock constituents (Speight and Ozum, 2002; Parkash, 2003; Hsu and Robinson, 2006; Gary et al., 2007; Chaudhuri, 2011; Speight, 2014a; Navarro et al., 2015).

The feedstock quality depends on several factors: (1) the quality and type of feedstock processed by the crude distillation units, (2) the quality and type of feedstock processed by other processing units, and (3) the complexity of the refinery. Products from other processes, such as delayed coking, coking, and visbreaking, and hydrotreating units (such as gas oil and residuum hydrotreaters), hydrocracking units, and deasphalting units. By-product streams from lubricant production may also be sent to the FCC unit.

One of the biggest advantages of the FCC unit is the flexibility to process all types of feedstock streams (and feedstock blends) that are complex mixtures in which the mixing processes are not always efficient to ensure completely homogeneous blends. Thus, the feedstocks to a FCC unit may be blends of (1) atmospheric gas oil and vacuum gas oil, (2) atmospheric residua, (3) coker gas oil and visbreaking gas oil, (4) hydrocracking residua, (5) hydrotreated gas oils and residua, and (6) blends of any of the aforementioned feedstocks. However, it is essential that there should be an availability of physical–chemical analytical programs for feedstock characterization to allow predicting the impact of feedstock quality on product yield, as well as operating conditions (heat balance) and product quality (Speight, 2014a, 2015; Navarro et al., 2015).

4.6.3 Catalysts

The catalyst, which may be an activated natural or synthetic material, is employed in bead, pellet, or microspherical form and can be used as a fixed bed, moving bed, or fluid bed. The fixed-bed process was the first process to be used commercially and uses a static bed of catalyst in several reactors, which allows a continuous flow of feedstock to be maintained. Thus, the cycle of operations consists of (1) flow of feedstock through the catalyst bed, (2) discontinuance of feedstock flow and removal of coke from the catalyst by burning, and (3) insertion of the reactor on-stream. The moving-bed process uses a reaction vessel (in which cracking takes place) and a kiln (in which the spent catalyst is regenerated), and catalyst movement between the vessels is provided by various means.

Natural clays have long been known to exert a catalytic influence on the cracking of oils, but it was not until about 1936 that the process using silica–alumina catalysts was developed sufficiently for commercial use. Since then, catalytic cracking has progressively supplanted thermal cracking as the most advantageous means of converting distillate oils into naphtha. The main reason for the wide adoption of catalytic cracking is the fact that a better yield of higher-octane naphtha can be obtained than by any known thermal operation. At the same time, the gas produced consists mostly of propane and butane with less methane and ethane. The production of heavy oils and tars, higher in molecular weight than the charge material, is also minimized, and both the naphtha and the uncracked cycle oil are more saturated than the products of thermal cracking.

The major innovations of the 20th century lie not only in reactor configuration and efficiency but also in catalyst development. There is probably not an oil company in the United States that does not have some research and development activity related to catalyst development. Much of the work is proprietary and, therefore, can only be addressed here in generalities.

The cracking of crude oil fractions occurs over many types of catalytic materials; however, high yields of desirable products are obtained with hydrated aluminum silicates. These may be either activated (acid-treated) natural clays of the bentonite type of synthesized silica–alumina or silica–magnesia preparations. Their activity to yield essentially the same products may be enhanced to some extent by the incorporation of small amounts of other materials such as the oxides of zirconium, boron (which has a tendency to volatilize away on use), and thorium. Natural and synthetic catalysts can be used as pellets or beads and also in the form of powder; in either case, replacements are necessary because of attrition and gradual loss of efficiency. It is essential that they are stable to withstand the physical impact of loading and thermal shocks, and that they can withstand the action of carbon dioxide, air, nitrogen compounds, and steam. They also should be resistant to sulfur and nitrogen compounds, and synthetic catalysts, or certain selected clays, appear to be better in this regard than average untreated natural catalysts.

The catalysts are porous and highly adsorptive, and their performance is affected markedly by the method of preparation. Two chemically identical catalysts having pores of different sizes and distributions may have different activity, selectivity, temperature coefficients of reaction rates, and responses to poisons. The intrinsic chemistry and catalytic action of a surface may be independent of pore size; however, small pores produce different effects because of the manner in which hydrocarbon vapors are transported into and out of the pore systems.

4.6.4 DESULFURIZATION DURING CATALYTIC CRACKING

Although catalytic cracking was originally designed to convert gas oil to naphtha, process modifications have allowed the feedstock types to include heavy oil feedstock. The manner in which a feedstock will crack depends on a number of process variables such as feedstock composition, boiling range, structural types present, reactor conditions, and catalyst type. Thus, it is not surprising that the heavier (high-heteroatom) feedstocks present special problems when used to feed a catalytic cracker. The susceptibility of the catalyst to deposited high-sulfur carbonaceous species and various nitrogen compounds, as well as to the metals (iron, nickel, vanadium, and copper), represents a major drawback to using heavy feedstocks for catalytic cracking units. For example, if the metals concentration (expressed in ppm) on the catalyst fulfills the following relation

$$4V + 14Ni + Fe + Cu > 1000$$

the catalyst is severely contaminated and may actually have to be replaced.

Nevertheless, it is possible to process the heavier feedstocks in catalytic crackers and bring about some degree of desulfurization in the process. The degree of desulfurization depends on the amount of sulfur in the feedstock (Table 4.16) as well as the absence of asphaltene constituents

TABLE 4.16
Fluid Catalytic Cracking of Residua from Hydrodesulfurization of Alaskan Crude Oils

	Sag River 375°F+ (190°C+)	Put River 375°F+ (190°C+)
Residuum Properties		
API gravity	21.5	21.1
Sulfur, wt.%	0.1	0.3
Aniline point, °F	199	192
Conradson carbon, wt.%	2.2	4.8
Nitrogen, wt.%	–	0.25
Conversion, vol.%	80.2	78.0
Yields		
C_4, vol.%	20.1	28.5
C_5–430°F gasoline, vol.%	60.8	58.9
430–650°F, vol.%	14.5	15.1
Decanted oil	5.3	6.9
Coke, wt.%	7.8	8.1
Gasoline Properties		
API gravity	57	56
Sulfur, wt.%	0.01	0.03
Distillate Properties		
API gravity	17.0	18.0
Sulfur, wt.%	0.24	0.50
Catalyst Makeup Rate		
lb/bbl fresh feed	0.5	1.4

Source: Speight, J.G. 2000. *The Desulfurization of Heavy Oils and Residua.* 2nd Edition. Marcel Dekker Inc., New York. Table 7.9, p. 279.

TABLE 4.17
Fluid Catalytic Cracking of Deasphalted Oils

	Virgin	Hydrotreated
Feedstock		
Specific gravity	0.970	0.931
API gravity	14.3	20.5
Aniline point, °C	96	101.5
Sulfur, wt.%	3.55	0.30
Nitrogen, wt.%	0.20	0.11
Distillation, °C		
10% v/v	530	480
30% v/v	570	545
50% v/v	590	580
Conversion, vol.%	70	70
Product Yields		
≤C_4, vol.%	22.1	21.6
Gasoline, C_5–204°C (ASTM)	58.5	61.0
EP, vol.%		
Light cycle oil, vol.%	24.0	24.0
Decanted oil, vol.%	6.0	6.0
Coke, wt.%	7.0	4.5
Product Properties		
Gasoline, C_5–204°C		
Specific gravity	0.759	0.745
API gravity	54.9	58.4
Sulfur, wt.%	0.39	0.02
Light Cycle Oil		
Specific gravity	0.959	0.929
API gravity	16.1	20.8
Sulfur, wt.%	3.9	0.02
Viscosity, cSt/210°F	0.8	0.7
Decanted Oil		
Specific gravity	1.058	1.013
API gravity	2.2	8.2
Sulfur, wt.%	9.7	0.49
Viscosity, cSt/210°F	21	16

Source: Speight, J.G. 2000. *The Desulfurization of Heavy Oils and Residua.* 2nd Edition. Marcel Dekker Inc., New York. Table 7.10, p. 280.

(Table 4.17). It is generally believed that to optimize use of a catalytic cracking unit, feedstocks should be treated to remove excess high molecular weight material and metals by processes such as visbreaking, coking, or deasphalting to prolong catalyst activity (Sadeghbeigi, 1995; Speight and Ozum, 2002; Parkash, 2003; Hsu and Robinson, 2006; Gary et al., 2007; Chaudhuri, 2011; Speight, 2014a). Furthermore, sulfur compounds are changed in such a way that the cumulative sulfur content of the liquid and nonvolatile products is lower than the sulfur content of the original feedstock. The decomposition of sulfur constituents into hydrocarbons and hydrogen sulfide (or other gaseous sulfur products) occurs.

4.7 HYDROPROCESSES

Hydroprocessing (variously referred to as *hydrotreating* and *hydrocracking*) is a group of refining processes in which the feedstock is heated with hydrogen under pressure (Table 4.18). The process objective is to remove contaminants (sulfur, nitrogen, metals) and saturate olefins and aromatics to produce a clean product for further processing or finished product sales. The primary process technique is hydrogenation (by means of a catalyst) to improve H/C atomic ratios and to remove sulfur, along with nitrogen, and metals. The outcome is the conversion of a variety of feedstocks to a range of products (Table 4.19) (Speight and Ozum, 2002; Parkash, 2003; Hsu and Robinson, 2006; Gary et al., 2007; Chaudhuri, 2011; Speight, 2011, 2014a).

For the purposes of this section, and to offer some relief from the confusion that exists in petroleum terminology (Speight, 2014a), a hydroprocess is a thermal conversion process in which hydrogen is used to accomplish the objectives of the refiner. Hydrotreating (nondestructive hydrogenation) is a process in which hydrogen is used to convert heteroatom constituents into their heteroatom hydrogen analogs and hydrocarbons:

$$R\text{-}S\text{-}R^1 + H_2 \rightarrow RH + R^1H + H_2S$$

The process has several variants: (1) naphtha hydrotreating—the primary objective is to remove sulfur contaminant for downstream processes, typically <1 ppm w/w; (2) hydrotreating gasoline blend stock, which removes sulfur from gasoline-blending components to meet recent clean fuels specifications; (3) mid-distillate hydrotreating, which involves sulfur removal from kerosene and conversion of kerosene to jet fuel by means mild aromatic saturation as well as sulfur removal from diesel for clean fuels; (4) satisfying ultra-low sulfur diesel requirements, leading to major unit revamps; and (5) pretreating or fluid catalytic cracker feedstock to remove sulfur removal for SO_x reduction in the flue gas and nitrogen removal for better FCC catalyst activity as well as of aromatics constituents to improve FCC feedstock processability and FCC efficiency.

Thus, hydrotreating is generally used for the purpose of improving product quality without appreciable alteration of the boiling range (Table 4.20). Mild processing conditions are employed so that only the more unstable materials are attacked. Thus, nitrogen, sulfur, and oxygen compounds undergo hydrogenolysis to split out ammonia, hydrogen sulfide, and water, respectively. Olefins are saturated, and unstable compounds, such as diolefins, which might lead to the formation of gums or insoluble materials, are converted to more stable compounds. Heavy metals present in the feedstock are also usually removed during hydrogen processing.

On the other hand, hydrocracking is a process in which thermal decomposition is extensive and the hydrogen assists in the removal of the heteroatoms as well as in mitigating the coke formation that usually accompanies thermal cracking of high molecular weight polar constituents.

4.7.1 HYDROTREATING

Distillate hydrotreating (Figure 4.10) is carried out by charging the feed to the reactor, together with hydrogen in the presence of catalysts such as tungsten–nickel sulfide, cobalt–molybdenum–alumina, nickel oxide–silica–alumina, and platinum–alumina. Most processes employ cobalt–molybdenum (Co_2O_3–MoO_2) catalysts, which generally contain about 10% of molybdenum oxide and <1% of cobalt oxide supported on alumina. The temperatures employed are in the range of 260–345°C (500–655°F), while the hydrogen pressures are about 500–1000 psi.

The reaction generally takes place in the vapor phase but, depending on the application, may be a mixed-phase reaction. Generally, it is more economical to hydrotreat high-sulfur feedstocks before

TABLE 4.18
Process Characteristics for Hydroconversion Processes

Feedstock	Hydrocracking	Aromatics Removal	Sulfur Removal	Nitrogen Removal	Metals Removal	Coke Mitigation	n-Paraffins Removal	Olefins Removal	Products
Naphtha	✓		✓	✓				✓	Reformer feedstock
									Liquefied petroleum gas (LPG)
Atmospheric	✓	✓			Gas oil				Diesel fuel
		✓					✓		Jet fuel
		✓							Petrochemical feedstock
									Naphtha
Vacuum	✓	✓	✓	✓	✓				Catalytic cracker feedstock
	✓	✓	✓						Diesel fuel
	✓	✓	✓						Kerosene
	✓		✓						Jet fuel
	✓								Naphtha
	✓	✓							LPG
									Lubricating oil
Residuum	✓		✓	✓	✓	✓			Catalytic cracker feedstock
			✓		✓	✓			Coker feedstock
									Diesel fuel (others)

Source: Speight, J.G. 2000. *The Desulfurization of Heavy Oils and Residua.* 2nd Edition. Marcel Dekker Inc., New York. Table 7.15, p. 290.

TABLE 4.19
Summary of Hydrocracking Processes

Conditions	Feedstocks	Products	Variations
Solid acid catalyst (silica–alumina with rare earth metals, various other options)	Refractory (aromatic) streams	Lower molecular weight paraffins	Fixed bed (suitable for liquid feedstocks
Temperature: 260–450°C (500–845°F; solid/liquid contact)	Coker oils	Some methane, ethane, propane, and butane	Ebullating bed (suitable for heavy feedstocks)
Pressure: 1000–6000 psi hydrogen	Cycle oils	Hydrocarbon distillates (full range depending on the feedstock)	
Frequent catalysts renewal for heavier feedstocks	Gas oils	Residual tar (recycle)	
Gas oil: catalyst life up to 3 years	Residua (as a full hydrocracking or hydrotreating option)	Contaminants (asphaltic constituents) deposited on the catalyst as coke or metals	
Heavy oil/tar sand bitumen: catalyst life <1 year	In some cases, asphaltic constituents (S, N, and metals) removed by deasphalting		

Source: Speight, J.G. 2014. *The Chemistry and Technology of Petroleum.* 5th Edition. CRC Press, Taylor & Francis Publishers, Boca Raton, FL. Table 15.4, p. 409.

TABLE 4.20
Process Parameter for Hydrotreating Various Feedstocks

Parameter	Naphtha	Residuum
Temperature, °C	300–400	340–425
Pressure, atm	35–70	55–170
LHSV	4.0–10.0	0.2–1.0
H_2 recycle rate, scf/bbl	400–1000	3000–5000
Catalyst life, years	3.0–10.0	0.5–1.0
Sulfur removal, %	99.9	85.0
Nitrogen removal, %	99.5	40.0

catalytic cracking than to hydrotreat the products from catalytic cracking. The advantages are that (1) sulfur is removed from the catalytic cracking feedstock, and corrosion is reduced in the cracking unit; (2) carbon formation during cracking is reduced so that higher conversions result; and (3) the cracking quality of the gas oil fraction is improved.

Hydrofining is a process that first went on-stream in the 1950s and is one example of the many hydroprocesses available. It can be applied to lubricating oils, naphtha, and gas oils. The feedstock is heated in a furnace and passed with hydrogen through a reactor containing a suitable metal oxide catalyst, such as cobalt and molybdenum oxides on alumina. Reactor operating conditions range from 205°C to 425°C (400–800°F) and from 50 to 800 psi, and depend on the kind of feedstock and the degree of treating required. Higher-boiling feedstocks, high sulfur content, and maximum sulfur removal require higher temperatures and pressures.

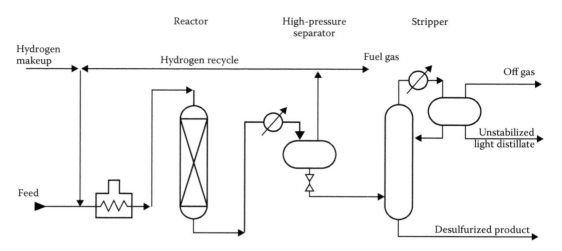

FIGURE 4.10 Distillate hydrotreater. (From OSHA Technical Manual, Section IV, Chapter 2: Petroleum Refining Processes. http://www.osha.gov/dts/osta/otm/otm_iv/otm_iv_2.html; From Speight, J.G. 2014. *The Chemistry and Technology of Petroleum*. 5th Edition. CRC Press, Taylor & Francis Publishers, Boca Raton, FL. Figure 15.10, p. 410.)

After passing through the reactor, the treated oil is cooled and separated from the excess hydrogen, which is recycled through the reactor. The treated oil is pumped to a stripper tower where hydrogen sulfide, formed by the hydrogenation reaction, is removed by steam, vacuum, or flue gas, and the finished product leaves the bottom of the stripper tower. The catalyst is not usually regenerated; it is replaced after about 1 year use.

4.7.2 HYDROCRACKING

Hydrocracking is similar to *catalytic cracking*, with hydrogenation superimposed and with the reactions taking place either simultaneously or sequentially. The process objective is to remove feedstock contaminants (nitrogen, sulfur, metals) and to convert low-value gas oils to valuable products (naphtha, middle distillates, and ultra-clean lube base stocks). Hydrocracking was initially used to upgrade low-value distillate feedstocks, such as cycle oils (highly aromatic products from a catalytic cracker that usually are not recycled to extinction for economic reasons), thermal and coker gas oils, and heavy-cracked and straight-run naphtha (Table 4.19). These feedstocks are difficult to process by either catalytic cracking or reforming, since they are usually characterized by a high polycyclic aromatic content and/or by high concentrations of the two principal catalyst poisons—sulfur and nitrogen constituents.

The choice of processing schemes for a given hydrocracking application depends on the nature of the feedstock as well as the product requirements (Suchanek and Moore, 1986). The process can be simply illustrated as a single-stage or as a two-stage operation (Figure 4.11).

Thus, hydrocracking is an extremely versatile process that can be utilized in many different ways, such as conversion of the high-boiling aromatic streams that are produced by catalytic cracking or by coking processes. To take full advantage of hydrocracking, the process must be integrated in the refinery with other process units.

The commercial processes for treating, or finishing, petroleum fractions with hydrogen all operate in essentially the same manner. The feedstock is heated and passed with hydrogen gas through a tower or reactor filled with catalyst pellets. The reactor is maintained at a temperature of 260–425°C (500–800°F) and at pressures from 100 to 1000 psi, depending on the particular process, the nature of the feedstock, and the degree of hydrogenation required. After leaving the reactor, excess hydrogen is separated from the treated product and recycled through the reactor after removal

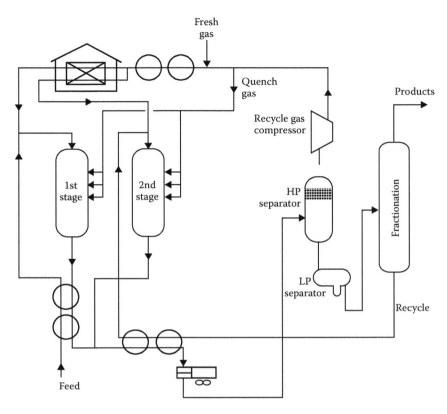

FIGURE 4.11 Single-stage or two-stage (optional) hydrocracking unit. (From OSHA Technical Manual, Section IV, Chapter 2: Petroleum Refining Processes. http://www.osha.gov/dts/osta/otm/otm_iv/otm_iv_2 .html; From Speight, J.G. 2014. *The Chemistry and Technology of Petroleum.* 5th Edition. CRC Press, Taylor & Francis Publishers, Boca Raton, FL. Figure 15.11, p. 411.)

of hydrogen sulfide. The liquid product is passed into a stripping tower where steam removes dissolved hydrogen and hydrogen sulfide and, after cooling, the product is taken to product storage or, in the case of feedstock preparation, pumped to the next processing unit.

4.7.3 Desulfurization during Hydroprocessing

A comparison of hydrocracking with hydrotreating is useful in assessing the parts played by these two processes in refinery operations. Hydrotreating of distillates may be defined simply as the removal of nitrogen–sulfur and oxygen-containing compounds by selective hydrogenation. The hydrotreating catalysts are usually in cobalt plus molybdenum or nickel plus molybdenum (in the sulfide) form impregnated on an alumina base. The hydrotreating operating conditions are such that appreciable hydrogenation of aromatics will not occur—1000 to 2000 psi hydrogen and about 370°C (700°F). The desulfurization reactions are usually accompanied by small amounts of hydrogenation and hydrocracking.

The reactions that are used to chemically define the processes (i.e., cracking and subsequent hydrogenation of the fragments, hydrogenation of unsaturated material, hydrodesulfurization, hydrodenitrogenation) may all occur. Hydrocracking a feedstock will, in all likelihood, be accompanied by hydrodesulfurization, thereby producing not only low-boiling products but also low-boiling products that are low in sulfur (Table 4.21).

Hydroprocesses offer direct desulfurization of heavy feedstocks and high conversion. These processes were not originally designed for heavy feedstocks, and yet the evolution of the various processes and catalysts has seen their application to heavy oils and residua. Application of a hydroprocess as a pretreatment process has some merit. Changing the character of the feedstock constituents can make for easier conversion in a later stage of the refining sequence. The advantages are as follows: (1) the products require less finishing; (2) sulfur is removed from the catalytic cracking feedstock, and corrosion is reduced in the cracking unit; (3) carbon formation during cracking is reduced and higher conversions result; and (4) the catalytic cracking quality of the gas oil fraction is improved.

The problems encountered in hydrocracking heavy feedstocks can be directly equated to the amount of complex, higher-boiling (heteroatom-containing) constituents (Figure 4.12) that may require pretreatment (Speight and Moschopedis, 1979; Reynolds and Beret, 1989). Processing these feedstocks is not merely a matter of applying know-how derived from refining conventional crude oils but requires knowledge of composition and behavior under various conditions (Speight, 2001, 2014a, 2015; Speight and Ozum, 2002; Parkash, 2003; Hsu and Robinson, 2006; Gary et al., 2007; Chaudhuri, 2011).

TABLE 4.21
Feedstock and Product Data for Hydrocracking Residua and Gas Oils

	Feedstock Properties				
Feed source	Kuwait	Khafji	Kuwait	Khafji	Tia Juana
Cut range, °F	620+	700+	700/1100	700/1100	825/1050
Gravity, °API	16.0	10.9	19.9	20.9	20.1
Sulfur, wt.%	3.94	4.25	3.25	3.05	1.80
Conradson carbon, wt.%	9.3	14.2	–	–	–
Metals, ppm					
V	43	99	–	–	–
Ni	13	32	–	–	–
Average Yields and Qualities					
Light Ends, wt.% on FF					
$\leq C_4$	3.7	4.1	3.5	3.5	2.5
Naphtha (C_5–400°F)					
Yield, vol.%	1.4	1.4	1.7	1.7	1.7
Gravity, °API	45.0	45.0	45.0	45.0	45.0
Sulfur, wt.%	0.02	0.02	0.02	0.02	0.02
Gas Oil (>400°F)					
Yield, vol.%	100.1	100.4	99.6	99.5	98.6
Gravity, °API	22.2	17.5	26.2	26.9	23.1
Sulfur, wt.%	1.00	1.00	0.50	0.48	0.18
H_2 consumption, scf/bbl	600	660	420	400	325

Source: Speight, J.G. 2000. *The Desulfurization of Heavy Oils and Residua.* 2nd Edition. Marcel Dekker Inc., New York. Table 7.16, p. 293.

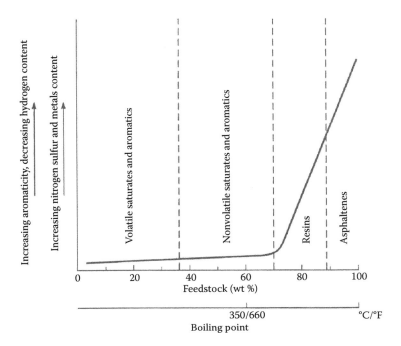

FIGURE 4.12 Relative distribution of heteroatoms in the various fractions. (From Speight, J.G. 2014. *The Chemistry and Technology of Petroleum*. 5th Edition. CRC Press, Taylor & Francis Publishers, Boca Raton, FL. Figure 15.19, p. 422.)

4.8 DEASPHALTING

The original object of *deasphalting* (*solvent extraction* or *solvent treatment*) was to remove unwanted constituents from lubricating oils. More recently, these processes have been applied to asphalt production from residua to enable the asphalt to meet the desired specifications by removing undesirable constituents from the charge material. Following from this, the deasphalting process has been employed to remove all, or a substantial portion, of the asphaltene constituents and resin constituents from feedstocks that are destined for further processing. The method involves use of a liquid hydrocarbon solvent, usually the lower-boiling hydrocarbons of the propane, butane(s), or pentane(s) group either singly or as mixtures.

In addition, the process is amenable to being combined with conversion processes, such as visbreaking, coking, and hydroconversion, where removal of asphaltene constituents and resin constituents affords a much better slate of higher-quality products than either the deasphalting process alone or the conversion processes. However, disposal of any carbonaceous by-product (deasphalter pitch or process coke) must also be accommodated.

4.8.1 DEASPHALTING PROCESSES

Solvent deasphalting processes are a major part of refinery operations (Speight and Ozum, 2002; Parkash, 2003; Hsu and Robinson, 2006; Gary et al., 2007; Chaudhuri, 2011; Speight, 2011, 2014a) and are not often appreciated for the tasks for which they are used. In the solvent deasphalting processes, an alkane is injected into the feedstock to disrupt the dispersion of components and causes the polar constituents to precipitate. Propane (or sometimes propane/butane mixtures) is extensively used for deasphalting and produces a deasphalted oil and propane deasphalted asphalt (or propane deasphalted tar) (Dunning and Moore, 1957). Propane has unique solvent properties; at lower temperatures (38–60°C, 100–140°C), paraffins are very soluble in propane and at higher temperatures (about 93°C, 200°F) all hydrocarbons are almost insoluble in propane. The process provides an extension to vacuum distillation and is a later addition to the petroleum refinery. Before its use, many processes capable of removing asphaltic materials from feedstocks were employed in the form of distillation (atmospheric and vacuum), as well as clay and sulfuric acid treatment.

The liquid hydrocarbon causes separation of the asphaltene constituents and/or the resin constituents from the feedstock, leaving deasphalted oil that contains substantially less sulfur and metals than the feedstock. On a commercial scale, propane is the most common solvent in the process (Figure 4.13) and the units are designed to operate at 40–80°C (105–175°F) and at pressures of 400–550 psi with the solvent/feedstock ratio usually falling in the range 5:1–13:1.

In the process, the heavy feedstock and 3–10 times its volume of liquefied propane are pumped together through a mixing device and then into a settling tank. The temperature is maintained between 27°C (80°F) and 71°C (160°F): the higher the temperature, the greater the tendency of asphaltic materials to separate. The propane is maintained in the liquid state by a pressure of about 200 psi. The asphalt settles in the setting tank and is pumped to an asphalt recovery unit, where propane is separated from the asphalt. The upper layer in the settling tank consists of deasphalted oil dissolves in a large amount of propane, and propane is separated from the oil–propane mixture in evaporators heated by steam. The last trace of propane is removed from the oil by steam in a stripper tower, and the propane is condensed and reused.

Many deasphalting plants use a countercurrent tower. Liquefied propane is pumped into the bottom of the tower to form a continuous phase, and lubricating stock or reduced crude (crude residuum) is pumped into the tower near the top. As the reduced crude descends, the oil components are dissolved and carried with the propane out the top of the tower, and the asphaltic components are pumped from the bottom of the tower. Thus, the extraction takes place countercurrently and some degree of selectivity of the solvent is possible since propane, contrary to the behavior of the majority of solvents, dissolves less asphaltic material at higher temperatures.

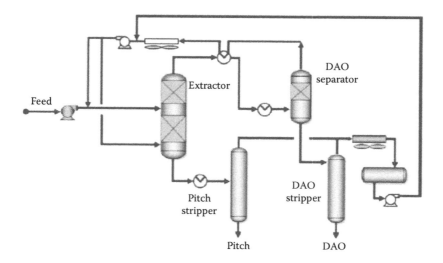

FIGURE 4.13   Propane deasphalting.

4.8.2 Desulfurization during Deasphalting

Deasphalting allows removal of sulfur compounds as well as nitrogen compounds and metallic constituents by balancing yield with the desired feedstock properties (Table 4.22). However, unless there are uses for the deasphalted oil as a fuel oil or as a blending stock for heavy fuel oils, there still remains the need to remove the remaining sulfur if the deasphalted oil is ultimately intended for use as one of the lighter liquid fuels.

TABLE 4.22
Feedstock and Product Data for the Deasphalting Process

	Murban	Light Arabian	Kuwait	Buzurgan	Boscan
Fxeedstock					
Yield on crude, wt.%	10	21.3	28.0	52.0	77.8
Specific gravity	0.982	1.003	1.021	1.051	1.037
API gravity	12.6	9.6	7.1	3.1	4.9
Viscosity at 210°F cSt	195	345	1105	3355	5250
Sulfur, wt.%	3.03	4.50	5.35	6.20	5.90
Nickel, ppm	17	19	33	76	133
Vanadium, ppm	26	61	87	233	1264
Solvent					
Deasphalted Oil	C_3	C_3	C_3	C_3	C_3
Yield on residuum, wt.%	54	45	29	31	33
Specific gravity	0.924	0.933	0.937	0.945	0.953
API gravity	21.6	20.2	19.5	18.2	17.0
Viscosity at 210°F cSt	38	34.9	46.0	30.0	33.0
Conradson carbon, wt.%	2.0	1.65	1.90	2.0	1.8
Asphaltenes, wt.%	<0.05	<0.05	<0.05	<0.05	<0.05
Nickel, ppm	<1	1.0	1.0	2.0	6.0
Vanadium, ppm	<1	1.4	2.0	4.0	10.0
Sulfur, wt.%	1.80	2.55	2.70	3.60	4.70
Nitrogen, wt.%	0.08	0.12	0.13	0.09	0.16
Solvent					
Deasphalted Oil	C_5	C_5	C_5	C_5	C_5
Yield on residuum, wt.%	93.0	85.5	78	64.2	64.0
Specific gravity	0.970	0.974	0.986	0.985	0.987
API gravity	14.4	13.8	12.0	12.2	11.9
Viscosity at 210°F cSt	124	105	201	78	135
Conradson carbon, wt.%	9	7.9	8.9	7.4	6.4
Asphaltenes, wt.%	<0.05	<0.05	<0.05	<0.05	<0.05
Nickel, ppm	8	7	12	12	36
Vanadium, ppm	9	15.5	28	26	270
Sulfur, wt.%	2.85	3.65	4.70	4.80	5.30
Nitrogen, wt.%	0.16	0.22	0.21	0.24	0.43

Source: Speight, J.G. 2000. *The Desulfurization of Heavy Oils and Residua.* 2nd Edition. Marcel Dekker Inc., New York. Table 7.17, p. 296.

TABLE 4.23

Distribution of Feedstock Components in Deasphalted Oils

	Sulfur			Nitrogen			Nickel			Vanadium		
Deasphalted oil yield, vol.%	70	80	90	70	80	90	70	80	90	70	80	90
Component left in oil, wt.%												
Gach Saran	50	68	83	33	48	70	17	32	57	11	26	55
Kuwait	57	68	83	38	54	74	6	15	38	7	19	44
Iranian Light	53	66	81	34	51	71	10	19	41	6	12	28
Tia Juana	50	64	80	34	50	71	5.5	14.5	38	3.5	10	33

Source: Speight, J.G. 2000. *The Desulfurization of Heavy Oils and Residua.* 2nd Edition. Marcel Dekker Inc., New York. Table 7.18, p. 297.

Although deasphalting offers a form of desulfurization of heavy feedstocks, there is (as might be expected) a downside to the application of the process in this manner or as a pretreatment step. Deasphalting removes potentially high-value material that might be otherwise converted to liquid products with the need to dispose of a product that might not be suitable for asphalt production.

Propane deasphalting (or modifications of this process) is most efficient in improving cracking feedstocks because the amount of coke forming (heteroatom constituents, including the sulfur-containing constituents and metal-containing constituents) is lower in the deasphalted oil than in the original feedstock (Table 4.23). Indeed, installation of a deasphalting unit just before a catalytic cracking (hydrocracking) unit may actually improve the yield and quality of the products from the cracker. There is some dispute over claims of this type, but there is no dispute that a deasphalting unit will improve the quality of the feedstock by removing the high molecular weight high-polarity constituents (Ditman, 1973; Mitchell and Speight, 1973; Long and Speight, 1997; Speight and Ozum, 2002; Parkash, 2003; Hsu and Robinson, 2006; Gary et al., 2007; Chaudhuri, 2011; Speight, 2014a).

4.8.3   DEWAXING PROCESSES

Paraffinic crude oils often contain paraffin wax constituents, and the crude oil may be treated with a solvent such as methyl ethyl ketone or methyl ethyl ketone–toluene mixtures to remove this wax before it is processed (Speight and Ozum, 2002; Parkash, 2003; Hsu and Robinson, 2006; Gary et al., 2007; Nimer et al., 2010; Chaudhuri, 2011; Speight, 2014a). This is not a common practice, however, and *solvent dewaxing processes* are designed to remove wax from lubricating oils to give the product good fluidity characteristics at low temperatures (e.g., low pour point) rather than from the whole crude oil. The mechanism of solvent dewaxing involves either the separation of wax as a solid that crystallizes from the oil solution at low temperature or the separation of wax as a liquid that is extracted at temperatures above the melting point of the wax through the preferential selectivity of the solvent. However, the former mechanism is the usual basis for commercial dewaxing processes.

There are several processes in use for solvent dewaxing (Figure 4.14), but all have the same general steps: (1) contacting the feedstock with the solvent, (2) precipitating the wax from the mixture by chilling, and (3) recovering the solvent from the wax and dewaxed oil for recycling. The processes use benzene–acetone (solvent dewaxing), propane (propane dewaxing), trichloroethylene (Separator–Nobel dewaxing), ethylene dichloride–benzene (Bari-Sol dewaxing), and urea (urea dewaxing), as well as liquid sulfur dioxide–benzene mixtures. The processes differ chiefly in the use of the solvent.

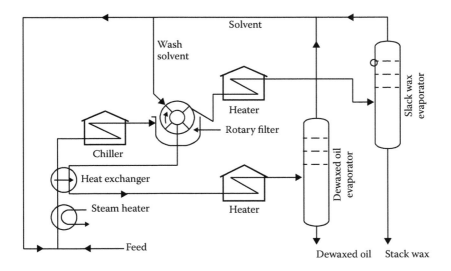

FIGURE 4.14 Solvent dewaxing. (From OSHA Technical Manual, Section IV, Chapter 2: Petroleum Refining Processes. http://www.osha.gov/dts/osta/otm/otm_iv/otm_iv_2.html; From Speight, J.G. 2014. *The Chemistry and Technology of Petroleum*. 5th Edition. CRC Press, Taylor & Francis Publishers, Boca Raton, FL. Figure 28.18, p. 819.)

Commercially used solvents were naphtha, propane, sulfur dioxide, acetone–benzene, trichloroethylene, ethylene dichloride–benzene (Bari-Sol), methyl ethyl ketone–benzene (benzol), methyl-*n*-butyl ketone, and methyl-*n*-propyl ketone. The process as now practiced involves mixing the feedstock with one to four times its volume of the ketone after which the mixture is then heated until the oil is in solution and the solution is chilled at a slow, controlled rate in double-pipe, scraped-surface exchangers. Cold solvent, such as filtrate from the filters, passes through the 2-in annular space between the inner and outer pipes and chills the waxy oil solution flowing through the inner 6-in pipe.

To prevent wax from depositing on the walls of the inner pipe, blades or scrapers extending the length of the pipe and fastened to a central rotating shaft scrape off the wax. Slow chilling reduces the temperature of the waxy oil solution to 2°C (35°F), and then faster chilling reduces the temperature to the approximate pour point required in the dewaxed oil. The waxy mixture is pumped to a filter case into which the bottom half of the drum of a rotary vacuum filter dips. The drum (8 ft in diameter, 14 ft long), covered with filter cloth, rotates continuously in the filter case. Vacuum within the drum sucks the solvent and the oil dissolved in the solvent through the filter cloth and into the drum. Wax crystals collect on the outside of the drum to form a wax cake, and as the drum rotates, the cake is brought above the surface of the liquid in the filter case and under sprays of ketone that wash oil out of the cake and into the drum. A knife-edge scrapes off the wax, and the cake falls into the conveyor and is moved from the filter by the rotating scroll.

The recovered wax is actually a mixture of wax crystals with a little ketone and oil, and the filtrate consists of the dewaxed oil dissolved in a large amount of ketone. Ketone is removed from both by distillation; however, before the wax is distilled, it is de-oiled, mixed with more cold ketone, and pumped to a pair of rotary filters in series, where further washing with cold ketone produces a wax cake that contains very little oil. The de-oiled wax is melted in heat exchangers and pumped to a distillation tower operated under vacuum, where a large part of the ketone is evaporated or flashed from the wax. The rest of the ketone is removed by heating the wax and passing it into a fractional distillation tower operated at atmospheric pressure and then into a stripper where steam removes the last traces of ketone.

An almost identical system of distillation is used to separate the filtrate into dewaxed oil and ketone. The ketone from both the filtrate and wax slurry is reused. Clay treatment or hydrotreating finishes the dewaxed oil as previously described. The wax (slack wax), even though it contains essentially no oil as compared with 50% in the slack wax obtained by cold pressing, is the raw material for either sweating or wax recrystallization, which subdivides the wax into a number of fractions with different melting points.

Solvent dewaxing can be applied to light, intermediate, and heavy lubricating oil distillates; however, each distillate produces a different kind of wax. Each of these waxes is actually a mixture of a number of waxes. For example, the wax obtained from light paraffin distillate consists of a series of paraffin waxes that have melting points in the range of 30–70°C (90–160°F), which are characterized by a tendency to harden into large crystals. However, heavy paraffin distillate yields a wax composed of a series of waxes with melting points in the range of 60–90°C (140–200°F), which harden into small crystals from which they derive the name of *microcrystalline wax* or *microwax*. On the other hand, intermediate paraffin distillates contain paraffin waxes and waxes intermediate in properties between paraffin and microwax.

Catalytic dewaxing is a hydrocracking process and is therefore operated at elevated temperatures (280–400°C, 550–750°F) and pressures (300–1500 psi). However, the conditions for a particular dewaxing operation depend on the nature of the feedstock and the product pour point required. The catalyst employed for the process is a mordenite-type catalyst that has the correct pore structure to be selective for *n*-paraffin cracking. Platinum on the catalyst serves to hydrogenate the reactive intermediates so that further paraffin degradation is limited to the initial thermal reactions. The process has been employed to successfully dewax a wide range of naphthenic feedstocks; however, it may not be suitable to replace solvent dewaxing in all cases. The process has the flexibility to fit into normal refinery operations and can be adapted for prolonged periods on-stream.

Other processes include the ExxonMobil distillate dewaxing (MDDW) process by which dewaxing is achieved by selective cracking in which the long paraffin chains are cracked to form shorter chains using a shape-selective zeolite that rejects ring compounds and isoparaffins. In a related process (MIDW process), the paraffins are selectively isomerized using low-pressure conditions. This process also uses a zeolite catalyst to convert low-quality gas oil into diesel fuel. In the process, the proprietary catalyst can be reactivated to fresh activity by relatively mild nonoxidative treatment. Of course, the time allowed between reactivation is a function of the feedstock; however, after numerous reactivations, it is possible that there will be coke buildup on the catalyst. The process can be used to dewax a full range of lubricating base stocks and, as such, has the potential to completely replace solvent dewaxing or can even be used in combination with solvent dewaxing. This latter option, of course, serves to de-bottleneck existing solvent dewaxing facilities.

4.8.4 DESULFURIZATION DURING DEWAXING

The effect of the solvent dewaxing process on the sulfur content of the feedstock is largely unknown. In some instances, when toluene is used as a cosolvent, it can be reasonably assumed that toluene will serve to maintain the high-sulfur molecular species constituents (resin and asphaltene constituent) in solution, and desulfurization will not occur. There is a slight potential for any long-chain paraffin-type compounds that contain sulfur within the molecule to be isolated with the wax constituents as part of the precipitate.

The catalytic dewaxing process not only provides a viable alternative to the solvent processes but also offers a means of desulfurizing the feedstock (Gorring and Shipman, 1975; Shen, 1983a,b; Oleck and Wilson, 1985; Chu and Shih, 1986; Kyan and Oswald, 1997). For example, in a catalytic dewaxing process, the feedstock with hydrogen is heated and mixed with hot recycle gas and introduced into a first reactor where sulfur-containing components and nitrogen-containing components are converted into hydrocarbons long with hydrogen sulfide and ammonia, respectively. The products stream (second reactor feedstock) is then fed to the second reactor containing a dewaxing catalyst where long-chain *n*-paraffins are converted to lower molecular weight products. The liquid is sent to a fractionation section, and hydrogen gas, after cooling, is fed to a cold high-pressure separator. For prevention of corrosion and deposition of ammonium salts, wash water can be supplied before cooling. In summary, the catalytic dewaxing process has the ability to remove any sulfur in the feedstocks as hydrogen sulfide.

4.9   FEEDSTOCK MODIFICATION

Refining conventional feedstocks presumes that the distillation is the first pretreatment process applied to a feedstock. Application of distillation as a separation process to heavy feedstocks is a moot point, often being considered economically unnecessary and of little benefit to the refining scenario. Processes that receive consideration to separate the residuum include deasphalting as well as thermal treatment. These options have not been accepted for wide use because of increased cost or because they recover low-value portions of the feedstock that must be used or disposed of at some stage of the refining operation. Nevertheless, some consideration is worthy of note here because of the potential for such concepts in the future.

However, for heavy feedstocks, in the current context, feedstock modification is the pretreatment of a heavy feedstock by application of a process that prepares the feedstock for further processing, i.e., desulfurization, and that has an economic benefit insofar as coke yields and catalyst costs are reduced.

One obvious method of cleaning the feed is to remove asphaltic material (asphaltene constituents plus resin constituents) using a solvent such as propane in a deasphalting unit. The resulting deasphalted oil has less metals than the original feedstock; however, coke formation and catalyst deactivation are not completely eliminated. The by-product stream is usually only acceptable as a raw material for asphalt manufacture. Even then, the asphaltic by-product may be unsuitable for specification-grade asphalt and require disposal by other means. Nevertheless, deasphalting is often considered as an alternative to hydrocracking; however, the economics of removing a portion of the feed stream and diverting it to asphalt production need careful consideration.

Visbreaking (or even hydrovisbreaking—i.e., visbreaking in an atmosphere of hydrogen or in the presence of a hydrogen donor material), the long ignored stepchild of the refining industry, may see a surge in use as a pretreatment process (Radovanović and Speight, 2011; Speight, 2011, 2012, 2014a). Management of the process to produce a liquid product that has been freed of the high potential for coke deposition (by taking the process parameters into the region where sediment forms) either in the absence or presence of (for example) a metal oxide scavenger could be valuable ally to catalyst cracking or hydrocracking units.

In addition, operating the catalytic cracking unit solely as a slurry riser cracker (without the presence of the main reactor) followed by separation of coke (sediment) would save the capital outlay required for a new catalytic cracker and might even show high conversion to valuable liquids. The quality (i.e., boiling range) of the distillate would be dependent on the residence time of the slurry in the pipe.

Scavenger additives such as metal oxides may also see a surge in use. As a simple example, a metal oxide (such as calcium oxide) has the ability to react with sulfur-containing feedstock to produce a hydrocarbon (and calcium sulfide):

$$\text{Feedstock[S]} + \text{CaO} \rightarrow \text{hydrocarbon product} + \text{CaS} + \text{H}_2\text{O}$$

Low-temperature hydroconversion has also been advocated as a pretreatment step for heavy feedstocks in reactors for residue conversion (Speight and Moschopedis, 1979; Reynolds and Beret, 1989; Beret and Reynolds, 1990; Speight and Ozum, 2002; Parkash, 2003; Hsu and Robinson, 2006; Gary et al., 2007; Chaudhuri, 2011; Speight, 2014a). The lower temperatures change the integrity of the feedstock without the accompanying disadvantage of coke formation. The mitigation of coke formation is often carried over to the hydroconversion step proper, at some significant saving to the catalyst and disposal costs. There may even be some savings in the overall cost of hydrogen (Beret and Reynolds, 1990).

A particular benefit of low-temperature hydrogenation is the ability to hydrogenate the high molecular weight components, which helps convert precursors to coke (Sanford and Chung, 1991) thereby preventing coke formation in the reactor, or to prevent precipitation of solids downstream of the reactor (Mochida et al., 1990). The protection against coking and precipitation was effective even at high conversion.

These benefits, coupled with possible savings in catalyst consumption, unwanted by-product production, and hydrogen costs suggest that two-stage processing may be attractive in individual cases.

In the long-term, new desulfurization technologies or evolution of the older technologies will reduce the need for hydrogen (Speight, 2011, 2014a). At the same time, refineries are constantly faced with challenges to reduce air pollution and other energy-related issues. Thus, traditional end-of-pipe air emission control technologies will lead to increased energy use and decreasing energy efficiency in the refinery. The petroleum refining industry will face many other challenges—climate change and new developments in automotive technology and biotechnology—which are poised to affect the future structure of refineries.

The increasing focus to reduce sulfur content in fuels will assure that the role of desulfurization in the refinery increases in importance (Babich and Moulijn, 2003). Currently, the process of choice is the hydrotreater, in which hydrogen is added to the fuel to remove the sulfur from the fuel. Some hydrogen may be lost to reduce the octane number of the fuel, which is undesirable.

Because of the increased attention for fuel desulfurization, various new process concepts are being developed with various claims of efficiency and effectiveness. The major developments in desulfurization have three main routes: (1) advanced hydrotreating (new catalysts, catalytic distillation, processing at mild conditions), (2) reactive adsorption (type of adsorbent used, process design), and (3) oxidative desulfurization (catalyst, process design).

However, residuum hydrotreating requires considerably different catalysts and process flows, depending on the specific operation so that efficient hydroconversion through uniform distribution of liquid, hydrogen-rich gas, and catalyst across the reactor is assured. In addition to an increase in guard bed use, the industry will see an increase in automated demetallization of fixed-bed systems as well as more units that operate as ebullating-bed hydrocrackers. In fact, for heavy feedstock upgrading, hydrotreating and hydrocracking technologies will be the processes of choice (Speight, 2011, 2014a). For cleaner transportation fuel production, the main task is the desulfurization of naphtha and kerosene. With the advent of various techniques, such as adsorption and biodesulfurization, the future development will still be centralized on hydrodesulfurization techniques.

Catalyst development will be key in the modification of processes and the development of new ones to make environmentally acceptable fuels (Rostrup-Nielsen, 2004). Conversion of crude oil is expected to remain the principal source of motor fuels for another 30–50 years, but it is likely that the production of fuel additives in large quantities along with conversion of natural gas will become significant (Sousa-Aguiar et al., 2005). Although crude oil conversion is expected to remain the principal source of fuels and petrochemicals in the future, natural gas reserves are emerging, and will continue to emerge, as a major hydrocarbon resource. This trend has already started to result in a shift toward use of natural gas (methane) as a significant feedstock for chemicals and also for fuels. As a result, deployment of technology for direct and indirect conversion of methane will probably displace much of the current production of liquefied natural gas.

REFERENCES

Abraham, H. 1945. *Asphalts and Allied Substances*, Vol. I. Van Nostrand, New York.

Babich, I.V., and Moulijn, J.A. 2003. Science and Technology of Novel Processes for Deep Desulfurization of Oil Refinery Streams: A Review. *Fuel* 82: 607–631.

Beret, S., and Reynolds, J.G. 1990. Effect of Prehydrogenation on the Hydroconversion of Maya Residuum. II. Hydrogen Incorporation. *Fuel Sci. Technol. Int.* 8: 191–220.

Bland, W.F., and Davidson, R.L. 1967. *Petroleum Processing Handbook*. McGraw-Hill, New York.

Carlson, C.S., Langer, A.W., Stewart, J., and Hill, R.M. 1958. Thermal Hydrogenation. Transfer of Hydrogen from Tetralin to Cracked Residua. *Ind. Eng. Chem.* 50: 1067–1070.

Chaudhuri, U.R. 2011. *Fundamentals of Petroleum and Petrochemical Engineering*. CRC Press, Taylor & Francis Group, Boca Raton, FL.

Chu, Y.F., and Shih, S.S. 1986. Process for Upgrading Petroleum Residua. United States Patent 4,592,828, June 3.

Ditman, J.G. 1973. Deasphalting Process. *Hydrocarbon Process.* 52(5): 110.

Dunning, H.N., and Moore, J.W. 1957. Propane Removes Asphalts from Crudes. *Petrol. Refiner.* 36(5): 247–250.

Evans, R.E., and Quinn, G.P. 1993. Fluid Catalytic Cracking: Science and Technology. *Stud. Surf. Sci. Catal.* 76: 563.

Forbes, R.J. 1958. *A History of Technology*, Vol. V. Oxford University Press, Oxford, UK.

Gary, J.H., Handwerk, G.E., and Kaiser, M.J. 2007. *Petroleum Refining: Technology and Economics*, 5th Edition. CRC Press, Taylor & Francis Group, Boca Raton, FL.

Gorring, R.L., and Shipman, G.F. 1975. Catalytic Dewaxing of Gas Oils. United States Patent 3,894,938, July 15.

Henderson, J.H., and Weber, L. 1965. Resid Conversion Kinetics. *J. Can. Pet. Tech.* 4: 206.

Hoiberg, A.J. 1960. *Bituminous Materials: Asphalts, Tars and Pitches*, Vols. I–II. Interscience, New York.

Hsu, C.S., and Robinson, P.R. (Editors) 2006. *Practical Advances in Petroleum Processing*, Vols. 1–2. Springer Science, New York.

Kubo, J., Higashi, H., Ohmoto, Y., and Arao, H. 1996. Heavy Oil Hydroprocessing with the Addition of Hydrogen-Donating Hydrocarbons Derived from Petroleum. *Energy Fuels* 10: 474–481.

Kyan, C.P., and Oswald, P.J. 1997. Hydrocarbon Upgrading Process. United States Patent 5,603,828, February 18.

Langer, A.W., Stewart, J., Thompson, C.E., White, H.Y., and Hill, R.M. 1962. Hydrogen Donor Diluent Visbreaking of Residua. *Ind. Eng. Chem. Proc. Design Dev.* 1: 309–312.

Long, R.B., and Speight, J.G. 1990. Studies in Petroleum Composition. III: The Distribution of Nitrogen Species, Metals, and Coke Precursors during High Vacuum Distillation. *Rev. Inst. Fr. Pétrol.* 45: 553.

MathPro. 2003. Evolution of Process Technology for FCC Naphtha Desulfurization: 1997–2003: An Example of Technical Progress Induced by Environmental Regulation. MathPro, West Bethesda, MD, March. http://www.mathproinc.com/pdf/2.1.3_FCCNaphDesulf.pdf; accessed September 5, 2014.

MathPro. 2011. An Introduction to Petroleum Refining and the Production of Ultra Low Sulfur Gasoline and Diesel Fuel. MathPro, West Bethesda, MD, October. http://www.theicct.org/sites/default/files/publica tions/ICCT05_Refining_Tutorial_FINAL_R1.pdf; accessed November 1, 2014.

McConaghy, J.R. 1987. Short Residence Time Hydrogen Donor Diluent Cracking Process. United States Patent 4,698,147, October 6.

Mitchell, D.L., and Speight, J.G. 1973. The Solubility of Asphaltenes in Hydrocarbon Solvents. *Fuel* 52: 149.

Mochida, I., Zhao, X.Z., and Sakanishi, K. 1990. Suppression of Sludge Formation by 2-Stage Hydrocracking of Vacuum Residue at High Conversion. *Ind. Eng. Chem. Res.* 29: 2324–2327.

Mushrush, G.W., and Speight, J.G. 1995. *Petroleum Products: Instability and Incompatibility*. Taylor & Francis, Washington, DC.

Navarro, U., Li, M., and Orlicki, D. 2015. FCC 101: How to Estimate Product Yields Cost-Effectively and Improve Operations. *Hydrocarbon Process.* 94(2): 41–49.

Nimer, A.A., Mohamed, A.A., and Rabah, A.A. 2010. Nile Blend Crude Oil: Wax Separation Using MEK Toluene Mixtures. *Arab. J. Sci. Eng.* 35(2): 17–24.

Oleck, S.M., and Wilson, R.C. Jr. 1985. Multi-Stage Process for Demetalation, Desulfurization, and Dewaxing of Petroleum Oils. United States Patent 4,508,615, April 2.

Parkash, S. 2003. *Refining Processes Handbook*. Gulf Professional Publishing, Elsevier, Amsterdam, the Netherlands.

Pellegrino, J.L. 1998. *Energy and Environmental Profile of the US Petroleum: Refining Industry*. Office of Industrial Technologies, United States Department of Energy, Washington, DC.

Radovanović, L.J., and Speight, J.G. 2011. Visbreaking: A Technology of the Future. In: *Proceedings. First International Conference—Process Technology and Environmental Protection (PTEP 2011)*. University of Novi Sad, Technical Faculty "Mihajlo Pupin," Zrenjanin, Republic of Serbia, December 7, pp. 335–338.

Reynolds, J.G., and Beret, S. 1989. Effect of Prehydrogenation on Hydroconversion of Maya Residuum I. Process Characterization. *Fuel Sci. Technol. Int.* 7: 165–186.

Rhoe, A., and de Blignieres, C. 1979. Visbreaking—A Flexible Process. *Hydrocarbon Process.* 58(1): 131.

Rostrup-Nielsen, J.R. 2004. Fuels and Energy for the Future: The Role of Catalysis. *Catal. Rev.* 46(3–4): 247–270.

Sadeghbeigi, R. 1995. *Fluid Catalytic Cracking: Design, Operation, and Troubleshooting of FCC Facilities*. Gulf Publishing Company, Houston, TX.

Sanford, E.C., and Chung, K.H. 1991. The Mechanism of Pitch Conversion during Coking, Hydrocracking, and Catalytic Hydrocracking of Athabasca Bitumen. *AOSTRA J. Res.* 7: 37–45.

Shen, R.C. 1983a. Catalytic Dewaxing Process. United States Patent 4,394,249, July 19.

Shen, R.C. 1983b. Cascade Catalytic Dewaxing/Hydrodewaxing Process. United States Patent 4,400,265, August 23.

Sousa-Aguiar, E.F., Appel, L.G., and Mota, C. 2005. Natural Gas Chemical Transformations: The Path to Refining in the Future. *Catal. Today* 10(1): 3–7.

Speight, J.G. 1992. A Chemical and Physical Explanation of Incompatibility during Refining Operations. *Proceedings. 4th International Conference on the Stability and Handling of Liquid Fuels*. US. Department of Energy (DOE/CONF-911102), p. 169.

Speight, J.G. 1996. Petroleum Refinery Processes. In *Kirk-Othmer Encyclopedia of Chemical Technology*, 4th edition. Wiley Interscience, New York, 18: 433.

Speight, J.G. 2000. *The Desulfurization of Heavy Oils and Residua*, 2nd Edition. Marcel Dekker Inc., New York.

Speight, J.G. 2001. *Handbook of Petroleum Analysis*. John Wiley & Sons Inc., Hoboken, NJ.

Speight, J.G. 2003. Thermal Cracking of Petroleum. In: *Natural and Laboratory-Simulated Thermal Geochemical Processes*, R. Ikan (Editor). Kluwer Academic Publishers Inc., Dordrecht, the Netherlands, Chapter 2.

Speight, J.G. 2011. *The Refinery of the Future*. Gulf Professional Publishing, Elsevier, Oxford, UK.

Speight, J.G. 2012. Visbreaking: A Technology of the Past and the Future. *Sci. Iran. C* 19(3): 569–573.

Speight, J.G. 2014a. *The Chemistry and Technology of Petroleum*, 5th Edition. CRC Press, Taylor & Francis Group, Boca Raton, FL.

Speight, J.G. 2014b. *Oil and Gas Corrosion Prevention*. Gulf Professional Publishing, Elsevier, Oxford, UK.

Speight, J.G. 2015. *Handbook of Petroleum Product Analysis*, 2nd Edition. John Wiley & Sons Inc., Hoboken, NJ.

Speight, J.G., and Francisco, M.A. 1990. Studies in Petroleum Composition IV: Changes in the Nature of Chemical Constituents during Crude Oil Distillation. *Rev. de Inst. Fr. Pétrol.* 45: 733.

Speight, J.G., and Moschopedis, S.E. 1979. The Production of Low-Sulphur Liquids and Coke from Athabasca Bitumen. *Fuel Process. Technol.* 2: 295.

Speight, J.G., and Ozum, B. 2002. *Petroleum Refining Processes*. Marcel Dekker Inc., New York.

Suchanek, A.J., and Moore, A.S. 1986. Efficient Carbon Rejection Upgrades Mexico's Maya Crude Oil. *Oil Gas J.* 84(31): 36–40.

Sullivan, D.W. 2011. New Continuous Coking Process. In: *Proceedings 14th Topical on Refinery Processing*. Spring Meeting AIChE. 2011 AIChE Spring Meeting and Global Congress on Process Safety, Chicago, March 13–17.

Vernon, L.W., Jacobs, F.E., and Bauman, R.F. 1984. Process for Converting Petroleum Residuals. United States Patent 4,425,224, January 10.

Yan, T.Y. 1991. Catalytic Visbreaking Process. United States Patent 5,057,204, October 15.

5 Upgrading Heavy Feedstocks

5.1 INTRODUCTION

Petroleum refining has entered a significant transition period as the industry moved into the 21st century, as the demands for petroleum products, particularly transportation fuels, have shown a sharp growth in recent years. Refinery operations have evolved to include a range of next-generation processes as the acceptance by refineries of heavy feedstocks became the norm.

Over the past three decades, crude oils available to refineries have generally decreased in API gravity (Speight, 2001, 2005, 2011, 2014, 2015; Speight and Ozum, 2002; Parkash, 2003; Houde and McGrath, 2006; Hsu and Robinson, 2006; Gary et al., 2007; Rana et al., 2007; Rispoli et al., 2009; Stratiev and Petkov, 2009; Stratiev et al., 2009; Motaghi et al., 2010a,b; Lucke et al., 2015). There is, nevertheless, a major focus in refineries (Figure 5.1) through a variety of conversion processes on the ways in which heavy feedstocks might be converted into low-boiling high-value products (Khan and Patmore, 1998). Simultaneously, the changing crude oil properties are reflected in changes such as an increase in asphaltene constituents and in sulfur, nitrogen, and metal contents. Pretreatment processes for removing these contaminants are playing a greater role than ever. However, the essential step required of refineries is the upgrading of heavy feedstocks (Dickenson et al., 1997; Speight, 2013). In fact, the increasing supply of heavy crude oils is a matter of serious concern for the petroleum industry.

Thus, one of the major tasks of the petroleum industry is the current and continuing production of gasoline and diesel fuels with ultra-low sulfur content from heavy feedstocks. Current legislation limits the maximum content of sulfur in gasoline and diesel fuels to parts per million (ppm) levels. Reducing the sulfur content in fuels is important in order to reduce their impact on the environment and protect public health. The effect of sulfur from fuels on the environment is mainly associated with its conversion into sulfur oxides that form during combustion in the engine, which then pollutes the environment and affects air quality. In addition, sulfur in fuels significantly decreases the efficiency and lifetime of emission gas treatment systems in vehicles.

Heavy feedstocks are generally characterized by low API gravity (high density) and high viscosity, high initial boiling point, high carbon residue, high nitrogen content, high sulfur content, and high metals content (Speight, 2001, 2005, 2011, 2014, 2015; Speight and Ozum, 2002; Parkash, 2003; Houde and McGrath, 2006; Hsu and Robinson, 2006; Gary et al., 2007; Rana et al., 2007; Rispoli et al., 2009; Stratiev and Petkov, 2009; Stratiev et al., 2009; Motaghi et al., 2010a,b). In addition to these properties, the heavy feedstocks also have an increased molecular weight and reduced hydrogen content (Figure 5.2). However, to adequately define the behavior of heavy feedstocks in refinery operations, some reference should also be made to composition and thermal characteristics of the constituents (Speight, 2001, 2005, 2011, 2014, 2015; Speight and Ozum, 2002; Parkash, 2003; Houde and McGrath, 2006; Hsu and Robinson, 2006; Gary et al., 2007; Rana et al., 2007; Rispoli et al., 2009; Stratiev and Petkov, 2009; Stratiev et al., 2009; Motaghi et al., 2010a,b).

When catalytic processes are employed, complex molecules (such as those that may be found in the original asphaltene fraction) or those formed during the process, are not sufficiently mobile (or are too strongly adsorbed by the catalyst) to be saturated by hydrogenation. The chemistry of the thermal reactions of some of these constituents dictates that certain reactions, once initiated, cannot be reversed and proceed to completion, usually resulting in the formation of coke that deposits on the catalyst, which deactivates the catalyst sites and eventually interfere with the hydroprocess (Speight, 2000, 2014; Ancheyta and Speight, 2007).

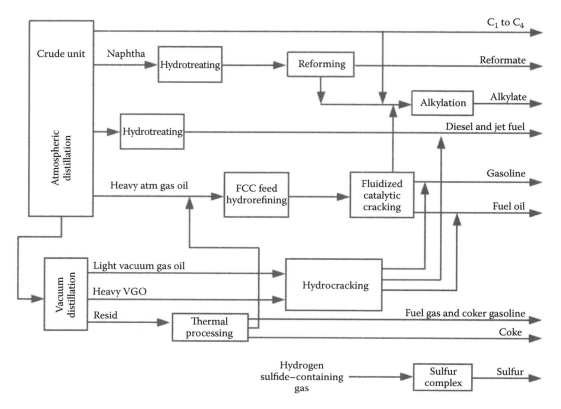

FIGURE 5.1 Schematic overview of a refinery. (From Speight, J.G. 2014. *The Chemistry and Technology of Petroleum.* 5th Edition. CRC Press, Taylor & Francis Publishers, Boca Raton, FL. Figure 15.1, p. 392.)

To satisfy the changing pattern of product demand, significant investments in refining conversion processes will be necessary to profitably utilize the heavy oils and residua. The most efficient and economical solution to this problem will depend to a large extent on individual refinery situations. However, the most promising technologies will likely focus on the conversion of vacuum residua and extra-heavy crude oils into useful low-boiling and middle-distillate products.

Upgrading heavy feedstocks began with the introduction of desulfurization processes (Speight, 1984, 2014). In the early days, the goal was desulfurization but, in later years, the processes were adapted to a 10–30% partial conversion operation, as intended to achieve desulfurization and obtain low-boiling fractions simultaneously, by increasing severity in operating conditions. Refinery evolution has seen the introduction of a variety of conversion processes for heavy feedstocks that are based on thermal cracking, catalytic cracking, and hydroconversion. Those processes are different from one another in terms of the method and product slates and will find employment in refineries according to their respective features.

New processes for the conversion of residua and heavy oils will eventually find use in place of visbreaking, the various coking options, catalytic cracking, and deasphalting that occur in current refining scenarios (Figure 5.1). It may also be opportune to use a degree of hydrocracking or hydrotreating as a means of *primary conversion* before other processes are applied. Such primary conversion processes may replace or augment the deasphalting units in many refineries. For example, the upgrading of bitumen from tar sands uses coking technology as the primary conversion (*primary upgrading*) (Speight, 2014). The bitumen is subjected to either delayed coking or

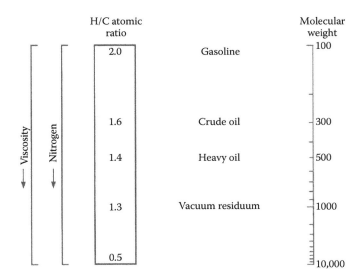

FIGURE 5.2 Relative hydrogen content (through the atomic H/C ratio) and molecular weight of various refinery feedstocks. (From Speight, J.G. 2014. *The Chemistry and Technology of Petroleum*. 5th Edition. CRC Press, Taylor & Francis Publishers, Boca Raton, FL. Figure 15.18, p. 421.)

fluid coking as the primary upgrading step (Figure 5.3) without prior distillation or topping. After primary upgrading, the product streams are hydrotreated and combined to form a synthetic crude oil that is shipped to a conventional refinery for further processing. Conceivably, a heavy feedstock could be upgraded in the same manner and, depending on the upgrading facility, upgraded further for sales. Such procedures are currently considered an exception to the usual method of refining but may become the preferred method for refining heavy oil and bitumen.

There are tried-and-true processes that are used in refineries and which were developed for conversion of residua as an option to asphalt production (Figure 5.1) (Speight, 2014), and these

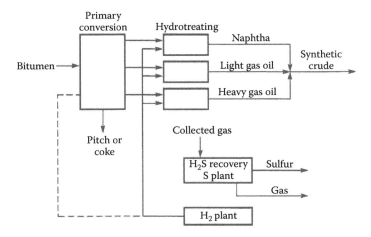

FIGURE 5.3 Processing sequence for tar sand bitumen. (From Speight, J.G. 2014. *The Chemistry and Technology of Petroleum*. 5th Edition. CRC Press, Taylor & Francis Publishers, Boca Raton, FL. Figure 14.20, p. 423.)

processes (visbreaking, coking, and deasphalting) helped balance the seasonal demands for variations in the product slate (Beeston, 2013). It is, therefore, opportune to include in this text a description of those processes that have emerged over the past two to four decades for upgrading heavy feedstocks, in addition to the most popular processes and that are now on-stream (Table 5.1). Again, as with the visbreaking, coking, and deasphalting processes, the prime motive for using these processes is conversion; however, desulfurization also occurs to varying extents.

Technologies for upgrading heavy feedstocks such as heavy oil, bitumen, and residua can be broadly divided into (1) carbon rejection and (2) hydrogen addition processes. *Carbon rejection* processes redistribute hydrogen among the various components, resulting in fractions with increased H/C atomic ratios and fractions with lower H/C atomic ratios (Speight and Ozum, 2002; Hsu and Robinson, 2006; Gary et al., 2007; Speight, 2014). On the other hand, *hydrogen addition* processes involve reaction of heavy crude oils with an external source of hydrogen and result in an overall increase in H/C ratio (Stanislaus and Cooper, 1994). Within these broad ranges, all upgrading technologies can be subdivided as follows: (1) *carbon rejection processes*, which include visbreaking, steam cracking, fluid catalytic cracking (FCC), and coking; (2) *separation processes*, such as deasphalting; and (3) *hydrogen addition processes*, which include catalytic hydroconversion (hydrocracking), fixed-bed catalytic hydroconversion, ebullated catalytic bed hydroconversion, thermal slurry hydroconversion (hydrocracking), hydrovisbreaking, hydropyrolysis, and donor solvent processes.

TABLE 5.1
Most-Used Processes for Heavy Feedstock Upgrading

Process	Type	Catalyst	General Product	Feed/Product Treating Requirement	Conversion
Delayed coking	Semibatch thermal cracking	No	Light gas, naphtha, light gas oil, heavy gas oil	Feed demetalization, product treatment: HDS, olefin/aromatic saturation	54–76%
Fluid coking	Continuous thermal cracking	No	Light gas, naphtha, light gas oil, heavy gas oil	Feed demetalization, product treatment: HDS, olefin/aromatic saturation	54–76%
Solvent deasphalting (SDA)	Solvent	No	Deasphalted oil	No	35–75%
Visbreaker	Mild thermal cracking	No	Light gas, naphtha, gas oil	No	4–30%
Catalytic cracking (RFCC)	Catalytic cracking	Regenerable	Light gas, diesel, gasoline, fuel oil	Feed or product hydrotreating, and product treatment; olefin/aromatic saturation	63–95%
H-oil	Catalytic in presence of hydrogen	Regenerable	Light gas, naphtha, kerosene, diesel fuel, vacuum gas oil, fuel oil	Product amine treating	
LC-fining	Catalytic in presence of hydrogen	Regenerable	Light gas, naphtha, kerosene, diesel fuel, vacuum gas oil, fuel oil	Product amine treating	
HRH	Catalytic in presence of hydrogen	Regenerable	Light gas, naphtha, kerosene, diesel	Product amine treating	95%

In the current context, *distillation* is excluded from the *separation processes* category and has been described elsewhere (Chapter 3) (Speight, 2011, 2014). However, a revamp of the vacuum distillation unit cutting deeper into the residue—to produce making incremental FCC unit feedstock or hydrocracker feedstock—is one of the attractive options available to the refiner. Over the past 20 years or so, the cut point of the vacuum gas oil (VGO) has increased from 510°C to approximately 590°C (950°F to approximately 110°F) or higher, as distillation column internals have been developed and the causes of coking in the vacuum column coking have become better understood. When such a revamp is planned, it is important to evaluate the impact on the quality of the VGO, especially if this gas oil forms part of the feed to downstream conversion units such as hydrocracking or FCC units. Typically, the specifications for feedstocks FCC and hydrocracking units will be on the order of

Property	FCC	Hydrocracker
Metals, ppm w/w	35	2
Conradson carbon residue, % w/w	2–10	1

The incremental yield of the gas oil (obtained by deeper distillation in the vacuum tower) is usually a relatively small proportion of total cracker feedstock, and (depending on the quality of the original petroleum feedstock) the quality of this stream can be poorer than the specification above. In addition, if a policy of increased catalyst consumption has been authorized, steps can be taken to mitigate the effect of high metals. For example, in the case of the FCC, a policy of increased catalyst consumption can be adopted or a higher proportion of demetallization catalyst can be utilized in the hydrocracker. Alternatively, a guard bed reactor may be inserted before both units—the guard bed reactor may be set at parameters in which the initial stages of coking commence and the metals are deposited with the coke.

These steps result in an increase in operating costs, which must be justified by the upgrading of the poorer-quality VGO to prime products. However, in refineries that are based on the presence of an FCC unit, the growing application of feed hydrotreatment allows processing much poorer-quality feed components in the FCC unit.

The mature and well-established processes such as visbreaking (Chapter 3), delayed coking (Chapter 3), fluid coking (Chapter 3), flexicoking (Chapter 3), and propane deasphalting (Chapter 3) have also been described elsewhere in this text. In the not too distant past, these processes were deemed adequate for upgrading heavy feedstocks. Now more options are sought and have become available. The *hydrogen addition processes* are also excluded from this chapter, having been described elsewhere (Chapter 7). However, the new generation of hydrogen addition processes are also described in detail elsewhere (Chapter 9). Even though the simplest means to cover the demand in low-boiling products is to increase the imports of light crude oils and low-boiling petroleum products (Chapter 1), the desired feedstocks are not always available and these actions are not the complete answer to the issues. Thus, new conversion processes have come on-stream.

The primary goal of these processes is to convert the heavy feedstocks to lower-boiling products, and during the conversion there is a *reduction* in the cumulative sulfur content of the liquid and nonvolatile products compared with the sulfur content of the original feedstock. Sulfur is eliminated as hydrogen sulfide or as other gaseous sulfur-containing products. Thus, even though these processes are not considered desulfurization processes in the strictest sense of the definition (Chapter 8), they need to be given some consideration because of the inclusion (or *integration*) into various refinery scenarios.

Thus, the current chapter will deal with those processes that are relatively latecomers to refinery scenarios. These processes have evolved during the last three decades, and were developed (and installed) to address the refining of the heavy feedstocks. Refining heavy feedstocks has become a major issue in modern refinery practice, and several process configurations have evolved to accommodate the heavy feedstocks. A detailed description of every process is beyond the scope of this chapter, and there are alternate sources of information that might be consulted that also provide references given for each process (RAROP, 1991; Shih and Oballa, 1991; Meyers, 1997). Instead, this chapter will discuss upgrading in terms of the type of technology involved.

5.2 THERMAL PROCESSES

Thermal cracking processes offer attractive methods of conversion of heavy feedstocks because they enable low operating pressure, while involving high operating temperature, without requiring expensive catalysts (Speight, 2003). Currently, the widest operated residuum conversion processes are visbreaking (Chapter 3) and delayed coking (Chapter 3). Furthermore, these are still attractive processes for refineries from an economic point of view (Dickenson et al., 1997).

In the current context, visbreaking is often employed to increase refinery net distillate yield through conversion of residuum (or residua) and/or by a reduction in the volume of blend stock for fuel oil production. Since the blend cutter stock is usually potential diesel blend stock, the economics are strongly driven by the price differential between diesel and fuel oil.

Briefly, there are two main variants of visbreaking: (1) the coil design, by which thermal cracking takes place only in the visbreaker heater, and (2) the soaker design, whereby heater products reside in soaker drums for the thermal reactions to continue. The soaker is, therefore, a lower-temperature, longer-residence-time design compared with coil visbreaking. Both versions are in use; however, in the coil, decoking is easier as no periodic decoking of drums is required, and process control is easier, especially if frequent changes in feed quality are required.

In the process, the degree of conversion is low (approximately 25% or less, depending on the properties of the feedstock) and is limited by sediment (coke) formation in the reactor due to the presence of asphaltene constituents in the feedstocks. In addition, the product may be unsuitable for blending with other fuel oil blend stocks. Recent evolution of the process has given attention to mitigating sediment (coke) formation and has also included optimization of the coil design by developing online spalling/decoking procedures, and introducing a heater design and proprietary installation valves, which allows for isolation and removal of one or more passes from operation. This permits steam-air decoking or pigging of the isolated pass without the need to shut down the entire visbreaker. By way of explanation, the term *spall* refers to flakes of a material that are broken off a larger solid body and can be produced by a variety of mechanisms. *Spalling* and *spallation* both describe the process of surface failure in which spall is shed.

5.2.1 ASPHALT COKING TECHNOLOGY (ASCOT) PROCESS

The ASCOT process is a residual oil upgrading process that integrates the delayed coking process and the deep solvent deasphalting process (low-energy deasphalting [LEDA]) (Bonilla and Elliott, 1987).

In the process, the vacuum residuum is brought to the desired extraction temperature and then sent to the extractor where solvent (straight-run naphtha, coker naphtha) flows upward, extracting soluble material from the down-flowing feedstock. The solvent deasphalting phase leaves the top of the extractor and flows to the solvent recovery system where the solvent is separated from the deasphalted oil and recycled to the extractor.

The deasphalted oil is sent to the delayed coker where it is combined with the heavy coker gas oil from the coker fractionator and sent to the heavy coker gas oil stripper, where low-boiling hydrocarbons are stripped off and returned to the fractionator. The stripped deasphalted oil/heavy coker gas oil mixture is removed from the bottom of the stripper and used to provide heat to the naphtha stabilizer–reboiler before being sent to battery limits as a cracking stock.

The raffinate phase containing the asphalt and some solvent flows at a controlled rate from the bottom of the extractor and is charged directly to the coking section.

The solvent contained in the asphalt and deasphalted oil is condensed in the fractionator overhead condensers, where it can be recovered and used as lean oil for a propane/butane recovery in the absorber, eliminating the need for lean oil recirculation from the naphtha stabilizer. The solvent introduced in the coker heater and coke drums results in a significant reduction in the partial pressure of asphalt feed, compared win a regular delayed coking unit. The low asphalt partial pressure results in low coke and high liquid yields in the coking reaction.

As an example, a feedstock (540°C+/1005°F Orinico vacuum residuum) having the properties of 2.8° API, 4.2% w/w sulfur, 1.0% w/w nitrogen, 198 ppm nickel plus vanadium, and 22.3% w/w carbon residue will produce 3.8% w/w liquefied petroleum gas (LPG) hydrocarbons, 7.7% w/w coker naphtha (54.7° API), 69.9% w/w coker gas oil plus deasphalted oil (13.4° API), and 25% w/w coke. The coke contains 5.8% w/w sulfur and 2.7% w/w nitrogen that represents 35% of the feedstock sulfur and 68% of the feedstock nitrogen.

5.2.2 CHERRY-P (COMPREHENSIVE HEAVY ENDS REFORMING REFINERY) PROCESS

The Cherry-P process is a process for the conversion of heavy crude oil or residuum into distillate and a cracked residuum.

In the process (Figure 5.4), the feedstock is mixed with coal powder in a slurry mixing vessel, heated in the furnace, and fed to the reactor where the feedstock undergoes thermal cracking reactions for several hours at a temperature >400°C (750°F) and under pressure. Gas and distillate from the reactor are sent to a fractionator, and the cracked residuum residue is extracted out of the system after distilling low-boiling fractions by the flash drum and vacuum flasher to adjust its softening point.

Product yields (Table 5.2) indicate that there is no coke. The distillates produced by this process are generally lower in the content of olefin hydrocarbons than the other thermal cracking process, comparatively easy to desulfurize in hydrotreating units, and compatible with straight-run distillates.

5.2.3 ET-II PROCESS

The ET-II process is a thermal cracking process for the production of distillates and cracked residuum for use as a metallurgical coke and is designed to accommodate feedstocks such as heavy oils, atmospheric residua, and vacuum residua. The distillate (referred to in the process as *cracked oil*)

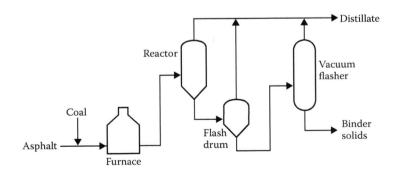

FIGURE 5.4 Cherry-P process. (From Speight, J.G. 2000. *The Desulfurization of Heavy Oils and Residua.* 2nd Edition. Marcel Dekker Inc., New York. Figure 8.4, p. 308.)

TABLE 5.2
Product Yields and Properties from the Cherry-P Process

	Feedstock	Naphtha	Kerosene	Gas Oil	Vacuum Gas Oil
Yield	100.0	21	21	41	17
Sp. gr.	1.026	0.743	0.823	0.863	0.931
°API	6.4	58.9	40.4	32.5	20.5
S, wt.%	3.1	0.57	0.83	1.35	2.46
Carbon residue	21.3	–	–	–	–

is suitable as a feedstock to hydrocracker and FCC. The basic technology of the ET-II process is derived from that of the original Eureka process.

In the process (Figure 5.5), the feedstock is heated up to 350°C (660°F) by passage through the preheater and fed into the bottom of the fractionator, where it is mixed with recycle oil, the high-boiling fraction of the cracked oil. The ratio of recycle oil to feedstock is within the range of 0.1–0.3 wt.%. The feedstock mixed with recycle oil is then pumped out and fed into the cracking heater, where the temperature is raised to approximately 490–495°C (915–925°F) and the outflow is fed to the stirred-tank reactor where it is subjected to further thermal cracking. Both cracking and condensation reactions take place in the reactor.

The heat required for the cracking reaction is brought in by the effluent itself from the cracking heater, as well as by the superheated steam, which is heated in the convection section of the cracking heater and blown into the reactor bottom. The superheated steam reduces the partial pressure of the hydrocarbons in the reactor and accelerates the stripping of volatile components from the cracked residuum. This residual product is discharged through a transfer pump and transferred to a cooling drum, where the thermal cracking reaction is terminated by quenching with a water spray, after which it is sent to the pitch water slurry preparation unit.

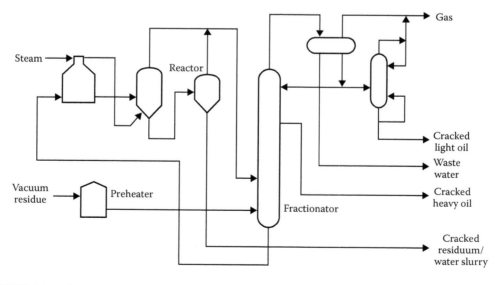

FIGURE 5.5   ET-II process. (From Speight, J.G. 2000. *The Desulfurization of Heavy Oils and Residua.* 2nd Edition. Marcel Dekker Inc., New York. Figure 8.5, p. 310.)

The cracked oil and gas products, together with steam from the top of the reactor, are introduced into the fractionator where the oil is separated into two fractions, *cracked light oil* and *cracked heavy oil*.

Commencing with a feedstock having and API gravity of 6.7°, 4.1 wt.% sulfur, and 0.6 wt.% nitrogen, the products are gas (6.3 wt.%), light oil (28.6 wt.%, 315°C⁻/600°F⁻), heavy oil (32.3 wt.%, 315°C⁻/600°F⁻–535°C/1000°F), and cracked residuum (32.8 wt.%, 535°C⁺/1000°F⁺). The cracked residuum (1.3° API) contains 6.25 wt.% sulfur (50% of the original sulfur) and 1.57 wt.% nitrogen (86% of the original nitrogen). The cracked residuum also contains 39.6 wt.% volatile matter as measured by a standard test for volatile matter.

5.2.4 EUREKA PROCESS

The Eureka process is a thermal cracking process to produce a cracked oil and aromatic residuum from heavy residual materials (Ohba et al., 2008).

In this process (Figure 5.6), the feedstock, usually a vacuum residuum, is fed to the preheater and then enters the bottom of the fractionator, where it is mixed with the recycle oil. The mixture is then fed to the reactor system that consists of a pair of reactors operating alternately. In the reactor, thermal cracking reaction occurs in the presence of superheated steam that is injected to strip the cracked products out of the reactor and supply a part of heat required for cracking reaction. At the end of the reaction, the bottom product is quenched.

The oil and gas products (Table 5.3) and steam pass from the top of the reactor to the lower section of the fractionator, where a small amount of entrained material is removed by a wash operation. The upper section is an ordinary fractionator, where the heavier fraction of cracked oil is drawn as a sidestream.

The original Eureka process uses two batch reactors, while the newer ET-II and the HSC process both employ continuous reactors.

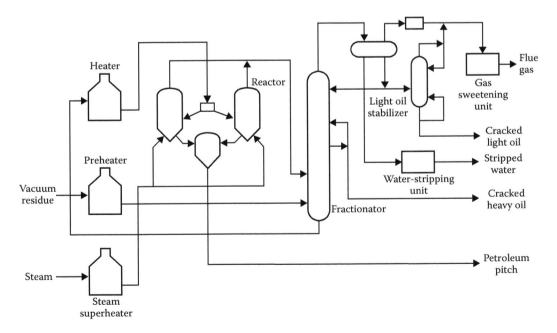

FIGURE 5.6 Eureka process. (From Speight, J.G. 2000. *The Desulfurization of Heavy Oils and Residua*. 2nd Edition. Marcel Dekker Inc., New York. Figure 8.6, p. 311.)

TABLE 5.3

Product Yields and Properties from the Eureka Process

Feedstock: Middle East Vacuum Residuum Mix[a]

API, °	7.6
Sulfur, wt.%	3.9
C7-asphaltene constituents, wt.%	5.7
Carbon residue, wt.%	20.0
Nickel, ppm	136.0
Vanadium, ppm	202.0

Products

Light Oil (C5–240°C, C5–465°F), wt.%	14.9
API, °	53.0
Sulfur, wt.%	1.1
Nitrogen, wt.%	<0.1
Gas Oil (240–540°C, 465–1000°F), wt.%	50.7
API	21.3
Sulfur, wt.%	2.7
Nitrogen, wt.%	0.3
Pitch (>540°C, >1000°F), wt.%	29.6
Sulfur, wt.%	5.7
Nitrogen, wt.%	1.2
Nickel, ppm	487.0
Vanadium, ppm	688.0

Source: Speight, J.G. 2014. *The Chemistry and Technology of Petroleum.*
           5th Edition. CRC Press, Taylor & Francis Publishers, Boca Raton,
           FL. Table 18.9, p. 504.
[a] >500°C, >930°F, not defined by name.

5.2.5 Fluid Thermal Cracking (FTC) Process

The FTC process is a heavy oil and residuum upgrading process in which the feedstock is thermally cracked to produce distillate and coke, which is gasified to fuel gas.

The feedstock, mixed with recycle stock from the fractionator, is injected into the cracker, immediately absorbed into the pores of the particles by capillary force, and is subjected to thermal cracking (Figure 5.7). In consequence, the surface of the noncatalytic particles is kept dry and good fluidity is maintained, allowing a good yield of, and selectivity for, middle-distillate products (Table 5.4). Hydrogen-containing gas from the fractionator is used for the fluidization in the cracker.

Excessive coke caused by the metals accumulated on the particle is suppressed in the presence of hydrogen. The particles with deposited coke from the cracker are sent to the gasifier, where the coke is gasified and converted into carbon monoxide (CO), hydrogen (H_2), carbon dioxide (CO_2), and hydrogen sulfide (H_2S) with steam and air. Regenerated hot particles are returned to the cracker.

5.2.6 High-Conversion Soaker Cracking (HSC) Process

The HSC process (Figure 5.8) is a cracking process to cover a moderate conversion—higher than visbreaking but lower than coking. The process features less gas make and a variety of distillate products (Table 5.5). The process can be used to convert a wide range of feedstocks with high sulfur and heavy metals, including heavy oils, oil sand bitumen, residua, and visbroken residua. As a note

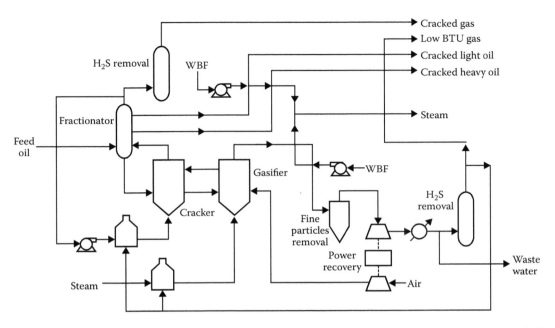

FIGURE 5.7 FTC process. (From Speight, J.G. 2000. *The Desulfurization of Heavy Oils and Residua.* 2nd Edition. Marcel Dekker Inc., New York. Figure 8.7, p. 313.)

TABLE 5.4
Product Yields and Properties from the FTC Process

	Feedstock	Gas	Naphtha	Middle Distillate	Heavy Distillate	Coke
Yield, wt.%	100	9.9	10.8	34.4	21.8	23.1
Sp. gr.	1.034	–	0.7504	0.8273	0.9421	–
API, °	5.3	–	57.1	39.5	18.7	–
Sulfur, wt.%	4.4					
Nitrogen, wt.%	0.5	–	0.01	0.05	0.19	–

of interest, the HSC process employs continuous reactors, whereas the original Eureka process uses two batch reactors.

The preheated feedstock enters the bottom of the fractionator, where it is mixed with the recycle oil. The mixture is pumped up to the charge heater and fed to the soaking drum (ca. atmospheric pressure, steam injection at the top and bottom), where sufficient residence time is provided to complete the thermal cracking. In the soaking drum, the feedstock and some product flows downward, passing through a number of perforated plates, while steam with cracked gas and distillate vapors flow through the perforated plates countercurrently.

The volatile products from the soaking drum enter the fractionator where the distillates are fractionated into desired product oil streams, including a heavy gas oil fraction. The cracked gas product is compressed and used as refinery fuel gas after sweetening. The cracked oil product after

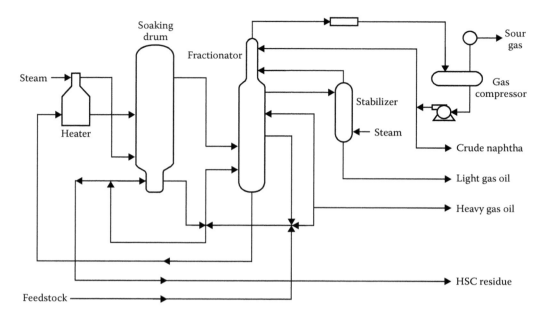

FIGURE 5.8 HSC process. (From Speight, J.G. 2000. *The Desulfurization of Heavy Oils and Residua.* 2nd Edition. Marcel Dekker Inc., New York. Figure 8.8, p. 313.)

hydrotreating is used as FCC or hydrocracker feedstock. The residuum is suitable for use as boiler fuel, road asphalt, binder for the coking industry, and as a feedstock for partial oxidation.

5.2.7 TERVAHL PROCESS

In the Tervahl T process (Peries et al., 1988) (Figure 5.9), the feedstock is heated to the desired temperature using the coil heater and heat recovered in the stabilization section, and held for a specified residence time in the soaking drum. The soaking drum effluent is quenched and sent to a conventional stabilizer or fractionator where the products are separated into the desired streams. The gas produced from the process is used for fuel.

In the Tervahl H process (Figure 5.9), the feedstock and hydrogen-rich stream are heated using heat recovery techniques and fired heater and held in the soak drum as in the Tervahl T process. The gas and oil from the soaking drum effluent are mixed with recycle hydrogen and separated in the hot separator where the gas is cooled and passed through a separator and recycled to the heater and soaking drum effluent. The liquids from the hot and cold separator are sent to the stabilizer section where purge gas and synthetic crude are separated. The gas is used as fuel, and the synthetic crude can now be transported or stored.

5.3 CATALYTIC CRACKING PROCESSES

The FCC process using VGO feedstock was introduced into refineries in the 1930s. In recent years, because of a trend for low-boiling products, most refineries perform the operation by partially blending residues into VGO. However, conventional FCC processes have limits in residue processing, so residue FCC processes have lately been employed one after another. Because the residue FCC process enables efficient naphtha production directly from residues, it will play the most important role as a residue cracking process, along with the residue hydroconversion process.

TABLE 5.5
Product Yields and Properties from the HSC Process

Feedstock	Iranian Heavy Vacuum Residuum	Maya Vacuum Residuum
API, °	5.7	2.2
Sulfur, wt.%	4.8	5.1
Nitrogen, wt.%	0.6	0.8
C7-asphaltene constituents, wt.%	11.3	19.0
Carbon residue, wt.%	22.6	28.7
Nickel, ppm	69.0	121.0
Vanadium, ppm	205.0	649.0
Products		
Naphtha (C5–200°C, C5–390°F), wt.%	6.3	3.8
API, °	54.4	51.5
Sulfur, wt.%	1.1	1.1
Light Gas Oil (200–350°C, 390–660°F), wt.%	15.0	13.3
API, °	30.2	29.5
Sulfur, wt.%	2.6	3.1
Nitrogen, wt.%	0.1	<0.1
Heavy Gas Oil (350–520°C, 660–970°F), wt.%	32.2	18.6
API, °	16.4	16.5
Sulfur, wt.%	3.5	3.7
Nitrogen, wt.%	0.3	0.3
Vacuum Residue (>520°C, >970°F), wt.%	43.2	62.0
Sulfur, wt.%	5.8	5.6
Nitrogen, wt.%	1.0	1.2
Carbon residue, wt.%	49.2	49.2
Nickel, ppm	148.0	148.0
Vanadium, ppm	453.0	453.0

Source: Speight, J.G. 2014. *The Chemistry and Technology of Petroleum.* 5th Edition. CRC Press, Taylor & Francis Publishers, Boca Raton, FL. Table 18.11, p. 506.

Another role of the residuum FCC process is to generate high-quality gasoline blending stock and petrochemical feedstock. Olefins (propene, butenes, and pentenes) serve as feed for alkylating processes, for polymer gasoline, as well as for additives for reformulated gasoline.

The processes described below are the evolutionary offspring of the FCC and the residuum catalytic cracking processes. Some of these newer processes use catalysts with different silica/alumina ratios as acid support of metals such as Mo, Co, Ni, and W. In general, the first catalyst used to remove metals from oils was the conventional hydrodesulfurization catalyst. Diverse natural minerals are also used as raw material for elaborating catalysts addressed to the upgrading of heavy fractions. Among these minerals are clays; manganese nodules; bauxite activated with vanadium (V), nickel (Ni), chromium (Cr), iron (Fe), and cobalt (Co), as well as and iron laterites, sepiolites; and mineral nickel and transition metal sulfides supported on silica and alumina. Other kinds of catalysts, such as vanadium sulfide, are generated *in situ*, possibly in colloidal states.

5.3.1 Asphalt Residual Treating (ART) Process

The ART process is a process for increasing the production of high-value transportation fuels and reduces heavy fuel oil production, without hydrocracking (Bartholic et al., 1992).

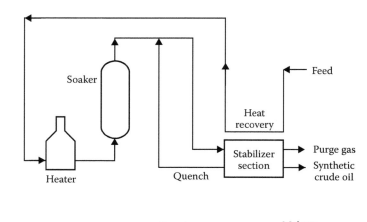

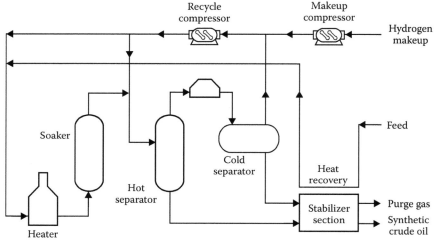

FIGURE 5.9 Tervahl processes. (From Speight, J.G. 2000. *The Desulfurization of Heavy Oils and Residua.* 2nd Edition. Marcel Dekker Inc., New York. Figure 8.9, p. 316.)

In the process (Figure 5.10), the preheated feedstock (which may be whole crude, atmospheric residuum, vacuum residuum, or bitumen) is injected into a stream of fluidized, hot catalyst (trade name: ArtCat). Complete mixing of the feedstock with the catalyst is achieved in the contactor, which is operated within a pressure–temperature envelope to ensure selective vaporization. The vapor and the contactor effluent are quickly and efficiently separated from each other, and entrained hydrocarbons are stripped from the contaminant (containing spent solid) in the stripping section. The contactor vapor effluent and vapor from the stripping section are combined and rapidly quenched in a quench drum to minimize product degradation. The cooled products are then transported to a conventional fractionator that is similar to that found in an FCC unit. Spent solid from the stripping section is transported to the combustor bottom zone for carbon burn-off.

In the combustor, coke is burned from the spent solid that is then separated from combustion gas in the surge vessel. The surge vessel circulates regenerated catalyst streams to the contactor inlet for feed vaporization, and to the combustor bottom zone for premixing.

The components of the combustion gases include carbon dioxide (CO_2), nitrogen (N_2), oxygen (O_2), sulfur oxides (SO_x), and nitrogen oxides (NO_x), which are released from the catalyst with the

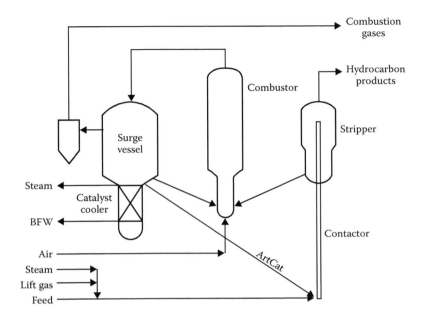

FIGURE 5.10 ART process. (From Speight, J.G. 2000. *The Desulfurization of Heavy Oils and Residua.* 2nd Edition. Marcel Dekker Inc., New York. Figure 8.10, p. 317.)

combustion of the coke in the combustor. The concentration of sulfur oxides in the combustion gas requires treatment for their removal.

5.3.2 RESIDUE FCC PROCESS

The residue FCC heavy oil cracking (HOC) process is a version of the FCC process that has been adapted to conversion of residua that contain high amounts of metal and asphaltene constituents (Feldman et al., 1992).

In the process, a residuum is desulfurized and the nonvolatile fraction from the hydrodesulfurizer is charged to the residuum FCC unit. The reaction system is an external vertical riser terminating in a closed cyclone system. Dispersion steam in amounts higher than that used for gas oils is used to assist in the vaporization of any volatile constituents of heavy feedstocks.

A two-stage stripper is used to remove hydrocarbons from the catalyst. Hot catalyst flows at low velocity in dense phase through the catalyst cooler and returns to the regenerator. Regenerated catalyst flows to the bottom of the riser to meet the feed.

The coke deposited on the catalyst is burned off in the regenerator along with the coke formed during the cracking of the gas oil fraction. If the feedstock contains high proportions of metals, control of the metals on the catalyst requires excessive amounts of catalyst withdrawal and fresh catalyst addition. This problem can be addressed by feedstock pretreatment.

5.3.3 HEAVY OIL TREATING (HOT) PROCESS

The HOT process is a catalytic cracking process for upgrading heavy feedstocks such as topped crude oils, vacuum residua, and solvent deasphalting bottoms using a fluidized bed of iron ore particles.

The main section of the process (Figure 5.11) consists of three fluidized reactors and separate reactions take place in each reactor (cracker, regenerator, and desulfurizer):

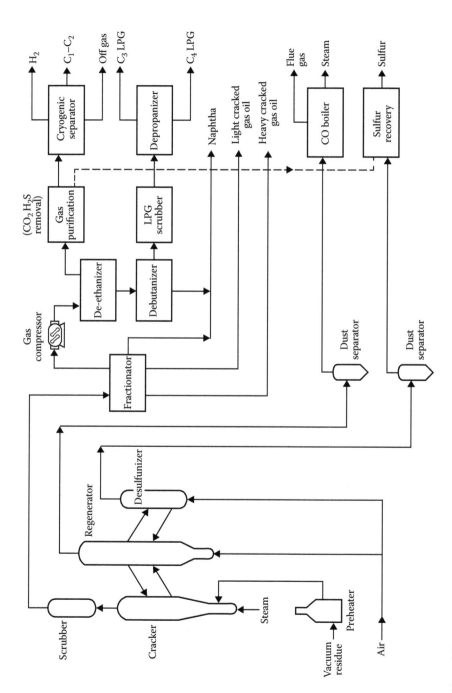

FIGURE 5.11 HOT process. (From Speight, J.G. 2000. *The Desulfurization of Heavy Oils and Residua*. 2nd Edition. Marcel Dekker Inc., New York. Figure 8.11, p. 319.)

Fe_3O_4 + asphaltene constituents → coke/Fe_3O_4 + oil + gas (in the cracker)

$3FeO + H_2O → Fe_3O_4 + H_2$ (in the cracker)

Coke/$Fe_3O_4 + O_2 → 3FeO + CO + CO_2$ (in the regenerator)

$FeO + SO_2 + 3CO → FeS + 3CO_2$ (in the regenerator)

$3FeS + 5O_2 → Fe_3O_4 + 3SO_2$ (in the desulfurizer)

In the cracker, heavy oil cracking and the steam–iron reaction take place simultaneously under the conditions similar to thermal cracking. Any unconverted feedstock is recycled to the cracker from the bottom of the scrubber. The scrubber effluent is separated into hydrogen gas, LPG, and liquid products that can be upgraded by conventional technologies to priority products.

In the regenerator, coke deposited on the catalyst is partially burned to form carbon monoxide in order to reduce iron tetroxide and to act as a heat supply. In the desulfurizer, sulfur in the solid catalyst is removed and recovered as molten sulfur in the final recovery stage.

5.3.4 R2R PROCESS

The R2R process is an FCC process for conversion of heavy feedstocks.

In the process (Figure 5.12), the feedstock is vaporized upon contacting the hot regenerated catalyst at the base of the riser and lifts the catalyst into the reactor vessel separation chamber, where rapid disengagement of the hydrocarbon vapors from the catalyst is accomplished by both a special solids separator and cyclones. The bulk of the cracking reactions takes place at the moment of contact and continues as the catalyst and hydrocarbons travel up the riser. The reaction products, along

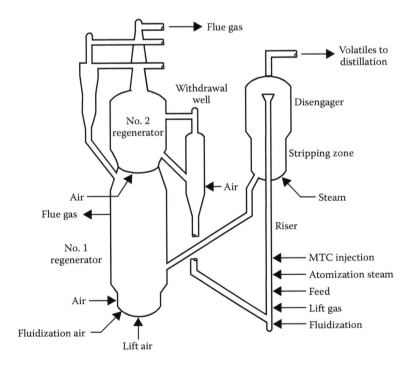

FIGURE 5.12 R2R process. (From Speight, J.G. 2000. *The Desulfurization of Heavy Oils and Residua.* 2nd Edition. Marcel Dekker Inc., New York. Figure 8.12, p. 320.)

with a minute amount of entrained catalyst, then flow to the fractionation column. The stripped spent catalyst, deactivated with coke, flows into the no. 1 regenerator.

Partially regenerated catalyst is pneumatically transferred via an air riser to the no. 2 regenerator, where the remaining carbon is completely burned in a dryer atmosphere. This regenerator is designed to minimize catalyst inventory and residence time at high temperature while optimizing the coke-burning rate. Flue gases pass through external cyclones to a waste heat recovery system. Regenerated catalyst flows into a withdrawal well and after stabilization is charged back to the oil riser.

5.3.5 Reduced Crude Oil Conversion (RCC) Process

In the RCC process (Figure 5.13), the clean regenerated catalyst enters the bottom of the reactor riser where it contacts low-boiling hydrocarbon lift gas, which accelerates the catalyst up the riser before feed injection. At the top of the lift gas zone, the feed is injected through a series of nozzles located around the circumference of the reactor riser.

The catalyst/oil disengaging system is designed to separate the catalyst from the reaction products and then rapidly remove the reaction products from the reactor vessel. Spent catalyst from the reaction zone is first steam stripped, to remove adsorbed hydrocarbon, and then routed to the regenerator. In the regenerator, all of the carbonaceous deposits are removed from the catalyst by combustion, restoring the catalyst to an active state with a very low carbon content. The catalyst is then returned to the bottom of the reactor riser at a controlled rate to achieve the desired conversion and selectivity to the primary products.

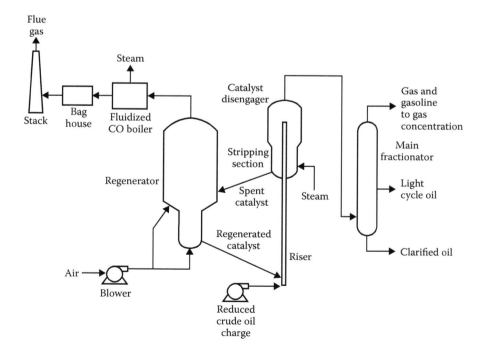

FIGURE 5.13 RCC process. (From Speight, J.G. 2000. *The Desulfurization of Heavy Oils and Residua*. 2nd Edition. Marcel Dekker Inc., New York. Figure 8.13, p. 321.)

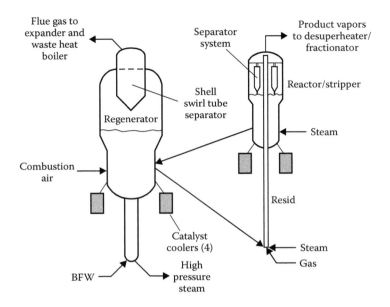

FIGURE 5.14 Shell FCC process. (From Speight, J.G. 2000. *The Desulfurization of Heavy Oils and Residua.* 2nd Edition. Marcel Dekker Inc., New York. Figure 8.14, p. 322.)

5.3.6 SHELL FCC PROCESS

In the Shell FCC process (Figure 5.14), the preheated feedstock (VGO, atmospheric residuum) is mixed with the hot regenerated catalyst (Khouw, 1990). After reaction in a riser, volatile materials and catalyst are separated after which the spent catalyst is immediately stripped of entrained and adsorbed hydrocarbons in a very effective multistage stripper. The stripped catalyst gravitates through a short standpipe into a single-vessel, simple, reliable, and yet efficient catalyst regenerator. Regenerative flue gas passes via a cyclone/swirl tube combination to a power recovery turbine. From the expander turbine, the heat in the flue gas is further recovered in a waste heat boiler.

Depending on the environmental conservation requirements, a $deNO_x$ing, $deSO_x$ing, and particulate emission control device can be included in the flue gas train.

Furthermore, hydrogenation pretreatment of bitumen before FCC (or for that matter, any catalytic cracking process) can result in enhanced yield of naphtha (Sato et al., 1992). It is suggested that mild hydrotreating be carried out upstream of an FCC unit to provide an increase in yield and quality of distillate products. This is in keeping with earlier work (Speight and Moschopedis, 1979) where mild hydrotreating of bitumen was reported to produce low-sulfur liquids that would be amenable to further catalytic processing.

5.3.7 S&W FCC PROCESS

In the S&W FCC process (Figure 5.15), the heavy feedstock is injected into a stabilized, upward flowing catalyst stream whereupon the feedstock–steam–catalyst mixture travels up the riser and is separated by a high-efficiency inertial separator. The product vapor goes overhead to the main fractionator.

The spent catalyst is immediately stripped in a staged, baffled stripper to minimize hydrocarbon carryover to the regenerator system. The first regenerator (650–700°C, 1200–1290°F) burns 50–70% of the coke in an incomplete carbon monoxide combustion mode running countercurrently. This

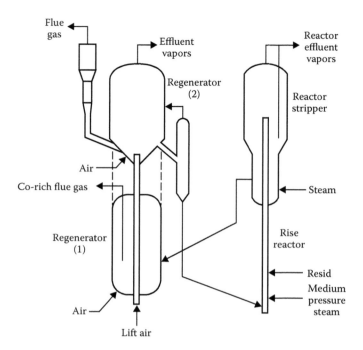

Flue gas

Effluent vapors

Regenerator (2)

Reactor effluent vapors

Reactor stripper

Air

Co-rich flue gas

Steam

Regenerator (1)

Rise reactor

Resid

Medium pressure steam

Air

Lift air

FIGURE 5.15 S&W FCC process. (From Speight, J.G. 2000. *The Desulfurization of Heavy Oils and Residua*. 2nd Edition. Marcel Dekker Inc., New York. Figure 8.15, p. 323.)

relatively mild, partial regeneration step minimizes the significant contribution of hydrothermal catalyst deactivation. The remaining coke is burned in the second regenerator (ca. 775°C, 1425°F) with an extremely low steam content. Hot clean catalyst enters a withdrawal well that stabilizes its fluid qualities before being returned to the reaction system.

5.3.8 Millisecond Catalytic Cracking (MSCC) Process

Short residence times for feedstock conversion are usually associated with high-severity processes such as *ultrapyrolysis* (high temperature and very short residence time). However, such processes are used commercially only for cracking ethane, propane, butane, and light-distillate feeds to produce ethylene and higher olefins.

However, the use of short residence times with the catalyst placed in a more optimal position to ensure better contact with the feedstock has resulted in the MSCC unit that is used to process residua. The unit is flexible in terms of feedstock changes, and the improved metals tolerance of the process allows the unit to handle a wide range of feedstocks.

5.3.9 Residuum Desulfurization (RDS) and Vacuum Residuum Desulfurization (VRDS) Processes

Fixed-bed residuum desulfurization technologies—RDS for atmospheric residuum and VRDS for vacuum residuum—provide a short, economical process path to higher-value products from difficult feeds. These hydrogen-efficient processes sufficiently saturate products so that further processing in conversion units is greatly enhanced, ultimately producing more and higher-value light products.

For example, by pretreating *residuum fluid catalytic cracking* feedstock in an RDS/VRDS reactor, refiners have more flexibility to choose less expensive crudes or process more residuum, while achieving higher product yields and higher on-stream efficiency. Adding an on-stream catalyst replacement (OCR) unit enables refiners to significantly increase capacity or improve product quality from the residuum desulfurization unit.

In the process, the feedstock (after passing through preheat exchangers) is combined with recycle gas and sent through additional exchangers and the feed furnace. After reaching a set inlet temperature, the combined feed enters the reactors from the top. The number of reactors, in parallel or series, is determined by the overall objectives and feed rate. From the reactor section, heat is recovered in the feed/effluent exchangers and then further cooled and flashed in the separation section. The recycle gas, bleed, and makeup gas are optimized to provide the highest-purity hydrogen for the reactor section at minimal cost. The fractionation section draws the liquid from the separation section and separates lower-boiling constituents into (1) off-gas, (2) distillates, and (3) treated atmospheric residual fractions as required.

Fresh catalyst is added at the top of the reactor, while residuum is fed into the bottom. The catalyst moves through the reactor in countercurrent flow, causing the dirtiest residuum to contact the oldest catalyst first. Spent catalyst is removed at the bottom of the reactor in a batch operation (typically once or twice per week), with no interruption to the process.

The countercurrent, moving-bed reactor can be integrated into the typical RDS reactor circuit. With the ability to replace spent catalyst on-line, compared with a standard RDS unit, refiners can increase feed throughput, process heavier feeds with higher levels of contaminant metals, or achieve deeper desulfurization when processing low-metal feeds. With the OCR addition, refiners can increase cycle lengths considerably as the life and efficiency of the catalyst is improved substantially.

5.4 SOLVENT PROCESSES

Solvent deasphalting processes have not realized their maximum potential and can play a major role in modern refineries. The technology can be used in a variety of ways for heavy feedstock upgrading (Houde and McGrath, 2006; Speight, 2011, 2014). With ongoing improvements in energy efficiency, such processes would display its effects in a combination with other processes. Solvent deasphalting allows removal of sulfur and nitrogen compounds as well as metallic constituents by balancing yield with the desired feedstock properties (Figure 5.16) (Ditman, 1973). Another version of the deasphalting process, *critical solvent deasphalting*, has been developed that uses critical state solvents for the best separation of lighter liquid products from all kinds of heavy residues (supercritical extraction), especially with coking or gasification for the elimination of solvent deasphalting bottoms (Dickenson et al., 2004), and the method is claimed to be energy efficient.

5.4.1 Deep Solvent Deasphalting Process

The deep solvent deasphalting process is an application of the LEDA process, which is used to extract high-quality lubricating oil bright stock or to prepare catalytic cracking feeds, hydrocracking feeds, hydrodesulfurizer feeds, and asphalt from vacuum residue materials. The LEDA process uses a low-boiling hydrocarbon solvent specifically formulated to ensure the most economical deasphalting design for each operation. For example, a propane solvent may be specified for a low deasphalted oil yield operation, while a solvent containing hydrocarbons such as hexane may be used to obtain a high deasphalted oil yield from a vacuum residuum. The deep deasphalting process can be integrated with a delayed coking operation (ASCOT process; in this case, the solvent can be a low-boiling naphtha).

Low-energy deasphalting operations are usually carried out in a rotating disc contractor, which provides more extraction stages than a mixer-settler or baffle-type column. Although not essential

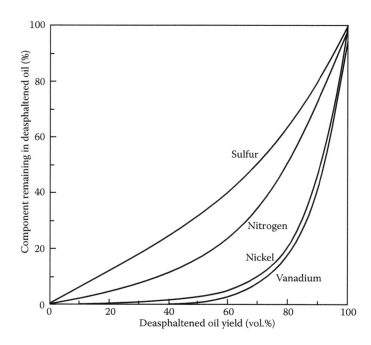

FIGURE 5.16 Relation of product properties and yield in a deasphalting process. (From Speight, J.G. 2000. *The Desulfurization of Heavy Oils and Residua.* 2nd Edition. Marcel Dekker Inc., New York. Figure 8.16, p. 324.)

to the process, the rotating disc contactor provides higher-quality deasphalted oil at the same yield, or higher yields at the same quality.

The low-energy solvent deasphalting process selectively extracts the more paraffinic components from vacuum residua while rejecting the condensed ring aromatics. As expected, deasphalted oil yields vary as a function of solvent type and quantity, and feedstock properties (Chapter 7).

In the process (Figure 5.17), vacuum residue feed is combined with a small quantity of solvent to reduce its viscosity and cooled to a specific extraction temperature before entering the rotating disc contactor. Recovered solvent from the high-pressure and low-pressure solvent receivers are combined, adjusted to a specific temperature by the solvent heater–cooler, and injected into the bottom section of the rotating disc contactor. Solvent flows upward, extracting the paraffinic hydrocarbons from the vacuum residuum, which is flowing downward through the rotating disc contactor.

Steam coils at the top of the tower maintain the specified temperature gradient across the rotating disc contactor. The higher temperature in the top section of the rotating disc contactor results in separation of the less soluble heavier material from the deasphalted oil mix and provides internal reflux, which improves the separation. The deasphalted oil mix leaves the top of the rotating disc contactor tower. It flows to an evaporator where it is heated to vaporize a portion of the solvent. It then flows into the high-pressure flash tower where high-pressure solvent vapors are taken overhead.

The deasphalted oil mix from the bottom of this tower flows to the pressure vapor heat exchanger where additional solvent is vaporized from the deasphalted oil mix by a condensing high-pressure flash. The high-pressure solvent, totally condensed, flows to the high-pressure solvent receiver. Partially vaporized, the deasphalted oil mix flows from the high-pressure vapor heater exchanger to the low-pressure flash tower where low-pressure solvent vapor is taken overhead, condensed, and collected in the low-pressure solvent receiver. The deasphalted oil mix flows down the low-pressure flash tower to the reboiler, where it is heated, and then to the deasphalted oil stripper, where the

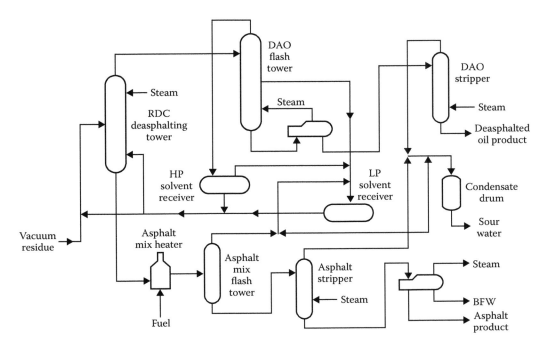

FIGURE 5.17 Deep solvent deasphalting process. (From Speight, J.G. 2000. *The Desulfurization of Heavy Oils and Residua*. 2nd Edition. Marcel Dekker Inc., New York. Figure 8.17, p. 326.)

remaining solvent is stripped overhead with superheated steam. The deasphalted oil product is pumped from the stripper bottom and is cooled, if required, before flowing to battery limits.

The raffinate phase containing asphalt and small amount of solvent flows from the bottom of the rotating disc contactor to the asphalt mix heater. The hot, two-phase asphalt mix from the heater is flashed in the asphalt mix flash tower where solvent vapor is taken overhead, condensed, and collected in the low-pressure solvent receiver. The remaining asphalt mix flows to the asphalt stripper where the remaining solvent is stripped overhead with superheated steam. The asphalt stripper overhead vapors are combined with the overhead from the deasphalted oil stripper, condensed, and collected in the stripper drum. The asphalt product is pumped from the stripper and is cooled by generating low-pressure steam.

5.4.2 Demex Process

The Demex process is a solvent extraction demetallizing process that separates high-metal vacuum residuum into demetallized oil of relatively low metal content and asphaltene of high metal content (Table 5.6). The asphaltene and condensed aromatic contents of the demetallized oil are very low. The demetallized oil is a desirable feedstock for fixed-bed hydrodesulfurization and, in cases where the metals and carbon residues are sufficiently low, is a desirable feedstock for FCC and hydrocracking units.

The Demex process is an extension of the propane deasphalting process and employs a less selective solvent to recover not only the high-quality oils but also higher molecular weight aromatics and other processable constituents present in the feedstock. Furthermore, the Demex process requires a much less solvent circulation in achieving its objectives, thus reducing the utility costs and unit size significantly.

TABLE 5.6

Product Yields and Properties from the Demex Process

		Vac. Resid	DMO		Pitch	
Vol.% of vac. resid		100	56	78	44	22
Sp. gr., 15/4°C	wt.%	1.02	0.959	0.98	1.10	1.16
Sulfur	wt.%	4.0	2.74	3.25	5.4	6.3
Nitrogen	wt.%	0.31	0.14	0.21	–	–
Carbon residue	wt.%	20.8	5.6	10.7	–	–
C_6 insols.	wt.%	10	0.05	0.05	–	–
Metal, V + Ni	ppm	98	6	19	201	341
UOP K factor		11.4	11.8	11.6	–	–
Softening pt.	°C	–	–	–	116	177

Source: Speight, J.G. 2000. *The Desulfurization of Heavy Oils and Residua.* 2nd Edition. Marcel Dekker Inc., New York. Table 8.5, p. 327.

The Demex process selectively rejects asphaltene constituents, metals, and high-molecular-weight aromatics from vacuum residues. The resulting demetallized oil can then be combined with VGO to give a greater availability of acceptable feed to subsequent conversion units.

In the process (Figure 5.18), the vacuum residuum feedstock, mixed with Demex solvent recycling from the second stage, is fed to the first-stage extractor. The pressure is kept high enough to maintain the solvent in liquid phase. The temperature is controlled by the degree of cooling of the recycle solvent. The solvent rate is set near the minimum required to ensure that the desired separation occurs.

Asphaltene constituents are rejected in the first stage. Some resin constituents are also rejected to maintain sufficient fluidity of the asphaltene for efficient solvent recovery. The asphaltene is heated and steam stripped to remove solvent.

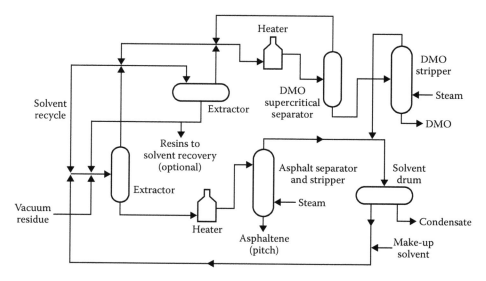

FIGURE 5.18 Demex process. (From Speight, J.G. 2000. *The Desulfurization of Heavy Oils and Residua.* 2nd Edition. Marcel Dekker Inc., New York. Figure 8.18, p. 328.)

The first-stage overhead is heated by an exchange with hot solvent. The increase in temperature decreases the solubility of resin constituents and high molecular weight aromatics. These precipitate in the second-stage extractor. The bottom stream of this second-stage extractor is recycled to the first stage. A portion of this stream can also be drawn as a separate product.

The overhead from the second stage is heated by an exchange with hot solvent. The fired heater further raises the temperature of the solvent/demetallized oil mixture to a point above the critical temperature of the solvent. This causes the demetallized oil to separate. It is then flashed and steam stripped to remove all traces of solvent. The vapor streams from the demetallized oil and asphalt strippers are condensed, dewatered, and pumped up to process pressure for recycle. The bulk of the solvent goes overhead in the supercritical separator. This hot solvent stream is then effectively used for process heat exchange. The subcritical solvent recovery techniques, including multiple effect systems, allow much less heat recovery. Most of the low-grade heat in the solvent vapors from the subcritical flash vaporization must be released to the atmosphere requiring additional heat input to the process.

5.4.3 MDS PROCESS

The MDS process is a technical improvement of the solvent deasphalting process, particularly effective for upgrading heavy feedstocks (Table 5.7). Combined with hydrodesulfurization, the process is fully applicable to the feed preparation for FCC and hydrocracking. The process is capable of using a variety of feedstocks, including atmospheric and vacuum residues derived from various crude oils, oil sand, visbroken tar, and so on.

TABLE 5.7
Product Yields and Properties from the MDS Process

Feedstock	Iranian Heavy Atmospheric Residuum	Kuwait Atmospheric Residuum	Khafji Vacuum Residuum
API, °	17.0	16.4	5.2
Sulfur, wt.%	2.7	3.7	5.2
Carbon residue, wt.%	9.1	9.4	21.9
Nickel, ppm	40.0	14.0	49.0
Vanadium, ppm	130.0	48.0	140.0
Products			
Deasphalted oil, vol.%	93.4	93.8	72.4
API, °	19.0	16.4	11.3
Sulfur, wt.%	2.4	3.7	4.3
Carbon residue, wt.%	5.9	9.4	10.9
Nickel, ppm	18.0	14.0	6.0
Vanadium, ppm	53.0	48.0	28.0
Asphalt, vol.%	6.6	6.2	27.6
API, °	<0.0	<0.0	<0.0
Sulfur, wt.%	5.4	7.2	7.3
Carbon residue, wt.%			49.3
Nickel, ppm	320.0	113.0	150.0
Vanadium, ppm	1010.0	425.0	400.0

Source: Speight, J.G. 2014. *The Chemistry and Technology of Petroleum.* 5th Edition. CRC Press, Taylor & Francis Publishers, Boca Raton, FL. Table 20.5, p. 552.

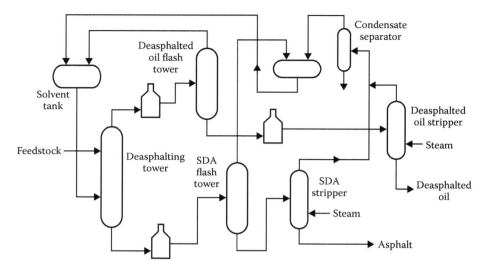

FIGURE 5.19 MDS process. (From Speight, J.G. 2000. *The Desulfurization of Heavy Oils and Residua*. 2nd Edition. Marcel Dekker Inc., New York. Figure 8.19, p. 330.)

In the process (Figure 5.19), the feed and the solvent are mixed and fed to the deasphalting tower. Deasphalting extraction proceeds in the upper half of the tower. After the removal of the asphalt, the mixture of deasphalted oil and solvent flows out of the tower through the tower top. Asphalt flows downward to come in contact with a countercurrent of rising solvent. The contact eliminates oil from the asphalt; the asphalt then accumulates on the bottom.

Deasphalted oil–containing solvent is heated through a heating furnace, and fed to the deasphalted oil flash tower where most of the solvent is separated under pressure. Deasphalted oil still containing a small amount of solvent is again heated and fed to the stripper, where the remaining solvent is completely removed.

Asphalt is withdrawn from the bottom of the extractor. Since this asphalt contains a small amount of solvent, it is heated through a furnace and fed to the flash tower to remove most of the solvent. The asphalt is then sent to the asphalt stripper, where the remaining portion of solvent is completely removed.

Solvent recovered from the deasphalted oil and asphalt flash towers is cooled and condensed into liquid, and sent to a solvent tank. The solvent vapor leaving both strippers is cooled to remove water and compressed for condensation. The condensed solvent is then sent to the solvent tank for further recycling.

5.4.4 Residuum Oil Supercritical Extraction (ROSE) Process

The ROSE process is a solvent deasphalting process with minimum energy consumption using a supercritical solvent recovery system. The process is of value in obtaining oils for further processing.

In the process (Figure 5.20), the residuum is mixed (M-l) with several-fold volume of a low-boiling hydrocarbon solvent and passed into the asphaltene separator vessel (V-l). Asphaltene constituents rejected by the solvent are separated from the bottom of the vessel and are further processed by heating (H-l) and steam stripping (T-l) to remove a small quantity of dissolved solvent. The solvent-free asphaltene constituents are pumped to fuel oil blending or further processing.

The main flow, solvent and extracted oil, passes overhead from the asphaltene separator (V-1) through a heat exchanger (E-1) and heater (H-2) into the oil separator (V-2). Here, the extracted oil is

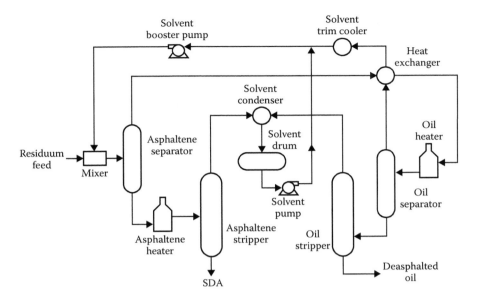

FIGURE 5.20 ROSE process. (From Speight, J.G. 2000. *The Desulfurization of Heavy Oils and Residua.* 2nd Edition. Marcel Dekker Inc., New York. Figure 8.20, p. 331.)

separated without solvent vaporization. The solvent, after heat exchange, is recycled to the process. The small amount of solvent contained in the oil is removed by steam stripping (T-2). The resulting vaporized solvent from the strippers is condensed (E-3) and returned to the process. Product oil is cooled by heat exchange before being pumped to storage or further processing.

5.4.5 SOLVAHL PROCESS

The Solvahl process is a solvent deasphalting process for application to vacuum residua (Peries et al., 1995).

The process (Figure 5.21) uses a higher-boiling solvent and a low solvent/feedstock ratio in addition to the conventional propane/butane/pentane deasphalting procedure, and was developed to give maximum yields of deasphalted oil. A high temperature and low vertical velocity enable the production of high-purity (no asphaltene constituents and low metals) deasphaltened oil. Deasphaltened oil is suitable for downstream processing such as catalytic cracking and hydrotreating. Solvent recovery is achieved under supercritical conditions.

5.5 FUTURE

The process alternatives for upgrading heavy feedstocks discussed above are only several of the currently available and future processes (Chavan et al., 2012; Castañeda et al., 2014; Speight 2014). Some of these processes can be used for partial upgrading of oil, while other process options will be adapted for total conversion upgrading with significant removal of sulfur and metals. Other emerging technologies are starting to evolve toward commercialization, while others remain at the concept stage and have yet to be proven to be suitable for commercial practice (Castañeda et al., 2014; Speight 2014).

However, there is no technology that offers the compete answer to heavy feedstock upgrading and desulfurization. The degree of conversion and the product slate are dependent on the properties

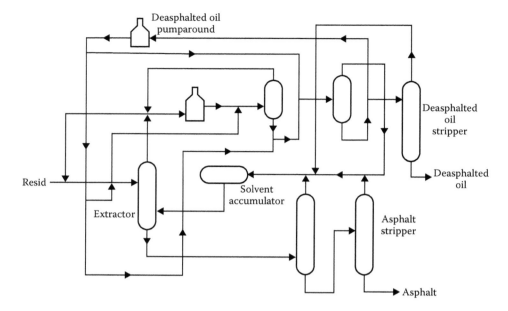

FIGURE 5.21 Solvahl process. (From Speight, J.G. 2000. *The Desulfurization of Heavy Oils and Residua.* 2nd Edition. Marcel Dekker Inc., New York. Figure 8.21, p. 332.)

of the feedstock as well as the process parameters. For example, in many of the processes, tar sand bitumen or heavy oil or resid will not process in the same manner even when the process parameters are adjusted.

Typically, hydrogen addition processes produce higher yields of higher-quality products but have high investment and operating costs. On the other hand, carbon rejection processes have more attractive investment and operating costs but produce lower yields of lower-quality products. At moderate severity, hydrogen addition can also be used for partial upgrading, after which the partially upgraded product can be further upgraded (or even used as fuel oil blend stock) under more favorable conditions and costs.

Therefore, each refinery will define its own scheme for heavy feedstock upgrading since there is no one single upgrading option that is suitable for all refineries. The successful selection of a technology for heavy feedstock upgrading will be determined by (1) the type and properties of the feedstock to be processed, (2) the degree of upgrading required (partial or full), (3) the yield and quality of the products, and (4) the complexity and flexibility of the process as well as the process parameters.

Finally, integration of two or more of the available technologies could also offer advantages in terms of product yields, product quality, elimination of low-value by-products, and reduction of impurities such as sulfur.

REFERENCES

Ancheyta, J., and Speight, J.G. 2007. *Hydroprocessing of Heavy Oils and Residua.* CRC Press, Taylor & Francis Group, Boca Raton, FL.

Bartholic, D.B., Center, A.M., Christian, B.R., and Suchanek, A.J. 1992. Petroleum Refinery of the Future. In: *Petroleum Processing Handbook,* J.J. McKetta (Editor). Marcel Dekker Inc., New York, pp. 108–129.

Beeston, S. 2013. Pathway to Maximizing Residue Upgrading Margins. In: *Proceedings. ICHE Meeting, Houston, TX.* American Institute of Chemical Engineers, New York, April 4.

Bonilla, J., and Elliott, J.D. 1987. Asphalt Coking Method. United States Patent 4,686,027, August 11.

Castañeda, L.C., Muñoz, J.A.D., and Ancheyta, J. 2014. Current Situation of Emerging Technologies for Upgrading of Heavy Oils. *Catal. Today* 220–222: 248–273.

Chavan, S., Kini, H., and Ghosal, R. 2012. Process for Sulfur Reduction from High Viscosity Petroleum Oils. *Int. J. Environ. Sci. Dev.* 3(3): 228–231.

Dickenson, R., Johnson, H., and Chang, E. 2004. Toward Higher Value Transportation Fuels from Residual Fuel Oil. In: *Proceedings. 3rd Bottom of the Barrel Technology Conference & Exhibition*. Antwerp, Belgium.

Dickenson, R.L., Biasca, F.E., Schulman, B.L., and Johnson, H.E. 1997. Refinery Options for Converting and Utilizing Heavy Fuel Oil. *Hydrocarbon Process.* 76(2): 57–62.

Ditman, J.G. 1973. Deasphalting Process. *Hydrocarbon Process.* 52(5): 110.

Feldman, J.A., Lutter, B.E., and Hair, R.L. 1992. Cracking, Catalytic: Optimization and Control. In: *Petroleum Processing Handbook*, J.J. McKetta (Editor). Marcel Dekker Inc., New York, pp. 516–526.

Gary, J.H., Handwerk, G.E., and Kaiser, M.J. 2007. *Petroleum Refining: Technology and Economics*, 5th Edition. CRC Press, Taylor & Francis Group, Boca Raton, FL.

Houde, E.J., and McGrath, M.J. 2006. Residue Upgrading. *Petrol. Technol. Q.* Q2: 1–9.

Hsu, C.S., and Robinson, P.R. (Editors) 2006. *Practical Advances in Petroleum Processing*, Vols. 1–2. Springer Science, New York.

Khan, M.R., and Patmore, D.J. 1998. Heavy Oil Upgrading Processes. In: *Petroleum Chemistry and Refining*, J.G. Speight (Editor). Taylor & Francis, Washington, DC, Chapter 6.

Khouw, F.H.H. 1990. Shell Residue FCC Technology. Annual Meeting. National Petroleum Refiners Association, March.

Lucke, E., Sloley, A.W., and Feyen, M. 2015. Consider the Pros and Cons When Importing Heavy Oil for Cokers. *Hydrocarbon Process.* 94(2): 59–64.

Meyers, R.A. (Editor) 1997. *Petroleum Refining Processes*, 2nd Edition. McGraw-Hill, New York.

Motaghi, M., Shree, K., and Krishnamurthy, S. 2010a. Consider New Methods for Bottom of the Barrel Processing—Part 1. *Hydrocarbon Process.* 89(2): 35–40.

Motaghi, M., Shree, K., and Krishnamurthy, S. 2010b. Consider New Methods for Bottom of the Barrel Processing—Part 2. *Hydrocarbon Process.* 89(2): 55–88.

Ohba, T., Shibutani, I., Watari, R., Inomata, J., and Nagata, H. 2008. The Advanced EUREKA Process: Environment Friendly Thermal Cracking Process. Paper No. WPC-19-2856. In: *Proceedings. 19th World Petroleum Congress*. Madrid, Spain, June 29–July 3.

Parkash, S. 2003. *Refining Processes Handbook*. Gulf Professional Publishing, Elsevier, Amsterdam.

Peries, J.P., Billon, A., Hennico, A., Morrison, E., and Morel, F. 1995. From Heavy Oils to Marketable Products: IFP Proposes New Schemes for Maximum Conversion. In: *Proceedings. 6th UNITAR International Conference on Heavy Crude and Tar Sand*, Vol. 2, pp. 229–244.

Peries, J.P., Quignard, A., Farjon, C., and Laborde, M. 1988. Thermal and Catalytic ASVAHL Processes under Hydrogen Pressure for Converting Heavy Crudes and Conventional Residues. *Rev. Inst. Fr. Pétrol.* 43(6): 847–853.

Rana, M.S., Sámano, V., Ancheyta, J., and Diaz, J.A.I. 2007. A Review of Recent Advances on Process Technologies for Upgrading of Heavy Oils and Residua. *Fuel* 86: 1216–1231.

RAROP. 1991. *Heavy Oil Processing Handbook*. Research Association for Residual Oil Processing. T. Noguchi (Chairman). Ministry of Trade and International Industry (MITI), Tokyo, Japan.

Rispoli, G., Sanfilippo, D., and Amoroso, A. 2009. Advanced Hydrocracking Technology Upgrades Extra Heavy Oil. *Hydrocarbon Process.* 88(12): 39–46.

Sato, Y., Yamamoto, Y., Kamo, T., and Miki, K. 1992. Fluid Catalytic Cracking of Alberta Tar Sand Bitumen. *Energy Fuels* 6: 821–825.

Shih, S.S., and Oballa, M.C. (Editors) 1991. *Tar Sand Upgrading Technology*. Symposium Series No. 282. American Institute for Chemical Engineers, New York.

Speight, J.G. 1984. Upgrading Heavy Oils and Residua: The Nature of the Problem. In: *Catalysis on the Energy Scene*, S. Kaliaguine and A. Mahay (Editors). Elsevier, Amsterdam.

Speight, J.G. 2000. *The Desulfurization of Heavy Oils and Residua*, 2nd Edition. Marcel Dekker Inc., New York.

Speight, J.G. 2001. *Handbook of Petroleum Analysis*. John Wiley & Sons Inc., Hoboken, NJ.

Speight, J.G. 2003. Thermal Cracking of Petroleum. In: *Natural and Laboratory-Simulated Thermal Geochemical Processes*, R. Ikan (Editor). Kluwer Academic Publishers Inc., Dordrecht, the Netherlands, Chapter 2.

Speight, J.G. 2005. Natural Bitumen (Tar Sands) and Heavy Oil. In Coal, Oil Shale, Natural Bitumen, Heavy Oil and Peat, from Encyclopedia of Life Support Systems (EOLSS), Developed under the Auspices of the UNESCO. EOLSS Publishers, Oxford, UK. http://www.eolss.net.

Speight, J.G. 2011. *The Refinery of the Future*. Gulf Professional Publishing, Elsevier, Oxford, UK.

Speight, J.G. 2013. *Heavy and Extra Heavy Oil Upgrading Technologies*. Gulf Professional Publishing, Elsevier, Oxford, UK.

Speight, J.G. 2014. *The Chemistry and Technology of Petroleum*, 5th Edition. CRC Press, Taylor & Francis Group, Boca Raton, FL.

Speight, J.G. 2015. *Handbook of Petroleum Product Analysis*, 2nd Edition. John Wiley & Sons Inc., Hoboken, NJ.

Speight, J.G., and Moschopedis, S.E. 1979. The Production of Low-Sulphur Liquids and Coke from Athabasca Bitumen. *Fuel Process. Technol.* 2: 295.

Speight, J.G., and Ozum, B. 2002. *Petroleum Refining Processes*. Marcel Dekker Inc., New York.

Stanislaus, A., and Cooper, B.H. 1994. Aromatic Hydrogenation Catalysis: A Review. *Catal. Rev. Sci. Eng.* 36(1): 75–123.

Stratiev, D., and Petkov, K. 2009. Residue Upgrading: Challenges and Perspectives. *Hydrocarbon Process.* 88(9): 93–96.

Stratiev, D., Tzingov, T., Shishkova, I., and Dermatova, P. 2009. Hydrotreating Units Chemical Hydrogen Consumption Analysis: A Tool for Improving Refinery Hydrogen Management. In: *Proceedings. 44th International Petroleum Conference*. Bratislava, Slovak Republic, September 21–22.

6 Refining Chemistry

6.1 INTRODUCTION

Understanding refining chemistry not only allows an explanation of the means by which these products can be formed from crude oil but also offers a chance of predictability. This is very necessary when the different types of crude oil accepted by refineries are considered. Moreover, the major processes by which these products are produced from crude oil constituents involve thermal decomposition. In addition, the key to understanding desulfurization is an understanding of the chemistry of refining, as illustrated by the various refining processes.

There are various theories relating to the thermal decomposition of organic molecules, and this area of petroleum technology has been the subject of study for several decades (Hurd, 1929; Fabuss et al., 1964; Fitzer et al., 1971). The relative reactivity of petroleum constituents can be assessed on the basis of bond energies; however, the thermal stability of an organic molecule is dependent on the bond strength of the weakest bond. Moreover, even though the use of bond energy data is a method for predicting the reactivity or the stability of specific bonds under designed conditions, the reactivity of a particular bond is also subject to its environment. Thus, it is not only the reactivity of the constituents of petroleum that are important in processing behavior; the stereochemistry of the constituents as they relate to one another is also of some importance (Speight, 2014). It must be appreciated that the stereochemistry of organic compounds is often a major factor in determining reactivity and properties (Eliel and Wilen, 1994).

In the present context, it is necessary to recognize that (*parum affinis* or not) most hydrocarbons decompose thermally at temperatures above about 650°F (340°C), so the high boiling points of many petroleum constituents cannot be measured directly and must be estimated from other measurements. In the present context, it is as well that hydrocarbons decompose at elevated temperatures. Thereby lies the route to many modern products. For example, in a petroleum refinery, the highest-value products are transportation fuels:

1. Gasoline (boiling range: 35–220°C, 95–425°F)
2. Jet fuel (boiling range 175–290°C, 350–550°F)
3. Diesel (175–370°C, 350–700°F)

The boiling ranges of these fuels are subject to variation and depend on the process used for their production. In winter, gasoline will typically (in cold regions) have butane added to the mix (to facilitate cold starting), thereby changing the boiling range to 0–220°C (32–425°F). The fuels are produced by thermal decomposition of a variety of hydrocarbons, high molecular weight paraffins included. Less than one-third of a typical crude oil distills in these ranges, and thus the goal of refining chemistry might be stated simply as the methods by which crude oil is converted to these fuels. It must be recognized that refining involves a wide variety of chemical reactions; however, the production of liquid fuels is the focus of a refinery.

Refining processes involve the use of various thermal and catalytic processes to convert molecules in the heavier fractions to smaller molecules in fractions distilling at these lower temperatures (Jones, 1995). This efficiency translates into a strong economic advantage, leading to widespread use of conversion processes in refineries today. However, in order to understand the principles of catalytic cracking, understanding the principles of adsorption and reaction on solid surfaces is valuable (Samorjai, 1994; Masel, 1995).

A refinery is a complex network of integrated unit processes for the purpose of producing a variety of products from crude oil. Refined products establish the order in which the individual refining units will be introduced, and the choice from among several types of units and the size of these units is dependent on economic factors. The trade-off among product types, quantity, and quality influences the choice of one kind of processing option over another.

Each refinery has its own range of preferred crude oil feedstock from which a desired distribution of products is obtained. Nevertheless, refinery processes can be divided into three major types: (1) *separation*—division of crude oil into various streams (or fractions) depending on the nature of the crude material; (2) *conversion*—production of salable materials from crude oil, usually by skeletal alteration, or even by alteration of the chemical type, of the crude oil constituents; and (3) *finishing*—purification of various product streams by a variety of processes that essentially remove impurities from the product; for convenience, processes that accomplish molecular alteration, such as *reforming*, are also included in this category.

The separation and finishing processes may involve distillation or even treatment with a wash solution, either to remove impurities or, in the case of distillation, to produce a material boiling over a narrower range, and the chemistry of these processes can be represented by simple equations, even to the disadvantage of oversimplification of the process (Speight, 2014). The inclusion of reforming processes in this category is purely for descriptive purposes rather than being representative of the chemistry involved. Reforming processes produce streams that allow the product to be "finished," as the term applies to product behavior and utility.

Conversion processes are, in essence, processes that change the number of carbon atoms per molecule, alter the molecular hydrogen-to-carbon ratio, or change the molecular structure of the material without affecting the number of carbon atoms per molecule. These latter processes (isomerization processes) essentially change the shape of the molecule(s) and are used to improve the quality of the product (Speight and Ozum, 2002; Hsu and Robinson, 2006; Gary et al., 2007; Speight, 2014).

Nevertheless, the chemistry of conversion process may be quite complex (King et al., 1973), and an understanding of the chemistry involved in the conversion of a crude oil to a variety of products is essential to an understanding of refinery operations. It is therefore the purpose of this chapter to serve as an introduction to the chemistry involved in these conversion processes so that the subsequent chapters dealing with desulfurization will be accepted in the correct perspective. However, as a word of caution, understanding refining chemistry from the behavior of model compounds under refining conditions is not as straightforward as it may appear because of the interactions of the individual constituents of complex mixtures versus the reaction of a single model compound (Ebert et al., 1987).

In fact, the complexity of the individual reactions occurring in an extremely complex mixture and the *interference* of the products with those from other components of the mixture is often unpredictable. Also, the interference of secondary and tertiary products with the course of a reaction and, hence, with the formation of primary products, may also be a cause for concern. Hence, caution is advised when applying the data from model compound studies to the behavior of petroleum, especially the molecularly complex heavy oils. These have few, if any, parallels in organic chemistry.

6.2 CRACKING

The term *cracking* applies to the decomposition of petroleum constituents that is induced by elevated temperatures (>350°C, >660°F), whereby the higher molecular weight constituents of petroleum are converted to lower molecular weight products. Cracking reactions involve carbon–carbon bond rupture and are thermodynamically favored at high temperature (Egloff, 1937).

6.2.1 THERMAL CRACKING

Cracking is a phenomenon by which higher-boiling (higher molecular weight) constituents in petroleum are converted into lower-boiling (lower molecular weight) products. However, certain products may interact with one another to yield products having higher molecular weights than the constituents of the original feedstock. Some of the products are expelled from the system as, say, gases, naphtha-range materials, kerosene-range materials, and the various intermediates that produce other products such as coke. Materials that have boiling ranges higher than naphtha and kerosene may (depending on the refining options) be referred to as *recycle* stock, which is recycled in the cracking equipment until conversion is complete.

There are two general types of reaction that occur during cracking:

1. The decomposition of large molecules into small molecules (primary reactions)

$$CH_3CH_2CH_2CH_3 \rightarrow CH_4 + CH_3CH = CH_2$$

$$\text{Butane} \qquad \text{methane} \quad \text{propene}$$

$$CH_3CH_2CH_2CH_3 \rightarrow CH_3CH_3 + CH_2 = CH_2$$

$$\text{Butane} \qquad \text{ethane} \quad \text{ethylene}$$

2. Reactions by which some of the primary products interact to form higher molecular weight materials (secondary reactions)

$$CH_2{=}CH_2 + CH_2{=}CH_2 \rightarrow CH_3CH_2CH{=}CH_2$$

$$RCH{=}CH_2 + R^1CH{=}CH_2 \rightarrow \text{cracked residuum, } + \text{ coke } + \text{ other products}$$

Thermal cracking is a free radical chain reaction; a free radical is an atom or group of atoms possessing an unpaired electron. Free radicals are very reactive, and it is their mode of reaction that actually determines the product distribution during thermal cracking. Free radical reacts with a hydrocarbon by abstracting a hydrogen atom to produce a stable end product and a new free radical. Free radical reactions are extremely complex, and it is hoped that these few reaction schemes illustrate potential reaction pathways. Any of the preceding reaction types are possible; however, it is generally recognized that the prevailing conditions and those reaction sequences that are thermodynamically favored determine the product distribution.

One of the significant features of hydrocarbon free radicals is their resistance to isomerization, for example, migration of an alkyl group; as a result, thermal cracking does not produce any degree of branching in the products other than that already present in the feedstock.

Data obtained from the thermal decomposition of pure compounds indicate certain decomposition characteristics that permit predictions to be made of the product types that arise from the thermal cracking of various feedstocks. For example, normal paraffins are believed to form, initially, a high molecular weight material, which subsequently decomposes as the reaction progresses. Other paraffinic materials and α (terminal) olefins are produced. An increase in pressure inhibits the formation of low molecular weight gaseous products, and therefore promotes the formation of higher molecular weight materials.

Branched paraffins react somewhat differently to the normal paraffins during cracking processes and produce substantially higher yields of olefins having one fewer carbon atom than the parent hydrocarbon. Cycloparaffins (naphthenes) react differently to their noncyclic counterparts and are somewhat more stable. For example, cyclohexane produces hydrogen, ethylene, butadiene, and benzene: alkyl-substituted cycloparaffins decompose by means of scission of the alkyl chain to produce an olefin and a methyl or ethyl cyclohexane.

The aromatic ring is considered fairly stable at moderate cracking temperatures (350–500°C, 660–930°F). Alkylated aromatics, like the alkylated naphthenes, are more prone to dealkylation than to ring destruction. However, ring destruction of the benzene derivatives occurs above 500°C (930°F), but condensed aromatics may undergo ring destruction at somewhat lower temperatures (450°C, 840°F).

6.2.2 Catalytic Cracking

Catalytic cracking is the thermal decomposition of petroleum constituents in the presence of a catalyst (Pines, 1981). Thermal cracking has been superseded by catalytic cracking as the process for gasoline manufacture. Indeed, naphtha produced by catalytic cracking is richer in branched paraffins, cycloparaffins, and aromatics, which all serve to increase the quality of the naphtha. Catalytic cracking also results in the production of the maximum amount of butene and butane derivatives (C_4H_8 and C_4H_{10}) rather than production of ethylene and ethane (C_2H_4 and C_2H_6).

Catalytic cracking processes evolved in the 1930s from research on petroleum and coal liquids. The petroleum work came to fruition with the invention of acid cracking. The work to produce liquid fuels from coal, most notably in Germany, resulted in metal sulfide hydrogenation catalysts. In the 1930s, a catalytic cracking catalyst for petroleum that used solid acids as catalysts was developed using acid-treated clays.

Clay minerals are a family of crystalline aluminosilicate solids, and the acid treatment develops acidic sites by removing aluminum from the structure. The acid sites also catalyze the formation of coke, and Houdry developed a moving-bed process that continuously removed the cooked beads from the reactor for regeneration by oxidation with air.

Although thermal cracking is a free radical (neutral) process, catalytic cracking is an ionic process involving carbonium ions, which are hydrocarbon ions having a positive charge on a carbon atom. The formation of carbonium ions during catalytic cracking can occur by (1) addition of a proton from an acid catalyst to an olefin and/or (2) abstraction of a hydride ion (H^-) from a hydrocarbon by the acid catalyst or by another carbonium ion. However, carbonium ions are not formed by cleavage of a carbon–carbon bond.

In essence, the use of a catalyst permits alternate routes for cracking reactions, usually by lowering the free energy of activation for the reaction. The acid catalysts first used in catalytic cracking were amorphous solids composed of approximately 87% silica (SiO_2) and 13% alumina (Al_2O_3), and were designated low-alumina catalysts. However, this type of catalyst is now being replaced by crystalline aluminosilicates (zeolites) or molecular sieves.

The first catalysts used for catalytic cracking were acid-treated clays, formed into beads. In fact, clays are still employed as catalysts in some cracking processes (Speight and Ozum, 2002; Hsu and Robinson, 2006; Gary et al., 2007; Speight, 2014). Clays are a family of crystalline aluminosilicate solids, and the acid treatment develops acidic sites by removing aluminum from the structure. The acid sites also catalyze the formation of coke, and the development of a moving-bed process that continuously removed the cooked beads from the reactor reduced the yield of coke; clay regeneration was achieved by oxidation with air.

Clays are natural compounds of silica and alumina, containing major amounts of the oxides of sodium, potassium, magnesium, calcium, and other alkali and alkaline earth metals. Iron and other transition metals are often found in natural clays, substituted for the aluminum cations. Oxides of virtually every metal are found as impurity deposits in clay minerals.

Clays are layered crystalline materials. They contain large amounts of water within and between the layers (Keller, 1985). Heating the clays above 100°C can drive out some or all of this water; at higher temperatures, the clay structures themselves can undergo complex solid-state reactions. Such behavior makes the chemistry of clays a fascinating field of study in its own right. Typical clays include kaolinite, montmorillonite, and illite (Keller, 1985). They are found in most natural soils and in large, relatively pure deposits, from which they are mined for applications ranging from adsorbents to paper making.

Once the carbonium ions are formed, the modes of interaction constitute an important means by which product formation occurs during catalytic cracking, for example, isomerization either by hydride ion shift or by methyl group shift, both of which occur readily. The trend is for stabilization of the carbonium ion by *movement* of the charged carbon atom toward the center of the molecule, which accounts for the isomerization of α-olefins to internal olefins when carbonium ions are produced. Cyclization can occur by internal addition of a carbonium ion to a double bond, which, by continuation of the sequence, can result in aromatization of the cyclic carbonium ion.

Like the paraffins, naphthenes do not appear to isomerize before cracking. However, the naphthenic hydrocarbons (from C_9 upward) produce considerable amounts of aromatic hydrocarbons during catalytic cracking. Reaction schemes similar to that outlined here provide possible routes for the conversion of naphthenes to aromatics. Alkylated benzenes undergo nearly quantitative dealkylation to benzene without apparent ring degradation below 500°C (930°F). However, polymethylbenzenes undergo disproportionation and isomerization with very little benzene formation.

Catalytic cracking can be represented by simple reaction schemes. However, questions have arisen about how the cracking of paraffins is initiated. Several hypotheses for the initiation step in catalytic cracking of paraffins have been proposed (Cumming and Wojciechowski, 1996). The Lewis site mechanism is the most obvious, as it proposes that a carbenium ion is formed by the abstraction of a hydride ion from a saturated hydrocarbon by a strong Lewis acid site: a tricoordinated aluminum species. On Brønsted sites, a carbenium ion may be readily formed from an olefin by the addition of a proton to the double bond or, more rarely, via the abstraction of a hydride ion from a paraffin by a strong Brønsted proton. This latter process requires the formation of hydrogen as an initial product. This concept was, for various reasons that are of uncertain foundation, often neglected.

It is therefore not surprising that the earliest cracking mechanisms postulated that the initial carbenium ions are formed only by the protonation of olefins generated either by thermal cracking or present in the feed as an impurity. For a number of reasons, this proposal was not convincing, and, in the continuing search for initiating reactions, it was even proposed that electrical fields associated with the cations in the zeolite are responsible for the polarization of reactant paraffins, thereby activating them for cracking. More recently, however, it has been convincingly shown that a pentacoordinated carbonium ion can be formed on the alkane itself by protonation, if a sufficiently strong Brønsted proton is available (Cumming and Wojciechowski, 1996).

Coke formation is considered, with just cause to a malignant side reaction of normal carbenium ions. However, while chain reactions dominate events occurring on the surface, and produce the majority of products, certain less desirable bimolecular events have a finite chance of involving the same carbenium ions in a bimolecular interaction with one another. Of these reactions, most will produce a paraffin and leave carbene/carboid-type species (Chapter 1) on the surface. This carbene/carboid-type species can produce other products, but the most damaging product will be one that remains on the catalyst surface and cannot be desorbed and results in the formation of coke, or remains in a noncoke form but effectively blocks the active sites of the catalyst.

A general reaction sequence for coke formation from paraffins involves oligomerization, cyclization, and dehydrogenation of small molecules at active sites within zeolite pores:

Alkanes → alkenes
Alkenes → oligomers
Oligomers → naphthenes
Naphthenes → aromatics
Aromatics → coke

Whether or not these are the true steps to coke formation can only be surmised. The problem with this reaction sequence is that it ignores sequential reactions in favor of consecutive reactions. Furthermore, it must be accepted that the chemistry leading up to coke formation is a complex process, consisting of many sequential and parallel reactions.

There is a complex and little understood relationship between coke content, catalyst activity, and the chemical nature of the coke. For instance, the atomic hydrogen-to-carbon ratio of coke depends on how the coke was formed; its exact value will vary from system to system (Cumming and Wojciechowski, 1996). Moreover, it seems that catalyst decay is not related in any simple way to the hydrogen-to-carbon atomic ratio of the coke, or to the total coke content of the catalyst, or any simple measure of coke properties. Moreover, despite many and varied attempts, there is currently no consensus about the detailed chemistry of coke formation. There is, however, much evidence and good reason to believe that catalytic coke is formed from carbenium ions that undergo addition, dehydrogenation, and cyclization, and elimination side reactions in addition to the main-line chain propagation processes (Cumming and Wojciechowski, 1996).

6.2.3 Dehydrogenation

The common primary reactions of pyrolysis are dehydrogenation and carbon bond scission. The extent of one or the other varies with the starting material and operating conditions; however, because of its practical importance, methods have been found to increase the extent of dehydrogenation and, in some cases, to render it almost the only reaction.

Dehydrogenation is essentially the removal of hydrogen from the parent molecule. For example, at 550°C (1025°F), n-butane loses hydrogen to produce butene-1 and butene-2. The development of selective catalysts, such as chromic oxide (chromia, Cr_2O_3) on alumina (Al_2O_3), has rendered the dehydrogenation of paraffins to olefins particularly effective, and the formation of higher molecular weight material is minimized.

Naphthenes are somewhat more difficult to dehydrogenate, and cyclopentane derivatives form only aromatics if a preliminary step to form the cyclohexane structure can occur. Alkyl derivatives of cyclohexane usually dehydrogenate at 480–500°C (895–930°F), and polycyclic naphthenes are also quite easy to dehydrogenate thermally. In the presence of catalysts, cyclohexane and its derivatives are readily converted into aromatics; reactions of this type are prevalent in catalytic cracking and reforming. Benzene and toluene are prepared by the catalytic dehydrogenation of cyclohexane and methylcyclohexane, respectively.

Polycyclic naphthenes can also be converted to the corresponding aromatics by heating at 450°C (840°F) in the presence of a chromia–alumina (Cr_2O_3–Al_2O_3) catalyst.

Alkylaromatics also dehydrogenate to various products. For example, styrene is prepared by the catalytic dehydrogenation of ethylbenzene. Other alkylbenzenes can be dehydrogenated similarly; isopropyl benzene yields α-methyl styrene.

6.2.4 Dehydrocyclization

Catalytic aromatization involving the loss of 1 mol hydrogen followed by ring formation and further loss of hydrogen has been demonstrated for a variety of paraffins (typically n-hexane and n-heptane). Thus, n-hexane can be converted to benzene, heptane is converted to toluene, and octane is converted to ethylbenzene and o-xylene. Conversion takes place at low pressures, even atmospheric, and at temperatures above 300°C (570°F), although 450–550°C (840–1020°F) is the preferred temperature range.

The catalysts are metals (or their oxides) of the titanium, vanadium, and tungsten groups and are generally supported on alumina; the mechanism is believed to be dehydrogenation of the paraffin to an olefin, which in turn is cyclized and dehydrogenated to the aromatic hydrocarbon. In support of this, olefins can be converted to aromatics much more easily than the corresponding paraffins.

6.3 HYDROGENATION

The purpose of hydrogenating petroleum constituents is (1) to improve existing petroleum products or develop new products or even new uses, (2) to convert inferior or low-grade materials into valuable products, and (3) to transform higher molecular weight constituents into liquid fuels.

The distinguishing feature of the hydrogenating processes is that, although the composition of the feedstock is relatively unknown and a variety of reactions may occur simultaneously, the final product may actually meet all the required specifications for its particular use (Furimsky, 1983; Speight, 2000).

Hydrogenation processes (Speight and Ozum, 2002; Hsu and Robinson, 2006; Gary et al., 2007; Speight, 2014) for the conversion of petroleum and petroleum products may be classified as *destructive* and *nondestructive*. The former (*hydrogenolysis* or *hydrocracking*) is characterized by the rupture of carbon–carbon bonds and is accompanied by hydrogen saturation of the fragments to produce lower-boiling products. Such treatment requires rather high temperatures and high hydrogen pressures, the latter to minimize coke formation. Many other reactions, such as isomerization, dehydrogenation, and cyclization, can occur under these conditions (Dolbear et al., 1987).

On the other hand, nondestructive, or simple, hydrogenation is generally used for the purpose of improving product (or even feedstock) quality without appreciable alteration of the boiling range. Treatment under such mild conditions is often referred to as *hydrotreating* or *hydrofining* and is essentially a means of eliminating nitrogen, oxygen, and sulfur as ammonia, water, and hydrogen sulfide, respectively.

6.3.1 HYDROCRACKING

Hydrocracking (Speight and Ozum, 2002; Hsu and Robinson, 2006; Gary et al., 2007; Speight, 2014) is a thermal process (>350°C, >660°F) in which hydrogenation accompanies cracking. Relatively high pressure (100–2000 psi) is employed, and the overall result is usually a change in the character or quality of the products.

The wide range of products possible from hydrocracking is the result of combining catalytic cracking reactions with hydrogenation (Dolbear, 1998; Hajji et al., 2010). The reactions are catalyzed by dual-function catalysts in which the cracking function is provided by silica–alumina (or zeolite) catalysts, and platinum, tungsten oxide, or nickel provides the hydrogenation function.

Essentially all the initial reactions of catalytic cracking occur; however, some of the secondary reactions are inhibited or stopped by the presence of hydrogen. For example, the yields of olefins and the secondary reactions that result from the presence of these materials are substantially diminished and branched-chain paraffins undergo demethanation. The methyl groups attached to secondary carbons are more easily removed than those attached to tertiary carbon atoms, whereas methyl groups attached to quaternary carbons are the most resistant to hydrocracking.

The effect of hydrogen on naphthenic hydrocarbons is mainly that of ring scission followed by immediate saturation of each end of the fragment produced. The ring is preferentially broken at favored positions, although generally all the carbon–carbon bond positions are attacked to some extent. For example, methyl-cyclopentane is converted (over a platinum–carbon catalyst) to 2-methylpentane, 3-methylpentane, and *n*-hexane.

Aromatic hydrocarbons are resistant to hydrogenation under mild conditions; however, under more severe conditions, the main reactions are conversion of the aromatic to naphthenic rings and scissions within the alkyl side chains. The naphthenes may also be converted to paraffins. However, polynuclear aromatics are more readily attacked than the single-ring compounds, the reaction proceeding by a stepwise process in which one ring at a time is saturated and then opened. For example, naphthalene is hydrocracked over a molybdenum oxide molecular catalyst to produce a variety of low-weight paraffins ($\leq C_6$).

6.3.2 HYDROTREATING

It is generally recognized that the higher the hydrogen content of a petroleum product, especially the fuel products, the better is the quality of the product. This knowledge has stimulated the use of hydrogen-adding processes in the refinery.

Thus, hydrotreating (i.e., hydrogenation without simultaneous cracking) (Speight and Ozum, 2002; Hsu and Robinson, 2006; Gary et al., 2007; Speight, 2014) is used for saturating olefins or for converting aromatics to naphthenes as well as for heteroatom removal. Under atmospheric pressure, olefins can be hydrogenated up to about 500°C (930°F); however, beyond this temperature, dehydrogenation commences. Application of pressure and the presence of catalysts make it possible to effect complete hydrogenation at room or even cooler temperature; the same influences are helpful in minimizing dehydrogenation at higher temperatures.

A wide variety of metals are active hydrogenation catalysts; those of most interest are nickel, palladium, platinum, cobalt, iron, nickel-promoted copper, and copper chromite. Special preparations of the first three are active at room temperature and atmospheric pressure. The metallic catalysts are easily poisoned by sulfur-containing and arsenic-containing compounds, and even by other metals. To avoid such poisoning, less effective but more resistant metal oxides or sulfides are frequently employed, generally those of tungsten, cobalt, chromium, or molybdenum.

Alternatively, catalyst poisoning can be minimized by mild hydrogenation to remove nitrogen, oxygen, and sulfur from feedstocks in the presence of more resistant catalysts, such as cobalt–molybdenum–alumina (Co–Mo–Al$_2$O$_3$). The reactions involved in nitrogen removal are somewhat analogous to those of the sulfur compounds and follow a stepwise mechanism to produce ammonia and the relevant substituted aromatic compound.

6.4   ISOMERIZATION

The importance of isomerization in petroleum-refining operations is twofold. First, the process is valuable in converting n-butane into isobutane, which can be alkylated to liquid hydrocarbons in the naphtha boiling range. Second, the process can be used to increase the octane number of the paraffins boiling in the naphtha boiling range by converting some of the n-paraffins present into isoparaffins.

The process involves contact of the hydrocarbon feedstock and a catalyst under conditions favorable to good product recovery (Speight and Ozum, 2002; Hsu and Robinson, 2006; Gary et al., 2007; Speight, 2014). The catalyst may be aluminum chloride promoted with hydrochloric acid or a platinum-containing catalyst. Both are very reactive and can lead to undesirable side reactions along with isomerization. These side reactions include disproportionation and cracking, which decrease the yield and produce olefinic fragments that may combine with the catalyst and shorten its life. These undesired reactions are controlled by such techniques as the addition of inhibitors to the hydrocarbon feed or by carrying out the reaction in the presence of hydrogen.

Paraffins are readily isomerized at room temperature, and the reaction is believed to occur by means of the formation and rearrangement of carbonium ions. The chain-initiating ion R$^+$ is formed by the addition of a proton from the acid catalyst to an olefin molecule, which may be added, present as an impurity, or formed by dehydrogenation of the paraffin.

Except for butane, the isomerization of paraffins is generally accompanied by side reactions involving carbon–carbon bond scissions when catalysts of the aluminum halide type are used. Products boiling both higher and lower than the starting material are formed, and the disproportionation reactions that occur with the pentanes and higher paraffins (>C$_5$) are caused by unpromoted aluminum halide. A substantial pressure of hydrogen tends to minimize these side reactions.

The ease of paraffin isomerization increases with molecular weight, but the extent of disproportionation reactions also increases. Conditions can be established under which isomerization takes place only with the butanes; however, this is difficult for the pentanes and higher hydrocarbons. At temperatures higher than 27°C (81°F), aluminum bromide (AlBr$_3$), the equilibrium mixture of n-pentane and isopentane, contains >70% of the branched isomer; at 0°C (32°F), approximately 90% of the branched isomer is present. Higher- and lower-boiling hydrocarbon products, hexanes, heptanes, and isobutane are also formed in side reactions even at 0°C (32°F) and in increased amounts when the temperature is raised. Although the thermodynamic conditions are favorable, neopentane [C(CH$_3$)$_4$] does not appear to isomerize under these conditions.

Olefins are readily isomerized; the reaction involves either movement of the position of the double bond (hydrogen-atom shift) or skeletal alteration (methyl group shift). The double-bond shift may also include a reorientation of the groups around the double bond to bring about a *cis–trans* isomerization. Thus, 1-butene is isomerized to a mixture of *cis-* and *trans*-2-butene. *Cis* (same side) and *trans* (opposite side) refer to the spatial arrangement of the methyl groups with respect to the double bond.

Olefins having a terminal double bond are the least stable. They isomerize more rapidly than those in which the double bond carries the maximum number of alkyl groups.

Naphthenes can isomerize in various ways; for example, in the case of cyclopropane (C_3H_6) and cyclobutane (C_4H_8), ring scission can occur to produce an olefin. Carbon–carbon rupture may also occur in any side chains to produce polymethyl derivatives, whereas cyclopentane (C_5H_{10}) and cyclohexane (C_6H_{12}) rings may expand and contract, respectively.

The isomerization of alkylaromatics may involve changes in the side-chain configuration, disproportionation of the substituent groups, or their migration about the nucleus. The conditions needed for isomerization within attached long side chains of alkylbenzenes and alkylnaphthalenes are also those for the scission of such groups from the ring. Such isomerization, therefore, does not take place unless the side chains are relatively short. The isomerization of ethylbenzene to xylenes, and the reverse reaction, occurs readily.

Disproportionation of attached side chains is also a common occurrence; higher and lower alkyl substitution products are formed. For example, xylenes disproportionate in the presence of hydrogen fluoride–boron trifluoride or aluminum chloride to form benzene, toluene, and higher alkylated products; ethylbenzene in the presence of boron trifluoride forms a mixture of benzene and 1,3-diethylbenzene.

6.5 ALKYLATION

Alkylation in the petroleum industry refers to a process for the production of high-octane motor fuel components by the combination of olefins and paraffins. The reaction of isobutane with olefins, using an aluminum chloride catalyst, is a typical alkylation reaction.

In acid-catalyzed alkylation reactions, only paraffins with tertiary carbon atoms, such as isobutane and isopentane, react with the olefin. Ethylene is slower to react than the higher olefins. Olefins higher than propene may complicate the products by engaging in hydrogen exchange reactions. Cycloparaffins, especially those containing tertiary carbon atoms, are alkylated with olefins in a manner similar to the isoparaffins; however, the yields are low because of the side reactions that also occur (Speight and Ozum, 2002; Hsu and Robinson, 2006; Gary et al., 2007; Speight, 2014).

Aromatic hydrocarbons are more easily alkylated than the isoparaffins by olefins. Cumene (isopropylbenzene) is prepared by alkylating benzene with propene over an acid catalyst. The alkylating agent is usually an olefin, although cyclopropane, alkyl halides, aliphatic alcohols, ethers, and esters may also be used. The alkylation of aromatic hydrocarbons is presumed to occur through the agency of the carbonium ion.

Thermal alkylation is also used in some plants; however, like thermal cracking, it is presumed to involve the transient formation of neutral free radicals and therefore tends to be less specific in production distribution.

6.6 POLYMERIZATION

Polymerization is a process in which a substance of low molecular weight is transformed into one of the same composition but of higher molecular weight while maintaining the atomic arrangement present in the basic molecules. It has also been described as the successive addition of one molecule to another by means of a functional group, such as that present in an aliphatic olefin.

In the petroleum industry, polymerization is used to indicate the production of, say, gasoline components that fall into a specific (and controlled) molecular weight range, hence the term *polymer gasoline*. Furthermore, it is not essential that only one type of monomer be involved:

$$CH_3CH=CH_2 + CH_2=CH_2 \rightarrow CH_3CH_2CH_2CH=CH_2$$

This type of reaction is correctly called *copolymerization*; however, polymerization in the true sense of the word is usually prevented, and all attempts are made to terminate the reaction at the dimer or trimer (three monomers joined together) stage. It is the 4- to 12-carbon compounds that are required as the constituents of liquid fuels. However, in the petrochemical section of the refinery, polymerization, which results in the production of, say, polyethylene, is allowed to proceed until materials of the required high molecular weight have been produced.

6.7　PROCESS CHEMISTRY

In a mixture as complex as petroleum, the reaction processes can only be generalized because of difficulties in analyzing not only the products but also the feedstock, as well as the intricate and complex nature of the molecules that make up the feedstock. The formation of coke from the higher molecular weight and polar constituents of a given feedstock is detrimental to process efficiency and to catalyst performance (Speight, 1987; Dolbear, 1998).

Refining the constituents of heavy oil and bitumen has become a major issue in modern refinery practice. The limitations of processing heavy oils and residua depend, to a large extent, on the amount of higher molecular weight constituents (i.e., asphaltene constituents) present in the feedstock (Ternan, 1983; Speight, 1984, 2000, 2004a; Schabron and Speight, 1997; Ancheyta et al., 2009) that are responsible for high yields of thermal and catalytic coke (Speight and Ozum, 2002; Hsu and Robinson, 2006; Gary et al., 2007; Speight, 2014).

6.7.1　THERMAL CHEMISTRY

When petroleum is heated to temperatures in excess of 350°C (660°F), the rate of thermal decomposition of the constituents increases significantly. The higher the temperature, the shorter the time to achieve a given conversion, and the severity of the process conditions is based on a combination of residence time of the crude oil constituents in the reactor and the temperature needed to achieve a given conversion.

Thermal conversion does not require the addition of a catalyst. This approach is the oldest technology available for residue conversion, and the severity of thermal processing determines the conversion and the product characteristics. As the temperature and residence time are increased, the primary products undergo further reaction to produce various secondary products, and so on, with the ultimate products (coke and methane) being formed at extreme temperatures of approximately 1000°C (1830°F).

The thermal decomposition of petroleum asphaltene constituents has received some attention (Magaril and Aksenova, 1967, 1968, 1970a,b, 1972; Magaril and Ramazaeva, 1969; Magaril et al., 1970, 1971; Schucker and Keweshan, 1980; Speight, 1990, 1998). Special attention has been given to the nature of the volatile products of asphaltene decomposition mainly because of the difficulty of characterizing the nonvolatile coke.

The organic nitrogen originally in the asphaltene constituents invariably undergoes thermal reaction to concentrate in the nonvolatile coke (Speight, 1970, 1989, 2004; Vercier, 1981). Thus, although asphaltene constituents produce high yields of thermal coke, little is known of the actual chemistry of coke formation. In a more general scheme, the chemistry of asphaltene coking has been suggested to involve the thermolysis of thermally labile bonds to form reactive species that then react with each other (condensation) to form coke. In addition, the highly aromatic and highly

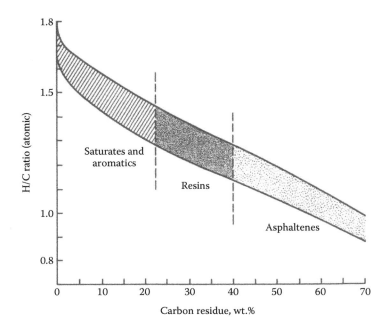

FIGURE 6.1 Yields of thermal coke for various petroleum fractions as determined by the Conradson carbon residue test. (From Speight, J.G. 2014. *The Chemistry and Technology of Petroleum*. 5th Edition. CRC Press, Taylor & Francis Publishers, Boca Raton, FL. Figure 16.1, p. 442.)

polar (refractory) products separate from the surrounding oil medium as an insoluble phase and proceed to form coke. It is also interesting to note that although the aromaticity of the asphaltene constituents is approximately equivalent to the yield of thermal coke (Figure 6.1), not all the original aromatic carbons in the asphaltene constituents form coke. Volatile aromatic species are eliminated during thermal decomposition, and it must be assumed that some of the original aliphatic carbons play a role in coke formation.

Various patterns of thermal behavior have been observed for the constituents of petroleum feedstocks (Table 6.1). Since the chemistry of thermal and catalytic cracking has been studied and well resolved, there has been a tendency to focus on the refractory (nonvolatile) constituents. These constituents of petroleum generally produce coke in yields varying from almost zero to >60% by weight (Figure 6.2). As an aside, it should also be noted that because of the differences in thermal behavior, the different subfractions of the asphaltene fraction detract from the concept of average structure. However, the focus of thermal studies has been, for obvious reasons, on the asphaltene constituents that produce thermal coke in amounts varying from approximately 35% by weight to approximately 65% by weight. Petroleum mapping techniques often show the nonvolatile constituents, specifically the asphaltene constituents and the resin constituents, producing coke while the volatile constituents produce distillates. It is often ignored that the asphaltene constituents also produce high yields (35–65% by weight) of volatile thermal products, which vary from gases to condensable liquids.

It has been generally thought that the chemistry of coke formation involves immediate condensation reactions to produce higher molecular weight, condensed aromatic species. There have been claims that coking is a bimolecular process but more recent approaches to the chemistry of coking render the bimolecular process debatable. The rate of decomposition will vary with the nature of the individual constituents, thereby giving rise to the perception of second-order or even multiorder kinetics. The initial reactions of asphaltene constituents involve thermolysis of pendant alkyl chains to form lower molecular weight higher polar species (carbenes and carboids), which then react

TABLE 6.1
General Indications of Feedstock Cracking

Feedstock Type	Characterization Factor, K	Naphtha Yield, % v/v	Coke Yield, % w/w	Relative Reactivity (Relative Crackability)
Aromatic	11.0(1)	35.0	13.5	Refractory
Aromatic	11.2(2)	49.6	12.5	Refractory
Aromatic	11.2(1)	37.0	11.5	Refractory
Aromatic–naphthenic	11.4(2)	47.0	9.1	Intermediate
Aromatic–naphthenic	11.4(1)	39.0	9.0	Intermediate
Naphthenic	11.6(2)	45.0	7.1	Intermediate
Naphthenic	11.6(1)	40.0	7.2	Intermediate
Naphthenic–paraffinic	11.8(2)	43.0	5.3	High
Naphthenic–paraffinic	11.8(1)	41.0	6.0	High
Naphthenic–paraffinic	12.0(2)	41.5	4.0	High
Naphthenic–paraffinic	12.0(1)	41.5	5.3	High
Paraffinic	12.2(2)	40.0	3.0	High

Source: Speight, J.G. 2014. *The Chemistry and Technology of Petroleum.* 5th Edition. CRC Press, Taylor & Francis Publishers, Boca Raton, FL. Table 16.1, p. 444.

Note: (1) Cycle oil/cracked feedstocks, 60% conversion; (2) straight-run/uncracked feedstocks, 60% conversion.

to form coke. Indeed, as opposed to the bimolecular approach, the initial reactions in the coking of petroleum feedstocks that contain asphaltene constituents appear to involve unimolecular thermolysis of asphaltene aromatic-alkyl systems to produce volatile species (paraffins and olefins) and nonvolatile species (aromatics) (Figure 6.3) (Speight, 1987; Schabron and Speight, 1997).

Thermal studies using model compounds confirm that volatility of the fragments has a major influence on carbon residue formation, and a pendant-core model for the high molecular weight

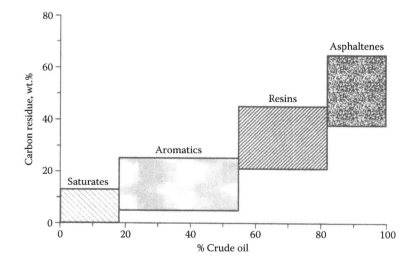

FIGURE 6.2 Illustration of the yields of thermal coke from fractions and subfractions of a specific crude oil as determined by the Conradson carbon residue test. (From Speight, J.G. 2014. *The Chemistry and Technology of Petroleum.* 5th Edition. CRC Press, Taylor & Francis Publishers, Boca Raton, FL. Figure 16.2, p. 444.)

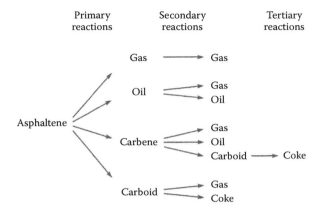

FIGURE 6.3 Multireaction sequence for the thermal decomposition of asphaltene constituents. (From Speight, J.G. 2014. *The Chemistry and Technology of Petroleum.* 5th Edition. CRC Press, Taylor & Francis Publishers, Boca Raton, FL. Figure 16.3, p. 445.)

constituents of petroleum has been proposed (Wiehe, 1994). In such a model, the scission of alkyl side chains occurs, thereby leaving a polar core of reduced volatility that commences to produce a carbon residue (Speight, 1994; Wiehe, 1994). In addition, the pendant-core model also suggests that even one-ring aromatic cores can produce a carbon residue if multiple bonds need to be broken before a core can volatilize (Wiehe, 1994).

In support of the participation of asphaltene constituents in sediment or coke formation, it has been reported that the formation of a coke-like substance during heavy oil upgrading is dependent on several factors (Storm et al., 1997):

1. The degree of polynuclear condensation in the feedstock
2. The average number of alkyl groups on the polynuclear aromatic systems
3. The ratio of heptane-insoluble material to the pentane-insoluble/heptane-soluble fraction
4. The hydrogen-to-carbon atomic ratio of the pentane-insoluble/heptane-soluble fraction

These findings correlate quite well with the proposed chemistry of coke or sediment formation during the processing of heavy feedstocks, and even offer some predictability since the characteristics of the whole feedstocks are evaluated.

Nitrogen species also appear to contribute to the pattern of the thermolysis. For example, the hydrogen or carbon–carbon bonds adjacent to ring nitrogen undergo thermolysis quite readily, as if promoted by the presence of the nitrogen atom (Fitzer et al., 1971; Speight, 1998). If it can be assumed that heterocyclic nitrogen plays a similar role in the thermolysis of asphaltene constituents, the initial reactions therefore involve thermolysis of aromatic-alkyl bonds that are enhanced by the presence of heterocyclic nitrogen. An ensuing series of secondary reactions, such as aromatization of naphthenic species and condensation of the aromatic ring systems, then leads to the production of coke. Thus, the initial step in the formation of coke from asphaltene constituents is the formation of volatile hydrocarbon fragments and nonvolatile heteroatom-containing systems.

It has been reported that as the temperature of a 1-methylnaphthalene is raised from 100°C (212°F) to 400°C (750°F), there is a progressive decrease in the size of the asphaltene constituent particles (Thiyagarajan et al., 1995). Furthermore, there is also the inference that the structural integrity of the asphaltene particle is compromised and that irreversible thermochemistry has occurred. Indeed, that is precisely what is predicted and expected from the thermal chemistry of asphaltene constituents and molecular weight studies of asphaltene constituents.

An additional corollary to this work is that conventional models of petroleum asphaltene constituents (which, despite evidence to the contrary, invoked the concept of a large polynuclear aromatic system) offer little, if any, explanation of the intimate events involved in the chemistry of coking. Models that invoke the concept of the asphaltene fractions as a complex solubility class with molecular entities composed of smaller polynuclear aromatic systems (Speight, 1994, 2014) are more in keeping with the present data. However, the concept of an average structure is not in keeping with the complexity of the fraction and the chemical or thermal reactions of the constituents (Speight, 1994, 2014; Ancheyta et al., 2009).

Little has been acknowledged about the role of low molecular weight polar species (resin constituents) in coke formation. However, it is worthy of note that the resin constituents are presumed to be lower molecular weight analogs of the asphaltene constituents. This being the case, similar reaction pathways may apply.

Thus, it is now considered more likely that molecular species within the asphaltene fraction, which contains nitrogen and other heteroatoms (and have lower volatility than the pure hydrocarbons), are the prime movers in the production of coke (Speight, 1987). Such species, containing various polynuclear aromatic systems, can be denuded of the attendant hydrocarbon moieties and are undoubtedly insoluble (Bjorseth, 1983; Dias, 1987, 1988) in the surrounding hydrocarbon medium. The next step is gradual carbonization of such entities to form coke (Cooper and Ballard, 1962; Magaril and Aksenova, 1967, 1968, 1970a,b, 1972; Magaril and Ramazaeva, 1969; Magaril et al., 1970, 1971).

Thermal processes (such as visbreaking and coking) are the oldest methods for crude oil conversion and are still used in modern refineries. The thermal chemistry of petroleum constituents has been investigated for more than five decades, and the precise chemistry of the lower molecular weight constituents has been well defined because of the bountiful supply of pure compounds. The major issue in determining the thermal chemistry of the nonvolatile constituents is, of course, their largely unknown chemical nature and, therefore, the inability to define their thermal chemistry with any degree of accuracy. Indeed, it is only recently that some light has been cast on the thermal chemistry of the nonvolatile constituents.

Thus, the challenges facing process chemistry and physics are determining (1) the means by which petroleum constituents thermally decompose, (2) the nature of the products of thermal decomposition, (3) the subsequent decomposition of the primary thermal products, (4) the interaction of the products with each other, (5) the interaction of the products with the original constituents, and (6) the influence of the products on the composition of the liquids.

When petroleum is heated to temperatures over approximately 410°C (770°F), the thermal or free radical reactions start to crack the mixture at significant rates. Thermal conversion does not require the addition of a catalyst; therefore, this approach is the oldest technology available for residue conversion. The severity of thermal processing determines the conversion and the product characteristics.

Asphaltene constituents are major components of residua, and heavy oils and their thermal decomposition have been the focus of much attention (Wiehe, 1993; Gray, 1994; Speight, 1994). The thermal decomposition not only produces high yields (40 wt.%) of coke but also, optimistically and realistically, produces equally high yields of volatile products (Speight, 1970). Thus, the challenge in studying the thermal decomposition of asphaltene constituents is to decrease the yields of coke and increase the yields of volatile products.

Several chemical models describe the thermal decomposition of asphaltene constituents (Wiehe, 1993; Gray, 1994; Speight, 1994). Using these available asphaltene models as a guide, the prevalent thinking is that the asphaltene nuclear fragments become progressively more polar as the paraffinic fragments are stripped from the ring systems by scission of the bonds (preferentially) between the carbon atoms α and β to the aromatic rings.

The higher-polarity polynuclear aromatic systems that have been denuded of the attendant hydrocarbon moieties are somewhat less soluble in the surrounding hydrocarbon medium than their parent systems (Bjorseth, 1983; Dias, 1987, 1988). Two factors are operative in determining the solubility of the polynuclear aromatic systems in the liquid product. The alkyl moieties that have a solubilizing effect have been removed, and there is also enrichment of the liquid medium in paraffinic constituents. Again, there is an analogy with the deasphalting process, except that the paraffinic material is a product of the thermal decomposition of the asphaltene molecules and is formed *in situ* rather than being added separately (Speight and Ozum, 2002; Hsu and Robinson, 2006; Gary et al., 2007; Speight, 2014).

The coke has a lower hydrogen-to-carbon atomic ratio than the hydrogen-to-carbon ratio of any of the constituents present in the original crude oil. The hydrocarbon products *may* have a higher hydrogen-to-carbon atomic ratio than the hydrogen-to-carbon ratio of any of the constituents present in the original crude oil, or hydrogen-to-carbon atomic ratios at least equal to those of many of the original constituents. It must also be recognized that the production of coke and volatile hydrocarbon products is accompanied by a shift in the hydrogen distribution.

Mild-severity and high-severity processes are frequently used for processing of residue fractions, whereas conditions similar to those of *ultrapyrolysis* (high temperature and very short residence time) are used commercially only for cracking ethane, propane, butane, and light-distillate feeds to produce ethylene and higher olefins.

The formation of solid sediments, or coke, during thermal processes is a major limitation on processing. Furthermore, the presence of different types of solids shows that solubility controls the formation of solids. Moreover, the tendency for solid formation changes in response to the relative amounts of the light ends, middle distillates, and residues and to their changing chemical composition during the process (Gray, 1994). In fact, the prime mover in the formation of incompatible products during the processing of feedstocks containing asphaltene constituents is the nature of the primary thermal decomposition products, particularly those designated as *carbenes* and *carboids* (Chapter 1) (Speight, 1987, 1992; Wiehe, 1992, 1993).

Coke formation during the thermal treatment of petroleum residua is postulated to occur by a mechanism that involves the liquid–liquid phase separation of reacted asphaltene constituents (which may be carbenes) to form a phase that is lean in abstractable hydrogen. The unreacted asphaltene constituents were found to be the fraction with the highest rate of thermal reaction but with the least extent of reaction. This not only described the appearance and disappearance of asphaltene constituents but also quantitatively described the variation in molecular weight and hydrogen content of the asphaltene constituents with reaction time. Thus, the main features of coke formation are (1) an induction period before coke formation, (2) a maximum concentration of asphaltene constituents in the reacting liquid, (3) a decrease in the asphaltene concentration that parallels the decrease in heptane-soluble material, and (4) high reactivity of the unconverted asphaltene constituents.

The induction period has been observed experimentally by many previous investigators (Levinter et al., 1966, 1967; Magaril and Aksenova, 1967, 1968, 1970a,b, 1972; Magaril and Ramazaeva, 1969; Magaril et al., 1970, 1971; Valyavin et al., 1979; Takatsuka et al., 1989a) and makes visbreaking and the Eureka processes possible. The postulation that coke formation is triggered by the phase separation of asphaltene constituents or reacted asphaltene constituents (Magaril and Aksenova, 1967, 1968, 1970a,b, 1972; Magaril and Ramazaeva, 1969; Magaril et al., 1970, 1971) led to the use of linear variations of the concentration of each fraction with reaction time, resulting in the assumption of zero-order kinetics rather than first-order kinetics. More recently (Yan, 1987), coke formation in visbreaking was described as resulting from a phase separation step; however, the phase-separation step was not included in the resulting kinetic model for coke formation.

This model represents the conversion of asphaltene constituents over the entire temperature range and of heptane-soluble materials in the coke induction period as first-order reactions. The data also show that the four reactions give simultaneously lower aromatic and higher aromatic products, on the basis of other evidence (Wiehe, 1992). Also, the previous work showed that residua fractions can be converted without completely changing solubility classes (Magaril and Aksenova, 1967, 1968, 1970a,b, 1972; Magaril and Ramazaeva, 1969; Magaril et al., 1970, 1971) and that coke formation is triggered by the phase separation of converted asphaltene constituents.

The maximum solubility of these product asphaltene constituents is proportional to the total heptane-soluble materials, as suggested by the observation that the decrease in asphaltene constituents parallels the decrease of heptane-soluble materials. Finally, the conversion of the insoluble product asphaltene constituents into toluene-insoluble coke is pictured as producing a heptane-soluble by-product, which provides a mechanism for the heptane-soluble conversion to deviate from first-order behavior once coke begins to form.

In support of this assumption, it is known (Langer et al., 1961) that partially hydrogenated refinery process streams provide abstractable hydrogen and, as a result, inhibit coke formation during residuum thermal conversion. Thus, the heptane-soluble fraction of a residuum, which contains naturally occurring partially hydrogenated aromatics, can provide abstractable hydrogen during thermal reactions.

As the conversion proceeds, the concentration of asphaltene cores continues to increase and the heptane-soluble fraction continues to decrease until the solubility limit, S_L, is reached. Beyond the solubility limit, the excess asphaltene cores, A_{ex}^*, phase separate to form a second liquid phase that is lean in abstractable hydrogen. In this new phase, asphaltene radical–asphaltene radical recombination is quite frequent, causing a rapid reaction to form solid coke and a by-product of a heptane-soluble core.

The asphaltene concentration varies little in the coke induction period (Wiehe, 1993) but then decreases once coke begins to form. Observing this, it might be concluded that asphaltene constituents are unreactive; however, it is the high reactivity of the asphaltene constituents down to the asphaltene core that offsets the generation of asphaltene cores from the heptane-soluble materials to keep the overall asphaltene concentration nearly constant.

Previously, it was demonstrated (Schucker and Keweshan, 1980; Savage et al., 1988) that the hydrogen-to-carbon atomic ratio of the asphaltene constituents decreases rapidly with reaction time for asphaltene thermolysis and then approaches an asymptotic limit at long reaction times, which provides qualitative evidence for asphaltene cracking down to a core.

The measurement of the molecular weight of petroleum asphaltene constituents is known to give different values depending on the technique, the solvent, and the temperature (Dickie and Yen, 1967; Moschopedis et al., 1976; Speight et al., 1985; Speight, 2014). As shown by small-angle x-ray (Kim and Long, 1979) and neutron (Overfield et al., 1989) scattering, this is because asphaltene constituents tend to self-associate and form aggregates.

Thus, coke formation is a complex process involving both chemical reactions and thermodynamic behavior. Reactions that contribute to this process are cracking of side chains from aromatic groups, dehydrogenation of naphthenes to form aromatics, condensation of aliphatic structures to form aromatics, condensation of aromatics to form higher fused-ring aromatics, and dimerization or oligomerization reactions. Loss of side chains always accompanies thermal cracking, and dehydrogenation and condensation reactions are favored by hydrogen-deficient conditions.

The importance of solvents in coking has been recognized for many years (e.g., Langer et al., 1961); however, their effects have often been ascribed to hydrogen donor reactions rather than to phase behavior. The separation of the phases depends on the solvent characteristics of the liquid. Addition of aromatic solvents suppresses phase separation, whereas paraffins enhance separation. Microscopic examination of coke particles often shows evidence for the presence of mesophase, spherical domains that exhibit the anisotropic optical characteristics of liquid crystals.

This phenomenon is consistent with the formation of a second liquid phase; the mesophase liquid is denser than the rest of the hydrocarbon, has a higher surface tension, and probably wets metal surfaces better than the rest of the liquid phase. The mesophase characteristic of coke diminishes as the liquid phase becomes more compatible with the aromatic material.

The phase separation phenomenon that is the prelude to coke formation can also be explained by use of the solubility parameter, δ, for petroleum fractions and for the solvents (Yen, 1984; Speight, 1994, 2014). As an extension of this concept, there is sufficient data to draw a correlation between the atomic hydrogen-to-carbon ratio and the solubility parameter for hydrocarbons and the constituents of the lower-boiling fractions of petroleum (Speight, 1994). It is recognized that hydrocarbon liquids can dissolve polynuclear hydrocarbons in which there is usually less than a three-point difference between the lower-solubility parameter of the solvent and the higher-solubility parameter of the solute. Thus, a parallel, or near-parallel, line can be assumed that allows the solubility parameter of the asphaltene constituents and resin constituents to be estimated.

By this means, the solubility parameter of asphaltene constituents can be estimated to fall in the range 9–12, which is in keeping with the asphaltene constituents being composed of a mixture of different compound types with an accompanying variation in polarity. Removal of alkyl side chains from the asphaltene constituents decreases the hydrogen-to-carbon atomic ratio (Wiehe, 1993; Gray, 1994) and increases the solubility parameter, thereby bringing about a concurrent decrease of the asphaltene product in the hydrocarbon solvent.

In fact, on the molecular weight polarity diagram for asphaltene constituents, carbenes and carboids can be shown as lower molecular weight, highly polar entities in keeping with molecular fragmentation models (Speight, 1994). If this increase in polarity and solubility parameter (Mitchell and Speight, 1973) is too drastic relative to the surrounding medium (Figure 6.4), phase separation will occur. Furthermore, the available evidence favors a multistep mechanism rather than a stepwise mechanism (Figure 6.5) as the means by which the thermal decomposition of petroleum constituents occurs (Speight, 1997).

Any chemical or physical interactions (especially thermal effects) that cause a change in the solubility parameter of the solute relative to that of the solvent will also cause *incompatibility* be it called *instability, phase separation, sediment formation,* or *sludge formation.*

Instability or incompatibility results in the separation of solids during refining, and can occur during a variety of processes, either by intent (such as in the deasphalting process) or inadvertently when the separation is detrimental to the process—the phenomenon is also manifested under certain conditions in petroleum products (Mushrush and Speight, 1995, 1998; Speight, 2014). Thus, separation of solids occurs whenever the solvent characteristics of the liquid phase are no longer adequate to maintain polar and/or high molecular weight material in solution. Examples of such occurrences are (1) asphaltene separation, which occurs when the paraffin content or character of the liquid medium increases; (2) wax separation, which occurs when there is a drop in temperature or the aromatic content or character of the liquid medium increases; (3) sludge or sediment formation in a reactor, which occurs when the solvent characteristics of the liquid medium change so that asphalt or wax materials separate; (4) coke formation, which occurs at high temperatures and commences when the solvent power of the liquid phase is not sufficient to maintain the coke precursors in solution; and (5) sludge or sediment formation in fuel products, which occurs because of the interplay of several chemical and physical factors (Mushrush and Speight, 1995, 1998; Speight and Ozum, 2002; Speight, 2014).

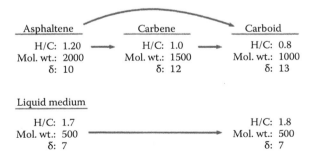

FIGURE 6.4 Illustration of the changes in the solubility parameter of the various fractions of petroleum during thermal treatment. (From Speight, J.G. 2014. *The Chemistry and Technology of Petroleum.* 5th Edition. CRC Press, Taylor & Francis Publishers, Boca Raton, FL. Figure 16.4, p. 449.)

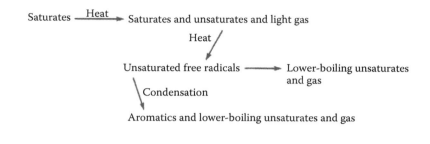

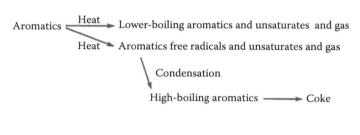

Resins and
asphaltenes } ——Heat——▸ Coke and lower-boiling aromatics and unsaturates
and light gas

FIGURE 6.5 Simplified schematic of the thermal decomposition of petroleum constituents. (From Speight, J.G. 2014. *The Chemistry and Technology of Petroleum*. 5th Edition. CRC Press, Taylor & Francis Publishers, Boca Raton, FL. Figure 16.5, p. 449.)

This mechanism also appears to be operable during residua hydroconversion, which has included a phase-separation step (the formation of *dry sludge*) in a kinetic model but was not included as a preliminary step to coke formation in a thermal cracking model (Takatsuka et al., 1989a,b; Speight, 2004a,b).

6.7.2 HYDROCONVERSION CHEMISTRY

There have also been many attempts to focus attention on the asphaltene constituents during hydrocracking studies. The focus has been on the macromolecular changes that occur by investigation of the changes to the generic fractions, i.e., the asphaltene constituents, the resin constituents, and the other fractions that make up such a feedstock (Ancheyta and Speight, 2007; Ancheyta et al., 2009). In terms of hydroprocessing, the means by which asphaltene constituents are desulfurized, as one step of a hydrocracking operation, is also suggested as part of the process. This concept can then be taken one step further to indicate that the thermal dealkylation of the various polynuclear aromatic systems is a definitive step in the hydrocracking process (Speight, 1970, 1987).

When catalytic processes are employed, complex molecules (such as those that may be found in the original asphaltene fraction or those formed during the process) are not sufficiently mobile (or are too strongly adsorbed by the catalyst) to be saturated by the hydrogenation components. Hence, these molecular species continue to condense and eventually degrade to coke. These deposits deactivate the catalyst sites and eventually interfere with the process.

Several noteworthy attempts have been made to focus attention on the asphaltene constituents during hydroprocessing studies. The focus has been on the macromolecular changes that occur by investigation of the changes in the generic fractions, i.e., the asphaltene constituents, the resin constituents, and the other fractions that make up such a feedstock. This option suggests that the overall pathway by which hydrotreating and hydrocracking of heavy oils and residua occur involves a stepwise mechanism:

Asphaltene constituents → resin-type constituents (polar aromatics)
Resin-type constituents → aromatics
Aromatics → saturates

A direct step from either the asphaltene constituents or the resin constituents to the saturates is not considered a predominant pathway for hydroprocessing.

The means by which asphaltene constituents are desulfurized, as one step of a hydrocracking operation, is also suggested as part of this process. This concept can then be taken one step further to show the dealkylation of the aromatic systems as a definitive step in the hydrocracking process (Speight, 1987). It is also likely that molecular species (within the asphaltene fraction) that contain nitrogen and other heteroatoms, and have lower volatility than their hydrocarbon analogs, are the prime movers in the production of coke (Speight, 1987).

When catalytic processes are employed, complex molecules such as those that may be found in the original asphaltene fraction or those formed during the process, are not sufficiently mobile (or are too strongly adsorbed by the catalyst) to be saturated by the hydrogenation components and, hence, continue to condense and eventually degrade to coke. These deposits deactivate the catalyst sites and eventually interfere with the hydroprocess.

A convenient means of understanding the influence of feedstock on the hydrocracking process is through a study of the hydrogen content (hydrogen-to-carbon atomic ratio) and molecular weight (carbon number) of the feedstocks and products. Such data show the extent to which the carbon number must be reduced and/or the relative amount of hydrogen that must be added to generate the desired lower molecular weight, hydrogenated products. In addition, it is possible to use data for hydrogen usage in residuum processing, where the relative amount of hydrogen consumed in the process can be shown to be dependent on the sulfur content of the feedstock.

6.7.3 Chemistry in the Refinery

Thermal cracking processes are commonly used to convert petroleum residua into distillable liquid products, although thermal cracking processes as used in the early refineries are no longer applied. Examples of modern thermal cracking processes are visbreaking and coking (delayed coking, fluid coking, and flexicoking) (Speight and Ozum, 2002; Hsu and Robinson, 2006; Gary et al., 2007; Speight, 2014). In all of these processes, the simultaneous formation of sediment or coke limits the conversion to usable liquid products. However, for the purposes of this section, the focus will be on the visbreaking and hydrocracking processes. The coking processes in which the reactions are taken to completion with the maximum yields of products are not a part of this discussion.

6.7.3.1 Visbreaking

To study the thermal chemistry of petroleum constituents, it is appropriate to select the visbreaking process (a carbon rejection process) and the hydrocracking process (a hydrogen addition process) as used in a modern refinery (Speight and Ozum, 2002; Hsu and Robinson, 2006; Gary et al., 2007; Speight, 2014). The processes operate under different conditions (Figure 6.6) and have different levels of conversion (Figure 6.7) and, although they do offer different avenues for conversion, these processes are illustrative of the thermal chemistry that occurs in refineries.

The visbreaking process is primarily a means of reducing the viscosity of heavy feedstocks by controlled thermal decomposition insofar as the hot products are quenched before complete conversion can occur (Speight and Ozum, 2002; Hsu and Robinson, 2006; Gary et al., 2007; Speight, 2014). However, the process is often plagued by sediment formation in the products. This sediment, or

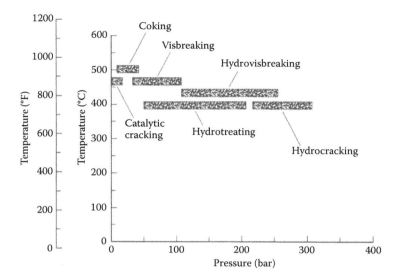

FIGURE 6.6 Temperature and pressure ranges for various processes. (From Speight, J.G. 2014. *The Chemistry and Technology of Petroleum*. 5th Edition. CRC Press, Taylor & Francis Publishers, Boca Raton, FL. Figure 16.6, p. 451.)

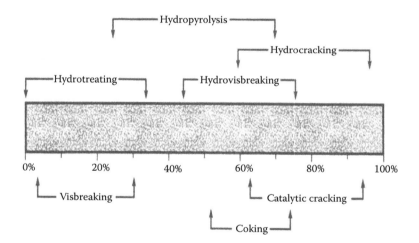

FIGURE 6.7 Feedstock conversion in various processes. (From Speight, J.G. 2014. *The Chemistry and Technology of Petroleum*. 5th Edition. CRC Press, Taylor & Francis Publishers, Boca Raton, FL. Figure 16.7, p. 452.)

sludge, must be removed if the products are to meet fuel oil specifications. The process (Figure 6.8) uses mild thermal cracking (partial conversion) as a relatively low-cost and low-severity approach to improving the viscosity characteristics of the residue without attempting significant conversion to distillates. Low residence times are required to avoid coking reactions, although additives can help suppress coke deposits on the tubes of the furnace (Allan et al., 1983).

A visbreaking unit consists of a reaction furnace, followed by quenching with a recycled oil, and fractionation of the product mixture. All of the reactions in this process occur as the oil flows through the tubes of the reaction furnace. The severity is controlled by the flow rate through the furnace and the temperature; typical conditions are 475–500°C (885–930°F) at the furnace exit

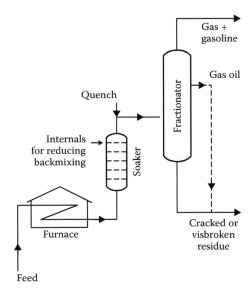

FIGURE 6.8 Visbreaking process using a soaker. (From OSHA Technical Manual, Section IV, Chapter 2: Petroleum Refining Processes. http://www.osha.gov/dts/osta/otm/otm_iv/otm_iv_2.html; Speight, J.G. 2014. *The Chemistry and Technology of Petroleum.* 5th Edition. CRC Press, Taylor & Francis Publishers, Boca Raton, FL. Figure 16.8, p. 452.)

with a residence time of 1–3 min, with operation for 3–6 months on-stream (continuous use) being possible before the furnace tubes must be cleaned and the coke removed. The operating pressure in the furnace tubes can range from 0.7 to 5 MPa depending on the degree of vaporization and the residence time desired. For a given furnace tube volume, a lower operating pressure will reduce the actual residence time of the liquid phase.

The reduction in viscosity of the unconverted residue tends to reach a limiting value with conversion, although the total product viscosity can continue to decrease (Figure 6.9). Conversion of residue in visbreaking follows first-order reaction kinetics (Henderson and Weber, 1965). The minimum viscosity of the unconverted residue can lie outside the range of allowable conversion if sediment begins

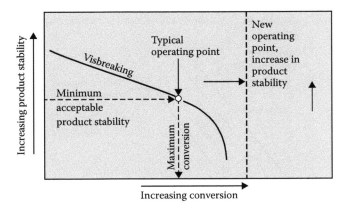

FIGURE 6.9 Representation of the break point above which maximum conversion is assured but product stability (inhibition of sediment formation) is less certain. (From Speight, J.G. 2014. *The Chemistry and Technology of Petroleum.* 5th Edition. CRC Press, Taylor & Francis Publishers, Boca Raton, FL. Figure 16.9, p. 453.)

to form (Rhoe and de Blignieres, 1979). When pipelining of the visbreaker product is a process objective, a diluent such as gas condensate can be added to achieve a further reduction in viscosity.

The high viscosity of the heavier feedstocks and residua is thought to be due to entanglement of the high molecular weight components of the oil and the formation of ordered structures in the liquid phase. Thermal cracking at low conversion can remove side chains from the asphaltene constituents and break bridging aliphatic linkages. A 5–10% conversion of atmospheric residue to naphtha is sufficient to reduce the entanglements and structures in the liquid phase and give at least a fivefold reduction in viscosity.

The stability of visbroken products is also an issue that might be addressed at this time. Using this simplified model, visbroken products might contain polar species that have been denuded of some of the alkyl chains and which, on the basis of solubility, might be more rightly called carbenes and carboids; however, an induction period is required for phase separation or agglomeration to occur. Such products might initially be soluble in the liquid phase but after the induction period, cooling, and/or diffusion of the products, incompatibility (phase separation, sludge formation, or agglomeration) occurs.

On occasion, higher temperatures are employed in various reactors as it is often assumed that if no side reactions occur, longer residence times at a lower temperature are equivalent to shorter residence times at a higher temperature. However, this assumption does not acknowledge the change in thermal chemistry that can occur at the higher temperatures, irrespective of the residence time. Thermal conditions can, indeed, induce a variety of different reactions in crude oil constituents, so that selectivity for a given product may change considerably with temperature. The onset of secondary, tertiary, and even quaternary reactions under the more extreme high-temperature conditions can convert higher molecular weight constituents of petroleum to low-boiling distillates, butane, propane, ethane, and (ultimately) methane. Caution is advised in the use of extreme temperatures.

Obviously, the temperature and residence time of the asphaltene constituents in the reactor are key to the successful operation of a visbreaker. Visbreakers must operate in temperature and residence time regimes that do not promote the formation of sediment (often referred to as coke). However, as already noted, there is a *break point* that might be increased, but then the possibility of sediment deposition also increases (Figure 6.9). At the temperatures and residence times outside of the most beneficial temperature and residence time regimes, thermal changes to the asphaltene constituents cause phase separation of a solid product that then progresses to coke. Furthermore, it is in such operations that models derived from average parameters can be ineffective and misleading.

For example, the amphoteric constituents of the asphaltene fraction are more reactive than the less polar constituents (Speight, 1994, 2014). The thermal products from the amphoteric constituents form first and will separate out from the reaction matrix before other products (Figure 6.10).

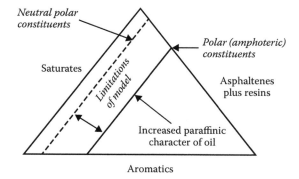

FIGURE 6.10 Limitations of the visbreaking process showing the points when phase separation occurs. (From Speight, J.G. 2014. *The Chemistry and Technology of Petroleum.* 5th Edition. CRC Press, Taylor & Francis Publishers, Boca Raton, FL. Figure 16.10, p. 454.)

Under such conditions, models based on average structural parameters or on average properties will not predict early-phase separation to the detriment of the product and the process as a whole.

Thus, knowledge of the actual nature and behavior (or thermal properties) of the subtypes of the asphaltene constituents is obviously beneficial and will allow steps to be taken to correct any such unpredictable occurrence. Indeed, the concept of hydrovisbreaking (visbreaking in the presence of hydrogen) could be of valuable assistance when high-asphaltene feedstocks are used.

6.7.3.2 Hydroprocessing

Hydrotreating is the (relatively) low-temperature removal of heteroatomic species by treatment of a feedstock or product in the presence of hydrogen (Speight and Ozum, 2002; Hsu and Robinson, 2006; Gary et al., 2007; Speight, 2014). On the other hand, hydrocracking (Figure 6.11) is the thermal decomposition of a feedstock in which carbon–carbon bonds are cleaved in addition to the removal of heteroatomic species—the presence of hydrogen changes the nature and yields of the products, especially decreasing coke yield (Speight and Ozum, 2002; Hsu and Robinson, 2006; Gary et al., 2007; Speight, 2014).

In the absence of hydrogen, the buildup of precursors occurs and these products are incompatible with the liquid medium and separate to form coke (Magaril and Aksenova, 1967, 1968, 1970a,b, 1972; Magaril and Ramazaeva, 1969; Magaril et al., 1970, 1971; Speight and Moschopedis, 1979). In fact, the chemistry involved in the reduction of asphaltene constituents to liquids using models in which the polynuclear aromatic system borders on graphitic is difficult to visualize, let alone

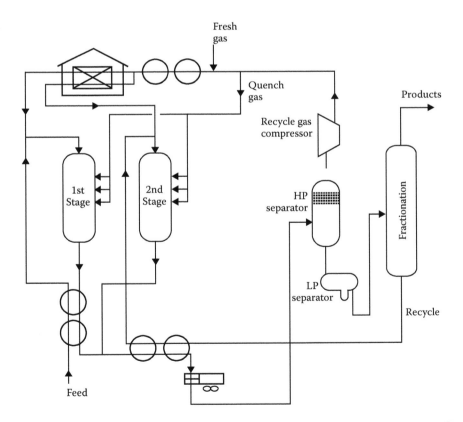

FIGURE 6.11 Two-stage hydrocracking unit. (From OSHA Technical Manual, Section IV, Chapter 2: Petroleum Refining Processes. http://www.osha.gov/dts/osta/otm/otm_iv/otm_iv_2.html; Speight, J.G. 2014. *The Chemistry and Technology of Petroleum.* 5th Edition. CRC Press, Taylor & Francis Publishers, Boca Raton, FL. Figure 16.11, p. 455.)

justify. However, the *paper chemistry* derived from the use of a molecularly designed model composed of smaller polynuclear aromatic systems is much easier to visualize (Speight, 1994, 2014). Nevertheless, the precise manner by which the asphaltene constituents interact with the catalysts is open to much more speculation.

In contrast to the visbreaking process, in which the general principle is the production of products for use as fuel oil, hydroprocessing is employed to produce a slate of products for use as liquid fuels. Nevertheless, the decomposition of asphaltene constituents is, again, an issue, and just as models consisting of large polynuclear aromatic systems are inadequate to explain the chemistry of visbreaking, they are also of little value for explaining the chemistry of hydrocracking.

Deposition of solids or incompatibility is still possible when asphaltene constituents interact with catalysts, especially acidic support catalysts, through the functional groups, e.g., the basic nitrogen species just as they interact with adsorbents. Moreover, there is a possibility for interaction of the asphaltene with the catalyst through the agency of a single functional group in which the remainder of the asphaltene molecule remains in the liquid phase. There is also a less desirable option in which the asphaltene reacts with the catalyst at several points of contact, causing immediate incompatibility on the catalyst surface.

There is evidence to show that during the early stages of the hydrotreating process, the chemistry of the asphaltene constituents follows the same routes as the thermal chemistry (Ancheyta et al., 2005). Thus, initially, there is an increase in the amount of asphaltene constituents followed by a decrease indicating that, in the early stages of the process, resin constituents are being converted to asphaltene material by aromatization and by some dealkylation. In addition, aromatization and dealkylation of the original asphaltene constituents yields asphaltene products that are of higher polarity and lower molecular weight than the original asphaltene constituents. Analogous to the thermal processes, this produces an overall asphaltene fraction that is more polar material and also of lower molecular weight. As the hydrotreating process proceeds, the amount of asphaltene constituents precipitated decreases due to conversion of the asphaltene constituents to products. At more prolonged on-stream times, there is a steady increase in the yield of the asphaltene constituents. This is accompanied by a general increase in the molecular weight of the precipitated material.

These observations are in keeping with observations for the thermal reactions of asphaltene constituents in the absence in hydrogen where the initial events are a reduction in the molecular weight of the asphaltene constituents, leading to lower molecular weight, more polar products that are derived from the asphaltene constituents but are often referred to as carbenes and carboids. As the reaction progresses, these derived products increase in molecular weight and eventually become insoluble in the reaction medium, deposit on the catalyst, and form coke.

As predicted from the chemistry of the thermal reactions of the asphaltene constituents, there is a steady increase in aromaticity (reflected as a decrease in the hydrogen-to-carbon atomic ratio) with on-stream time. This is due to (1) aromatization of naphthene ring system that are present in asphaltene constituents, (2) cyclodehydrogenation of alkyl chains to form other naphthene ring systems, (3) dehydrogenation of the new naphthene ring systems to form more aromatic rings, and (4) dealkylation of aromatic ring systems.

As the reaction progresses, the aromatic carbon atoms in the asphaltene constituents show a general increase and the degree of substitution of the aromatic rings decreases. Again, this is in keeping with the formation of products from the original asphaltene constituents (carbenes, carboids, and eventually coke) that have an increased aromaticity and decreased number of alkyl chains as well as a decrease in the alkyl chain length. Thus, as the reaction progresses with increased on-stream time, new asphaltene constituents are formed that, relative to the original asphaltene constituents, have increased aromaticity coupled with a lesser number of alkyl chains that are shorter than the original alkyl chains.

It may be that the chemistry of hydrocracking has to be given serious reconsideration insofar as the data show that the initial reactions of the asphaltene constituents appear to be the same as the reactions under thermal conditions where hydrogen is not present. Rethinking of the process conditions and the potential destruction of the catalyst by the deposition of carbenes and carboids require further investigation of the chemistry of asphaltene hydrocracking.

If these effects are prevalent during hydrocracking high-asphaltene feedstocks, the option may be to hydrotreat the feedstock first and then to hydrocrack the hydrotreated feedstock. There are indications that such hydrotreatment can (at some obvious cost) act beneficially in the overall conversion of the feedstocks to liquid products.

REFERENCES

Allan, D.E., Martinez, C.H., Eng, C.C., and Barton, W.J. 1983. Visbreaking Gains Renewed Interest. *Chem. Eng. Progr.* 79(1): 85–89.

Ancheyta, J., Centeno, G., Trejo, F., Betancourt, G., and Speight, J.G. 2005. Asphaltene Characterization as a Function of Time On-stream during Hydroprocessing of Maya Crude. *Catal. Today* 109: 162.

Ancheyta, J., and Speight, J.G. 2007. Feedstock Evaluation and Composition. In: *Hydroprocessing of Heavy Oils and Residua*, J. Ancheyta and J.G. Speight (Editors). CRC Press, Taylor & Francis Group, Boca Raton, FL, Chapter 2.

Ancheyta, J., Trejo, F., and Rana, M.S. 2009. *Asphaltenes: Chemical Transformations during Hydroprocessing Heavy Oils*. CRC Press, Taylor & Francis Group, Boca Raton, FL.

Bjorseth, A. 1983. *Handbook of Polycyclic Aromatic Hydrocarbons*. Marcel Dekker Inc., New York.

Cooper, T.A., and Ballard, W.P. 1962. *Advances in Petroleum Chemistry and Refining*, Vol. 6, K.A. Kobe and J.J. McKetta (Editors). Interscience, New York, Chapter 4.

Cumming, K.A., and Wojciechowski, B.W. 1996. *Catal. Rev. Sci. Eng.* 38: 101–157.

Dias, J.R. 1987. *Handbook of Polycyclic Hydrocarbons. Part A. Benzenoid Hydrocarbons*. Elsevier, New York.

Dias, J.R. 1988. *Handbook of Polycyclic Hydrocarbons. Part B. Polycyclic Isomers and Heteroatom Analogs of Benzenoid Hydrocarbons*. Elsevier, New York.

Dickie, J.P., and Yen, T.F. 1967. Macrostructures of the Asphaltic Fractions by Various Instrumental Methods. *Anal. Chem.* 39: 1847–1852.

Dolbear, G.E. 1998. Hydrocracking: Reactions, Catalysts, and Processes. In: *Petroleum Chemistry and Refining*, J.G. Speight (Editor). Taylor & Francis Publishers, Washington, DC, Chapter 7.

Dolbear, G.E., Tang, A., and Moorehead, E.L. 1987. Upgrading Studies with California, Mexican, and Middle Eastern Heavy Oils. In: *Metal Complexes in Fossil Fuels*, R.H. Filby and J.F. Branthaver (Editors). American Chemical Society, Washington, DC, pp. 220–232.

Ebert, L.B., Mills, D.R., and Scanlon, J.C. 1987. *Prepr. Div. Petrol. Chem. Am. Chem. Soc.* 32(2): 419.

Egloff, G. 1937. *The Reactions of Pure Hydrocarbons*. Reinhold, New York.

Eliel, E., and Wilen, S. 1994. *Stereochemistry of Organic Compounds*. John Wiley & Sons Inc., New York.

Fabuss, B.M., Smith, J.O., and Satterfield, C.N. 1964. Thermal Cracking of Pure Saturated Hydrocarbons. In: *Advances in Petroleum Chemistry and Refining*, Vol. 3, J.J. McKetta (Editor). John Wiley & Sons, New York, pp. 156–201.

Fitzer, E., Mueller, K., and Schaefer, W. 1971. The Chemistry of the Pyrolytic Conversion of Organic Compounds to Carbon. *Chem. Phys. Carbon* 7: 237–383.

Furimsky, E. 1983. Thermochemical and Mechanistic Aspects of Removal of Sulfur, Nitrogen and Oxygen from Petroleum. *Erdol. Kohle* 36: 518.

Gary, J.G., Handwerk, G.E., and Kaiser, M.J. 2007. *Petroleum Refining: Technology and Economics*, 5th Edition. CRC Press, Taylor & Francis Group, Boca Raton, FL.

Gray, M.R. 1994. *Upgrading Petroleum Residues and Heavy Oils*. Marcel Dekker Inc., New York.

Hajji, A.A., Muller, H., and Koseoglu, O.R. 2010. Molecular Details of Hydrocracking Feedstocks. *Saudi Aramco J. Technol.* Spring: 1–12.

Henderson, J.H., and Weber, L. 1965. Physical Upgrading of Heavy Oils by the Application of Heat. *J. Can. Pet. Tech.* 4: 206–212.

Hsu, C.S., and Robinson, P.R. (Editors) 2006. *Practical Advances in Petroleum Processing*, Vols. 1–2. Springer Science, New York.

Hurd, C.D. 1929. *The Pyrolysis of Carbon Compounds*. The Chemical Catalog Company Inc., New York.

Jones, D.S.J. 1995. *Elements of Petroleum Processing*. John Wiley & Sons Inc., New York.

Keller, W.D. 1985. Clays. In: *Kirk Othmer Concise Encyclopedia of Chemical Technology*, M. Grayson (Editor). Wiley Interscience, New York, p. 283.

Kim, H., and Long, R.B. 1979. *Ind. Eng. Chem. Fundam.* 18: 60.

King, P.J., Morton, F., and Sagarra, A. 1973. In: *Modern Petroleum Technology*, G.D. Hobson and W. Pohl (Editors). Applied Science Publishers, Barking, Essex, UK.

Langer, A.W., Stewart, J., Thompson, C.E., White, H.T., and Hill, R.M. 1961. *Ind. Eng. Chem.* 53: 27.

Levinter, M.E., Medvedeva, M.I., Panchenkov, G.M., Agapov, G.I., Galiakbarov, M.F., and Galikeev, R.K. 1967. *Khim. Tekhol. Topl. Masel.* 4: 20.

Levinter, M.E., Medvedeva, M.I., Panchenkov, G.M., Aseev, Y.G., Nedoshivin, Y.N., Finkelshtein, G.B., and Galiakbarov, M.F. 1966. *Khim. Tekhnol. Topl. Masel.* 9: 31.

Magaril, R.Z., and Aksenova, E.L. 1967. Mechanism of Coke Formation during the Cracking of Petroleum Tars. *Izv. Vyssh. Zaved. Neft Gaz.* 10(11): 134–136.

Magaril, R.Z., and Aksenova, E.L. 1968. Study of the Mechanism of Coke Formation in the Cracking of Petroleum Resin Constituents. *Int. Chem. Eng.* 8: 727–729 [first published in *Vyssh. Ucheb. Zaved. Neft Gaz.* 1967, 11: 134–136].

Magaril, R.Z., and Aksenova, E.I. 1970a. Kinetics and Mechanism of Coking Asphaltene Constituents. *Khim. Izv. Vyssh. Ucheb. Zaved. Neft Gaz.* 13(5): 47–53.

Magaril, R.Z., and Aksenova, E.I. 1970b. Mechanism of Coke Formation in the Thermal Decomposition of Asphaltene Constituents. *Khim. Tekhnol. Topl. Masel.* 15(7): 22–24.

Magaril, R.Z., and Aksenova, E.I. 1972. Coking Kinetics and Mechanism of the Thermal Decomposition of Asphaltene Constituents. *Khim. Tekhnol. Tr. Tyumen. Ind. Inst.* 169–172.

Magaril, R.Z., and Ramazaeva, L.F. 1969. Study of Coke Formation in the Thermal Decomposition of Asphaltene Constituents in Solution. *Izv. Vyssh. Ucheb. Zaved. Neft Gaz.* 12(1): 61–64.

Magaril, R.Z., Ramazaeva, L.F., and Askenova, E.I. 1970. Kinetics of Coke Formation in the Thermal Processing of Petroleum. *Khim. Tekhnol. Topliv Masel.* 15(3): 15–16.

Magaril, R.Z., Ramazeava, L.F., and Aksenova, E.I. 1971. Kinetics of Coke Formation in the Thermal Processing of Crude Oil. *Int. Chem. Eng.* 11: 250.

Masel, R.I. 1995. *Principles of Adsorption and Reaction on Solid Surfaces*. John Wiley & Sons Inc., New York.

Mitchell, D.L., and Speight, J.G. 1973. The Solubility of Asphaltene Constituents in Hydrocarbon Solvents. *Fuel* 52: 149.

Moschopedis, S.E., Fryer, J.F., and Speight, J.G. 1976. An Investigation of Asphaltene Molecular Weights. *Fuel* 55: 227.

Mushrush, G.W., and Speight, J.G. 1995. *Petroleum Products: Instability and Incompatibility*. Taylor & Francis Publishers, Philadelphia, PA.

Mushrush, G.W., and Speight, J.G. 1998. Instability and Incompatibility of Petroleum Products. In: *Petroleum Chemistry and Refining*, J.G. Speight (Editor). Taylor & Francis, Washington, DC, Chapter 8.

Overfield, R.E., Sheu, E.Y., Sinha, S.K., and Liang, K.S. 1989. SANS Study of Asphaltene Aggregation. *Fuel Sci. Technol. Int.* 7: 611.

Pines, H. 1981. *The Chemistry of Catalytic Hydrocarbon Conversions*. Academic Press, New York.

Rhoe, A., and de Blignieres, C. 1979. *Hydrocarbon Process.* 58(1): 131–136.

Samorjai, G.A. 1994. *Introduction to Surface Chemistry and Catalysis*. John Wiley & Sons Inc., New York.

Savage, P.E., Klein, M.T., and Kukes, S.G. 1988. Asphaltene Reaction Pathways 3. Effect of Reaction Environment. *Energy Fuels* 2: 619–628.

Schabron, J.F., and Speight, J.G. 1997. An Evaluation of the Delayed Coking Product Yield of Heavy Feedstocks using Asphaltene Content and Carbon Residue. *Rév. Inst. Fr. Pét.* 52(1): 73–85.

Schucker, R.C., and Keweshan, C.F. 1980. Reactivity of Cold Lake Asphaltene constituents. *Prepr. Div. Fuel Chem. Am. Chem. Soc.* 25: 155.

Speight, J.G. 1970. Thermal Cracking of Athabasca Bitumen, Athabasca Asphaltene constituents, and Athabasca Deasphalted Heavy Oil. *Fuel* 49: 134.

Speight, J.G. 1984. Upgrading Heavy Oils and Residua: The Nature of the Problem. In: *Catalysis on the Energy Scene*, S. Kaliaguine and A. Mahay (Editors). Elsevier, Amsterdam, p. 515.

Speight, J.G. 1987. Preprints. Initial Reactions in the Coking of Residua. *Div. Petrol. Chem. Am. Chem. Soc.* 32(2): 413.

Speight, J.G. 1989. Thermal Decomposition of Asphaltene Constituents. *Neftekhimiya* 29: 732.

Speight, J.G. 1990. The Chemistry of the Thermal Degradation of Petroleum Asphaltene Constituents. *Acta Pet. Sin. (Pet. Process. Sect.)* 6(1): 29.

Speight, J.G. 1992. A Chemical and Physical Explanation of Incompatibility during Refining Operations. In: *Proceedings 4th International Conference on the Stability and Handling of Liquid Fuels*. US Department of Energy (DOE/CONF-911102), p. 169.

Speight, J.G. 1994. Chemical and Physical Studies of Petroleum Asphaltene Constituents. In: *Asphalts and Asphaltene Constituents*, Vol. 1, T.F. Yen and G.V. Chilingarian (Editors). Elsevier, Amsterdam, Chapter 2.

Speight, J.G. 1997. *Petroleum Chemistry and Refining*, J.G. Speight (Editor) Taylor & Francis Publishers, Washington, DC, Chapter 5.

Speight, J.G. 1998. Thermal Chemistry of Petroleum Constituents. *Petroleum Chemistry and Refining*, J.G. Speight (Editor). Taylor & Francis, Washington, DC, Chapter 5.

Speight, J.G. 2000. *The Desulfurization of Heavy Oils and Residua*, 2nd Edition. Marcel Dekker Inc., New York.

Speight, J.G. 2004a. New Approaches to Hydroprocessing. *Catal. Today* 98(1–2): 55–60.

Speight, J.G. 2004b. Petroleum Asphaltene Constituents Part 2: The Effect of Asphaltene and Resin Constituents on Recovery and Refining Processes. *Rev. Inst. Fr. Pét. Oil Gas Sci. Technol.* 59(5): 479–488.

Speight, J.G. 2014. *The Chemistry and Technology of Petroleum*, 5th Edition. CRC Press, Taylor & Francis Group, Boca Raton, FL.

Speight, J.G., and Moschopedis, S.E. 1979. The Production of Low-Sulfur Liquids and Coke From Athabasca Bitumen. *Fuel Process. Technol.* 2: 295.

Speight, J.G., and Ozum, B. 2002. *Petroleum Refining Processes*. Marcel Dekker Inc., New York.

Speight, J.G., Wernick, D.L., Gould, K.A., Overfield, R.E., Rao, B.M.L., and Savage, D.W. 1985. Molecular Weights and Association of Asphaltene Constituents: A Critical Review. *Rev. Inst. Fr. Pet.* 40: 51.

Storm, D.A., Decanio, S.J., Edwards, J.C., and Sheu, E.Y. 1997. Sediment Formation during Heavy Oil Upgrading. *Pet. Sci. Technol.* 15: 77.

Takatsuka, T., Kajiyama, R., Hashimoto, H., Matsuo, I., and Miwa, S.A. 1989a. *J. Chem. Eng. Jpn.* 22: 304.

Takatsuka, T., Wada, Y., Hirohama, S., and Fukui, Y.A. 1989b. *J. Chem. Eng. Jpn.* 22: 298.

Ternan, M. 1983. Catalysis, Molecular Weight Change and Fossil Fuels. *Can. J. Chem. Eng.* 61(2): 133–147.

Thiyagarajan, P., Hunt, J.E., Winans, R.E., Anderson, K.B., and Miller, J.T. 1995. Temperature Dependent Structural Changes of Asphaltene Constituents in 1-Methylnaphthalene. *Energy Fuels* 9: 629.

Valyavin, G.G., Fryazinov, V.V., Gimaev, R.H., Syunyaev, Z.I., Vyatkin, Y.L., and Mulyukov, S.F. 1979. *Khim. Tekhol. Topl. Masel.* 8: 8.

Vercier, P. 1981. *The Chemistry of Asphaltenes*, J.W. Bunger and N.C. Li (Editors). Advances in Chemistry Series No. 195. American Chemical Society, Washington, DC.

Wiehe, I.A. 1992. A Solvent-Resid Phase Diagram for Tracking Resid Conversion. *Ind. Eng. Chem. Res.* 31: 530–536.

Wiehe, I.A. 1993. A Phase-Separation Kinetic Model for Coke Formation. *Ind. Eng. Chem. Res.* 32: 2447–2454.

Wiehe, I.A. 1994. The Pendant-Core Building Block Model of Petroleum Residua. *Energy Fuels* 8: 536–544.

Yan, T.Y. 1987. Coker Formation in the Visbreaking Process. *Prepr. Div. Petrol. Chem. Am. Chem. Soc.* 32: 490.

Yen, T.F. 1984. *The Future of Heavy Crude Oil and Tar Sands*, R.F. Meyer, J.C. Wynn, and J.C. Olson (Editors). McGraw-Hill, New York.

7 Influence of Feedstock

7.1 INTRODUCTION

The problems encountered in processing heavy feedstocks can be directly equated to the amount of complex, higher-boiling constituents (Speight, 2001, 2014, 2015). Processing these feedstocks is not just a matter of applying know-how derived from refining conventional crude oils that often use hydrogen-to-carbon (H/C) atomic ratios and/or API gravity to determine the refining sequences, but requires knowledge of composition as well as the properties of the residua (Table 7.1) where the majority of the difficult-to-desulfurize constituents are found (Speight, 2000, 2014).

A wide choice of commercial processes is available for the catalytic hydrodesulfurization of heavy feedstocks (Chapter 13). The suitability of any particular process depends not only on the nature of the feedstock but also on the degree of desulfurization that is required. There is also a dependence on the relative amounts of the lower-boiling products that are to be produced as feedstocks for further refining and generation of liquid fuels.

The constituents of residua are not only complex in terms of the carbon number and boiling points but also because a large part of this *envelope* falls into a range about which very little is known about model compounds (Speight, 2000, 2014). It is also established that the majority of the higher molecular weight materials produce coke (with some liquids), while the majority of the lower molecular weight constituents produce liquids (with some coke). To mitigate the latter process trend—the formation of coke (Figure 7.1)—hydrocracking is aimed.

It is the physical and chemical composition of a feedstock that plays a large part not only in determining the nature of the products that arise from refining operations but also in determining the precise manner by which a particular feedstock should be processed. Furthermore, it is apparent that the conversion of heavy feedstocks requires new lines of thought to develop suitable processing scenarios. Indeed, the use of thermal (*carbon rejection*) and hydrothermal (*hydrogen addition*) processes that were inherent in the refineries designed to process lighter feedstocks has been a particular cause for concern and has brought about the evolution of processing schemes that will accommodate the heavier feedstocks. However, processes based on carbon rejection are not chemically efficient since they degrade usable portions of the feedstock to coke (Speight and Ozum, 2002; Parkash, 2003; Hsu and Robinson, 2006; Gary et al., 2007; Speight, 2013, 2014).

Thus, there is the potential for the application of more efficient conversion processes to heavy feedstock refining. Hydrocracking is probably the most versatile of the petroleum refining processes because of its applicability to a wide range of feedstocks. In fact, hydrocracking can be applied to the conversion of the heavier feedstocks, and there are a variety of processes that are designed specifically for this particular use (Speight and Ozum, 2002; Parkash, 2003; Hsu and Robinson, 2006; Gary et al., 2007; Speight, 2013, 2014). It is worthy of note at this point that there is often very little effort (it is often difficult) to differentiate between a hydrocracking process and a hydrotreating process.

In actual practice, the reactions that are used to chemically define the processes i.e., cracking and subsequent hydrogenation of the fragments, hydrogenation of unsaturated material, as well as hydrodesulfurization and hydrodenitrogenation, can all occur (or be encouraged to occur) in any one particular process. Thus, hydrocracking will, in all likelihood, be accompanied by hydrodesulfurization, thereby producing not only low-boiling products but also low-boiling products that are low in sulfur. Thus, the choice of processing schemes for a given hydrocracking application depends on the nature of the feedstock and the product requirements (Speight, 2000, 2014; Speight and Ozum, 2002; Parkash, 2003; Hsu and Robinson, 2006; Ancheyta and Speight, 2007;

TABLE 7.1

Properties of Different Residua

Feedstock	Gravity API	Sulfur wt.%	Nitrogen wt.%	Nickel ppm	Vanadium ppm	Asphaltene Constituents (Heptane) wt.%	Carbon Residue (Conradson) wt.%
Arabian light, >650°F	17.7	3.0	0.2	10.0	26.0	1.8	7.5
Arabian light, >1050°F	8.5	4.4	0.5	24.0	66.0	4.3	14.2
Arabian heavy, >650°F	11.9	4.4	0.3	27.0	103.0	8.0	14.0
Arabian heavy, >1050°F	7.3	5.1	0.3	40.0	174.0	10.0	19.0
Alaska North Slope, >650°F	15.2	1.6	0.4	18.0	30.0	2.0	8.5
Alaska North Slope, >1050°F	8.2	2.2	0.6	47.0	82.0	4.0	18.0
Lloydminster (Canada), >650°F	10.3	4.1	0.3	65.0	141.0	14.0	12.1
Lloydminster (Canada), >1050°F	8.5	4.4	0.6	115.0	252.0	18.0	21.4
Kuwait, >650°F	13.9	4.4	0.3	14.0	50.0	2.4	12.2
Kuwait, >1050°F	5.5	5.5	0.4	32.0	102.0	7.1	23.1
Tia Juana, >650°F	17.3	1.8	0.3	25.0	185.0		9.3
Tia Juana, >1050°F	7.1	2.6	0.6	64.0	450.0		21.6
Taching, >650°F	27.3	0.2	0.2	5.0	1.0	4.4	3.8
Taching, >1050°F	21.5	0.3	0.4	9.0	2.0	7.6	7.9
Maya, >650°F	10.5	4.4	0.5	70.0	370.0	16.0	15.0

Source: Speight, J.G. 2014. *The Chemistry and Technology of Petroleum.* 5th Edition. CRC Press, Taylor & Francis Publishers, Boca Raton, FL. Table 26.6, p. 757.

Gary et al., 2007). The process can be simply illustrated as a single-stage or as a two-stage operation (Figure 7.2).

The single-stage process can be used to produce naphtha, but is more often used to produce middle distillate from heavy vacuum gas oils. The two-stage process was developed primarily to produce high yields of naphtha from straight-run gas oils, and the first stage may actually be a purification step to remove sulfur-containing and nitrogen-containing organic materials. Both processes use an extinction/recycle technique to maximize the yields of the desired product. Significant conversion of heavy feedstocks can be accomplished by hydrocracking at high severity (Ancheyta and Speight, 2007). For some applications, the products boiling up to 340°C (650°F) can be blended to give the desired final product.

As a general rule for a given feedstock, an increase in the conversion level of the vacuum residuum (i.e., the material boiling 565°C+/1050°F+) will lead to an increase in the quantity of the distillate; however, there will be a decline in the quality of the distillate. If the unit is followed by vacuum distillation, the gas oil fraction can be taken to cracking units and converted into naphtha and kerosene. The quality of these fractions also depends on cracking severity in the unit.

Attention must also be given to the coke mitigation aspects. For example, in the hydrogen addition options, particular attention must be given to hydrogen management, thereby promoting asphaltene fragmentation to lighter products rather than to the production of coke. The presence of a material with good solvating power, to diminish the possibility of coke formation, is preferred. In

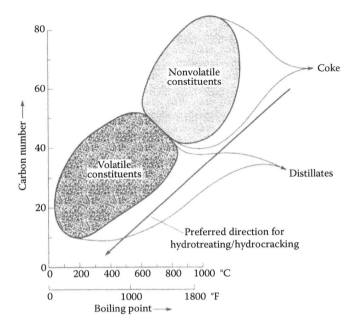

FIGURE 7.1   Simplified representation of the formation of thermal products from petroleum. (From Speight, J.G. 2000. *The Desulfurization of Heavy Oils and Residua*. 2nd Edition. Marcel Dekker Inc., New York. Figure 7.2, p. 211.)

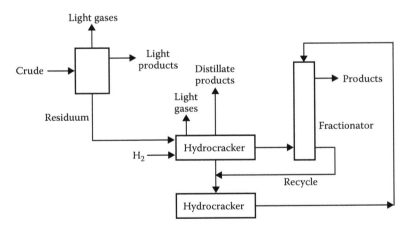

FIGURE 7.2   Single-stage and two-stage hydrocracking. (From Speight, J.G. 2000. *The Desulfurization of Heavy Oils and Residua*. 2nd Edition. Marcel Dekker Inc., New York. Figure 7.3, p. 211.)

this respect, it is worth noting the reappearance of donor solvent processing of heavy oil (Vernon et al., 1984; McConaghy, 1987), which has its roots in the older hydrogen donor diluent visbreaking process (Carlson et al., 1958; Langer et al., 1961, 1962; Bland and Davidson, 1967).

In reality, no single bottom-of-the-barrel processing scheme is always the best choice. Refiners must consider the potential of proven processes, evaluate the promise of newer ones, and choose based on the situation. The best selection will always depend on the kind of crude oil, the market for products, and financial and environmental consideration. Although there are no simple solutions, the available established processes and the growing number of new ones under development offer some reasonable choices. The issue then becomes how to most effectively handle the asphaltene fraction of the feedstock at the most reasonable cost. Solutions to this processing issue can be separated into two broad categories: (1) conversion of asphaltene constituents into another, salable product and (2) use of the asphaltene constituents by concentration into a marketable, or useable, product such as asphalt.

The hydrodesulfurization process variables (Chapter 10) usually require some modification to accommodate the various feedstocks that are submitted for this particular aspect of refinery processing. The main point of the text is to outline the hydrodesulfurization process with particular reference to the heavier oils and residua. However, some reference to the lighter feedstocks is warranted. This will serve as a base point to indicate the necessary requirements for heavy feedstock hydrodesulfurization (Table 7.2).

One particular aspect of the hydrodesulfurization process that needs careful monitoring, with respect to feedstock type, is the exothermic nature of the reaction. The heat of the reaction is proportional to the hydrogen consumption, and with the more saturated lower-boiling feedstocks where hydrocracking may be virtually eliminated, the overall heat production during the reaction may be small, leading to a more controllable temperature profile. However, with the heavier feedstocks where hydrogen consumption is appreciable (either by virtue of the hydrocracking that is necessary to produce a usable product or by virtue of the extensive hydrodesulfurization that must occur), it may be desirable to provide internal cooling of the reactor (Speight, 2013). This can be accomplished by introducing cold recycle gas to the catalyst bed to compensate for excessive heat. One other generalization may apply to the lower-boiling feedstocks in the hydrodesulfurization process. The process may actually have very little effect on the properties of the feedstock (assuming that hydrocracking reactions are negligible)—removal of sulfur will cause some drop in specific gravity that could give rise to volume recoveries approaching (or even above) 100%. Furthermore, with the assumption that cracking reactions are minimal, there may be a slight lowering of the boiling range due to sulfur removal from the feedstock constituents. However, the production of lighter fractions is usually small and may only amount to some 1–5% by weight of the products boiling below the initial boiling point of the feedstock.

One consideration for the heavier feedstocks is that it may be more economical to hydrotreat and desulfurize high-sulfur feedstocks before catalytic cracking than to hydrotreat the products from catalytic cracking. This approach (Speight and Moschopedis, 1979; Decroocq, 1984; Speight, 2013, 2014) has the potential for several advantages: (1) the products require less finishing; (2) sulfur

TABLE 7.2
Lower (Naphtha Desulfurization) and Upper (Residua Desulfurization) Extremes or the Hydrodesulfurization Process Parameters

Parameter	Naphtha	Residuum
Temperature, °C	300–400	340–425
Pressure, atm	35–70	55–170
LHSV	4.0–10.0	0.2–1.0
H_2 recycle rate, scf/bbl	400–1000	3000–5000
Catalysts life, years	3.0–10.0	0.5–1.0
Sulfur removal, %	99.9	85.0
Nitrogen removal, %	99.5	40.0

Source: Speight, J.G. 2014. *The Chemistry and Technology of Petroleum.* 5th Edition. CRC Press, Taylor & Francis Publishers, Boca Raton, FL. Table 21.1, p. 569.

is removed from the catalytic cracking feedstock, and corrosion is reduced in the cracking unit; (3) coke formation is reduced; (4) higher feedstock conversions; and (5) the potential for better-quality products. The downside is that many of the heavier feedstocks act as hydrogen sinks in terms of their ability to interact with the expensive hydrogen. A balance of the economic advantages/disadvantages must be struck on an individual feedstock basis.

Thus, the purpose of this chapter is to provide an overview of the methods that are available to study the composition of heavy feedstocks (heavy oil, extra heavy oil, tar sand bitumen, and residua) in terms of their physical and, to a lesser extent, chemical composition. It is also the purpose of the present section to relate the physical properties (or evaluation) of these feedstocks to their performance in the hydrodesulfurization process, with special reference to any parameters that adversely affect process efficiency.

7.2 CHEMICAL COMPOSITION

As already noted (Chapter 3), strategies for upgrading petroleum emphasize the difference in their properties that, in turn, influences the choice of methods or combinations thereof for conversion of petroleum to various products (Speight and Ozum, 2002; Parkash, 2003; Hsu and Robinson, 2006; Gary et al., 2007; Speight, 2013, 2014). Similar principles are applied to heavy feedstocks, and the availability of processes that can be employed to convert these feedstocks to usable products has increased significantly in recent years and will continue to increase (Chapter 4) (Speight, 2011, 2014).

Refining petroleum involves subjecting the feedstock to a series of physical and chemical processes (Chapter 3) (Figure 7.2), as a result of which a variety of products are generated. In some of the processes, e.g., distillation, the constituents of the feedstock are isolated unchanged, whereas in other processes, e.g., cracking, considerable changes are brought about to the constituents (Speight and Ozum, 2002; Parkash, 2003; Hsu and Robinson, 2006; Gary et al., 2007; Speight, 2014).

Standard analyses on whole heavy crude oil or residua, such as determinations of elemental compositions and various physical property tests, have served to provide some indications of processability and may give an indication of the feedstock behavior (Chapter 2) (Speight, 2001, 2014, 2015). However, there is some question of the reliability of the tests when applied to the heavier feedstocks. For example, it might be wondered if the carbon residue tests (ASTM D189, ASTM D524, and ASTM D4530) are really indicative of the yields of coke formed under process conditions. Moreover, for the heavier feedstocks, it must be emphasized that to proceed from the raw evaluation data to full-scale production, insofar as the heavy feedstock is immediately used in the refinery, is to proceed without caution. The thermal chemistry of the feedstock constituents will remain an unknown until the feedstock is used on-stream and the compatibility of the feedstock and the products with other feedstocks and products will also be unknown. Further evaluation of the processability of the feedstock is usually necessary. Predicting processability is a matter of correlating the test data (Chapter 2) with an understanding of the chemical and physical composition of the feedstock as it relates to properties, refining behavior, and product yields (Speight, 1992). The molecular complexity of the heavy feedstocks offers disadvantages to the reliance on the use of bulk properties as the sole means of predicting behavior (Dolbear et al. 1987).

Recognition that refinery behavior is related to the composition of the feedstock has led to multiple attempts to establish petroleum and its fractions as compositions of matter. As a result, various analytical techniques have been developed for the identification and quantification of every molecule in the lower-boiling fractions of petroleum. It is now generally recognized that the name *petroleum* does not describe a composition of matter but rather a mixture of various organic compounds that includes a wide range of molecular weights and molecular types that exist in balance with each other (Speight, 1994; Long and Speight, 1998). There must also be some questions of the advisability (perhaps *futility* is a better word) of attempting to describe *every molecule* in petroleum. The true focus should be to what ends these molecules can be used.

The problems encountered in processing the heavy feedstocks can be directly equated to the amount of complex, higher-boiling constituents in the residual portion of the oil. Refining these feedstocks is not just a matter of applying know-how derived from refining conventional crude oils but requires knowledge of the composition of these more complex feedstocks.

Thus, investigations of the character of petroleum need to be focused on the influence of its character on refining operations and the nature of the products that will be produced. Furthermore, one means by which the character of petroleum has been studied is through its fractional composition. However, the fractional composition of petroleum varies markedly with the method of isolation or separation, thereby leading to potential complications (especially in the case of the heavier feedstocks) in the choice of suitable processing schemes for these feedstocks.

Knowledge of the composition of petroleum allows the geologist to answer questions of precursor–product relationships and conversion/maturation mechanisms. Knowledge of the composition of petroleum allows the refiner to optimize the conversion of raw petroleum into high-value products. Petroleum is now the world's main source of energy and petrochemical feedstock. Originally, petroleum was distilled and sold as fractions (such as kerosene) with desirable physical properties. The present demand is for gasoline, solvents, diesel fuel, jet fuel, heating oil, lubricant oils, and asphalt, leading to a multiplicity of products (Speight and Ozum, 2002; Parkash, 2003; Hsu and Robinson, 2006; Gary et al., 2007; Speight, 2014). However, all are not available from raw petroleum as they once were, and a variety of refining and finishing processes are required (Speight, 2014). Alternatively, petroleum is converted to petrochemical feedstocks such as ethylene, propylene, butene(s), butadiene, and isoprene. These feedstocks are important, for they form the basis for plastics, elastomers, and artificial fibers, and other industrial products.

7.2.1 HYDROCARBON COMPOUNDS

The identification of hydrocarbon constituents, especially those in the higher-boiling weight fractions, is an important aspect of petroleum science. Determination of the skeletal structures of the hydrocarbons gives indications of the types of locations in which the heteroatoms might be found. However, because of the complexity of the structures in heavy feedstocks, it is difficult on the basis of the data obtained from synthesized hydrocarbons to determine the identity or even the similarity of the synthetic hydrocarbons to those that constitute many of the higher-boiling fractions of petroleum. Nevertheless, it has been well established that the hydrocarbon components of petroleum are composed of paraffinic, naphthenic, and aromatic groups. Olefinic groups ($>C=C<$) are not usually found in crude oils, and acetylene-type hydrocarbons ($-C\equiv C-$) are very rare indeed.

Thus, methane is the main hydrocarbon component of petroleum gases with lesser amounts of ethane, propane, butane, isobutane, and some C_5^+ light hydrocarbons. Other gases, such as hydrogen, carbon dioxide, hydrogen sulfide, and carbonyl sulfide, are also present.

The *naphtha fraction* is dominated by the saturates with lesser amounts of mono- and di-aromatics. Within the saturate fraction, every possible paraffin from the simplest C_1 hydrocarbon (methane) to n-C_{10} (normal decane) is present. Depending on the source, one of these low-boiling paraffins may be the most abundant compound in a crude oil reaching several percent. The isoparaffins begin at C_4 with isobutane, and the number of isomers grows rapidly with carbon number. The saturate fractions also contain cycloalkanes (naphthenes) with mainly five or six carbons in the ring. Fused ring dicycloalkane derivatives such as *cis*-decahytronapthalene and *trans*-decahytronapthalene (decalin) are also common. The numerous aromatics in naphtha begin with benzene, but the C_1 to C_3 alkylated derivatives generally are present in larger amounts. Benzenes with fused cycloparaffin rings (naphthene aromatics), such as indane and tetralin, have been isolated along with a number of their respective methyl derivatives. Naphthalene is included in this fraction, while the 1-methylnaphthalene and 2-methylnaphthalene and higher homologues of fused two-ring aromatics appear in the mid-distillate fraction.

The saturates remain the major component in the mid-distillate fractions of petroleum; however, aromatics, which now include simple compounds with up to three aromatic rings, and heterocyclic compounds are present and represent a larger portion of the total. Kerosene, jet fuel, and diesel fuel are all derived from middle-distillate fractions and can also be obtained from cracked and hydro-processed refinery streams.

Within the middle-distillate fractions, the concentration of *n*-paraffins decreases regularly from C_{11} to C_{20} and since the number of possible C_{15} isomers exceeds 4000, it is not surprising that few additional isoparaffins have been identified. Mono- and di-cycloparaffins with five or six carbons per ring constitute the bulk of the naphthenes in the middle-distillate boiling range, decreasing in concentration as the carbon number increases; substituted three-ring naphthenes are also present.

The most abundant aromatics in the *mid-distillate fractions* are di- and tri-methyl naphthalene derivatives. Other one and two ring aromatics are undoubtedly present in small quantities as either naphthene aromatics or alkyl homologues in the C_{11}–C_{20} range. In addition to these homologues of alkylbenzenes, tetralin, and naphthalene, the mid-distillate contains some fluorene and phen-anthrene derivatives. The phenanthrene structure appears to be favored over that of anthracene structure (Tissot and Welte, 1978; Speight, 2014).

Saturated constituents contribute less to the vacuum gas oil than the aromatics but more than the polar constituents that are now present at percentage rather than trace levels. The vacuum gas oil itself is occasionally used as heating oil; however, most commonly, it is processed by catalytic cracking to produce naphtha or by extraction to yield lubricating oil. Within the vacuum gas oil saturates, the distribution of paraffins, isoparaffins, and naphthenes is highly dependent on the petroleum source. The bulk of the vacuum gas oil–saturated constituents consist of isoparaffins and naphthenes. The naphthenes contain from one to more than six fused rings and have alkyl substitu-ents. For mono- and di-aromatics, the alkyl substitution typically involves one long side chain and several short methyl and ethyl substituents.

The aromatics in vacuum gas oil may contain one to six fused aromatic rings that may bear additional naphthene rings and alkyl substituents in keeping with their boiling range. Mono- and di-aromatics account for about 50% of the aromatics in vacuum gas oil samples.

Within petroleum, certain aromatic structures appear to be favored. For example, alkyl phenanthrene derivatives outnumber alkyl anthracene derivatives by as much as 100:1. In addition, despite the bias in the separation methods, the alkyl derivatives appear to be more abundant than the parent ring compounds.

The abundance of the different members of the same homologous series varies considerably in absolute and relative values. However, in any particular crude oil or crude oil fraction, there may be a small number of constituents forming the greater part of the fraction, and these have been referred to as the *predominant constituents*. This generality may also apply to other constituents and is very dependent on the nature of the source material as well as the relative amounts of the individual source materials prevailing during maturation conditions (Speight, 2014).

Paraffins usually occur as vestiges in many heavy feedstocks and appear as alkyl side chains on aromatic and naphthenic systems. Furthermore, these alkyl chains can contain 40 or more carbon atoms (Speight, 2014). The cycloparaffins occur as polycyclic structures containing two to six rings per molecule. There is also the premise that the naphthene ring systems carry alkyl chains that are generally shorter than the alkyl substituents carried by aromatic systems. This can be reflected on hypothetical structures for the asphaltene constituents (Speight, 2014).

In the asphaltene fraction, free condensed naphthenic ring systems may occur; however, gen-eral observations favor the occurrence of combined aromatic–naphthenic systems that are vari-ously substituted by alkyl systems. There is also general evidence that the aromatic systems are responsible for the polarity of the asphaltene constituents. Components with two aromatic rings are presumed to be naphthalene derivatives and those with three aromatic rings may be phenanthrene derivatives. Currently, and because of the consideration of the natural product origins of petroleum, phenanthrene derivatives are favored over anthracene derivatives. In addition, trace amounts of peri-condensed polycyclic aromatic hydrocarbons, such as methyl chrysene, methyl- and dimethyl perylene derivatives, and benzofluorene derivatives, have been identified in crude oil. Chrysene and benzofluorene homologues seem to predominate over those of pyrene.

The polycyclic aromatic systems in the asphaltene fraction are complex molecules that fall into a molecular weight and boiling range where very little is known about model compounds (Speight, 1994, 2014). There has not been much success in determining the nature of such systems in the higher-boiling constituents of petroleum, i.e., the residua or nonvolatile constituents. In fact, it has been generally assumed that as the boiling point of a petroleum fraction increases, so does the number of condensed rings in a polycyclic aromatic system. To an extent, this is true, but the simplicities of such assumptions cause an omission of other important structural constituents of the petroleum matrix, the alkyl substituents, the heteroatoms, and any polycyclic systems that are linked by alkyl chains or by heteroatoms.

The active principle is that petroleum is a continuum (Long and Speight, 1989; Speight, 2014) and has natural product origins (Eglinton and Murphy, 1969; Sakarnen and Ludwig, 1971; Tissot and Welte, 1978; Durand, 1980; Weiss and Edwards, 1980; Brooks and Welte, 1984; Speight, 2014). As such, it might be anticipated that there is a continuum of aromatic systems throughout petroleum that might differ from volatile to nonvolatile fractions but which, in fact, are based on natural product systems. It might also be argued that substitution patterns of the aromatic nucleus that are identified in the volatile fractions, or in any natural product counterparts, also apply to the nonvolatile fractions.

One method that had provided valuable information about the aromatic systems in the nonvolatile fractions of crude oil is ultraviolet spectroscopy.

Typically, the ultraviolet spectrum of an asphaltene shows two major regions with very little fine structure. Interpretation of such a spectrum can only be made in general terms. However, the technique can add valuable information about the degree of condensation of polycyclic aromatic ring systems through the auspices of high-performance liquid chromatography (Speight, 2014). Indeed, when this approach is taken, the technique not only confirms the complex nature of the asphaltene fraction but also allows further detailed identifications to be made of the individual functional constituents of the nonvolatile fractions. The amphoteric and basic nitrogen subfractions contain polycyclic aromatic systems having two-to-six rings per system. The acid subfractions (phenolic/carboxylic functions) and neutral polar subfractions (amides/imino functions) contain few, if any, polycyclic aromatic systems having more than three rings per system. In all cases, the evidence favored the preponderance of the smaller (one-to-four) ring systems.

7.2.2 SULFUR COMPOUNDS

Sulfur compounds are perhaps the most important nonhydrocarbon constituents of petroleum and occur as a variety of structures (Chapters 1 and 6). During the refining sequences involved to convert crude oils to salable products, a great number of the sulfur compounds that occur in any particular petroleum are concentrated in the residua and other heavy fractions.

The relative importance attached to sulfur compounds in petroleum may, at first, seem unwarranted, but the presence of sulfur compounds in any crude oil can only result in harmful effects. For example, the presence of sulfur compounds in finished petroleum products such as gasoline will cause corrosion of engine parts, especially under winter conditions when water containing sulfur dioxide (from the internal combustion) may collect in the crankcase. On the other hand, mercaptans cause the corrosion of copper and brass in the presence of air and also have an adverse effect on the color stability of gasoline and other liquid fuels.

The distribution of sulfur compounds in crude oils has been studied extensively since the 1890s, and it has become possible to note various generalities. For example, the proportion of sulfur will increase with the boiling point of the crude oil fraction (Speight, 2001, 2014, 2015). If the distillation is allowed to proceed at too high a temperature, thermal decomposition of the high molecular weight sulfur compounds will ensue. Hence, the middle fractions will contain more sulfur compounds than the higher-boiling fractions. The distribution of the various types of sulfur compounds varies markedly among crude oils of diverse origin. It is difficult to assign specific trends to the occurrence of compound types within the different crude oils other than that an increase in boiling point of fractions from a particular crude oil is accompanied by an increase in sulfur content.

Sulfur is usually the only heteroatom to be found in the naphtha fraction, and then only at trace levels and can be in the form of mercaptans (thiols, RSH), thiophenol derivatives (C_6H_5SH), sulfide derivatives (R-S-R[1]), alkyl sulfide derivatives, and five- or six-ring cyclic (thiacyclane) structures, and to a lesser extent, disulfide derivatives (RSSR[1]). In general, the sulfur structure distribution mimics the hydrocarbons, i.e., naphthenic oils with high cyloalkane derivatives have high thiacyclane content.

The sulfur–heterocyclic compounds in the mid-distillate range are primarily the thiacyclane, benzothiophene, and dibenzothiophene derivatives, with lesser amounts of dialkyl–, diaryl–, and aryl–alkyl sulfides. Because of the ring structures, sulfur compounds are significant contributors to the vacuum gas oil fraction. The major sulfur species are alkyl benzothiophene, dibenzothiophene, benzonaphthene–thiophene, and phenanthro–thiophene derivatives.

7.2.3 NITROGEN COMPOUNDS

The presence of nitrogen in petroleum is of much greater significance in refinery operations than might be expected from the small amounts present. Nitrogen compounds can be responsible for the poisoning of cracking catalysts, and they also contribute to gum formation in such products as domestic fuel oil. The trend in recent years toward cutting deeper into the crude to obtain stocks for catalytic cracking has accentuated the harmful effects of the nitrogen compounds that are concentrated largely in the higher-boiling portions.

In general, the nitrogen content of crude oil is low and falls within the range 0.1–0.9% w/w, although early work indicates that some crude oils may contain up to 2% w/w nitrogen. However, crude oils with no detectable nitrogen or even trace amounts are not uncommon; however, in general, the more asphaltic the oil, the higher its nitrogen content and the greater the tendency of the nitrogen to exist in a homologue of pyridine (Mitra-Kirtley et al., 1993; Speight, 2014). Insofar as an approximate correlation exists between the sulfur content and API gravity of crude oils, there also exists a correlation between nitrogen content and the API gravity of crude oil (Speight, 2014). It also follows that there is an approximate correlation between the nitrogen content and the carbon residue—the higher the carbon residue, the higher the nitrogen content.

Nitrogen in petroleum may be classed arbitrarily as *basic* and *nonbasic*. The basic nitrogen compounds, which are composed mainly of pyridine homologues and occur throughout the boiling ranges, have a decided tendency to exist in the higher-boiling fractions and residua (Speight and Ozum, 2002; Parkash, 2003; Hsu and Robinson, 2006; Gary et al., 2007; Speight, 2014). The nonbasic nitrogen compounds, which are usually of the pyrrole, indole, and carbazole types, also occur in the higher-boiling fractions and residua.

For example, although only trace amounts (ppm levels) of nitrogen are found in the mid-distillates, both neutral and basic nitrogen compounds have been isolated and identified in fractions boiling below 345°C (650°F). Pyrrole and indole derivatives account for the about two-thirds of the nitrogen, while the remainder is found in the basic pyridine and quinoline compounds. Most of these compounds carry alkyl chains.

In the vacuum gas oil range, the nitrogen-containing compounds include higher molecular weight pyridine derivatives, quinoline derivatives, benzoquinoline derivatives, amides, indole derivatives, and carbazole derivatives, and molecules with two nitrogen atoms (diaza compounds) with three and four aromatic rings are especially prevalent. Typically, about one-third of the compounds are basic (pyridine and pyridine benzologs), while the remainder are present as neutral species (amides and carbazole derivatives). Although benzo- and di-benzoquinoline derivatives found in petroleum are rich in sterically hindered structures, hindered and unhindered structures have also been found.

Porphyrins (nitrogen–metal complexes) are also constituents of petroleum and usually occur in the nonbasic portion of the nitrogen-containing concentrate (Reynolds, 1998). Pyrrole, the chief constituent of the porphyrin molecule, is marked by high stability due to its aromatic character. As a result of aromatization arising from the conjugated bond system, pyrrole is not strongly basic even though it is a secondary amine. Pyrrole, like other heterocyclic molecules, differs from homocyclic aromatic compounds in that it is quite reactive at the position α to the nitrogen atom and tends to form dimers, trimers, and higher condensation products in which the fundamental pyrrole structure is preserved to a marked degree.

The simplest porphyrin is porphine and consists of four pyrrole molecules joined by methine (–CH=) bridges (Speight, 2014). The methine bridges (=CH–) establish conjugated linkages between the component pyrrole nuclei, forming a more extended resonance system. Although the resulting structure retains much of the inherent character of the pyrrole components, the larger conjugated system gives increased aromatic character to the porphine molecule. The reactivity of the pyrrole nuclei in porphine is greatly reduced, partially because of the increased aromatic character but primarily because the reactive α positions are occupied. Two closely related classes of porphyrin-type compounds are the dihydroporphyrin derivatives and the tetrahydroporphyrin derivatives. The dihydroporphyrin derivatives, commonly known as *chlorins*, include two additional hydrogen atoms on one of the pyrrole rings and the tetrahydroporphyrin derivatives carry four extra hydrogen atoms.

The most common metal complexes of porphyrins are those formed by replacement of the two nitrogen-bound hydrogen atoms with a relatively small cation. Indeed, metal complexes of porphyrins are widespread in petroleum. Their concentrations are quite low, and in some oils porphyrins do not appear to be present (Reynolds, 1998).

The presence of vanadium and nickel in crude oils, especially as metal porphyrin complexes, has focused much attention in the petroleum refining industry on the occurrence of these metals in feedstocks (Reynolds, 1998). Only a part of the total nickel and vanadium in crude oil is recognized to occur in porphyrin structure. In general, it is assumed that about 10% w/w of the total metal in a crude oil is accommodated as porphyrin complexes, although as much as 40% of the vanadium and nickel may be present as metal porphyrin complexes in petroleum.

The levels of nitrogen plus oxygen may begin to approach the concentration of sulfur in certain residua. These two elements consistently concentrate in the most polar fractions of heavy feedstocks to the extent that every molecule contains more than two heteroatoms. At this point, precise structural identification cannot be accomplished and other methods of structural evaluation may be attempted (Speight, 2014).

7.2.4 OXYGEN COMPOUNDS

Oxygen in organic compounds can occur in a variety of forms (Speight, 2014), and it is not surprising that the more common of the oxygen-containing compounds occur in petroleum. The total oxygen content of petroleum is usually <2% w/w, although larger amounts have been reported; however, in cases where the oxygen content is phenomenally high, it may be that the oil has suffered prolonged exposure to the atmosphere either during or after production. However, the oxygen content of petroleum does increase with the boiling point of the fractions examined; in fact, the nonvolatile residua may have oxygen contents up to 8% w/w. Although these high molecular weight compounds contain most of the oxygen in petroleum, little is known concerning their structure, but those of lower molecular weight have been investigated with considerably more success and have been shown to contain carboxylic acids ($R\text{-}CO_2H$) and phenols (Ar-OH, where Ar is an aromatic moiety).

The presence of acid substances in petroleum first appears to have been reported in 1874, and it was established 9 years later that these substances contained carboxyl groups and were carboxylic acids. These were termed *naphthenic acids*. Although alicyclic (naphthenic) acids appear to be the more prevalent, it is now well known that aliphatic acids are also present. In addition to the carboxylic acids, alkaline extracts from petroleum contain phenols.

It has generally been concluded that the carboxylic acids in petroleum with less than eight carbon atoms per molecule are almost entirely aliphatic in nature; monocyclic acids begin at C_6 and predominate above C_{14}. This indicates that the structures of the carboxylic acids correspond with those of the hydrocarbons with which they are associated in the crude oil. Thus, in the range where paraffins are the prevailing type of hydrocarbon, the aliphatic acids may be expected to predominate; similarly, in the ranges where the monocycloparaffin derivatives and dicycloparaffin derivatives prevail, one may expect to find principally monocyclic and dicyclic acids, respectively.

In addition to the carboxylic acids and phenolic compounds, the presence of ketones, esters, ethers, and anhydrides has been claimed for a variety of crude oils. However, the precise identification of these compounds is difficult, as most of them occur in the higher molecular weight nonvolatile residua. They are claimed to be products of the air blowing of the residua, and their existence in virgin petroleum may yet need to be substantiated.

Oxygen levels in the vacuum gas oil parallel the nitrogen content. The most commonly identified oxygen compounds are the carboxylic acids and phenols, collectively called naphthenic acids. Among the different structures, a number of specific steroid carboxylic acids have been identified.

Although comparisons are frequently made between the sulfur and nitrogen contents and physical properties such as the API gravity, it is not the same with the oxygen content of crude oil. It is possible to postulate, and show, that such relationships exist. However, the ease with which some of the crude oil constituents can react with oxygen (aerial or dissolved) to incorporate oxygen functions into their molecular structure often renders the exercise somewhat futile if meaningful deductions are to be made.

7.2.5 METALLIC COMPOUNDS

The occurrence of metallic constituents in crude oil is of considerably greater interest to the petroleum industry than might be expected from the very small amounts present. Even minute amounts of iron, copper, and particularly nickel and vanadium in the charging stocks for catalytic cracking affect the activity of the catalyst and result in increased gas and coke formation and reduced yields of naphtha. In high-temperature power generators, such as oil-fired gas turbines, the presence of metallic constituents, particularly vanadium in the fuel, may lead to ash deposits on the turbine rotors, thus reducing clearances and disturbing their balance. More particularly, damage by corrosion may be very severe. The ash resulting from the combustion of fuels containing sodium and especially vanadium reacts with refractory furnace linings to lower their fusion points, and thus cause their deterioration.

Thus, the ash residue left after burning of a crude oil is due to the presence of these metallic constituents, part of which occur as inorganic water-soluble salts (mainly chlorides and sulfates of sodium, potassium, magnesium, and calcium) and which occur in the water phase of crude oil emulsions. These are removed in the desalting operations, either by evaporation of the water and subsequent water washing, or by breaking the emulsion, thereby causing the original mineral content of the crude to be substantially reduced. Other metals are present in the form of oil-soluble organometallic compounds either as complexes, metallic soaps, or in the form of colloidal suspensions, and the total ash from desalted crude oils is of the order of 0.1–100 mg/L.

Two groups of elements appear in significant concentrations in the original crude oil, associated with well-defined types of compounds. Zinc, titanium, calcium, and magnesium appear in the form of organometallic soaps with surface-active properties, adsorbed in the water–oil interfaces, and act as emulsion stabilizers. However, vanadium, copper, nickel, and part of the iron found in crude oils seem to be in a different class and are present as oil-soluble compounds (Reynolds, 1998). These metals, capable of complexing with pyrrole pigment compounds derived from chlorophyll and hemoglobin, are almost certain to have been present in plant and animal source materials. It is easy to surmise that the metals in question are present in such form, ending in the ash content. Evidence for the presence of several other metals in oil-soluble form has been produced, and, thus, zinc, titanium, calcium, and magnesium compounds have been identified in addition to vanadium, nickel, iron, and copper (Speight, 2014). Examination of the analyses of a number of crude oils for iron, nickel, vanadium, and copper indicates a relatively high vanadium content that usually exceeds that of nickel, although the reverse can also occur.

Distillation concentrates the metallic constituents in the residua (Chapter 3)—some can appear in the higher-boiling distillates (due to, for example, entrainment) but the latter may, in part, be due to entrainment. Nevertheless, there is evidence that a portion of the metallic constituents may occur in the distillates by volatilization of the organometallic compounds present in the petroleum. In fact, as the percentage overhead obtained by vacuum distillation of reduced crude is increased, the amount of metallic constituents in the overhead oil is also increased. The majority of the vanadium, nickel, iron, and copper in residual stocks may be precipitated along with the asphaltene constituents by low-boiling alkane hydrocarbon solvents. Thus, removal of the asphaltene constituents with *n*-pentane reduces the vanadium content of the oil by up to 95% with substantial reductions in the amounts of iron and nickel.

7.3 PHYSICAL COMPOSITION

The term *physical composition* (or *bulk composition*) refers to the composition of crude oil as determined by various physical techniques. For example, the separation of petroleum using solvents and adsorbents into various bulk fractions determines the physical composition of crude oil (Speight, 2014). However, in many instances, the physical composition may not be equivalent to the chemical composition. These methods of separation are not always related to chemical properties, and the terminology applied to the resulting fractions is often a terminology of convenience.

For a variety of reasons, it is often necessary to define a feedstock in terms of its physical composition. Furthermore, the physical composition of heavy feedstocks vary markedly with the method of isolation or separation, thereby leading to further complications in the choice of suitable processing schemes for these feedstocks (Speight and Ozum, 2002; Parkash, 2003; Hsu and Robinson, 2006; Gary et al., 2007; Speight, 2014). However, in the simplest sense, petroleum and heavy feedstocks can be considered to be composites of four major fractions—saturates, aromatics, resin, and asphaltene fractions (Figure 4.3) and, in the current context, much of the focus has been on the resin and asphaltene fractions because of the high sulfur content and high coke-forming propensity of the constituents (Speight, 1994, 2014). The nomenclature of these fractions lies within the historical development of petroleum science, and in that the fraction names are operational and are released more to the general characteristics than to the identification of specific compound types. Nevertheless, once a convenient fractionation technique has been established, it is possible to compare a variety of different feedstocks and the relevant choices of refinery options (Speight and Ozum, 2002; Parkash, 2003; Hsu and Robinson, 2006; Gary et al., 2007; Speight, 2014).

It is noteworthy that, throughout the history of studies related to petroleum composition, there has been considerable attention paid to the asphaltic constituents (i.e., the asphaltene and resin constituents). This is due, in no small part, to the tendency of the asphaltene constituents to be responsible for high yields of thermal coke and also for shortened catalyst lifetimes in refinery operations (Speight, 2014). In fact, it is the unknown character of the asphaltene constituents that has also been responsible for drawing the attention of investigators for the last five decades. In addition, residua contain the majority of all of the potential coke-forming constituents and catalyst poisons that were originally in the feedstock. The distillation process is essentially a concentration process, and most of the coke formers and catalyst–poisons are nonvolatile. Furthermore, the asphaltene fraction contains most of the coke-forming constituents and catalyst poisons that are originally present in a heavy feedstock.

There are also two other operational definitions that should be noted at this point, and these are the terms *carbenes* and *carboids*. Both such fractions are, by definition, insoluble in benzene (or toluene); however, the carbenes are soluble in carbon disulfide, whereas the carboids are insoluble in carbon disulfide. Only traces of these materials occur in natural petroleum and heavy feedstocks (with the exception of cracked residua), and the quantities are not sufficient to alter the character of the asphaltene constituents if the benzene treatment of the feedstock is omitted. On the other hand, feedstocks that have received some thermal treatment (such as visbroken feedstocks) may have considerable quantities of these materials present as they are also considered to be precursors to coke (Speight, 2014).

Heavy feedstock, especially vacuum residua (*vacuum bottoms*), are the most complex feedstocks in the refinery. Few molecules are free of heteroatoms, and the molecular weight of the constituents extends from 400 to >2000 and, at the upper end of this molecular weight range, characterization of individual species is virtually impossible. Separations by group type become blurred by the shear frequency of substitution and by the occurrence of multiple functionality in single molecules. However, like conventional petroleum, heavy feedstocks can also be separated into a variety of fractions using a myriad of different techniques that have been used since the beginning of petroleum science (Speight, 2014). However, the evolution of these techniques has been accompanied by subtle interlaboratory (and even intralaboratory) variations to an extent that many of the nuclear fractionation procedures appear to bear very little relationship to one another.

In the present context, the deposition of coke on a desulfurization catalyst will seriously affect catalyst activity with a marked decrease in the rate of desulfurization (Speight, 2000; Ancheyta and Speight, 2007). In fact, it has been noted that even with a deasphalted feedstock, i.e., a heavy feedstock from which the asphaltene constituents have previously been removed, the accumulation of carbonaceous deposits on the catalyst is still substantial. It has been suggested that this deposition of carbonaceous material is due to the condensation reactions that are an integral part of any thermal (even hydrocracking) process in which heavy feedstocks are involved.

It appears that the high molecular weight species originally present in the feedstock (or formed during the process) are not sufficiently mobile (or are too strongly adsorbed by the catalyst) to be saturated by the hydrogenation components and, hence, continue to condense and eventually degrade to coke. These deposits deactivate the catalyst sites and eventually interfere with the hydrodesulfurization process. Thus, the deposition of coke and, hence, the rate of catalyst deactivation, is subject to variations in the asphaltene content and the resin content of the feedstock, as well as the adsorptive properties of the catalyst for the heavier molecules.

Furthermore, the organometallic compounds (of which nickel and vanadium are the principal constituents) that are present to varying degrees in all residua and in the majority of heavy feedstocks cause catalyst deterioration—in fact, when refining heavy feedstock using catalytic processes, it is usual if catalyst deterioration does not occur. Deposition of these metals in any form onto the catalyst leads to catalyst deactivation; however, the exact mechanism of deactivation is still subject to speculation. Nickel tends to be deposited throughout the catalyst, whereas vanadium is more usually concentrated in the outer layers of the catalyst. In either case, catalyst deactivation is certain whether it be by physical blockage of the pores or destruction of reactive sites.

7.3.1 Asphaltene Separation

The suggestion arose in 1914 that the systematic separation of petroleum is effected by treatment with solvents. If chosen carefully, solvents effect a separation between the constituents of residua, bituminous materials, and virgin petroleum according to differences in molecular weight and aromatic character. The nature and the quantity of the components separated depend on the conditions of the experiment, namely, the degree of dilution temperature and the nature of the solvent.

By definition, the asphaltene fraction is that portion of the feedstock that is precipitated when a large excess (40 volumes) of a low-boiling liquid hydrocarbon (e.g., *n*-pentane or *n*-heptane) is added to (1 volume) of the crude oil (Speight, 1994, 2014). *n*-Heptane is the preferred hydrocarbon with *n*-pentane still being used (Speight et al., 1984; Speight, 1994), although hexane is used on occasion but not regularly in the petroleum field (Yan et al., 1997), which makes hexane-separated asphaltene fractions of little relevance for comparison with asphaltene fractions separated by the standard test methods.

Thus, although the use of both *n*-pentane and *n*-heptane has been widely advocated, and although *n*-heptane is becoming the deasphalting liquid of choice, this is by no means a hard-and-fast rule. Moreover, it must be recognized that large volumes of solvent may be required to effect a *qualitative* and *quantitative* reproducible separation. In addition, whether *n*-pentane or *n*-heptane is employed, the method effects a separation of the chemical components with the most complex structures from the mixture, and this fraction should be correctly identified as *n*-pentane asphaltene constituents or as *n*-heptane asphaltene constituents.

Method	Deasphalting Liquid	Volume, mL/g
ASTM D893	*n*-Pentane	10
ASTM D2007	*n*-Pentane	10
ASTM D3279	*n*-Heptane	100
ASTM D4124	*n*-Heptane	100

However, it must be recognized that some of these methods were developed for use with the more conventional feedstocks, and adjustments are necessary when applied to the separation of heavy feedstocks (Speight, 2001, 2014, 2015).

Although *n*-pentane and *n*-heptane are the solvents of choice in the laboratory, other solvents can be used (Speight, 1979) and cause the separation of asphaltene constituents as brown-to-black powdery materials. In the refinery, supercritical low molecular weight hydrocarbons (e.g., liquid propane, liquid butane, or mixtures of both) are the solvents of choice, and the product is a semi-solid (tacky) to solid asphalt. The amount of asphalt that settles out of the paraffin/residuum mixture depends on the size of the paraffin, the temperature, and the paraffin-to-feedstock ratio (Mitchell and Speight, 1973a; Corbett and Petrossi, 1978; Speight et al., 1984; Speight, 2014). At constant temperature, the quantity of precipitate first increases with increasing ratio of solvent to feedstock, and then reaches a maximum. In fact, there are indications that when the proportion of solvent in the mix is <35%, little or no asphaltene constituents are precipitated. When pentane and the lower molecular weight hydrocarbon solvents are used in large excess, the quantity of precipitate and the composition of the precipitate changes with increasing temperature (Mitchell and Speight, 1973a,b; Andersen, 1994; Speight, 2014).

Contact time between the hydrocarbon and the feedstock also plays an important role in asphaltene separation. Yields of the asphaltene fraction reach a maximum after approximately 8 h, which may be ascribed to the time required for the asphaltene particles to agglomerate into particles of a filterable size as well as the diffusion-controlled nature of the process. Heavier feedstocks also need time for the hydrocarbon to penetrate their mass.

For example, if the precipitation method (deasphalting) involves the use of a solvent and a heavy feedstock, the process is essentially leaching of soluble constituents from the insoluble residue; this process may be referred to as *extraction*. However, under the prevailing conditions now in laboratory use, the term *precipitation* is perhaps more correct and descriptive of the method. Variation of solvent type also causes significant changes in asphaltene yield. Thus, the contact time between the feedstock and the hydrocarbon liquid can have an important influence on the yield and character of the asphaltene fraction.

7.3.2 Fractionation

After removal of the asphaltene fraction, further fractionation of petroleum is also possible by variation of the hydrocarbon solvent. For example, liquefied gases, such as propane and butane, precipitate as much as 50% by weight of the residuum or bitumen. The precipitate is a black, tacky, semisolid material, in contrast to the pentane-precipitated asphaltene constituents, which are usually brown, amorphous solids. Treatment of the propane precipitate with pentane then yields the insoluble brown, amorphous asphaltene constituents and soluble, near-black, semisolid resin constituents, which are, as near as can be determined, equivalent to the resin constituents isolated by adsorption techniques (Speight, 2014, 2015).

Separation by adsorption chromatography essentially commences with the preparation of a porous bed of finely divided solid—the adsorbent. The adsorbent is usually contained in an open tube (column chromatography); the sample is introduced at one end of the adsorbent bed and induced to flow through the bed by means of a suitable solvent. As the sample moves through the bed, the various components are held (adsorbed) to a greater or lesser extent depending on the chemical nature of the component. Thus, those molecules that are strongly adsorbed spend considerable time on the adsorbent surface rather than in the moving (solvent) phase; however, components that are slightly adsorbed move through the bed comparatively rapidly. There are many procedures that have received attention over the years (Speight, 2014); however, for the purposes of this text, this section will focus on the standard (ASTM) methods of fractionation.

There are three ASTM methods that provide for the separation of a feedstock into four or five constituent fractions—it is interesting to note that as the methods have evolved, there has been a change from the use of pentane (ASTM D2007) to heptane (ASTM D4124) to separate the asphaltene fraction, while the use of hexane has largely been ignored in petroleum science in favor of standard test methods that employ *n*-pentane or *n*-heptane, with the latter liquid being the most favored separation solvent (Speight, 2014). Furthermore, the use of heptane as the precipitation (or separation) medium is in keeping with the production of a more consistent fraction that represents these higher molecular weight, more complex constituents of petroleum (Speight et al., 1984; Speight, 2014). Two methods (ASTM D2007 and D4124) use adsorbents to fractionate the deasphalted oil.

Obviously, there are precautions that must be taken when attempting to separate heavy feedstocks or polar feedstocks into constituent fractions. The disadvantages in using ill-defined adsorbents are that adsorbent performance differs with the same feed and, in certain instances, may even cause chemical and physical modification of the feed constituents. The use of a chemical reactant like sulfuric acid should only be advocated with caution since feeds react differently and may even cause irreversible chemical changes and/or emulsion formation. These advantages may be of little consequence when it is not, for various reasons, the intention to recover the various product fractions *in toto* or in the original state; however, in terms of the compositional evaluation of different feedstocks, the disadvantages are very real.

In general terms, group-type analysis of petroleum is often identified by the acronyms for the names: PONA (paraffins, olefins, naphthenes, and aromatics), PIONA (paraffins, isoparaffins, olefins, naphthenes, and aromatics), PNA (paraffins, naphthenes, and aromatics), PINA (paraffins, isoparaffins, naphthenes, and aromatics), or SARA (saturates, aromatics, resin constituents, and asphaltene constituents). However, it must be recognized that the fractions produced by the use of different adsorbents will differ in content and will also be different from fractions produced by solvent separation techniques.

The variety of fractions isolated by these methods and the potential for the differences in composition of the fractions makes it even more essential that the method is described accurately and that it is reproducible not only in any one laboratory but also between various laboratories.

7.4 FEEDSTOCK TYPES

To understand the composition of heavy feedstocks, it is necessary to present a very brief description of the constituents of the lower-boiling fractions of petroleum. Acceptance that petroleum is a continuum of molecular types that continues from the low-boiling fractions to the nonvolatile fractions (Speight, 2014) is an aid to understanding the chemical nature of the heavy feedstocks.

The chemical composition of heavy feedstocks is, in spite of the large volume of work performed in this area, largely speculative (Altgelt and Boduszynski, 1994; Speight, 2014). Indeed, the simpler crude oils are extremely complex mixtures of organic compounds. In fact, the composition of petroleum can vary with the location and age of the field, in addition to any variations that occur with the depth of the individual well. Two adjacent wells are more than likely to produce petroleum with very different characteristics.

On a molecular basis, petroleum contains hydrocarbons as well as the organic compounds of nitrogen, oxygen, and sulfur; metallic constituents may also be present, but only to a minor extent. While the hydrocarbon content of petroleum may be as high as 97% (as, for example, in the lighter paraffinic crude oils), it is nevertheless the nonhydrocarbon (i.e., nitrogen, oxygen, and sulfur) constituents that play a large part in determining the nature and, hence, the processability of the feedstock.

The elemental composition (ultimate analysis) of petroleum, regardless of the origin of the particular petroleum, varies only slightly over very narrow limits (Chapter 1):

Carbon, % w/w	83.0–87.0
Hydrogen, % w/w	10.0–14.0
Sulfur, % w/w	0.05–6.0
Nitrogen, % w/w	0.1–2.0
Oxygen, % w/w	0.05–1.5
Metals, ppm	10–1000

The high proportions of carbon and hydrogen in petroleum indicate that hydrocarbons are the major constituents and may actually account for >75% of the constituents of many crude oils. The inclusion of organic compounds of nitrogen, oxygen, and sulfur only serves to present crude oils as even more complex mixtures, and the appearance of appreciable amounts of these nonhydrocarbon compounds causes some concern in the refining of crude oils. Even though the concentration of nonhydrocarbon constituents (i.e., those organic compounds containing one or more nitrogen, oxygen, or sulfur atom) in certain fractions may be quite small, they do, however, tend to concentrate in the higher-boiling fractions of petroleum. Indeed, their influence on the processability of the petroleum is important irrespective of their molecular size and the fraction in which they occur.

Thus, the presence of traces of nonhydrocarbon compounds can impart objectionable characteristics to finished products, leading to discoloration and/or lack of stability during storage. On the other hand, catalyst poisoning and corrosion will be the most noticeable effects during refining sequences when these compounds are present. It is, therefore, not surprising that considerable attention must be given to the nonhydrocarbon constituents of petroleum, as the trend in the refining industry, of late, has been to process the heavier crude oils and residua that contain substantial proportions of these nonhydrocarbon materials.

7.4.1 Low-Boiling Distillates

The hydrodesulfurization of light (low-boiling) distillate (naphtha) is one of the more common catalytic hydrodesulfurization processes since it is usually used as a pretreatment of such feedstocks before catalytic reforming (Table 7.3). Hydrodesulfurization of such feedstocks is required because sulfur compounds poison the precious metal catalysts used in reforming, and desulfurization can be achieved under relatively mild conditions and is near quantitative. If the feedstock arises from a cracking operation, hydrodesulfurization will be accompanied by some degree of saturation resulting in increased hydrogen consumption.

TABLE 7.3

Hydrodesulfurization of Straight-Run Naphtha

Feedstock	Boiling Range		Sulfur, wt.%	Desulfurization,[a] %
	°C	°F		
Visbreaker naphtha	65–220	150–430	1.00	90
Visbreaker–coker naphtha	65–220	150–430	1.03	85
Straight-run naphtha	85–170	185–340	0.04	99
Catalytic naphtha (light)	95–175	200–350	0.18	89
Catalytic naphtha (heavy)	120–225	250–440	0.24	71
Thermal naphtha (heavy)	150–230	300–450	0.28	57

Source: Speight, J.G. 2014. *The Chemistry and Technology of Petroleum.* 5th Edition. CRC Press, Taylor & Francis Publishers, Boca Raton, FL. Table 21.1, p. 569.

[a] Process conditions: Co–Mo on alumina, 260–370°C/500–700°F, 200–500 psi hydrogen.

The hydrodesulfurization of low-boiling (naphtha) feedstocks is usually a gas-phase reaction and may employ the catalyst in fixed beds, and (with all of the reactants in the gaseous phase) only minimal diffusion problems are encountered within the catalyst pore system. It is, however, important that the feedstock be completely volatile before entering the reactor, as there may be pressure variations (leading to less satisfactory results) if some of the feedstock enters the reactor in the liquid phase and is vaporized within the reactor.

In applications of this type, the sulfur content of the feedstock may vary from 100 ppm to 1%, and the necessary degree of desulfurization to be effected by the treatment may vary from as little as 50% to >99%. If the sulfur content of the feedstock is particularly low, it will be necessary to presulfide the catalyst. For example, if the feedstock only has 100–200 ppm sulfur, several days may be required to sulfide the catalyst as an integral part of the desulfurization process even with complete reaction of all of the feedstock sulfur to, say, cobalt and molybdenum (catalyst) sulfides. In such a case, presulfiding can be conveniently achieved by the addition of sulfur compounds to the feedstock or by addition of hydrogen sulfide to the hydrogen.

Generally, hydrodesulfurization of naphtha feedstocks to produce catalytic reforming feedstocks is carried to the point where the desulfurized feedstock contains <20 ppm sulfur. The net hydrogen produced by the reforming operation may actually be sufficient to provide the hydrogen consumed in the desulfurization process.

The hydrodesulfurization of middle distillates is also an efficient process, and applications include predominantly the desulfurization of kerosene, diesel fuel, jet fuel, and heating oils, which boil over the general range 250–400°C (480–750°F). However, with this type of feedstock, hydrogenation of the higher-boiling catalytic cracking feedstocks has become increasingly important where hydrodesulfurization is accomplished alongside the saturation of condensed-ring aromatic compounds as an aid to subsequent processing.

Under the relatively mild processing conditions used for the hydrodesulfurization of these particular feedstocks, it is difficult to achieve complete vaporization of the feed. Process conditions may dictate that only part of the feedstock is actually in the vapor phase and that sufficient liquid phase is maintained in the catalyst bed to carry the larger molecular constituents of the feedstock through the bed. If the amount of liquid phase is insufficient for this purpose, molecular stagnation (leading to carbon deposition on the catalyst) will occur.

Hydrodesulfurization of middle distillates causes a more marked change in the specific gravity of the feedstock, and the amount of low-boiling material is much more significant when compared with the naphtha-type feedstock. In addition, the somewhat more severe reaction conditions (leading to see degree of hydrocracking) also leads to an overall increase in hydrogen consumption when middle distillates are employed as feedstocks in place of the naphtha.

7.4.2 High-Boiling Distillates

The high-boiling distillates, such as the atmospheric and vacuum gas oils, are not usually produced as a refinery product but merely serve as feedstocks to other processes for conversion to lower-boiling materials. For example, gas oils can be desulfurized to remove >80% of the sulfur originally in the gas oil with some conversion of the gas oil to lower-boiling materials (Speight, 2000, 2014). The treated gas oil (which has a reduced carbon residue as well as lower sulfur and nitrogen contents relative to the untreated material) can then be converted to lower-boiling products in, say, a catalytic cracker where an improved catalyst life and volumetric yield may be noted.

The conditions used for the hydrodesulfurization of a gas oil may be somewhat more severe than the conditions employed for the hydrodesulfurization of middle distillates with, of course, the feedstock in the liquid phase.

In summary, the hydrodesulfurization of the low-, middle-, and high-boiling distillates can be achieved quite conveniently using a variety of processes. One major advantage of this type of feedstock is that the catalyst does not become poisoned by metal contaminants in the feedstock since only negligible amounts of these contaminants will be present. Thus, the catalyst may be regenerated several times, and on-stream times between catalyst regeneration (while varying with the process conditions and application) may be of the order of 3–4 years (Table 7.2).

7.4.3 Heavy Feedstocks

Heavy feedstocks are those feedstocks that were considered to be less desirable for conversion to liquid fuels; however, the depletion of the reserves of conventional petroleum has brought about an interest in the conversion of the heavier feedstocks. It has also been realized that any degree of self-sufficiency would necessitate the use of heavy feedstocks in refineries. The changing nature of refinery feedstocks (Chapter 1) has required that the industry evolve to accommodate the new refining chemistry. However, it has been asserted that more needs to be done in correlating analytical data obtained for heavy feedstocks with processability (Speight, 1984, 2014). In particular, measurements of properties are needed that reflect the uniqueness of individual heavy oils and processes such as hydrodesulfurization and hydrodemetallization.

Modern refineries use a combination of heat, catalyst, and hydrogen to convert the petroleum constituents into these products. Conversion processes include coking, catalytic cracking, and hydrocracking to convert the higher molecular weight constituents into lower molecular weight products, and reduce the heteroatom content to create environmentally acceptable products (Speight and Ozum, 2002; Parkash, 2003; Hsu and Robinson, 2006; Gary et al., 2007; Speight, 2013, 2014).

An understanding of the chemical types (or composition) of any feedstock can lead to an understanding of the chemical aspects of processing the feedstock. Processability is not only a matter of knowing the elemental composition of a feedstock; it is also a matter of understanding the bulk properties as they relate to the chemical or physical composition of the material. For example, it is difficult to understand *a priori* the process chemistry of various feedstocks from the elemental composition alone. It might be surmised that the major difference between a heavy feedstock and a more conventional material is the H/C atomic ratio alone. This property indicates that the heavier crude oils (having a lower H/C atomic ratio and being more atomic in character) would require more hydrogen for upgrading to liquid fuels. This is, indeed, true; however, much more information is necessary to understand the processability of the feedstock.

The chemical composition of the heavier feedstocks is complex. Physical methods of fractionation usually indicate high proportions of asphaltene constituents and resin constituents even in amounts up to 50% (or higher) of the residuum. In addition, the presence of ash-forming metallic constituents, including such organometallic compounds as those of vanadium and nickel, is also a distinguishing feature of residua and the heavier oils. Furthermore, the deeper the cut into the crude oil, the greater is the concentration of sulfur and metals in the residuum and the greater the deterioration in physical properties (Figure 4.2) (Speight, 2014).

Throughout the history of studies related to petroleum composition, there has been considerable attention paid to the asphaltic constituents (i.e., the asphaltene constituents and resin constituents). This is due, in no small part, to the tendency of the asphaltene constituents to be responsible for high yields of thermal coke and for shortened catalyst lifetimes in refinery operations (Speight, 1994, 2013, 2014; Speight and Ozum, 2002; Parkash, 2003; Hsu and Robinson, 2006; Gary et al., 2007). In fact, it is the unknown character of the asphaltene constituents that has also been responsible for drawing the attention of investigators for the last five decades (Speight, 1994, 2014). Residua, because of the concentrating nature of the distillation process, contain the majority of the potential coke-forming constituents and catalyst poisons that were originally in the feedstock. The majority of the coke-forming constituents and catalyst poisons are nonvolatile. Thus, the asphaltene fraction contains most of the coke-forming constituents and catalyst poisons that are originally present in heavy feedstocks. As a result, there have been several attempts to focus attention on the asphaltene constituents during hydroprocessing studies. The focus has been on the macromolecular changes that occur by investigation of the changes to the generic fractions, i.e., the asphaltene constituents, the resin constituents, and other fractions that make up such a feedstock (Chapter 3).

When catalytic processes are employed, complex molecules (such as those that may be found in the original asphaltene fraction), or those formed during the process, are not sufficiently mobile. They are also too strongly adsorbed by the catalyst to be saturated by the hydrogenation component and, hence, continue to react and eventually degrade to coke. These deposits deactivate the catalyst sites and eventually interfere with the hydroprocess.

The asphaltene (or asphaltic) fraction is particularly important because as the proportion of this fraction in a crude oil increases, there is concomitant increase in thermal coke and a decrease in the relative rate of hydrodesulfurization (Speight, 2000, 2014; Ancheyta and Speight, 2007). Thus, it is no surprise that as refineries have to accept more of the heavier crude oils for processing, there are serious attempts to define the chemical nature of petroleum asphaltene constituents. Such knowledge is seen to be not only an aid to understanding the routes by which asphaltene constituents form thermal coke, but also assisting in the development of suitable processing options to mitigate coke formation.

The reactivity of the feedstock to coking is related to the number of substituents attached to the aromatic hydrocarbons. Feedstocks with moderate aromaticity and having a high percentage of substituents on the aromatic nucleus are much more reactive than feedstocks with high aromaticity and a low percentage of substituents. A balance between these two structures is what is required for optimum carbonization (Rodriguez et al., 1991; Speight, 1998, 2014, 2015). Usually, aromatic feeds produce more coke than do aliphatic feeds. For example, the vacuum gas oil from western shale oil having an API gravity of 17.0 and an H/C atomic ratio of 1.76 produce 19.5 wt.% coke (Thomas and Hunter, 1989). The vacuum gas oil from a coal-derived liquid having an API gravity of −3.8 and an H/C atomic ratio of 1.12 produce 29.2 wt.% coke (Thomas and Hunter, 1989). The Conradson carbon residues for the original whole oils for these studies are 1.2 and 4.3 wt.%, respectively.

Thus, the availability of detailed structural information regarding the feedstock and the produced coke is necessary for the optimization of various thermal upgrading processes.

It is apparent that the conversion of heavy feedstocks requires new lines of thought to develop suitable processing scenarios. Indeed, the use of thermal and hydrothermal processes that were inherent in the refineries designed to process lighter feedstocks have been a particular cause for concern and has brought about the evolution of processing schemes that will accommodate the heavier feedstocks. It follows that these feedstocks require more severe hydrodesulfurization conditions to produce low-sulfur liquid product streams that can then, as is often now desired, be employed as feedstocks for other refining operations. Hydrodesulfurization of the heavier feedstocks is normally accompanied by a high degree of hydrocracking, and thus the process conditions required to achieve 70–90% desulfurization will also effect substantial conversion of the feedstock to lower-boiling products.

The extent of the hydrocracking is, like the hydrodesulfurization reaction, dependent on the temperature, and both reaction rates increase with increase in temperature. However, the rate of hydrocracking tends to show more marked increases with temperature than the rate of hydrodesulfurization. The overall effect of the increase in the rate of the hydrocracking reaction is to increase the rate of carbon deposition on the catalyst. This adversely affects the rate of hydrodesulfurization; hydrocracking reactions are not usually affected by carbon deposition on the catalyst since they are more dependent on the noncatalytic scission of covalent bonds brought about by the applied thermal energy.

In contrast to the lighter feedstocks that may be subjected to the hydrodesulfurization operation, the heavy feedstocks may require pretreatment. For example, the process catalysts are usually susceptible to poisoning by nitrogen (and oxygen) compounds and metallic salts (in addition to the various sulfur compound types) that tend to be concentrated in residua (Chapter 3) or exist as an integral part of the heavy feedstock matrix. Thus, any processing sequence devised to hydrodesulfurize heavy feedstocks must be capable of accommodating the constituents that adversely affect the ability of the catalyst to function in the most efficient manner possible.

The conditions employed for the hydrodesulfurization of the heavier feedstocks may be similar to those applied to the hydrodesulfurization of gas oil fractions but with the tendency to increased pressures. However, carbon deposition on, and metal contamination of, the catalyst is much greater when heavy feedstocks are employed and, unless a low level of desulfurization is acceptable, frequent catalyst regeneration is necessary.

7.5 FEEDSTOCK COMPOSITION

There is one aspect of feedstock properties that has not yet been discussed fully, and that is feedstock composition. This particular aspect of the nature of the feedstock is, in fact, related to the previous section where the influence of various feedstock types on the hydrodesulfurization process was noted, but it is especially relevant when heavy feedstocks from various sources are to be desulfurized.

The composition of the various feedstocks may, at first sight, seem to be of minor importance when the problem of hydrodesulfurization of heavy feedstocks comes under consideration. However, consideration of the variation in process conditions that are required for different feedstocks (where the feedstocks are relatively well-defined boiling fractions of petroleum) presents some indication of the problems that may be encountered where the feedstocks are less well defined. Molecular composition is as important as molecular weight (or boiling range). Such is the nature of the problem when dealing with various heavy feedstocks, which are (to say the least) unknown in terms of their chemical composition. In fact, the complexity of these particular materials (Chapter 3) has allowed little more than speculation about the molecular structure of the constituents.

One of the major drawbacks to defining the influence of the feedstock on the process is that the research with respect to feedstocks has been fragmented. In every case, a conventional catalyst has been used, and the results obtained are only valid for the operating conditions, reactor system, and catalyst used. More rigorous correlation is required, and there is a need to determine the optimum temperature for each type of sulfur compound. To obtain a useful model, the intrinsic kinetics of the reaction for a given catalyst should also be known. In addition, other factors that influence the desulfurization process, such as (1) catalyst inhibition or deactivation by hydrogen sulfide, (2) effect of nitrogen compounds, and (3) the effects of various solvents, should also be included to obtain a comprehensive model that is independent of the feedstock. The efficacy of other catalytic systems on various feedstocks also needs to be evaluated.

For example, the cracking and desulfurization activity of constituents containing the polynuclear aromatic nucleus may be sluggish and more refractory to desulfurization (Gates et al., 1979; Ma et al., 1994). Moreover, alkyldibenzothiophenes (in particular, 4,9-dimethyldibenzothiophene) are very resistant to desulfurization. This unreactivity, or refractory behavior, may inhibit the reactivity of paraffinic and naphthenic constituents. In addition, organometallic compounds produce metallic products that influence the reactivity of the catalyst by deposition on its surface. The transfer of metal from the feedstock to the catalyst constitutes an irreversible poisoning of the catalyst. Such observations have serious consequences for processing heavy feedstocks where aromatic and polynuclear aromatic systems as well as sterically hindered benzothiophenes can be major constituents in some of the fractions. However, this is not the whole answer to hydrodesulfurization. The inter- and intra-molecular interactions of the feedstock constituents are very rarely taken into consideration.

Nevertheless, there have been some successful attempts to define the behavior of heavy feedstocks during the hydrodesulfurization process in terms of physical composition, which has also led to the development of process modifications to suit various heavy feedstocks. This line of investigation arose because of the tendency, over the years, to classify all heavy feedstocks as useless (really, too expensive) for further processing. The only exception is the production of asphalt from certain residua. It was only when these so-called useless materials entered the refinery operation as feedstocks for the production of additional liquid fuels that it was realized that each particular feedstock may have to be considered for processing on the basis of an examination of several fundamental properties.

7.5.1 Asphaltene and Resin Content

Heavy feedstocks, like conventional petroleum, can be fractionated by a variety of techniques (Speight, 2014, 2015) to provide broad general fractions termed asphaltene constituents, resin constituents, aromatics, and saturates (Figure 7.3). By convention, the asphaltene and resin fractions are often referred to as the asphaltic fraction because of their insolubility in liquid propane and subsequent separation from a liquid propane solution of residua as asphalt.

In addition, thermally cracked residua will also contain two other fractions: the carbenes and the carboids (Figure 7.3). These fractions are also defined by solubility and are precursors to coke (Speight, 2014, 2015). Both fractions are, by definition, insoluble in benzene (or toluene), but the

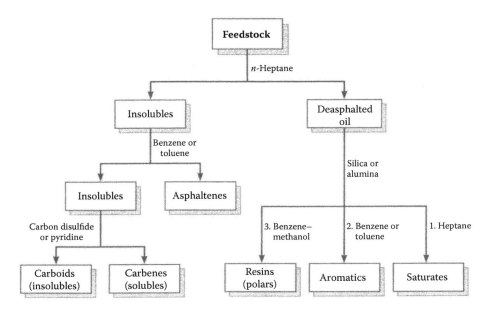

FIGURE 7.3 General scheme for feedstock fractionation. (From Speight, J.G. 2014. *The Chemistry and Technology of Petroleum*. 5th Edition. CRC Press, Taylor & Francis Publishers, Boca Raton, FL. Figure 9.1, p. 212.)

carbenes are soluble in carbon disulfide (or pyridine) whereas the carboids are insoluble in carbon disulfide (or pyridine). Only traces of these materials occur in conventional petroleum and heavy feedstocks, with the exception of cracked residua, which may contain substantial amounts of carbenes and carboids. Any such quantities are not sufficient to alter the character of the asphaltene constituents if the benzene treatment of the feedstock is omitted. On the other hand, feedstocks that have received some thermal treatment (such as visbroken feedstocks and cracked residua) may have considerable quantities of these materials present, as they are also considered to be precursors to coke.

Briefly, the asphaltene fraction of crude oil is that fraction that is precipitated by the addition of a large excess of a low-boiling liquid hydrocarbon (usually *n*-heptane) (Chapter 2). On the other hand, resin constituents are those materials that remain soluble in the pentane but will be adsorbed by a surface-active material such as fuller's earth, while the oils fraction is soluble in pentane but is not adsorbed from the pentane solution by any surface-active material. The asphaltic fraction is usually a combination of the asphaltene constituents and resin constituents and, in many instances, may constitute a large portion of a heavy feedstock.

The asphaltene and resin fractions contain the higher molecular weight materials and higher proportions of sulfur, nitrogen, and oxygen than the aromatics and saturate fractions. The H/C atomic ratios will vary from about 1.0 in the asphaltene constituents to about 1.3 in the resin constituents, indicating significant amounts of aromatic centers, while the more aliphatic (i.e., the asphaltene constituents plus the resin constituents) fraction will have an atomic H/C ratio of the order of 1.6. In any investigations on the asphaltic constituents of residua, attention has usually been devoted to the asphaltene fraction, with the resin fraction receiving considerably less attention. Nevertheless, the resin fraction of heavy feedstocks is just as important as the asphaltene fraction since the resin constituents have the potential to be converted into asphaltene constituents under remarkably mild conditions, either through the application of heat or through exposure to aerial oxygen.

Therefore, in any assessment of the influence of the asphaltene fraction on the processability of a residuum (or similar material), attention should also be paid to the proportion of the resin constituents. The resin constituents have the potential to act as a precursor to the asphaltene constituents and, eventually, to coke.

High amounts of asphaltene constituents and resin constituents require high hydrogen partial pressures and may actually limit the maximum level of hydrodesulfurization, or final traces of sulfur in the residuum may only be eliminated under extremely severe reaction conditions where hydrocracking is the predominant reaction in the process. High asphaltene and resin contents are also responsible for high viscosity (Speight, 2014), which may increase the resistance to mass transfer of the reactants to the catalyst surface and thus decrease the rate of sulfur removal. In such an instance, it may be deemed satisfactory by the refiner to simply blend the heavy feedstock with a lighter feedstock, thereby alleviating the problem for that particular process. Similarly, a high asphaltic content of a feedstock usually indicates high sulfur content (Speight, 2000, 2014) of that particular feedstock, which will require the higher partial pressure of hydrogen as well as the more severe reaction conditions indicated in the earlier part of this section. The sulfur content of a feedstock may also be expressed in terms of the asphaltene content alone (Speight, 2000, 2014), there seemingly being a relationship between the asphaltene constituents and the asphaltene sulfur.

It is generally true that the sulfur content of heavy feedstocks has a marked influence on the characteristics of the desulfurization process. Thus, for a series of residual feedstocks, the relative rate constants for the desulfurization process decrease with increasing sulfur content (Speight, 2000, 2014) under standard process conditions. With the general proviso that the majority of the sulfur in a feedstock is located in the asphaltene and resin fractions, it is not surprising that there have been many suggestions to improve the desulfurization process by prior removal of these extremely complex fractions. There are also other problems that arise by virtue of the presence of these fractions in a feedstock; for example, the asphaltic constituents of feedstocks also pose the problem of coke deposition. The coke-forming tendencies of most feedstocks can be directly estimated from (and are, in fact, related to) the carbon residue of that feedstock for that particular process, and the higher the asphaltic content of a feedstock, the higher the carbon residue (Speight, 2000, 2014). It has been postulated that the asphaltene fraction is the sole source of coke deposited during processing various feedstocks, but there is some evidence to indicate that not only the resin constituents but also some of the heavier constituents of the gas oil also contribute to coke formation. However, coke formation from the asphaltene constituents is presumed to be the more rapid reaction rate. Again, as with the problem of sulfur removal from the asphaltic constituents, some alleviation of the problem of carbon formation on the catalyst can be achieved by increasing the partial pressure of the hydrogen in the process (Beuther and Schmid, 1963).

One suggested method of facilitating hydrodesulfurization of heavy feedstocks is to remove completely, or partially, all of the asphaltic from the stock (Mitchell and Speight, 1973b). In the first instance, the asphaltic constituents can be completely, and conveniently, removed from a heavy stock by treatment with liquid propane under pressure whereupon the asphaltene constituents and the resin constituents are deposited as a heavy, thick, black asphalt leaving the lower molecular weight oil fraction in the liquid phase. Alternatively, use of other hydrocarbon solvents allows adjustment of the yield of precipitate in the form of partial removal of the asphaltene constituents and resin constituents, thereby retaining some, or all, of the resin constituents in the stock to serve as potential liquid-forming materials.

Thus, although the concept of deasphalting has been discussed elsewhere as a potential desulfurizing technique (Chapter 3), it has also been advocated as a method of feedstock pretreatment to remove the objectionable asphaltic constituents from feedstocks that are destined for catalytic hydrodesulfurization. For example, removal of asphaltic material (thereby removing macromolecular species containing high proportions of sulfur, nitrogen, and oxygen) may allow a several-fold increase in space velocity. It will be possible to maintain a comparable level of desulfurization and substantially improve the life span of the hydrodesulfurization catalyst, or allow substantial increases in the rate of hydrodesulfurization (Speight, 2000, 2014). However, in such an instance, the yield of naphtha may remain the same for the deasphalted material as for the whole residuum, but the overall quality of the residual fraction is considerably improved when the deasphalted material is used as the feedstock (Speight, 2000, 2014).

The asphaltene and resin constituents of petroleum bear a molecular relationship to each other (Speight, 1999). Moreover, it is because of this relationship that the resin constituents confer molecular and conformational stability on the asphaltene constituents in the oil. Heating destroys this stability by altering the stable conformations leading to destabilization of the structure of the petroleum.

In summary, the presence of large proportions of asphaltic materials in hydrodesulfurization feedstocks can adversely affect the process in many ways, not the least of which are reduced catalyst life and hydrogen requirements. It appears that there may be a case for deasphalting the feedstock before the hydrodesulfurization step; however, obviously, each feedstock should be evaluated on its own merits, and the process economics need careful examination before any decision is made on a particular refining sequence.

7.5.2 Metal Content

Heavy feedstocks contain impurities other than sulfur, nitrogen, and oxygen, and the most troublesome of these impurities are the organometallic compounds of nickel and vanadium. The metal content of oil can vary from several parts per million to >1000 per million (Speight, 2000, 2014), and there does seem to be a more than chance relationship between the metal content of a feedstock and its physical properties (Reynolds, 1998; Speight, 2014). In the hydrodesulfurization of the heavier feedstocks, the metals (nickel plus vanadium) are an important factor since large amounts (>150 ppm) will cause rapid deterioration of the catalyst. The free metals, or the sulfides, deposit on the surface of the catalyst and within the pores of the catalyst, thereby poisoning the catalyst by making the more active catalyst sites inaccessible to the feedstock and the hydrogen. This results in frequent replacement of an expensive process commodity unless there are adequate means by which the catalyst can be regenerated.

The problem of metal deposition on the hydrodesulfurization catalysts has generally been addressed using any one of three methods: (1) suppressing deposition of the metals on the catalyst, (2) development of a catalyst that will accept the metals and can tolerate high levels of metals without marked reduction in the hydrodesulfurization capabilities of the catalyst, and (3) removal of the metal contaminants before the hydrodesulfurization step. The first two methods involve a careful and deliberate choice of the process catalyst and operating conditions. However, these methods may only be viable for feedstocks with <150 ppm total metals since the decrease in the desulfurizing activity of the catalyst is directly proportional to the metal content of the feedstock. There are, however, catalysts that can tolerate substantial proportions of metals within their porous structure before the desulfurizing capability drops to an unsatisfactory level. Unfortunately, data on such catalysts are extremely limited because of their proprietary nature, and details are not always available; however, tolerance levels for metals that are equivalent to 15–65% by weight of the catalyst have been quoted.

The third method may be especially applicable to feedstocks with a high metal content and requires a separate demetallization step just before the hydrodesulfurization reactor. Such a step might involve passage of the feedstock through a demetallization chamber that contains a catalyst with a high selectivity for metals but whose activity for sulfur removal is low. Nevertheless, demetallization applied as a separate process can be used to generate low-metal feedstocks and will allow a more active, and stable, desulfurization system so that a high degree of desulfurization can be achieved on high-metal feedstocks with an acceptable duration of operation (Speight, 2000, 2014).

7.6   PRODUCT DISTRIBUTION

Hydrocracking is an extremely versatile process that can be utilized in many different ways. One of the advantages of hydrocracking is its ability to break down high-boiling aromatic feedstocks that are produced by catalytic cracking or coking. This is particularly desirable where maximum naphtha and minimum fuel oil must be made. For example, in the early days of the process, one particular type of feedstock was used to provide limited distribution of lower-boiling products. However, it must not be forgotten that product distribution and quality will vary considerably depending on the nature of the feedstock (Speight, 2000, 2014) as well as on the process. In modern refineries, hydrocracking is one of several process options that can be applied to the production of liquid fuels from the heavier feedstocks (Speight, 2014). A most important aspect of the modern refinery operation is the desired product slate that dictates the matching of a process with any particular feedstock to overcome differences in feedstock composition.

An example of how feedstock composition can influence the variation in product distribution and quality comes from application of the ABC (asphaltene bottom cracking) hydrocracking process to different feedstocks (Komatsu et al., 1986; Takeuchi et al., 1986). A further example of variations in product distributions from different feedstocks comes from the mild resid hydrocracking process. In addition, different processes will produce variations in the product slate from any one particular feedstock, and the feedstock recycle option adds another dimension to variations in product slate (Speight and Ozum, 2002; Parkash, 2003; Hsu and Robinson, 2006; Gary et al., 2007; Speight, 2014).

Hydrogen consumption is also a parameter that varies with feedstock composition and is dependent on factors other than the theoretical amount of hydrogen required to remove sulfur (Speight, 2000; Ancheyta and Speight, 2007). This indicates the need for a thorough understanding of the feedstock constituents if the process is to be employed to maximum efficiency. Hydrogen consumption increases with the extent of the conversion of the higher-boiling constituents (Speight, 2000) and is dependent on whether the feedstock is to be hydrogenated and/or hydrocracked (Speight,

2000). In addition, conversion is enhanced by the use of higher hydrogen pressures (Beuther and Schmid, 1963). One of the major advantages of hydrocracking is that it may be used to process the higher-boiling refractory feedstocks that may be produced by catalytic cracking or by any of the coking processes. However, it must be recognized that hydrocracking does involve the use, and cost, of hydrogen.

A very convenient means of understanding the influence of feedstock on the hydrocracking process is through a study of the hydrogen content (H/C atomic ratio), feedstock gravity, and molecular weight (carbon number) of the various feedstocks/products (Scott and Bridge, 1971; Speight, 2000, 2013, 2014; Speight and Ozum, 2002; Parkash, 2003; Hsu and Robinson, 2006; Ancheyta and Speight, 2007; Gary et al., 2007). Such data show the extent to which the carbon number must be reduced and/or the relative amount of hydrogen that must be added to generate the desired lower molecular weight, hydrogenated products. In addition, it is also possible to use data for hydrogen usage in residuum processing where the relative amount of hydrogen consumed in the process can be shown to be dependent on the sulfur content of the feedstock (Scott and Bridge, 1971; Speight, 2000, 2014; Ancheyta and Speight, 2007). In addition, the direct hydrodesulfurization of various feedstocks has received some attention, and the concept may be applicable directly to the conversion of petroleum residua (Speight, 2000, 2013, 2014; Speight and Ozum, 2002; Parkash, 2003; Hsu and Robinson, 2006; Gary et al., 2007).

The heavier feedstocks add a new dimension to hydrocracking chemistry by virtue of their highly complex nature. During the early evolution of the hydrocracking process, when gas oils were the feedstock of choice, prediction of product distribution was possible from a general knowledge of the composition of the gas oil. There is the need, therefore, to understand the nature of the constituents of the heavier feedstocks in more detail in order to be able to predict product yield and product distribution. The characterization techniques currently at hand are an aid to accomplishing this goal (Chapter 2). However, it is only when a full understanding of the character of the heavier feedstocks is available that a confident predictability will emerge.

7.7 USE OF THE DATA

The use of composition data to model feedstock behavior during refining is becoming increasingly important in refinery operations (Speight, 2014). For example, molecular models have been used for some time with limited success, and the use of the analytical data requires somewhat more than reduction of the data to a mental paper exercise. The trend of the use of compositional data for process modeling has progressed beyond this stage, and compositional models use (1) analytical data, (2) representation of the molecular structure of a large number of components, and (3) molecular structure property relationships. Currently, such techniques have been reported as successful when applied to lube oil technology and show promise for application to heavier feedstocks.

Nevertheless, in the simplest sense, petroleum can be considered a composite of four major operational fractions. However, it must never be forgotten that the nomenclature of these fractions lies within the historical development of petroleum science, and that the fraction names are operational and are related more to the general characteristics than to the identification of specific compound types. Nevertheless, once a convenient fractionation technique has been established, it is possible to compare a variety of different feedstocks varying from conventional petroleum to propane asphalt (Corbett and Petrossi, 1978). Later studies have focused not only on the composition of petroleum and its major operational fractions but also on further fractionation, which allows different feedstocks to be compared on a relative basis to provide a very simple but convenient feedstock map.

Thus, by careful selection of an appropriate technique, it is possible to obtain an overview of petroleum composition that can be used for behavioral predictions. By taking the approach one step further and by assiduous collection of various subfractions, it becomes possible to develop the petroleum map and add an extra dimension to compositional studies (Figure 4.4). Petroleum and heavy feedstocks then appear more as a continuum than as four specific fractions. Such a map does not give any indication of the complex interrelationships of the various fractions (Koots and Speight, 1975), although the prediction of feedstock behavior is possible using such data. It is necessary to take the composition studies one step further using subfractionation of the major fractions to obtain a more representative indication of petroleum composition.

Further development of this concept (Long and Speight, 1998) involved the construction of a different type of compositional map using the molecular weight distribution and the molecular type distribution as coordinates. The separation involved the use of an adsorbent such as clay, and the fractions were characterized by using *solubility parameter* as a measure of the polarity of the molecular types. The molecular weight distribution can be determined by gel permeation chromatography. Using these two distributions, a map of composition can be prepared using molecular weight and solubility parameter as the coordinates for plotting the two distributions. Such a composition map can provide insights into the many separation and conversion processes used in petroleum refining.

The molecular type was characterized by the polarity of the molecules, as measured by the increasing adsorption strength on an adsorbent. At the time of the original concept, it was unclear how to characterize the continuum in molecular type or polarity. For this reason, the molecular type coordinate of their first maps was the yield of the molecular types ranked in order of increasing polarity. However, this type of map can be somewhat misleading because the areas are not related to the amounts of material in a given type. The horizontal distance on the plot is a measure of the yield, and there is no continuous variation in polarity for the horizontal coordinate. It was suggested that the solubility parameter of the different fractions could be used to characterize both polarity and adsorption strength.

To attempt to remove some of these potential ambiguities, more recent developments of this concept have focused on the solubility parameter. The simplest map that can be derived using the solubility parameter is produced with the solubility parameters of the solvents used in solvent separation procedures, and equating these parameters to the various fractions (Speight, 1994, 2014). However, the solubility parameter boundaries determined by the values for the eluting solvents that remove the fractions from the adsorbent offer a further step in the evolution of petroleum maps (Long and Speight, 1998; Speight, 2014).

Measuring the overall solubility parameter of a petroleum fraction is a time-consuming chore. Therefore, it is desirable to have a simpler, less time-consuming measurement that can be made on petroleum fractions that will correlate with the solubility parameter and thus give an alternative continuum in polarity. In fact, the hydrogen-to-carbon atomic ratio and other properties of petroleum fractions that can be correlated with the solubility parameter also provide correlation for the behavior of crude oils (Speight, 1994, 2014; Wiehe, 1994, 2008).

REFERENCES

Altgelt, K.H., and Boduszynski, M.M. 1994. *Compositional Analysis of Heavy Petroleum Fractions*. Marcel Dekker Inc., New York.

Ancheyta, J., and Speight, J.G. (Editors) 2007. *Hydroprocessing Heavy Oils and Residua*. CRC Press, Taylor & Francis Group, Boca Raton, FL.

Andersen, S.I. 1994. Dissolution of Solid Boscan Asphaltene Constituents in Mixed Solvents. *Fuel Sci. Technol. Int.* 12: 51.

ASTM D189. 2014. Standard Test Method for the Conradson Carbon Residue of Petroleum Products. In: *Annual Book of Standards*. ASTM International, West Conshohocken, PA.

ASTM D524. 2014. Standard Test Method for the Ramsbottom Conradson Carbon Residue of Petroleum Products. In: *Annual Book of Standards*. ASTM International, West Conshohocken, PA.

ASTM D2007. 2014. Standard Test Method for Characteristic Groups in Rubber Extender and Processing Oils and Other Petroleum-Derived Oils by the Clay–Gel Absorption Chromatographic Method. In: *Annual Book of Standards*. ASTM International, West Conshohocken, PA.

ASTM D4124. 2014. Standard Test Method for Separation of Asphalt into Four Fractions. In: *Annual Book of Standards*. ASTM International, West Conshohocken, PA.

ASTM D4530. 2014. Standard Test Method for Determination of Carbon Residue (Micro Method). In: *Annual Book of Standards*. ASTM International, West Conshohocken, PA.

Beuther, H., and Schmid, B.K. 1963. Reaction Mechanisms and Rates in Residua Hydrodesulfurization. Paper No. 20, Section III. In: *Proceedings. 6th World Petroleum Congress*, Vol. 3, p. 297.

Bland, W.F., and Davidson, R.L. 1967. *Petroleum Processing Handbook*. McGraw-Hill, New York.

Brooks, J., and Welte, D.H. 1984. *Advances in Petroleum Geochemistry*. Academic Press Inc., New York.

Carlson, C.S., Langer, A.W., Stewart, J., and Hill, R.M. 1958. Thermal Hydrogenation. Transfer of Hydrogen from Tetralin to Cracked Residua. *Ind. Eng. Chem.* 50: 1067–1070.

Corbett, L.W., and Petrossi, U. 1978. Differences in Distillation and Solvent Separated Asphalt Residua. *Ind. Eng. Chem. Prod. Res. Dev.* 17: 342–346.

Decroocq, D. 1984. *Catalytic Cracking of Heavy Petroleum Fractions*. Editions Technip, Paris, France.

Dolbear, G.E., Tang, A., and Moorehead, E.L. 1987. Upgrading Studies with California, Mexican, and Middle Eastern Heavy Oils. In: *Metal Complexes in Fossil Fuels*, R.H. Filby and J.F. Branthaver (Editors). Symposium Series No. 344. American Chemical Society, Washington, DC, pp. 220–232.

Durand, B. (Editor) 1980. *Kerogen: Insoluble Organic Matter from Sedimentary Rocks*. Editions Technip, Paris, France.

Eglinton, G., and Murphy, B. 1969. *Organic Geochemistry: Methods and Results*. Springer-Verlag, New York.

Gary, J.H., Handwerk, G.E., and Kaiser, M.J. 2007. *Petroleum Refining: Technology and Economics*, 5th Edition. CRC Press, Taylor & Francis Group, Boca Raton, FL.

Gates, B.C., Katzer, J.R., and Schuit, G.C.A. 1979. *Chemistry of Catalytic Processes*. McGraw-Hill, New York.

Hsu, C.S., and Robinson, P.R. (Editors) 2006. *Practical Advances in Petroleum Processing*, Vols. 1–2. Springer Science, New York.

Komatsu, S., Hori, Y., and Shimizu, S. 1986. The Asphaltene Bottoms Conversion Cracking. In: *Proceedings 51st Midyear Meeting*. American Petroleum Institute, Washington, DC.

Koots, J.A., and Speight, J.G. 1975. The Relation of Petroleum Resin Constituents to Asphaltene Constituents. *Fuel* 54: 179.

Langer, A.W., Stewart, J., Thompson, C.E., White, H.T., and Hill, R.M. 1961. Thermal Hydrogenation of Crude Residua. *Ind. Eng. Chem.* 53: 27–30.

Langer, A.W., Stewart, J., Thompson, C.E., White, H.T., and Hill, R.M. 1962. Hydrogen Donor Diluent Visbreaking of Residua. *Ind. Eng. Chem. Proc. Design Dev.* 1: 309–312.

Long, R.B., and Speight, J.G. 1989. Studies in Petroleum Composition. I: Development of a Compositional Map for Various Feedstocks. *Rev. Inst. Fr. Pét.* 44: 205.

Long, R.B., and Speight, J.G. 1998. The Composition of Petroleum. In: *Petroleum Chemistry and Refining*, J.G. Speight (Editor). Taylor & Francis Publishers, Washington, DC, Chapter 1.

Ma, X., Sakanishi, K., and Mochida, I. 1994. Hydrodesulfurization Reactivities of Various Sulfur Compounds in Diesel Fuel. *Ind. Eng. Chem. Res.* 33: 218–222.

McConaghy, J.R. 1987. Short Residence Time Hydrogen Donor Diluent Cracking Process. United States Patent 4,698,147, October 6.

Mitchell. D.L., and Speight, J.G. 1973a. The Solubility of Asphaltene Constituents in Hydrocarbon Solvents. *Fuel* 52: 149.

Mitchell, D.L., and Speight, J.G. 1973b. Preparation of Mineral-Free Asphaltene Constituents. United States Patent 3,779,902, December 18.

Mitra-Kirtley, S., Mullins, O.C., Branthaver, J.F., and Cramer, S.P. 1993. Nitrogen Chemistry of Kerogens and Bitumens from X-ray Absorption Near-Edge Spectroscopy. *Energy Fuels* 7: 1128–1134.

Parkash, S. 2003. *Refining Processes Handbook*. Gulf Professional Publishing, Elsevier, Amsterdam.

Reynolds, J.G. 1998. Metals and Heteroatoms in Heavy Crude Oils. In: *Petroleum Chemistry and Refining*, J.G. Speight (Editor). Taylor & Francis Publishers, Philadelphia, PA, Chapter 3.

Rodriguez, J., Tierney, J.W., and Wender, I. 1991. *Prepr. Div. Fuel Chem. Am. Chem. Soc.* 36(3): 1295.

Sakarnen, K.V., and Ludwig, C.H. 1971. *Lignins: Occurrence, Formation, Structure and Reactions*. John Wiley & Sons Inc., New York.

Scott, W., and Bridge, A.G. 1971. The Continuing Development of Hydrocracking. In: *Origin and Refining of Petroleum*, H.G. McGrath and M.E. Charles (Editors). Advances in Chemistry Series No. 103. American Chemical Society, Washington, DC, Chapter 6, pp. 113–129.

Speight, J.G. 1979. *Studies on Bitumen Fractionation—(A) Fractionation by a Cryoscopic Method. (B) Effect of Solvent Type on Asphaltene Solubility*. Information Series No. 84. Alberta Research Council, Edmonton, Alberta, Canada.

Speight, J.G. 1984. The Chemical Nature of Petroleum Asphaltene Constituents. In: *Characterization of Heavy Crude Oils and Petroleum Residues*, S. Kaliaguine and A. Mahay (Editors). Elsevier, Amsterdam, p. 515.

Speight, J.G. 1992. A Chemical and Physical Explanation of Incompatibility during Refining Operations. In: *Proceedings. Eastern Oil Shale Symposium*. IMMR, Lexington, KY, p. 17.

Speight, J.G. 1994. Chemical and Physical Studies of Petroleum Asphaltene Constituents. In: *Asphaltene Constituents and Asphalts. I. Developments in Petroleum Science*, Vol. 40, T.F. Yen and G.V. Chilingarian (Editors). Elsevier, Amsterdam, Chapter 2.

Speight, J.G. 1998. Thermal Chemistry of Petroleum Constituents. In: *Petroleum Chemistry and Refining*, J.G. Speight (Editor). Taylor & Francis Publishers, Philadelphia, PA, Chapter 5.

Speight, J.G. 2000. *The Desulfurization of Heavy Oils and Residua*, 2nd Edition. Marcel Dekker Inc., New York.

Speight, J.G. 2001. *Handbook of Petroleum Analysis*. John Wiley & Sons Inc., Hoboken, NJ.

Speight, J.G. 2011. *The Refinery of the Future*. Gulf Professional Publishing, Elsevier, Oxford, UK.

Speight, J.G. 2013. *Heavy and Extra Heavy Oil Upgrading Technologies*. Gulf Professional Publishing, Elsevier, Oxford, UK.

Speight, J.G. 2014. *The Chemistry and Technology of Petroleum*, 5th Edition. CRC Press, Taylor & Francis Group, Boca Raton, FL.

Speight, J.G. 2015. *Handbook of Petroleum Product Analysis*, 2nd Edition. John Wiley & Sons Inc., Hoboken, NJ.

Speight, J.G., Long, R.B., and Trowbridge, T.D. 1984. On the Definition of Asphaltene Constituents. *Fuel* 63: 616.

Speight, J.G., and Moschopedis, S.E. 1979. The Production of Low-Sulphur Liquids and Coke from Athabasca. *Fuel Process. Technol.* 2: 295.

Speight, J.G., and Ozum, B. 2002. *Petroleum Refining Processes*. Marcel Dekker Inc., New York.

Takeuchi, C., Fukui, Y., Nakamura, M., and Shiroto, Y. 1986. The Asphaltene Bottoms Cracking Process. *Ind. Eng. Chem. Process. Des. Dev.* 22: 236.

Thomas, K.P., and Hunter, D.E. 1989. The Evaluation of a Coal-Derived Liquid as a Feedstock for the Production of High-Density Aviation Turbine Fuel. DOE Report DOE/MC/11079.2993. United States Department of Energy, Washington, DC.

Tissot, B.P., and Welte, D.H. 1978. *Petroleum Formation and Occurrence*. Springer-Verlag, New York.

Vernon, L.W., Jacobs, F.E., and Bauman, R.F. 1984. Process for Converting Petroleum Residuals. United States Patent 4,425,224, January 10.

Weiss, V., and Edwards, J.M. 1980. *The Biosynthesis of Aromatic Compounds*. John Wiley & Sons, Inc., New York.

Wiehe, I.A. 1994. The Pendant-Core Building Block Model of Petroleum Residua. *Energy Fuels* 8: 536–544.

Wiehe, I.A. 2008. *Process Chemistry of Petroleum Macromolecules*. CRC Press, Taylor & Francis Group, Boca Raton, FL.

Yan, J., Plancher, H., and Morrow, N.R. 1997. Paper No. SPE 37232. SPE International Symposium on Oilfield Chemistry, Houston, TX.

8 Desulfurization Methods

8.1 INTRODUCTION

The increasing use of fossil fuels is polluting the environment at a constantly increasing rate. Besides carbon dioxide, a greenhouse gas, the use of petroleum and its fractions generates significant quantities of sulfur oxides (SO_x), which play a major role in the formation of acid depositions. Thus, refineries are facing many challenges, including the influx of heavy crude oils, increased fuel quality standards in terms of a severely diminished and regulated sulfur content, and the need to reduce all forms of emissions to meet air pollution regulations. In fact, environmental regulations in every country worldwide are becoming more stringent in order to reduce all forms of emissions, including that of sulfur oxides. Furthermore, as the global community moves toward zero-sulfur fuels, the only differences are the starting point of sulfur level (feedstock sulfur) and rate reduction of sulfur content during refinery operations. To meet these more stringent regulations, high-quality crude oil with low sulfur content is preferred; however, when not available, the only alternative option is to remove sulfur during the refining process.

For example, in the United States, the Environmental Protection Agency (US EPA) has established stringent sulfur control programs for gasoline and diesel fuel (Lappinen and Higgins, 2013). Three broad alternatives are available for meeting these sulfur standards: (1) posttreating—hydrotreat or otherwise treat part or all of the fluid catalytic cracking (FCC) naphtha, and hydrotreat straight-run kerosene and distillate, coker distillate, and FCC light cycle oil; (2) conventional pretreating—hydrotreat FCC feedstock to reduce sulfur content as well as nitrogen and metal content, hydrofinish all or part of the FCC naphtha, and hydrotreat straight-run kerosene and distillate, coker distillate, and light cycle oil (but under certain conditions, treating the FCC naphtha would be unnecessary because its sulfur content would be sufficiently low without posttreatment); and (3) partial conversion hydrocracking—partially hydrocrack FCC feed (to reduce sulfur content, as well as nitrogen and metal content, and, importantly, to produce some low-sulfur naphtha and distillate), and hydrotreat straight-run kerosene and distillate, coker distillate, and light cycle oil.

With partial conversion hydrotreating, treating the resulting FCC naphtha would not be necessary. Posttreating processes alone are sufficient for meeting these standards, and they entail lower investment and operating costs than the other approaches. Thus far, posttreating has been the approach of choice for most refineries in the United States, except in California where pretreating is practiced for a variety of reasons that are unique to state regulations.

Generally, of the desulfurization methods (Figure 8.1), hydrodesulfurization is the most common technology used by refineries to remove sulfur from intermediate product streams and product blending stock; however, the chemistry can be complex and catalyst deactivation ever present if the correct precautions are not taken (Gates and Topsøe, 1997; Furimsky and Massoth, 1999; Inoue et al., 2000). However, hydrodesulfurization has several disadvantages, in that it is energy intensive, costly to install and to operate, and does not work well on refractory organosulfur compounds. Recent research has therefore focused on improving hydrodesulfurization catalysts and processes, and also on the development of alternative technologies. Among the new technologies, one possible approach is biocatalytic desulfurization, which has the advantage that it can be operated in conditions that require less energy and hydrogen. In fact, biocatalytic desulfurization operates at ambient temperature and pressure with high selectivity, resulting in decreased energy costs, low emission, and no generation of undesirable side products.

It is the purpose of this chapter to present to the reader an update of the various concepts that are currently being investigated as a means of sulfur removal. Indeed, it is likely that many of these

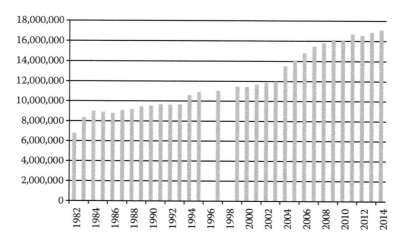

FIGURE 8.1 US refinery desulfurization (including catalytic hydrotreating) downstream charge capacity as of January 1 (barrels per stream day). (From US Energy Information Administration: http://www.eia.gov /dnav/pet/hist/LeafHandler.ashx?n=PET&s=8_NA_8CD0_NUS_5&f=A.)

concepts (that have not already been tested) will be developed for commercial installation as sulfur removal processes in refineries (Speight, 2011, 2014).

8.2 METHODS FOR SULFUR REMOVAL

There are various reported desulfurization methods to remove sulfur from fossil fuels (Speight, 2000, 2014; Speight and Ozum, 2002; Parkash, 2003; Hsu and Robinson, 2006; Ancheyta and Speight, 2007; Gary et al., 2007). Among these, hydrodesulfurization is currently considered as the most important one. Desulfurization by ionic liquids, selective adsorption, oxidative desulfurization, and biological methods have also shown good potential to be a plausible substitution for hydrodesulfurization technology or to be used in line with hydrodesulfurization technology in future refining systems (Speight, 2011).

The conventional desulfurization process that is widely practiced for crude oil and the distillate products (naphtha, kerosene, and gas oil) is hydrodesulfurization, which is a technology that converts organic sulfur compounds to hydrogen sulfide and other inorganic sulfides, under high temperature (290–455°C, 555–850°F) and high pressure (150–3000 psi), and uses hydrogen gas in the presence of metal catalysts, e.g., $CoMo/Al_2O_3$ or $NiMo/Al_2O_3$. The produced hydrogen sulfide is then catalytically air oxidized to elemental sulfur (Speight, 2014). Oil refiners depend on such a costly, extreme chemical process to treat approximately 20 million barrels of crude oil per day (Gupta et al., 2005). Although hydrodesulfurization can easily remove the inorganic sulfur or simple organic sulfur compounds, it is not effective for removing complicated polycyclic aromatic heterocyclic compounds, such as dibenzothiophene derivatives and alkylated dibenzothiophene derivatives (Cx-DBTs) (Babich and Moulijn, 2003; Bustos-Jaimes et al., 2003; Franchi et al., 2003). To reach lower concentrations of sulfur (<15 ppm), higher temperature and pressure are required. There are low-temperature and low-pressure processes being developed that do not rely on hydrotreating, such as biodesulfurization and chemical oxidation. Sulfur can be removed via these processes up front in the refinery, such as from crude oil, before being processed in the refinery into fuels. Or, sulfur can be removed from those refinery streams that are to be blended directly into fuel products.

However, there is a no universal approach to classify desulfurization processes. The processes can be categorized by the fate of the organosulfur compounds during desulfurization, the role of

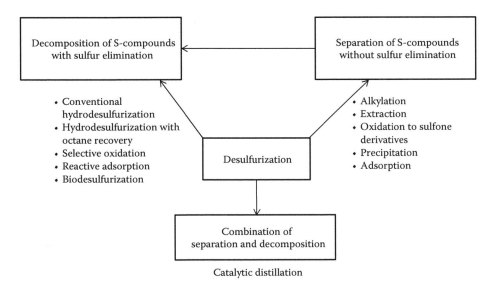

FIGURE 8.2 Classification of desulfurization processes based on the transformation of organosulfur compounds.

hydrogen, or the nature of the process used (chemical and/or physical). However, based on the way in which the organosulfur compounds are transformed, the processes can be divided into three groups depending on whether the sulfur compounds are decomposed separated from refinery stream without decomposition, or separated and then decomposed (Figure 8.2). When organosulfur compounds are decomposed, gaseous or solid sulfur products are formed and the hydrocarbon part is recovered and remains in the refinery streams. Conventional hydrodesulfurization technology is the most typical example of this type of process. In other processes, the organosulfur compounds are simply separated from the refinery streams. Some processes of this type first transform the organosulfur compounds into other compounds that are easier to separate from the refinery streams. When streams are desulfurized by separation, some desired product can be lost and disposal of the retained organosulfur molecules is still a problem. In the third type of process, organosulfur compounds are separated from the streams and simultaneously decomposed in a single reactor unit rather than in a series of reaction and separation vessels. These combined processes, which provide the basis for many technologies currently proposed for industrial application, may prove very promising for producing ultra-low-sulfur fuels. Desulfurization by catalytic distillation is the fascinating example of this type of process.

Desulfurization processes can be also classified in two groups, hydrodesulfurization-based technology and non-hydrodesulfurization-based technology, depending on the role of hydrogen in removing sulfur. In hydrodesulfurization-based processes, hydrogen is used to decompose organosulfur compounds and eliminate sulfur from refinery streams, while non-hydrodesulfurization technology–based processes do not require hydrogen. Different combinations of refinery streams pre- or postdistilling treatments with hydrotreating to maintain desired fuel specifications can also be assigned as hydrodesulfurization-based processes since hydrodesulfurization treatment is one of the key steps. These two classifications overlap to some extent. Most sulfur elimination processes, with the exception of selective oxidation, are hydrodesulfurization based. The organosulfur compound separation processes are usually non-hydrodesulfurization technology based since they do not require hydrogen if concentrated sulfur-rich streams are not subsequently hydrotreated.

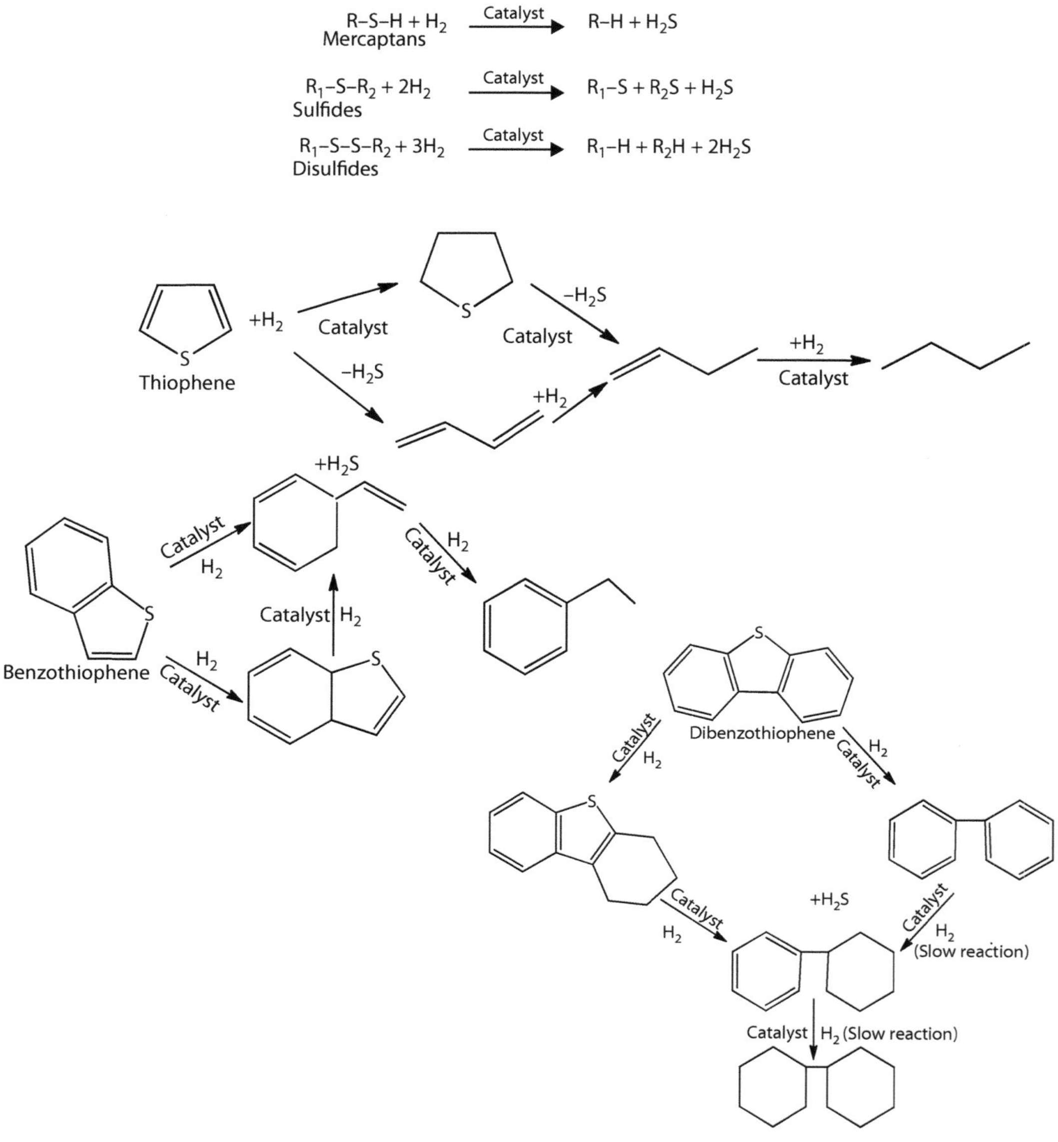

FIGURE 8.3 Typical organosulfur compounds and their hydrotreating pathway. Reaction pathway for alkylated thiophene, benzothiophene, and dibenzothiophene is similar to the reaction of nonalkylated counterparts.

Finally, desulfurization processes can be classified on the basis of the nature of the key physicochemical process used for sulfur removal (Figure 8.3). The most developed and commercialized technologies are those that catalytically convert organosulfur compounds with sulfur elimination. Such catalytic conversion technologies include (1) conventional hydrotreating, (2) hydrotreating with advanced catalysts and/or reactor design, and (3) a combination of hydrotreating with some additional chemical processes to maintain fuel specifications. The main feature of the technologies of the second type is the application of physicochemical processes different in nature from catalytic hydrodesulfurization to separate and/or to transform organosulfur compounds from refinery streams. Such technologies include as a key step distillation, alkylation, oxidation, extraction, adsorption, or combination of these processes, but typically by oxidation followed by extraction (Tam et al., 1990a,b; Zamikos et al., 1995; Otsuki et al., 2000).

8.2.1 Hydrodesulfurization

The reactivity of organosulfur compounds varies widely depending on their structure and local sulfur atom environment. The low-boiling crude oil fraction contains mainly the aliphatic organosulfur compounds: mercaptans, sulfides, and disulfides. They are very reactive in a conventional hydrotreating process and can easily be completely removed from the fuel. For higher-boiling crude oil fractions such as heavy straight-run naphtha, straight-run diesel, and light FCC naphtha, the organosulfur compounds predominantly contain thiophene rings. These compounds include thiophene and benzothiophene derivatives. Thiophene-containing compounds are more difficult to convert via hydrotreating than mercaptans and sulfides. The heaviest fractions blended to the gasoline and diesel pools—bottom FCC naphtha, coker naphtha, FCC, and coker diesel—contain mainly alkyl-benzothiophene derivatives, dibenzothiophene derivatives, and polynuclear organic sulfur compounds, i.e., the least reactive S-containing molecules in the hydrodesulfurization reaction.

One of the current technologies to reduce sulfur in middle-distillate/diesel fuels is known as hydrodesulfurization. In hydrodesulfurization, the sulfur atom in sulfur compounds is reduced to hydrogen sulfide (H_2S) on $CoMo/Al_2O_3$ or $NiMo/Al_2O_3$ catalyst in the presence of hydrogen gas. The hydrogen sulfide is then catalytically air oxidized to elemental sulfur. Depending on the hydrocarbon type and degree of desulfurization, hydrodesulfurization may occur at 200–425°C (390–795°F) and 150–250 psi hydrogen (Gupta et al., 2005). To reach a lower concentration of sulfur (<15 ppm) in the product, higher temperature and pressure are required. By this means, a variety of some of the organosulfur compounds (Figure 8.3)—namely, mercaptans, sulfides, disulfides, thiophene derivatives, and benzothiophene derivatives—are converted to hydrocarbon products (Speight, 2000, 2014; Babich and Moulijn, 2003). The reactivity of sulfur compounds in the hydrodesulfurization reaction follows the order from most to least reactive (Babich and Moulijn, 2003):

Thiophene > alkylated thiophene > benzothiophene > alkylated benzothiophene[1]
Alkylated benzothiophene > dibenzothiophene and alkylated benzothiophene
Dibenzothiophene and alkylated dibenzothiophene > alkylated dibenzothiophene[2]
Alkylated dibenzothiophene[2] > alkylated dibenzothiophene[3]
Alkylated dibenzothiophene[3] > dibenzothiophene and alkylated dibenzothiophene[4]
Dibenzothiophene and alkylated dibenzothiophene[4] > alkylated dibenzothiophene[5] > alkylated dibenzothiophene[5] > alkylated dibenzothiophene[6]

1: Without substituents at the 4 and 6 positions
2: With one substituent at either the 4 or 6 position
3: With alkyl substituents at the 4 and 6 positions
4: Without substituents at the 4 and 6 positions
5: With one substituent at either the 4 or 6 position
6: With alkyl substituents at the 4 and 6 positions

Although the concentrations of benzothiophene and dibenzothiophene derivatives are considerably decreased by hydrodesulfurization (Monticello, 1998) and the process has been commercially used for a long time, it has several issues that must be overcome: (1) hydrodesulfurization of diesel feedstock for a low-sulfur (or ultra-low-sulfur) product requires a larger reactor volume, longer processing times, and substantial hydrogen and energy inputs; (2) for refractory sulfur compounds, hydrodesulfurization requires higher temperatures, higher hydrogen pressures, and longer residence time, which make the process costly owing to the requirement of stronger reaction vessels and facilities (McHale, 1981); (3) the application of extreme conditions to desulfurize refractory compounds results in the deposition of carbonaceous coke on the catalysts; (4) exposure of crude oil fractions to severe conditions, including temperatures above about 360°C (680°F), initiates cracking reactions (Speight, 2000, 2014), and the products of such reactions decrease the fuel value of the treated product; (5) deep hydrodesulfurization processes—which implies that more of the least reactive sulfur compounds must be converted—requires large capital investments for equipment as well as the accompanying operating costs; (6) the generated hydrogen sulfide poisons the catalysts and shortens their useful life; (7) deep hydrodesulfurization is affected by components in the reaction mixture, such as organic hetero-compounds and polynuclear aromatic hydrocarbons (Egorova, 2003); (8) for older units, which are not sufficiently efficient to produce products that meet the new sulfur removal levels, erection of new hydrodesulfurization facilities and heavy load of capital cost is inevitable; (9) hydrodesulfurization removes paraffinic sulfur compounds such as thiol derivatives, sulfide derivatives, and disulfide derivatives effectively, but some aromatic sulfur-containing compounds such as 4- and 4,6-substituted dibenzothiophene derivatives and polynuclear aromatic sulfur heterocycles are resistant to hydrodesulfurization and form the most abundant organosulfur compounds after hydrodesulfurization (Ma et al., 1994; Monticello, 1998); (10) the hydrogen atmosphere in hydrodesulfurization results in the hydrogenation of olefin derivatives and reduces the calorific value of fuel—to increase the calorific value, the hydrodesulfurization-treated stream is sent to the FCC unit, which adds to the process cost (Hernández-Maldonado and Yang, 2004a,b); and (11) although hydrodesulfurization is considered a cost-effective method for fossil fuel desulfurization, the cost of sulfur removal from refractory compounds by hydrodesulfurization is high (Atlas et al., 2001)—for example, reducing the cost of lowering the sulfur content from 200 to 50 ppm has been estimated to be four times higher than reducing the sulfur content of a product from 500 to 200 ppm. Nevertheless, further reduction of sulfur concentration by hydrodesulfurization to <1 ppm will be necessary for future fuels, and remains a challenging research and economic target.

Recently, the refining industry has made a great deal of progress toward developing more active catalysts and more economical processes to remove sulfur from gasoline and diesel fuel (oil) (Okada et al., 2002a,b; Egorova and Prins, 2004; La Paz Zavala and Rodriguez, 2004). For example, California refineries are producing gasoline that contains 29 ppm sulfur on average (US EPA, 2000), despite the high sulfur content of California and Alaska crude oils used as feedstocks up to 11,000 ppm. Owing to incentives and regulations, 10 ppm sulfur diesel has been commercially available in Sweden for several years. However, although this process tends to improve diesel quality by raising cetane number, it decreases naphtha quality by lowering the octane number (Babich and Moulijn, 2003; Song and Ma, 2003).

Additional investments are required to achieve low or near-zero sulfur products to meet the international environmental regulations for producing ultraclean transportation fuels, which will need higher temperatures and pressures with larger consumption of energy as well as new catalysts to desulfurize the most recalcitrant molecules, leading to increased operations and capital costs as well as more carbon dioxide emissions (Castorena et al., 2002). In addition to desulfurization, this will cause demetallization and a reduction in carbon residue reduction, some denitrogenation, hydrocracking, and coking (Song, 2003).

Transportation fuels, such as gasoline, jet fuel, and diesel, are ideal fuels owing to their high energy density, ease of storage and transportation, and established distribution network. However, their sulfur concentration must be <10 ppm to protect the deactivation of catalysts in a reforming process and electrodes in a fuel cell system (Wild et al., 2006). Accordingly, refineries started to establish new complementary routes in addition to the hydrodesulfurization process.

8.2.2 Extraction

Desulfurization by extraction relies on the higher solubility of organic sulfur compounds in an appropriate solvent than other hydrocarbons present in a petroleum fraction. The organic sulfur compounds are removed from the feed into the solvent, after which the mixture of sulfur-rich solvent and feed is separated. The organic sulfur compounds are removed from the solvent by distillation and the solvent is recycled (Abinaya et al., 2013; Mužic and Sertić-Bionda, 2013).

By way of introduction to extraction processes, the UOP Merox extraction process for removing mercaptans is used for liquefied gases and for all liquid fuel raw fractions (Speight, 2014). In the process, mercaptans are extracted with caustic solution, after which the process catalyst and dissolved mercaptans are oxidized with air to disulfides.

$$2RSH \rightarrow RSSR$$

The separated disulfide oil can be hydrotreated or sold as a special product.

Paraffinic sulfur compounds, i.e., mercaptans (RSH), sulfide derivatives (R^1SR^2), and disulfide derivatives (R^1SSR^2), are readily hydrodesulfurized. However, cyclic and sterically hindered aromatic S-compounds, e.g., alkylated dibenzothiophene derivatives, are less reactive to hydrodesulfurization. For example, the rate of hydrodesulfurization of 4-methyldibenzothiophene and 4,6-dimethyldibenzothiophene decrease by factors of approximately 2 and 10, respectively, relative to dibenzothiophene (Bösmann et al., 2002). In such cases, extractive desulfurization can be applied as an alternative or supplemental technology for the desulfurization of refinery streams.

Extractive desulfurization is based on the fact that organosulfur compounds are more soluble than hydrocarbons in an appropriate solvent. The general process flow (Figure 8.4) involves a mixing tank in which the sulfur compounds are transferred from the fuel oil into the solvent due to their higher solubility in the solvent. Subsequently, the solvent–fuel mixture is fed into a separator in which hydrocarbons are separated from the solvent. The desulfurized hydrocarbon stream is used either as a component to be blended into the final product or as a feed for further transformations. The organosulfur compounds are separated by distillation and the solvent is recycled. The most attractive feature of the extractive desulfurization is the applicability at ambient conditions. The process does not change the chemical structure of the fuel oil components. As the equipment used is rather conventional without special requirements, the process can be easily integrated into the refinery. Dispersed phase contractors can be used for desulfurization of hydrocarbons; however, it requires high solvent holdup, which is consequently associated with emulsion formation, foaming, unloading, and flooding. Liquid–liquid extraction, using intermixed solvents, suffers from slow phase separation and stable emulsions (Yahiya et al., 2013).

The efficiency of extractive desulfurization is mainly limited by the solubility of the organic sulfur compounds in the solvent. The solvent must have a boiling temperature different from that of the sulfur-containing compounds, and it must be inexpensive to ensure the economic feasibility of the process. Solvents of different nature have been tried, among which acetone, ethanol, polyethylene glycols, and nitrogen-containing solvents showed a reasonable level of desulfurization of 50–90% w/w sulfur removal, depending on the number of extraction cycles (Babich and Moulijn, 2003). Preparation of such a *solvent cocktail* is challenging and intrinsically nonefficient since the composition of the solvent mix depends on the spectrum of the organosulfur compounds present in the process feedstock. Solubility can also be enhanced by transforming the organic sulfur compounds to increase their solubility in a polar solvent. One way to do this is by selectively oxidizing

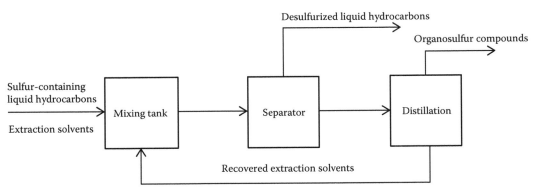

FIGURE 8.4   General process flow for extractive desulfurization.

the organic sulfur compounds (thiophene derivatives, benzothiophene derivatives, and dibenzo-thiophene derivatives) to sulfone derivatives possessing higher polarity with a greater tendency to solubilize in various polar solvents.

Desulfurization of all naphtha streams from FCC can be easily achieved through the conventional hydrodesulfurization process, because sulfide derivatives, disulfide derivatives, and thiophene derivatives are relatively reactive for hydrodesulfurization (Speight, 2000, 2014; Speight and Ozum, 2002; Parkash, 2003; Hsu and Robinson, 2006; Ancheyta and Speight, 2007; Gary et al., 2007). However, hydrotreating leads to olefin saturation and loss in octane number. The GT-Desulf process is an example of desulfurization technology based on organosulfur compound extraction (Chapados, 2000). This process separates the organosulfur compounds and aromatics from naphtha (from the FCC unit) by extractive distillation using a blend of solvents. A desulfurized/dearomatized olefin-rich naphtha stream and an aromatic stream containing the sulfur compounds are formed after treatment in a GT-Desulf reactor. The first stream is directly used as a gasoline blend stock. Unfortunately, available literature does not contain any information on the level of sulfur removal from the treated stream. The aromatics fraction with the sulfur compounds is sent to a hydrodesulfurization reactor. After treatment in the hydrodesulfurization reactor, aromatics recovery is proposed as an additional option to increase the economic efficiency of the process. The GT-Desulf process may be economically suitable due to an integrated approach to the refinery processing (segregated sulfur removal and aromatics recovery) and lower hydrogen consumption since less naphtha is treated in the hydrodesulfurization reactor.

The biggest advantage of extractive desulfurization is the possibility of conducting it at relatively low temperatures and pressures, while the hydrocarbon compounds in the feed remain mainly intact. The process equipment necessary for carrying out this process is conventional and can easily be integrated into refineries. However, in order for the process to be efficient, several conditions must be met: organic sulfur compounds must be fully soluble in the solvent, the solvent must have a different boiling point in relation to the removed organic sulfur compounds, and the solvent must be nontoxic and relatively cheap so the process can be economically viable. The efficiency of extractive desulfurization is mainly restricted by (1) the solubility of the organic sulfur compounds in the applied solvent and (2) the choice of the solvent, which must take into account the nature of the organic sulfur compounds to be extracted (Mužic and Sertić-Bionda, 2013).

Photochemical extractive desulfurization combines photochemical reactions with extraction of the organic sulfur compounds with an appropriate (usually aqueous) solvent (Shiraishi et al., 1999). The organic sulfur compounds are suspended in an aqueous-soluble solvent and irradiated by ultraviolet or visible light in a specially designed photoreactor—as anticipated, the outcome is the oxidation of the sulfur compounds. The polar compounds formed are rejected by the nonpolar hydrocarbon phase and are concentrated in the solvent. The photochemical reaction is assisted by a photosensitizer species such as 9,10-dicyanoanthracene. Acetonitrile, which provides relatively high solubility of initial and oxidized sulfur compounds, is a suitable solvent. After photooxidation, the solvent and the hydrocarbon phases are separated, as in extractive desulfurization.

In addition, the aromatics from the solvent as well as the photosensitizer from the solvent and desulfurized hydrocarbon stream must be recovered to increase product yield and economic efficiency. Aromatics are usually recovered by liquid–liquid extraction using light paraffinic solvents subsequently blended into the desulfurized fuel stream. The 9,10-dicyanoanthracene is removed by adsorption, using a silica gel as an adsorbent. It can be returned to the process after desorption with aqueous solution of acetonitrile. All of these processes are rather common refinery processes (although not all of the chemicals are common) that can be easily integrated into the refinery and do not require special equipment or conditions. This photooxidation process exhibits high selectivity to the removal of organic sulfur compounds from light oils, catalytic-cracked naphtha, and vacuum gas oils (Mužic and Sertić-Bionda, 2013).

Extractive distillation is done in the presence of a miscible, high-boiling, relatively nonvolatile component (the solvent) that does not form an azeotrope with the other components in the mixture.

The method is used for the separation of mixtures that cannot be separated by simple distillation, because the volatility of the two components in the mixture is nearly the same, causing them to evaporate at nearly the same temperature at a similar rate, making normal distillation impractical.

The process has been suggested to be suitable for removing organic sulfur compounds from FCC naphtha without octane number loss and without increasing hydrogen consumption (Mužic and Sertić-Bionda, 2013).

8.2.3 DESULFURIZATION BY IONIC LIQUIDS

By definition, ionic liquids are a new class of green solvents that have the ability to incorporate catalytic reactions within the liquids (Sheldon, 2001; Zhang and Zhang, 2002); intensive research is currently under way on the removal of thiophene-type sulfur species (e.g., dibenzothiophene) from fuels because of the limitation of the traditional hydrodesulfurization method in removing these species. Ionic liquids have the ability to extract aromatic sulfur-containing compounds at ambient conditions without hydrogen consumption. In addition, ionic liquids are immiscible with fuel, and the used ionic liquids can be regenerated and recycled by solvent washing or distillation (Zhang and Zhang, 2002; Esser et al., 2004; Planeta et al., 2006; Nie et al., 2007; Li et al., 2009). Moreover, desulfurization using ionic liquids has received growing attention (Bösmann et al., 2001; Zhang and Zhang, 2002; Holbrey et al., 2003; Zhu et al., 2003, 2011; Huang et al., 2004; Liu et al., 2006a,b; Lu et al., 2006; Nie et al., 2006; Planeta et al., 2006; Zhang et al., 2007, 2011; Alonso et al., 2008; Jiang et al., 2008; Zhao et al., 2008a,b; Qiu et al., 2009; Kulkarni and Alonso, 2010).

Ionic liquids are organic salts that are in liquid state at temperatures below 100°C (212°F) and which have been predicted to take the place of organic solvents, because they have no measurable vapor pressure below their decomposition temperature and can be designed to have different properties depending on their structure (Ramroop Singh and Speight, 2011). Furthermore, ionic liquids have been recognized as novel designable solvents, i.e., their properties can be tuned by altering their ionic structures to meet specific demands, by tailoring their cationic and anionic structures to optimize their physico–chemical properties. Ionic liquids as a kind of extraction solvent do not remain in the organic phase, which is very convenient for separation after the desulfurization process (Zhang et al., 2004; Dai et al., 2008; Kuhlmann et al., 2009; Dharaskar, 2012).

The use of ionic liquids for selective extraction of sulfur compounds from diesel fuel is a proposal that was first developed in the 1990s (Bösmann et al., 2001). The ionic agents that produced the most success were aluminum trichloride-1-butyl-3-methylimidazolium chloride (BMIMCl/AlCl$_3$). Deep desulfurization using a chlorine-free ionic liquid (anion is octyl sulfate and cation N-octyl-N-methylimidazolium) has also been reported; however, this agent is not as sensitive to water as aluminum complexes (Welton, 1999; Sheldon, 2001).

Desulfurization by ionic liquids is based on extraction theories and is a mild solvent-based process, and organic ions in ionic liquids can be designed in numerous varieties and combined together to make practically unlimited number of ionic liquids (Figure 8.5) (Freemantle, 2001). Among these, imidazolium-based ionic liquids, such as [BMIM][PF$_6$], [EMIM][BF$_4$], [BMIM][MeSO$_4$], [BMIM][AlCl$_4$], and [BMIM][OcSO$_4$], have demonstrated a high selective partitioning for heterocyclic sulfur-containing molecules such as dibenzothiophene derivatives, single β-, and di-β-methylated dibenzothiophene derivatives. Selection of ions for ionic liquids used in organic sulfur removal from fuel oils is very important. Some of the chlorometallate ionic liquids—such as those with [BMIM] [AlCl$_4$] structures—show good selectivity for sulfur removal; however, they are very sensitive to air and moisture and may cause alkene polymerization in fuels (Huang et al., 2004). The size of the anions in ionic liquids is also important for the extraction of dibenzothiophene from an oil phase—larger anions tend to extract dibenzothiophene derivatives more effectively than smaller anions (Bösmann et al., 2001). Esser et al. (2004) have reported that imidazolium ions with larger alkyl substitution groups are better solvents for dibenzothiophene removal. However, they found that alkyl groups beyond a certain size lowered the selectivity.

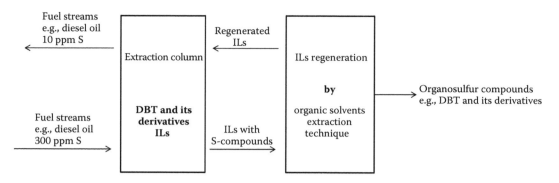

FIGURE 8.5   Desulfurization using ionic liquids (ILs).

Although [BMIM][OcSO$_4$] demonstrated a high partitioning for dibenzothiophene derivatives, it could also considerably dissolve nonsulfur organic molecules, especially cycloalkanes and aromatics. Therefore, it was not considered a selective ionic liquid for dibenzothiophene derivative compounds. There is an increasing trend on desulfurization of fossil fuels by ionic liquids with the purpose of the research on the use of ionic liquids in future refineries to economize desulfurization energy requirements, and to decrease carbon dioxide production that is associated with other desulfurization processes such as hydrodesulfurization. The recovery and recycling of ionic liquids during desulfurization process is difficult (Earle et al., 2006). Organic solvent extraction techniques can be used to recycle or recover ionic liquids, but solvent loss during the extraction process is undesirable.

It is the properties of ionic liquids that render them suitable for the process. For example, ionic liquids have very high electric conductivity and very low vapor pressure; they are inflammable and have high thermal stability, a wide range of liquidity, and the ability to separate several components. The miscibility of ionic liquids with water and organic solvents varies with the length of substituted chain on the cation, and with the choice of anion—ionic liquids can be formulated to behave as acids, bases, or ligands. Despite the low vapor pressure, some ionic liquids can be distilled under vacuum at temperatures of around 300°C (570°F). Ionic liquids have exhibited low affinity toward paraffins and olefins, somewhat higher affinity to aromatics, while their affinity toward thiophene derivatives and methylthiophene derivatives is very high. The structure and size of cations and anions of ionic liquids influence the ability to remove particular components. The affinity of ionic liquids toward aromatics increases with the electron density. However, ionic liquids are capable of removing organic sulfur compounds without the simultaneous removal of aromatics, which is desirable because when the feed is naphtha, this does not cause the reduction in octane number (Mužic and Sertić-Bionda, 2013).

In summary, the extraction of sulfur-containing compounds (nitrogen-containing compounds) from distillates by ionic liquids indicates that such a process could be an alternative to common hydrodesulfurization for deep desulfurization down to values of 10 ppm sulfur or even lower. The results show the selective extraction properties of ionic liquids, especially with regard to those S-compounds that are difficult to remove by hydrodesulfurization, such as dibenzothiophene derivatives in which the sulfur is shielded by adjacent alkyl groups present in middle distillate fractions. The application of mild process conditions (ambient pressure and temperature) and the fact that no hydrogen is needed are additional advantages compared with hydrodesulfurization. However, although extraction of sulfur-containing constituents (especially benzothiophene derivatives and dibenzothiophene derivatives) with ionic liquids is relatively straightforward, a major drawback is the loss in octane number by olefin saturation, which favors extraction with ionic liquids (Esser et al., 2004; Kuhlmann et al., 2009).

8.2.4 ALKYLATION

The alkylation technique is based on the concept that, when the boiling temperature of organosulfur compounds is shifted to a higher value, they can be removed from light fractions by distillation and concentrated in the heavy boiling part of the refinery streams (Javadli and De Klerk, 2012). The process employs alkylation of thiophene compounds via reaction with olefins present in the stream:

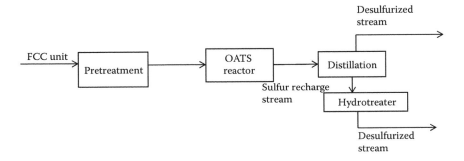

As a result of the alkylation, the boiling temperature (or boiling range) of the sulfur-containing hydrocarbon compounds increases. In comparison with thiophene (boiling point: 84°C, 183°F), alkylated thiophene derivatives such as 3-hexylthiophene and/or 2-octylthiophene have a much higher boiling point (221°C and 259°C, or 430°F and 498°F, respectively). This enables the alkylated derivatives to be easily separated from the main naphtha stream (boiling range 0–200°C, 32–390°F) by distillation. The high-boiling compounds produced can be blended into the diesel pool and desulfurized by conventional hydrotreating as the octane number is not important for diesel.

For example, the application of reactive distillation in naphtha alkylation desulfurization has been reported using a conceptual design and a rigorous steady-state simulation (Guo and Li, 2012). The conceptual design considered three aspects: (1) a thermodynamic analysis of the reactive system, (2) benefits and constraints of the reactive distillation, and (3) computation of the reactive residue curve maps. The simulation results illustrated that reactive distillation is feasible for naphtha alkylation desulfurization; would produce naphtha with sulfur content <1 ppm w/w; and the process has no constraints and has great potential for capital savings, improved conversion, reduced by-product formation, improved catalyst stability, and avoidance of reactor hotspots.

The OATS (olefin alkylation of thiophene sulfur) process technology consists of a pretreatment section, a fixed-bed reactor, and a product separation unit (Figure 8.6). Thiophene sulfur derivatives are alkylated in a reactor employing acidic catalysts such as boron trifluoride (BF_3), aluminum chloride ($AlCl_3$), zinc chloride ($ZnCl_2$), or antimony pentachloride ($SbCl_5$) deposited on silica (SiO_2), alumina (Al_2O_3), or silica–alumina (SiO_2–Al_2O_3) supports. From the reactor, the feed is sent to a conventional distillation column where it is separated into light sulfur-free naphtha and a higher-boiling sulfur-rich stream. The light naphtha is directly sent to the gasoline pool and the heavy stream is preferably hydrotreated. The hydrotreater is not an essential part of the technology,

FIGURE 8.6 OATS process.

but its application after the fractionators increases the product yield. Another advantage of the process is that less hydrogen is consumed since only a relatively low volume of the naphtha stream is hydrotreated. The efficiency of the process can be limited by competing processes—alkylation of aromatic hydrocarbons and olefin polymerization. Fortunately, under the conditions employed, alkylation of the sulfur-containing compounds occurs more rapidly than that of aromatics. One of the disadvantages of the process is that the alkylated sulfur compounds produced require more severe hydrotreating conditions to eliminate sulfur (http://www.axens.net/product/technology-licensing/11002/oats.html).

In the process, thiophene sulfur and light sulfur species are converted to higher-boiling sulfur compounds by alkylation with olefins, thus producing a sulfur-free low-boiling fraction without the need for hydrogen. A higher fraction is produced that is sent to a reduced-size selective hydrodesulfurization unit for octane retention. When operated in the desulfurization mode, the OATS process severity is adjusted to meet the required sulfur target in the light cracked naphtha while minimizing oligomerization side reactions.

8.2.5 Desulfurization by Precipitation

Desulfurization by precipitation is based on the formation and subsequent removal of insoluble charge-transfer complexes (Figure 8.7). For example, the methylation of dibenzothiophene by methyl iodide (CH_3I) in the presence of silver tetrafluoroborate ($AgBF_4$) at room temperature has been reported to produce the corresponding highly polarized water-soluble S-methylthiophenium tetrafluoroborate, and suggested its application for the removal of benzothiophene derivatives and dibenzothiophene derivatives from the light gas oil (Acheson and Harrison, 1969).

Preliminary experiments were reported for a model organosulfur compound (4,6-dimethyl-dibenzothiophene), hexane, and gas oil, using 2,4,5,7-tetranitro-9-fluorene as the most efficient π-acceptor (Milenkovic et al., 1999). A suspension of the π-acceptor and sulfur-containing gas oil was stirred in a batch reactor where insoluble charge-transfer complexes between π-acceptor and dibenzothiophene derivatives formed. The consecutive steps include filtration to remove the formed complex from gas oil and the recovery of the π-acceptor excess using a solid adsorbent. Currently, the efficiency is very low—one treatment results in the removal of only 20% w/w of the sulfur. Moreover, there is a competition in complex formation between dibenzothiophene derivatives and other nonsulfur aromatic derivatives that results in low selectivity for the removal of dibenzothiophene derivatives. The experimental results reported are not very informative because the role of other compounds that might form π-complexes (aromatics, n-compounds) has not been studied. Moreover, a large overstoichiometric amount of 2,4,5,7-tetranitro-9-fluorene is used to provide good complexing, and its excess should be removed from the oil stream afterward. It seems interesting to introduce a complexing agent into a solid organic or inorganic matrix. This would simplify the process since the filtration and π-acceptor recovery steps are avoided.

It has also been reported (Shiraishi et al., 2001a) that the desulfurization of light gas oil can be accelerated by increasing the quantity of the alkylating agents added in the desulfurization reactor,

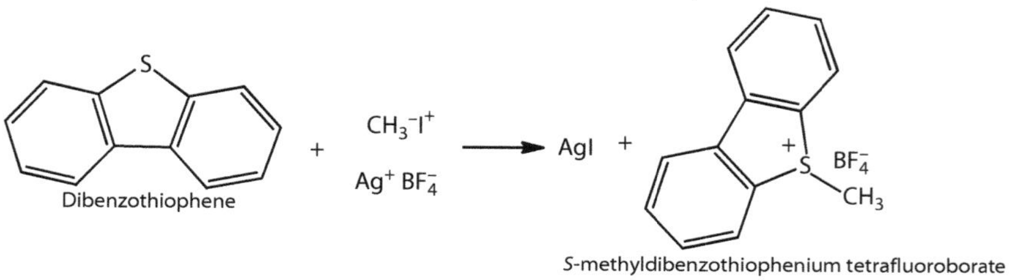

FIGURE 8.7 Desulfurization by precipitation.

such that the sulfur content of the light oils was decreased successfully to <0.005% w/w. In addition, the desulfurization of catalytic-cracked naphtha by a precipitation technique using alkylating agents (methyl iodide and silver tetrafluoroborate) under mild conditions at room temperature has also been reported (Shiraishi et al., 2001b). The sulfur is removed as the corresponding alkyl sulfonium salts of thiols, disulfide derivatives, tetrahydrothiophene derivatives, thiophene derivatives, and benzothiophene derivatives (which are the most difficult compounds to remove in the desulfurization process). The sulfur content of the catalytic-cracked naphtha was decreased from 100 to <30 ppm. Although the olefin concentration was decreased significantly following desulfurization, the resulting catalytic-cracked naphtha showed as high an octane number as the inlet feedstock.

8.2.6 SELECTIVE ADSORPTION

Conventional hydrodesulfurization technology as it is currently being used in the petroleum industry involves the use of (1) specialized reactors, (2) purified hydrogen, (3) high temperature, and (4) high pressure (Speight, 2000, 2014; Speight and Ozum, 2002; Parkash, 2003; Hsu and Robinson, 2006; Ancheyta and Speight, 2007; Gary et al., 2007). An alternative to the hydrodesulfurization process is desulfurization by means of adsorption, wherein sulfur compounds are selectively removed through adsorption on the solid adsorbent, leaving behind sulfur-free fuel (Adeyi and Aberuagba, 2012).

Adsorptive desulfurization is a promising approach to produce fuel cell grade gasoline and diesel at relatively low temperature and low pressure without using hydrogen gas, and advantageous compared with the conventional hydrodesulfurization method that uses high temperature and high hydrogen pressure. In fact, adsorptive desulfurization of liquid fuels has been achieved using a variety of procedures (Salem, 1994; Gamil and Nasser, 1997; Ma et al., 2002, 2005; Song, 2002; Song and Ma, 2003; Velu et al., 2003a,b; Chavan et al., 2012; Javadli and De Klerk, 2012)—including photodegradation adsorption (Faghihian and Sadeghinia, 2014)—and a variety of adsorbents have been used for this purpose, such as modified composite oxides (Seredych and Bandosz, 2010), activated carbon (Marin-Rosas et al., 2010), and mesoporous and microporous zeolites such as Faujasite (Salem, 1994; McKinley and Angelici, 2003), 5-A, 13-X, ZSM-5, and Y-Zeolite (Weitkamp et al., 1991; Salem and Hamid, 1997; Velu et al., 2003a).

Thus, adsorption can become a key separation technique in industry, particularly in the oil and gas industry (Mikhail et al., 2002; Hernández-Maldonado and Yang, 2003, 2004a,b; Zaki et al., 2005; Wang et al., 2006, 2008; Wang and Yang, 2007; Shakirullah et al., 2009; Majid and Seyedeyn-Azad, 2010). Adsorbents used industrially are generally synthetic microporous solids: activated carbon, molecular sieve carbon, activated alumina, silica gel, zeolites, and bleaching clay. The adsorbents are usually agglomerated with binders in the form of beads, extrudates, and pellets of a size consistent with the application that is considered (Al Zubaidy et al., 2013; Gawande and Kaware, 2014). The sulfur compounds can be removed from commercial fuels either via reactive adsorption by chemisorption π-complexation, or van der Waals forces and electrostatic interactions. In fact, the adsorptive desulfurization process is an easy and rapid method to remove sulfur from fuel (such as diesel fuel). In the process, the addition of activated carbon can cause reduction in sulfur content by >54% w/w of the original sulfur content (Al Zubaidy et al., 2013).

Activated carbon and carbon fibers are well-known multipurpose adsorbents used for treating different gases and liquids. Their surface structure and other surface properties can be adjusted to improve adsorption efficiency (Mužic and Sertić-Bionda, 2013). Specific application of activated carbon as adsorbent depends on the molecules that are being targeted for adsorption. Adsorption on activated carbon can be physical, which depends on the size and volume of the pores, or chemical, which depends on the chemical properties of the surface, i.e., surface properties that are conducive to chemisorption. Adsorption on activated carbon can be carried out in several types of reactors such as batch vessel reactor, continuously stirred reactor, and in fixed- or fluidized-bed reactors.

The advantage of the fluidized-bed column reactors is its ability to achieve high mixing efficiency. However, after some time, increased wearing and erosion of the adsorbent particles occurs. These drawbacks of the fluidized-bed adsorption column as well as the high prices of powdered activated carbon, mean that adsorptive desulfurization could be competitively carried out in fixed-bed columns with granulated activated carbon. The saturated activated carbon must be regenerated for it to be used again. The most commonly used method of regeneration is thermal treatment; however, chemical and biological treatments as well as ultrasound could also be used (Mei et al., 2003; Javadli and De Klerk, 2012; Mužic and Sertić-Bionda, 2013; Hosseini and Hamidi, 2014).

Solvent extraction is another method used to regenerate the desulfurization adsorbents. There are some disadvantages in these two methods. For the solvent extraction method, it is difficult to separate sulfur compounds from the organic solvents and to reuse these solvents. For the calcination method, sulfur compounds and aromatics are burned out, which can lose the heat value of fuels.

Reactive adsorption desulfurization is a process where the metal-based adsorbent bonds the sulfur on the surface and forms disulfide—the S-Zorb process was developed for removing sulfur from naphtha and kerosene by reactive adsorption and is based on fluidized-bed technology at relatively severe conditions: temperatures between 340°C and 400°C (645°F and 750°F) and pressures up to 300 psi. The sulfur atom in the organic sulfur compound is first adsorbed on the surface, after which it reacts with the adsorbent and the newly formed hydrocarbon is released into the main stream. The used adsorbent is continuously removed from the reactor and transported into the regeneration chamber. The sulfur is removed from the surface of the adsorbent by burning, and the formed sulfur dioxide is sent to the sulfur plant. The adsorbent is then reduced with hydrogen and recycled back to the reactor (Mužic and Sertić-Bionda, 2013).

Desulfurization by adsorption is a green technology and is characterized by easy operation, low cost, and no pollution, and leads to deep desulfurization. The sulfur compounds can be removed from commercial fuels either via reactive adsorption by chemisorption; π-complexation, i.e., activated adsorption through the interaction between adsorbent and adsorbate, or physical adsorption; and van der Waals and electrostatic interactions, i.e., adsorption through intermolecular forces of attraction between molecules of the adsorbent and the adsorbate (Yang et al., 2014).

The modified y-type zeolite was popularly used as an adsorbent to remove sulfur from fuels via π-complexation and copper (Cu^+), and silver-exchanged y-type zeolites were effective to remove sulfur compounds from naphtha (Yang et al., 2001; Takahashi et al., 2002; Hernández-Maldonado and Yang, 2004a). Selective removal of dibenzothiophene and 4,6-dimethyl dibenzothiophene can be removed from naphtha-type feedstocks, kerosene feedstocks, and crude oil is also possible by adsorptive desulfurization (McKinley and Angelici, 2003; Velu et al., 2005; Shimizu et al., 2007; Adeyi and Aberuagba, 2012; Al Zubaidy et al., 2013). Carbon prepared from rice husks have also been used as adsorbent (Ania and Bandosz, 2005), as has adsorbents composed of metal–organic framework compounds (Blanco-Brieva et al., 2010), and activated carbon (Opara et al., 2013).

In addition, special-morphology zinc oxide materials with smaller crystallite sizes were synthesized by the hydrothermal homogeneous precipitation method, and the corresponding Ni/ZnO adsorbents were prepared by an incipient impregnation method (Zhang et al., 2013). Reactive adsorption desulfurization of model gasoline using thiophene as model sulfur-containing compounds over Ni/ZnO adsorbent was carried out in a fixed-bed reactor in the presence of hydrogen. The results showed that the Ni/ZnO adsorbent using ZnO with a larger surface area and smaller crystal gains as active component shows higher desulfurization activity and stability. During the adsorption process of thiophene on the Ni/ZnO adsorbent, sulfur is trapped by zinc oxide and converted to zinc sulfide. Furthermore, in a comparative study of the desulfurization potential of two metal oxides—activated manganese dioxide and activated zinc oxide—the indications were that significant sulfur depletion occurred with activated manganese dioxide in 5–6 h, and the kinetics of the process was best described as pseudo-second order (Adeyi and Aberuagba, 2012).

However, adsorption desulfurization (Figure 8.8) has some problems to be solved. There are high proportions of aromatics compared with the amounts of sulfur compounds in middle distillate such

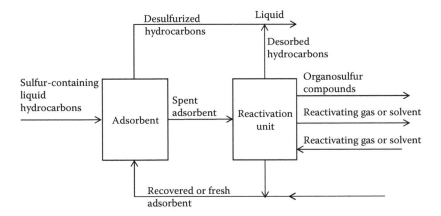

FIGURE 8.8   Desulfurization by adsorption.

as kerosene and diesel, and the aromatics can also be adsorbed on the desulfurization adsorbents. Thus, the adsorbents should be well designed to achieve suitable selectivity; when the selectivity is low, the adsorbents are easily regenerated. However, this can lead to heat loss because of the comparative adsorption. As the selectivity increases, the spent adsorbents become more and more difficult to be regenerated (Hernández-Maldonado and Yang, 2004a,b). For example, S-Zorb desulfurization can decrease sulfur in feedstock containing >2 mg/mL sulfur to a very low level of <0.005 mg/mL in the presence of hydrogen within an operating temperature and pressure of 300–400°C (570–750°F) and 275–500 MPa, respectively. This would affect the hydrocarbon skeleton of dibenzothiophene and its derivatives (Stanislaus et al., 2010). Consequently, the major challenge in adsorptive desulfurization is to design an adsorbent material that can selectively adsorb the sulfur compounds from fuel without altering the aromatic content.

Solvent extraction and oxidation in the air are two methods to regenerate the desulfurization adsorbents; however, there are disadvantages inherent in these two methods. The solvent extraction method suffers from the disadvantage that it is difficult to separate sulfur compounds from the organic solvents and to reuse these solvents. On the other hand, in the calcination method, sulfur compounds and aromatics are burned out, which can lose the heat value of fuels.

In summary, adsorption can be used for desulfurization of petroleum fractions based on the ability of solid adsorbent material to selectively adsorb organic sulfur compounds. There are two types of adsorptive desulfurization: adsorptive desulfurization during which physical and/or chemical adsorption of organic sulfur compounds takes place on the surface of the adsorbent, and reactive adsorption desulfurization during which organic sulfur compounds react with chemical species on the surface of the adsorbent and the sulfur is chemically bound, usually in the form of sulfide, while the newly formed hydrocarbon compound is released into the stream. The efficiency of adsorptive desulfurization is, for the most part, influenced by the adsorbent properties: capacity, selectivity, stability, and ability to regenerate. Materials used as adsorbents for desulfurization include different forms of activated carbon, zeolite, metallic oxides, and porous metals.

8.2.7   OXIDATIVE DESULFURIZATION

There are several types of organic sulfur compounds in petroleum (Chapter 2), and of particular interest are the aromatic-type sulfur compounds such as benzothiophene or dibenzothiophene, which are chemically stable; moreover, carbon–sulfur bond fission is not an easy process. Traditionally, catalytic hydrogenation (Chapter 4) has been used for such bond cleavage, which has several advantages for industrial purposes, but also some disadvantages. The use of hydrogen gas and catalyst necessitated taking more and more drastic reaction conditions (higher pressure and higher temperature) in order to attain a higher degree of the sulfur removal. This requires continuous development and use of highly active catalysts that are sufficiently efficient (and expensive) to achieve the long-term target of sulfur removal.

Oxidative desulfurization is an innovative technology that can be used to reduce the cost of producing ultra-low-sulfur diesel, and is a process during which organic sulfur compounds are oxidized to sulfoxides and sulfone (Jiang et al., 2011; Javadli and De Klerk, 2012):

Suflur compound   Sulfoxide   Sulfone

The oxidation products are subsequently removed from the feed by a separation method

Feedstock → oxidation → phase separation → sulfone separation → clean fuel

The methods used for separating oxidized sulfur compounds from treated fuel feedstocks include extraction, adsorption, distillation, and thermal decomposition.

During oxidative desulfurization, the refractory sulfur compounds, such as dibenzothiophene derivatives, are oxidized to form sulfones. Sulfones have higher polarity than the initial sulfur compounds, which is why they are more easily removed from petroleum fractions. In practice, the oxidative desulfurization process has encountered some technological and economic problems. When the process includes extraction, an additional problem of product loss occurs. In addition, the oxidative desulfurization process is a source of new, so-called, sulfonic waste, which requires special treatment.

The introduction of regulations stipulating ultra-low sulfur content in fuels caused the peroxides to become the most used oxidizing agents, such as hydrogen peroxide (H_2O_2) (Jiang et al., 2011; Da Silva and Dos Santos, 2013; Mamaghani et al., 2013; Joskić et al., 2014) and molecular oxygen (Murata et al., 2003). The use of molecular oxygen may be appealing to refineries that already have the infrastructure for an oxidation facility to prepare blown asphalt.

Other oxidizing agents include *in situ* formed per-acids, organic acids, phosphate acid and heteropolyphosphate acids, Fe-tetra amido macrocyclic ligand (Fe-TAML), Fenton and Fenton-like compounds, as well as solid catalysts such as those based on titanium-silica, W-V-TiO$_2$, solid bases such as magnesium and lanthanum metal oxides or hydro-talcite compounds, iron oxides and oxidizing catalysts based on monoliths, and *tert*-butyl-hydroperoxide with catalyst support. These compounds can effectively oxidize organic sulfur compounds into sulfones with less residue formation (Abdul Jalil and Falah Hasan, 2012; Mužic and Sertić-Bionda, 2013).

Oxidative desulfurization is envisaged as a milder approach for deep desulfurization in which sulfur compounds (sulfide derivatives) are oxidized to more polar compounds (such as sulfone derivatives and sulfoxide derivatives) using an oxidizing reagent (such as aqueous hydrogen peroxide, *tert*-butyl hydroperoxide, oxygen/aldehyde, and potassium ferrate). The reaction requires the presence of homogenous catalysts such as formic acid, cobalt–manganese–nickel acetates, and acetic acid, or heterogeneous catalysts such as tungsten/zirconia, titanium/mesoporous silica, peroxy-carboxylic acid, functionalized hexagonal mesoporous silica, molybdenum–vanadium oxides supported on alumina, titania, ceria, niobia, silica, and cobalt–manganese–nickel oxides supported on alumina, followed by their separation from the hydrocarbon part by a suitable separation process (extraction or adsorption) (Nanoti et al., 2009; Jiang et al., 2011). However, the efficiency and economics of an oxidative desulfurization process is strongly dependent on the methods used for oxidizing the sulfur compounds, and successively the methods used for separating the sulfone derivative and/or the S-sulfoxide derivatives from the oxidized fuels.

Oxidative desulfurization of petroleum is based on the polarity of sulfur compounds. It is well known that the electronegativity of sulfur is similar to that of carbon. Consequently, the carbon–sulfur bond is relatively nonpolar, and the organo-sulfur compounds exhibit properties quite similar to their corresponding organic compounds, so their solubility in polar and nonpolar solvents are nearly the same. Upon oxidation of these sulfur compounds, the polarity of the produced sulfoxides/sulfones increases, and consequently their solubility in polar solvents increase, which can be separated by extraction or adsorption (Ito and Van Veen, 2006). However, it is known that homogeneous catalysts are difficult to separate from the reaction products, and this limits their recycling. Thus, the preparation of new supported catalysts is the most desirable improvement of the oxidative desulfurization process (García-Gutiérrez et al., 2008). Nowadays, different catalysts (Wang and Yu, 2013; Kadijani et al., 2014; Mjalli et al., 2014) and ultrasound oxidative desulfurization (Wu and Ondruschka, 2010; Zhao and Wang, 2013; Liu et al., 2014; Wittanadecha et al., 2014) are applied to improve reaction efficiency. For example, ultrasonic-assisted catalytic ozonation combined with extraction process exhibits high catalytic efficiency for the removal of dibenzothiophene from simulated diesel oil (Zhao and Wang, 2013).

Compared with hydrodesulfurization, oxidative desulfurization has several advantages: the refractory sulfur compounds, such as alkylated dibenzothiophene derivatives, are easily oxidized under low operating temperature and pressure; also, it does not require use of expensive hydrogen, so the process is safer and can be applied in small and medium-sized refineries, isolated ones, and those located away from hydrogen pipelines. The overall capital cost and requirements for oxidative desulfurization units are significantly less than that for a deep hydrodesulfurization unit (Gore, 2001; Guo et al., 2011).

In recent years, oxidative desulfurization has gained importance and is considered an excellent option after the hydrodesulfurization process, since the alkyl-substituted dibenzothiophene derivatives are easily oxidized under low-temperature and low-pressure conditions to form the corresponding sulfoxide derivatives and sulfone derivatives. While sulfur compounds such as disulfide derivatives are easy to hydrodesulfurize, they oxidize slowly. For this reason, oxidative desulfurization can be used as a second stage after existing hydrodesulfurization units, taking a low sulfur diesel (500 ppm) down to ultra-low sulfur diesel (<10 ppm) levels. The efficiency and economics of an oxidative desulfurization process is strongly dependent on the methods used for oxidizing the sulfur compounds and, successively, on the methods used for separating the sulfoxide derivatives and sulfone derivatives from the oxidized fuels.

Solvent extraction of the sulfone–sulfoxide derivatives using γ-butyrolactone, n-methyl pyrrolidone, methanol, dimethylformamide, acetonitrile, and furfural has been reported (Liu et al., 2008). In general, one of the major drawbacks of the solvent extraction method is the appreciable solubility of hydrocarbon fuels in polar solvents, which leads to significant losses of usable hydrocarbon fuel. Such a loss is completely unacceptable on a commercial basis. Beside this, sulfone derivatives are polar compounds and form strong bonding with polar solvents, and it is difficult to remove them from the solvents to <10 ppm w/w. Hence, there will be buildup of sulfone derivatives in the solvent during solvent recovery (Nanoti et al., 2009).

On the other hand, the removal of sulfone–sulfoxide derivatives could be achieved using an adsorption technique (using adsorbents such as silica gel, activated carbon, bauxite, clay, coke, alumina, silicalite [polymorph of silica], ZSM-5, zeolite β, zeolite x, and zeolite y) in addition to the mesoporous oxide-based materials. These have attracted much attention in the recent years due to their large pore sizes and controlled pore size distribution, which may be beneficial in allowing accessibility of large-molecular-size sulfone derivatives to the surface active sites. However, one of the major drawbacks of the adsorption technique is that the amount of oil treated per unit weight of adsorbent is low (Babich and Moulijn, 2003).

In 1996, Petro Star Inc. combined conversion and extraction desulfurization to remove sulfur from diesel fuel (Chapados et al., 2000); briefly, the diesel fuel is oxidized by mixing with per-oxyacetic acid, i.e., H_2O_2/acetic acid at low temperature (<100°C) and under atmospheric pressure. Then, liquid–liquid extraction takes place, producing low-sulfur-content diesel oil followed by adsorption treatment to yield ultra-low-sulfur diesel oil. Recycling of extract solvent for reuse takes place, and the concentrated extract is further processed to remove sulfur. British Petroleum used hydrogen peroxide, phosphotungstic acid as a catalyst, and tetraoctylammoniumbromide as the phase-transfer agent in a mixture of water and toluene, and applied this oxidative desulfurization process after the hydrodesulfurization process (Collins et al., 1997). Lyondell Chemical Company used t-butyl hydroperoxide in a fixed-bed reactor for a cost-effective oxidative desulfurization process (Liotta and Han, 2003).

The advantages of oxidative desulfurization—in comparison with conventional hydrodesulfurization—include (1) the requirement of rather moderate reaction conditions, (2) no need for expensive hydrogen, and (3) higher reactivity of aromatic sulfur compounds in oxidation reactions since the electrophilic reaction with sulfur atom is enhanced by the higher electron density of aromatic rings (Javadli and De Klerk, 2012). Alkyl groups attached to the aromatic ring increase the electron density on the sulfur atom even more; however, the reactivity of the feedstock constituents to the process is also dependent on the type of catalyst.

Oxidative desulfurization technology offers a non-hydrogen-consuming, lower-capital-cost, more sustainable alternative to conventional hydrodesulfurization technology. The technology can be applied as part of an effective strategy for an alternative to revamping intermediate- and low-pressure hydrodesulfurization units to produce ultra-low-sulfur diesel, or as a means to reduce the operating cost of low-pressure units that have already been revamped. Oxidative desulfurization can also be considered as an attractive revamp opportunity for any existing unit where construction of new high-pressure hydrodesulfurization units is the alternative. Thus, the key to the successful implementation of this technology in most refinery applications is effectively integrating the oxidative desulfurization unit with the existing diesel hydrotreating unit in a revamp situation (Gatan et al., 2004).

8.2.8 BIOCATALYTIC DESULFURIZATION

Biocatalytic desulfurization is often considered as a potential alternative to the conventional deep hydrodesulfurization processes used in refineries. In this process, bacteria remove organosulfur from petroleum fractions without degrading the carbon skeleton of the organosulfur compounds. During the process, alkylated dibenzothiophene derivatives are converted to nonsulfur compounds—for example, dibenzothiophene is converted to 2-hydroxybiphenyl. Biocatalytic desulfurization offers mild processing conditions and reduces the need for hydrogen. Both these features would lead to high energy savings in the refinery—however, the by-products may require hydrogen treatment to complete the conversion to hydrocarbons. Furthermore, significant reductions in greenhouse gas emissions have also been predicted if biocatalytic desulfurization is used.

As already mentioned, hydrodesulfurization is not equally effective in desulfurizing all classes of sulfur compounds present in fossil fuels. On the other hand, the biocatalytic desulfurization process is effective regardless of the position of alkyl substituents (Pacheco, 1999). However, the hydrodesulfurization process conditions are sufficient not only to desulfurize sensitive (labile) organosulfur compounds, but also to (1) remove nitrogen and metals from organic compounds, (2) induce saturation of at least some carbon–carbon double bonds, (3) remove substances having an unpleasant smell or color, (4) clarify the product by drying it, and (5) improve the cracking characteristics of the material (Swaty, 2005). Therefore, in consideration of such advantages, placing a biocatalytic desulfurization unit downstream of a hydrodesulfurization unit as a complementary technology (rather than as a replacement technology) to achieve ultra-deep desulfurization is a realistic consideration giving a multistage process for desulfurization of fossil fuels (Monticello et al., 1996; Pacheco, 1999; Fang et al., 2006).

8.2.9 MEMBRANE SEPARATION

Membrane separation has been used for desulfurization of refined hydrocarbons, upstream distillation products, and other hydrocarbon products of cracking processes and/or other refining operations, e.g., naphtha (Lin, 2009) and FCC naphtha (Kong et al., 2008). A dense layer of polymeric membrane is needed to improve the selectivity of the membrane toward sulfur compounds, which consequently requires extensive energy consumption to pass the fuel feed through the membrane and/or requires the use of a transport agent to stimulate the transportation rate of sulfur compounds through the dense layer of the polymeric membrane. Pervaporation, where liquid fuels are transformed to gaseous phase by heating and/or vacuum using high energy, is required to increase the permeate transport rate throughout the membrane desulfurization processes, and combined techniques can be employed (Yahiya et al., 2013).

In the combined techniques, the membranes are either hosting liquid within its pores to provide selective media for the separation, or providing a good contact with an extractive or selective liquid within its pores. The membranes are not contributing to the selectivity of the separation, but they are still contributing to the overall mass transfer of the fuel stream. This depends on the porosity, pore size distribution, membrane thickness, and pore tortuosity. The sulfur-containing liquid hydrocarbons pass through the retentate side of the membrane, and extractive liquid passes along the permeate side of the membrane. The membrane provides a controlled interface to allow the extractive liquid to draw sulfur-containing compounds from the hydrocarbon liquid. However, this process faces a big challenge due to the formation of fouling material on the membrane surface, which consequently would slow the mass transfer. In this proposed technique, light naphtha is found to be more promising than ionic liquids or furfural as an extractive liquid because it transfers under the osmotic pressure difference to the whole crude oil. Thus, it prevents fouling and consequently promotes faster diffusion of sulfur-containing species.

8.2.10 OTHER METHODS

8.2.10.1 Ambient or Mild Conditions without Hydrogen

In this process, selective adsorption for the removal of sulfur compounds is carried out and followed by hydrodesulfurization of concentrated sulfur compounds using high-activity catalysts such as cobalt–molybdenum (CoMo/MCM-41) (Song, 2003). Such a technique is much easier than conventional hydrodesulfurization of diesel streams for two reasons: (1) its reactants are more concentrated and thus reactor utilization is more efficient and (2) the rate of hydrodesulfurization reaction is faster in relation to the removal of aromatics that inhibit the hydrodesulfurization by competitive adsorption in the hydrogenation sites.

8.2.10.2 Elevated Temperatures under Hydrogen without Hydrogenation of Aromatics

Desulfurization takes place in the presence of hydrogen in order to accelerate the reaction between the sulfur compounds and the adsorbing agent; however, less hydrogen is consumed than is the case with hydrodesulfurization processes. Research and development work has focused on different methods: fluidized bed, moving bed, and slurry for conveying the adsorbing agent through the reaction column and regeneration column.

The S-Zorb process uses a sorbent that is based on the reduced metal that reacts with sulfur in treated distillate containing 500 ppm w/w sulfur to become metal sulfide in fluidized bed. The spent sorbent is continuously withdrawn from the reactor and transferred to the regenerator section. The cleansed sorbent is further reduced by hydrogen and then recycled back to the reactor for removing more sulfur (Hernández-Maldonado and Yang, 2004a).

The Irvad process uses a slurry catalyst and consumes very little (if any) hydrogen. An alumina-based selective adsorbent is used to contact the feedstock in a countercurrent operation at temperatures below 240°C (465°F) in a multistage adsorber. The adsorbent is regenerated in a continuous cross-flow reactivator using heated gas. The process operates at lower pressure and does not consume hydrogen or saturate olefin constituents of the feedstocks. The adsorption mechanism is based on the polarity of sulfur compounds (Sano et al., 2005).

8.3   MOLECULAR IMPRINTING TECHNOLOGY

Deep desulfurization of liquid fuels by molecular imprinting technology is expected to find wide application since the technology is based on the application of biosensors, separation media, and affinity supports for the recognition of target molecules (Liu et al., 2006b). Molecular imprinting technology has a relationship to the more recent catalyst design studies in which catalysts are designed with specific surface and pore features to accommodate specific feedstock molecules.

Thus, molecular imprinting technology has a unique predeterminative characteristic, specificity, and practicability since it is based on the creation of specific molecular recognition sites in polymers to identify template molecules. The prepared polymer is known as a molecularly imprinted polymer because it is complimentary to the template in space structure and binding sites. Consequently, based on the selectivity mechanism of molecularly imprinted polymer toward the sulfur compounds in fuel oils, the prepared molecularly imprinted polymer using benzothiophene, dibenzothiophene, and dibenzothiophene sulfone as templates would remove thiophene derivative from fuel oils.

As an example, the preparation of molecularly imprinted polymers by bulk polymerization has been reported using dibenzothiophene as the template, 4-vinyl pyridine as the functional monomer, ethylene glycol dimethacrylate as the cross-linker, and toluene as the solvent, with maximum binding capacity of 48.3 mg/g at 20°C (68°F) and good removing capacity toward benzothiophene derivatives and dibenzothiophene derivatives (Chang et al., 2003). In addition, the synthesis of imprinted chitosan hydrogels by bulk polymerization has also been described using glutaric dialdehyde as cross-linker, dibenzothiophene as the template, and acetonitrile/water as the solvent; the adsorption capacity for dibenzothiophene was on the order of 16.54 mg/g at 25°C (77°F) and increased to a maximum 27.5 mg/g at 50°C (122°F) (Aburto and Borgne, 2004). Also, chitosan molecularly imprinted polymer was prepared by dispersion polymerization using dibenzothiophene as the template, paraffin as the dispersed phase, Span 80 as the surfactant, and glutaric dialdehyde as the cross-linking agent (Chang et al., 2010). Adsorptive desulfurization of naphtha at 25°C (77°F) showed a capacity of 3.52 mg/g that remained unchanged after 10 adsorption–regeneration cycles.

Surface molecular imprinting technology is very promising for deep desulfurization of liquid fuel oils since molecular imprinting polymers can be prepared with high selectivity, fast adsorption, and good mechanical and thermal stability. The surface molecularly imprinting polymer can be prepared by grafting a thin imprinted polymer film with a large amount of binding sites onto inorganic support—such as silica gel, titanium dioxide (TiO_2), potassium titanate ($K_2Ti_4O_9$), or carbon microspheres. Silica gel is usually preferred because of the high porosity, large surface area, good computability, mechanical property, and stability (Yang et al., 2014). It is also possible to accommodate graft polymerization using benzothiophene as template, silica gel modified by KH-550 as support, methacrylic acid as the monomer, ethylene glycol dimethacrylate as cross-linker, azo-iso-butyrontrile as the initiator, and toluene as solvent (Hu et al., 2010). This product had an adsorptive desulfurization capacity on the order of 57.4 mg/g in naphtha at 25°C (77°F) and remained unchanged after 30 cycles.

8.4 FUTURE

Alternative desulfurization processes, including advanced hydrodesulfurization processes based on separation, adsorption, and extraction; oxidation; and biodesulfurization processes are necessary because the conventional hydrodesulfurization process is experiencing difficulties in producing fuels in accordance with latest regulations, i.e., environmentally friendly fuels (Campos-Martin et al., 2010). Furthermore, biodesulfurization (Chapter 9), which is based on the application of microorganisms that selectively remove sulfur atoms from organosulfur compounds, appears to be a viable technology to complement the traditional hydrodesulfurization of fuels. Enzymes in bacteria selectively oxidize the sulfur, and then cleave carbon–sulfur bonds. Biodesulfurization will operate at ambient temperatures and atmospheric pressure, and thus will require substantially less energy than conventional hydrodesulfurization methods to achieve sulfur levels below those required by current regulatory standards. Biodesulfurization generates a fraction of the carbon dioxide that is generated in association with hydrodesulfurization, and it does not require hydrogen. Additionally, biodesulfurization can effectively remove some key sulfur-containing compounds that are among the most difficult for hydrodesulfurization to treat. Biodesulfurization can be used instead of, or complementary with, hydrodesulfurization (Piddington et al., 1995; Gray et al., 1996, 2003; Monticello, 2000; Noda et al., 2003; Martín et al., 2004; Gupta et al., 2005; Martín et al., 2005; Kilbane, 2006; Lee et al., 2006; Li et al., 2006; Xu et al., 2006; Soleimani et al., 2007).

From the technological standpoint, it is prudent to first try to improve the current hydrodesulfurization catalysts and reactors; however, it is clear that if significant advancement were to be made, new types of hydrodesulfurization catalysts and reactor designs have to be developed. In fact, several commercial processes have been developed and successfully installed in refineries, including the reactive adsorption S-Zorb process and extractive desulfurization Sulf-X process. However, for the most part, these alternative desulfurization processes need further development, although there are indications of economic viability that has already started on the road to widespread commercial application (Mužic and Sertić-Bionda, 2013).

In view of stringent environmental regulations, utilization of sulfur-containing fuel oils has severe limitations regarding emission of sulfur dioxide. Technology for reduction of sulfur in diesel fuel to 15 ppm is currently available, and new technologies that could reduce the cost of desulfurization are under development. Chemical oxidation in conjunction with ionic liquid extraction can sharply increase the removal of sulfur. Ionic liquids have the ability to extract aromatic sulfur-containing compounds at ambient conditions without H_2 consumption. The cations, anions, structure, and size of ionic liquids are important parameters affecting the extracting ability. In addition, ionic liquids are immiscible with fuel, and the used ionic liquids can be regenerated and recycled by solvent washing or distillation.

REFERENCES

Abdul Jalil, T., and Falah Hasan, L. 2012. Oxidative Desulfurization of Gas Oil Using Improving Selectivity for Active Carbon from Rice Husk. *Diyala J. Pure Sci.* 8(3): 68–81.

Abinaya, K., Sivalingam, A., and Kannadasan, T. 2013. Desulfurization of Liquid Fuels by Selective Extraction Method: A Review. *Int. J. Sci. Res.* 2(6): 172–175.

Aburto, J., and Borgne, S.L. 2004. Selective Adsorption for Dibenzothiophene Sulfone by an Imprinted and Stimuli-Responsive Chitosan Hydrogel. *Macromolecules* 37: 2938–2943.

Acheson, R.M., and Harrison, D.R. 1969. S-Alkylthiophenium Salts. *J. Chem. Soc. D Chem. Commun.* 724.

Adeyi, A.A., and Aberuagba, F. 2012. Comparative Analysis of Adsorptive Desulphurization of Crude Oil by Manganese Dioxide and Zinc Oxide. *Res. J. Chem. Sci.* 2(8): 14–20.

Alonso, L., Arce, A., Francisco, M., and Soto, A. 2008. Solvent Extraction of Thiophene from *n*-Alkanes (C7, C12, and C16) Using the Ionic Liquid [C8mim][BF4]. *J. Chem. Thermodynam.* 40(6): 966–972.

Al Zubaidy, I.A.H., Tarsh, F.B., Darwish, N.N., Majeed, B.S.S.A., Sharafi, A.A., and Chacra, L.A. 2013. Adsorption Process of Sulfur Removal from Diesel Oil Using Sorbent Materials. *J. Clean Energy Technol.* 1(1): 66–68.

Ancheyta, J., and Speight, J.G. (Editors) 2007. *Hydroprocessing Heavy Oils and Residua.* CRC Press, Taylor & Francis Group, Boca Raton, FL.

Ania, C.O., and Bandosz, T.J. 2005. Importance of Structural and Chemical Heterogeneity of Activated Carbon Surfaces for Adsorption of Dibenzothiophene. *Langmuir* 21: 7752–7759.

Atlas, R.M., Boron, D.J., Deever, W.R., Johnson, A.R., McFarland, B.L., and Meyer, J.A. 2001. Method for Removing Organic Sulfur from Heterocyclic Sulfur Containing Organic Compounds. US Patent H1,986, August 7.

Babich, I.V., and Moulijn, J.A. 2003. Science and Technology of Novel Processes for Deep Desulfurization of Oil Refinery Streams: A Review. *Fuel* 82: 607–631.

Blanco-Brieva, G., Campos-Martin, J.M., Al-Zahrani, S.M., and Fierro, J.L.G. 2010. Removal of Refractory Organic Sulfur Compounds in Fossil Fuels Using MOF Sorbents. *Glob. NEST J.* 12(3): 296–304.

Bösman, A., Datsevich, L., Jess, A., Lauter, A., Schmitz, C., and Wasserscheid, P. 2001. Deep Desulfurization of Diesel Fuel by Extraction with Ionic Liquids. *Chem. Commun.* 7(23): 2494–2495.

Bustos-Jaimes, I., Amador, G., Castorena, G., and Le Borgne, S. 2003. Genotypic Characterization of Sulfur-Oxidative Desulfurizing Bacterial Strains Isolated from Mexican Refineries. *Oil Gas Sci. Technol. Rev. Inst. Fr. Pet.* 58(4): 521–526.

Campos-Martin, J.M., Capel-Sanchez, M.C., Perez-Presas, P., and Fierro, J.L.G. 2010. Oxidative Processes for Desulfurization of Liquid Fuels. *J. Chem. Technol. Biotechnol.* 85(7): 879–890.

Castorena, G., Suarez, C., Valdez, I., Amador, G., Fernandez, L., and Le Borgne, S. 2002. Sulfur-Selective Desulfurization of Dibenzothiophene and Diesel Oil by Newly Isolated *Rhodococcus* Sp. Strains. *FEMS Microbiol. Lett.* 215(1): 157–161.

Chang, Y.H., Liu, B., Ying, H.J., and He, M.F. 2003. Solid-Phase Extraction Sorbent for Organosulfur Compounds Present in Fuels Made by Molecular Imprinting. *Ion Exchange Adsorp.* 19(5): 450–456.

Chang, Y.H., Zhang, L., Ying, H.J., Li, Z.J., Lv, H., and Ouyang, P.K. 2010. Desulfurization of Gasoline Using Molecularly Imprinted Chitosan as Selective Adsorbents. *Appl. Biochem. Biotechnol.* 160: 593–603.

Chapados, D., Bonde, S.E., Chapados, D., Gore, W.L., Dolbear, G., and Skov, E. 2000. Desulfurization by Selective Oxidation and Extract of Sulfur-Containing Compounds to Economically Achieve Ultra-Low Proposed Diesel Fuel Sulfur Requirements. In: *Proceedings. NPRA Annual Meeting AM-00-25.* San Antonio, TX, March 26–28.

Chavan, S., Kini, H., and Ghosal, R. 2012. Process for Sulfur Reduction from High Viscosity Petroleum Oils. *Int. J. Environ. Sci. Dev.* 3(3): 228–231.

Collins, F.M., Lucy, A.R., and Sharp, C. 1997. Oxidative Desulphurisation of Oils via Hydrogen Peroxide and Heteropolyanion Catalysis. *J. Mol. Catal.* A 117: 397–403.

Dai, W., Zhou, Y., Wang, S., Su, W., Sun, Y., and Zhou, L. 2008. Desulfurization of Transportation Fuels Targeting at Removal of Thiophene/Benzothiophene. *Fuel Process. Technol.* 89(8): 749–755.

Da Silva, M.J., and Dos Santos, L.F. 2013. Novel Oxidative Desulfurization of a Model Fuel with H_2O_2 Catalyzed by $AlPMo_{12}O_{40}$ under Phase Transfer Catalyst-Free Conditions. *J. Appl. Chem.* 2013: 147945. http://dx.doi.org/10.1155/2013/147945; accessed October 14, 2014.

Dharaskar, S.A. 2012. Ionic Liquids (A Review): The Green Solvents for Petroleum and Hydrocarbon Industries. *Res. J. Chem. Sci.* 2(8): 80–85.

Earle, M.J., Esperança, J.M.S.S., Gilea, M.A., Lopes, J.N.C., Rebelo, L.P.N., and Magee, J.W. 2006. The distillation and volatility of ionic liquids. *Nature* 439: 831–834.

Egorova, M. 2003. Study of the Aspects of Deep Hydrodesulfurization by Means of Model Reactions. PhD Thesis. Swiss Federal Institute of Technology, Zurich, Switzerland.

Egorova, M., and Prins, R. 2004. Mutual Influence of the HDS of Dibenzothiophene and HDN of 2-Methylpyridine. *J. Catal.* 221: 11–19.

Esser, J., Wasserscheid, P., and Jess, A. 2004. Deep Desulfurization of Oil Refinery Streams by Extraction with Ionic Liquids. *Green Chem.* 6: 316–322.

Faghihian, H., and Sadeghinia, R. 2014. Photo Degradation–Adsorption Process as a Novel Desulfurization Method. *Adv. Chem. Eng. Res.* 3: 18–26.

Fang, X.X., Zhang, Y.L., Luo, L.L., Xu P., Chen, Y.L., Zhou, H., and Hai, L. 2006. Organic Sulfur Removal from Catalytic Diesel Oil by Hydrodesulfurization Combined with Biodesulfurization. *Mod. Chem. Ind.* 26: 234–238 (Chinese journal; Abstract in English).

Franchi, E., Rodriguez, F., Serbolisca, L., and Ferra, F. 2003. Vector Development Isolation of New Promoters Enhancement of Catalytic Activity of DSZ Enzyme Complex in *Rhodococcus* sp. Strain. *Oil Gas Sci. Technol. Rev. Inst. Fr. Pet.* 58(4): 515–520.

Freemantle, M. 2001. New horizons for ionic liquids. *Chem. Eng. News* 79: 21–25.

Furimsky, E., and Massoth, F.E. 1999. Deactivation of Hydroprocessing Catalysts. *Catal. Today* 52: 381–495.

Gamil, M.H.R., and Nasser, M.M. 1997. Desulphurization of Um Al Nar Refinery Straight Run Kerosene and Gas Oil Using Pal Fruit Kernel Activated Charcoal: A Locally Made Adsorbent. *Adsorp. Sci. Technol.* 15(4): 311–321.

García-Gutiérrez, J.L., Fuentes, G.A., Hernandez-Teran, M.E., Garcıa, P., Murrieta-Guevara, F., and Jimenez-Cruz, F. 2008. Ultra-Deep Oxidative Desulfurization of Diesel Fuel by the Mo/Al$_2$O$_3$-H$_2$O$_2$ System: The Effect of System Parameters on Catalytic Activity. *Appl. Catal. A Gen.* 334: 366–373.

Gary, J.H., Handwerk, G.E., and Kaiser, M.J. 2007. *Petroleum Refining: Technology and Economics*, 5th Edition. CRC Press, Taylor & Francis Group, Boca Raton, FL.

Gatan, R., Barger, P., Gembicki, V., Cavanna, A., and Molinari, D. 2004. Oxidative Desulfurization: A New Technology for ULSD. *Prepr. Div. Fuel Chem. Am. Chem. Soc.* 49(2): 577–579.

Gates, B.C., and Topsøe, H. 1997. Reactivities in Deep Catalytic Hydrodesulfurization: Challenges, Opportunities, and the Importance of 4-Methyldibenzothiophene and 4,6-Dimethyldibenzothiophene. *Polyhedron* 16(18): 3213–3217.

Gawande, P.R., and Kaware, J. 2014. A Review on Desulphurization of Liquid Fuel by Adsorption. *Int. J. Sci. Res.* 3(7): 2255–2259.

Gore, W. 2001. Method of Desulfurization of Hydrocarbons. United States Patent 6,274,785, August 14.

Gray, K.A., Mrachko, C.T., and Squires, C.H. 2003. Biodesulfurization of Fossil Fuels. *Curr. Opin. Microbiol.* 6(3): 229–235.

Gray, K.A., Pogrebinsky, O.S., Mrachko, G.T., Xi, L., Monticello, D.J., and Squires, C.H. 1996. Molecular Mechanisms of Biocatalytic Desulfurization of Fossil Fuels. *Nat. Biotechnol.* 14(13): 1705.

Guo, B., and Li, Y. 2012. Analysis and Simulation of Reactive Distillation for Gasoline Alkylation. *Chem. Eng. Sci.* 72: 115–125.

Guo, W., Wang, C., Lin, P., and Lu, X. 2011. Oxidative Desulfurization of Diesel with TBHP/Isobutyl Aldehyde/Air Oxidation System. *Appl. Energy* 88: 175–179.

Gupta, N., Roychoudhury, K., and Deb, J.K. 2005. Biotechnology of Desulfurization of Diesel: Prospects and Challenges. *Appl. Microbiol. Biotechnol.* 66(4): 356–366.

Hernández-Maldonado, A.J., and Yang, R.T. 2003. Desulfurization of Liquid Fuels by Adsorption via P Complexation with Cu(I)Y and AgY Zeolites. *Ind. Eng. Chem. Res.* 42: 123.

Hernández-Maldonado, A.J., and Yang, R.T. 2004a. Desulfurization of Diesel Fuels by Adsorption via p-Complexation with Vapor Phase Exchanged (VPIE) Cu(I)-Y Zeolites. *J. Am. Chem. Soc.* 126: 992–993.

Hernández-Maldonado, A.J., and Yang, R.T. 2004b. New Sorbents for Desulfurization of Diesel Fuels via π-Complexation. *AIChE J.* 50(4): 791–801.

Holbrey, J.D., Reichert, W.M., Nieuwenhuyzen, M., Sheppard, O., Hardacre, C., and Rogers, R.D. 2003. Liquid Clathrate Formation in Ionic Liquid–Aromatic Mixtures. *Chem. Commun.* 4: 476–477.

Hosseini, H., and Hamidi, A. 2014. Sulfur Removal of Crude Oil by Ultrasound-Assisted Oxidative Method. In: *Proceedings. International Conference on Biological, Civil and Environmental Engineering (BCEE-2014)*. Dubai, United Arab Emirates (UAE), March 17–18. http://iicbe.org/siteadmin/upload /9150C0314090.pdf; accessed October 12, 2014.

Hsu, C.S., and Robinson, P.R. (Editors) 2006. *Practical Advances in Petroleum Processing*, Vols. 1–2. Springer Science, New York.

Hu, T.P., Zhang, Y.M., Zheng, L.H., and Fan, G.Z. 2010. Molecular Recognition and Adsorption Performance of Benzothiophene Imprinted Polymer on Silica Gel Surface. *J. Fuel Chem. Technol.* 38(6): 722–729.

Huang, C., Chen, B., Zhang, J., Liu, Z., and Li, Y. 2004. Desulfurization of Gasoline by Extraction with New Ionic Liquids. *Energy Fuels* 18: 1862–1864.

Inoue, S., Takatsaku, T., Wada, Y., Hirohama, S., and Ushida, T. 2000. Distribution Function Model for Deep Desulfurization of Diesel Fuel. *Fuel* 79(7): 843–849.

Ito, E., and Van Veen, R.J.A. 2006. On Novel Processes for Removing Sulphur from Refinery Streams. *Catal. Today* 116: 446–460.

Javadli, J., and De Klerk, A. 2012. Desulfurization of Heavy Oil. *Appl. Petrochem. Res.* 1(1–4): 3–19.

Jiang, X.C., Nie, Y., Li, C.X., and Wang, Z.H. 2008. Imidazolium-Based Alkylphosphate Ionic Liquids: A Potential Solvent for Extractive Desulfurization of Fuel. *Fuel* 87(1): 79–84.

Jiang, Z., Lu, H., Zhang, Y., and Li, C. 2011. Oxidative Desulfurization of Fuel Oils. *Chin. J. Catal.* 32: 707–715.

Joskić, R., Dunja, M., and Sertić-Bionda, K. 2014. Oxidative Desulfurization of Model Diesel Fuel with Hydrogen Peroxide. *Goriva Maziva* 53(1): 11–18 (in English).

Kadijani, J.A., Narimani, E., and Kadijani, H.A. 2014. Oxidative Desulfurization of Organic Sulfur Compounds in the Presence of Molybdenum Complex and Acetone as Catalysts. *Petrol. Coal* 56(1): 116–123.

Kilbane, J.J. 2006. Microbial Biocatalyst Developments to Upgrade Fossil Fuels. *Curr. Opin. Microbiol.* 17(3): 1–10.

Kong, Y., Lin, L., Zhang, Y., Lu, F., Xie, K., and Liu, R. 2008. Desulfurization of Naphtha using a Membrane. *Eur. Polym. J.* 44: 3335–3343.

Kuhlmann, E., Haumann, M., Jess, A., Seeberger, A., and Wasserscheid, P. 2009. Ionic Liquids in Refinery Desulfurization: Comparison between Biphasic and Supported Ionic Liquid Phase Suspension Processes. *ChemSusChem* 2(10): 969–977.

Kulkarni, P.S., and Alonso, C.A.M. 2010. Deep Desulfurization of Diesel Fuel Using Ionic Liquids: Current Status and Future Challenges. *Green Chem.* 2: 139–149.

La Paz Zavala, C., and Rodriguez, J.E. 2004. Practical Applications of a Process Simulator of Middle Distillates Hydrodesulfurization. *Petrol. Sci. Technol.* 22(1/2): 61–71.

Lappinen, M., and Higgins, T. 2013. *Review of EPA Proposed Tier 3 Motor Gasoline Refinery Cost Model.* Hart Energy: Research and Consulting, McLean, VA. Available at http://nepis.epa.gov/Exe/ZyNET.exe /P100ISWM.TXT?ZyActionD=ZyDocument&Client=EPA&Index=2011+Thru+2015&Docs=&Query =&Time=&EndTime=&SearchMethod=1&TocRestrict=n&Toc=&TocEntry=&QField=&QFieldYear =&QFieldMonth=&QFieldDay=&IntQFieldOp=0&ExtQFieldOp=0&XmlQuery=&File=D%3A%5Czyfil es%5CIndex%20Data%5C11thru15%5CTxt%5C00000010%5CP100ISWM.txt&User=ANONYMOU S&Password=anonymous&SortMethod=h%7C-&MaximumDocuments=1&FuzzyDegree=0&ImageQ uality=r75g8/r75g8/x150y150g16/i425&Display=p%7Cf&DefSeekPage=x&SearchBack=ZyActionL& Back=ZyActionS&BackDesc=Results%20page&MaximumPages=1&ZyEntry=1&SeekPage=x&ZyP URL; accessed September 4, 2014.

Lee, W.C., Ohshiro, T., Matsubara, T., Izumi, Y., and Tanokura, M. 2006. Crystal Structure and Desulfurization Mechanism of 2-Hydroxybiphenyl-2-Sulfinic Acid Desulfinase. *J. Biol. Chem.* 281(43): 32534–32539.

Li, F.T., Liu, R.H., Wen, J.H., Zhao, D.S., Sun, Z.M., and Liu, Y. 2009. Desulfurization of Dibenzothiophene by Chemical Oxidation and Solvent Extraction with e3NCH$_2$C6H$_5$Cl·2ZnCl$_2$ Ionic Liquid. *Green Chem.* 11: 883–888.

Li, G., Ma, T., Li, J., Liang, F., and Liu, R. 2006. Desulfurization of Dibenzothiophene by *Bacillus subtilis* Recombinants Carrying dszABC and dszD Genes. *Biotechnol. Lett.* 28(14): 1095–1100.

Lin, L. 2009. FCC Gasoline Desulphurization by Pervaporation. *Prepr. Am. Chem. Soc. Div. Fuel Chem.* 54(1): 188.

Liotta, F.J., and Han, Y.Z. 2003. Production of Ultra-Low Sulfur Fuels by Selective Hydroperoxide Oxidation. In: *Proceedings of NPRA Annual Meeting AM-03-23.* San Antonio, TX, March 26–28.

Liu, L., Zhang, Y., and Tan, W. 2014. Ultrasound-Assisted Oxidation of Dibenzothiophene with Phosphotungstic Acid Supported on Activated Carbon. *Ultrason. Sonochem.* 21(3): 970–974.

Liu, S., Wang, B., Cui, B., and Sun, L. 2008. Deep Desulfurization of Diesel Oil Oxidized by Fe(VI) Systems. *Fuel* 87: 422–428.

Liu, X.J., Ouyang, C.B., Zhao, R., Shangguan, D.H., Chen, Y., and Liu, G.Q. 2006a. Monolithic Molecularly Imprinted Polymer for Sulfamethoxazole and Molecular Recognition Properties in Aqueous Mobile Phase. *Anal. Chim. Acta* 571(2): 235–241.

Liu, Z.C., Hu, J.R., and Gao, J.S. 2006b. FCC Naphtha Desulfurization via Alkylation Process over Ionic Liquid Catalyst. *Petrol. Proc. Petrochem.* 37(10): 22–26.

Lu, H.Y., Gao, J.B., Jiang, Z.X., Jing, F., Yang, Y.X., Wang, G., and Li, C. 2006. Ultra-Deep Desulfurization of Diesel by Selective Oxidation with [C$_{18}$H$_{37}$N(CH$_3$)$_3$]$_4$[H$_2$NaPW$_{10}$O$_{36}$] Catalyst Assembled in Emulsion Droplets. *J. Catal.* 239(2): 369–375.

Ma, X., Sakanishi, K., and Mochida, I. 1994. Hydrodesulfurization Reactivities of Various Sulfur Compounds in Diesel Fuel. *Ind. Eng. Chem. Res.* 33: 218–222.

Ma, X., Sun, L., and Song, C. 2002. A New Approach to Deep Desulfurization of Gasoline, Diesel Fuel and Jet Fuel by Selective Adsorption for Ultra-Clean Fuels and for Fuel Cell Applications. *Catal. Today* 77: 107–116.

Ma, X.L., Sprague, M., and Song, C. 2005. Deep Desulphurization of Gasoline by Selective Adsorption over Nickel-Based Adsorbent for Fuel Cell Application. *Ind. Eng. Chem. Res.* 44(15): 5768–5775.

Majid, D., and Seyedeyn-Azad, F. 2010. Desulfurization of Gasoline over Nanoporous Nickel-Loaded Y-Type Zeolite at Ambient Conditions. *Ind. Eng. Chem. Res.* 49: 11254–11259.

Mamaghani, A.H., Fatemi, S., and Asgari, M. 2013. Investigation of Influential Parameters in Deep Oxidative Desulfurization of Dibenzothiophene with Hydrogen Peroxide and Formic Acid. *Int. J. Chem. Eng.* 2013: 951045. http://dx.doi.org/10.1155/2013/951045; accessed June 25, 2014.

Marin-Rosas, C., Ramírez-Verduzco, L.F., Murrieta-Guevara, F.R., Hernández-Tapia, G., and Rodriguez-Otal, L.M. 2010. Desulfurization of Low Sulfur Diesel by Adsorption Using Activated Carbon: Adsorption Isotherms. *Ind. Eng. Chem. Res.* 49: 4372–4376.

Martín, A.B., Alcón, A., Santos, V.E., and Garcia-Ochoa, F. 2004. Production of a Biocatalyst of *Pseudomonas putida* CECT5279 for Dibenzothiophene (DBT Biodesulfurization for Different Media Composition. *Energy Fuels* 18(3): 851–857.

Martín, A.B., Alcón, A., Santos, V.E., and Garcia-Ochoa, F. 2005. Production of a Biocatalyst of *Pseudomonas putida* CECT5279 for Dibenzothiophene (DBT): Influence of the Operational Conditions. *Energy Fuels* 19: 775–782.

McHale, W.D. 1981. Process for Removing Sulfur from Petroleum Oils. United States Patent 4,283,270.

McKinley, S.G., and Angelici, R.J. 2003. Deep Desulfurization by Selective Adsorption of Dibenzothiophenes on Ag^+/SBA-15 and $Ag^{+/}SiO_2$. *Chem. Commun.* 20: 2620–2621.

Mei, H., Mei, B.M., and Yen, T.F. 2003. A New Method for Obtaining Ultra-Low Sulfur Diesel Fuel via Ultrasound Assisted Oxidative Desulfurization. *Fuel* 82(4): 405–414.

Mikhail, S., Zaki, T., and Khalil, L. 2002. Desulphurization by Economically Adsorption Technique. *Appl. Catal. A Gen.* 227: 265–278.

Milenkovic, A., Shulz, E., Meille, V., Lofferdo, D., Forissier, M., Vriant, M., Saultet, P., and Lemair, M. 1999. Selective Elimination of Alkyldibenzothiophenes from Gas Oil by Formation of Insoluble Charge-Transfer Complexes. *Energy Fuels* 13: 881.

Mjalli, F.S., Ahmed, A.U., Al-Wahaibi, T., Al-Wahaibi, Y., and AlNashef, I.M. 2014. Deep Oxidative Desulfurization of Liquid Fuels. *Rev. Chem. Eng.* 30(4): 337–378.

Monticello, D.J., Haney, I.I.I., and William, M. 1996. Biocatalytic Process for Reduction of Petroleum Viscosity. US Pat. 5,529,930.

Monticello, D.J. 1998. Biodesulfurization of Diesel Fuels. *Chem. Technol.* 28(7): 38–45.

Monticello, D.J. 2000. Biodesulfurization and the Upgrading of Petroleum Distillates. *Curr. Opin. Biotechnol.* 11(6): 540–546.

Murata, S., Murata, K., Kidena, K., and Nomura, M. 2003. Oxidative Desulfurization of Diesel Fuels by Molecular Oxygen. *Prepr. Div. Fuel Chem., Am. Chem. Soc.* 48(2): 531.

Mužic, M., and Sertić-Bionda, K. 2013. Alternative Processes for Removing Organic Sulfur Compounds from Petroleum Fractions. *Chem. Biochem. Eng. Q.* 27(1): 101–108.

Nanoti, A., Dasgupta, S., Goswami, A.N., Nautiyal, B.R., Rao, T.V., Sain, B., Sharma, Y.K., Nanoti, S.M., Garg, M.O., and Gupta, P. 2009. Mesoporous Silica as Selective Sorbents for Removal of Sulfones from Oxidized Diesel Fuel. *Micropor. Mesopor. Mater.* 124: 94–99.

Nie, Y., Li, C.X., Sun, A.J., Meng, H., and Wang, Z. 2006. Extractive Desulfurization of Gasoline Using Imidazolium-Based Phosphoric Ionic Liquids. *Energy Fuels* 20(5): 2083.

Nie, Y., Li, C.X., and Wang, Z.H. 2007. Extractive Desulfurization of Fuel Oil Using Alkylimidazole and Its Mixture with Dialkylphosphate Ionic Liquids. *Ind. Eng. Chem. Res.* 46(15): 5108–5112.

Noda, K.I., Watanabe, K., and Maruhashi, K. 2003. Recombinant *Pseudomonas putida* Carrying Both the *dsz* and *hcu* Genes Can Desulfurize Dibenzothiophene in *n*-Tetradecane. *Biotechnol. Lett.* 25(14): 147–150.

Okada, H., Nomura, N., Nakahara, T., and Maruhashi, K. 2002a. Analyses of Substrate Specificity of the Desulfurizing Bacterium *Mycobacterium* sp. G3. *J. Biosci. Bioeng.* 93(2): 228–233.

Okada, H., Nomura, N., Nakahara, T., and Maruhashi, K. 2002b. Analysis of Dibenzothiophene Metabolic Pathway in *Mycobacterium* strain G3. *J. Biosci. Bioeng.* 93(5): 491–497.

Opara, C.C., Oyom, A.I., and Okonkwo, M.C. 2013. Deodorization of Kerosene Using Activated Carbon as Adsorbent. *Greener J. Phys. Sci.* 3(2): 70–75.

Otsuki, S., Nowaka, T., Takashima, N., Qian, W., Ishihara, A., Imai, T., and Kabe, T. 2000. Oxidative Desulfurization of Light Oil and Vacuum Gas Oil by Oxidation and Solvent Extraction. *Energy Fuels* 14: 1232–1239.

Pacheco, M.A. 1999. Recent Advances in Biodesulfurization (BDS) of Diesel Fuel. Paper presented at the NPRA Annual Meeting, San Antonio, TX, March 21–23, 1999.

Parkash, S. 2003. *Refining Processes Handbook*. Gulf Professional Publishing, Elsevier, Amsterdam.

Piddington, C.S., Kovacevich, B.R., and Rambosek, J. 1995. Sequence and Molecular Characterization of a DNA Region Encoding the Dibenzothiophene Desulfurization Operon of *Rhodococcus* sp. Strain IGTS8. *Appl. Environ. Microbiol.* 61(2): 468–475.

Planeta, J., Karásek, P., and Roth, M. 2006. Distribution of Sulfur-Containing Aromatics Between [hmim] [Tf2N] and Supercritical CO_2: A Case Study for Deep Desulfurization of Oil Refinery Streams by Extraction With Ionic Liquids. *Green Chem.* 8: 70–77.

Qiu, J., Wang, G., Zeng, D., Tang, D.Y., Wang, M., and Li, Y. 2009. Oxidative Desulfurization of Diesel Fuel Using Amphiphilic Quaternary Ammonium Phosphomolybdate Catalysts. *Fuel Process. Technol.* 90(12): 1538–1542.

Ramroop Singh, N., and Speight, J.G. 2011. Applications of Ionic Liquids in Industry. *Chem. Technol.* 6(2): 114–112.

Salem, A.B.S.H. 1994. Naphtha Desulfurization by Adsorption. *Ind. Eng. Chem. Res.* 33: 336–340.

Salem, A.B.S.H., and Hamid, H.S. 1997. Removal of Sulfur Compounds from Naphtha Solutions Using Solid Adsorbents. *Chem. Eng. Technol.* 20: 342–347.

Sano, Y., Kazomi, S., Ki-Hyonk, C., Yozo, K., and Isao, M. 2005. Two-Step Adsorption Process for Deep Desulfurization of Diesel Oil. *Fuel* 84: 903–910.

Seredych, M., and Bandosz, T.J. 2010. Adsorption of Dibenzothiophenes on Nanoporous Carbons: Identification of Specific Adsorption Sites Governing Capacity and Selectivity. *Energy Fuels* 24: 3352–3360.

Shakirullah, M., Ahmad, I., Ishaq, M., and Ahmad, W. 2009. Study on the Role of Metal Oxides in Desulphurization of Some Petroleum Fractions. *J. Chin. Chem. Soc.* 56: 107–114.

Sheldon, R. 2001. Catalytic Reactions in Ionic Liquids. *Chem. Commun.* 23: 2399–2407.

Shimizu, Y., Kumagai, S., Takeda, K., and Enda, Y. 2007. Adsorptive Removal of Sulfur Compounds in Kerosene by Using Rice Husk Activated Carbon. http://acs.omnibooksonline.com/data/papers/2007 _P035.pdf; accessed October 29, 2014.

Shiraishi, Y., Hara, H., Hirai, T., and Komasawa, I. 1999. A Deep Desulfurization Process for Light Oil by Photosensitized Oxidation using a Triplet Photosensitizer and Hydrogen Peroxide in an Oil/Water Two-Phase Liquid–Liquid Extraction System. *Ind. Eng. Chem. Res.* 38(4): 1589–1595.

Shiraishi, Y., Tachibana, K., Taki, Y., Hirai, T., and Komasawa, I. 2001b. A Novel Desulfurization Process for Fuel Oils Based on the Formation and Subsequent Precipitation of *S*-Alkylsulfonium Salts. 2. Catalytic-Cracked Gasoline. *Ind. Eng. Chem. Res.* 40(4): 1225–1233.

Shiraishi, Y., Taki, Y., Hirai, T., and Komasawa, I. 2001a. A Novel Desulfurization Process for Fuel Oils Based on the Formation and Subsequent Precipitation of *S*-Alkylsulfonium Salts. 1. Light Oil Feedstocks. *Ind. Eng. Chem. Res.* 40(4): 1213–1224.

Soleimani, M., Bassi, A., and Margaritis, A. 2007. Biodesulfurization of Refractory Organic Sulfur Compounds in Fossil Fuels. *Biotechnol. Adv.* 25(6): 570–596.

Song, C. 2002. Fuel Processing for Low-Temperature and High-Temperature Fuel Cells: Challenges, and Opportunities for Sustainable Development in the 21st Century. *Catal. Today* 77: 17–49.

Song, C.S. 2003. An Overview of New Approaches to Deep Desulfurization for Ultra-Clean Gasoline, Diesel Fuel, and Jet Fuel. *Catal. Today* 86: 211–263.

Song, C., and Ma, X. 2003. New Design Approaches to Ultra-Clean Diesel Fuels by Deep Desulfurization and Deep De-aromatization. *Appl. Cat. B Environ.* 41(1–2): 207–238.

Speight, J.G. 2000. *The Desulfurization of Heavy Oils and Residua*, 2nd Edition. Marcel Dekker Inc., New York.

Speight, J.G. 2011. *The Refinery of the Future*. Gulf Professional Publishing, Elsevier, Oxford, UK.

Speight, J.G. 2014. *The Chemistry and Technology of Petroleum*, 5th Edition. CRC Press, Taylor & Francis Group, Boca Raton, FL.

Speight, J.G., and Ozum. B. 2002. *Petroleum Refining Processes*. Marcel Dekker Inc., New York.

Stanislaus, A., Marafi, A., and Rana, M.S. 2010. Recent Advances in the Science and Technology of Ultra Low Sulfur Diesel (ULSD) Production. *Catal. Today* 153: 1–68.

Swaty, T.E. 2005. Global Refining Industry Trends: The Present and Future. *Hydrocarbon Process.* September 2005: 35–46.

Takahashi, A., Yang, F.H., and Yang, R.T. 2002. New Sorbents for Desulfurization by Complexation: Thiophene/Benzene Adsorption. *Ind. Eng. Chem. Res.* 41: 2487–2496.

Tam, P.S., Kittrell, J.R., and Eldridge, J.W. 1990a. Desulfurization of Fuel Oil by Oxidation and Extraction. 1. Enhancement of Extraction Oil Yield. *Ind. Eng. Chem. Res.* 29(3): 321–324.

Tam, P.S., Kittrell, J.R., and Eldridge, J.W. 1990b. Desulfurization of Fuel Oil by Oxidation and Extraction. 2. Kinetic Modeling of Oxidation Reaction. *Ind. Eng. Chem. Res.* 29(3): 324–329.

US EPA. 2000. Heavy-Duty Engine and Vehicle Standards and Highway Diesel Fuel Sulfur Control Requirements. Report No. EPA 420-F-00-057. US Environmental Protection Agency, Washington, DC.

Velu, S., Ma, X., and Song, C. 2003a. Mechanistic Investigations on the Adsorption of Organic Sulfur Compounds over Solid Adsorbents in the Adsorptive Desulfurization of Transportation Fuels. *Prepr. Div. Fuel Chem., Am. Chem. Soc.* 48(2): 693–694.

Velu, S., Ma, X., and Song, C. 2003b. Selective Adsorption for Removing Sulfur from Jet Fuel over Zeolite-Based Adsorbents. *Ind. Eng. Chem. Res.* 42: 5293–5304.

Velu, S., Ma, X., and Song, C. 2005. Desulfurization of JP-8 Jet Fuel by Selective Adsorption over a Ni-Based Adsorbent for Micro Solid Oxide Fuel Cells. *Energy Fuels* 19: 1116–1125.

Wang, Y., and Yang, R.T. 2007. Desulfurization of Liquid Fuels by Adsorption on Carbon-Based Sorbents and Ultrasound-Assisted Sorbent Regeneration. *Langmuir* 23: 3825–3831.

Wang, R., and Yu, F. 2013. Deep Oxidative Desulfurization of Dibenzothiophene in Simulated Oil and Real Diesel Using Heteropolyanion-Substituted Hydrotalcite-Like Compounds as Catalysts. *Molecules* 18(11): 13691–13704.

Wang, Y., Yang, F.H., Yang, R.T., Heinzel, J.M., and Nickens, A.D. 2006. Desulfurization of High-Sulfur Jet Fuel by P-Complexation with Copper and Palladium Halide Sorbents. *Ind. Eng. Chem. Res.* 45: 7649–7655.

Wang, Y., Yang, R.T., and Heinzel, J.M. 2008. Desulfurization of Jet Fuel by Complexation Adsorption with Metal Halides Supported on MCM-41 and SBA-15 Mesoporous Materials. *Chem. Eng. Sci.* 63: 356–365.

Weitkamp, J., Schwark, M., and Ernst, S. 1991. Removal of Thiophene Impurities from Benzene by Selective Adsorption on Zeolite ZSM-5. *J. Chem. Soc. Chem. Commun.* 16: 1133–1134.

Welton, T. 1999. Room-Temperature Ionic Liquids. Solvents for Synthesis and Catalysis. *Chem. Rev.* 99(8): 2071–2083.

Wild, P.J., Nyqvist, R.G., Bruijn, F.A., and Stobbe, E.R. 2006. Removal of Sulphur-Containing Odorants from Fuel Gases for Fuel Cell-Based Combined Heat and Power Applications. *J. Power Sources* 159: 995–1002.

Wittanadecha, W., Laosiripojana, N., Ketcong, A., Ningnuek, N., Praserthdam, P., and Assabumrungrat, S. 2014. Synthesis of Au/C Catalysts by Ultrasonic-Assisted Technique for Vinyl Chloride Monomer Production. *Eng. J.* 8(3): 65–71.

Wu, Z., and Ondruschka, B. 2010. Ultrasound-Assisted Oxidative Desulfurization of Liquid Fuels and Its Industrial Application. *Ultrason. Sonochem.* 17(6): 1027–1032.

Xu, P., Yu, B., Li, F.L., and Cai, X.F. 2006. Microbial Degradation of Sulfur, Nitrogen and Oxygen Heterocycles. *Trends Microbiol.* 14: 398–405.

Yahiya, G.O., Hamad, F., Bahamdan, A., Tammana, V.V.R., and Hamad, E.Z. 2013. Supported Ionic Liquid Membrane and Liquid–Liquid Extraction Using Membrane for Removal of Sulfur Compounds from Diesel/Crude Oil. *Fuel Process. Technol.* 113: 123–129.

Yang, R.T., Hernandez-Maldonado, A.J., and Yang, F.H. 2001. New Sorbents for Desulfurization of Liquid Fuels by π-Complexation. *Ind. Eng. Chem. Res.* 40: 6236–6239.

Yang, Y.Z., Liu, X.G., and Xu, B.S. 2014. Recent Advances in Molecular Imprinting Technology for the Deep Desulfurization of Fuel Oils. *New Carbon Mater.* 29(1): 1–14.

Zaki, T., Riad, M., Saad, L., and Mikhail, S. 2005. Selected Oxide Materials for Sulfur Removal. *Chem. Eng. J.* 113: 41–46.

Zamikos, F., Lois, E., and Stournas, S. 1995. Desulfurization of Petroleum Fractions by Oxidation and Solvent Extraction. *Fuel Proc. Technol.* 42: 35–45.

Zhang, J., Huang, C.P., Chen, B.H., Li, Y.X., and Qiao, C.Z. 2007. Extractive Desulfurization from Gasoline by [BMIM][Cu_2Cl_3]. *J. Fuel Chem. Tech.* 33(4): 431–434.

Zhang, J., Nan, J., Liu, Z., Yu, H., Geng, S., Shi, Y., Qu, X., Liu, C., and Liu, H. 2013. Synthesis of Special Morphology ZnO Materials and the Performance for Reaction Adsorption Desulfurization of Thiophene. *Appl. Phys. Front.* 1(1): 9–15.

Zhang, J., Wang, A.J., Li, X., and Ma, X.H. 2011. Oxidative Desulfurization of Dibenzothiophene and Diesel Over [Bmim]3$PMo_{12}O_{40}$. *J. Catal.* 279(2): 269–275.

Zhang, S.G., and Zhang, Z.C. 2002. Novel Properties of Ionic Liquids in Selective Sulfur Removal from Fuels at Room Temperature. *Green Chem.* 4: 376–379.

Zhang, S.G., Zhang, Q.L., and Zhang, Z.C. 2004. Extractive Desulfurization and Denitrogenation of Fuels Using Ionic Liquids. *Ind. Eng. Chem. Res.* 43: 614–622.

Zhao, D.S., Liu, R., Wang, J.L., and Liu, B.Y. 2008a. Photochemical Oxidation and Ionic Liquid Extraction Coupling Technique in Deep Desulphurization of Light Oil. *Energy Fuels* 22(2): 1100.

Zhao, Y., and Wang, R. 2013. Deep Desulfurization of Diesel Oil by Ultrasound-Assisted Catalytic Ozonation Combined with Extraction Process. *Petrol. Coal* 55(1): 62–67.

Zhao, D.S., Sun, Z.M., Li, F.T., Liu, R., and Shan, H.D. 2008b. Oxidative Desulfurization of Thiophene Catalyzed by (C_4H_9)$_4$NBr 2C_6H_{11}NO Coordinated Ionic Liquid. *Energy Fuels* 22(5): 3065–3069.

Zhu, H.P., Yang, F., Tang, J., and He, M.Y. 2003. Brønsted Acidic Ionic Liquid 1-Methylimidazolium Tetrafluoroborate: A Green Catalyst and Recyclable Medium for Esterification. *Green Chem.* 5: 38–39.

Zhu, W.S., Li, H.M., Gu, Q.Q., Wu, P.W., Zhu, G.P., Yan, S., and Chen, Y. 2011. Kinetics and Mechanism for Oxidative Desulfurization of Fuels Catalyzed by Peroxomolybdenum Amino Acid Complexes in Water-Immiscible Ionic Liquids. *J. Mol. Catal. A Chem.* 336(1–2): 16–22.

9 Biocatalytic Desulfurization

9.1 INTRODUCTION

The high sulfur, nitrogen, and metal content (and the accompanying high viscosity) in petroleum cause expensive processing problems in the refinery. Conventional technology does not exist to economically remove these contaminants from crude oil, so the problem is left for the refiners to handle downstream at a high cost.

Sulfur is the major concern for producers and refiners, and has long been a key determinant of the value of crude oils for several reasons: (1) sulfur presents a processing problem for refiners—desulfurization offers refiners the opportunity to reduce the sulfur of their crude feedstocks before they ever enter the refinery system, minimizing downstream desulfurization costs; (2) the amount of sulfur in many finished products such as gasoline and diesel fuel is limited—the regulations restricting allowable levels of sulfur in end products continues to become increasingly stringent, and this creates an ever more challenging technical and economic situation for refiners as the sulfur levels in available crude oils continue to increase, and creates a market disadvantage for producers of high-sulfur crudes. Low-sulfur crudes continue to command a premium price in the market, while higher sulfur crude oils sell at a discount. Desulfurization would offer producers the opportunity to economically upgrade their resources.

Metals in petroleum lead to two major problems for the industry. Combustion of these fuels leads to the formation of ash with high concentrations of metal oxides, leading to undesirable waste disposal issues. Also, when crude oil is refined, the metals are concentrated in the residual fraction, which is then further processed by coking or more often by catalytic cracking where metals from the oil deposit on the cracking catalyst, resulting in catalyst poisoning, thereby decreasing catalyst selectivity and activity. In addition, nitrogen in crude oil also leads to poisoning of the refinery catalysts and results in increased nitrogen oxide emissions upon combustion in car engines.

High viscosity significantly hampers the pumping, transportation, refining, and handling of petroleum. Common methods used to overcome problems associated with high viscosity include heating (perhaps even visbreaking), dilution with a suitable solvent, and use of chemical additives. As a result, the need for a safe, economical, and effective method for reducing viscosity has long been a goal of the industry. Biocatalytic processes for addressing these problems offer the petroleum industry potentially great rewards. Some process options focus on the removal of sulfur from crude oil and refinery streams by a microbial process (biocatalytic desulfurization). Furthermore, biocatalytic approaches to viscosity reduction, as well as the removal of metals and nitrogen as additional approaches to fuel upgrading, are also options.

Generally, biological processing of petroleum feedstocks offers an attractive alternative to conventional thermochemical treatment due to the mild operating conditions and greater reaction specificity afforded by the nature of biocatalysis. Efforts in microbial screening and development have identified microorganisms capable of petroleum desulfurization, denitrogenation, and demetallization. Biological desulfurization of petroleum may occur either oxidatively or reductively. In the oxidative approach, organic sulfur is converted to sulfate and may be removed in process water. This route is attractive because it would not require further processing of the sulfur and may be amenable for use at the wellhead where process water may then be reinjected. In the reductive desulfurization scheme, organic sulfur is converted into hydrogen sulfide, which may then be catalytically converted into elemental sulfur, an approach of utility at the refinery. Regardless of the mode of biodesulfurization, key factors affecting the economic viability of such processes are biocatalyst activity and cost, differentials in product selling price, sale or disposal of coproducts or wastes from the treatment process, and the capital and operating costs of unit operations in the treatment scheme.

However, first and by way of definition, *petroleum biotechnology* is based on biotransformation processes. *Petroleum microbiology* research is advancing on many fronts, spurred on most recently by new knowledge of cellular structure and function gained through molecular and protein engineering techniques, combined with more conventional microbial methods. *Petroleum bioremediation* refers specifically to the cleanup of spills of petroleum and petroleum products using microorganisms (Speight and Arjoon, 2012). Furthermore, *biodegradation (biotic degradation, biotic decomposition)* is the chemical degradation of contaminants by bacteria or other biological means. Organic material can be degraded aerobically (in the presence of oxygen) or anaerobically (in the absence of oxygen). Most bioremediation systems operate under aerobic conditions; however, a system under anaerobic conditions may permit microbial organisms to degrade chemical species that are otherwise nonresponsive to aerobic treatment, and vice versa.

Thus, bioremediation—the use of living organisms to reduce or eliminate environmental hazards resulting from accumulations of toxic chemicals and other hazardous wastes—is an option that offers the possibility to destroy or render harmless various contaminants using natural biological activity (Gibson and Sayler, 1992). In addition, bioremediation can also be used in conjunction with a wide range of traditional physical and chemical technology to enhance their effectiveness (Vidali, 2001). However, the lessons learned from the application of microbes as bioremedial agents can be applied to the use of microorganisms in petroleum recovery and refining (Speight, 2014).

Biotechnology is now accepted as an attractive means of improving the efficiency of any industrial process, and resolving serious environmental problems. One of the reasons for this is the extraordinary metabolic capability that exists within the bacterial world. Microbial enzymes are capable of biotransforming a wide range of compounds, and the worldwide increase in attention being paid to this concept can be attributed to several factors, including the presence of a wide variety of catabolic enzymes and the ability of many microbial enzymes to transform a broad range of unnatural compounds (xenobiotic compounds) as well as natural compounds. Biotransformation processes have several advantages compared with chemical processes, including the following: (1) microbial enzyme reactions are often more selective; (2) biotransformation processes are often more energy efficient; (3) microbial enzymes are active under mild conditions; and (4) microbial enzymes are environmentally friendly biocatalysts. Although many biotransformation processes have been described, only a few of these have been used as part of an industrial process, and opportunities exist for biorefining of petroleum (Mohebali and Ball, 2008). Of particular interest in this connection is the phenomenon of biodesulfurization (biological desulfurization, microbial desulfurization) in which microorganisms are used to oxidize sulfur compounds in crude oil, ultimately resulting in desulfurization. This represents the ability of microbial species to "desulfurize compounds that are recalcitrant to the current standard technology in the oil industry" (Abin-Fuentes et al., 2013).

Current applied research on petroleum microbiology encompasses oil spill remediation, fermenter- and wetland-based hydrocarbon treatment, biofiltration of volatile hydrocarbons, enhanced oil recovery, oil and fuel biorefining, fine-chemical production, and microbial community-based site assessment (Van Hamme et al., 2003). From this work, it is evident that biorefining is a possible alternative to some of the current oil-refining processes (Figures 9.1 and 9.2) (Speight, 2000, 2014; Speight and Ozum, 2002; Parkash, 2003; Hsu and Robinson, 2006; Ancheyta and Speight, 2007; Gary et al., 2007). The major potential applications of biorefining are biodesulfurization, biodenitrogenation, biodemetallization, and biotransformation of heavy crude oils into lighter crude oils. The most advanced area is biodesulfurization for which pilot plants exist (Le Borgne and Quintero, 2003; Bachmann et al., 2014).

In fact, the application of biotechnology to petroleum refining (biorefining) is a possible alternative to some of the current refining processes. The major potential applications of biorefining are biodesulfurization, biodenitrogenation, biodemetallization, and biotransformation of heavy crude oils into lighter crude oils (Le Borgne and Quintero, 2003).

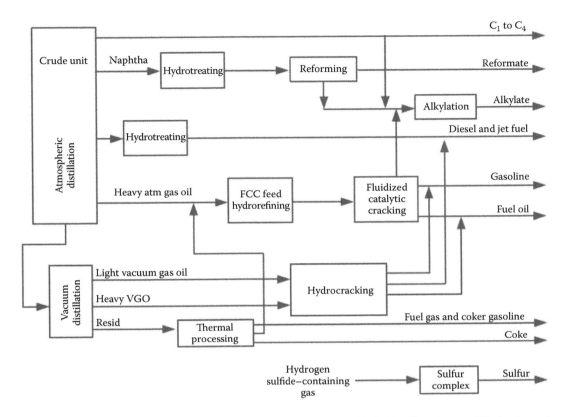

FIGURE 9.1 Schematic overview of a refinery. (From Speight, J.G. 2014. *The Chemistry and Technology of Petroleum*. 5th Edition. CRC Press, Taylor & Francis Publishers, Boca Raton, FL. Figure 15.1, p. 392.)

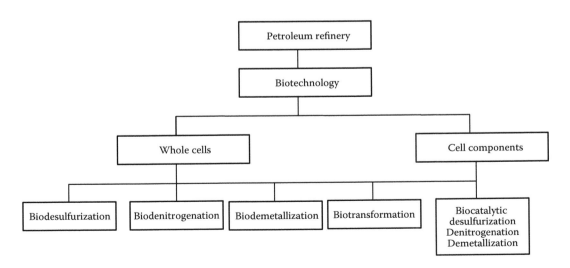

FIGURE 9.2 Potential applications of biotechnology in a petroleum refinery.

Biodesulfurization is a process that removes organic sulfur compounds from fossil fuels using enzyme-catalyzed reactions (also known as *biocatalyzed desulfurization*). Biocatalytic sulfur removal from fuels has applicability for producing low-sulfur naphtha and kerosene. Certain microbial biocatalysts have been identified that can biotransform sulfur compounds found in fuels, including ones that selectively remove sulfur from dibenzothiophene derivatives (Abbad-Andaloussi et al., 2003; Mužic and Sertić-Bionda, 2013).

Anaerobic biodesulfurization reactions proceed more slowly than aerobic reactions, but generate the same products as the conventional hydrodesulfurization technology: hydrogen sulfide and desulfurized oil. Although anaerobic biodesulfurization would be attractive because it avoids costs associated with aeration, it has the advantage of liberating sulfur as a gas and does not liberate sulfate as a by-product that must be disposed by some appropriate treatment. However, owing to the low reaction rates, safety and cost concerns, and the lack of identification of specific enzymes and genes responsible for anaerobic desulfurization, effective anaerobic microorganisms for practical petroleum desulfurization have not been found yet. Consequently, aerobic biodesulfurization has been the focus of the majority of research in biodesulfurization.

Biodesulfurization is often considered as a potential alternative to the conventional deep hydrodesulfurization processes used in refineries. In this process, microorganisms, their enzymes, or cellular extracts as catalysts remove organosulfur from petroleum fractions without degrading the carbon skeleton of the organosulfur compounds. During a biodesulfurization process, alkylated dibenzothiophene derivatives (Cx-DBTs) are converted to nonsulfur compounds, e.g., 2-hydroxybiphenyl (2-HBP), and sulfate. A hypothetical oxidative desulfurization pathway has been proposed (Kilbane, 1989) that, if ever existed in nature, could specifically remove sulfur from dibenzothiophene. The pathway was named as the 4S pathway (Figure 9.3) and implied consecutive oxidation of dibenzothiophene sulfur to sulfoxide (DBTO), sulfone (DBTO$_2$), sulfinate (HPBSi), and/or sulfonate (HPBSo), and then finally to the oil-soluble product 2-HBP or 2,2'-dihydroxybiphenyl, which finds its way back into the petroleum fractions while retaining the fuel value. The flavin-dependent monooxygenase DszC catalyzes the first two steps, and the monooxygenase DszA catalyzes the third step. The third enzyme, desulfinase (DszB), catalyzes the last step in the 4S route. The use of this pathway has been proposed for the desulfurization of petroleum in production fields and also in refineries (Soleimani et al., 2007).

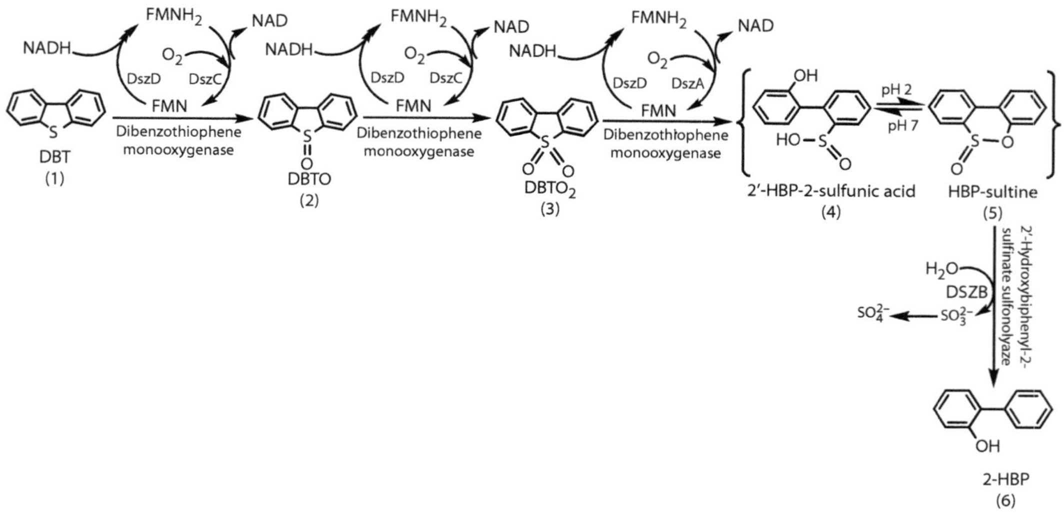

FIGURE 9.3 4S desulfurization pathway.

Biodesulfurization offers specificity of enzymes, relatively lower capital and operating costs, and mild processing conditions, as well as reduces the need for hydrogen. Both these features would lead to high energy savings in the refinery (Dinamarca et al., 2010). Furthermore, significant reductions in greenhouse gas emissions have also been predicted if biodesulfurization is used (Calzada et al., 2011). Biodesulfurization is a complementary technology; as already mentioned, hydrodesulfurization is not equally effective in desulfurizing all classes of sulfur compounds present in fossil fuels. The biodesulfurization process, on the other hand, is effective regardless of the position of alkyl substitutions (Pacheco, 1999). However, the hydrodesulfurization process conditions are sufficient not only to desulfurize sensitive (labile) organosulfur compounds, but also to (1) remove nitrogen and metals from organic compounds, (2) induce saturation of at least some carbon–carbon double bonds, (3) remove substances having an unpleasant smell or color, (4) clarify the product by drying it, and (5) improve the cracking characteristics of the material (Swaty, 2005). Therefore, with respect to these advantages, placing the biodesulfurization unit downstream of a hydrodesulfurization unit as a complementary technology to achieve ultra-deep desulfurization, rather than as a replacement, should also be considered. Monticello et al. (1996) suggested a multistage process for desulfurization of fossil fuels. This method was based on subjecting vacuum gas oil to hydrodesulfurization before biodesulfurization in defined conditions. Pacheco (1999) reported that the Energy Biosystems Corporation (EBC) used biodesulfurization downstream of hydrodesulfurization.

Biodesulfurization is often considered as a potential alternative to the conventional deep hydrodesulfurization processes used in refineries. In this process, microorganisms, their enzymes, or cellular extracts as catalysts remove organosulfur from petroleum fractions without degrading the carbon skeleton of the organosulfur compounds. During a biodesulfurization process, alkylated dibenzothiophene derivatives (Cx-DBTs) are converted to nonsulfur compounds, e.g., 2-HBP, and sulfate. The 4S pathway (Figure 9.3) (Kilbane, 1989) implies that consecutive oxidation of dibenzothiophene sulfur to sulfoxide (DBTO), sulfone (DBTO$_2$), sulfinate (HPBSi), and/or sulfonate (HPBSo), and then finally to the oil-soluble product 2-HBP or 2,2'-dihydroxybiphenyl, is possible, which finds its way back into the petroleum fractions while retaining the fuel value. There are four genes responsible for the 4S-pathway: the flavin-dependent monooxygenase *dszC* catalyzes the first two steps and the monooxygenase *dszA* catalyzes the third step. The third enzyme, desulfinase (*dszB*), catalyzes the last step in the 4S route, and another enzyme (*dszD*) encodes an NADH-flavin mononucleotide oxidoreductase (NADH is the reduced form of nicotinamide adenine dinucleotide). The use of this pathway has been proposed for the desulfurization of petroleum in production fields (before application of recovery operations) and also in refinery operations (Soleimani et al., 2007).

Biodesulfurization offers specificity of enzymes, relatively lower capital and operating costs, and mild processing conditions, as well as reduces the need for hydrogen. Both these features would lead to high energy savings in the refinery (Dinamarca et al., 2010). Furthermore, significant reductions in greenhouse gas emissions have also been predicted if biodesulfurization is used (Calzada et al., 2011). Biodesulfurization is a complementary technology; as already mentioned, hydrodesulfurization is not equally effective in desulfurizing all classes of sulfur compounds present in fossil fuels (Speight, 2000, 2014; Ancheyta and Speight, 2007). The biodesulfurization process, on the other hand, is effective regardless of the position of alkyl substitutions (Pacheco, 1999). However, the hydrodesulfurization process conditions are sufficient not only to desulfurize sensitive (labile) organosulfur compounds, but also to (1) remove nitrogen and metals from organic compounds, (2) induce saturation of at least some carbon–carbon double bonds, (3) remove substances having an unpleasant smell or color, (4) clarify the product by drying it, and (5) improve the cracking characteristics of the material (Swaty, 2005).

Therefore, with respect to the above-listed advantages, placing the biodesulfurization unit downstream of a hydrodesulfurization unit as a complementary technology (rather than as a replacement technology) to achieve ultra-deep desulfurization and then production of ultra low-sulfur products should also be considered. Monticello et al. (1996) suggested a multistage process for desulfurization of fossil fuels. This method was based on subjecting vacuum gas oil to hydrodesulfurization before biodesulfurization in defined conditions. Pacheco (1999) reported that the EBC used biodesulfurization downstream of hydrodesulfurization. It has also been reported (Fang et al., 2006) that a combination of hydrodesulfurization and biodesulfurization could reduce the sulfur content of catalytic diesel oil from 3358 to <20 ppm, and the desulfurization of diesel fuel after hydrodesulfurization treatment by *Rhodococcus erythropolis* FSD-2 can achieve 97% desulfurization (Zhang et al., 2007).

9.2 SCALE-UP OF THE BIODESULFURIZATION TECHNIQUE

Development of biodesulfurization from laboratory-scale or pilot-scale units to commercial-scale units requires improvement not only in the process itself, i.e., reactor design, optimization of operation conditions, and downstream operation, but also in the characteristics of the biocatalyst (Denome et al., 1993, 1994; Piddington et al., 1995; Li et al., 1996; Gallardo et al., 1997; Larose et al., 1997; Oldfield et al., 1997; Maghsoudi et al., 2000, 2001).

Biodesulfurization is preferred to occur under thermophilic conditions because it enhances the biodesulfurization rate, and the operating temperature would be closer to fluid catalytic cracking (FCC) or hydrodesulfurization outlet streams. Since distillate fractions are often treated at high temperatures, there may be some cost savings through the use of moderate thermophiles if biodesulfurization is integrated with hydrodesulfurization during refining without cooling the stock to 30°C (86°F). Moreover, the desulfurization activity will also be enhanced owing to the higher mass transfer rate at high temperatures. For practical biodesulfurization, it is useful to obtain microorganisms that exhibit much higher dibenzothiophene and benzothiophene desulfurization activities at high temperatures.

A number of mesophilic and thermophilic dibenzothiophene-desulfurizing microorganisms have been isolated. However, most of them belong to the *Rhodococcus* genus, mainly *Rhodococcus rhodochrous* IGTS8 (ATCC 53968), *Rhodococcus* FMF (R. FMF) native bacterium, and *Bacillus sphaericus* IGTS9 (ATCC 53969) (Monticello and Kilbane, 1994; Akbarzadeh et al., 2003). IGTS8 has been licensed by EBC, and it was reported to have a very low desulfurization activity (Naito et al., 2001).

The R. FMF native bacterium was isolated from soil contaminated with oil in Tabriz Refinery. This bacterium carries three SO_x genes (*dszA, B,* and *C*) on its genomic DNA. Preliminary studies have proved that the R. FMF strain possesses desulfurization activity, and that the microbe is capable of desulfurizing dibenzothiophene (Akbarzadeh et al., 2003).

A number of mesophilic and thermophilic DBT-desulfurizing microorganisms has been isolated; *Corynebacterium, Arthrobacter, Pseudomonas,* and *Gordonia* species (Izumi et al., 1994; Lee et al., 1995; Li et al., 1996, 2006; Serbolisca et al., 1999; Monticello, 2000; Luo et al., 2003); *R. erythropolis* D-1 (Izumi et al., 1994); *R. erythropolis* Ni-36 (Wang and Krawiec, 1994); *R. erythropolis* H-2 (Ohshiro et al., 1996a,b); *R. erythropolis* I-19 (Folsom et al., 1999); *R. erythropolis* KA-2-5-1 (Kishimoto et al., 2000; Naito et al., 2001); *Rhodococcus* sp. strain P32C1 (Maghsoudi et al., 2001); *Rhodococcus* sp. strain WU-K2R (Kirimura et al., 2002); *R. erythropolis* X309 (Bustos-Jaimes et al., 2003); *Rhodococcus* sp. strain ECRD-1 (Prince and Grossman, 2003); and *R. erythropolis* SHT87 (Davoodi-Dehaghani et al., 2010).

It has been reported (Gunam et al., 2006) that the majority of the sulfur (59% w/w) can be removed from light gas oil and from Liaoning crude oil within 36 and 72 h, respectively, using *Shingomonas subarctica* T7B. In addition, the isolation of *Lysinibacillus sphaericus* DMT-7 from diesel-contaminated soil for its ability to desulfurize the sulfur compounds benzothiophene, dibenzothiophene, 3,4-benzodibenzothiophene, 4,6-dibenzothiophene, and 4,6-dibutyldibenzothiophene at 37°C (100°F) has also been reported (Bahuguna et al., 2011), and the procedure was recommend for application to biodesulfurization of diesel and crude oil. *Paenibacillus* sp. A11-2, which desulfurizes dibenzothiophene at 60°C (140°F), has been isolated and the strain proved to have a different desulfurization gene cluster (tdsABC), which was 73%, 61%, and 52% homologous with *dszABC* genes (Ishii et al., 2000). *Bacillus subtilis* WU-S2B and *Mycobacterium phlei* WU-FI are also thermophilic strains that can specifically cleave the carbon–sulfur bond of dibenzothiophene and its alkylated homologues up to 52°C (126°F). The dibenzothiophene desulfurization gene cluster in WU-S2B was identified and named as biodesulfurizationABC. The DNA and amino acid sequencing of biodesulfurization genes with the genes of IGTS8 showed 61% homology (Kirimura et al., 2004).

Mycobacterium pheli WU-F1 (Furuya et al., 2001), which could desulfurize dibenzothiophene and its derivatives over a wide temperature range of 20–50°C (68–122°F) with the highest level at 45–50°C (113–122°F), and another *Bacillus* sp., which desulfurizes dibenzothiophene at 45°C (113°F), were isolated (Hosseini et al., 2006). In addition, thermophilic and hydrocarbon-tolerant *Mycobacterium goodii* X7B, which had been primarily isolated as a bacterial strain, is capable of desulfurizing dibenzothiophene to produce 2-HBP via the 4S pathway and was also found to desulfurize benzothiophene to *O*-hydroxystyrene at 40°C (104°F) (Li et al., 2005). This strain appeared to have the ability to remove organic sulfur from a broad range of sulfur species in naphtha at 40°C (104°F) and 1:9 oil–water phase ratio.

Furthermore, the hydrocarbon-tolerant and thermophilic dibenzothiophene-desulfurizing bacterium *M. phlei* WU-F1, which grew in a medium with hydrodesulfurized light gas oil as the sole source of sulfur, has been shown to exhibit high desulfurizing activity toward light gas oil between 30°C and 50°C (86°F and 122°F) (Furuya et al., 2003). When WU-F1 was cultivated at 45°C with B-light gas oil (390 ppm S), F-light gas oil (120 ppm S), or X-light gas oil (34 ppm S) as the sole sulfur source, biodesulfurization was around 60–70% for all three types of hydrodesulfurized light gas oils (Furuya et al., 2003). When resting cells were incubated at 45°C (113°F) with hydrodesulfurized light gas oils in the reaction mixtures containing 50% (v/v) oils, biodesulfurization reduced the sulfur content from 390 to 100 ppm of B-light gas oil, from 120 to 42 ppm of F-light gas oil, and from 34 to 15 ppm of X-light gas oil. The fungus *Rhodosporidium toruloides* strain DBVPG6662 was found to be able to utilize dibenzothiophene as a sulfur source, producing 2,2′-dihydroxybiphenyl. When it was grown on glucose in the presence of commercial emulsion of bitumen (Orimulsion), 68% of the benzo- and dibenzothiophene derivatives were removed after 15 days of incubation; not only this, but it was also able to utilize the organic sulfur in a large variety of thiophene derivatives that occur extensively in commercial fuel oils by physically adhering to the organic sulfur source (Baldi et al., 2003). *Stachybotrys* sp. WS4 was reported to accomplish 76% and 65% desulfurization of heavy crude oil from the Soroush and Kunhemond oil fields within 72 and 144 h, respectively (Torkamani et al., 2009).

Biodesulfurization with high hydrocarbon phase tolerance is considered an advantage because a less amount of water is required for biodesulfurization. *Rhodococcus* sp. IMP-S02 could remove 60% w/w the sulfur from diesel oil when incubated for 7 days at 30°C (86°F) (Castorena et al., 2002). In another study, the sulfur content of straight-run diesel oil was reduced from 1807 to 741 ppm by resting cells of *Nocardia globerula* R-9 (Luo et al., 2003), and it has been shown (Ishii et al., 2005) that, with growing cells of *M. phlei* WU-0103, the total sulfur content in 12-fold-diluted straight-run light gas oil was reduced from 1000 to 475 ppm sulfur at 45°C (113°F). *Mycobacterium goodii* X7B and *R. erythropolis* XP have been demonstrated to remove 47.2–62.3% w/w of sulfur from crude oil after treatment for 2 h at 30°C (86°F) (Yu et al., 2006a; Li et al., 2007), and whole cells of *R. erythropolis* XP was able to decrease the sulfur content of fluid catalytic cracker naphtha and straight-run naphtha by 30% and 85% w/w, respectively, indicating the tolerance of the microorganisms to the inhibitory effect of naphtha constituents (Yu et al. 2006b). Furthermore, *Desulfobacterium indolicum* isolated from oil-contaminated soil exhibited very high desulfurizing ability toward kerosene at 30°C in 1:9 oil–water phase ratio, which resulted in reduction of sulfur from 48.68 to 13.76 ppm over a period of 72 h with significant decrease in thiophene and 2,5-dimethyl thiophene (Aribike et al., 2008). Later work (Aribike et al., 2009) demonstrated that *Desulfobacterium anilini* isolated from petroleum products–polluted soil causes a significant decrease of benzothiophene and dibenzothiophene in diesel with 82% w/w removal of total sulfur after 72 h at 30°C (86°F) with 1/9 oil–water phase ratio. *Pantoea agglomerans* D23W3 were found to remove 26.38–71.42% of sulfur from different petroleum oils with the highest sulfur removal from light crude oil at 1/9 oil–water phase ratio (Bhatia and Sharma, 2010a,b). The native fungus that has been identified as *Stachybotrys* sp. is able to remove sulfur and nitrogen from heavy crude oil selectively at 30°C (86°F) (Torkamani et al., 2009). This fungus strain, which has been isolated as a part of the heavy crude oil biodesulfurization project initiated by Petroleum Engineering Development Company, a subsidiary of National Iranian Oil Company, is able to remove 76% and 64.8% w/w of the sulfur from heavy crude oil of Soroush oil field and the Kuhemond oil field—the heavy oils had an initial sulfur content of 5% w/w and 7.6% w/w, respectively—in 72 and 144 h, respectively.

Whether the commercialization of the biocatalytic desulfurization process will be realistic in the near future remains questionable because of the low desulfurization rate of the known microorganisms. There are various reports for biodesulfurization efficiencies of different distillate fractions: 30–70% for mid-distillates (Grossman et al., 1999; Pacheco, 1999), 65–70% for partially hydrodesulfurization-treated mid-distillates (Folsom et al., 1999), 90% for extensively hydrodesulfurization-treated mid-distillates, 20–60% for light gas oils (Chang et al., 1998; Pacheco, 1999; Noda et al., 2003), 75–90% for cracked stocks (Pacheco, 1999), and 25–60% for crude oils (Premuzic and Lin, 1999).

As promising as these data may seem, the desulfurization level is still insufficient to meet the required ultra-low sulfur levels for all fuels. Hence, the future development will depend on either genetically modifying the currently available bacteria or identifying novel biodesulfurization agents. Moreover, the known biodesulfurization agents have an approximately 500-fold lower desulfurization rate than what is required in industrial processes (Bhatia and Sharma, 2010a,b). An improvement in the uptake of sulfur compounds in oil fractions should be effective in enhancing the biodesulfurization activity. Watanabe et al. (2003) transferred the *dsz* gene cluster from *R. erythropolis* KA2-5-1 into *R. erythropolis* MC1109, which was unable to desulfurize light gas oil. Resting cells of the resultant recombinant strain, named MC0203, decreased the sulfur concentration of light gas oil from 120 to 70 ppm in 2 h. The desulfurization activity of this strain was about twice that of strain KA2-5-1. Moreover, *R. erythropolis* strain KA2-5-1 is unable to desulfurize 4,6-dipropyl dibenzothiophene in the oil phase (Noda et al., 2003). The *dsz* desulfurization gene cluster from *R. erythropolis* strain KA2-5-1 was transferred into 22 rhodococcal and mycobacterial strains using a transposon–transposase complex. The recombinant strain MR65, from *Mycobacterium* sp. NCIMB10403, was able to grow on a minimal medium supplemented with 1.0 mM 4,6-dipropyl dibenzothiophene in *n*-tetradecane (50%, v/v) as the sole sulfur source. The concentration of sulfur in the light gas oil was reduced by strain MR65 from 126 to 58 ppm and by strain KA2-5-1 from 126 to 80 ppm, within 24 h. Strain MR65 had about a 1.5-fold higher desulfurization activity for light gas oil than *R. erythropolis* strain KA2-5-1. The application of a recombinant, which is able to utilize 4,6-dipropyl dibenzothiophene in the oil phase, was effective in enhancing light gas oil biodesulfurization, and the desulfurization activity for light gas oil was likely to be higher than that for 4,6-dipropyl dibenzothiophene in *n*-tetradecane, since other sulfur sources are found in light gas oil: 4,6-dimethyldibenzothiophene or 4,6-diethyldibenzothiophene in addition to 4,6-dipropyl dibenzothiophene.

To develop an efficient biocatalyst, many investigators have constructed recombinant biocatalysts. For example, the *dsz* genes from *R. erythropolis* DS-3 were successfully integrated into the chromosomes of *Bacillus subtilis* ATCC 21332 and UV1, yielding two recombinant strains, *B. subtilis* M29 and M28, in which the integrated *dsz* genes were expressed efficiently under control of the promoter Pspac. The dibenzothiophene desulfurization efficiency of M29 was significantly higher than that of *R. erythropolis* DS-3, and also showed no product inhibition (Ma et al., 2006). It has been reported (Ishii et al., 2005) that 52% desulfurization of a 12-fold diluted straight-run light gas oil fraction by *M. pheli* WU-0103 is possible. Using mixtures of microbial cells of different ages would add advantage to the biodesulfurization process. Also, a genetically modified microorganism, *Pseudomonas putida* CECT5279, can be employed as a desulfurizing biocatalyst (Calzada et al., 2011). Experimental results show the possibility of optimizing a biocatalyst by mixing two-age cells of *P. putida* CECT5279 for complete transformation of dibenzothiophene in a minimized reaction time by mixing different biomass concentration of both cells with higher dibenzothiophene removal activity and cells with higher HBP production activity. Combining cells with 5 and 23 h growth time, in 0.7 and 1.4 g DCW (dry cell weight)/L, respectively, was found to be the best option. This biocatalyst formulation attains 100% dibenzothiophene conversion while reducing biodesulfurization time. In addition, this cell combination achieves higher initial biodesulfurization elimination rate than 23-h cells used alone.

In desulfurization of petroleum, there are many different compounds (solvents) that have an inhibitory effect on desulfurization-competent strains. *Pseudomonas* sp. was found to be an ideal candidate for biodesulfurization in petroleum, because they are organic solvent tolerant and have a high growth rate. With the properties noted, *dszABC* genes from *R. erythropolis* XP were cloned into *P. putida* to construct a solvent-tolerant, desulfurizing *P. putida* A4. This strain, when contacted with sulfur refractory compounds dissolved in hydrocarbon solvent, maintained the same substrate desulfurization traits as observed in *R. erythropolis* XP. Resting cells of *P. putida* A4 could desulfurize 86% of dibenzothiophene in 10% (v/v) *p*-xylene in 6 h. In the first 2 h, the desulfurization occurred with a rate of 1.29 mM dibenzothiophene/g DCW/h. No dibenzothiophene reduction was noticed with *R. erythropolis* or *P. putida* at identical conditions (Tao et al., 2006). The design of a recombinant microorganism to remove the highest amount of sulfur compounds in fossil fuels has also been reported (Raheb et al., 2009). Three genes (*dszA, B,* and *C*) from the dsz operon responsible for the 4S pathway (biodesulfurization pathway) in *R. erythropolis* IGTS8 were inserted into the chromosome of a novel indigenous *P. putida*. The reaction catalyzed by products of *dszABC* genes requires $FMNH_2$ supplied by dszD enzyme. Thus, pVLT31 vector harboring *dszD* gene was transferred into this recombinant strain. This new indigenous bacterium is an ideal biocatalyst for a desulfurizing enzyme system owing to the solvent-tolerant characteristic and optimum growth temperature at 40°C (104°F), which is suitable for the industrial biodesulfurization process. In addition, this strain produces a rhamnolipid biosurfactant that accelerates the two-phase separation step in the biodesulfurization process through increasing emulsification. Moreover, it has a high growth rate, which causes removal of sulfur compounds faster than *R. erythropolis* IGTS8, and has the highest biodesulfurization activity in the shortest time.

The biodesulfurization activities of recombinant indigenous *P. putida* and *R. erythropolis* IGTS8 were compared, and the results showed that the 2-HBP production of recombinant *P. putida* was more than that of *R. erythropolis* IGTS8 in the primary (1–20 h), while the 2-HBP production of *R. erythropolis* IGTS8 increased after a long cultivation time (approximately 22 h). Therefore, engineered *P. putida* could be a promising candidate for industrial and environmental application in biodesulfurization owing to its removal of higher sulfur amounts from oil in the shortest time. In addition to a higher optimal growth temperature, ability to produce rhamnolipid biosurfactant and solvent toleration were the other privileges of this recombinant strain that are applicable in biodesulfurization processes. Furthermore, *P. putida* DS23 has been developed using one of the organic solvent-responsive expression vectors newly constructed for biocatalysts, in which gene expression could be regulated in an organic solvent-dependent fashion (Tao et al., 2011). The biodesulfurization activity of *P. putida* DS23 induced by *n*-hexane achieved 56% biodesulfurization of 0.5 mM dibenzothiophene in 12 h in a biphasic reaction with 33.3% (v/v) *n*-hexane. This activity decreased to 26% when the strain was induced by isopropyl β-D-1-thiogalactopyranoside. However, a disadvantage to the use of certain microbes (especially those that produce the HBP derivatives) is that cell growth and desulfurization activity are inhibited by the end products of dibenzothiophene desulfurization: 2-HBP and/or 2,2′-dihydroxybiphenyl and sulfate are known to be severe (Ohshiro et al., 1996b; Okada et al., 2003; Kim et al., 2004).

Several investigators have reported that the desulfurization activity in various bacteria was completely repressed, and the production of dibenzothiophene desulfurizing enzymes has been shown to be inhibited by sulfate or other readily bioavailable sulfur sources, including methionine, cysteine, taurine, methanesulfonic acid, and casamino acids (a mixture of amino acids and some low molecular weight peptides obtained from acid hydrolysis of casein) (Kayser et al., 1993; Ohshiro et al., 1995, 1996b; Rhee et al., 1998; Kertesz, 2000; Chang et al., 2001; Matsui et al., 2002; Noda et al., 2002; Kim et al., 2004; Gunam et al., 2006).

The inhibition of dibenzothiophene desulfurization activity by sulfate is considered to be a gene-level regulation. The expression of *dsz* genes that are involved in desulfurization is strongly repressed by sulfate derivatives. The removal of this feedback regulation is of great importance from a process viewpoint (Kim et al., 2004). Therefore, for efficient desulfurization of fossil fuels, it would be advantageous to develop a new strain that is not susceptible to sulfate repression. To alleviate the sulfate repression and enhance the desulfurization activity, several recombinant biocatalysts, including recombinant *Escherichia coli*, *Pseudomonas aeruginosa*, and *Rhodococcus* strains, have been developed (Kilbane, 2006).

Recently, *Gordonia alkanivorans* RIPI90A has been reported as a desulfurizing strain for dibenzothiophene with special emulsion stabilization properties that could be of special interest in biodesulfurization process design (Mohebali et al., 2007). By cloning and sequencing of the *dszABC* genes of *G. alkanivorans* RIPI90A, it was possible to identify a suitable vector system for the *Gordonia* genus and to self-clone the *dszABC* genes in a *Gordonia* isolate (Shavandi et al., 2009). The recombinant strain was able to desulfurize dibenzothiophene in the presence of inorganic sulfate and sulfur-containing amino acids, while the native strain could not. The maximum desulfurization activity by recombinant resting cells (131.8 μM 2-HBP/g DCW/h) was increased 2.67-fold in comparison to the highest desulfurization activity of native resting cells.

As already reported on the repression of growth and desulfurization activity by 2-HBP and 2,2′-dihydroxybiphenyl (Ohshiro et al., 1996b; Setti et al., 1999; Okada et al., 2003; Kim et al., 2004), other workers (Chen et al., 2008) have studied the effect of 2-HBP, the end product of dibenzothiophene desulfurization via the 4S pathway, on the cell growth and desulfurization activity of *Microbacterium* sp. ZD-M2. The data indicated that 2-HBP would inhibit the desulfurization activity. However, if 2-HBP is added to the reaction media, the dibenzothiophene degradation rate would decrease along with the increase of the addition of 2-HBP. By contrast, cell growth would be promoted in the addition of 2-HBP at a low concentration (<0.1 mM). At high concentration of 2-HBP, inhibition of cell growth occurred. Meanwhile, the inhibitory effect of 2-HBP on dibenzothiophene desulfurization activity was tested both in the oil/aqueous two-phase system and the aqueous system. On the other hand, higher dibenzothiophene degradation activity and lower 2-HBP inhibitory effect in the oil/aqueous two-phase system than that in the sole aqueous system were obtained.

There are two reasons for these possibilities. First, the addition of an organic phase would enhance the transfer of dibenzothiophene to the cell (Okada et al., 2002; Luo et al., 2003). Second, 2-HBP is very soluble in the oil phase and, in practice, finds its way back to the petroleum fraction (Monticello, 2000). Thus, the inhibition by 2-HBP of dibenzothiophene desulfurization might be lower in the oil/aqueous two-phase system than that in the sole aqueous phase because of most of the 2-HBP would turn into the oil (*n*-hexadecane) phase and minimize the toxicity of 2-HBP to the desulfurization enzyme. By comparing the influence of 2-HBP on biodesulfurization in aqueous and oil/aqueous systems, it is advised that the removal of 2-HBP in the reaction medium is necessary for biodesulfurization, to decrease the inhibition of 2-HBP on the enzyme.

Also, overcoming the effect of toxic by-products like phenolic compounds can be done by adding surfactants (Feng et al., 2006). Biodesulfurization of hydrodesulfurized oil by *R. erythropolis* lawq was enhanced by adding the surfactant Tween 80. Tween 80 was shown to decrease the product concentration associated with the cells, reducing product inhibition.

The problems involved in the biodesulfurization of fuels are inhibition of the biocatalyst by the by-products and slow diffusion between the organic and aqueous phases. Inverse-phase-transfer biocatalysis uses supramolecular receptors (modified cyclodextrins like hydroxypropyl-β-cyclodextrin) that selectively pick up the sulfur aromatic compounds in the organic phase and transfers them into the water phase, which contains the biocatalyst. The inverse-phase-transfer biocatalysis approach can increase the mass transfer of water-insoluble substrates between the aqueous and organic phases, and eliminate or reduce feedback inhibition of the biocatalyst due to accumulation of the by-products in the water phase. It has been reported that 2-HBP inhibits *R. rhodochrous* IGTS8—in the presence of 10 mM cyclodextrin, the growth of IGTS8 decreases from 100% to only 80% (Setti et al., 2003). This indicates that cyclodextrin probably picks up the HBP in solution as well as directly from the interphase of the cellular biomembrane, thus protecting the microorganisms from the irreversible inhibition effect of this phenol. It was also observed that hydroxypropyl cyclodextrin can improve the mass transfer of water-insoluble substrates such as dibenzothiophene in *n*-hexadecane and the aqueous phase. The specific rate of the dibenzothiophene converted by IGTS8 (i.e., the parts per million of dibenzothiophene converted per hour per gram of dry cell) at a dibenzothiophene concentration of 120 ppm in *n*-hexadecane increased from 2.9 to 4.3 in the absence and presence of 3.14 mM hydroxypropyl cyclodextrin, respectively. Furthermore, to increase the availability of the organosulfur compounds to the microorganisms, the use of surfactant (Li and Jiang, 2013) and immobilized cells (Derikvand et al., 2014) is considered to be a promising approach. In addition, bacterial immobilization by adsorption is better than the cell entrapment method, as it reduces mass transfer limitation and any steric hindrance effects.

All the biodesulfurization processes reported hitherto are triphasic systems composed of cells, water, and oil. There are numerous reports on the treatment of diesel oils or model oil mixtures by using suspensions of growing or resting cells (Chang et al., 2001; Noda et al., 2003; Labana et al., 2005; Yu et al., 2006a,b). Treatment of petroleum oils using free cells has some limitations, such as high cost of the biocatalyst and low volumetric ratio between the organic phase and the aqueous one. In these cases, oil is mixed together with the cells as a suspension, which produces a sort of surfactant, emulsifying the oil. It seems to be very difficult to separate oil and water from the emulsified oil. Furthermore, the recovery of the cells is also difficult. On the other hand, it has been reported (Yu et al., 1998) that separation of oil, water, and cells from the emulsified oil is possible by using a cyclone. However, this concept of separation is based on centrifugal force, and it seems to be difficult to separate the components completely.

Cell immobilization was considered to be one of the most promising approaches. Compared with cell suspension, biodesulfurization with immobilized cells has some advantages: ease of biocatalyst separation from the treated fuels, low risk of contamination, relatively high oil–water volumetric ratios, high stability, and long lifetime of the biocatalyst (Chang et al., 2000; Hou et al., 2005). Nevertheless, physical interactions between bacterial cells and sulfur substrates require further studies to upscale biodesulfurization (Yang et al., 2007), in order to resolve problems associated with the limited access of microorganisms to organic substrates (Tao et al., 2006). In this context, surfactants and immobilized cells are considered promising solutions to the problem of low solubility. Biomodification of inorganic supports using cell immobilization increases the interaction between the reactants present in two-phase systems, thus avoiding the need to use expensive surfactants (Feng et al., 2006).

Generally, entrapment and adsorption are the preferred methods for cell immobilization. In entrapment, living cells are enclosed in a polymeric matrix that is porous enough to allow diffusion of substrates to the cells and of products away from the cells. The materials used for entrapment of cells are mainly natural polymers, such as alginate, carrageenan, gelatin, and chitosan. They may also be synthetic polymers such as polysaccharides, photo-cross-linkable resins, polyurethane, polyvinyl alcohol, polyacrylamide, and so on. The major drawbacks of an entrapment technique are diffusional limitations and steric hindrance, especially when diffusion of macromolecular substrates, such as starch and proteins, is involved. Mass transfer involved in the diffusion of a substrate to a reaction site and in the removal of inhibitory or toxic products from the environment may be impeded. Cell immobilization by adsorption is currently gaining considerable importance because of a major advantage, namely, reducing or eliminating the mass transfer problems associated with the common entrapment methods. However, the adsorption technique is generally limited by biomass loading, strength of adhesion, biocatalytic activity, and operational stability. This occurs because immobilization by adsorption involves attachment of cells to the surface of an adsorbent such as *celite*. Adsorption is a simple physical process in which the forces involved in cell attachment are so weak that cells that are several micrometers across are not strongly adsorbed and are readily lost from the surface of the adsorbent (Shan et al., 2005a).

ENT-4000 has been selected (Naito et al., 2001) as a suitable gel material and has succeeded in constructing a biphasic biodesulfurization system (immobilized *R. erythropolis* KA2-5-1 cells and oil) with good desulfurization activity and without leakage of cells from the support. Furthermore, ENT-4000-immobilized cells catalyzed biodesulfurization repeatedly in this system for >900 h with reactivation, and the recovery of both the biocatalyst and the desulfurized model oil was easy. This study would give a solution to the problems in biodesulfurization, such as the troublesome process of recovering desulfurized oil and the short lifetime of biodesulfurization biocatalysts.

Immobilization of *M. goodii* X7B cells by entrapment with calcium alginate, carrageenan, agar, polyvinyl alcohol, polyacrylamide, and gelatin–glutaraldehyde has been attempted (Li et al., 2005), and it was found that calcium alginate–immobilized cells had the highest dibenzothiophene desulfurization activity. When immobilized *M. goodii* X7B cells were incubated at 40°C (104°F) with Dushanzi straight-run naphtha (DSRG227) in a reaction mixture containing 10% v/v oil for 24 h, the sulfur content of the naphtha decreased from 275 to 121 ppm, after which the desulfurization reaction was repeated by exchanging the used immobilized cells for fresh ones and the sulfur content further decreased to 54 ppm, corresponding to a reduction of 81%.

It has been observed (Klein et al., 1993) that the small microbead size is important for minimizing the mass transfer resistance problem normally associated with immobilized cell culture involved in diffusion of a substrate to a reaction site and in removal of inhibitory or toxic products from the environment. To minimize these effects, it is necessary to minimize the diffusional distance through a reduction in bead size (Cassidy et al., 1996). Furthermore, immobilization of resting cells of *Pseudomonas delafieldii* R-8 by entrapment in calcium alginate is possible using a new gas jet extrusion technique, which is a rapid and simple method (Li et al., 2008a). In the method, the resultant slurry is extruded through a cone-shaped needle into a stirred 0.1 M $CaCl_2$ gelling solution. The slurry is intruded as discrete droplets so as to form calcium alginate beads with normal size (2.5 and 4 mm in diameter). To prepare smaller beads than 2 mm diameter, nitrogen gas was introduced around the tip of the needle to blow off the droplets. By adjusting the gas flow rate to 0.5 L/min, the size of the beads can be controlled at 1.5 mm in diameter. Song et al. (2005) have employed Tween 20 to improve the permeability of the entrapment–encapsulation hybrid membrane. The immobilized beads without Tween 20 would rupture because of the formation of carbon dioxide and nitrogen as a result of respiration and denitrification (Song et al., 2005). During cell immobilization, the addition of nontoxic and nonionic surfactants, including Span 20, Span 80, Tween 20, and Tween 80, greatly enhanced the desulfurization rate compared with the control (Li et al., 2008a). Span 80 showed the highest effect on desulfurization activity. In 24 h, the desulfurization rate with the addition of 0.5% Span 80 was 1.8-fold higher than that of without Span 80.

A study of the resting cells of *Gordonia* sp. WQ-01A, a dibenzothiophene-desulfurizing strain, showed that the cells can be immobilized by calcium alginate (Peng and Wen, 2010). Batch dibenzothiophene biodesulfurization experiments using immobilized cells and *n*-dodecane as the oil phase were conducted in a fermenter under varying operating conditions, such as initial dibenzothiophene concentration, bead loading, and the oil phase volume fraction. When the initial dibenzothiophene concentration is 0.5, 1, and 5 mM, the dibenzothiophene concentration dropped to almost zero after 40, 60, and 100 h, respectively. The influence of bead loading and the oil-phase volume fraction on the dibenzothiophene biodesulfurization was small. Furthermore, a mathematical model was proposed to simulate the batch dibenzothiophene biodesulfurization process in an oil–water immobilization system, which took into account the internal and external mass transfer resistances of dibenzothiophene and oxygen, and the intrinsic kinetics of bacteria. As with most immobilization systems, the diffusion rate of substrates and products within the bead often limits productivity. According to the results, it was concluded that (1) the rate-limiting step in the oil–water immobilization system is not mass transfer resistance but bioconversion, and (2) compared with the effect of dibenzothiophene, oxygen concentration is not an important factor affecting the dibenzothiophene biodesulfurization in the immobilized system. That work allowed an understanding of the dynamic behavior of the immobilized system, and may be generally applicable to other areas of biocatalytic and biotransformation processes. In a study of the biodesulfurization of model oil with a sulfur content of 300 mg/kg, 1:1 thiophene (Th)/dibenzothiophene using *P. delafieldii* R-8 cells immobilized in calcium alginate beads recorded 40% and 25% removal of Th and dibenzothiophene within 10 h in 1:2 oil/water ratio, and was reused over 15 batch cycles with only 25% decrease in desulfurization activity from the first batch (Huang et al., 2012).

The key of entrapment method is the selection of materials and methods for preparation. The use of polyvinyl alcohol for cell entrapment has been investigated (Lozinsky and Plieva, 1998). Polyvinyl alcohol is low cost and nontoxic. It does not have adverse effects on cells and is becoming one of the most promising materials for entrapment (Hashimoto and Furukawa, 1987). Although easily operated, the saturated boric acid solution used to cross-link the polyvinyl alcohol is highly acidic (pH, ~4) and could cause difficulty in maintaining cell viability. In addition, the sphering speed of polyvinyl alcohol is very slow, and it results in agglomeration of the polyvinyl alcohol beads.

A successful attempt to quicken the sphering speed of polyvinyl alcohol and prevent the agglomeration problem of the polyvinyl alcohol beads was done by adding a small amount of calcium alginate to saturated boric acid (Wu and Wisecarver, 1992). However, calcium alginate may be easily damaged by a salt of phosphoric acid and tends to be eroded or dissolved when used in reaction systems (Fernandes et al., 2002). To overcome the difficulties, a freezing–thawing technique was used to prepare polyvinyl alcohol beads (Giuliano et al., 2003). The beads have excellent water resistance, elasticity, and flexibility, and shows high safety to living bodies because no chemical agent is used for gel formation. In addition, its high water content and porous structure make it ideal for use as a carrier for the culture and propagation of the immobilized microorganisms (Freeman and Lilly, 1998).

Bacterial immobilization by adsorption is an improvement over the cell entrapment method that reduces mass transference and the steric effect. Bacterial cell adsorption involves the use of inorganic compounds as ideal biosupports. These materials must have controlled porosity and a specific area. They must also be inert to biological attack, insoluble in the growth media, and nontoxic to microbial cells. Moreover, adsorbed cells on inorganic supports should be able to maintain the metabolic activity required for the biodesulfurization process. Few studies have evaluated the influence of inorganic supports, with different physicochemical properties, on the biodesulfurization activity of dibenzothiophene or gas oil (Hwan et al., 2000; Zhang et al., 2007)—alumina and celite are selected as the most common supports used in these bioprocesses.

Celite beads are used as filter aids in pharmaceutical and beverage processing, as bulk filters for food and planes. Concerning chemical composition, celite is virtually inert for biological attack, insoluble in culture media, and nontoxic to microbial cells. These properties justified the choice of celite beads as a support material for immobilization. Desulfurization of a model oil (hexadecane containing dibenzothiophene) and diesel oil by using immobilized dibenzothiophene-desulfurizing bacterial strains, *Gordona* sp. CYKS1 and *Nocardia* sp. CYKS2, where celite beads were used as a biosupport for cell immobilization, has been reported (Chang et al., 2000). Immobilized cells were used for eight cycles each of 24-h duration, and the desulfurization rate of diesel oil was approximately four to seven times higher than that of model oil since model oil contained only dibenzothiophene, a recalcitrant compound, while diesel oil contained various readily desulfurizable compounds, such as thiol derivatives and sulfide derivatives.

The efficiency of the adsorption of *Pseudomonas stutzeri* on silica (Si), alumina (Al), sepiolite (Sep), and titania (Ti), and their influence on the biodesulfurization of gas oil, with emphasis on the interaction of inorganic supports on metabolic activity, have been studied (Dinamarca et al., 2010). The highest interaction was observed in the *P. stutzeri*/Si and *P. stutzeri*/Sep biocatalysts, and a direct relation between biodesulfurization activity and the adsorption capacity of the bacterial cells was observed at the adsorption/desorption equilibrium level. The biomodification of inorganic supports generates dynamic biostructures that facilitate the interaction with insoluble organic substrates, improving the biodesulfurization of gas oil in comparison to whole nonadsorbed cells. It was concluded that immobilization by adsorption of bacterial cells is a simple and effective method that can be applied in biodesulfurization reactions of gas oil. The addition of biosurfactant or Tween 80 improves the biodesulfurization efficiency of dibenzothiophene and gas oil in free and immobilized cell systems of *R. rhodochrous* IGTS8, using adsorption on silica and alumina, and an immobilizing technique (Dinamarca et al., 2014).

The technological issues of biodesulfurization include good reactor design, product recovery, and oil–water separations. Generally, batch-stirred tank reactors have been used because of the absence of immobilization technologies. A continuous two-phase (organic/aqueous) bioreactor can effect 12% sulfur removal from diesel oil by *Rhodococcus globerulus* DAQ3 (Yang et al., 2007). Multistaged airlift reactors can also be used to overcome poor reaction kinetics at low sulfur concentrations and to reduce mixing costs. This would enhance the concept of continuous growth and regeneration of the biocatalyst in the reactions system rather than in separate, external tanks.

A typical process consists of charging the biocatalyst, oil, air, and a small amount of water into a batch reactor (Figure 9.4). In the reactor, as the polynuclear aromatic sulfur species are oxidized to water-soluble products, the sulfur segregates into the aqueous phase. The oil–water–biocatalyst–sulfur by-product emulsion from the reactor effluent is separated into two streams, namely, the oil (which is further processed and returned to the refinery) and the water–biocatalyst–sulfur–by-product stream. A second separation is needed to allow most of the water and biocatalyst to return to the reactor for reuse. Airlift bioreactors with different microorganisms and operating conditions have also been employed and achieved 50–100% w/w sulfur removal (Nandi, 2010; Irani et al., 2011).

Effective oil–cell–water contact and mixing is essential for good mass transfer. Unfortunately, a tight emulsion is usually formed, and it must be broken in order to recover the desulfurized oil, recycle the cells, and separate the by-products. The phases are usually separated by liquid–liquid hydrocyclones. Another approach is to separate two immiscible liquids of varying densities by using a settling tank, where the liquid mixture is given enough residence time for them to form two layers, which are then drained.

Bacteria usually partition to the oil–water interface and move with the discontinuous phase into a two-phase emulsion. In a water-in-oil emulsion, cells associate with the water droplets. A small

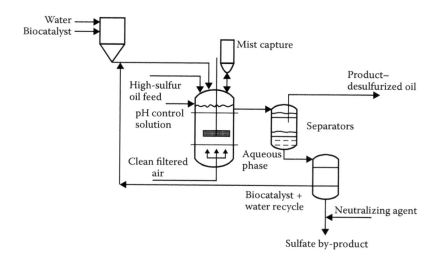

FIGURE 9.4   General process flow sheet for biocatalytic desulfurization.

amount of fresh oil can then be added to create an oil-in-water emulsion so that the cells will stick to the oil droplets. Passage of the emulsion through a hydrocyclone will yield a clean water phase and a concentrated cell and oil mixture that can be recycled to the reactor. By manipulating the nature of these emulsions, relatively clean oil and water can be separated from the mixture without resorting to high-energy separations. This causes a significant reduction in capital and operating costs (US Department of Energy, 2003).

There are very few reports on biodesulfurization process designs and cost analysis. To ensure that the capital and operating costs for biodesulfurization will be lower than those for hydrodesulfurization, it is necessary to design a suitable biocatalytic process (Monticello, 2000). The cost of building a bioreactor can be reduced by changing from a mechanically agitated reactor to airlift designs to minimize energy costs (Pacheco, 1999; Monticello, 2000). However, specific details about the process and the results achieved were not published (Kilbane and Le Borgne, 2004). A new type of airlift reactor with immobilized *Gordonia nitida* CYKS1 cells on a fibrous support was designed (Lee et al., 2005) and used for the biocatalytic desulfurization (biodesulfurization) of diesel oil. It was shown that cells immobilized on nylon fibers well sustained their growth accompanied by desulfurization activity during a series of repeated batch runs over an extended period of time. The advantages of easy separation of the biocatalyst from the treated fuels, and the high stability and long lifetime of the biocatalyst using immobilized cells were concluded. Sanchez et al. (2008) reported biodesulfurization of 50:50 water–kerosene emulsions carried out at 100-mL scale and in a 0.01-m^3 airlift reactor with resting cells of the reference strain ATCC 39327 and *Pseudomonas* native strains No. 02, 05, and 06. The reactor conditions were 30°C, pH 8.0, and 0.34 m^3/h air flow. After 7 culture days, the mean sulfur removal for the strains No. 06 and ATCC 39327 was 64% and 53%, respectively, with a mean calorific power loss of 4.5% for both strains. The use of the native strain No. 06 and the designed airlift reactor is shown as an alternative for biodesulfurization process and constitute a first step for its scale-up to pilot plant.

Enhanced dibenzothiophene biodesulfurization has been accomplished using a microchannel reactor (Noda et al., 2008). The bacterial cell suspension and *n*-tetradecane containing 1 mM dibenzothiophene were introduced separately into the double-Y channel microfluidic device (Figure 9.5), both at 0.2–4 μL/min. It was confirmed that a stable *n*-tetradecane/cell suspension interface was formed in the microchannel and the two liquids were separated almost completely at the exit of the microchannel. An emulsion was generated in the *n*-tetradecane/cell batch reaction, which decreased the recovery rate in the *n*-tetradecane phase. By contrast, the emulsion was not observed in the microchannel, and the *n*-tetradecane phase and cell suspension separated completely. The rate of biodesulfurization in the oil–water phase of the microchannel reaction was more than 9-fold that

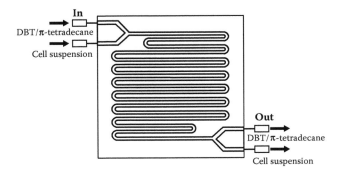

FIGURE 9.5 Double-Y channel microchannel reactor. (From Noda, K., Kogure, T., Irisa, S., Murakami, Y., Sakata, M., and Kuroda, A., *Biotechnol. Lett.*, 30, 451, 2008.)

in a batch (control) reaction. In addition, the microchannel reaction system using a bacterial cell suspension degraded the alkylated dibenzothiophene that was not degraded by the batch reaction system. This work provides a foundation for the application of a microchannel reactor system consisting of biological catalysts using an oil–water phase reaction.

Rhodococcus sp. NCIM 2891 has shown high activity to reduce sulfur level in diesel (Mukhopadhyay et al., 2006). The initial sulfur concentration was varied over the range 200–540 ppm. A trickle bed reactor (diameter, 0.066 m; height, 0.6 m) in continuous mode was studied with liquid flow rate and inlet sulfur concentration as parameters. Pith balls have been used as the immobilization matrix for the microorganisms with a constant bed porosity of 0.6. Sulfur conversion up to 99% has been achieved.

Other workers (Li et al., 2009a) proposed a method to produce ultra-low-sulfur diesel by adsorption and biodesulfurization in which the adsorbents were regenerated adsorbents with microbial cells. The adsorption and bioregeneration properties of the adsorbents were studied with *P. delafieldii* R-8 and different types of adsorbents. The regeneration system contained *n*-octane, aqueous phase, lyophilized cells, and spent adsorbents. All reactions were carried out in 100 mL flasks at 30°C on a rotary shaker operated at 200 rpm. Adsorption–bioregeneration properties were tested in an *in situ* adsorption–bioregeneration system, which can conveniently be divided into two parts: adsorption and bioregeneration (Figure 9.6). After the saturation of adsorbents, the adsorption system is shut down and the adsorption reactor was connected with the bioreactor. Then, the desorbed sulfur compounds were converted by R-8 cells. Finally, the desorbed adsorbents were treated with air at 550°C to remove water from the adsorption system. The integrated system is able to efficiently desulfurize dibenzothiophene, and the adsorption property of bioregenerated adsorbents is similar to fresh ones.

Immobilized cells have been used for increased volumetric reaction rate and lower operating costs for biodesulfurization (Lee et al., 2005; Dinamarca et al., 2010) due to the utilization of high cell concentrations of the biocatalyst and increased transport rate of organosulfur compounds to

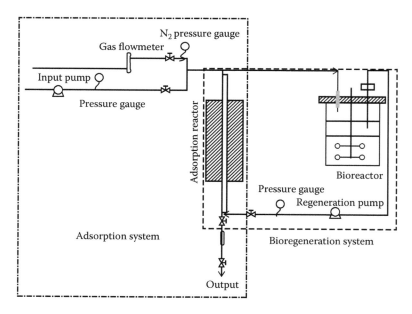

FIGURE 9.6 *In situ* adsorption–bioregeneration system. (From Li, W., Tang, H., Liu, Q., Xing, J., Li, Q., Wang, D., Yang, M., Li, X., and Liu, H., *Biochem. Eng. J.*, 44, 297, 2009.)

the biocatalyst. In most of biodesulfurization studies, two separate stages are employed—one for cell growth and the second for induction and biodesulfurization activity. To simplify the process and lower the operating cost, a single vertical rotating immobilized cell reactor using *R. erythropolis* IGTS8, glucose as a carbon source, and polyurethane foam for immobilization to desulfurize model oil has also been used and contained 11.68 mM dibenzothiophene/hexadecane in a ratio of 1:6 oil/water, with a feed rate of 320 mL/h (Amine, 2011). The highest biodesulfurization activity of 132 mM 2-HBP/kg dry cells/h was recorded within 15 h and then decreased dramatically. By applying 25 g/L each of glucose and ethanol as a carbon source and 11.68 mM dibenzothiophene in hexadecane as a sulfur source after 35 h, the activity of the reactor was restored and sulfate concentration was decreased and 2-HBP was detected in bioreactor effluent at 40 h and steadily increased, reaching a steady state at 55 h with complete removal of sulfur (of 22% and specific desulfurization of 166.86 mM 2-HBP/kg dry cells/h for 120 h). The vertical rotating immobilized cell reactor method has longevity of operation, final biomass concentration, and volumetric biodesulfurization activity that are 1.4, 3.5, and 7.3 times higher, respectively—this is an important aspect for the industrial implementation of such a process.

From a cost perspective, biodesulfurization has favorable features: (1) operation at low temperature and pressure; (2) biodesulfurization is estimated to have 70–80% lower carbon dioxide emissions; (3) in the case of reaching adequate biodesulfurization efficiency level, the capital cost required for an industrial biodesulfurization process is predicted to be two-third of that for a hydrodesulfurization process; (4) cost-effective—for biodesulfurization, the capital and operating costs are 50% and 10–15% less than that for hydrodesulfurization, respectively (Pacheco, 1999); Atlas et al. (2001) estimated the hydrodesulfurization process cost to decrease the sulfur content from 500 to 200 ppm to be approximately 0.26 cent/L, and desulfurization cost would increase by a factor of 4 if the sulfur content is further lowered to 50 ppm; and (5) flexible nature—the process can be applied to many process streams: crude feeds, FCC naphtha, and middle distillates.

However, the application of large-scale biodesulfurization technology is still only at the pilot scale. US-EBC developed a pilot plant with a working capacity of 5 barrels of oil per day. In the process, the biocatalyst and fossil fuel are mixed in a bioreactor where biodesulfurization takes place, then passed through series of filters to a container for disposal of sewage. Finally, the mixture is added to a basic aqueous solution for neutralization and removal of sodium sulfate in wastewater treatment (Figure 9.7) (Monticello, 2000). Since the degradation rate and metabolism of currently available microorganisms are still low, three bioreactors are needed to reach low sulfur concentration. Renewing microbial biomass during the biodesulfurization process would make the process commercially viable.

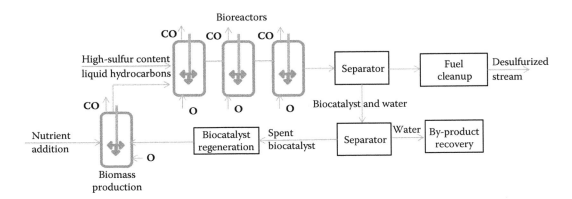

FIGURE 9.7 Biodesulfurization pilot plant. (From Monticello, D.J., *Curr. Opin. Biotechnol.*, 11, 540, 2000.)

9.3   NANO-BIOTECHNOLOGY AND BIODESULFURIZATION

Compared with biodesulfurization, adsorptive desulfurization has a much faster reaction rate (Song and Ma, 2003). Adsorbent preparation is the key of adsorptive desulfurization. Recently, most adsorbents for desulfurization were based on π-complexation (Shan et al., 2005b) or formation of metal–sulfur bonds (such as Ni–S, La–S) (Tian et al., 2006). Adsorbents based on π-complexation are easy to regenerate, but their selectivity is very low, resulting in a loss of fuel quality. Meanwhile, adsorbents that form metal–sulfur bonds with sulfur have high selectivity but are difficult to regenerate. Hence, adsorptive desulfurization technology also has a long way to go before being industrialized. If a desulfurization technology has both the high reaction rate of adsorptive desulfurization and the high selectivity of biodesulfurization, it can increase the desulfurization rate without damaging fuel quality.

The adsorbent is an important factor of this coupling technology because different adsorbents have different interactions with the organic sulfur compounds and cells, which would affect their assembly onto the cell surfaces and the desorption behavior of organic sulfur from them. Because the adsorbents are assembled on the cell surfaces, the property of cell surface is another factor that affects the coupling technology. Moreover, desulfurization conditions, such as temperature and volume ratio of the oil-to-water phase, would also affect the *in situ* coupling technology.

In situ coupling of adsorptive desulfurization and biodesulfurization is a new desulfurization technology for fossil oil. It has the merits of high selectivity of biodesulfurization and high rate of adsorptive desulfurization. It is carried out by assembling nanoadsorbents onto surfaces of microbial cells. For example, the combination of adsorptive desulfurization and biodesulfurization, an *in situ* coupling technology, has been proposed (Guobin et al., 2005). γ-Al$_2$O$_3$ nanosorbents, which have the ability to selectively adsorb dibenzothiophene from organic phase, were assembled on the surfaces of *P. delafieldii* R-8 cells, a desulfurization strain. γ-Al$_2$O$_3$ nanosorbents have the ability to adsorb dibenzothiophene from oil phase, and the rate of adsorption was significantly higher than that of biodesulfurization (Guobin et al., 2005). Thus, dibenzothiophene can be quickly transferred to the biocatalyst surface where nanosorbents were located, which quickened dibenzothiophene transfer from the organic phase to the biocatalyst surface, and resulted in the increase of biodesulfurization rate. The desulfurization rate of the cells assembled with nanosorbents was approximately 2.5-fold higher than that of original cells. An improved *in situ* coupling technology of adsorptive desulfurization and biodesulfurization by *P. delafieldii* R-8 cells has also been reported (Zhang et al., 2007). γ-Alumina (Al$_2$O$_3$) nanoparticles were synthesized and modified using gum Arabic to avoid agglomeration in aqueous solutions, and its effect on adsorptive desulfurization and biodesulfurization was also evaluated. Results showed that γ-Al$_2$O$_3$ nanoparticles dispersed well in aqueous solutions after modification with gum Arabic. The adsorptive desulfurization capacity of γ-alumina nanoparticles was increased from 0.56 mmol S/g (Al$_2$O$_3$) to 0.81 mmol S/g (Al$_2$O$_3$) after treatment with gum Arabic. Compared with unmodified γ-alumina nanoparticles, the biodesulfurization rate by adsorbing gum Arabic-modified γ-alumina nanoparticles onto the surfaces of R-8 cells was increased from 17.8 to 25.7 mmol/kg/h, which may be due to improvement in the dispersion and biocompatibility of γ-alumina nanoparticles after modification.

Other workers (Zhang et al., 2008) used different kinds of widely used adsorbents (alumina, molecular sieves, and active carbon) in the *in situ* coupling technology of adsorptive desulfurization and biodesulfurization. The procedure was carried out by assembling nanoadsorbents onto surfaces of *P. delafieldii* R-8 cells. The data showed that Na-Y molecular sieves restrain the activity of R-8 cells, and active carbon cannot desorb the substrate dibenzothiophene. Thus, they are not applicable to *in situ* coupling desulfurization technology. γ-Alumina can adsorb dibenzothiophene from the oil phase quickly, and then desorb and transfer it to R-8 cells for biodegradation, thus increasing desulfurization rate. It was also found that nano-sized γ-alumina increases desulfurization rate more than regular-sized γ-alumina. Therefore, nano-γ-alumina is regarded as a better adsorbent for this *in situ* coupling desulfurization technology.

In the last two decades, numerous studies have been carried out on biodesulfurization using whole cells (Maghsoudi et al., 2001; Tao et al., 2006; Yang et al., 2007; Caro et al., 2008) or isolated enzymes (Monticello and Kilbane, 1994) in the free or immobilized form. Biodesulfurization of dibenzothiophene occurs via a multienzyme system that requires cofactors (e.g., NADH—the reduced form of nicotinamide adenine dinucleotide). The use of enzymes is disadvantageous since extraction and purification of the enzyme is costly and, frequently, enzymes catalyzing reactions require cofactors that must be regenerated (Setti et al., 1997). Therefore, biodesulfurization can often be designed by using whole-cell biotransformation rather than that of the enzyme. However, there are still some bottlenecks limiting the commercialization of the biodesulfurization process.

One of the challenges is to improve the current biodesulfurization rate by about 500-fold, assuming the target industrial process is 1.2–3 mmol/g DCW/h (Kilbane, 2006). When free cells are used for petroleum biodesulfurization, deactivation of the biocatalyst and troublesome oil–water–biocatalyst separation are significant barriers (Konishi et al., 2005; Yang et al., 2007). Cell immobilization may give a solution to the problems, providing advantages such as repeated or continuous use, enhanced stability, and easy separation.

For biodesulfurization, cells need to be harvested from the culture medium, and several separation schemes had been evaluated, including settling tanks (Schilling et al., 2002), hydrocyclones (Yu et al., 1998), and centrifuges (Monticello, 2000). However, these procedures are time consuming and costly. Magnetic separation technology provides a quick, easy, and convenient alternative over traditional methods in biological systems (Haukanes and Kvam, 1993). Superparamagnetic nanoparticles are increasingly used to achieve affinity separation of high-value cells and biomolecules (Molday et al., 1977). Magnetic supports for cell immobilization offer several advantages, such as the ease of magnetic collection. The magnetic supports present further options in continuous reactor systems when used in a magnetically stabilized, fluidized bed. In addition, the mass transfer resistance can be reduced by the spinning of magnetic beads under a revolving magnetic field (Sada et al., 1981).

There is a report (Shan et al., 2003) on the modification of magnetite (Fe^{2+}–Fe^{3+} oxide, Fe_3O_4) magnetic fluid preparation with the coprecipitation method (Liu et al., 2003) to produce a hydrophilic magnetic fluid. Addition of nitrogen gas during preparation prevents oxidation of ferrous ion in the aqueous solution, prevents blackening, and controls particle size. Magnetic particles prepared by the coprecipitation method have a large number of hydroxyl groups on their surface in contact with the aqueous phase. The -OH groups on the surface of mixed iron oxide nanoparticles (with an average particle diameter on the order of 8 nm) react readily with carboxylic acid head groups of the added oleic acid, after which the excess oleic acid will be adsorbed to the first layer of oleic acid to form a hydrophilic shell. When this magnetic fluid was added to aqueous ammonium hydroxide, the outer layer of oleic acid on the magnetite surface is transformed into ammonium oleate and the hydrophilic magnetic fluid is produced. Magnetic fluid was directly mixed into hydrophilic support liquids such as polyvinyl alcohol and sodium alginate with dried cells of *P. delafieldii* R-8. Immobilized beads were formed by extruding the aforementioned mixture through a syringe into a gelling solution of 0.1 M calcium chloride ($CaCl_2$) saturated with boric acid, and solidified for 24 h. The immobilized beads formed were washed with saline, and then freeze dried for 48 h under vacuum. The magnetic fluid is mainly composed of magnetite particles that not only provide the magnetic property of support but also improve the mechanical strength of the supports, which are superparamagnetic. Therefore, the magnetic immobilized supports could be easily separated and recycled by an external magnetic field, and the recovered magnetic supports could be redispersed by gentle shaking with the removal of the external magnetic field. Compared with nonmagnetic immobilized cells, the beads of magnetic immobilized cells showed higher reaction activity of desulfurization and also higher strength against swelling, longer-term stability, and the possibility to be reused for seven cycles of reaction, while the nonmagnetic immobilized cells could be used for only five cycles. Also, the magnetic immobilized cells can be easily separated from the reaction medium, stored, and reused to give consistent results. The support is relatively cheap, easy to prepare, and good for large-scale industrial applications.

Other workers have attempted to increase the efficiency of cells and to decrease the cost of operations in a biodesulfurization process (Shan et al., 2005c). In their work, magnetic polyvinyl alcohol beads were prepared by a freezing–thawing technique in liquid nitrogen, and the beads have distinct superparamagnetic properties. The desulfurization rate of the immobilized cells could reach 40.2 mmol/kg/h, twice that of free cells. The heat resistance of the cells apparently increased when the cells were entrapped in magnetic polyvinyl alcohol beads. The cells immobilized in magnetic polyvinyl alcohol beads could be stably stored and be repeatedly used over 12 times for biodesulfurization. The immobilized cells could be easily separated by magnetic field. To understand cell distribution in magnetic polyvinyl alcohol beads, the sections of the beads after being repeatedly used for six times were observed by scanning electron microscopy. A highly macroporous structure is found in the beads in favor of diffusion of substrates and dissolved gas. On average, the size of the beads was about 3 mm. It is evident that the R8 cells mainly covered the edges and submarginal sections of the bead, while there was no cell in the center of the bead because of insufficiencies of oxygen and nutrients and, gradually, autolysis.

A new technique in which magnetic nanoparticles are used to coat the cells could successfully overcome the difficulties of conventional cell immobilization, such as mass transfer problems, cell loss, and separation of carrier with adsorbed cells from the reaction mixture at the end of a desulfurization treatment. A coating layer of nanoparticles does not change the hydrophilic nature of the cell surface. The coating layer has negligible effect on mass transfer because the structure of the layer is looser than that of the cell wall and does not interfere with mass transfer of dibenzothiophene. The coated cells have good stability and can be reused. This new technique has the advantage of magnetic separation and is convenient as well as easy to perform, so it offers the potential to be suitable for large-scale industrial applications.

In addition, a report (Shan et al., 2005a) of a technique described a process in which microbial cells of *P. delafieldii* R-8 were coated with magnetic nanoparticles (Fe_3O_4) and then immobilized by external application of a magnetic field. The nanoparticles were strongly adsorbed on the cell surfaces because of their high specific surface area and high surface energy. It was possible to concentrate the dispersed coated cells by application of a magnetic field for reuse, and when dispersed, the coated cells experienced minimal mass transfer problems. Thus, this technique has advantages over conventional immobilization by adsorption to carrier materials such as celite. Furthermore, the method can overcome drawbacks such as limitations in biomass loading and in the loss of cells from the carrier associated with conventional immobilization by adsorption. The nanoparticles were synthesized by a coprecipitation method followed by modification with ammonium oleate. The surface-modified magnetite nanoparticles were monodispersed in an aqueous solution and did not precipitate over 18 months. Using transmission electron microscopy, the average size of the magnetic particles was found to be in the range from 10 to 15 nm (Figure 9.8). Transmission electron microscopic cross-section analysis of the cells showed further that the magnetite nanoparticles were, for the most part, strongly absorbed by the surfaces of the cells and coated the cells. The coated cells had distinct superparamagnetic properties. The coated cells not only had the same desulfurizing activity as free cells but could also be reused more than five times, while the free cells could be used only once. Compared with cells immobilized on celite, the cells coated with magnetite nanoparticles had greater desulfurizing activity and operational stability insofar as the coated cells did not experience a mass transfer problem.

The *in situ* cell separation and immobilization of bacterial cells for biodesulfurization, which were developed by using superparamagnetic magnetite nanoparticles, have also been reported (Li et al., 2009b). The magnetite nanoparticles were synthesized by coprecipitation followed by modification with ammonium oleate. The surface-modified nanoparticles were monodispersed, and the particle size was on the order of 13 nm. After adding the magnetic fluids to the culture broth, *R. erythropolis* LSSE8-1 cells were immobilized by adsorption and then separated with an externally magnetic field. Analysis showed that the nanoparticles were strongly adsorbed to the surface and coated the cells. Compared with free cells, the coated cells not only had the same desulfurizing

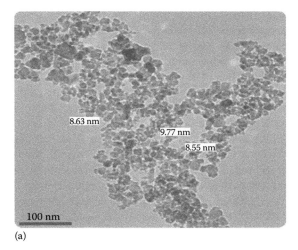

(a)

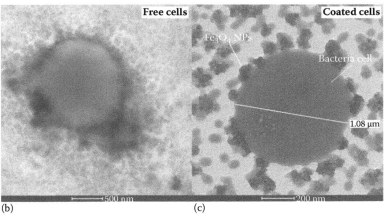

(b) (c)

FIGURE 9.8 Transmission electron microscopy images of (a) synthesized Fe_3O_4 nanoparticles and (b) free and (c) coated cells.

activity but could also be easily separated from fermentation broth by magnetic force. It was believed that oleate-modified magnetite nanoparticles adsorbed bacterial cells mainly because of the nano-size effect and hydrophobic interaction.

From the point of commercial application, the *Rhodococcus* strains possess several properties favorable for desulfurization over *Pseudomonas* in an oil–water system. First, the hydrophobic nature of *Rhodococcus* makes them access preferentially Cx-DBT derivatives from the oil, resulting in little mass transfer limitation (Le Borgne and Quintero, 2003). Moreover, the *Rhodococcus* bacteria are more resistant to solvents than *Pseudomonas* (Bouchez-Naitali et al., 2004). Therefore, there has been an attempt to develop a simple and effective technique by integrating the advantages of magnetic separation and cell immobilization for biodesulfurization process with gram-positive *R. erythropolis* LSSE8-1 and gram-negative *P. delafieldii* R-8 (Li et al., 2009b). Cells were grown to the late exponential phase (the phase where microbes exhibit exponential growth and some have also called the logarithmic phase) and the culture was transferred into Erlenmeyer flasks. A volume of magnetic fluids was added and mixed thoroughly—the microbial cells were coated by adsorbing the magnetic nanoparticles. The ammonium oleate-modified magnetite nanoparticles formed a stable suspension in distilled water, and the magnetic fluid did not settle during 8 months of storage at room temperature. The transmission electron microscope image of magnetite nanoparticles showed that the particles have an approximately spherical morphology with an average diameter of about 13 nm. The magnetite nanoparticles on the cell surface were not washed out by deionized water, ethanol, saline water (0.85 wt.%), or phosphate buffer (0.1 M, pH 7). Thus, there is little cell loss or decrease in biomass loading when cells were coated with magnetic nanoparticles. This outcome was different from that obtained with cells immobilized by traditional adsorption to a carrier. The cells coated with magnetite nanoparticles were superparamagnetic. Therefore, the cell–nanoparticle aggregates in aqueous suspension could be easily separated with an externally magnetic field and redispersed by gentle shaking after the removal of magnetic field. For magnetic separation, a permanent magnet can be placed at the side of the vessel. After several minutes (3–5 min), the coated cells can be concentrated and separated from the suspension medium by decantation (Figure 9.9).

An adsorption mechanism between the magnetite nanoparticles and desulfurizing cells has also been proposed (Li et al., 2009b). The large specific surface area and the high surface energy of the magnetite nanoparticles (i.e., the nano-size effect) ensure that the magnetite nanoparticles are strongly adsorbed on the surfaces of microbial cells. Furthermore, the hydrophobic interaction between the bacterial cell wall and the hydrophobic tail of oleate-modified magnetite nanoparticles may play another important role in cell adsorption. The suspension of oleate-modified magnetite nanoparticles was considered a bilayer surfactant–stabilized aqueous magnetic fluid (Liu et al., 2006). The iron oxide nanocrystals were first chemically coated with oleic acid molecule, after which the excess oleic acid was weakly adsorbed on the primary layer through the hydrophobic interaction between the subsequent molecule and the hydrophobic tail of oleate. Since the bacterial cell wall is composed of proteins, carbohydrates, and other substances (such as peptidoglycan, lipopolysaccharide, and mycolic acid), the extracellular matrix can form a hydrophobic interaction with the hydrophobic tail of the oleate of nanoparticles.

Previous work (Shan et al., 2005a) had utilized resting cells of *P. delafieldii* R-8 coated with magnetite nanoparticles for biodesulfurization; however, the process was more complicated because centrifugation was necessary in the preparation of the resting cells. However, the development of magnetically separated/immobilized desulfurizing bacterial cells of the two gram types from their original culture fermentation broth, namely *in situ* magnetic separation and immobilization of bacteria, has been reported (Li et al., 2009b). This one-step technology dramatically optimized the biodesulfurization process flow, and much fewer magnetite nanoparticles were needed.

FIGURE 9.9 Photograph of coated cell suspension collected by application of an external magnetic field.

A microbial method has also been used to regenerate desulfurization adsorbents (Li et al., 2006). Most of the sulfur compounds can be desorbed and removed, and heat losses during the bioregeneration process are markedly reduced. The particle size of cells is similar to that of desulfurization adsorbents, which is about several microns. Therefore, it is difficult to separate regenerated adsorbents and cells. Superparamagnetism is an efficient method to separate small particles. To solve the problem of separation of cells and adsorbents, magnetite nanoparticle–modified *P. delafieldii* R-8 cells were used in the bioregeneration of adsorbents. Biodesulfurization with *P. delafieldii* R-8 strains coated with magnetite nanoparticles has been reported previously (Shan et al., 2005a), and cells coated with magnetite nanoparticles can be separated and reused several times.

There is also the report (Li et al., 2008b) on the bioregeneration of the desulfurization adsorbent silver–yttrium (AgY) with magnetic cells. Superparamagnetic magnetite nanoparticles are prepared by the coprecipitation method followed by modification with ammonia oleate. Magnetic *P. delafieldii* R-8 cells can be prepared by mixing the cells with magnetite nanoparticles. When magnetic cells were used in the bioregeneration of the desulfurization adsorbent AgY, the concentration of dibenzothiophene and 2-HBP with free cells is a little higher that that with magnetic cells. The adsorption capacity of the regenerated adsorbent is 93% that of the fresh one after being desorbed with magnetic *P. delafieldii* R-8, dried at 100°C for 24 h, and calcined in the air at 500°C for 4 h. The magnetic cells can be separated from the adsorbent bioregeneration system after desulfurization with an external magnetic field and, thus, can be reused.

9.4 FUTURE

Microbial desulfurization in nature is different from other more common biotechnological processes, and the process has several limitations that prevent it from being applied in a modern refinery (Gupta et al., 2005). The metabolism of sulfur compounds is typically slow compared with chemical reactions employed in a refinery, and generally the rate of metabolism limits the rate of the process, although mass transfer resistance from the oil–water interface to the microbe is also slow compared with the rate of transfer of the sulfur compound to the oil–water interface. Large amounts of biomass are needed (typically 2.5 g biomass per gram sulfur), and biological systems must be kept alive to function, which can be difficult under the variable input conditions found in refineries. The rate of desulfurization depends strongly on pH, temperature, and dissolved oxygen concentration. Separation of the cells from the oil can also be difficult, and immobilized cells often have lower activity and limited lifetime.

Furthermore, the delicate enzymes of desulfurization-competent cells have to obtain part of their substrates from a different phase in which they cannot survive. Since the first time that a specific sulfur removal was introduced, research in this field has steadily continued. Many other strains have been isolated or cloned, and different methods have been tried. The most serious problem in the implementation of biodesulfurization as an alternative industrial approach to produce ultra-low sulfur content lies in the isolation or design of a microbial strain with higher efficiency. Any small success that provides the possibility to remove sulfur at a higher temperature, with a higher rate, or longer stability of desulfurization activity is considered a significant step toward industry-level biodesulfurization. Moreover, process development and any unexpected problems that might occur as a result of upscaling the operations must also be considered.

Once success has been achieved, the process may operate in a line after hydrodesulfurization unit. Total desulfurization of fossil fuel by the microbial approach is not expected to be on-stream in the near future, and more research is needed to design a recombinant strain with a broader range of target sulfur compounds or to use successive desulfurizing microbial systems with high potency. Most researches on desulfurization of refractory compounds have been performed with a simple model fuel to understand the nature of desulfurization. Dealing with genuine fossil fuel will open up new challenges to solve.

Adsorption, biodesulfurization, and hydrodesulfurization are more likely to be employed to be complementary to each other. However, to develop the biodesulfurization process as a complementary process, a multidisciplinary approach, as well as the participation of scientists and engineers from the fields of biotechnology, biochemistry, refining processes, and engineering, is essential. In the long run, applying nanotechnology to desulfurization of petroleum and petroleum products will lead to a process capable of producing ultra-low-sulfur to no-sulfur products on an economical and cost-effective basis.

Furthermore, desulfurizing biocatalysts can be modified to remove specific sulfur structures or broader classes of sulfur compounds. Biocatalysts that can upgrade oils modify specific organic structures like cleaving or opening aromatic rings. The advantage of this selectivity is that a process user can reliably predict the chemical changes that will occur. Unfortunately, crude oil being a complex natural product is variable in composition; therefore, for crude oil applications, the biocatalysts will have to be customized for each crude oil and the total process may involve multiple biocatalysts.

However, if these hurdles can be overcome, the bioprocessed oil can be processed at a refinery using current technology at a lower cost. For sulfur, hydrotreating is very effective for removing mercaptans and straight-chain sulfides, whereas biodesulfurization is more effective with organic ring sulfides like dibenzothiophene. The hydrotreating conditions are less severe (lower cost) for straight-chain sulfur compounds, so an economic advantage is achieved when both hydrotreating and biodesulfurization processes are used. Moreover, bioprocessing can have a beneficial effect on oil properties, especially properties that affect the handling of the bulk oil—improvements such as (1) viscosity reduction, (2) shifts to lower molecular weight distribution in the oil, or (3) lower asphaltene content would reduce fluid piping and transportation problems and costs. Indeed, a higher-grade and cleaner oil (low sulfur content and high API gravity) could well be the outcome of bioprocessing crude oil at the wellhead.

For example, there have been significant developments in the area of defining a truly efficient and economical biodesulfurization process. Critical aspects of such a process include the (1) suitability of the microbes, (2) reaction rate, (3) product recovery, and (4) reactor design. In addition, other technical issues that need to be fully resolved include molecular mechanisms, optimized microbial strains for commercial use, mass transfer, and reactivity of heavy feedstocks (without having a detrimental effect on the biocatalyst), where viscosity and density play an important role in feedstock processing (Gray et al., 1996; Monticello, 2000; Bachmann et al., 2014).

As a corollary, one aspect of biodesulfurization that should also be explored is the application of the process at the wellhead. Many of the biocatalysts that can be applied to crude oil upgrading and desulfurization are water soluble, and in an oil field operation, water is coproduced with the oil; thus, the process of water and oil separation is a routine field process, and the only added process step would be agitation of the oil and water mixture with the biocatalyst. The process would generate a wastewater stream, which must be handled whether it is generated in the oil field or at the refinery. In a refinery, this new waste stream becomes an added problem. However, in the field, the water containing the formed salts can be diluted and reinjected as part of the field water flood program, which, if allowable under environmental regulations, may have minimal effect on oil field operation.

One advantage of such a concept is the potential simplicity of the process—the crude oil is mixed with a water-soluble biocatalyst (either the microorganism or the enzyme) and with air. After the reaction, the formed water–oil emulsion is separated to recover the upgraded oil—the biocatalyst remains with the water and is (potentially) available for reuse. The only added feature (that may not always be available at the wellhead) is a mixing reactor before separation.

However, the issue of biodegradation of asphaltene constituents remains a significant issue, and feedstocks containing asphaltene constituents may be difficult to process. The qualitative and quantitative differences in the hydrocarbon and the nonhydrocarbon content of petroleum (Speight, 2014) influence the susceptibility of petroleum and certain petroleum products to biocatalysis. This must be acknowledged as a major consideration in determining the potential for biocatalytic desulfurization of asphaltene-containing feedstocks (Speight and Arjoon, 2012).

In summary, significant progress has been made toward the commercialization of the biodesulfurization of petroleum, which includes (1) characterization of feedstock candidates for the process, (2) improved biocatalyst performance that directly relates to crude oil biodesulfurization, and (3) development of an analytical method that can be employed in the development of biodesulfurization process concepts.

REFERENCES

Abbad-Andaloussi, S., Warzywoda, M., and Monot, F. 2003. Microbial Desulfurization of Diesel Oils by Selected Bacterial Strains. *Oil Gas Sci. Technol. Rév. Inst. Fr. Pétrol.* 58(4): 505–513.

Abin-Fuentes, A., Mohamed, M.E.S., Wang, E.I.C., and Prather, K.L.J. 2013. Exploring the Mechanism of Biocatalyst Inhibition in Microbial Desulfurization. *Appl. Environ. Microbiol.* 79(24): 7807–7817.

Akbarzadeh, S., Raheb, J., Aghaei, A., and Karkhane, A.A. 2003. Study of Desulfurization Rate in *Rhodococcus* FMF Native Bacterium. *Iranian J. Biotechnol.* 1(1): 36–40.

Amine, G.A. 2011. Integrated Two-Stage Process for Biodesulfurization of Model Oil by Vertical Rotating Immobilized Cell Reactor with the Bacterium *Rhodococcus erythropolis*. *J. Pet. Environ. Biotechnol.* 2:107. doi:10.4172/2157-7463.1000107.

Ancheyta, J., and Speight, J.G. (Editors) 2007. *Hydroprocessing Heavy Oils and Residua*. CRC Press, Taylor & Francis Group, Boca Raton, FL.

Aribike, D.S., Susu, A.A., Nwachukwu, S.C.U., and Kareem, S.A. 2008. Biodesulfurization of Kerosene by *Desulfobacterium indolicum*. *Nat. Sci.* 1(4): 55–63.

Aribike, D.S., Susu, A.A., Nwachukwu, S.C.U., and Kareem, S.A. 2009. Microbial Desulfurization of Diesel by *Desulfobacterium anilini*. *Acad. Arena* 1(4): 11–17.

Atlas, R.M., Boron, D.J., Deever, W.R., Johnson, A.R., McFarland, B.L., and Meyer, J.A. 2001. Method for Removing Organic Sulfur from Heterocyclic Sulfur Containing Organic Compounds. US Patent H1,986.

Bachmann, R.T., Johnson, A.C., and Edyvean, R.G.J. 2014. Biotechnology in the Petroleum Industry: An Overview. *Int. Biodeter. Biodegrad.* 86: 225–237.

Bahuguna, A., Lily, M.K., Munjal, A., Singh, R.N., and Dangwal, K. 2011. Desulfurization of Dibenzothiophene (DBT) by a Novel Strain *Lysinibacillus sphaericus* DMT-7 Isolated from Diesel Contaminated Soil. *J. Environ. Sci.* 23(6): 975–982.

Baldi, F., Pepi, M., and Fava, F. 2003. Growth of *Rhodosporidium toruloides* Strain DBVPG 6662 on Dibenzothiophene Crystals and Orimulsion. *Appl. Environ. Microbiol.* 69(8): 4689–4696.

Bhatia, S., and Sharma, D.K. 2010a. Biodesulfurization of Dibenzothiophene, Its Alkylated Derivatives, and Crude Oil by a Newly Isolated Strain of *Pantoea agglomerans* D23W3. *Biochem. Eng. J.* 50(3): 104–109.

Bhatia, S., and Sharma, D.K. 2010b. Mining of Genomic Databases to Identify Novel Biodesulfurizing Microorganisms. *J. Ind. Microbiol. Biotechnol.* 37: 425–429.

Bouchez-Naitali, M., Abbad-Andaloussi, S., Warzywoda, M., and Monot, F. 2004. Relation between Bacterial Strain Resistance to Solvents and Biodesulfurization Activity in Organic Medium. *Appl. Microbiol. Biotechnol.* 65: 440–445.

Bustos-Jaimes, I., Amador, G., Castorena, G., and Le Borgne, S. 2003. Genotypic Characterization of Sulfur-Oxidative Desulfurizing Bacterial Strains Isolated from Mexican Refineries. *Oil Gas Sci. Technol. Rev. Inst. Fr. Petrol.* 58(4): 521–526.

Calzada, J., Alcon, A., Santos, V.E., and Garcia-Ochoa, F. 2011. Mixtures of *Pseudomonas putida* CECT 5279 Cells of Different Ages: Optimization as Biodesulfurization Catalyst. *Process Biochem.* 46: 1323–1328.

Caro, A., Boltes, K., Letn, P., and Garca-Calvo, E. 2008. Biodesulfurization of Dibenzothiophene by Growing Cells of *Pseudomonas putida* CECT 5279 in Biphasic Media. *Chemosphere* 73(5): 663–669.

Cassidy, M.B., Lee, H., and Trevors, J.T. 1996. Environmental Applications of Immobilised Cells: A Review. *J. Ind. Microbiol. Biotechnol.* 16: 79–101.

Castorena, G., Suarez, C., Valdez, I., Amador, G., Fernandez, L., and Le Borgne, S. 2002. Sulfur-Selective Desulfurization of Dibenzothiophene and Diesel Oil by Newly Isolated *Rhodococcus* sp. Strains. *FEMS Microbiol. Lett.* 215(1): 157–161.

Chang, J.H., Chang, Y.K., Cho, K.S., and Chang, H.N. 2000. Desulfurization of Model and Diesel Oils by Resting Cells of *Gordona* sp. *Biotechnol. Lett.* 22: 193–196.

Chang, J.H., Kim, Y.J., Lee, B.H., Cho, K.S., Ryu, H.W., Chang, Y.K., and Chang, H.N. 2001. Production of a Desulfurization Biocatalyst by Two-Stage Fermentation and Its Application for the Treatment of Model and Diesel Oils. *Biotechnol. Progress* 17(5): 876–880.

Chang, J.H., Rhee, S.K., Chang, Y.K., and Chang, H.N. 1998. Desulfurization of Diesel Oils by a Newly Isolated Dibenzothiophene-Degrading *Nocardia* sp. Strain CYKS2. *Biotechnol. Prog.* 14: 851–855.

Chen, H., Zhang, W.J., Cai, Y.B., Zhang, Y., and Li, W. 2008. Elucidation of 2-Hydroxybiphenyl Effect on Dibenzothiophene Desulfurization by *Microbacterium* sp. Strain ZD-M2. *Bioresour. Technol.* 99: 6928–6933.

Davoodi-Dehaghani, F., Vosoughi, M., and Ziaee, A.A. 2010. Biodesulfurization of Dibenzothiophene by A Newly Isolated *Rhodococcus erythropolis* Strain. *Bioresour. Technol.* 101: 1102–1105.

Denome, S.A., Oldfield, C., and Nash, L.J. 1994. Characterization of the Desulfurization Genes from *Rhodococcus* sp. Strain IGTS8. *J. Bacteriol.* 176(21): 6707–6716.

Denome, S.A., Olson, E.S., and Young, K. 1993. Identification and Cloning of Genes Involved in Specific Desulfurization of Dibenzothiophene by *Rhodococcus* sp. Strain IGTS8. *Appl. Environ. Microbiol.* 59(9): 2837–2843.

Derikvand, P., Etemadifar, Z., and Biria, D. 2014. Taguchi Optimization of Dibenzothiophene Biodesulfurization by *Rhodococcus erythropolis* R1 Immobilized Cells in a Biphasic System. *Int. Biodeter. Biodegr.* 86: 343–348.

Dinamarca, M.A., Ibacache-Quiroga, C., Baeza, P., Galvez, S., Villarroel, M., Olivero, P., and Ojeda, J. 2010. Biodesulfurization of Gas Oil Using Inorganic Supports Biomodified with Metabolically Active Cells Immobilized by Adsorption. *Bioresour. Technol.* 101: 2375–2378.

Dinamarca, M.A., Rojas, A., Baeza, P., Espinoza, G., Ibacache-Quiroga, C.I., and Ojeda, J. 2014. Optimizing the Biodesulfurization of Gas Oil by Adding Surfactants to Immobilized Cell Systems. *Fuel* 116: 237–241.

Fang, X.X., Zhang, Y.L., Luo, L.L., Xu, P., Chen, Y.L., Zhou, H., and Hai, L. 2006. Organic Sulfur Removal from Catalytic Diesel Oil by Hydrodesulfurization Combined with Biodesulfurization. *Mod. Chem. Ind.* 26: 234–238 (Chinese journal; Abstract in English).

Feng, J., Zeng, Y., Ma, C., Cai, X., Zhang, Q., Tong, M., Yu, B., and Xu, P. 2006. The Surfactant Tween 80 Enhances Biodesulfurization. *Appl. Environ. Microbiol.* 72(11): 7390–7393.

Fernandes, P., Vidinha, P., Ferreira, T., Silvestre, H., Cabral, J.M.S., and Prazeres, D.M.F. 2002. Use of Free and Immobilized *Pseudomonas putida* Cells for the Reduction of a Thiophene Derivative in Organic Media. *J. Mol. Catal. B Enz.* 9–20: 353–361.

Folsom, B.R., Schieche, D.R., and Digrazia, P. 1999. Microbial Desulfurization of Alkylated Dibenzothiophenes from a Hydrodesulfurized Middle Distillate by *Rhodococcus erythropolis* I-19. *Appl. Environ. Microbiol.* 65: 4967–4972.

Freeman, A., and Lilly, M.D. 1998. Effect of Processing Parameters on the Feasibility and Operational Stability of Immobilized Viable Microbial Cells. *Enz. Microbiol. Technol.* 23: 335–345.

Furuya, T., Ishii, Y., Noda, K., Kino, K., and Kirimura, K. 2003. Thermophilic Biodesulfurization of Hydrodesulfurized Light Gas Oils by *Mycobacterium phlei* WU-F1. *FEMS Microbiol. Lett.* 221(1): 137–142.

Furuya, T., Kirimura, K., Kino, K., and Usami, S. 2001. Thermophilic Biodesulfurization of Dibenzothiophene and Its Derivatives by *Mycobacterium phlei* WU-F1. *FEMS Microbiol. Lett.* 204: 129–133.

Gallardo, M.E., Ferrandez, A., and Lorenzo, V.D. 1997. Designing Recombinant *Pseudomonas* Strain to Enhance Biodesulfurization. *J. Bacteriol.* 179(22): 7156–7160.

Gary, J.H., Handwerk, G.E., and Kaiser, M.J. 2007. *Petroleum Refining: Technology and Economics*, 5th Edition. CRC Press, Taylor & Francis Group, Boca Raton, FL.

Gibson, D.T., and Sayler, G.S. 1992. *Scientific Foundation for Bioremediation: Current Status and Future Needs.* American Academy of Microbiology, Washington, DC.

Giuliano, M., Schiraldi, C., Maresca, C., Esposito, V., and Rosa, M.D. 2003. Immobilized *Proteus mirabilis* in Poly(Vinyl Alcohol) Cryogels For L(−)-Carnitine Production. *Enz. Microb. Technol.* 32: 507–512.

Gray, K.A., Pogrebinsky, O., Mrachko, G., Xi, L., Monticello, D.J., and Squires, C. 1996. Molecular Mechanisms of Biocatalytic Desulfurization of Fossil Fuels. *Nat. Biotechnol.* 14: 1705–1709.

Grossman, M.J., Lee, M.K., Prince, R.C., Garrett, K.K., George, G.N., and Pickering, I.J. 1999. Microbial Desulfurization of a Crude Oil Middle-Distillate Fraction: Analysis of the Extent of Sulfur Removal and the Effect of Removal on Remaining Sulfur. *Appl. Environ. Microbiol.* 65: 181–188.

Gunam, I.B.W., Yaku, Y., Hirano, M., Yamamura, K., Tomita, F., and Sone, T. 2006. Biodesulfurization of Alkylated Forms of Dibenzothiophene and Benzothiophene by *Sphingomonas subarctica* T7b. *J. Biosci. Bioeng.* 101: 322–327.

Guobin, S., Huaiying, Z., Weiquan, C., Jianmin, X., and Huizhou, L. 2005. Improvement of Biodesulfurization Rate by Assembling Nanosorbents on the Surfaces of Microbial Cells. *Biophys. J.* 89(6): L58–L60.

Gupta, N., Roychoudhury, P.K., and Deb, J.K. 2005. Biotechnology of Desulfurization of Diesel: Prospects and Challenges. *Appl. Microbiol. Biotechnol.* 66(4): 356–366.

Hashimoto, S., and Furukawa, K. 1987. Immobilization of Activated Sludge by the PVA–Boric Acid Method. *Biotechnol. Bioeng.* 30(1): 52–59.

Haukanes, B.I., and Kvam, C. 1993. Application of Magnetic Beads in Bioassays. *Nat. Biotechnol.* 11: 60–63.

Hosseini, S.A., Yaghmaei, S., Mousavi, S.M., and Jadidi, A. 2006. Biodesulfurization of Dibenzothiophene by a Newly Isolated Thermophilic Bacteria Strain. *Iran J. Chem. Chem. Eng.* 25(3): 65–71.

Hou, Y., Kong, Y., Yang, J., Zhang, J., Shi, D., and Xin, W. 2005. Biodesulfurization of Dibenzothiophene by Immobilized Cells of *Pseudomonas stutzeri* UP-1. *Fuel* 84: 1975–1979.

Hsu, C.S., and Robinson, P.R. (Editors) 2006. *Practical Advances in Petroleum Processing*, Vols. 1–2. Springer Science, New York.

Huang, T., Qiang, L., Zelong, W., Daojiang, Y., and Jianmin, X. 2012. Simultaneous Removal of Thiophene and Dibenzothiophene by Immobilized *Pseudomonas delafieldii* R-8 Cells. *Chin. J. Chem. Eng.* 20(1): 47–51.

Hwan, J., Keun, Y., Wook, H., and Nam, H. 2000. Desulfurization of Light Gas Oil in Immobilized-Cell Systems of *Gordona* sp. CYKS1 and *Nocardia* sp. CYKS2. *FEMS Microbiol. Lett.* 182: 309–312.

Irani, Z.A., Mehrnia, M.R., Yazdian, F., Soheily, M., Mohebali, G., and Rasekh, B. 2011. Analysis of Petroleum Biodesulfurization in an Airlift Bioreactor Using Response Surface Methodology. *Bioresour. Technol.* 102: 10585–10591.

Ishii, Y., Konishi, J., Okada, H., Hirasawa, K., Onaka, T., and Suzuki, M. 2000. Operon Structure and Functional Analysis of the Genes Encoding Thermophilic Desulfurizing Enzymes of *Paenibacillus* sp. A11-2. *Biochem. Biophys. Res. Commun.* 270: 81–88.

Ishii, Y., Kozaki, S., Furuya, T., Kino, K., and Kirimura, K. 2005. Thermophilic Biodesulfurization of Various Heterocyclic Sulfur Compounds and Crude Straight-Run Light Gas Oil Fraction by a Newly Isolated Strain *Mycobacterium phlei* WU-0103. *Curr. Microbiol.* 50: 63–70.

Izumi, Y., Ohshiro, T., Ogino, H., Hine, Y., and Shimo, M. 1994. Selective Desulfurization of Dibenzothiophene by *Rhodococcus erythropolis* D-1. *Appl. Environ. Microbiol.* 60: 223–226.

Kayser, K.J., Bielaga-Jones, B.A., Jackowski, K., Odusan, O., and Kilbane, J.J. 1993. Utilization of Organo-sulphur Compounds by Axenic and Mixed Cultures of *Rhodococcus rhodochrous* IGTS8. *J. Gen. Microbiol.* 139: 3123–3129.

Kertesz, M.A. 2000. Riding the Sulfur Cycle—Metabolism of Sulfonates and Sulfate Esters in Gram-Negative Bacteria. *FEMS Microbiol. Rev.* 24: 135–175.

Kilbane, J.J. 1989. Desulfurization of Coal: The Microbial Solution. *Trends Biotechnol.* 7: 97–101.

Kilbane, J.J. 2006. Microbial Biocatalyst Development to Upgrade Fossil Fuels. *Curr. Opin. Biotechnol.* 17: 305–314.

Kilbane, J.J., and Le Borgne, S. 2004. Petroleum Biorefining: The Selective Removal of Sulfur, Nitrogen, and Metals. In: *Petroleum Biotechnology, Developments and Perspectives*, R. Vazquez-Duhalt and R. Quintero-Ramirez (Editors). Elsevier, Amsterdam, pp. 29–65.

Kim, Y.J., Chang, J.H., Cho, K.S., Ryu, H.W., and Chang, Y.K. 2004. A Physiological Study on Growth and Dibenzothiophene (DBT) Desulfurization Characteristics of *Gordonia* sp. CYKS1. *Korean J. Chem. Eng.* 21(2): 436–441.

Kirimura, K., Furuya, T., Sato, R., Yoshitaka, I., Kino, K., and Usami, S. 2002. Biodesulfurization of Naphthothiophene and Benzothiophene through Selective Cleavage of Carbon–Sulfur Bonds by *Rhodococcus* sp. Strain WU-K2R. *Appl. Environ. Microbiol.* 68(8): 3867–3872.

Kirimura, K., Harada, K., Iwasawa, H., Tanaka, T., Iwasaki, Y., Furuya, T., Ishii, Y., and Kino, K. 2004. Identification and Functional Analysis of the Genes Encoding Dibenzothiophene-Desulfurizing Enzymes from Thermophilic Bacteria. *Appl. Microbiol. Biotechnol.* 65: 703–713.

Kishimoto, M., Inui, M., Omasa, T., Katakura, Y., Suga, K., and Okumura, K. 2000. Efficient Production of Desulfurizing Cells with the Aid of Expert System. *Biochem. Eng. J.* 5: 143–147.

Klein, J., Stock, J., and Vorlop, D.K. 1993. Pore Size and Properties of Spherical Calcium Alginate Biocatalysts. *Eur. J. Appl. Microb. Biotechnol.* 18: 86–91.

Konishi, M., Kishimoto, M., Tamesui, N., Omasa, I., Shioya, S., and Ohtake, H. 2005. The Separation of Oil from an Oil–Water–Bacteria Mixture Using a Hydrophobic Tubular Membrane. *Biochem. Eng. J.* 24: 49–54.

Labana, S., Pandey, G., and Jain, R.K. 2005. Desulphurization of Dibenzothiophene and Diesel Oils by Bacteria. *Lett. Appl. Microbiol.* 40(3): 159–163.

Larose, C.D., Labbe, D., and Bergeron, H. 1997. Conservation of Plasmid Encoded Dibenzothiophene Desulfurization Genes in Several Rhodococci. *Appl. Environ. Microbiol.* 63(7): 2915–2919.

Le Borgne, S., and Quintero, R. 2003. Biotechnological Processes for the Refining of Petroleum. *Fuel Process. Technol.* 81: 155–169.

Lee, I.S., Bae, H., Ryu, H.W., Cho, K., and Chang, Y.K. 2005. Biocatalytic Desulfurization of Diesel Oil in an Air-Lift Reactor with Immobilized *Gordonia nitida* CYKS1 Cells. *Biotechnol. Prog.* 21: 781–785.

Lee, M.K., Senius, J.D., and Grossman, M.J. 1995. Sulfur-Specific Microbial Desulfurization of Sterically Hindered Analogs of Dibenzothiophene. *Appl. Environ. Microbiol.* 61: 4362–4366.

Li, F., Xu, P., Feng, J., Meng, L., Zheng, Y., Luo, L., and Ma, C. 2005. Microbial Desulfurization of Gasoline in a *Mycobacterium goodii* X7B Immobilized-Cell System. *Appl. Environ. Microbiol.* 71(1): 276–281.

Li, M.Z., Squires, C.H., and Monticello, D.J. 1996. Genetic Analysis of the Dsz Promoter and Associated Regulatory Regions of *Rhodococcus erythropolis* IGTS8. *J. Bacteriol.* 178: 6409–6418.

Li, W., and Jiang, X. 2013. Enhancement of Bunker Oil Biodesulfurization by Adding Surfactant. *World J. Microbiol. Biotechnol.* 29(1): 103–108.

Li, W., Tang, H., Liu, Q., Xing, J., Li, Q., Wang, D., Yang, M., Li, X., and Liu, H. 2009a. Deep Desulfurization of Diesel by Integrating Adsorption and Microbial Method. *Biochem. Eng. J.* 44: 297–301.

Li, W., Xing, J., Li, Y., Xiong, X., Li, X., and Liu, H. 2006. Feasibility Study on the Integration of Adsorption/ Bioregeneration of π-Complexation Adsorbent for Desulfurization. *Ind. Eng. Chem. Res.* 45(8): 2845–2849.

Li, W., Xing, J., Li, Y., Xiong, X., Li, X., and Liu, H. 2008b. Desulfurization and Bio-regeneration of Adsorbents with Magnetic *P. delafieldii* R-8 Cells. *Catal. Commun.* 9: 376–380.

Li, Y.G., Gao, H.S., Li, W.L., Xing, J.M., and Liu, H.Z. 2009b. *In situ* Magnetic Separation and Immobilization of Dibenzothiophene-Desulfurizing Bacteria. *Bioresour. Technol.* 100: 5092–5096.

Li, Y.G., Ma, J., Zhang, Q.Q., Wang, C.S., and Chen, Q. 2007. Sulfur-Selective Desulfurization of Dibenzothiophene and Diesel Oil by Newly Isolated *Rhodococcus erythropolis* NCC-1. *Chin. J. Org. Chem.* 25: 400–405.

Li, Y.G., Xing, J.M., Xiong, X.C., Li, W.L., Gao, S., and Liu, H.Z. 2008a. Improvement of Biodesulfurization Activity of Alginate Immobilized Cells in Biphasic Systems. *J. Ind. Microbiol. Biotechnol.* 35: 145–150.

Liu, X., Kaminski, M.D., Guan, Y., Chen, H., Liu, H., and Rosengart, A.J. 2006. Preparation and Characterization of Hydrophobic Superparamagnetic Gel. *J. Magn. Magn. Mater.* 306: 248–253.

Liu, X.Q., Liu, H.Z., Xing, J.M., Guan, Y.P., Ma, Z.Y., Shan, G.B., and Yang, C.L. 2003. Preparation and Characterization of Superparamagnetic Functional Polymeric Microparticles. *Chin. Particul.* 1: 76–79.

Lozinsky, V.I., and Plieva, F.M. 1998. Poly (Vinyl Alcohol) Cryogels Employed as Matrices for Cell Immobilization. 3. Overview of Recent Research and Development. *Enz. Microb. Technol.* 23: 227–242.

Luo, M.F., Xing, J.M., Gou, Z.X., Li, S., Liu, H.Z., and Chen, J.Y. 2003. Desulfurization of Dibenzothiophene by Lyophilized Cells of *Pseudomonas delafieldii* R-8 in the Presence of Dodecane. *Biochem. Eng. J.* 13: 1–6.

Ma, T., Li, G., Li, J., Liang, F., and Liu, R. 2006. Desulfurization of Dibenzothiophene by *Bacillus subtilis* Recombinants Carrying *dszABC* and *dszD* Genes. *Biotechnol. Lett.* 28: 1095–1100.

Maghsoudi, S., Kheirolomoom, A., and Vossoughi, M. 2000. Selective Desulfurization of Dibenzothiophene by Newly Isolated *Corynebacterium* sp. Strain P32C1. *Biochem. Eng. J.* 5: 11–16.

Maghsoudi, S., Vossoughi, M., Kheirolomoom, A., Tanaka, E., and Katoh, S. 2001. Biodesulfurization of Hydrocarbons and Diesel Fuels by *Rhodococcus* sp. strain P32C1. *Biochem. Eng. J.* 8: 151–156.

Matsui, T., Noda, K., Tanaka, Y., Maruhashi, K., and Kurane, R. 2002. Recombinant *Rhodococcus* sp. Strain T09 Can Desulfurize Dibenzothiophene in the Presence of Inorganic Sulfate. *Curr. Microbiol.* 45: 240–244.

Mohebali, G., and Ball, A.S. 2008. Biocatalytic Desulfurization (Biodesulfurization) of Petrodiesel Fuels. *Microbiology* 154: 2169–2183.

Mohebali, G., Ball, A.S., Keytash, A., and Rasekh, B. 2007. Stabilization of Water/Gas Oil Emulsions by Desulfurizing Cells of *Gordonia alkanivorans* RIPI90A. *Microbiology* 153: 1573–1581.

Molday, R.S., Yen, S.P., and Rembaum, A. 1977. Application of Magnetic Microspheres in Labelling and Separation of Cell. *Nature* 268: 437–438.

Monticello, D.J. 2000. Biodesulfurization and Upgrading of Petroleum Distillates. *Curr. Opin. Biotechnol.* 11: 540–546.

Monticello, D.J., Haney, I.I.I., and William, M. 1996. Biocatalytic Process for Reduction of Petroleum Viscosity. United States Patent 5,529,930.

Monticello, D.J., and Kilbane, J.J. 1994. Emulsification of Petroleum Oils and Aqueous Bacterial Enzyme, Then Incubation, for Selective Cleavage of Carbon–Sulfur Bonds. United States Patent 5,358,870.

Mukhopadhyay, M., Chowdhury, R., and Bhattacharya, P. 2006. Biodesulfurization of Hydrodesulfurized diesel in a Tickle Bed Reactor—Experiments and Modelling. *J. Sci. Ind. Res.* 65: 432–436.

Mužic, M., and Sertić-Bionda, K. 2013. Alternative Processes for Removing Organic Sulfur Compounds from Petroleum Fractions. *Chem. Biochem. Eng. Q.* 27(1): 101–108.

Naito, M., Kawamoto, T., Fujino, K., Kobayashi, M., Maruhashi, K., and Tanaka, A. 2001. Long-Term Repeated Biodesulfurization by Immobilized *Rhodococcus*. *Appl. Microbiol. Biotechnol.* 55: 374–378.

Nandi, S. 2010. Biodesulfurization of Hydro-desulfurized Diesel in Airlift Reactor. *J. Sci. Ind. Res.* 69: 543–547.

Noda, K., Kogure, T., Irisa, S., Murakami, Y., Sakata, M., and Kuroda, A. 2008. Enhanced Dibenzothiophene Biodesulfurization in a Microchannel Reactor. *Biotechnol. Lett.* 30: 451–454.

Noda, K., Watanabe, K., and Maruhashi, K. 2002. Cloning of *Rhodococcus* Promoter Using a Transposon for Dibenzothiophene Biodesulfurization. *Biotechnol. Lett.* 25: 1875–1888.

Noda, K.I., Watanabe, K., and Maruhashi, K. 2003. Isolation of A Recombinant Desulfurizing 4,6-Dipropyl Dibenzothiophene in *N*-Tetradecane. *Biosci. Bioeng.* 95(4): 354–360.

Ohshiro, T., Hirata, T., Hashimoto, I., and Izumi, Y. 1996b. Regulation of Dibenzothiophene Degrading Enzyme Activity of *Rhodococcus erythropolis* D-1. *J. Ferment. Bioeng.* 82(6): 610–612.

Ohshiro, T., Hirata, T., and Izumi, Y. 1995. Microbial Desulfurization of Dibenzothiophene in the Presence of Hydrocarbon. *Appl. Micorbiol. Biotechnol.* 44: 249–252.

Ohshiro, T., Hirata, T., and Izumi, Y. 1996a. Desulfurization of Dibenzothiophene Derivatives by Whole Cells of *Rhodococcus erythropolis* H-2. *FEMS Microbiol. Lett.* 142(1): 65–70.

Okada, H., Nomura, N., Nakahara, T., and Maruhashi, K. 2002. Analysis of Dibenzothiophene Metabolic Pathway in *Mycobacterium* sp. G3. *J. Biosci. Bioeng.* 93(5): 491–497.

Okada, H., Nomura, N., Nakahara, T., Saitoh, K., Uchiyama, H., and Maruhashi, K. 2003. Analyses of Microbial Desulfurization Reaction of Alkylated Dibenzothiophenes Dissolved in the Oil Phase. *Biotechnol. Bioeng.* 83(4): 489–497.

Oldfield, C., Poogrebinsky, O., and Simmonds, J. 1997. Elucidation of the Metabolic Pathway for Dibenzothiophene Desulphurization by *Rhodococcus* sp. Strain IGTS8 (ATCC53968). *Microbiology* 143: 2961–2973.

Pacheco, M.A. 1999. Recent Advances in Biodesulfurization (Biodesulfurization) of Diesel Fuel. In: *Proceedings. NPRA Annual Meeting.* San Antonio, TX, March 21–23.

Parkash, S. 2003. *Refining Processes Handbook.* Gulf Professional Publishing, Elsevier, Amsterdam.

Peng, Y., and Wen, J. 2010. Modeling of DBT Biodesulfurization by Resting Cells of *Gordonia* sp. *Chem. Biochem. Eng.* 24(1): 85–94.

Piddington, C.S., Kovacevich, B.R., and Rambosek, J. 1995. Sequence and Molecular Characterization of a DNA Region Encoding the Dibenzothiophene Desulfurization Operon of *Rhodococcus* sp. strain IGTS8. *Appl. Environ. Microbiol.* 61: 468–475.

Premuzic, E.T., and Lin, M.S. 1999. Induced Biochemical Conversions of Heavy Crude Oils. *J. Petrol. Sci.* 22: 171–180.

Prince, R.C., and Grossman, M.J. 2003. Substrate Preferences in Biodesulfurization of Diesel Range Fuels by *Rhodococcus* sp. Strain ECRD-1. *Appl. Environ. Microbiol.* 69(10): 5833–5838.

Raheb, J., Hajipour, M.J., Saadati, M., Rasekh, B., and Memari, B. 2009. The Enhancement of Biodesulfurization Activity in a Novel Indigenous Engineered *Pseudomonas putida*. *Iran. Biomed. J.* 13(4): 141–147.

Rhee, S.K., Chang, J.H., Chang, Y.K., and Chang, H.N. 1998. Desulfurization of Dibenzothiophene and Diesel Oils by a Newly Isolated *Gordona* Strain CYKS. *Appl. Environ. Microbiol.* 64(6): 2327–2331.

Sada, E., Katon, S., and Terashima, M. 1981. Enhancement of Oxygen Absorption by Magnetite-Containing Beads of Immobilized Glucose Oxidase. *Biotechnol. Bioeng.* 21: 1037–1044.

Sanchez, O.F., Almeciga-Diaz, C.J., Silva, E., Cruz, J.C., Valderrama, J.D., and Caicedo, L.A. 2008. Reduction of Sulfur Levels in Kerosene by *Pseudomonas* sp. Strain in an Airlift Reactor. *Latin Am. Appl. Res.* 38: 329–335.

Schilling, B.M., Alvarez, L.M., Wang, D.I.C., and Cooney, C.L. 2002. Continuous Desulfurization of Dibenzothiophene with *Rhodococcus rhodochrous* IGTS8 (ATCC 53968). *Biotechnol. Prog.* 18: 1207–1213.

Serbolisca, L., Ferra, F., and Megabit, I. 1999. Manipulation of the DNA Coding for the Desulphurizing Activity in a New Isolate of *Arthrobacter* sp. *Appl. Microbiol. Biotechnol.* 52: 122–126.

Setti, L., Bonoli, S., Badiali, E., and Giuliani, S. 2003. Inverse Phase Transfer Biocatalysis for a Biodesulfurization Process of Middle Distillates. *Bull. Moscow Univ. Ser. 2 Chem.* 44(1): 80–83.

Setti, L., Farinelli, P., Martino, S., Frassinetti, S., Lanzarini, G., and Pifferi, P.G. 1999. Developments in Destructive and Non-destructive Pathways for Selective Desulfurizations in Developments in Destructive and Non-destructive Pathways for Selective Desulfurization in Oil-Biorefining Processes. *Appl. Microbiol. Biotechnol.* 52: 111–117.

Setti, L., Lanzarini, G., and Pifferi, P.G. 1997. Whole Cell Biocatalysis for an Oil Desulfurization Process. *Fuel Process. Technol.* 52: 145–153.

Shan, G.B., Xing, J.M., Guo, C., Liu, H.Z., and Chen, J.Y. 2005c. Biodesulfurization Using *Pseudomonas delafieldii* in Magnetic Polyvinyl Alcohol Beads. *Lett. Appl. Microbiol.* 40: 30–36.

Shan, G.B., Xing, J.M., Luo, M.F., Liu, H.Z., and Chen, J.Y. 2003. Immobilization of *Pseudomonas delafieldii* with Magnetic Polyvinyl Alcohol Beads and Its Application in Biodesulfurization. *Biotechnol. Lett.* 25: 1977–1983.

Shan, G.B., Xing, J.M., Zhang, H., and Liu, H.Z. 2005a. Biodesulfurization of Dibenzothiophene by Microbial Cells Coated with Magnetite Nanoparticles. *Appl. Environ. Microbiol.* 71(8): 4497–4502.

Shan, G.B., Zhang, H., Liu, H., and Xing, J.M. 2005b. π-Complexation Studied by Fluorescence Technique: Application in Desulfurization of Petroleum Product using Magnetic π-Complexation Sorbents. *Sep. Sci. Technol.* 40(14): 2987–2999.

Shavandi, M., Sadeghizadeh, M., Zomorodipour, A., and Khajeh, K. 2009. Biodesulfurization of Dibenzo-thiophene by Recombinant *Gordonia alkanivorans* RIPI90A. *Bioresour. Technol.* 100: 475–479.

Soleimani, M., Bassi, A., and Margaritis, A. 2007. Biodesulfurization of Refractory Organic Sulfur Compounds in Fossil Fuels. *Biotechnol. Adv.* 25: 570–596.

Song, C.S., and Ma, X. 2003. New Design Approaches to Ultra-Clean Diesel Fuels by Deep Desulfurization and Deep Deamortization. *Appl. Catal. B. Environ.* 41: 207–238.

Song, S.H., Choi, S.S., Park, K., and Yoo, Y.J. 2005. Novel Hybrid Immobilization of Microorganisms and Its Applications to Biological Denitrification. *Enz. Microb. Technol.* 37: 567–573.

Speight, J.G. 2000. *The Desulfurization of Heavy Oil and Residua*, 2nd Edition. Marcel Dekker Inc., New York.

Speight, J.G. 2014. *The Chemistry and Technology of Petroleum*, 5th Edition. CRC Press, Taylor & Francis Group, Boca Raton, FL.

Speight, J.G., and Ozum, B. 2002. *Petroleum Refining Processes*. Marcel Dekker Inc., New York.

Speight, J.G., and Arjoon, K.K. 2012. *Bioremediation of Petroleum and Petroleum Products*. Scrivener Publishing, Beverly, MA.

Swaty, T.E. 2005. Global Refining Industry Trends: The Present and Future. *Hydrocarbon Process.* September: 35–46.

Tao, F., Liu, Y., Luo, Q., Su, F., Xu, Y., Li, F., Yu, B., Ma, C., and Xu, P. 2011. Novel Organic Solvent-Responsive Expression Vectors for Biocatalysis: Application for Development of an Organic Solvent-Tolerant Biodesulfurization Strain. *Bioresour. Technol.* 102: 9380–9387.

Tao, F., Yu, B., Xu, P., and Ma, C.O. 2006. Biodesulfurization in Biphasic Systems Containing Organic Solvents. *Appl. Environ. Microbiol.* 72: 4604–4609.

Tian, F., Wu, J.Z., Liang, C., Yang, Y., Ying, P., Sun, X., Cai, T., and Li, C. 2006. The Study of Thiophene Adsorption onto La(III)-Exchanged Zeolite NaY By FTIR Spectroscopy. *J. Colloid Interface Sci.* 301: 395–401.

Torkamani, S., Shayegan, J., Yaghmaei, S., and Alemzadeh, I. 2009. *Heavy Crude Oil Biodesulfurization Project. Annual Report*. Petroleum Engineering Development Company (PEDEC), National Iranian Oil Company, Tehran, Iran.

US Department of Energy. 2003. *Gasoline Biodesulfurization*. Office of Energy Efficiency and Renewable Energy, May. United States Department of Energy, Washington, DC. http://www.eere.energy.gov/industrial.

Van Hamme, J.D., Singh, A., and Ward, O.P. 2003. Recent Advances in Petroleum Microbiology. *Microbiol. Mol. Biol. Rev.* 67: 503–549.

Vidali, M. 2001. Bioremediation: An Overview. *Pure Appl. Chem.* 73(7): 1163–1172.

Wang, P., and Krawiec, S. 1994. Desulfurization of Dibenzothiophene to 2-Hydroxybiphenyl by Some Newly Isolated Bacterial Strains. *Arch. Microbiol.* 161: 266–271.

Watanabe, K., Noda, K., Konishi, J., and Maruhashi, K. 2003. Desulfurization of 2,4,6,8-Tetraethyl Dibenzothiophene by Recombinant *Mycobacterium* sp. Strain MR65. *Biotechnol. Lett.* 25: 1451–1456.

Wu, K.Y.A., and Wisecarver, K.D. 1992. Cell Immobilization Using PVA Cross-Linked with Boric Acid. *Biotechnol. Bioeng.* 39(4): 447–449.

Yang, J., Hu, Y., Zhao, D., Wang, S., Lau, P.C.K., and Marison, I.W. 2007. Two-Layer Continuous-Process Design for the Biodesulfurization of Diesel Oils under Bacterial Growth Conditions. *Biochem. Eng. J.* 37(2): 212–218.

Yu, B., Ma, C., Zhou, W., Wang, Y., Cai, X., Tao, F., Zhang, Q., Tong, M., Qu, J., and Xu, P. 2006b. Microbial Desulfurization of Gasoline by Free Whole-Cells of *Rhodococcus erythropolis* XP. *FEMS Microbiol. Lett.* 258: 284–289.

Yu, B., Xu, P., Shi, Q., and Ma, C. 2006a. Deep Desulfurization of Diesel Oil and Crude Oils by a Newly Isolated *Rhodococcus erythropolis* Strain. *Appl. Environ. Microbiol.* 72: 54–78.

Yu, L., Meyer, T.A., and Folsom, B.R. 1998. Oil/Water/Biocatalyst Three Phase Separation Process. United States Patent 5,772,901.

Zhang, H., Liu, Q.F., Li, Y., Li, W., Xiong, X., Xing, J., and Liu, H. 2008. Selection of Adsorbents for *In situ* Coupling Technology of Adsorptive Desulfurization and Biodesulfurization. *Sci. China Ser. B-Chem.* 51(1): 69–77.

Zhang, H., Shan, G., Liu, H., and, Xing, J. 2007. Surface Modification of γ-Al$_2$O$_3$ Nanoparticles with Gum Arabic and Its Applications in Adsorption and Biodesulfurization. *Surf. Coat. Technol.* 201: 6917–6921.

10 Hydrodesulfurization

10.1 INTRODUCTION

The crude oils that are being refined for the production of transportation fuels have become heavier and heavier. Because of this, many topping refineries have shut down because of their inability to process these heavier crude oils. In comparison, the total capacity of those processes that is intended for upgrading high-boiling distillates (such as residua) and heavy crude oils has increased (Speight and Ozum, 2002; Parkash, 2003; Hsu and Robinson, 2006; Gary et al., 2007; Speight, 2013, 2014). Furthermore, the growing dependence on high-heteroatom heavy feedstocks has emerged as a result of the continuing decreasing availability of conventional crude oil through the depletion of reserves in various parts of the world. Thus, the ever growing tendency to convert as much as possible of lower-grade feedstocks to liquid products is causing an increase in the total sulfur content in refined products. Refiners must, therefore, continue to remove substantial portions of sulfur from the lighter products; however, residua and heavy crude oil poses a particularly difficult problem. Indeed, it is now clear that there are other problems involved in the processing of the heavier feedstocks and that these heavier feedstocks, which are gradually emerging as the liquid fuel supply of the future, need special attention.

The trend of processing more heavy and sour crudes, the shift in demand away from naphtha toward more distillate, and the more stringent fuel qualities criteria change the fuel and hydrogen balances in most refineries (Speight and Ozum, 2002; Parkash, 2003; Ancheyta et al., 2005; Hsu and Robinson, 2006; Gary et al., 2007; Liu et al., 2009; Speight, 2011). Finding the right solution to managing fuel gas and hydrogen requirements has become essential for refineries to remain profitable. The refinery gas value is higher for generating hydrogen than for generating power, and slightly lower than the value for using hydrogen as chemical feedstock.

Hydrogenation processes are the principal processes used in the manufacture of naphtha (Brunet et al., 2005). In fact, the use of hydrogen in thermal processes is perhaps the single most significant advancement in refining technology during the 20th century (Dolbear, 1998). Indeed, with the influx of heavier feedstocks into refineries, hydroprocessing (Table 10.1) will assume a greater role in the refinery of the future for the production of low-sulfur and low-contaminant products (Speight and Ozum, 2002; Parkash, 2003; Hsu and Robinson, 2006; Gary et al., 2007; Speight, 2014). The process uses the principle that the presence of hydrogen during a thermal reaction of a petroleum feedstock terminates many of the coke-forming reactions and enhances the yields of the lower-boiling components, such as naphtha, kerosene, and jet fuel.

In addition, the hydrogen requirement for product improvement, which is the hydrotreatment of petroleum products, to ensure that they meet utility and performance specifications, is also increasing. Product improvement can involve not only hydrotreatment but also changes in molecular shape (*reforming* and *isomerization*) or molecular size (*alkylation* and *polymerization*), and hydrotreating can play a major role in product improvement (Speight and Ozum, 2002; Parkash, 2003; Hsu and Robinson, 2006; Gary et al., 2007; Speight, 2014).

Typically, in an integrated refinery, the high-boiling fractions of crude oils are subjected to either coking or hydrogenation processes to convert them to streams more easily processed for the production of transportation fuels. In the former case, up to about 20% of the original feed to the coker can be lost as a product of low economic value, especially if the coke contains large amounts of metals and heteroatoms. In the latter case, the addition of hydrogen to this heavy feed is very expensive because of the impact of metals and heteroatoms on catalyst lifetimes, as well as the cost of the hydrogen required to remove the heteroatoms and metals and to saturate aromatic rings.

TABLE 10.1

Outcome of Hydroprocesses during Refining

Reaction	Feedstock	Purpose
HDS[a]	Catalytic reformer feedstocks	Reduce catalyst poisoning
	Diesel fuel	Environmental specifications
	Distillate fuel oil	Environmental specifications
	Hydrocracker feedstocks	Reduce catalyst poisoning
	Coker feedstocks	Reduce sulfur content of coke
HDN[b]	Lubricating oil	Improve stability
	Catalytic cracking feedstocks	Reduce catalyst poisoning
	Hydrocracker feedstocks	Reduce catalyst poisoning
HDM[c]	Catalytic cracking feedstocks	Avoid metal deposition
		Avoid coke buildup
		Avoid catalyst destruction
	Hydrocracker feedstocks	Avoid metal deposition
		Avoid coke buildup
		Avoid catalyst destruction
CRR[d]	Catalytic cracker feedstocks	Reduce coke buildup on catalyst
	Residua	Reduce coke yield
	Heavy oils	Reduce coke yield

Source: Speight, J.G. 2000. *The Desulfurization of Heavy Oils and Residua.* 2nd Edition. Marcel Dekker Inc., New York. Table 5.1, p. 169.

[a] HDS, hydrodesulfurization.

[b] HDN, hydrodenitrogenation.

[c] HDM, hydrodemetallization.

[d] CRR, carbon residue reduction.

Hydrodesulfurization processes are used at several places in virtually every refinery, both to protect catalysts (since the catalyst represents a significant fraction of the costs of operating a hydrodesulfurization unit) and to meet product specifications. Thus, several types of chemistry might be anticipated as occurring during hydrodesulfurization that depend on the desired product and the nature of the feedstock (Speight and Ozum, 2002; Parkash, 2003; Hsu and Robinson, 2006; Gary et al., 2007; Speight, 2013, 2014). Similarly, hydrodenitrogenation is commonly used only in conjunction with hydrocracking, to protect catalysts (Ho, 1988; Kressmann et al., 2004). Other hydrotreating processes are used to saturate olefins and aromatics to meet product specifications or to remove metals from residual oils.

Thus, although hydrodesulfurization is widely used in the petroleum industries, it has two disadvantages: (1) the process is expensive to perform and (2) some of the compounds containing organic sulfur in protected molecular locales, such as dibenzothiophenes and their derivatives are not always desulfurized by this method, and other methods are necessary by changing the process parameters (Speight, 2000, 2014) or, in some cases, by using a different approach such as applying microorganisms for sulfur removal (Chapter 9) (Li et al., 1996; Maghsoudi et al., 2000; Akbarzadeh et al., 2003; Mohebali and Ball, 2008).

Virtually all of these processes rely on promoted molybdenum sulfide (MoS_2) catalysts. Hydrodenitrogenation catalysts are usually nickel-promoted molybdenum sulfide (MoS_2), supported on alumina (Al_2O_3) (Ho, 1988; Topsøe et al., 1996). These nickel–molybdenum catalysts are more active for hydrogenation than the corresponding cobalt catalysts. Nickel–molybdenum hydrodenitrogenation catalysts are generally good hydrodesulfurization catalysts.

As with many processes involving the use of hydrogen, the hydrodesulfurization process is actually a specific hydrogenation process and, as employed in petroleum refining, can be classified as *nondestructive* or *destructive*. Nondestructive, or simple, hydrogenation is generally used for the purpose of improving product quality without appreciable alteration of the boiling range. Mild processing conditions are employed so that only the more unstable materials are attacked and the sulfur, nitrogen, and oxygen compounds undergo hydrogenolysis to split hydrogen sulfide (H_2S), ammonia (NH_3), and water (H_2O), respectively. An example is the hydrodesulfurization of naphtha with temperatures on the order of 330–170°C (625–695°F) and hydrogen partial pressures of 100–500 psi.

On the other hand, destructive hydrogenation (hydrocracking) is characterized by the cleavage of carbon–carbon bonds with concurrent addition of hydrogen to the fragments to produce saturated lower-boiling products. Such treatment requires severe reaction temperatures (usually on the order of 360–410°C [680–770°F]) and high hydrogen pressures to minimize reactions that lead to the formation of undesirable products such as coke.

Destructive hydrogenation is employed in hydrocracking processes as a means of converting the heavier feedstocks (Kobayashi et al., 1987a,b). In such cases, desulfurization is still a major objective of the process; however, an additional factor, the conversion of the feedstock to lower-boiling products, may also be required. The tendency to form coke (from the higher-molecular weight constituents of the feedstock) is very strong. Using higher pressures of hydrogen (on the order of 750–2000 psi) than used for naphtha minimizes this tendency. Higher temperatures (360–410°C, 680–770°F) are required to convert the high molecular weight components of the feedstock to lower-boiling hydrogenated products. In essence, the feedstock plays an important part in determining the hydrodesulfurization conditions (Chapter 7).

One of the goals of any hydrogenation process is the removal of heteroatoms and metals, specifically nitrogen, sulfur, vanadium, and nickel. Another is the saturation of aromatic structures, primarily through hydrotreating. However, for residua, the conversion of high-boiling materials to low-boiling materials is an additional important goal that is accomplished by hydrocracking. The processes used include a variety of catalytic cracking processes. The capacity for these types of processes has increased (and will continue to increase) significantly in recent years due to the increase in the amount of heavy crude oils being processed. Examples include various fluid catalytic cracking (FCC) processes, the unicracking/hydrodesulfurization process, the H-oil process, and the heavy oil cracking process (Speight, 2011, 2014).

The feeds to these types of units are usually atmospheric and vacuum residua. The products include feeds for the production of transportation fuels, fuel oils, olefins, etc. However, the operating conditions of the reactor, whether it is an FCC unit or a fixed-bed unit, is dependent on the desired product slate and the properties of the feed.

Hydrodesulfurization, through the application of hydroprocesses, is linked to the relevant environmental regulations as well as to the desired distribution of products. Achieving these goals is compounded when heavy feedstocks are used because of the molecular complexity of the constituents.

Heavy feedstocks contain many thousands of different compounds that range in molecular weight from about 100 to >2000, and greatly influence the processes for sulfur removal (Chapters 2 and 7) (Speight, 2014). This broad range in molecular weight results in a boiling range from approximately 200°C (390°F) to approximately 1100°C (2000°F). The mixtures are so complex and so dependent on the history of the feedstock (Speight, 2014) that the patterns of thermal decomposition can be significantly different to warrant different process conditions. For example, the cracking and desulfurization activity of constituents containing the polynuclear aromatic nucleus may be sluggish and may inhibit the reactivity of paraffinic and naphthenic constituents. In addition, organometallic compounds produce metallic products that influence the reactivity of the catalyst by deposition on its surface.

Thus, on a chemical basis, it is possible to define the hydrodesulfurization process in terms of the feedstock type and the required products. In practice, it may actually be difficult, if not impossible, to carry out nondestructive hydrogenation and completely eliminate carbon–carbon bond scission, as in the case of the desulfurization of naphtha. For heavy feedstocks, it is usually preferable to promote hydrocracking as an integral part of the desulfurization process.

10.2 PROCESS DESCRIPTION

Hydrotreating is carried out by charging the feed to the reactor, together with a portion of all the hydrogen produced in the catalytic reformer. Suitable catalysts are tungsten–nickel sulfide, cobalt–molybdenum–alumina, nickel oxide–silica–alumina, and platinum–alumina (Kressmann et al., 2004). Most processes employ cobalt–molybdenum catalysts, which generally contain about 10% by weight molybdenum oxide and <1% by weight cobalt oxide supported on alumina. The temperatures employed are in the range of 300–345°C (570–850°F), and the hydrogen pressures are about 500–1000 psi.

The reaction generally takes place in the vapor phase; however, depending on the application, a mixed-phase reaction may occur. The reaction products are cooled in a heat exchanger and led to a high-pressure separator where hydrogen gas is separated for recycling. Liquid products from the high-pressure separator flow to a low-pressure separator (stabilizer) where dissolved light gases are removed. The product may then be fed to a reforming or cracking unit if desired.

Traditionally, hydrodesulfurization reactors are cocurrent in nature, in which hydrogen is mixed together with the feedstock at the entrance to the reactor and flow through the reactor together. Because the reaction is exothermic, heat must be removed periodically, which is often achieved through the introduction of fresh hydrogen and feedstock fuel in the middle of the reactor. The advantage of the cocurrent design is practical as it eases the control of gas–liquid mixing and contact with the catalyst. The disadvantage is that the concentration of hydrogen is the highest at the front of the reactor where the easiest to remove sulfur is, and lowest at the outlet where the hardest to remove sulfur remains. The opposite is true for the concentration of hydrogen sulfide, which increases the difficulty of achieving extremely low sulfur levels owing to the low hydrogen concentration and high hydrogen sulfide concentration at the end of the reactor.

In addition, naphtha (a gasoline blend stock or solvent) being fed to the refinery reformer should always be examined for sulfur content; if sulfur is present, the naphtha will need to be hydrotreated to remove nearly all sulfur, nitrogen, and metal contaminants, which would deactivate the noble metal catalyst used in the reforming process. Similarly, feed to the FCC unit can also be hydrotreated to remove most of the sulfur, nitrogen, and metal contaminants to improve the yield and quality of high-value products, such as naphtha and middle distillate, from the unit.

Generally, it is more economical to hydrotreat high-sulfur feedstocks before catalytic cracking than to hydrotreat the products from catalytic cracking. The advantages are as follows: (1) the products require less finishing; (2) sulfur is removed from the catalytic cracking feedstock, and corrosion is reduced in the cracking unit; and (3) coke formation during cracking is reduced and higher conversions result, and the catalytic cracking quality of the gas oil fraction is improved.

Although hydrocracking will occur during hydrotreating, attempts are made to minimize such effects; however, the degree of cracking is dependent on the nature of the feedstock. For example, decalin (decahydronaphthalene) cracks more readily than the corresponding paraffin analogue, n-decane $[CH_3(CH_2)_8CH_3]$, to give higher isoparaffin to n-paraffin product ratios than those obtained from the paraffin. A large yield of single-ring naphthenes is also produced, and these are resistant to further hydrocracking and contain a higher-than-equilibrium ratio of methylcyclopentane to cyclohexane.

There are several valid reasons for removing heteroatoms from petroleum fractions. These include (1) reduction, or elimination, of corrosion during refining, handling, or use of the various products; (2) production of products having an acceptable odor and specification; (3) increasing the performance and stability of naphtha; (4) decreasing smoke formation in kerosene; and (5) reduction of heteroatom content in fuel oil to a level that improves burning characteristics and is environmentally acceptable.

Heteroatom removal, as practiced in various refineries, can take several forms (Speight, 2000), such as concentration in residua during distillation, concentration in coke during coking, or chemical removal (acid treating, caustic treating, i.e., sweetening or finishing processes) (Speight, 2014). Nevertheless, the heteroatom removal from petroleum feedstocks is almost universally accomplished by the catalytic reaction of hydrogen with the feedstock constituents. However, there are other refinery processes that are adaptable to heavy feedstocks and that may be effective for reducing, but not necessarily effective for complete removal of the heteroatom-containing constituents (Speight and Ozum, 2002; Parkash, 2003; Hsu and Robinson, 2006; Gary et al., 2007; Speight, 2014).

The major differences between hydrotreating and hydrocracking are the time at which the feedstock remains at reaction temperature, and the extent of the decomposition of the non-heteroatom constituents and products. The upper limits of hydrotreating conditions may overlap with the lower limits of hydrocracking conditions. Furthermore, where the reaction conditions overlap, feedstocks to be hydrotreated will generally be exposed to the reactor temperature for shorter periods, hence the reason why hydrotreating conditions may be referred to as mild. All is relative.

The usual goal of hydrotreating is to hydrogenate olefins and to remove heteroatoms, such as sulfur, and to saturate aromatic compounds and olefins (Meyers, 1997; Speight, 2000). On the other hand, hydrocracking is a process in which thermal decomposition is extensive and the hydrogen assists in the removal of the heteroatoms as well as mitigates the coke formation that usually accompanies thermal cracking of high molecular weight polar constituents.

Thus, catalytic hydrotreating is a hydrogenation process used to remove about 90% of contaminants such as nitrogen, sulfur, oxygen, and metals from liquid petroleum fractions. These contaminants, if not removed from the petroleum fractions as they travel through the refinery processing units, can have detrimental effects on the equipment, the catalysts, and the quality of the finished product. Typically, hydrotreating is done before processes such as catalytic reforming so that the catalyst is not contaminated by untreated feedstock. Hydrotreating is also used before catalytic cracking to reduce sulfur and improve product yields, and to upgrade middle-distillate petroleum fractions into finished kerosene, diesel fuel, and heating fuel oils. In addition, hydrotreating converts olefins and aromatics to saturated compounds.

Hydrodesulfurization is one of the refinery hydrotreating processes (variously referred to as *hydroprocessing*) in which the feedstock is treated with hydrogen at temperature and pressure at which hydrocracking (thermal decomposition in the presence of hydrogen) is minimized (Figures 10.1 and 10.2). Organic sulfur compounds like thiol derivatives, sulfide derivatives, and thiophene derivatives can be readily removed from petroleum fractions using conventional hydrodesulfurization catalysts (sulfided $CoMo/Al_2O_3$ and $NiMo/Al_2O_3$ compounds) and conventional reactors. However, the efficiency toward the removal of so-called refractory sulfur compounds, such as dibenzothiophene and 4,6-dimethyldibenzothiophene, is greatly reduced. The efficiency of the hydrodesulfurization process can be increased by developing new catalysts and new reactors.

In a very simplified process for distillate hydrotreating (Figure 10.3), the feedstock is first pressurized to a pressure that is a little higher than that of the reactor section, mixed with hot recycle gas, and preheated to the temperature of the reactor inlet. The hot feedstock (and the recycle gas) is then introduced to the catalyst in the reactor where temperatures on the order of 290–455°C (550–850°F) and pressures in the range 150–3000 psi prevail.

Heat exchangers are employed to cool the reactor effluent, and the desulfurized liquid product is separated from the recycle gas at a pressure somewhat lower than that of the reactor section. Hydrogen sulfide and any light hydrocarbon gases are removed from the recycle gas, which is then mixed with fresh (makeup) hydrogen, compressed, and mixed with further hydrocarbon feedstock.

The use of a recycle gas technique in the hydrodesulfurization process minimizes losses of hydrogen—a very expensive commodity in petroleum refining. The hydrodesulfurization process requires high partial pressures of hydrogen to promote high desulfurization reaction rates and to diminish coke (carbon) deposition on the catalyst. To maintain high partial pressures of hydrogen,

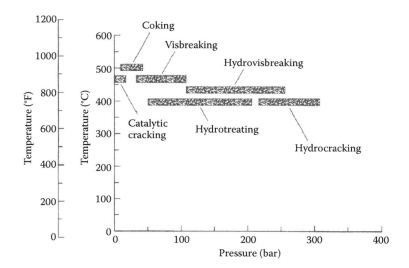

FIGURE 10.1 Temperature and pressure ranges for refinery processes. (From Speight, J.G. 2007. *The Chemistry and Technology of Petroleum*. 4th Edition. CRC Press, Taylor & Francis Publishers, Boca Raton, FL. Figure 21.2, p. 563.)

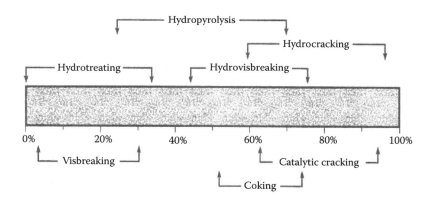

FIGURE 10.2 Feedstock conversion to liquids by refinery processes. (From Speight, J.G. 2007. *The Chemistry and Technology of Petroleum*. 4th Edition. CRC Press, Taylor & Francis Publishers, Boca Raton, FL. Figure 21.3, p. 564.)

it is necessary to introduce hydrogen into the reactor at several times the rate of hydrogen consumption, and it is possible to recover most of the unused hydrogen in the separator, after which it is then recycled for further use. Indeed, if it were not possible to recover most of the unused hydrogen, the process economics would probably be very questionable. However, tolerable amounts of hydrogen may be lost as a result of the solubility of the gas in the liquid hydrocarbon product or even during the removal of the hydrogen sulfide (and the light hydrocarbon gases) from the recycle gas.

Hydrogen for the hydrodesulfurization process has to be produced on site and would usually come from a hydrogen plant, which may utilize the concept of steam–methane reforming or a similar process involving the generation of hydrogen from a low molecular weight hydrocarbon

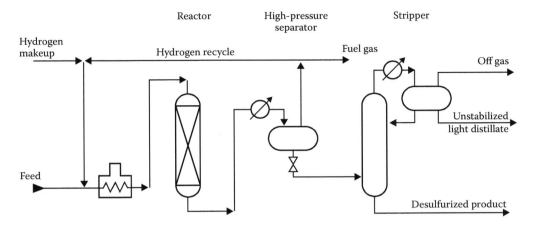

FIGURE 10.3 Hydrodesulfurization process for petroleum distillates. (From Speight, J.G. 2000. *The Desulfurization of Heavy Oils and Residua.* 2nd Edition. Marcel Dekker Inc., New York. Figure 5.1, p. 171.)

(Chapter 10). Alternatively, the hydrogen could be produced as a by-product from one of the many catalytic reforming processes that are available to refinery operators. These are the processes by which high-octane naphtha is produced from low-octane naphtha.

The reactions involved in this type of process are extremely complex. The overall effect of a reforming process is that the change in the boiling point of the feedstock passed through a reforming unit is relatively small. It is not the intent of the process to induce severe thermal degradation of the components of the charge stock but to rearrange the structure of the components and thus produce higher-octane naphtha. During the reforming process, dehydrogenation of certain molecular types occurs with the production of hydrogen gas that is then available for any one or more of the processes that require extraneous hydrogen. However, depending on the needs of any particular refinery, it may be necessary to have a hydrogen production plant to produce hydrogen in addition to that produced by a reforming unit.

Hydrogen requirements for the hydrodesulfurization process depend on the nature of the feedstock as well as on the extent of the desulfurization (Speight and Ozum, 2002; Parkash, 2003; Hsu and Robinson, 2006; Gary et al., 2007; Speight, 2013, 2014). For example, the heavier feedstocks require substantially more hydrogen to produce a given product than the lower-boiling feedstocks or to produce a product with a predetermined amount of sulfur (Speight and Ozum, 2002; Parkash, 2003; Hsu and Robinson, 2006; Gary et al., 2007; Speight, 2013, 2014). The theoretical hydrogen requirements often differ markedly from the experimental values because of the nitrogen and oxygen contents of the feedstock. These atoms are removed during the process as their respective hydrogen analogues—ammonia and water. There is also the occurrence of hydrocracking during the desulfurization. The metals content of the feedstock may also influence the hydrogen consumption by altering the characteristics of the catalyst. It is conceivable that metal deposition onto the catalyst may serve to increase the hydrocracking activity of the catalyst to the detriment of the hydrotreating activity, thereby promoting higher consumption of hydrogen in the process (Speight and Ozum, 2002; Parkash, 2003; Hsu and Robinson, 2006; Gary et al., 2007; Speight, 2013, 2014).

Sulfur removal by a hydrodesulfurization process is usually good to excellent and may even be on the order of 90–95%. The products formed by the removal of sulfur from the various molecular types that may make up any of a variety of feedstocks usually are considerably more volatile than the parent sulfur compounds. Certain amounts of these low-boiling compounds may have to be removed from the product mix to maintain an acceptable volatility. This is especially true when the feedstock to be desulfurized is a naphtha or middle distillate (light fuel oil). In the case of the naphtha feedstock, it may be preferable that the finished gasoline has a small proportion of low-boiling hydrocarbons present to promote ignition of the gasoline. The presence of these same hydrocarbons in a fuel oil might increase the flash point to such an extent that the fuel oil is classified as unstable and unfit for the specified use.

On the other hand, desulfurization may increase the aromatics content of the product stream relative to the feedstock. This is a very desirable property where an increase in the octane rating is sought. In short, the desulfurization process may need careful monitoring when feedstock type can have a marked influence on the properties of the product and downstream treatments may be necessary to ensure that product specifications are met.

In the same way, the desulfurization of heavy feedstocks may also require careful monitoring of the product properties. However, the nature of the feedstock may preclude the production of highly volatile materials if the sole purpose of the process is to reduce the sulfur content of the feedstock, thereby producing the equivalent of a heavy gas oil feedstock. If, however, the process conditions are such that hydrocracking becomes one of the major chemical reaction types, a high proportion of low-boiling material may be found in the product mix and have to be removed accordingly before any further processing of the product stream.

Perhaps the major advantage of the hydrodesulfurization process is the fact that the sulfur is ultimately removed as hydrogen sulfide, which may, by virtue of its gaseous nature, be readily and completely removed from the hydrocarbon product streams. The manner in which the hydrogen sulfide is removed (as a process waste or as a precursor for elemental sulfur manufacture) is in direct contrast to the problems that may arise when other methods of desulfurization are employed. For example, desulfurization of a feedstock by any of the thermal methods where extraneous hydrogen is not employed (Chapters 3 and 8) merely tends to concentrate the sulfur in the coke (or, if the process is distillation, in the residuum) that is formed, thereby merely postponing, rather than dispelling, the problem. Indeed, the sulfur still remains in one of the process products and will eventually have to be removed if the product is to satisfy specifications. Furthermore, it is doubtful if any one of these alternate methods will ever reach a satisfactory level of desulfurization compared with the hydrodesulfurization process where desulfurization levels of at least 80% are desirable, and, in the majority of cases, are often recorded.

The hydrodesulfurization of high-boiling feedstocks that have sulfur contents up to approximately 5% w/w are similar to those employed for gas oil. However, the tendency for coke (carbon) deposition on the catalyst and contamination of the catalyst by the metals in the feedstock is much higher. Consequently, unless a low level of desulfurization is acceptable, it is necessary to regenerate the catalyst at frequent intervals. Furthermore, these heavier feedstocks contain appreciable amounts of asphaltene and resin constituents, and the sulfur compounds are a significant proportion of the total compounds present.

When applied to residua, hydrotreating can be used for processes such as (1) fuel oil desulfurization, and (2) residuum hydrogenation, which is accompanied by hydrodesulfurization, hydrodenitrogenation, and partial conversion to produce products suitable as feedstocks for other processes, such as catalytic cracking.

One of the chief problems with hydroprocessing residua is the deposition of metals, in particular vanadium, on the catalyst. It is not possible to remove vanadium from the catalyst, which must therefore be replaced when deactivated, and the time taken for catalyst replacement can significantly reduce the unit time efficiency. Fixed-bed catalysts tend to plug owing to solids in the feed or carbon deposits when processing residual feeds. As mentioned previously, the highly exothermic reaction at high conversion causes difficult-to-solve reactor design problems in heat removal and temperature control.

The problems encountered in hydrotreating heavy feedstocks can be directly equated to the amount of complex, higher-boiling constituents that may require pretreatment (Speight and Moschopedis, 1979; Reynolds and Beret, 1989). Processing these feedstocks is not merely a matter of applying know-how derived from refining conventional crude oils but also requires knowledge of the composition (Chapters 2 and 7). The materials are not only complex in terms of the carbon number and boiling point ranges but also because a large part of this *envelope* falls into a range of model compounds and very little is known about the properties. It is also established that the majority of the higher molecular weight materials produce coke (with some liquids) but the majority of the lower molecular weight constituents produce liquids (with some coke). It is to both of these trends that hydrocracking is aimed.

It is the physical and chemical composition of a feedstock that plays a large part not only in determining the nature of the products that arise from refining operations but also in determining the precise manner by which a particular feedstock should be processed (Speight, 1986). Furthermore, it is apparent that the conversion of heavy feedstocks requires new lines of thought to develop suitable processing scenarios (Speight, 2011). Indeed, the use of thermal (*carbon rejection*) processes and of hydrothermal (*hydrogen addition*) processes, which were inherent in the refineries designed to process lighter feedstocks, has been a particular cause for concern. This has brought about the evolution of processing schemes that accommodate the heavier feedstocks (Khan and Patmore, 1998; Speight, 2011).

The choice of processing schemes for a given hydrotreating application depends on the nature of the feedstock and on the product requirements (Suchanek and Moore, 1986). For higher-boiling feedstocks, the process is usually hydrocracking and can be simply illustrated as a single-stage or a two-stage operation (Figure 10.4). Variations to the process are feedstocks dependent.

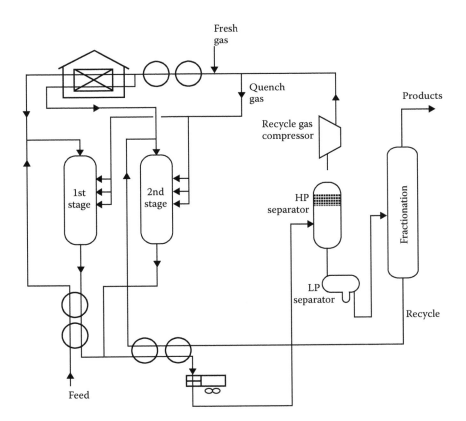

FIGURE 10.4 Single-stage or two-stage (optional) hydrocracking unit. (From OSHA Technical Manual, Section IV, Chapter 2: Petroleum Refining Processes. http://www.osha.gov/dts/osta/otm/otm_iv/otm_iv_2 .html.)

For example, the single-stage process can be used to produce naphtha but is more often used to produce middle distillate from heavy vacuum gas oils. The two-stage process was developed primarily to produce high yields of naphtha from straight-run gas oil, and the first stage may actually be a purification step to remove sulfur-containing (as well as nitrogen-containing) organic materials. Both processes use an extinction-recycling technique to maximize the yields of the desired product. Significant conversion of heavy feedstocks can be accomplished by hydrocracking at high severity (Howell et al., 1985). For some applications, the products boiling up to 340°C (650°F) can be blended to give the desired final product.

For lower-boiling feedstocks, the commercial processes for treating or finishing petroleum fractions with hydrogen all operate in essentially the same manner. The feedstock is heated and passed with hydrogen gas through a tower or reactor filled with catalyst pellets. The reactor is maintained at a temperature of 260–425°C (500–800°F) at pressures from 100 to 1000 psi, depending on the particular process, the nature of the feedstock, and the degree of hydrogenation required. After leaving the reactor, excess hydrogen is separated from the treated product and recycled through the reactor after the removal of hydrogen sulfide. The liquid product is passed into a stripping tower, where steam removes dissolved hydrogen and hydrogen sulfide, and after cooling the product is run to finished product storage or, in the case of feedstock preparation, pumped to the next processing unit.

Excessive contact time and/or temperature will create coking. Precautions need to be taken when unloading the coked catalyst from the unit to prevent iron sulfide fires. The coked catalyst should be cooled to below 49°C (<120°F) before removal, or dumped into nitrogen-blanketed bins where it can be cooled before further handling. Antifoam additives may be used to prevent catalyst poisoning from silicone carryover in the coker feedstock. There is a potential for exposure to hydrogen sulfide or hydrogen gas in the event of a release, or to ammonia should a sour-water leak or spill occur. Phenol also may be present if high-boiling-point feedstocks are processed.

Attention must also be given to the coke mitigation aspects of hydrotreating as a preliminary treatment option of feedstocks for other processes, especially heavier feedstocks. Although the visbreaking process (Chapter 3) reduces the viscosity of residua and partially converts the residue to lighter hydrocarbons and coke (Chapter 3) (Radovanović and Speight, 2011; Speight, 2014), the process can also be used to remove the undesirable higher molecular weight polar constituents before sending the visbroken feedstock to a catalytic cracking unit. The solvent deasphalting process separates the higher-value liquid product (deasphalted oil) using a low-boiling paraffinic solvent from low-value asphaltene-rich insoluble stream. Moreover, various residuum hydrotreating and/ or hydrocracking processes in which the feedstock is processed under high temperature and pressure, using a robust catalyst to remove sulfur, metals, condensed aromatic, or nitrogen, and increase the residue's hydrogen content to a desired degree, are also available (Speight and Ozum, 2002; Parkash, 2003; Hsu and Robinson, 2006; Gary et al., 2007; Speight, 2014). However, increasing number of options are becoming available in which the residuum is first hydrotreated (under milder conditions to remove heteroatoms and mitigate the effects of the asphaltene and resin constituents) before sending the hydrotreated product to, for example, an FCC unit. However, there is also the potential for hydrotreating to change the stability of the asphaltene constituents in the oil (Bartholdy and Andersen, 2000).

In such cases, it is even more important that particular attention is given to hydrogen management and promoting hydrodesulfurization and hydrodenitrogenation (even fragmentation) of asphaltene and resin constituents, thereby producing a product that may be suitable as a feedstock for catalytic cracking with reduced catalyst destruction. The presence of a material with good solvating power to assist in the hydrotreating process is preferred. In this respect, it is worth noting the reappearance of donor solvent processing of heavy feedstocks (Vernon et al., 1984; Bakshi and Lutz, 1987) that has its roots in the older hydrogen donor diluent visbreaking process (Bland and Davidson, 1967).

However, it must not be forgotten that product distribution and quality vary considerably depending on the nature of the feedstock constituents and on the process. In modern refineries, hydrocracking is one of several process options that can be applied to the production of liquid fuels from the heavier feedstocks (Speight and Ozum, 2002; Parkash, 2003; Hsu and Robinson, 2006; Gary et al., 2007; Speight, 2014). A most important aspect of the modern refinery operation is the desired product slate, which dictates the matching of a process with any particular feedstock to overcome differences in feedstock composition.

Hydrogen consumption is also a parameter that varies with feedstock composition (Dolbear, 1998; Speight and Ozum, 2002; Parkash, 2003; Hsu and Robinson, 2006; Gary et al., 2007; Speight, 2014), thereby indicating the need for a thorough understanding of the feedstock constituents if the process is to be employed to maximum efficiency. A convenient means of understanding the influence of feedstock on the hydrotreating process is through a study of the hydrogen content (H/C atomic ratio) and molecular weight (carbon number) of the various feedstocks or products. It is also possible to use data for hydrogen usage in residuum processing where the relative amount of hydrogen consumed in the process can be shown to be dependent on the sulfur content of the feedstock.

Hydrotreating processes differ depending on the feedstock available and catalysts used. Hydrotreating can be used to improve the burning characteristics of distillates such as kerosene. Hydrotreatment of a kerosene fraction can convert aromatics into naphthenes, which are cleaner-burning compounds.

Lube oil hydrotreating uses catalytic treatment of the oil with hydrogen to improve product quality. The objectives in mild lube hydrotreating include saturation of olefins and improvements in color, odor, and acid nature of the oil. Mild lube hydrotreating also may be used following solvent processing. Operating temperatures are usually below 315°C (600°F) and operating pressures below 800 psi. Severe lube hydrotreating, at temperatures in the 315–400°C (600–750°F) range and hydrogen pressures up to 3000 psi, is capable of saturating aromatic rings, along with sulfur and nitrogen removal, to impart specific properties not achieved at mild conditions.

Hydrotreating (Brunet et al., 2005) also can be employed to improve the quality of pyrolysis naphtha, a by-product from the manufacture of ethylene. Traditionally, the outlet for pyrolysis-produced naphtha has been motor gasoline blending, a suitable route in view of its high octane number. However, only small portions can be blended untreated because of the unacceptable odor, color, and gum-forming tendencies of this material. The quality of pyrolysis naphtha, which is high in olefin content, can be satisfactorily improved by hydrotreating, whereby conversion of olefins into mono-olefins provides an acceptable product for motor gas blending (Brunet et al., 2005).

10.3 REACTOR DESIGN

Reactor configurations within petroleum hydroprocessing units may include catalyst beds that are fixed or moving (Speight, 2000; Ancheyta, 2007). Most hydroprocessing reactors are fixed-bed reactors. Hydroprocessing units with fixed-bed reactors must be shut down to remove the spent catalyst when catalyst activity declines below an acceptable level (due to the accumulation of coke, metals, and other contaminants). There are also hydroprocessing reactors with moving or ebullating catalyst beds.

As the trend toward utilizing heavier petroleum feedstocks continues, the hydrotreating processes used to upgrade such stocks become increasingly important. Difficulties are encountered in the development of catalysts with high resistance to deactivation. Another important challenge is that of designing three-phase reactors capable of processing large quantities at high temperatures and pressures. The desirable features of such reactors are low-pressure drop, in the presence of deposits, and low mass transfer resistance between gas–liquid and liquid–solid. The monolithic reactor offers a viable alternative in which the monolith is typically 1 mm or a few millimeters in diameter. Each channel is bounded by either a porous wall or a solid wall onto which a porous washcoat may be applied. In these narrow channels, gas and liquid flow concurrently.

Reactor designs for hydrodesulfurization of various feedstocks vary in the way in which the feedstock is introduced into the reactor and in the arrangement, as well as the physical nature, of the catalyst bed. The conditions under which the hydrodesulfurization process operates (i.e., high temperatures and high pressures) dictate the required wall thickness (determined by the pressure/

temperature/strength ratio). In addition, resistance of the reactor walls to the corrosive attack by hydrogen sulfide and hydrogen (to name only two of the potential corrosive agents of all of the constituents in, or arising from, the feedstock) can be a problem. Precautions should be taken to ensure that wall thickness and composition yield maximum use and safety.

With these criteria in mind, various reactors have been designed to satisfy the needs of the hydroprocesses, including hydrodesulfurization (Ancheyta, 2007; Speight, 2014). Thus, reactors may vary from as little as 4 ft to as much as 20 ft in diameter and have a wall thickness anywhere from 4.5 to 10 in or so. These vessels may weigh from 150 tons to as much as 1000 tons. Obviously, before selecting a suitable reactor, shipping and handling requirements (in addition to the more conventional process economics) must be given serious consideration.

The hydrodesulfurization process operates using high hydrogen pressure, typically 1500–2500 psi, and temperatures on the order of 290–370°C (550–700°F). Several process configurations are used, depending on the feed and the design criteria. All include provisions for addition of cold hydrogen at several points in the hydrocracking reactor to control reactor temperatures, since a great amount of heat is released by hydrogenation. Reactor internals provided for this function are complex mechanical devices.

Finally, and before a discussion of the various reactor-bed types used in hydrodesulfurization, a note that the once-popular once-through reactors, where the incompletely converted or unconverted fraction of the feedstock is separated from the lower-boiling products, are being replaced by recycle reactors. In these reactors, any unconverted feedstock is sent back (recycled) to the reactor for further processing. In such a case, the volume flow of the combined (fresh and unconverted) feedstock is the sum of the inputs of the fresh feedstock and the recycled feedstock:

$$F_T = F_F + F_R$$

where F_T is the total feedstock into the unit, F_F is the fresh feedstock, and F_R is the recycled feedstock. Thus, the recycle ratio, τ, is defined as the ratio of the recycled feedstock to the fresh feedstock:

$$\tau = F_R/F_F$$

The recycle ratio may also be expressed as a percentage in some texts or references.

However, the recycled feedstock, having been through the reactor at least once, will have a lower reactivity than the original feedstock, thereby reducing the reactivity of the feedstock in each recycle event. However, the overall conversion is increased.

10.3.1 DOWNFLOW FIXED-BED REACTOR

The reactor design commonly used in hydrodesulfurization of distillates is the fixed-bed reactor design in which the feedstock enters at the top of the reactor and the product leaves at the bottom of the reactor (Salmi et al., 2011). The catalyst remains in a stationary position (fixed bed) with hydrogen and petroleum feedstock passing in a downflow direction through the catalyst bed. The hydrodesulfurization reaction is exothermic, and the temperature increases from the inlet to the outlet of each catalyst bed. With a high hydrogen consumption and subsequent large temperature increase, the reaction mixture can be quenched with cold recycled gas at intermediate points in the reactor system. This is achieved by dividing the catalyst charge into a series of catalyst beds, and the effluent from each catalyst bed is quenched to the inlet temperature of the next catalyst bed.

The extent of desulfurization is controlled by increasing the inlet temperature in each catalyst bed to maintain constant catalyst activity over the course of the process. Fixed-bed reactors are mathematically modeled as plug-flow reactors with very little back mixing in the catalyst beds.

The first catalyst bed is poisoned with vanadium and nickel at the inlet to the bed and may be a cheaper catalyst (guard bed). As the catalyst is poisoned in the front of the bed, the temperature exotherm moves down the bed and the activity of the entire catalyst charge declines, thus requiring an increase in the reactor temperature over the course of the process sequence. After catalyst regeneration, the reactors are opened and inspected, and the high metal content catalyst layer at the inlet to the first bed may be discarded and replaced with a fresh catalyst. The catalyst loses activity after a series of regenerations and, consequently, thereafter, it is necessary to replace the complete catalyst charge. In the case of feedstocks with very high metal content (such as residua), it is often necessary to replace the entire catalyst charge rather than to regenerate it. This is because the metal contaminants cannot be removed by economical means during rapid regeneration, and the metals have been reported to interfere with the combustion of carbon and sulfur, catalyzing the conversion of sulfur dioxide (SO_2) to sulfate $\left(SO_4^{2-}\right)$, which has a permanent poisoning effect on the catalyst.

Fixed-bed hydrodesulfurization units are generally used for distillate hydrodesulfurization; it may also be used for residuum hydrodesulfurization but require special precautions in processing. The residuum must undergo a two-stage electrostatic desalting so that salt deposits do not plug the inlet to the first catalyst bed and the residuum must be low in vanadium and nickel content to avoid plugging the beds with metal deposits, hence the need for a guard bed in residuum hydrodesulfurization reactors.

During the operation of a fixed-bed reactor, contaminants entering with fresh feed are filtered out and fill the voids between catalyst particles in the bed. The buildup of contaminants in the bed can result in the channeling of reactants through the bed and reducing the hydrodesulfurization efficiency. As the flow pattern becomes distorted or restricted, the pressure drop throughout the catalyst bed increases. If the pressure drop becomes high enough, physical damage to the reactor internals can result. When high-pressure drops are observed throughout any portion of the reactor, the unit is shut down and the catalyst bed is skimmed and refilled.

With fixed-bed reactors, a balance must be reached between reaction rate and pressure drop across the catalyst bed. As catalyst particle size is decreased, the desulfurization reaction rate increases but so does the pressure drop across the catalyst bed. Expanded-bed reactors do not have this limitation, and small 1/32-in (0.8-mm) extrudate catalysts or fine catalysts may be used without increasing the pressure drop.

The downflow fixed-bed reactor has been used widely for hydrodesulfurization processes and is so called because of the feedstock entry at the top of the reactor while the product stream is discharged from the base of the reactor (Figure 10.5). The catalyst is contained in the reactor as stationary beds with the feedstock and hydrogen passing through the bed in a downward direction. The exothermic nature of the reaction and the subsequent marked temperature increase from the inlet to the outlet of each catalyst bed require that the reaction mix be quenched by cold recycle gas at various points in the reactor, hence the incorporation of separate catalyst beds as part of the reactor design.

To combat the inevitable loss in desulfurizing activity of the catalyst, which must be presumed to occur with time under any predetermined set of reaction conditions, the bed inlet temperature may be increased slowly, thereby increasing the overall temperature of the catalyst bed and thus maintaining constant catalyst activity. Thus, depending on the nature of the feedstock, there may be a considerable difference between the start-of-run temperature and the end-of-run temperature.

In fixed-bed reactors, the catalyst may be poisoned (deactivated) progressively. For example, the first catalyst bed will most likely be poisoned by vanadium and nickel deposition initially at the inlet to the bed, and then progressively through the bed as the active zone is gradually pushed downward into the bed. Once catalyst poisoning has progressed through the bed, the catalyst may have to be discarded or regenerated. However, catalyst regeneration may only suffice for a limited time and the catalyst may have to be completely replaced—this is especially true in the case of heavy feedstocks when coke and metal deposition on the catalyst is a process constant.

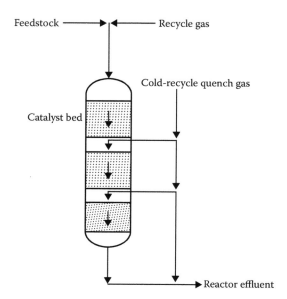

FIGURE 10.5   Fixed-bed downflow reactor. (From Speight, J.G. 2000. *The Desulfurization of Heavy Oils and Residua*. 2nd Edition. Marcel Dekker Inc., New York. Figure 5.6, p. 180.)

Removal of the metal contaminants is not usually economical, or efficient, during rapid regeneration. In fact, the deposited metals are believed to form sulfates during the removal of carbon and sulfur compounds by combustion that produce a permanent poisoning effect. Thus, if fixed-bed reactors are to be used for heavy feedstock hydrodesulfurization (in place of the more usual distillate hydrodesulfurization), it may be necessary to first process the heavier feedstocks to remove the metals (especially vanadium and nickel) and thus decrease the extent of catalyst bed plugging. Precautions should also be taken to ensure that plugging of the bed does not lead to the formation of channels within the catalyst bed, which will also reduce the efficiency of the process and may even lead to pressure variances within the reactor because of the distorted flow patterns with eventual damage.

10.3.2   RADIAL-FLOW FIXED-BED REACTOR

The radial-flow fixed-bed hydrodesulfurization reactor (Figure 10.6) is a variance of the downflow fixed-bed reactor (Salmi et al., 2011). Again, the feedstock enters the top of the reactor, but instead of flowing downward through the catalyst bed, the feedstock is encouraged to flow through the bed in a radial direction and then out through the base of the reactor. There are certain advantages of this type of reactor, not the least of which is a low pressure drop through the catalyst bed; in addition, a radial bed reactor has a larger catalyst cross-sectional area as well as a shorter bed depth than the corresponding downflow reactor. It is this latter property that gives rise to the smaller pressure drop across the catalyst bed.

However, there are more chances of localized heating in the catalyst bed, and (in addition to the more expensive reactor design per unit volume of catalyst bed) it may be more difficult to remove contaminants from the bed as part of the catalyst regeneration sequence. For this reason alone, it is preferable that this type of reactor is limited to hydrodesulfurization of low-boiling feedstocks such as naphtha and kerosene, and application to the higher-boiling feedstocks is usually not recommended.

In summary, fixed-bed processes have advantages in ease of scale-up and operation. The reactors operate in a downflow mode, with liquid feed trickling downward over the solid catalyst concurrent

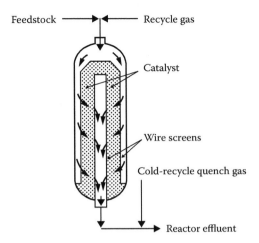

FIGURE 10.6 Fixed-bed radial flow reactor. (From Speight, J.G. 2000. *The Desulfurization of Heavy Oils and Residua*. 2nd Edition. Marcel Dekker Inc., New York. Figure 5.7, p. 182.)

with the hydrogen gas. The usual catalyst is cobalt/molybdenum (Co/Mo) or nickel/molybdenum (Ni/Mo) on alumina (Al_2O_3) and contain 11–14% molybdenum and 2–3% of the promoter nickel or cobalt. The alumina typically has a pore volume of 0.5 mL/g. The catalyst is formed into pellets by extrusion, in shapes such as cylinders (ca. 2 mm diameter), lobed cylinders, or rings.

These trickle bed reactors normally operate in the downflow configuration and have a number of operational problems, including poor distribution of liquid and pulsing operation at high liquid and gas loading. Scale-up of these liquid–gas–solid reactors is much more difficult than a gas–solid or gas–liquid reactor. Nevertheless, the downflow system is convenient when the bed is filled with small catalyst particles. Furthermore, because the catalyst particles are small, these reactors are quite effective as filters of the incoming feed. Any suspended fine solids, such as fine clays from production operations, accumulate at the front end of the bed. Eventually, this will lead to high-pressure differentials between the inlet and outlet end of the reactor.

The main limitation of this type of reactor is the gradual accumulation of metals when heavy feedstocks are processed. The metals accumulate in the pores of the catalyst and gradually block access for hydrogenation and desulfurization. The length of operation is then dictated by the metal-holding capacity of the catalyst and the nickel and vanadium content of the feed. As the catalyst deactivates, the reactor feed temperature is gradually increased to maintain conversion. Toward the end of a run, this mode of operation leads to accumulation of carbonaceous deposits on the catalyst, further reducing the activity.

At the end of a processing cycle, the reactor must be shut down and the catalyst bed replaced. This is an expensive operation, and such units are usually designed for infrequent changes of the catalyst. Many feedstocks are not suited to fixed-bed operation because of plugging and metal accumulation.

10.3.3 UPFLOW EXPANDED-BED REACTOR (PARTICULATE FLUIDIZED-BED REACTOR)

Expanded-bed reactors (Salmi et al., 2011) are applicable to distillates, but are commercially used for very heavy, high-metal-content, and/or dirty feedstocks having extraneous fine solid material. They operate in such a way that the catalyst is in an expanded state so that the extraneous solids pass through the catalyst bed without plugging. They are isothermal, which conveniently handles the high-temperature exotherms associated with high hydrogen consumptions. Since the catalyst is in an expanded state of motion, it is possible to treat the catalyst as a fluid and to withdraw and add catalyst during operation.

Expanded beds of catalyst are referred to as particulate fluidized insofar as the feedstock and hydrogen flow upward through an expanded bed of catalyst with each catalyst particle in independent motion. Thus, the catalyst migrates throughout the entire reactor bed. Expanded-bed reactors are mathematically modeled as back-mix reactors with the entire catalyst bed at one uniform temperature. Spent catalyst may be withdrawn and replaced with fresh catalyst on a daily basis. Daily catalyst addition and withdrawal eliminate the need for costly shutdowns to change out catalyst, and also result in a constant equilibrium catalyst activity and product quality. The catalyst is withdrawn daily and has a vanadium, nickel, and carbon content that is representative on a macro scale of what is found throughout the entire reactor. On a micro scale, individual catalyst particles have ages from that of fresh catalyst to as old as the initial catalyst charge to the unit; however, the catalyst particles of each age group are so well dispersed in the reactor that the reactor contents appear uniform.

In the unit, the feedstock and hydrogen recycle gas enter the bottom of the reactor, pass up through the expanded catalyst bed, and leave from the top of the reactor. Commercial expanded-bed reactors normally operate with 1/32-in (0.8-mm) extrudate catalysts that provide a higher rate of desulfurization than the larger catalyst particles used in fixed-bed reactors. With extrudate catalysts of this size, the upward liquid velocity based on fresh feedstock is not sufficient to keep the catalyst particles in an expanded state. Therefore, for each part of the fresh feed, several parts of product oil are taken from the top of the reactor, recycled internally through a large vertical pipe to the bottom of the reactor, and pumped back up through the expanded catalyst bed. The amount of catalyst bed expansion is controlled by the recycle of product oil back up through the catalyst bed.

The expansion and turbulence of gas and oil passing upward through the expanded catalyst bed are sufficient to cause almost complete random motion in the bed (particulate fluidized). This effect produces the isothermal operation. It also causes almost complete back mixing. Consequently, to effect near-complete sulfur removal (>75%), it is necessary to operate with two or more reactors in series. The ability to operate at a single temperature throughout the reactor or reactors, and to operate at a selected optimum temperature rather than an increasing temperature from the start to the end of the run, results in more effective use of the reactor and catalyst contents. When all these factors are put together, i.e., use of a smaller catalyst particle size, isothermal, fixed temperature throughout run, back mixing, daily catalyst addition, and constant product quality, the reactor size required for an expanded bed is often smaller than that required for a fixed bed to achieve the same product goals. This is generally true when the feeds have high initial boiling points and/or the hydrogen consumption is very high.

The upflow expanded-bed reactor (particulate fluidized-bed reactor) (Figure 10.7) operates in such a way that the catalyst remains loosely packed and is less susceptible to plugging, and they are therefore more suitable for the heavier feedstocks as well as for feedstocks that may contain considerable amounts of suspended solid material. Because of the nature of the catalyst bed, such suspended material will pass through the bed without causing frequent plugging problems. Furthermore, the expanded state of motion of the catalyst allows frequent withdrawal from, or addition to, the catalyst bed during operation of the reactor without the necessity of shutdown of the unit for catalyst replacement. This property alone makes the ebullated reactor ideally suited for the high-metal feedstocks that rapidly poison a catalyst, resulting in frequent catalyst replacement.

In the reactor, the feedstock and hydrogen are caused to flow upward through the catalyst bed in which each catalyst particle is reputed to have independent motion and can therefore (in theory) migrate throughout the entire catalyst bed. The heat of the reaction (which could be a problem in other reactors if it were not controlled) is dissipated in increasing the temperature of fresh feedstock to that existing in the reactor. Spent catalyst can be withdrawn from the reactor and replaced with fresh catalyst on a daily basis if necessary, and the need for time wasting and costly shutdowns can thus be eliminated.

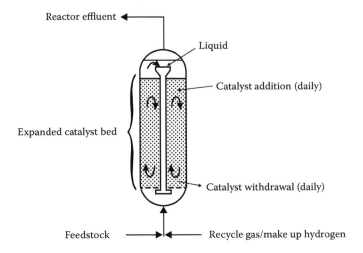

FIGURE 10.7 Expanded-bed reactor. (From Speight, J.G. 2000. *The Desulfurization of Heavy Oils and Residua.* 2nd Edition. Marcel Dekker Inc., New York. Figure 5.8, p. 183.)

Expanded-bed reactors normally operate with 1/32-in extruded catalysts that require partial recycle of the liquid products to maintain the catalyst particles in an expanded and fluidized state.

The near completely random motion of the catalyst bed virtually ensures an isothermal operation, but the efficiency of the hydrodesulfurization reaction tends to suffer because of the back mixing of the product and feedstock. Hence, to effect sulfur removal at >75% efficiency, it may be necessary to operate with two or more reactors in series. The need for two or more of these units to effectively desulfurize a feedstock may be cited as a disadvantage of the reactor; however, the ability of the reactor to operate under isothermal conditions as well as the on-stream catalyst addition–withdrawal system and the fact that the reactor size required for an expanded catalyst bed is often smaller than that required for a fixed bed can be cited in support of such a unit.

It is also possible to use finely divided catalyst in the expanded-bed reactor. If the catalyst is a suitable size (50–200 μm, i.e., 50×10^{-4} to 200×10^{-4} cm), it is possible to operate the expanded-bed reactor without recycling the liquid products to maintain the catalyst in the fluidized state. In addition, the finely divided catalyst has a relatively larger number of external pores on the surface than the extruded catalyst, and is less likely to have metal contaminants plugging up these pores because of their size. The overall effect of a finely divided catalyst in this reactor is more efficient sulfur removal for a given set of process conditions.

10.3.4 EBULLATING-BED REACTOR

The ebullated-bed units (Salmi et al., 2011) are typically used to replace fixed-bed units and moving-bed units—especially for heavy feedstocks—because hydroprocessing units with fixed-bed reactors must be shut down to remove the spent catalyst when catalyst activity declines below an acceptable level (due to the accumulation of coke, metals, and other contaminants deposited on the catalyst). In ebullated-bed units, the catalyst within the reactor bed is not fixed. In such a process, the hydrocarbon feed stream enters the bottom of the reactor and flows upward through the catalyst; the catalyst is kept in suspension by the pressure of the fluid feed. Ebullating-bed reactors are capable of converting the heavy feedstocks to lower-boiling products while simultaneously removing contaminants. The function of the catalyst is to remove contaminants such as sulfur and nitrogen heteroatoms, which accelerate the deactivation of the catalyst, while cracking (converting) the feed to lighter products (Eccles, 1993; Speight and Ozum, 2002; Parkash, 2003; Hsu and Robinson, 2006; Gary et al., 2007; Speight, 2014).

In such a process, the hydrocarbon feed stream enters the bottom of the reactor and flows upward through the catalyst—the catalyst is kept in suspension by the pressure of the fluid feed. Ebullating-bed reactors are capable of converting the most problematic feeds, such as atmospheric residua, vacuum residua, heavy oil, extra heavy oil, and tar sand bitumen (all of which have a high content of asphaltene constituents as well as metal constituents, sulfur constituents, and constituents ready to form sediment) to lower-boiling, more valuable products while simultaneously removing the contaminants.

In the unit, the hydrogen-rich recycle gas is bubbled up through a mixture of oil and catalyst particles, which provides three-phase turbulent mixing that is needed to ensure a uniform temperature distribution. At the top of the reactor, the catalyst is disengaged from the process fluids, which are separated in downstream flash drums. Some of the catalyst is withdrawn and replaced with fresh catalyst, while the majority of the catalyst is returned to the reactor.

The function of the catalyst is to remove contaminants such as sulfur and nitrogen heteroatoms, which accelerate the deactivation of the catalyst, while cracking (converting) the feed to lighter products. Because ebullating-bed reactors perform both hydrotreating and hydrocracking functions, they are considered to be dual-purpose reactors. Ebullating-bed catalysts are made of pellets that are <1 mm in size to facilitate suspension by the liquid phase in the reactor.

In contrast to fixed-bed hydrocracking units for vacuum gas oil, ebullating-bed units are suitable for processing residua and other heavy feedstocks, such as tar sand bitumen. The main advantages are (1) high conversion of atmospheric residue, up to 90% v/v; (2) better product quality than many other residue conversion processes, especially delayed coking; and (3) long run length—the catalyst life does not limit these units since fresh catalyst is added and spent catalyst is removed continuously and (subject only to mechanical issues) the units can typically run for a much longer time than fixed-bed units for the same heavy feedstock.

Examples of processes that use ebullated-bed reactors are the (1) H-oil process, (2) LC-Fining process, and (3) T-Star process.

10.3.5 DEMETALLIZATION REACTOR

The demetallization reactor (guard reactor, guard bed reactor) is a reactor that is placed in front of hydrocracking reactors to remove contaminants, particularly metals, before hydrocracking (Speight, 2000, 2014). Such reactors may employ an inexpensive catalyst to remove metals from an expanded-bed feed. Spent demetallization catalyst can be loaded to >30% vanadium. A catalyst support having large pores is preferentially demetallized with a low degree of desulfurization.

Feedstocks that have relatively high metal contents (>300 ppm) substantially increase catalyst consumption because the metals poison the catalyst, thereby requiring frequent catalyst replacement. The usual desulfurization catalysts are relatively expensive for these consumption rates; however, there are catalysts that are relatively inexpensive and can be used in the first reactor to remove a large percentage of the metals. Subsequent reactors downstream of the first reactor would use normal hydrodesulfurization catalysts. Since the catalyst materials are proprietary, it is not possible to identify them here. However, it is understood that such catalysts contain little or no metal promoters, i.e., nickel, cobalt, or molybdenum. Metals removal on the order of 90% has been observed with these materials.

Thus, one method of controlling demetallization is to employ separate smaller guard reactors just ahead of the fixed-bed hydrodesulfurization reactor section. The preheated feed and hydrogen pass through the guard reactors that are filled with an appropriate catalyst for demetallization that is often the same as the catalyst used in the hydrodesulfurization section. The advantage of this system is that it enables replacement of the most contaminated catalyst (guard bed), where pressure drop is highest, without having to replace the entire inventory or shut down the unit. The feedstock is alternated between guard reactors while the catalyst in the idle guard reactor is being replaced.

When the expanded-bed design is used, the first reactor could employ a low-cost catalyst (5% of the cost of Co/Mo catalyst) to remove the metals, and subsequent reactors can use the more selective hydrodesulfurization catalyst. The demetallization catalyst can be added continuously without taking the reactor out of service and the spent demetallization catalyst can be loaded to >30% vanadium, which makes it a valuable source of vanadium.

10.3.6 REACTOR OPTIONS

One of the ways to improve the hydrodesulfurization process is to change the direction of reaction streams, i.e., conduct the hydrodesulfurization countercurrently, which can lead to some satisfactory reactor concentration profiles (Salmi et al., 2011; Mužic and Sertić-Bionda, 2013). The feed is introduced at the top of the reactor and the hydrogen at the bottom. Hydrogen sulfide is removed from the reactor top, avoiding possible recombination reactions with olefins in the exit mixture.

An example of an advanced hydrodesulfurization reactor is the reactor where the hydrodesulfurization reactions are taking place in an ebullated catalyst bed. This process is used for treating heavier petroleum fractions that cause very fast deactivation of conventional hydrodesulfurization catalysts because of excessive coke formation. During this type of operations, the heat transfer efficiency is high, overheating of the catalyst carrier is minimized, and the formation of coke is decreased.

In the ebullated-bed process, the catalyst particles, the feed, and the hydrogen are in fluidized state, which enables very good mixing. The clogging and erosion of the catalyst particles is minimized and the reactor operates in almost isothermal conditions with constant, rather small, pressure drop. Also, the ebullated-bed reactor enables control of the catalyst activity by allowing constant addition and removal of catalyst particles (Babich and Moulijn, 2003; Song, 2003).

Reactor designs and configurations for deep hydrodesulfurization of gas oil preferentially involve both single-stage and two-stage desulfurization processes (Sie, 1999). Hydrogen sulfide strongly suppresses the activity of the catalyst for converting the refractory sulfur compounds, which should occur in the major downstream part of a cocurrent trickle-bed reactor during deep desulfurization, which is not the optimal technology for deep desulfurization—a second reactor can be (or should be) used, particularly to meet the lower sulfur levels. Both desulfurization and hydrogenation in the second reactor can be improved by removing hydrogen sulfide and ammonia from the exit gas of the first reactor before the stream enters the second reactor.

Another option for reactor design is to have two or more catalyst beds, which are normally placed in separate reactors, within a single reactor shell and have both cocurrent and countercurrent flows. The hydrogen is mixed with the feedstock (usually a distillate) at the entrance to the reactor, and the mixture flows through the reactor. The advantage of cocurrent design is ease of control of gas–liquid mixing and contact with the catalyst. The disadvantage is that the concentration of hydrogen is the highest in the front of the reactor and lowest at the outlet. The opposite is true for the concentration of hydrogen sulfide. The solution to this issues problem is to design a countercurrent reactor, where the fresh hydrogen is introduced from the bottom of the reactor and the liquid distillate from the top.

10.4 CATALYSTS

The increasing importance of hydrodesulfurization in petroleum processing to produce clean-burning fuels has led to a surge of research on the chemistry and engineering of process catalysts. There has been a growing need to develop catalysts that can carry out deep hydrodesulfurization (Vasudevan and Fierro, 1996; Turaga et al., 2003; Song et al., 2006). Indeed, catalysts are considered to have a major role in solving future problems related to fossil fuel conversion in a more efficient manner. This need has become even more pressing in view of recent environmental regulations limiting the amount of sulfur in fuels and gas oils to <0.05 wt.%.

The process dictates that sulfur-resistant noble metal catalysts are necessary (Cooper and Donnis, 1996; Song, 2002). However, the design of sulfur-resistant noble metal catalysts for low-temperature hydrotreating of sulfur-containing distillates to produce clean distillate fuels has been of prime importance to produce fuels such as diesel fuels and jet fuels. In addition, achieving low levels of sulfur is not a simple task because the sulfur compounds that remain after hydrodesulfurization are highly refractory. Steric factors often render the sulfur atom inaccessible to the catalyst and the hydrogen. The development of a new generation of catalysts to achieve this objective of low sulfur levels in the processing of different feedstocks presents a challenge. Indeed, the problem of sulfur emission from fuels is serious, and environmental regulations are likely to become even more stringent in the future. There is a growing demand for the development of better catalysts for hydrotreating that limit not only the amount of sulfur but also that of nitrogen and aromatics.

To achieve the goal of reducing sulfur levels in fuels, there is a clear need for understanding the mechanism of the reaction (Chapter 5) in conjunction with the nature of the catalyst and support. Most of the work has been carried out with the traditional cobalt–molybdenum catalyst supported on alumina. This system is time tested and effective.

More efficient conversion of the heavy feedstocks also requires consideration of catalyst types (Fischer and Angevine, 1986) as well as the development of various process options to respond to market depends. Some generalities of the behavior of different catalysts can be included here. In short, catalyst properties are the key to effective hydroprocessing of heavy feedstocks. Proper catalyst selection and application can make hydroprocessing an attractive process route to lower-boiling products, allowing process flexibility for a wide range of different feedstocks.

The hydrotreating catalysts are usually cobalt plus molybdenum or nickel plus molybdenum (in the sulfide) forms, impregnated on an alumina base. The hydrotreating operating conditions (1000–2000 psi hydrogen and about 370°C [700°F]) are such that appreciable hydrogenation of aromatics does not occur. The desulfurization reactions are invariably accompanied by small amounts of hydrogenation and hydrocracking, the extent of which depends on the nature of the feedstock and the severity of desulfurization.

One of the problems in the processing of high-sulfur and high-nitrogen feeds is the large quantity of hydrogen sulfide (H_2S) and ammonia (NH_3) that are produced. Substantial removal of both compounds from the recycle gas can be achieved by the injection of water in which, under the high-pressure conditions employed, both hydrogen sulfide and ammonia are very soluble compared with hydrogen and hydrocarbon gases. The solution is processed in a separate unit for the recovery of anhydrous ammonia and hydrogen sulfide.

The reactions of hydrocracking require a dual-function catalyst that posses both acidic (cracking) and metallic (hydrogenation) components. The acidic cracking component is usually an amorphous silica–alumina support, a crystalline zeolite material, alumina, or acid-treated clay. The metallic hydrogenation component is usually a metal sulfide incorporating metals such as nickel, tungsten, platinum, palladium, or cobalt, and which is finely dispersed on the support material. The relative strengths of the catalyst components influence the course of the various reactions that can occur during hydrocracking. Thus, they affect the resulting product yield and structure.

Hydrocracking catalysts are very sensitive to nitrogen compounds in the feed, which break down under the conditions of reaction to give ammonia and neutralize the acid sites. As many heavy feedstocks contain substantial amounts of nitrogen (Chapter 7), a pretreatment stage is frequently required. For gas oils and liquid feedstocks, this could involve an acid wash to remove the basic nitrogen; however, for heavy feedstocks, this involves a primary stage in which the feedstock is hydrotreated followed by the hydrocracking stage. In the primary stage, denitrogenation and desulfurization can be carried out using cobalt/molybdenum or nickel/cobalt/molybdenum on alumina or silica–alumina.

Catalysts containing platinum or palladium (~0.5 wt.%) on a zeolite base appear to be somewhat less sensitive to nitrogen than are nickel catalysts, and successful operation has been achieved with feedstocks containing 40 ppm nitrogen. This catalyst is also more tolerant of sulfur in the feedstock that acts as a temporary poison. The catalyst recovers its activity when the sulfur content of the feed is reduced.

Many of the catalysts for the hydrodesulfurization process are produced by composting a transition metal (or its salt) with a solid support. The metal constituent is the active catalyst (Dolbear, 1998). The most commonly used materials for supports are alumina, silica, silica–alumina, kieselguhr, magnesia (and other metal oxides), and the zeolites. The support can be manufactured in a variety of shapes or may even be crushed to particles of the desired size. The metal constituent can then be added by contact of the support with an aqueous solution of the metal salt. The whole is then subjected to further treatment that will dictate the final form of the metal on the support (i.e., the metal oxide, metal sulfide, or even the metal itself).

Molybdenum sulfide (MoS_2), usually supported on tailored alumina, is widely used in petroleum processes for hydrogenation reactions. It is a layered structure that can be made much more active by the addition of cobalt or nickel (Topsøe et al., 1996). When promoted with cobalt sulfide (CoS), making what is called cobalt–molybdenum catalysts, it is widely used in hydrodesulfurization processes. The nickel sulfide (NiS)-promoted version is used for hydrodenitrogenation as well as hydrodesulfurization. The closely related tungsten compound (WS_2) is used in commercial hydrocracking catalysts. Other sulfides (iron sulfide [FeS], chromium sulfide [Cr_2S_3], and vanadium sulfide [V_2S_5]) are also effective and used in some catalysts. A valuable alternative to the base metal sulfides is palladium sulfide (PdS). Although it is expensive, palladium sulfide forms the basis for several very active catalysts. Zeolites loaded with transition metal sulfides have also been used as hydrodesulfurization catalysts (Keville et al., 1995; Pawelec et al., 1997).

The surface area of the catalyst is usually large (200–300 m^2/g), but almost this entire surface is contained within the pore space of the alumina. Cobalt and molybdenum are two of the most common metals that are used as hydrodesulfurization catalysts and, as such, are dispersed in a thin layer within the pore system of the alumina. When these metals are used together as a hydrodesulfurization catalyst, the catalyst is more tolerant to poisoning agents and is usually considered suitable for a wide variety of feedstocks but more particularly to the heavy feedstocks. Other metals may be used such as a combination of nickel and molybdenum; however, this catalyst is a more active hydrogenation catalyst than the cobalt–molybdenum catalyst and consumes more hydrogen per mole of sulfur removed. However, the nickel–molybdenum catalyst is useful for the hydrodesulfurization of catalytic cracking feedstocks (where maximum hydrogen consumption is desirable). The catalyst is more selective for nitrogen removal from feedstocks and presumably can tolerate higher nitrogen feedstocks without losing as much activity in a given time than the cobalt–molybdenum catalyst. Another catalyst that has received some attention is nickel–tungsten on alumina, which is a very active hydrogenation catalyst and displays a high activity in hydrocracking reactions.

Hydrodesulfurization catalysts are normally used as extrudates or as porous pellets, but the particle size and pore geometry have an important influence on process design, especially for the heavier feedstocks. The reaction rates of hydrodesulfurization catalysts are limited by the diffusion of the reactants into, and the products out of, the catalyst pore systems. Thus, as the catalyst particle size is decreased, the rate of desulfurization is increased (Frost and Cottingham, 1971); however, the pressure differential across the catalyst bed also diminishes, and a balance must be reached between reaction rate and pressure drop across the bed.

Hydrodesulfurization catalysts are usually more active in the sulfide form with the external sulfur being applied either during the preparation from a sulfur-containing source or during the initial contact of the catalyst with the sulfur-containing feedstock. Chemically, the sulfiding process is a reduction in the oxide form of the metal (cobalt, nickel, etc.) on the alumina support by conversion to the metal sulfide. This can be achieved, in the absence of a sulfur-containing feedstock, by injecting hydrogen sulfide or any other low-boiling sulfide (carbon disulfide, dimethyl sulfide, and the like) directly into the recirculating hydrogen stream. If the feedstock is present and contains >1% sulfur, addition of an external source is not necessary. Temperatures for the sulfiding process are usually on the order of 290–315°C (550–600°F), while minimum pressures of about 150 psi are also recommended.

Caution is advised against using temperatures in excess of 315°C (600°F) since the hot hydrogen gas can reduce a substantial portion of the metal oxides in the catalyst. In addition, with respect to the sulfiding reaction, the reactivity of the reduced metals is lower than the reactivity of the metal oxides. The sulfiding process is continued until the sulfur content of the catalyst has reached a predetermined level, usually estimated by the level of hydrogen sulfide in the tail gas.

Catalyst consumption is a major aspect of the hydrodesulfurization process, and the costs of the process increase markedly with the high-metal feedstocks (Dolbear, 1998; Speight and Ozum, 2002; Parkash, 2003; Hsu and Robinson, 2006; Gary et al., 2007; Speight, 2014). The ease with which the catalyst can be replaced depends, to a large extent, on the bed type; with the high-metal feedstocks, it is inevitable that frequent catalyst replacement will occur. Attempts have been made to produce a correlation that will indicate the catalyst efficiency for various groups of metal-containing feedstocks. However, as with all estimations of this type, they should be used only to present an indication of the trend in the hydrodesulfurization rather than as absolute data.

New types of CoMo and NiMo catalysts with increased hydrogenation activity and selectivity have been developed for hydrodesulfurization by applying new techniques and technologies. New reactor designs have also been developed, such as reactors with multiple levels or reactors with different catalyst structures. It is expected that the best results will be achieved by combining the improved activity catalysts with the new design reactors. Furthermore, hydrodesulfurization catalysts with excellent desulfurization efficiency have been developed by combining new types of catalytically active species, such as noble metals, and new carrier materials such as those based on improved amorphous silica–alumina. However, the use of noble metals in hydrodesulfurization processing is limited owing to their susceptibility to sulfur poisoning, which is why they are used only after most of the organic sulfur compounds and hydrogen sulfide are removed from the feeds. To increase the catalysts' resistance to sulfur poisoning, new types of bifunctional hydrodesulfurization catalysts have been developed. The hydrodesulfurization process that uses catalysts based on noble metals is conducted in two or more steps, which include multilevel catalyst systems in order to achieve deep desulfurization and hydrogenation (Mužic and Sertić-Bionda, 2013).

10.5   CATALYST BED PLUGGING

Catalyst bed plugging can arise in a variety of ways; however, the overall effect of bed plugging is always the same: expensive shutdowns and possibly complete renewal of the expensive catalyst. Thus, the deposition of rust, coke, or metal salts (e.g., sodium chloride) from heavier and dirtier feedstock may all contribute to the plugging of a catalyst bed. Vanadium and nickel may also be deposited onto the surface of the catalyst as well as into the pore system. Asphaltene deposition is also a potential means of bed plugging—coagulation of the asphaltene constituents becomes appreciable at temperatures above 420°C (790°F) with the formation of coke on the catalyst.

The exothermal nature of the reaction may also contribute indirectly to catalyst plugging. For example, lack of proper control over the heat liberated during the hydrodesulfurization process may lead to the formation of localized hotspots in the catalyst, which can then initiate asphaltene coagulation at these points. In a similar manner, adsorption of the lower-boiling constituents from a whole feedstock or a from residuum feedstock could leave carbonaceous materials within the void space of the catalyst, thereby leading to the eventual denaturation of the catalyst with a concurrent loss in catalyst efficiency. Finally, another cause of bed plugging may be due to the physical nature of the catalyst itself. Agglomeration of catalyst particles may lead to the formation of catalyst lumps within the bed that again leads to catalyst inefficiency, presumably through formation of localized hotspots within the bed. The resulting end point of any of the above deficiencies will be an increased pressure drop through the catalyst bed that eventually requires process shutdown.

The catalysts used for processing heavy feedstocks inevitably deactivate with time due to the accumulation of nickel and vanadium sulfides and carbonaceous residues (coke) on the catalyst. The former deposits have been studied intensely, in part because the metal deposits tend to accumulate near the surface of catalyst pellets, rendering the interior ineffective. Both metal sulfides and coke may contribute to loss of activity.

The nickel molybdenum (Ni/Mo) or cobalt molybdenum (Co/Mo) on alumina catalysts for hydroprocessing give rapid accumulation of coke followed by constant concentrations (Ancheyta, 2007; Speight, 2014). The amount of coke deposited on catalysts during heavy feedstock processing ranges from 15–35 wt.% on a carbon basis (Thakur and Thomas, 1985). A significant loss in activity is associated with this rapid accumulation (Mosby et al., 1986).

Coke deposits are thought to develop via surface adsorption of coke precursors, followed by a combination of oligomerization and aromatization reactions. Since adsorption is the first step in the process, decreases in volatility (higher molecular weight) and increases in polarity would enhance the first step of the process. Residue fractions, therefore, give more coke deposition than an equivalent gas oil fraction due to the decrease in volatility of the reacting molecules. Although asphaltene constituents are commonly considered to be the heaviest, most polar components in residue, the asphaltene content of the feed does not correlate with coke deposition (Furimsky, 1979). For example, deasphalting a residue only reduced the coke levels from 30 wt.% of the catalyst to 20 wt.% (Thakur and Thomas, 1985).

Operating conditions are also very important. An increase in hydrogen pressure reduces coke accumulation by suppressing the oligomerization and by hydrogenating the adsorbed species. Higher temperatures produce higher yields of coke because the dehydrogenation reactions become more favored, even though adsorption would be reduced at a higher temperature. Since increased temperature also speeds up the coking reactions, these catalysts are more likely to have a large coke concentration at the outer edge of the catalyst pellets. Similarly, larger pellet sizes favor an uneven distribution of coke within a pellet. In fixed reactors, larger extrudates are used to control pressure drop, and the reactors are operated in an increasing-temperature mode to maintain constant catalyst activity. Both of these factors contribute to a higher concentration of coke at the outer edge of the catalyst pellets. In contrast, the smaller ebullated-bed catalysts that are exposed to isothermal conditions gave a flat profile of coke with pellet radius (Thakur and Thomas, 1985). A mean pore diameter of at least 10–15 nm in macropores is recommended to minimize diffusion resistance.

Metals accumulate more slowly on the catalyst surfaces because the inlet concentrations of metals are lower than for coke precursors. The accumulation of metals can be even greater than coke, for example, the vanadium concentration can reach 30–50 wt.% of the catalyst on a fresh catalyst basis (Thakur and Thomas, 1985). Demetallization reactions can be considered autocatalytic in the sense that once the surface of the catalyst is covered with metal sulfides, the catalyst remains quite active and continues to accumulate metal sulfides. The final loss of catalyst activity is usually associated with filling of pore mouths in the catalyst by metal sulfide deposits.

As in the case of coke deposition, larger catalyst pellets and higher operating temperatures increase the tendency for metal sulfides to deposit at the periphery of the pellet. Higher hydrogen pressure also increases the rate of metal conversion, which increases the tendency toward high peripheral concentrations. The problem of maintaining good catalyst activity in the presence of metal sulfide deposits, therefore, is a challenging combination of chemistry and transport, and has led to the development of catalysts with improved pore networks to sustain deposition of metals for longer periods of time.

Catalyst bed plugging (and a subsequent expensive shutdown) may be minimized by using corrosion-resistant alloys throughout the entire system of feed lines, heaters, reactors, etc., thereby reducing the risk of rust accumulation in the feedstock. Removal of debris that has accumulated in the feedstock from transportation and storage may be achieved by mechanical filtering. Every attempt should be made to desalt the crude oil in order to remove any inorganic materials (sodium chloride, etc.) that may originate from the brines that are associated with crude oils as they are pumped to the surface of the earth. In the processing of the heavier feedstocks, it may be advisable to suspend the top layer of catalyst in basket-type devices so that this layer, when plugged, can be conveniently removed and replaced. It may be necessary to use two or more reactors in parallel rather than one large reactor—this would not only lead to more catalyst surface area for collecting debris, but also, if shutdown is inevitable, one of the reactors could be left on-stream while the other reactor is being cleaned. Alternatively, upflow reactors would create the tendency for any debris in the feedstock to fall out of the catalyst, and a circulating stream of liquid below the catalyst bed could be channeled through a filter to remove any solid materials. In a similar manner, prefiltering of the feedstock through a bed of bauxite or similar material will facilitate removal of debris and, in the presence of hydrogen, may also encourage the removal of vanadium.

In summary, protection of the catalyst and the reactor is essential if shutdowns are to be minimized, and this is particularly the case when the heavier feedstocks are used in the hydrodesulfurization process. Some measure of protection from the asphaltene and the resin constituents can be achieved by the removal of these materials by means of a deasphalting step (Chapter 3). The overall relative merits of any particular total processing scheme should be assessed in terms of feedstock properties, product yields, process variables, and last but not least, process economics (Speight and Ozum, 2002; Parkash, 2003; Hsu and Robinson, 2006; Gary et al., 2007; Speight, 2014).

10.6 CATALYST POISONING

Hydrocarbons, especially aromatic hydrocarbons, can undergo multicondensation reactions in the presence of catalysts to form coke. This coke is a complex polynuclear aromatic material that is low in hydrogen. Coke can deposit on the surface of a catalyst, blocking access to the active sites and reducing the activity of the catalyst. Poisoning by coke deposits is a major problem in FCC catalysts where catalysts containing deposited coke are circulated to a fluidized-bed combustor to be regenerated. In hydrocracking, coke deposition is virtually eliminated by the catalyst's hydrogenation function. However, the product referred to as *coke* is not a single material. The first products deposited are tarry deposits that can, with time and temperature, continue to condense to a solid deposit.

In a hydrodesulfurization system, the hydrogenation function adds hydrogen to the tarry deposits. This reduces the concentration of coke precursors on the surface. There is, however, a slow accumulation of coke that reduces activity over a 1–2 year period. Refiners respond to this slow reduction in activity by raising the average temperature of the catalyst bed to maintain conversions. Eventually, however, an upper limit to the allowable temperature is reached, and the catalyst must be removed and regenerated.

Either metallurgy of the reactor, product quality considerations, or catalyst selectivity sets upper temperature limits. Metallurgy considerations result from decreases in wall strength at higher temperatures. Product quality can be compromised at higher temperatures because thermal cracking reactions begin to compete with acid cracking as temperatures approach 400°C (750°F). Selectivity changes result from all the possible side reactions that become more important as the temperatures increase.

Burning off the accumulated coke can regenerate catalysts containing deposited coke. This is achieved using rotary or similar kilns rather than leaving catalysts in the hydrocracking reactor, where the reactions could damage the metals in the walls. Removing the catalysts also allows inspection and repair of the complex and expensive reactor internals, discussed below. Regeneration of a large catalyst charge can take weeks or months, so refiners may own two catalyst loads: one in the reactor and one regenerated and ready for reload.

In addition, organometallic compounds produce metallic products that influence the reactivity of the catalyst by deposition on its surface. The transfer of metal from the feedstock to the catalyst constitutes an irreversible poisoning of the catalyst. After combustion to remove the carbonaceous deposits, the catalysts are treated to redisperse active metals.

10.7 PROCESS VARIABLES

The efficiency of the hydrodesulfurization process is measured by the degree of sulfur removal or, in other words, by the yields of sulfur-free products. However, there are several process variables

TABLE 10.2

Typical Process Conditions for Desulfurization and Hydrocracking

	Desulfurization		Hydrocracking	
	Middle Distillate	**Vacuum Gas Oil**	**MPHC**	**High Pressure**
Pressure, psi	400–800	500–1000	<1500	1500+
Space velocity, vol/hr/vol	2–4	1–2	0.4–1.0	0.4–1.0
Avg. reactor temp.				
°C	315–370	360–415	385–425	315–400
°F	600–700	675–775	725–800	600–750
Hydrogen rate, scf/bbl	100–300	300–500	400–1000	1500–3500
Conversion				
% to naphtha	1	1	5–15	100
% to distillate	–	10–20	20–50	–

Source: Speight, J.G. 2000. *The Desulfurization of Heavy Oils and Residua.* 2nd Edition. Marcel Dekker Inc., New York. Table 5.8, p. 199.

(Table 10.2) that need special attention, as any one of these variables can have a marked influence on the course and efficiency of the hydrodesulfurization process.

The major process variables are (1) reactor temperature, (2) hydrogen pressure, (3) liquid hourly space velocity, (4) hydrogen recycle rate, and (5) influence of feedstock properties—which has been discussed in Chapter 7.

10.7.1 REACTOR TEMPERATURE

A higher reaction temperature increases the rate of desulfurization at constant feed rate, and the start-of-run temperature is set by the design desulfurization level, space velocity, and hydrogen partial pressure. The capability to increase temperature as the catalyst deactivates is built into most process or unit designs. Temperatures of 415°C (780°F) and above result in excessive coking reactions and higher-than-normal catalyst aging rates. Therefore, units are designed to avoid the use of such temperatures for any significant part of the cycle life.

The temperature in the hydrodesulfurization reactor is often considered to be the primary means by which the process is controlled. For example, at stabilized reactor conditions, an increase of 10°C (18°F) in the reaction temperature will substantially increase, and may even double, the reaction rate. Generally, an increase in the temperature (from 360°C to 380°C, i.e., from 680°F to 715°F) will produce a noticeable increase in conversion, or for a fixed conversion of about 90% enables the quantity of catalyst necessary for the process to be halved.

In the same manner as in hydrocracking (Dolbear, 1998), hydrogen is added at intermediate points in hydrodesulfurization reactors. This is important for control of reactor temperatures. The mechanical devices in the reactor, called *reactor internals*, that accomplish this step are very important to successful processes. If redistribution is not efficient, some areas of the catalyst bed will have more contact with the feedstock. This can lead to three levels of problems:

1. *Poor selectivity*: Ratios of hydrogen, oil, and catalyst outside design ranges will change the yield structures. Some parts of the bed will be hotter than other parts. Some fractions of the feedstock will be cracked to undesirable low molecular weight (light hydrocarbon) products and conversion will be lower.

2. *Rapid catalyst aging*: Higher than desirable hydrogenation can increase local reactor temperatures markedly. Catalysts can sinter, losing surface area and activity and shortening run length.

3. *Hotspots*: When local reactor temperatures are well above 400°C (750°F), thermal cracking can become important. Thermal cracking produces olefins, which add hydrogen, releasing heat. This increases the temperatures further, and thermal cracking rates go up. These hotspots can easily reach temperatures higher than the safe upper limits for the reactor walls, and results can be catastrophic.

There are, however, limits to which the temperature can be increased without adversely affecting process efficiency; at temperatures above 410°C (770°F), thermal cracking of the hydrocarbon constituents becomes the predominant process that can lead to the formation of considerable amounts of low molecular weight hydrocarbon liquids and gases. In addition, increasing the partial pressure of the hydrogen cannot diminish these high-temperature cracking reactions. In addition, excessively high temperatures (above 400°C or 750°F) lead to deactivation of the catalyst much more quickly than lower temperatures.

10.7.2 Hydrogen Pressure

The important effect of hydrogen partial pressure is the minimization of coking reactions. If the hydrogen pressure is too low for the required duty at any position within the reaction system, premature aging of the remaining portion of catalyst will be encountered. In addition, the effect of hydrogen pressure on desulfurization varies with feed boiling range. For a given feed, there exists a threshold level above which hydrogen pressure is beneficial to the desired desulfurization reaction. Below this level, desulfurization drops off rapidly as hydrogen pressure is reduced.

The overall effect of increasing the partial pressure of the hydrogen is to increase the extent of the conversion through an increase in catalyst activity (Frost and Cottingham, 1971; Speight and Ozum, 2002; Parkash, 2003; Hsu and Robinson, 2006; Gary et al., 2007; Speight, 2014). This is to be expected since the essential function of the catalyst is to serve as a means by which the reactants are brought together, thereby promoting interaction between the feedstock constituents and the hydrogen. As with the temperature variable, there are also limitations to increasing the partial pressure of the hydrogen. Use of excessively high partial pressures (for conventional feedstocks: >1000 psi; for heavy feedstocks, this figure could be on the order of 2000 psi) may only serve to saturate the catalyst, and any increase in the partial pressure of the hydrogen will affect the conversion only slightly.

There are, however, two other instances where the partial pressure of the hydrogen will influence the course of the reaction. The nature of the hydrodesulfurization process is such that the rate will increase (subject to the above limitations) with increasing partial pressure of the hydrogen. However, feedstock conversion decreases with increasing ammonia (a product of the reaction of hydrogen with nitrogen-containing organic constituents of the feedstock) partial pressure. Caution should be exercised to ensure that an increase in the hydrogen partial pressure does not adversely affect the reaction by maintaining a relatively high concentration of ammonia in the vicinity of the catalyst.

It is necessary to exercise caution to ensure that high concentrations of hydrogen sulfide are not allowed to build up at any point in the reactor. Initially, hydrogen sulfide produced by the hydrodesulfurization process may be used beneficially, i.e., to sulfide the catalyst; however, high concentrations of this gas will only serve to cause corrosion of the equipment as well as adversely affect the activity of the catalyst.

10.7.3 Liquid Hourly Space Velocity

As the space velocity is increased, desulfurization is decreased; however, increasing the hydrogen partial pressure and/or the reactor temperature can offset the detrimental effect of increasing space velocity.

The liquid hourly space velocity is the ratio of the hourly volume flow of liquid in, say, barrels to the catalyst volume in barrels, and the reciprocal of the liquid hourly space velocity gives the contact time. Since the catalyst volume for the process will be constant, the space velocity will vary directly with the feed rate. A decrease in the liquid hourly space velocity (or, alternatively, an increase in the contact time) will usually bring about an increase in the efficiency (or extent) of the hydrodesulfurization process (Frost and Cottingham, 1971). To maintain a fixed rate of hydrodesulfurization when the feed rate is increased, it may be necessary to increase the temperature.

Since the liquid hourly space velocity is a ratio that involves the use of the volume of the catalyst in the reactor, it is also possible to use the data to estimate the additional catalyst required to increase the efficiency of the process. This is especially relevant when the activity of different catalysts is to be compared, and such comparisons are only meaningful if estimations can be made of the quantities of catalyst required to obtain a predetermined performance. For example, to intimate that one catalyst is twice as active as another catalyst signifies that one unit of the first catalyst will function as efficiently as two units of the second catalyst. Such differences in catalyst performance can, however, only be determined accurately through estimations of the liquid hourly space velocity and how this parameter varies with the degree of hydrodesulfurization.

10.7.4 Hydrogen Recycle Rate

The optimum use of hydrodesulfurization catalysts requires a relatively high hydrogen partial pressure, and it is therefore necessary to introduce with the feedstock quantities of hydrogen that are considerably greater than required on the basis of stoichiometric chemical consumption. In all cases, process economics dictate that unused hydrogen should be recycled after it has been partially (or completely) freed from hydrogen sulfide that was produced in the previous pass.

The overall hydrogen consumption is a summation of several processes:

1. Removal of the sulfur, nitrogen, and oxygen in the feedstock as their hydrogenated analogues, i.e., hydrogen sulfide, ammonia, and water, respectively
2. Addition of hydrogen to unsaturated (olefinic) functions in the products as will occur in the prevailing conditions of the hydrodesulfurization process
3. Destruction, by saturation (i.e., addition of hydrogen), of certain aromatic compound types
4. Stabilization of unsaturated short-lived organic intermediates that exist during hydrocracking

Thus, if the hydrogen is not recycled, the process economics are unfavorable and, in addition, the efficiency of the hydrodesulfurization reaction may be adversely affected because of the possible competing reactions outlined above.

10.7.5 Catalyst Life

Catalyst life depends on the charge stock properties and the degree of desulfurization desired. The only permanent poisons to the catalyst are metals in the feedstock that deposit on the catalyst, usually quantitatively, causing permanent deactivation as they accumulate. However, this is usually of little concern except when deasphalted oils are used as feedstocks since most distillate feedstocks contain low amounts of metals. Nitrogen compounds are a temporary poison to the catalyst, but there is essentially no effect on catalyst aging except that caused by a higher temperature requirement to achieve the desired desulfurization. Hydrogen sulfide can be a temporary poison in the reactor gas, and recycle gas scrubbing is employed to counteract this condition.

Providing that pressure drop buildup is avoided, cycles of 1 year or more and ultimate catalyst life of 3 years or more can be expected. The catalyst employed can be regenerated by normal steam–air or recycle combustion gas–air procedures. The catalyst is restored to near fresh activity by regeneration during the early part of its ultimate life. However, permanent deactivation of the catalyst occurs slowly during usage and repeated regenerations, so replacement becomes necessary.

10.7.6 FEEDSTOCK EFFECTS

The different types of streams that can undergo hydroprocessing range from heavy feedstocks of resid and vacuum gas oil to lighter feedstocks of naphtha and distillate. Naphtha is hydroprocessed to remove contaminants such as sulfur, which are harmful to downstream operations (such as precious metal–reforming catalyst). Diesel hydroprocessing removes sulfur to meet fuel requirements, and saturates aromatics. The purpose of resid and vacuum gas oil hydroprocessing is to remove metals, sulfur, and nitrogen (e.g., hydrotreating), as well as to convert high molecular weight hydrocarbons into lower molecular weight hydrocarbons (e.g., hydrocracking).

Thus, it is not surprising that the character of the feedstock properties, especially the feed boiling range, has a definite effect on the ultimate design of the desulfurization unit and process flow. In agreement, there is a definite relationship between the percent by weight sulfur in the feedstock and the hydrogen requirements.

In addition, the reaction rate constant in the kinetic relationships decreases rapidly with increasing average boiling point in the kerosene and light gas oil range, but much more slowly in the heavy gas oil range. This is attributed to the difficulty in removing sulfur from ring structures present in the entire heavy gas oil boding range.

The hydrodesulfurization of light (low-boiling) distillate (naphtha or kerosene) is one of the more common catalytic hydrodesulfurization processes since it is usually used as a pretreatment of such feedstocks before deep hydrodesulfurization or before catalytic reforming. This is similar to the concept of pretreating residua before cracking to improve the quality of the products (Speight and Ozum, 2002; Parkash, 2003; Hsu and Robinson, 2006; Gary et al., 2007; Speight, 2014). Hydrodesulfurization of such feedstocks is required because sulfur compounds poison the precious-metal catalysts used in reforming, and desulfurization can be achieved under relatively mild conditions and is near quantitative (Table 10.3). If the feedstock arises from a cracking operation, hydrodesulfurization will be accompanied by some degree of saturation, resulting in increased hydrogen consumption.

The hydrodesulfurization of low-boiling (naphtha) feedstocks is usually a gas-phase reaction and may employ the catalyst in fixed beds and (with all of the reactants in the gaseous phase) only

TABLE 10.3
Hydrodesulfurization of Various Naphtha Fractions

| Feedstock | Boiling Range | | Sulfur, wt.% | Desulfurization,[a] % |
	°C	°F		
Visbreaker naphtha	65–220	150–430	1.00	90
Visbreaker–coker naphtha	65–220	150–430	1.03	85
Straight-run naphtha	85–170	185–340	0.04	99
Catalytic naphtha (light)	95–175	200–350	0.18	89
Catalytic naphtha (heavy)	120–225	250–440	0.24	71
Thermal naphtha (heavy)	150–230	300–450	0.28	57

Source: Speight, J.G. 2007. *The Chemistry and Technology of Petroleum.* 5th Edition. CRC Press, Taylor & Francis Publishers, Boca Raton, FL. Table 21.1, p. 571.

[a] Process conditions: Co/Mo on alumina, 260–370°C (500–700°F), 200–500 psi hydrogen.

minimal diffusion problems are encountered within the catalyst pore system. It is, however, important that the feedstock be completely volatile before entering the reactor, as there may be the possibility of pressure variations (leading to less satisfactory results) if some of the feedstock enters the reactor in the liquid phase and is vaporized within the reactor.

In applications of this type, the sulfur content of the feedstock may vary from 100 ppm to 1%, and the necessary degree of desulfurization to be effected by the treatment may vary from as little as 50% to >99%. If the sulfur content of the feedstock is particularly low, it will be necessary to presulfide the catalyst. For example, if the feedstock only has 100–200 ppm sulfur, several days may be required to sulfide the catalyst as an integral part of the desulfurization process even with complete reaction of all of the feedstock sulfur to, say, cobalt and molybdenum (catalyst) sulfides. In such a case, presulfiding can be conveniently achieved by addition of sulfur compounds to the feedstock or by addition of hydrogen sulfide to the hydrogen.

Generally, hydrodesulfurization of naphtha feedstocks to produce catalytic reforming feedstocks is carried to the point where the desulfurized feedstock contains <20 ppm sulfur. The net hydrogen produced by the reforming operation may actually be sufficient to provide the hydrogen consumed in the desulfurization process.

The hydrodesulfurization of middle distillates is also an efficient process, and applications include predominantly the desulfurization of kerosene, diesel fuel, jet fuel, and heating oils that boil over the general range 250–400°C (480–750°F). However, with this type of feedstock, hydrogenation of the higher-boiling catalytic cracking feedstocks has become increasingly important where hydrodesulfurization is accomplished alongside the saturation of condensed-ring aromatic compounds as an aid to subsequent processing.

Under the relatively mild processing conditions used for the hydrodesulfurization of these particular feedstocks, it is difficult to achieve complete vaporization of the feed. Process conditions may dictate that only part of the feedstock is actually in the vapor phase and that sufficient liquid phase is maintained in the catalyst bed to carry the larger molecular constituents of the feedstock through the bed. If the amount of liquid phase is insufficient for this purpose, molecular stagnation (leading to carbon deposition on the catalyst) will occur.

Hydrodesulfurization of middle distillates causes a more marked change in the specific gravity of the feedstock, and the amount of low-boiling material is much more significant when compared with the naphtha-type feedstock. In addition, the somewhat more severe reaction conditions (leading to a designated degree of hydrocracking) also cause an overall increase in hydrogen consumption when middle distillates are employed as feedstocks in place of the naphtha.

High-boiling distillates, such as the atmospheric and vacuum gas oils, are not usually produced as a refinery product but merely serve as feedstocks to other processes for conversion to lower-boiling materials. For example, gas oils can be desulfurized to remove >80% of the sulfur originally in the gas oil with some conversion of the gas oil to lower-boiling materials. The treated gas oil (which has a reduced carbon residue as well as lower sulfur and nitrogen contents relative to the untreated material) can then be converted to lower-boiling products in, say, a catalytic cracker where an improved catalyst life and volumetric yield may be noted.

The conditions used for the hydrodesulfurization of gas oil may be somewhat more severe than the conditions employed for the hydrodesulfurization of middle distillates with, of course, the feedstock in the liquid phase.

In summary, the hydrodesulfurization of the low-, middle-, and high-boiling distillates can be achieved quite conveniently using a variety of processes. One major advantage of this type of feedstock is that the catalyst does not become poisoned by metal contaminants in the feedstock since only negligible amounts of these contaminants will be present. Thus, the catalyst may be regenerated several times, and on-stream times between catalyst regeneration (while varying with the process conditions and application) may be on the order of 3–4 years.

REFERENCES

Akbarzadeh, S., Raheb, J., Aghaei, A., and Karkhane, A.A. 2003. Study of Desulfurization Rate in *Rhodococcus FMF* Native Bacterium. *Iran. J. Biotechnol.* 1(1): 36–40.

Ancheyta, J. 2007. Reactors for Hydroprocessing. In: *Hydroprocessing of Heavy Oils and Residua*, J. Ancheyta and J.G. Speight (Editors). CRC Press, Taylor & Francis Group, Boca Raton, FL.

Ancheyta, J., Rana, N.S., and Furimsky, E. 2005. Hydroprocessing of Heavy Petroleum Feeds. *Catal. Today* 109: 3–15.

Babich, I.V., and Moulijn, J.A. 2003. Science and Technology of Novel Processes for Deep Desulfurization of Oil Refinery Streams: A Review. *Fuel* 82(2003): 607–631.

Bakshi, A., and Lutz, I. 1987. Adding Hydrogen Donor to Visbreaking Improves Distillate Yields. *Oil Gas J.* 85(28): 84–87.

Bartholdy, J., and Andersen, S.I. 2000. Changes in Asphaltene Stability during Hydrotreating. *Energy Fuels* 14: 52–55.

Bland, W.F., and Davidson, R.L. 1967. *Petroleum Processing Handbook*. McGraw-Hill, New York.

Brunet, S., Mey, D., Guy Pérot, G., Bouchy, C., and Diehl, F. 2005. On the Hydrodesulfurization of FCC Gasoline: A Review. *Appl. Catal. A* 278: 143–172.

Cooper, B.H., and Donnis, B.B.L. 1996. Aromatic Saturation of Distillates: An Overview. *Appl. Catal. A.* 137: 203–223.

Dolbear, G.E. 1998. Hydrocracking: Reactions, Catalysts, and Processes. In: *Petroleum Chemistry and Refining*, J.G. Speight (Editor). Taylor & Francis, Washington, DC, Chapter 7, pp. 175–198.

Eccles, R.M. 1993. Residue Hydroprocessing using Ebullated-Bed Reactors. *Fuel Process. Technol.* 35: 21–38.

Fischer, R.H., and Angevine, P.V. 1986. Dependence of Resid Processing Selectivity on Catalyst Pore Size Distribution. *Appl. Catal.* 27: 275–283.

Frost, C.M., and Cottingham, P.L. 1971. Hydrodesulfurization of Venezuelan Residual Fuel Oils. Report of Investigations RI 7557. US. Bureau of Mines, Washington, DC.

Gary, J.H., Handwerk, G.E., and Kaiser, M.J. 2007. *Petroleum Refining: Technology and Economics*, 5th Edition. CRC Press, Taylor & Francis Group, Boca Raton, FL.

Ho, T.C. 1988. Hydrodenitrogenation Catalysis. *Catal. Rev. Sci. Eng.* 30: 117–160.

Howell, R.L., Hung, C., Gibson, K.R., and Chen, H.C. 1985. Catalyst Selection Important for Residuum Hydrotreating. *Oil Gas J.* 83(30): 121–128.

Hsu, C.S., and Robinson, P.R. (Editors) 2006. *Practical Advances in Petroleum Processing*, Vols. 1–2. Springer Science, New York.

Keville, K.M., Timken, H.K.C., and Ware, R.A. 1995. Method for Preparing Catalysts Comprising Zeolites. United States Patent 5,378,671, January 3.

Khan, M.R., and Patmore, D.J. 1998. Heavy Oil Upgrading Processes. In: *Petroleum Chemistry and Refining*, J.G. Speight (Editor). Taylor & Francis, Washington, DC, Chapter 6.

Kobayashi, S., Kushiyama, S., Aizawa, R., Koinuma, Y., Inoue, K., Shmizu, Y., and Egi, K. 1987a. Kinetic Study on the Hydrotreating of Heavy Oil. 1. Effect of Catalyst Pellet Size in Relation to Pore Size. *Ind. Eng. Chem. Res.* 26: 2241–2245.

Kobayashi, S., Kushiyama, S., Aizawa, R., Koinuma, Y., Inoue, K., Shmizu, Y., and Egi, K. 1987b. Kinetic Study on the Hydrotreating of Heavy Oil. 2. Effect of Catalyst Pore Size. *Ind. Eng. Chem. Res.* 26: 2245–2250.

Kressmann, S., Guillaume, D., Roy, M., and Plain, C. 2004. A New Generation of Hydroconversion and Hydrodesulfurization Catalysts. In: *Proceedings. 14th Annual Symposium, Catalysis in Petroleum Refining and Petrochemicals, King Fahd University of Petroleum and Minerals*. Dhahran, Saudi Arabia, December 5–6.

Li, M.Z., Squires, C.H., and Monticello, D.J. 1996. Genetic Analysis of the Dsz Promoter and Associated Regulatory Regions of *Rhodococcus erythropolis* IGTS8. *J. Bacteriol.* 175(22): 6409–6418.

Liu, Y., Gao, L., Wen, L., and Zong, B. 2009. Recent Advances in Heavy Oil Hydroprocessing Technologies. *Rec. Patents Chem. Eng.* 2: 22–36.

Maghsoudi, S., Kheirolomoom, A., and Vossoughi, M. 2000. Selective Desulfurization of Dibenzothiophene by Newly Isolated *Corynebacterium* sp. strain P32C1. *Biochem. Eng. J.* 5: 11–16.

Meyers, R.A. (Editor) 1997. *Handbook of Petroleum Refining Processes*, 2nd Edition. McGraw-Hill, New York.

Mohebali, G., and Ball, A.S. 2008. Biocatalytic Desulfurization (Biodesulfurization) of Petrodiesel Fuels. *Microbiology* 154: 2169–2183.

Mosby, J.F., Buttke, R.D., Cox, J.A., and Nikolaides, C. 1986. Process Characterization of Expanded-Bed Reactors in Series. *Chem. Eng. Sci.* 41: 989–995.

Mužic, M., and Sertić-Bionda, K. 2013. Alternative Processes for Removing Organic Sulfur Compounds from Petroleum Fractions. *Chem. Biochem. Eng. Q.* 27(1): 101–108.

Parkash, S. 2003. *Refining Processes Handbook*. Gulf Professional Publishing, Elsevier, Amsterdam.

Pawelec, B., Navarro, R., Fierro, J.L.G., Cambra, J.F., Zugazaga, F., Guemez, M.B., and Arias, P.L. 1997. Hydrodesulfurization over PdMo/HY Zeolite Catalysts. *Fuel* 76: 61–71.

Radovanović, L.J., and Speight, J.G. 2011. Visbreaking: A Technology of the Future. In: *Proceedings. First International Conference—Process Technology and Environmental Protection (PTEP 2011)*. University of Novi Sad, Technical Faculty "Mihajlo Pupin," Zrenjanin, Republic of Serbia, December 7, pp. 335–338.

Reynolds, J.G., and Beret, S. 1989. Effect of Prehydrogenation on Hydroconversion of Maya Residuum. *Fuel Sci. Technol. Int.* 7: 165–186.

Salmi, T.O., Mikkola, J.-P., and Wärnå, J.P. 2011. *Chemical Reaction Engineering and Reactor Technology*. CRC Press, Taylor & Francis Group, Boca Raton, FL.

Sie, S.T. 1999. Reaction Order and Role of Hydrogen Sulfide in Deep Hydrodesulfurization of Gas Oil: Consequences for Industrial Reactor Configuration. *Fuel Proc. Technol.* 61: 149–171.

Song, C. 2002. New Approaches to Deep Desulfurization for Ultra-Clean Gasoline and Diesel Fuels: An Overview. *Prepr. Div. Fuel. Chem. Am. Chem. Soc.* 47(2): 438–444.

Song, C. 2003. An Overview of New Approaches to Deep Desulfurization for Ultra-Clean Gasoline, Diesel Fuel and Jet Fuel. *Catal. Today* 86: 211–263.

Song, C., Turaga, U.T., and Ma, X. 2006. Desulfurization. In: *Encyclopedia of Chemical Processing*, S. Lee (Editor). CRC Press, Taylor & Francis Group, Boca Raton, FL. pp. 651–661.

Speight, J.G. 1986. Upgrading Heavy Feedstocks. *Annu. Rev. Energ.* 11: 253.

Speight, J.G. 2000. *The Desulfurization of Heavy Oils and Residua*, 2nd Edition. Marcel Dekker Inc., New York.

Speight, J.G. 2011. *The Refinery of the Future*. Gulf Professional Publishing, Elsevier, Oxford, UK.

Speight, J.G. 2013. *Heavy and Extra Heavy Oil Upgrading Technologies*. Gulf Professional Publishing, Elsevier, Oxford, UK.

Speight, J.G. 2014. *The Chemistry and Technology of Petroleum*, 5th Edition. CRC Press, Taylor & Francis Group, Boca Raton, FL.

Speight, J.G., and Moschopedis, S.E. 1979. The Production of Low-Sulfur Liquids and Coke from Athabasca Bitumen. *Fuel Process. Technol.* 2: 295.

Speight, J.G., and Ozum, B. 2002. *Petroleum Refining Processes*. Marcel Dekker Inc., New York.

Suchanek, A.J., and Moore, A.S. 1986. Efficient Carbon Rejection Upgrades Mexico's Maya Crude Oil. *Oil Gas J.* 84(31): 36–40.

Thakur, D.S., and Thomas, M.G. 1985. Catalyst Deactivation in Heavy Petroleum and Synthetic Crude Processing: A Review. *Appl. Catal.* 15: 197–225.

Topsøe, H., Clausen, B.S., and Massoth, F.E. 1996. *Hydrotreating Catalysis*. Springer-Verlag, Berlin, Germany.

Turaga, U.T., Song, C., Turaga, T., and Song, C. 2003. MCM-41-Supported Co–Mo Catalysts for Deep Hydrodesulfurization of Light Cycle Oil. *Catal. Today* 86(1–4): 129–140.

Vasudevan, P.T., and Fierro, J.L.G. 1996. A Review of Deep Hydrodesulfurization Catalysis. *Catal. Rev. Sci. Eng.* 38: 161–188.

Vernon, L.W., Jacobs, F.E., and Bauman, R.F. 1984. Process for Converting Petroleum Residuals. United States Patent 4,425,224, June 10.

11 Desulfurization Processes—Gases

11.1 INTRODUCTION

Gas processing consists of separating all of the various hydrocarbons and fluids from pure natural gas (Mokhatab et al., 2006; Speight, 2007, 2014). Major transportation pipelines usually impose restrictions on the makeup of the natural gas that is allowed into the pipeline—before the natural gas can be transported, the level of contaminants must be markedly reduced. This is no easy task when the variable nature of the gas composition must be taken into consideration (Table 11.1).

Thus, while ethane, propane, butane, and pentanes must be removed from natural gas, this does not mean that they are all waste products. Gas processing is necessary to ensure that the natural gas intended for use is as clean and pure as possible, making it the clean burning and environmentally sound energy choice. Thus, natural gas, as it is used by consumers, is much different from the natural gas that is brought from underground up to the wellhead. Although the processing of natural gas is, in many respects, less complicated than the processing and refining of crude oil, it is equally as necessary before its use by end users. The natural gas used by consumers is composed almost entirely of methane. However, natural gas found at the wellhead, although still composed primarily of methane, is by no means as pure.

Raw natural gas comes from three types of wells: oil wells, gas wells, and condensate wells. Natural gas that comes from oil wells is typically termed *associated gas*. This gas can exist separate from oil in the formation (free gas), or dissolved in the crude oil (dissolved gas). Natural gas from gas and condensate wells, in which there is little or no crude oil, is termed *non-associated gas*. Gas wells typically produce raw natural gas by itself, while condensate wells produce free natural gas along with a semiliquid hydrocarbon condensate. Whatever the source of the natural gas, once separated from crude oil (if present), it commonly exists in mixtures with other hydrocarbons, principally ethane, propane, butane, and pentanes. In addition, raw natural gas contains water vapor, hydrogen sulfide (H_2S), carbon dioxide, helium, nitrogen, and other compounds. In fact, associated hydrocarbons known as *natural gas liquids* (NGLs) can be very valuable by-products of natural gas processing. Natural gas liquids include ethane, propane, butane, isobutane, and natural gasoline that are sold separately and have a variety of different uses, including enhancing oil recovery in oil wells, providing raw materials for oil refineries or petrochemical plants, and as sources of energy.

Petroleum refining as it is currently known will continue at least for the next three decades. Various political differences have caused fluctuations in petroleum imports, and imports have continued to increase over the past several years (Speight, 2011). It is also predictable that use of petroleum for the transportation sector will increase as increases in travel offset increased efficiency. As a consequence of this increase in use, petroleum will be the largest single source of carbon emissions from fuel. Acid gases corrode refining equipment, harm catalysts, pollute the atmosphere, and prevent the use of hydrocarbon components in petrochemical manufacture. When the amount of hydrogen sulfide is high, it may be removed from a gas stream and converted to sulfur or sulfuric acid. Some natural gases contain sufficient carbon dioxide to warrant recovery as dry ice (Bartoo, 1985; Mokhatab et al., 2006).

TABLE 11.1

Constituents of Natural Gas

Name	Formula	% v/v
Methane	CH_4	65–85
Ethane	C_2H_6	4–9
Propane	C_3H_8	2–5
Butane	C_4H_{10}	2–3
Pentane+	$C_5H_{12}^+$	2–5
Carbon dioxide	CO_2	2–5
Hydrogen sulfide	H_2S	1–3
Nitrogen	N_2	2–5
Helium	He	<0.5

Source: Speight, J.G. 2007. *The Chemistry and Technology of Petroleum.* 4th
            Edition. CRC Press, Taylor & Francis Publishers, Boca Raton, FL.
            Table 11.1, p. 708.

Note: Pentane+, pentane and higher molecular weight hydrocarbons, including
         benzene and toluene.

11.2 GAS STREAMS

Furthermore, off-gases from various refinery unit operations can be converted into valuable chemicals such as hydrogen. Off-gases contain large concentrations of sulfur and other impurities such as
arsenic that must be removed. Sulfur is a potent poison for the catalysts used in the conversion of
refinery off-gas to hydrogen. Traditionally, hydrodesulfurization is used for deep desulfurization in
a two-step process: (1) hydrogenation of sulfur compounds and (2) subsequent removal of hydrogen
sulfide with an expendable chemical absorbent. For higher sulfur levels, the cost of hydrodesulfurization is very high and also results in hydrogenation of olefins. High-capacity sorbents are needed
that can remove sulfur (both hydrogen sulfide and organic sulfur species) directly by adsorption
from refinery off-gases.

The actual practice of processing gas streams to pipeline dry gas quality levels can be quite
complex, but usually involves four main processes to remove the various impurities. Gas streams
produced during petroleum and natural gas refining, while ostensibly being hydrocarbon in nature,
may contain large amounts of acid gases such as hydrogen sulfide and carbon dioxide. Most commercial plants employ hydrogenation to convert organic sulfur compounds into hydrogen sulfide.
Hydrogenation is effected by means of recycled hydrogen-containing gases or external hydrogen
over a nickel molybdate or cobalt molybdate catalyst.

In summary, refinery process gas (which includes still gas), in addition to hydrocarbons, may
contain other contaminants, such as carbon oxides (CO_x, where $x = 1$ and/or 2), sulfur oxides (SO_x,
where $x = 2$ and/or 3), as well as ammonia (NH_3), mercaptans (R-SH), and carbonyl sulfide (COS)
(Table 11.2). The presence of the various impurities may eliminate some of the sweetening processes, since some processes remove large amounts of acid gas but not to a sufficiently low concentration. On the other hand, there are those processes not designed to remove (or incapable of
removing) large amounts of acid gases whereas they are capable of removing the acid gas impurities
to very low levels when the acid gases are present only in low to medium concentration in the gas
(Katz, 1959).

The processes that have been developed to accomplish gas purification vary from a simple once-
through wash operation to complex multistep recycling systems (Speight, 2007). In many cases,
the process complexities arise because of the need for recovery of the materials used to remove the

TABLE 11.2

Approximate Compositional Ranges for the Constituents of Process Gas[a]

Constituents (Listed Alphabetically)	Range (% v/v)
Butane(s)	<5.0
Butylene(s)	<2.0
Carbon dioxide	0–10.0
Ethane	<15.0
Ethylene	<5.0
Hexane(s)	<2.0
Hydrogen	40.0–50.0
Hydrogen sulfide	10.0–50.0
Methane	20.0–30.0
Nitrogen	0–10.0
Pentane(s)	<5.0
Pentylene(s) [amylene(s)]	<2.0
Propane	<5.0
Propylene	<2.0

[a] Also may contain traces of carbonyl sulfide (COS), dimethyl sulfide (CH_3SCH_3), ethyl mercaptan (C_2H_5SH), methyl mercaptan (CH_3SH), and propyl mercaptan (C_3H_8SH).

contaminants or even recovery of the contaminants in the original, or altered, form (Katz, 1959; Kohl and Riesenfeld, 1985; Newman, 1985).

There are many variables in treating refinery gas or natural gas. The precise area of application of a given process is difficult to define. Several factors must be considered:

1. Types of contaminants in the gas
2. Concentrations of contaminants in the gas
3. Degree of contaminant removal desired
4. Selectivity of acid gas removal required
5. Temperature of the gas to be processed
6. Pressure of the gas to be processed
7. Volume of the gas to be processed
8. Composition of the gas to be processed
9. Carbon dioxide–hydrogen sulfide ratio in the gas
10. Desirability of sulfur recovery due to process economics or environmental issues

Hydrogen sulfide is a highly toxic, flammable, and corrosive gas that dissolves in hydrocarbon and water streams, and it is present in the vapor phase above these streams. Partitioning of hydrogen sulfide to the oil, water, and vapor phases is influenced by temperature, pH, and pressure. Hydrogen sulfide is heavier than air and will, therefore, collect in low places, such as the bottom of storage or shipping vessels. The human odor-detection limit ranges from 3 parts per billion (ppb) to 20 ppb—the gas may be present long before it reaches a hazardous level.

In addition to hydrogen sulfide (H_2S) and carbon dioxide (CO_2), gas may contain other contaminants, such as mercaptans (RSH) and carbonyl sulfide (COS). The presence of these impurities may eliminate some of the sweetening processes since some processes remove large amounts of acid gas but not to a sufficiently low concentration. On the other hand, there are those processes that are not designed to remove (or are incapable of removing) large amounts of acid gases. However, these processes are also capable of removing the acid gas impurities to very low levels when the acid gases are present in low to medium concentrations in the gas.

Process selectivity indicates the preference with which the process removes one acid gas component relative to (or in preference to) another. For example, some processes remove both hydrogen sulfide and carbon dioxide; other processes are designed to remove hydrogen sulfide only. It is important to consider the process selectivity for, say, hydrogen sulfide removal compared with carbon dioxide removal that ensures minimal concentrations of these components in the product, thus the need for consideration of the carbon dioxide to hydrogen sulfide in the gas stream.

Gas processing involves the use of several different types of processes; however, there is always overlap between the various processing concepts. In addition, the terminology used for gas processing can often be confusing and/or misleading because of the overlap (Nonhebel, 1964; Curry, 1981; Maddox, 1982).

There are four general processes used for emission control (often referred to in another, more specific context as flue gas desulfurization): (1) adsorption, (2) absorption, (3) catalytic oxidation, and (4) thermal oxidation (Soud and Takeshita, 1994; Svoboda et al., 1994).

Adsorption is a physical–chemical phenomenon in which the gas is concentrated on the surface of a solid or liquid to remove impurities. Usually, carbon is the adsorbing medium, which can be regenerated upon *desorption* (Fulker, 1972; Mokhatab et al., 2006; Speight, 2007). The quantity of material adsorbed is proportional to the surface area of the solid and, consequently, adsorbents are usually granular solids with a large surface area per unit mass. Subsequently, the captured gas can be desorbed with hot air or steam either for recovery or for thermal destruction.

Adsorbers are widely used to increase a low gas concentration before incineration unless the gas concentration is very high in the inlet air stream. Adsorption also is employed to reduce problem odors from gases. There are several limitations to the use of adsorption systems; however, it is generally felt that the major one is the requirement for minimization of particulate matter and/or condensation of liquids (e.g., water vapor) that could mask the adsorption surface and drastically reduce the efficiency of the process (Mokhatab et al., 2006; Speight, 2007, 2014).

Absorption differs from *adsorption*, in that it is not a physical–chemical surface phenomenon, but an approach in which the absorbed gas is ultimately distributed throughout the absorbent (liquid). The process depends only on physical solubility and may include chemical reactions in the liquid phase (*chemisorption*). Common absorbing media used are water, aqueous amine solutions, caustic, sodium carbonate, and nonvolatile hydrocarbon oils, depending on the type of gas to be absorbed. Usually, the gas–liquid contactor designs that are employed are plate columns or packed beds (Mokhatab et al., 2006; Speight, 2007, 2014).

Absorption is achieved by dissolution (a physical phenomenon) or by reaction (a chemical phenomenon) (Barbouteau and Dalaud, 1972; Ward, 1972; Mokhatab et al., 2006; Speight, 2007). Chemical adsorption processes adsorb sulfur dioxide onto a carbon surface where it is oxidized (by oxygen in the flue gas) and absorbs moisture to give sulfuric acid impregnated into and on the adsorbent.

Liquid absorption processes (which usually employ temperatures below 50°C [120°F]) are classified either as *physical solvent processes* or *chemical solvent processes*. The physical solvent processes employ an organic solvent, and absorption is enhanced by low temperatures or high pressure, or both. Regeneration of the solvent is often accomplished readily (Staton et al., 1985). In chemical solvent processes, absorption of the acid gases is achieved mainly by use of alkaline solutions such as amines or carbonates (Kohl and Riesenfeld, 1985). Regeneration (desorption) can be brought about by use of reduced pressures and/or high temperatures, whereby the acid gases are stripped from the solvent.

Solvents used for emission control processes should have (Mokhatab et al., 2006; Speight, 2007)

1. High capacity for acid gas
2. Low tendency to dissolve hydrogen

3. Low tendency to dissolve low molecular weight hydrocarbons
4. Low vapor pressure at operating temperatures to minimize solvent losses
5. Low viscosity
6. Low thermal stability
7. Absence of reactivity toward gas components
8. Low tendency for fouling
9. Low tendency for corrosion
10. Economic acceptability

Amine washing of gas emissions involves chemical reaction of the amine with any acid gases, with the liberation of an appreciable amount of heat, and it is necessary to compensate for the absorption of heat. Amine derivatives such as ethanolamine (monoethanolamine), diethanolamine, triethanolamine, methyldiethanolamine, diisopropanolamine, and diglycolamine have been used in commercial applications (Table 11.3) (Katz, 1959; Jou et al., 1985; Kohl and Riesenfeld, 1985; Maddox et al., 1985; Polasek and Bullin, 1985; Pitsinigos and Lygeros, 1989; Speight, 1993).

The chemistry can be represented by simple equations for low partial pressures of the acid gases:

$$2RNH_2 + H_2S \rightarrow (RNH_3)_2S$$

$$2RHN_2 + CO_2 + H_2O \rightarrow (RNH_3)_2CO_3$$

At high acid gas partial pressure, the reactions will lead to the formation of other products:

$$(RNH_3)_2S + H_2S \rightarrow 2RNH_3HS$$

$$(RNH_3)_2CO_3 + H_2O \rightarrow 2RNH_3HCO_3$$

The reaction is extremely fast, with the absorption of hydrogen sulfide being limited only by mass transfer; this is not so for carbon dioxide.

Regeneration of the solution leads to near-complete desorption of carbon dioxide and hydrogen sulfide. A comparison between monoethanolamine, diethanolamine, and diisopropanolamine shows that monoethanolamine is the cheapest of the three but shows the highest heat of reaction and corrosion; the reverse is true for diisopropanolamine.

Carbonate washing is a mild alkali process for emission control by the removal of acid gases (such as carbon dioxide and hydrogen sulfide) from gas streams (Speight, 1993), and uses the principle that the rate of absorption of carbon dioxide by potassium carbonate increases with temperature. It has been demonstrated that the process works best near the temperature of reversibility of the reactions:

$$K_2CO_3 + CO_2 + H_2O \rightarrow 2KHCO_3$$

$$K_2CO_3 + H_2S \rightarrow KHS + KHCO_3$$

Water washing, in terms of the outcome, is analogous to washing with potassium carbonate (Kohl and Riesenfeld, 1985), and it is also possible to carry out the desorption step by pressure reduction. The absorption is purely physical, and there is also a relatively high absorption of hydrocarbons, which are liberated at the same time as the acid gases.

In *chemical conversion processes*, contaminants in gas emissions are converted to compounds that are not objectionable or that can be removed from the stream with greater ease than the original constituents. For example, a number of processes have been developed that remove hydrogen sulfide and sulfur dioxide from gas streams by absorption in an alkaline solution.

TABLE 11.3
Olamines Used for Gas Processing

Olamine	Formula	Derived Name	Molecular Weight	Specific Gravity	Melting Point (°C)	Boiling Point (°C)	Flash Point (°C)	Relative Capacity (%)
Ethanolamine (monoethanolamine)	$HOC_2H_4NH_2$	MEA	61.08	1.01	10	170	85	100
Diethanolamine	$(HOC_2H_4)_2NH$	DEA	105.14	1.097	27	217	169	58
Triethanolamine	$(HOC_2H_4)_3NH$	TEA	148.19	1.124	18	335, d	185	41
Diglycolamine (hydroxyethanolamine)	$H(OC_2H_4)_2NH_2$	DGA	105.14	1.057	−11	223	127	58
Diisopropanolamine	$(HOC_3H_6)_2NH$	DIPA	133.19	0.99	42	248	127	46
Methyldiethanolamine	$(HOC_2H_4)_2NCH_3$	MDEA	119.17	1.03	−21	247	127	51

Source: Speight, J.G. 2007. *The Chemistry and Technology of Petroleum.* 4th Edition. CRC Press, Taylor & Francis Publishers, Boca Raton, FL. Table 11.4, p. 716.
Note: d, with decomposition.

Catalytic oxidation is a chemical conversion process that is used predominantly for destruction of volatile organic compounds and carbon monoxide. These systems operate in a temperature regime of 205–595°C (400–1100°F) in the presence of a catalyst. Without the catalyst, the system would require higher temperatures. Typically, the catalysts used are a combination of noble metals deposited on a ceramic base in a variety of configurations (e.g., honeycomb shaped) to enhance good surface contact. A photocatalytic oxidative method is also available (Song et al., 2012).

Catalytic systems are usually classified on the basis of bed types such as *fixed bed* (or *packed bed*) and *fluid bed* (*fluidized bed*). These systems generally have very high destruction efficiencies for most volatile organic compounds, resulting in the formation of carbon dioxide, water, and varying amounts of hydrogen chloride (from halogenated hydrocarbons). The presence in emissions of chemicals such as heavy metals, phosphorus, sulfur, chlorine, and most halogens in the incoming air stream act as poison to the system and can foul up the catalyst.

Thermal oxidation systems, without the use of catalysts, also involve chemical conversion (more correctly, chemical destruction) and operate at temperatures in excess of 815°C (1500°F), or 220–610°C (395–1100°F) higher than catalytic systems.

Historically, *particulate matter control* (*dust control*) (Mody and Jakhete, 1988) has been one of the primary concerns of industries, since the emission of particulate matter is readily observed through the deposition of fly ash and soot as well as in impairment of visibility. Differing ranges of control can be achieved by use of various types of equipment. Upon proper characterization of the particulate matter emitted by a specific process, the appropriate piece of equipment can be selected, sized, installed, and performance tested. The general classes of control devices for particulate matter are as follows.

Cyclone collectors are the most common of the inertial collector class. Cyclones are effective in removing coarser fractions of particulate matter. The particle-laden gas stream enters an upper cylindrical section tangentially and proceeds downward through a conical section. Particles migrate by centrifugal force generated by providing a path for the carrier gas to be subjected to a vortex-like spin. The particles are forced to the wall and are removed through a seal at the apex of the inverted cone. A reverse-direction vortex moves upward through the cyclone and discharges through a top center opening. Cyclones are often used as primary collectors because of their relatively low efficiency (50–90% is usual). Some small-diameter high-efficiency cyclones are utilized. The equipment can be arranged either in parallel or in series to both increase efficiency and decrease pressure drop. However, there are disadvantages that must be recognized (Mokhatab et al., 2006; Speight, 2007, 2014). These units for particulate matter operate by contacting the particles in the gas stream with a liquid. In principle, the particles are incorporated in a liquid bath or in liquid particles, which are much larger and therefore more easily collected.

Fabric filters are typically designed with nondisposable filter bags. As the dusty emissions flow through the filter media (typically cotton, polypropylene, Teflon, or fiberglass), particulate matter is collected on the bag surface as a dust cake. Fabric filters are generally classified on the basis of the filter bag cleaning mechanism employed. Fabric filters operate with collection efficiencies up to 99.9%, although other advantages are evident (Mokhatab et al., 2006; Speight, 2007, 2014).

Wet scrubbers are devices in which a countercurrent spray liquid is used to remove particles from an air stream. Device configurations include plate scrubbers, packed beds, orifice scrubbers, venturi scrubbers, and spray towers, individually or in various combinations (Mokhatab et al., 2006; Speight, 2007, 2014).

Other methods include use of high-energy-input *venturi scrubbers* or electrostatic scrubbers where particles or water droplets are charged, and flux force/condensation scrubbers where a hot humid gas is contacted with cooled liquid or where steam is injected into saturated gas. In the latter scrubber, the movement of water vapor toward the cold water surface carries the particles with it (*diffusiophoresis*), while the condensation of water vapor on the particles causes the particle size to increase, thus facilitating collection of fine particles.

The *foam scrubber* is a modification of a wet scrubber in which the particle-laden gas is passed through a foam generator, where the gas and particles are enclosed by small bubbles of foam.

Electrostatic precipitators (Mokhatab et al., 2006; Speight, 2007, 2014) operate on the principle of imparting an electric charge to particles in the incoming air stream, which are then collected on an oppositely charged plate across a high-voltage field. Particles of high resistivity create the most difficulty in collection. Conditioning agents such as sulfur trioxide (SO_3) have been used to lower resistivity.

Important parameters include design of electrodes, spacing of collection plates, minimization of air channeling, and collection-electrode rapping techniques (used to dislodge particles). Techniques under study include the use of high-voltage pulse energy to enhance particle charging, electron-beam ionization, and wide plate spacing. Electrical precipitators are capable of efficiencies >99% under optimum conditions; however, performance is still difficult to predict in new situations.

11.2.1   Gas Streams from Crude Oil

To process and transport associated dissolved natural gas, it must be separated from the oil in which it is dissolved. This separation of natural gas from oil is most often done using equipment installed at or near the wellhead.

The actual process used to separate oil from natural gas, as well as the equipment that is used, can vary widely. Although dry pipeline-quality natural gas is virtually identical across different geographic areas, raw natural gas from different regions will vary in composition (Table 11.1), and therefore the requirements for gas separation may emphasize or deemphasize the optional separation processes. In many instances, natural gas is dissolved in oil underground primarily due to the formation pressure. When this natural gas and oil is produced, it is possible that it will separate on its own, simply due to decreased pressure, much like opening a can of soda pop allows the release of dissolved carbon dioxide. In these cases, separation of oil and gas is relatively easy, and the two hydrocarbons are sent separate ways for further processing. The most basic type of separator is known as a conventional separator. It consists of a simple closed tank, where the force of gravity serves to separate the heavier liquids like oil and the lighter gases like natural gas.

In certain instances, however, specialized equipment is necessary to separate oil and natural gas. An example of this type of equipment is the low-temperature separator. This is most often used for wells producing high-pressure gas along with light crude oil or condensate. These separators use pressure differentials to cool the wet natural gas and separate the oil and condensate. Wet gas enters the separator, being cooled slightly by a heat exchanger. The gas then travels through a high-pressure liquid "knockout," which serves to remove any liquids into a low-temperature separator. The gas then flows into this low-temperature separator through a choke mechanism, which expands the gas as it enters the separator. This rapid expansion of the gas allows for the lowering of the temperature in the separator. After liquid removal, the dry gas then travels back through the heat exchanger and is warmed by the incoming wet gas. By varying the pressure of the gas in various sections of the separator, it is possible to vary the temperature, which causes the oil and some water to be condensed out of the wet gas stream. This basic pressure–temperature relationship can work in reverse as well, to extract gas from a liquid oil stream.

On the other hand, petroleum refining produces gas streams that contain substantial amounts of acid gases such as hydrogen sulfide and carbon dioxide. These gas streams are produced during initial distillation of the crude oil and during the various conversion processes. Of particular interest is the hydrogen sulfide (H_2S) that arises from the hydrodesulfurization of feedstocks that contain organic sulfur:

$$[S]_{feedstock} + H_2 \rightarrow H_2S + \text{hydrocarbons}$$

The terms *refinery gas* (including still gas) and *process gas* are also often used to include all of the gaseous products and by-products that emanate from a variety of refinery processes. There are also components of the gaseous products that must be removed before release of the gases to the atmosphere or before use of the gas in another part of the refinery, i.e., as a fuel gas or as a process feedstock.

Petroleum refining typically involves, with the exception of the heavy feedstocks, *primary distillation* that results in separation into fractions differing in carbon number, volatility, specific gravity, and other characteristics (Chapter 3) (Speight and Ozum, 2002; Parkash, 2003; Hsu and Robinson, 2006; Gary et al., 2007; Speight, 2014). The most volatile fraction, which contains most of the gases that are generally dissolved in the crude, is referred to as *pipestill gas* or *pipestill light ends* and consists essentially of hydrocarbon gases ranging from methane to butane(s), or sometimes pentane(s).

The gas varies in composition and volume, depending on crude origin and on any additions to the crude made at the loading point. It is not uncommon to reinject light hydrocarbons such as propane and butane into the crude oil before dispatch by tanker or pipeline. This results in a higher vapor pressure of the crude; however, it allows increasing the quantity of light products obtained at the refinery. Since light ends in most petroleum markets command a premium, while in the oil field itself propane and butane may have to be reinjected or flared, the practice of *spiking* crude oil with liquefied petroleum gas is becoming fairly common.

In addition to the gases obtained by distillation of petroleum, more highly volatile products result from the subsequent processing of naphtha and middle distillate to produce gasoline. Hydrogen sulfide is produced in the desulfurization processes involving hydrogen treatment of naphtha, distillate, and residual fuel, and from the coking or similar thermal treatments of vacuum gas oils and residua (Speight and Ozum, 2002; Parkash, 2003; Hsu and Robinson, 2006; Gary et al., 2007; Speight, 2014). The most common processing step in the production of gasoline blend stock is the catalytic reforming of hydrocarbon fractions in the heptane (C_7) to decane (C_{10}) range.

Additional gases are produced in *thermal cracking processes*, such as the coking or visbreaking processes for the processing of heavy feedstocks (Speight and Ozum, 2002; Parkash, 2003; Hsu and Robinson, 2006; Gary et al., 2007; Speight, 2014). In the visbreaking process, fuel oil is passed through externally fired tubes and undergoes liquid-phase cracking reactions, which results in the formation of lighter fuel oil components. Oil viscosity is thereby reduced, and some gases, mainly hydrogen, methane, and ethane, are formed. Substantial quantities of both gas and carbon are also formed in coking (both fluid coking and delayed coking), in addition to the middle distillate and naphtha. When coking a residual fuel oil or heavy gas oil, the feedstock is preheated and contacted with hot carbon (coke), which causes extensive cracking of the feedstock constituents of higher molecular weight to produce lower molecular weight products ranging from methane, via liquefied petroleum gas(es) and naphtha, to gas oil and heating oil. Products from coking processes tend to be unsaturated, and olefin components predominate in the tail gases from coking processes.

Another group of refining operations that contributes to gas production is that of the *catalytic cracking processes* (Speight and Ozum, 2002; Parkash, 2003; Hsu and Robinson, 2006; Gary et al., 2007; Speight, 2014). These consist of fluid-bed catalytic cracking, and there are many process variants in which heavy feedstocks are converted into cracked gas, liquefied petroleum gas, catalytic naphtha, fuel oil, and coke by contacting the heavy hydrocarbon with the hot catalyst. Both catalytic and thermal cracking processes, the latter being now largely used for the production of chemical raw materials, result in the formation of unsaturated hydrocarbons, particularly ethylene ($CH_2=CH_2$), but also propylene (propene, $CH_3CH=CH_2$), isobutylene [isobutene, $(CH_3)_2C=CH_2$], and the *n*-butenes ($CH_3CH_2CH=CH_2$ and $CH_3CH=CHCH_3$), in addition to hydrogen (H_2), methane (CH_4), and smaller quantities of ethane (CH_3CH_3), propane ($CH_3CH_2CH_3$), and butanes [$CH_3CH_2CH_2CH_3$, $(CH_3)_3CH$]. Diolefins such as butadiene ($CH_2=CHCH=CH_2$) are also present.

A further source of refinery gas is *hydrocracking*, a catalytic high-pressure pyrolysis process in the presence of fresh and recycled hydrogen (Speight and Ozum, 2002; Parkash, 2003; Hsu and Robinson, 2006; Gary et al., 2007; Speight, 2014). The feedstock is again heavy gas oil or residual fuel oil, and the process is mainly directed at the production of additional naphtha and middle distillates. Since hydrogen is to be recycled, the gases produced in this process again have to be separated into lighter and heavier streams; any surplus recycle gas and the liquefied petroleum gas from the hydrocracking process are both saturated.

In a series of *reforming processes* (Speight and Ozum, 2002; Parkash, 2003; Hsu and Robinson, 2006; Gary et al., 2007; Speight, 2014), commercialized under names such as *platforming*, paraffin, and naphthene (cyclic nonaromatic), hydrocarbons are converted in the presence of hydrogen and a catalyst into aromatics, or isomerized to more highly branched hydrocarbons. Catalytic reforming processes thus not only result in the formation of a liquid product of higher octane number, but also produce substantial quantities of gases. The latter are rich in hydrogen, but also contain hydrocarbons from methane to butanes, with a preponderance of propane ($CH_3CH_2CH_3$), n-butane ($CH_3CH_2CH_2CH_3$), and isobutane [(CH_3)$_3CH$].

The composition of the process gas varies in accordance with reforming severity and reformer feedstock. All catalytic reforming processes require substantial recycling of a hydrogen stream. Therefore, it is normal to separate reformer gas into a propane ($CH_3CH_2CH_3$) and/or a butane stream [$CH_3CH_2CH_2CH_3$ plus (CH_3)$_3CH$], which becomes part of the refinery liquefied petroleum gas production, and a lighter gas fraction, part of which is recycled. In view of the excess of hydrogen in the gas, all products of catalytic reforming are saturated, and there are usually no olefin gases present in either gas stream.

Both hydrocracker gases and catalytic reformer gases are commonly used in catalytic desulfurization processes. In the latter, feedstocks ranging from light to vacuum gas oils are passed at pressures of 500–1000 psi with hydrogen over a hydrofining catalyst. This results mainly in the conversion of organic sulfur compounds to hydrogen sulfide

$$[S]_{feedstock} + H_2 \rightarrow H_2S + \text{hydrocarbons}$$

This process also produces some light hydrocarbons by hydrocracking.

Thus, refinery gas streams, while ostensibly being hydrocarbon in nature, may contain large amounts of acid gases such as hydrogen sulfide and carbon dioxide. Most commercial plants employ hydrogenation to convert organic sulfur compounds into hydrogen sulfide. Hydrogenation is effected by means of recycled hydrogen-containing gases or external hydrogen over a nickel molybdate or cobalt molybdate catalyst.

The presence of impurities in gas streams may eliminate some of the sweetening processes, since some processes remove large amounts of acid gas but not to a sufficiently low concentration. On the other hand, there are those processes not designed to remove (or incapable of removing) large amounts of acid gases but are capable of removing the acid gas impurities to very low levels when the acid gases are present only in low to medium concentration in the gas.

The processes that have been developed to accomplish gas purification vary from a simple once-through wash operation to complex multistep recycling systems. In many cases, the process complexities arise because of the need for recovery of the materials used to remove the contaminants or even recovery of the contaminants in the original, or altered, form (Katz, 1959; Kohl and Riesenfeld, 1985; Newman, 1985).

From an environmental viewpoint, it is not the means by which these gases can be utilized but it is the effects of these gases on the environment when they are introduced into the atmosphere.

In addition to the corrosion of equipment of acid gases, the escape into the atmosphere of sulfur-containing gases can eventually lead to the formation of the constituents of acid rain, i.e., the oxides of sulfur (SO_2 and SO_3). Similarly, the nitrogen-containing gases can also lead to nitrous and nitric acids (through the formation of the oxides NO_x, where $x = 1$ or 2), which are the other major contributors to acid rain. The release of carbon dioxide and hydrocarbons as constituents of refinery effluents can also influence the behavior and integrity of the ozone layer.

Finally, another acid gas, hydrogen chloride (HCl), although not usually considered to be a major emission, is produced from mineral matter and the brine that often accompany petroleum during production, and is gaining increasing recognition as a contributor to acid rain. However, hydrogen chloride may exert severe local effects because it does not need to participate in any further chemical reaction to become an acid. Under atmospheric conditions that favor a buildup of stack emissions in the areas where hydrogen chloride is produced, the amount of hydrochloric acid in rainwater could be quite high.

In summary, refinery process gas, in addition to hydrocarbons, may contain other contaminants, such as carbon oxides (CO_x, where $x = 1$ and/or 2), sulfur oxides (SO_x, where $x = 2$ and/or 3), as well as ammonia (NH_3), mercaptans (R-SH), and carbonyl sulfide (COS). From an environmental viewpoint, petroleum processing can result in a variety of gaseous emissions. It is a question of degree insofar as the composition of the gaseous emissions may vary from process to process but the constituents are, in the majority of cases, the same.

11.2.2 Gas Streams from Natural Gas

Natural gas is also capable of producing emissions that are detrimental to the environment. While the major constituent of natural gas is methane, there are components such as carbon dioxide (CO_2), hydrogen sulfide (H_2S), and mercaptans (thiols; R-SH), as well as trace amounts of sundry other emissions such as carbonyl sulfide (COS). The fact that methane has a foreseen and valuable end-use makes it a desirable product; however, in several other situations, it is considered a pollutant, having been identified as a greenhouse gas.

A sulfur removal process must be very precise, since natural gas contains only a small quantity of sulfur-containing compounds that must be reduced several orders of magnitude. Most consumers of natural gas require <4 ppm in the gas.

A characteristic feature of natural gas that contains hydrogen sulfide is the presence of carbon dioxide (generally in the range of 1–4% by volume). In cases where the natural gas does not contain hydrogen sulfide, there may also be a relative lack of carbon dioxide.

In practice, heaters and scrubbers are installed, usually at or near the wellhead. The scrubbers serve primarily to remove sand and other large-particle impurities, and the heaters ensure that the temperature of the gas does not drop too low. With natural gas that contains even low quantities of water, natural gas hydrates have a tendency to form when temperatures drop. These hydrates are solid or semisolid compounds, resembling ice-like crystals. If the hydrates accumulate, they can impede the passage of natural gas through valves and gathering systems (Zhang et al., 2007). To reduce the occurrence of hydrates, small natural gas–fired heating units are typically installed along the gathering pipe wherever it is likely that hydrates may form.

Natural gas hydrates are usually considered as possible nuisances in the development of oil and gas fields, mainly in deep-water drilling operations and if multiphase production and transportation technologies are to be examined. On the other hand, hydrates can be used for the safe and economic storage of natural gas, mainly in cold countries. In remote offshore areas, the use of hydrates for natural gas transportation is also presently considered as an economic alternative to the processes based either on liquefaction or on compression (Lachet and Béhar, 2000).

11.3 WATER REMOVAL

Water is a common impurity in gas streams, and removal of water is necessary to prevent condensation of the water and the formation of ice or gas hydrates ($C_nH_{2n+2} \cdot xH_2O$). Water in the liquid phase causes corrosion or erosion problems in pipelines and equipment, particularly when carbon dioxide and hydrogen sulfide are present in the gas. The simplest method of water removal (refrigeration or cryogenic separation) is to cool the gas to a temperature at least equal to or (preferentially) below the dew point (Mokhatab et al., 2006; Speight, 2007, 2014).

In addition to separating petroleum and some condensate from the wet gas stream, it is necessary to remove most of the associated water. Most of the liquid, free water associated with extracted natural gas is removed by simple separation methods at or near the wellhead. However, the removal of the water vapor that exists in solution in natural gas requires a more complex treatment. This treatment consists of "dehydrating" the natural gas, which usually involves one of two processes: either absorption or adsorption.

Absorption occurs when the water vapor is taken out by a dehydrating agent. Adsorption occurs when the water vapor is condensed and collected on the surface.

In a majority of cases, cooling alone is insufficient and, for the most part, impractical for use in field operations. Other more convenient water removal options use (1) hygroscopic liquids (e.g., diethylene glycol or triethylene glycol) and (2) solid adsorbents or desiccants (e.g., alumina, silica gel, and molecular sieves). Ethylene glycol can be directly injected into the gas stream in refrigeration plants.

11.3.1 ABSORPTION

An example of absorption dehydration is known as *glycol dehydration*, and diethylene glycol, the principal agent in this process, has a chemical affinity for water and removes water from the gas stream. In this process, a liquid desiccant dehydrator serves to absorb water vapor from the gas stream. Essentially, glycol dehydration involves using a glycol solution, usually either diethylene glycol or triethylene glycol, which is brought into contact with the wet gas stream in a contactor. The glycol solution will absorb water from the wet gas and, once absorbed, the glycol particles become heavier and sink to the bottom of the contactor where they are removed. The natural gas, having been stripped of most of its water content, is then transported out of the dehydrator. The glycol solution, bearing all of the water stripped from the natural gas, is put through a specialized boiler designed to vaporize only the water out of the solution. The boiling point differential between water (100°C, 212°F) and glycol (204°C, 400°F) makes it relatively easy to remove water from the glycol solution, allowing it be reused in the dehydration process.

As well as absorbing water from the wet gas stream, the glycol solution occasionally carries with it small amounts of methane and other compounds found in the wet gas. In the past, this methane was simply vented out of the boiler. In addition to losing a portion of the natural gas that was extracted, this venting contributes to air pollution and the greenhouse effect. To decrease the amount of methane and other compounds that are lost, flash tank separator–condensers work to remove these compounds before the glycol solution reaches the boiler. Essentially, a flash tank separator consists of a device that reduces the pressure of the glycol solution stream, allowing the methane and other hydrocarbons to vaporize (*flash*). The glycol solution then travels to the boiler, which may also be fitted with air- or water-cooled condensers, which serve to capture any remaining organic compounds that may remain in the glycol solution. The regeneration (stripping) of the glycol is limited by temperature: diethylene glycol and triethylene glycol decompose at or before their respective boiling points. Such techniques as stripping of hot triethylene glycol with dry gas (e.g., heavy hydrocarbon vapors, the *Drizo process*) or vacuum distillation are recommended.

In practice, absorption systems recover 90–99% by volume of methane that would otherwise be flared into the atmosphere.

The *Rectisol process* is a gas purification process for removing carbon dioxide and hydrogen sulfide/carbonyl sulfide from gas streams. The process uses methanol as a wash solvent, and the wash unit operates under favorable at temperatures below 0°C (32°F). To lower the temperature of the feed gas, it is cooled against the cold-product streams, before entering the absorber tower. At the absorber tower, carbon dioxide and hydrogen sulfide (with carbonyl sulfide) are removed. By use of an intermediate flash, coabsorbed products such as hydrogen and carbon monoxide are recovered, thus increasing the product recovery rate. To reduce the required energy demand for the carbon dioxide compressor, the carbon dioxide product is recovered in two different pressure steps (medium pressure and lower pressure). The carbon dioxide product is essentially sulfur free (H_2S-free, COS-free) and water free. The carbon dioxide products can be used for enhanced oil recovery and/or sequestration or as pure carbon dioxide for other processes.

11.3.2 ADSORPTION

Solid-adsorbent or solid-desiccant dehydration is the primary form of dehydrating natural gas using adsorption, and usually consists of two or more adsorption towers that are filled with a solid desiccant. Typical desiccants include activated alumina or a granular silica gel material. Wet natural gas is passed through these towers, from top to bottom. As the wet gas passes around the particles of desiccant material, water is retained on the surface of these desiccant particles. Passing through the entire desiccant bed, almost all of the water is adsorbed onto the desiccant material, leaving the dry gas to exit the bottom of the tower.

There are several solid desiccants that possess the physical characteristic to adsorb water from natural gas. These desiccants are generally used in dehydration systems consisting of two or more towers and associated regeneration equipment.

Molecular sieves—a class of aluminosilicates that produce the lowest water dew points, and that can be used to simultaneously sweeten dry gases and liquids (Maple and Williams, 2008)—are commonly used in dehydrators ahead of plants designed to recover ethane and other NGLs. These plants operate at very cold temperatures and require very dry feed gas to prevent formation of hydrates. Dehydration to $-100°C$ ($-148°F$) dew point is possible with molecular sieves. Water dew points less than $-100°C$ ($-148°F$) can be accomplished with special design and definitive operating parameters (Mokhatab et al., 2006).

Solid-adsorbent dehydrators are typically more effective than glycol dehydrators, and are usually installed as a type of straddle system along natural gas pipelines. These types of dehydration systems are best suited for large volumes of gas under very high pressure, and are thus usually located on a pipeline downstream of a compressor station. Two or more towers are required because after a certain period of use, the desiccant in a particular tower becomes saturated with water. To "regenerate" the desiccant, a high-temperature heater is used to heat gas to a very high temperature. Passing this heated gas through a saturated desiccant bed vaporizes the water in the desiccant tower, leaving it dry and allowing for further natural gas dehydration.

Although two-bed adsorbent treaters have become more common (while one bed is removing water from the gas, the other undergoes alternate heating and cooling), on occasion, a three-bed system is used: one bed adsorbs, one is being heated, and one is being cooled. An additional advantage of the three-bed system is the facile conversion of a two-bed system so that the third bed can be maintained or replaced, thereby ensuring continuity of the operations and reducing the risk of a costly plant shutdown.

Silica gel (SiO_2) and alumina (Al_2O_3) have good capacities for water adsorption (up to 8% by weight). Bauxite (crude alumina, Al_2O_3) adsorbs up to 6% by weight water, and molecular sieves adsorb up to 15% by weight water. Silica is usually selected for dehydration of sour gas because of its high tolerance to hydrogen sulfide and to protect molecular sieve beds from plugging by sulfur. Alumina guard beds (which serve as protectors by the act of attrition and may be referred to as an *attrition catalyst*) (Speight, 2000) may be placed ahead of the molecular sieves to remove the sulfur compounds. Downflow reactors are commonly used for adsorption processes, with an upward flow regeneration of the adsorbent and cooling in the same direction as adsorption.

Solid-desiccant units generally cost more to buy and operate than glycol units. Therefore, their use is typically limited to applications such as gases having a high hydrogen sulfide content, very low water dew point requirements, simultaneous control of water, and hydrocarbon dew points. In processes where cryogenic temperatures are encountered, solid-desiccant dehydration is usually preferred over conventional methanol injection to prevent hydrate and ice formation (Kindlay and Parrish, 2006).

11.3.3 USE OF MEMBRANES

Membrane separation processes are very versatile and are designed to process a wide range of feedstocks and offer a simple solution for the removal and recovery of higher-boiling hydrocarbons

(NGLs) from natural gas (Foglietta, 2004; Mokhatab et al., 2006; Speight, 2007, 2014). The separation process is based on high-flux membranes that selectively permeate higher-boiling hydrocarbons (compared to methane) and are recovered as a liquid after recompression and condensation. The residue stream from the membrane is partially depleted of higher-boiling hydrocarbons, and is then sent to sales gas stream. Gas permeation membranes are usually made of vitreous polymers that exhibit good selectivity; however, to be effective, the membrane must be very permeable with respect to the separation process.

11.4 LIQUID REMOVAL

Natural gas coming directly from a well contains many NGLs that are commonly removed. In most instances, NGLs have a higher value as separate products, and it is thus economical to remove them from the gas stream. The removal of NGLs usually takes place in a relatively centralized processing plant, and uses techniques similar to those used to dehydrate natural gas. Recovery of the liquid hydrocarbons can be justified either because it is necessary to make the gas salable or because economics dictate this course of action. The justification for building a liquid recovery (or a liquid removal) plant depends on the price differential between the enriched gas (containing the higher molecular weight hydrocarbons) and lean gas with the added value of the extracted liquid.

There are two basic steps to the treatment of NGLs in the natural gas stream. First, the liquids must be extracted from the natural gas. Second, these NGLs must be separated themselves, down to their base components. These two processes account for approximately 90% of the total production of NGLs.

11.4.1 Extraction

There are two principal techniques for removing NGLs from the natural gas stream: the absorption method and the cryogenic expander process (Mokhatab et al., 2006; Speight, 2007, 2014).

In the process, a turboexpander is used to produce the necessary refrigeration and very low temperatures, and high recovery of light components, such as ethane and propane, can be attained. The natural gas is first dehydrated using a molecular sieve followed by cooling. The separated liquid containing most of the heavy fractions is then demethanized, and the cold gases are expanded through a turbine that produces the desired cooling for the process. The expander outlet is a two-phase stream that is fed to the top of the demethanizer column. This serves as a separator in which (1) the liquid is used as the column reflux and the separator vapors combined with vapors stripped in the demethanizer are exchanged with the feed gas, and (2) the heated gas, which is partially recompressed by the expander compressor, is further recompressed to the desired distribution pressure in a separate compressor. This process allows for the recovery of about 90–95% by volume of the ethane originally in the gas stream. In addition, the expansion turbine is able to convert some of the energy released when the natural gas stream is expanded into recompressing the gaseous methane effluent, thus saving energy costs associated with extracting ethane.

The extraction of NGLs from the natural gas stream produces both cleaner, purer natural gas, as well as the valuable hydrocarbons that are the NGLs themselves.

11.4.2 Absorption

The absorption method of extraction is very similar to using absorption for dehydration. The main difference is that, in the absorption of NGLs, absorbing oil is used as opposed to glycol. This absorbing oil has an affinity for NGLs in much the same manner as glycol has an affinity for water. Before the oil has picked up any NGLs, it is termed *lean* absorption oil.

The *oil absorption process* involves the countercurrent contact of the lean (or stripped) oil with the incoming wet gas with the temperature and pressure conditions programmed to maximize the dissolution of the liquefiable components in the oil. The *rich* absorption oil (sometimes referred to as *fat* oil), containing NGLs, exits the absorption tower through the bottom. It is now a mixture of absorption oil, propane, butanes, pentanes, and other higher-boiling hydrocarbons. The rich oil is fed into lean oil stills, where the mixture is heated to a temperature above the boiling point of the NGLs but below that of the oil. This process allows for the recovery of around 75% by volume of the butanes, and 85–90% by volume of the pentanes and higher-boiling constituents from the natural gas stream.

The basic absorption process above can be modified to improve its effectiveness, or to target the extraction of specific NGLs. In the refrigerated oil absorption method, where the lean oil is cooled through refrigeration, propane recovery can be upward of 90% by volume and approximately 40% by volume of the ethane can be extracted from the natural gas stream. Extraction of the other, higher-boiling NGLs can be close to 100% by volume using this process.

11.4.3 Fractionation of Natural Gas Liquids

After NGLs have been removed from the natural gas stream, they must be broken down into their base components to be useful. That is, the mixed stream of different NGLs must be separated. The process used to accomplish this task is called fractionation. Fractionation works based on the different boiling points of the different hydrocarbons in the NGLs stream. Essentially, fractionation occurs in stages consisting of the boiling off of hydrocarbons one by one. The name of a particular fractionator gives an idea about its purpose, as it is conventionally named for the hydrocarbon that is boiled off. The entire fractionation process is broken down into steps, starting with the removal of the lighter NGLs from the stream. The particular fractionators are used in the following order: (1) deethanizer that separates the ethane from the stream of NGLs; (2) depropanizer that separates the propane from the deethanized stream; (3) debutanizer that separates the butanes, leaving the pentanes and higher-boiling hydrocarbons in the stream; and (4) the butane splitter or de-isobutanizer that separates the isobutane and *n*-butane.

11.5 NITROGEN REMOVAL

Nitrogen may often occur in sufficient quantities in natural gas and, consequently, lower the heating value of the gas. Thus, several plants for nitrogen removal from natural gas have been built; however, it must be recognized that nitrogen removal requires liquefaction and fractionation of the entire gas stream, which may affect process economics. In many cases, the nitrogen-containing natural gas is blended with a gas having a higher heating value and sold at a reduced price depending on the thermal value (Btu/ft^3).

11.6 ACID GAS REMOVAL

In addition to water and NGLs removal, one of the most important parts of gas processing involves the removal of hydrogen sulfide and carbon dioxide. Natural gas from some wells contains significant amounts of hydrogen sulfide and carbon dioxide and is usually referred to as *sour gas*. Sour gas is undesirable because the sulfur compounds it contains can be extremely harmful, even lethal, to breathe, and the gas can also be extremely corrosive. The process for removing hydrogen sulfide from sour gas is commonly referred to as *sweetening* of the gas.

The primary process for sweetening sour natural gas (desulfurizing natural gas) is quite similar to the processes of glycol dehydration and removal of NGLs by absorption (Figure 11.1) (Mokhatab et al., 2006; Speight, 2007, 2014). In this case, however, amine (olamine) solutions are used to remove the hydrogen sulfide (the amine process). The sour gas is run through a tower that contains the olamine solution. There are two principal amine solutions used: monoethanolamine and diethanolamine. Either of these compounds, in liquid form, will absorb sulfur compounds from natural

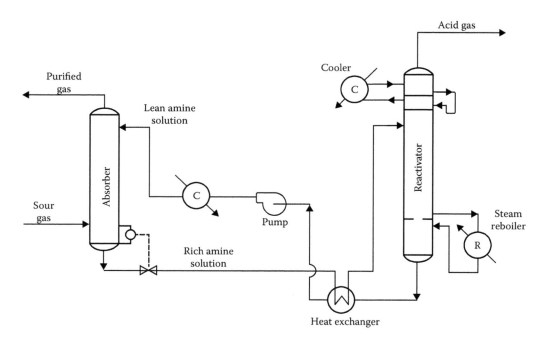

FIGURE 11.1 Amine or olamine process. (From Speight, J.G. 2007. *The Chemistry and Technology of Petroleum*. 4th Edition. CRC Press, Taylor & Francis Publishers, Boca Raton, FL. Figure 11.6, p. 725.)

gas as it passes through. The effluent gas is virtually free of sulfur compounds, and thus loses its sour gas status. Like the process for the extraction of NGLs and glycol dehydration, the amine solution used can be regenerated for reuse. Although most sour gas sweetening processes involve amine absorption, it is also possible to use solid desiccants like iron sponge (q.v.) to remove hydrogen sulfide and carbon dioxide.

Treatment of gas to remove the acid gas constituents (hydrogen sulfide and carbon dioxide) is most often accomplished by contact of the natural gas with an alkaline solution. The most commonly used treating solutions are aqueous solutions of ethanolamine (Table 11.3) or alkali carbonates, although a considerable number of other treating agents have been developed in recent years (Mokhatab et al., 2006; Speight, 2007, 2014). Most of these newer treating agents rely on physical absorption and chemical reaction. When only carbon dioxide is to be removed in large quantities or when only partial removal is necessary, a hot carbonate solution or one of the physical solvents is the most economical selection. The most well-known hydrogen sulfide removal process is based on the reaction of hydrogen sulfide with iron oxide (often also called the iron sponge process or the dry box method), in which the gas is passed through a bed of wood chips impregnated with iron oxide.

The iron oxide process is the oldest and still the most widely used batch process for sweetening natural gas and NGLs (Zapffe, 1963; Duckworth and Geddes, 1965; Anerousis and Whitman, 1984; Mokhatab et al., 2006; Speight, 2007, 2014). The process was implemented during the 19th century. In the process (Figure 11.2), the sour gas is passed down through the bed. In the case where continuous regeneration is to be utilized, a small concentration of air is added to the sour gas before it is processed. This air serves to continuously regenerate the iron oxide that has reacted with hydrogen sulfide, which serves to extend the on-stream life of a given tower but probably serves to decrease the total amount of sulfur that a given weight of bed will remove.

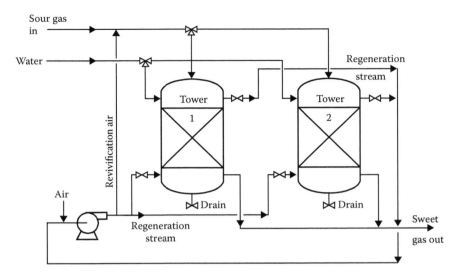

FIGURE 11.2 Iron oxide process. (From Speight, J.G. 2007. *The Chemistry and Technology of Petroleum.* 4th Edition. CRC Press, Taylor & Francis Publishers, Boca Raton, FL. Figure 25.7, p. 726.)

The process is usually best applied to gases containing low to medium concentrations (300 ppm) of hydrogen sulfide or mercaptans. This process tends to be highly selective and does not normally remove significant quantities of carbon dioxide. As a result, the hydrogen sulfide stream from the process is usually of high purity. The use of the iron sponge process for sweetening sour gas is based on the adsorption of the acid gases on the surface of the solid sweetening agent followed by the chemical reaction of ferric oxide (Fe_2O_3) with hydrogen sulfide:

$$2Fe_2O_3 + 6H_2S \rightarrow 2Fe_2S_3 + 6H_2O$$

The reaction requires the presence of slightly alkaline water and a temperature below 43°C (110°F), and bed alkalinity (pH + 8 to 10) should be checked regularly, usually on a daily basis. The pH level is maintained through the injection of caustic soda with the water. If the gas does not contain sufficient water vapor, water may need to be injected into the inlet gas stream.

The ferric sulfide produced by the reaction of hydrogen sulfide with ferric oxide can be oxidized with air to produce sulfur and regenerate the ferric oxide:

$$2Fe_2S_3 + 3O_2 \rightarrow 2Fe_2O_3 + 6S$$

$$S_2 + 2O_2 \rightarrow 2SO_2$$

The regeneration step is exothermic, and air must be introduced slowly so the heat of reaction can be dissipated. If air is introduced quickly, the heat of reaction may ignite the bed. Some of the elemental sulfur produced in the regeneration step remains in the bed. After several cycles, this sulfur will cake over the ferric oxide, decreasing the reactivity of the bed. Typically, after 10 cycles, the bed must be removed and a new bed introduced into the vessel.

The iron oxide process is one of several metal oxide-based processes that scavenge hydrogen sulfide and organic sulfur compounds (mercaptans) from gas streams through reactions with the solid-based chemical adsorbent (Kohl and Riesenfeld, 1985). They are typically nonregenerable, although some are partially regenerable, losing activity upon each regeneration cycle. Most of the processes are governed by the reaction of a metal oxide with hydrogen sulfide to form the metal sulfide. For regeneration, the metal oxide is reacted with oxygen to produce elemental sulfur and the regenerated metal oxide. In addition to iron oxide, the primary metal oxide used for dry sorption processes is zinc oxide.

In the zinc oxide process, the zinc oxide media particles are extruded cylinders 3–4 mm in diameter and 4–8 mm in length (Kohl and Nielsen, 1997) and react readily with the hydrogen sulfide:

$$ZnO + H_2S \rightarrow ZnS + H_2O$$

At increased temperatures (205–370°C, 400–700°F), zinc oxide has a rapid reaction rate, therefore providing a short mass transfer zone, resulting in a short length of unused bed and improved efficiency.

Removal of larger amounts of hydrogen sulfide from gas streams requires a continuous process, such as the *Ferrox process* or the *Stretford process*. The Ferrox process is based on the same chemistry as the iron oxide process except that it is fluid and continuous. The Stretford process employs a solution containing vanadium salts and anthraquinone disulfonic acid (Maddox, 1974). Most hydrogen sulfide removal processes return the hydrogen sulfide unchanged; however, if the quantity involved does not justify installation of a sulfur recovery plant (usually a Claus plant), it is necessary to select a process that directly produces elemental sulfur.

The processes using ethanolamine and potassium phosphate are now widely used. The ethanolamine process, known as the *Girbotol process*, removes acid gases (hydrogen sulfide and carbon dioxide) from liquid hydrocarbons as well as from natural and from refinery gases. The Girbotol process uses an aqueous solution of ethanolamine ($H_2NCH_2CH_2OH$) that reacts with hydrogen sulfide at low temperatures and releases hydrogen sulfide at high temperatures. The ethanolamine solution fills a tower called an absorber through which the sour gas is bubbled. Purified gas leaves the top of the tower, and the ethanolamine solution leaves the bottom of the tower with the absorbed acid gases. The ethanolamine solution enters a reactivator tower where heat drives the acid gases from the solution. Ethanolamine solution, restored to its original condition, leaves the bottom of the reactivator tower to go to the top of the absorber tower, and acid gases are released from the top of the reactivator.

The process using potassium phosphate is known as phosphate desulfurization, and it is used in the same way as the Girbotol process to remove acid gases from liquid hydrocarbons as well as from gas streams. The treatment solution is a water solution of tripotassium phosphate (K_3PO_4), which is circulated through an absorber tower and a reactivator tower in much the same way as the ethanolamine is circulated in the Girbotol process; the solution is regenerated thermally.

Moisture may be removed from hydrocarbon gases at the same time as hydrogen sulfide is removed. Moisture removal is necessary to prevent harm to anhydrous catalysts and to prevent the formation of hydrocarbon hydrates (e.g., $C_3H_8 \cdot 18H_2O$) at low temperatures. A widely used dehydration and desulfurization process is the glycolamine process, in which the treatment solution is a mixture of ethanolamine and a large amount of glycol. The mixture is circulated through an absorber and a reactivator in the same way as ethanolamine is circulated in the Girbotol process. The glycol absorbs moisture from the hydrocarbon gas passing up the absorber; the ethanolamine absorbs hydrogen sulfide and carbon dioxide. The treated gas leaves the top of the absorber; the spent ethanolamine–glycol mixture enters the reactivator tower, where heat drives off the absorbed acid gases and water.

Other processes include the *Alkazid process* for removal of hydrogen sulfide and carbon dioxide using concentrated aqueous solutions of amino acids. The hot potassium carbonate process decreases the acid content of natural and refinery gas from as much as 50% to as low as 0.5%, and operates in a unit similar to that used for amine treating. The *Giammarco–Vetrocoke process* is used for hydrogen sulfide and/or carbon dioxide removal. In the hydrogen sulfide removal section, the reagent consists of sodium or potassium carbonates containing a mixture of arsenites and arsenates; the carbon dioxide removal section utilizes hot aqueous alkali carbonate solution activated by arsenic trioxide or selenous acid or tellurous acid.

Molecular sieves are highly selective for the removal of hydrogen sulfide (as well as other sulfur compounds) from gas streams and offer continuously high absorption efficiency. They are also an effective means of water removal and thus offer a process for the simultaneous dehydration and desulfurization of gas. Gas that has excessively high water content may require upstream dehydration, however (Mokhatab et al., 2006; Speight, 2007).

In the *Lo–Cat process*, hydrogen sulfide is converted to elemental sulfur by an environmentally safe, chelated iron catalyst:

$$2H_2S + O_2–Fe\ catalyst \rightarrow H_2O + S$$

The chelated iron is capable of oxidizing the sulfide ions to elemental sulfur by means of a redox reaction in which the iron is reduced from the ferric state to the ferrous state and then is regenerated back to the ferric state by reacting with oxygen supplied from atmospheric air. In the process, the primary chemical consumptions are replacement of chelated iron lost in the sulfur removal process, replacement of chelating agents that oxidize over time, and a small caustic addition required to maintain the pH of the operating solution in the mildly alkaline range.

If liquid entrainment in the sour gas is anticipated, the first two unit operations are a knockout pot and a coalescing filter. The sour gas is routed to the static mixer absorbers, where it contacts the oxidized catalyst solution. A two-phase stream consisting of sweet gas and reduced solution exits the static mixer absorber, and then enters the absorber separator, where the gas and liquid separate. The treated gas exits the unit after passing through a mist eliminator, after which the gas is cooled downstream of the unit to remove water, then compressed to 950–1000 psi in the third-stage compressor.

The reduced LO-CAT solution from the separator passes through a pressure-reducing valve and flash drum, then gravity drains to the oxidizer where the iron is regenerated for reuse in the absorber. A slipstream of this oxidized solution is then sent to the settler where sulfur is allowed to settle and is subsequently transferred as slurry to the belt filter. The filtrate from the filter is returned to the oxidizer, while the sulfur cake is discharged after washing. A small quantity of flash gas is vented to the flare header.

The *molecular sieve process* is similar to the iron oxide process. Regeneration of the bed is achieved by passing heated clean gas over the bed. As the temperature of the bed increases, it releases the adsorbed hydrogen sulfide into the regeneration gas stream. The sour effluent regeneration gas is sent to a flare stack, and up to 2% v/v of the gas seated can be lost in the regeneration process. A portion of the natural gas may also be lost by the adsorption of hydrocarbon components by the sieve (Mokhatab et al., 2006; Speight, 2007).

In this process, unsaturated hydrocarbon components, such as olefins and aromatics, tend to be strongly adsorbed by the molecular sieve. Molecular sieves are susceptible to poisoning by such chemicals as glycols and require thorough gas cleaning methods before the adsorption step. Alternatively, the sieve can be offered some degree of protection by the use of guard beds in which a less expensive catalyst is placed in the gas stream before contact of the gas with the sieve, thereby protecting the catalyst from poisoning. This concept is analogous to the use of guard beds or attrition catalysts in the petroleum industry (Speight, 2000).

Another option for hydrogen sulfide removal is to use hydrogen peroxide, which can control hydrogen sulfide and other sulfides in two ways, depending on the application: (1) destruction, in which is converted to the sulfide to elemental sulfur or sulfate ion; and (2) prevention, in which dissolved oxygen inhibits the septic conditions that lead to biological sulfide formation.

Under neutral or slightly acid conditions, the reaction proceeds according to

$$2S + H_2O_2 \rightarrow S + 2H_2O$$

The product of the oxidation is predominately elemental sulfur. On the other hand, under alkaline conditions, the reaction proceeds according to

$$S^{2-} + 4H_2O_2 \rightarrow SO_4^{2-} + 4H_2O$$

The above reaction predominates at pH > 9.2 and yields soluble sulfate as the reaction product.

Finally, in moving from pH 7 to 9, both of the above reactions may occur with the following results: (1) the reaction products transition from elemental sulfur to sulfate; (2) the hydrogen peroxide requirement increases from 1:1 to 4.25:1; and (3) the rate of the reaction is increased. Catalysts, such as iron, favor sulfate formation and may be used to economize hydrogen peroxide use or to produce a clear effluent. In both cases, the speed of reaction is greatly accelerated.

The SNOX process (developed by Haldor Topsøe) is a process designed for the removal of sulfur dioxide, nitrogen oxides, and particulates from flue gases. The sulfur is recovered as concentrated sulfuric acid, and the nitrogen oxides are reduced to free nitrogen. The process is based on catalytic reactions and does not consume water or absorbents and includes the following steps: (1) dust removal, (2) catalytic reduction of NO_x by adding ammonia (NH_3) to the gas upstream the DeNO$_x$ reactor, (3) catalytic oxidation of sulfur dioxide (SO_2) to sulfur trioxide (SO_3) in the oxidation reactor, and (4) cooling of the gas to about approximately 100°C (212°F) whereby the sulfuric acid condenses and can be withdrawn as concentrated sulfuric acid product.

The SNOX technology is especially suitable for cleaning flue gases from combustion of high-sulfur fuels in refineries, and the process represents an energy-efficient way to convert the nitrogen oxides (NO_x) in the flue gas into nitrogen (N_2) and the sulfur oxides (SO_x) into concentrated sulfuric acid of commercial quality without using any absorbents and without producing waste products or wastewater. Along with the flue gases, other sulfurous waste streams from a refinery can be treated, such as hydrogen sulfide, Claus tail gas, and elemental sulfur (Mokhatab et al., 2006; Speight, 2007, 2014).

Finally, several classes of chemicals effectively lessen the hazards associated with hydrogen sulfide. Selection of an appropriate chemical hydrogen sulfide scavenger should include (1) ease of handling and use, (2) efficiency of reaction and selectivity for hydrogen sulfide, and (3) reaction irreversibility. Selecting an unsuitable scavenger can result in adverse downstream impacts—if a metal salt is used in a fuel oil, it may result in the specification for the mineral content (reflected as mineral ash) limit being exceeded and require the finished fuel to be reprocessed.

Typically, natural gas is desulfurized in amine absorbers; however, where this is not possible, scavengers may be needed. Triazine derivatives have been widely used hydrogen sulfide scavengers (a triazine derivative is a reaction product of an amine and formaldehyde). Amines, such as methyl amine or monoethanolamine, are used to produce water-based triazine derivatives, whereas higher molecular weight amines, such as methoxypropylamine, are used to produce a water-free (oil-soluble) version. The primary reaction product is dithiazine—1 mol triazine will react with 2 mol hydrogen sulfide and liberate 2 mol amine. Triazine derivatives are generally more effective at higher pH ranges and, because triazine derivatives release amines that increase the pH of a system, the use of triazine derivatives in oil production can negatively affect the performance of scale inhibitors by decreasing the solubility of calcium carbonate.

The production of heavier high-sulfur sour crude oils results in increased scavenger levels in the crude oil, which may increase negative downstream impacts. Triazine derivatives have been successfully used to reduce the concentration of hydrogen sulfide in crude oil; however, because of the adverse downstream effects, the use of non-amine scavengers can help avoid many of these problems (Kenreck, 2014). Chemical hydrogen sulfide scavengers have also been advocated for use at the wellhead (Bowers and Cash, 2014), thereby allowing (1) a reduced need for downstream hydrogen sulfide monitoring, and (2) reduced downstream hydrogen sulfide contamination and the reduced potential for fouling.

11.7 ENRICHMENT

The purpose of enrichment is to produce natural gas for sale and enriched tank oil. The tank oil contains more light hydrocarbon liquids than natural petroleum, and the residue gas is drier (leaner, i.e., has lesser amounts of the higher molecular weight hydrocarbons). Therefore, the process concept is essentially the separation of hydrocarbon liquids from the methane to produce a lean, dry gas.

Crude oil enrichment is used when there is no separate market for light hydrocarbon liquids or when the increase in API gravity of the crude provides a substantial increase in the price per unit volume as well as volume of the stock tank oil. A very convenient method of enrichment involves manipulation of the number and operating pressures of the gas–oil separators (traps). However, it must be recognized that alteration or manipulation of the separator pressure affects the gas compression operation as well as influences other processing steps.

One method of removing light ends involves the use of a pressure reduction (vacuum) system. Generally, stripping of light ends is achieved at low pressure, after which the pressure of the stripped crude oil is elevated so that the oil acts as an absorbent. The crude oil, which becomes enriched by this procedure, is then reduced to atmospheric pressure in stages or using fractionation (rectification).

11.8 FRACTIONATION

Fractionation processes are very similar to those processes classed as liquid removal processes, but often appear to be more specific in terms of the objectives, hence the need to place the fractionation processes into a separate category. The fractionation processes are those processes that are used (1) to remove the more significant product stream first, or (2) to remove any unwanted light ends from the heavier liquid products.

In the general practice of natural gas processing, the first unit is a deethanizer followed by a depropanizer then by a debutanizer and, finally, a butane fractionator. Thus, each column can operate at a successively lower pressure, thereby allowing the different gas streams to flow from column to column by virtue of the pressure gradient, without necessarily the use of pumps.

The purification of hydrocarbon gases by any of these processes is an important part of refinery operations, especially in regard to the production of liquefied petroleum gas. This is actually a mixture of propane and butane, which is an important domestic fuel, as well as an intermediate material in the manufacture of petrochemicals (Speight and Ozum, 2002; Parkash, 2003; Hsu and Robinson, 2006; Gary et al., 2007; Speight, 2014). The presence of ethane in liquefied petroleum gas must be avoided because of the inability of this lighter hydrocarbon to liquefy under pressure at ambient temperatures and its tendency to register abnormally high pressures in the liquefied petroleum gas containers. On the other hand, the presence of pentane in liquefied petroleum gas must also be avoided, since this particular hydrocarbon (a liquid at ambient temperatures and pressures) may separate into a liquid state in the gas lines.

11.9 CLAUS PROCESS

The Claus process is the most significant gas desulfurizing process, recovering elemental sulfur from gaseous hydrogen sulfide, a toxic gas that can originate from physical and chemical gas treatment units (Selexol, Rectisol, Purisol, and amine scrubbers) in refineries, natural gas processing plants, gasification plants, or synthesis gas plants. The multistep process recovers sulfur from the gaseous hydrogen sulfide found in raw natural gas and from the by-product gases containing hydrogen sulfide derived from refining crude oil and other industrial processes. These by-product gases may also contain hydrogen cyanide, hydrocarbons, sulfur dioxide, or ammonia.

More specifically, in refineries, hydrogen sulfide originates in crude oils and is also produced in the coking, catalytic cracking, hydrotreating, and hydrocracking processes, and is an issue with many refiners. Burning hydrogen sulfide as a fuel gas component or as a flare gas component is precluded by safety and environmental considerations since one of the combustion products is the highly toxic sulfur dioxide (SO_2). As described above, hydrogen sulfide is typically removed from the refinery light ends gas streams through an olamine process, after which application of heat regenerates the olamine and forms an acid gas stream. Thereafter, the acid gas stream is treated to convert the hydrogen sulfide elemental sulfur and water. The conversion process utilized in most modern refineries is the Claus process, or a variant thereof.

The removal of sulfur dioxide from flue gas is an essential part of desulfurization and various flue gas desulfurization technologies (Mokhatab et al., 2006; Speight, 2007; Li et al., 2010). Both dry and semidry flue gas desulfurization systems use jetting calcium powder as a sorbent to react with sulfur (Z.W. Sun et al., 2010). The wet flue gas desulfurization process has been widely commercialized to achieve sulfur dioxide removal rates in excess of 95%. However, many of the current technologies generate substantial amounts of wet solid waste, which requires extensive wastewater treatment (Maina and Mbarawa, 2012). Although sulfur dioxide is a precursor to acid rain, it is important and useful resource of sulfur fertilizer if it is reasonably captured and transformed (Z.W. Sun et al., 2010).

Absorption processes employing organic solvents are promising routes for flue gas purification because of the selectivity in the absorption allowed, and also because of the possible regeneration of both the absorption medium and the absorbed species under mild conditions and without producing unwanted side products (Hong et al., 2014). Furthermore, the choice of the appropriate organic solvents can lead to high selectivity for sulfur dioxide, while the selectivity to other gaseous components is zero or very low (Van Dam et al., 1997). However, currently available technologies are often energy-intensive processes and the production of by-products is a disadvantage (Huang et al., 2014).

N-formylmorpholine, which is readily miscible with water and of low partial pressure and toxicity, has been used as a physical solvent for recovery of high-purity aromatics and for the treatment of subquality natural gas (Lee and Rangaiah, 2009; Miller et al., 2011). The liquid-phase reaction after sulfur dioxide absorption in N-formylmorpholine is very rapid, possibly instantaneous (Nagel et al., 2002). On the other hand, carbon dioxide is barely soluble in N-formylmorpholine at low pressure (Jou et al., 1989), raising the potential for the wide use of this solvent as a means of flue gas desulfurization by the capture of sulfur dioxide where the carbon dioxide content is several orders of magnitude higher than that the sulfur dioxide content.

The most common sulfur dioxide utilization process is the Claus process (Figure 11.3), which involves combustion of approximately one-third of the hydrogen sulfide to sulfur dioxide and then reaction of the sulfur dioxide with the remaining hydrogen sulfide in the presence of a fixed bed of activated alumina and cobalt molybdenum catalyst, resulting in the formation of elemental sulfur:

$$2H_2S + 3O_2 \rightarrow 2SO_2 + 2H_2O$$

$$2H_2S + SO_2 \rightarrow 3S + 2H_2O$$

Different process flow configurations are in use to achieve the correct hydrogen sulfide/sulfur dioxide ratio in the conversion reactors.

In a split-flow configuration, one-third split of the acid gas stream is completely combusted, and the combustion products are then combined with the noncombusted acid gas upstream of the conversion reactors. In a once-through configuration, the acid gas stream is partially combusted by only providing sufficient oxygen in the combustion chamber to combust one-third of the acid gas. Two or three conversion reactors may be needed depending on the level of hydrogen sulfide conversion required. Each additional stage provides incrementally less conversion than the previous stage.

Overall, a conversion of 96–97% of the hydrogen sulfide to elemental sulfur is achievable in a Claus process. If this is insufficient to meet air quality regulations, a Claus process tail gas treater is utilized to remove essentially the entire remaining hydrogen sulfide in the tail gas from the Claus unit. The tail gas treater employs a proprietary solution to absorb the hydrogen sulfide followed by conversion to elemental sulfur.

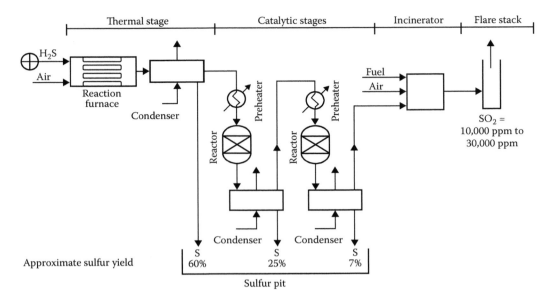

FIGURE 11.3 Claus process. (From Maddox, R.N. 1974. *Gas and Liquid Sweetening.* 2nd Edition. Campbell Publishing Co., Norman, OK; Speight, J.G. 2007. *The Chemistry and Technology of Petroleum.* 4th Edition. CRC Press, Taylor & Francis Publishers, Boca Raton, FL. Figure 11.8, p. 730.)

The Shell Claus off-gas treating (SCOT®) unit is the most common type of tail gas unit and uses a hydrotreating reactor followed by amine scrubbing to recover and recycle sulfur, in the form of hydrogen, to the Claus unit (Nederland, 2004).

In the process (Figure 11.4), tail gas (containing hydrogen sulfide and sulfur dioxide) is contacted with hydrogen and reduced in a hydrotreating reactor to form hydrogen sulfide and water. The catalyst is typically cobalt/molybdenum on alumina. The gas is then cooled in a water contractor. The hydrogen sulfide–containing gas enters an amine absorber, which is typically in a system segregated from the other refinery amine systems. The purpose of segregation is twofold: (1) the tail gas treater frequently uses a different amine than the rest of the plant, and (2) the tail gas is frequently cleaner than the refinery fuel gas (in regard to contaminants) and segregation of the systems reduces the maintenance requirements for the SCOT unit. Amines chosen for use in the tail gas system tend to be more selective for hydrogen sulfide and are not affected by the high levels of carbon dioxide in the off-gas.

The hydrotreating reactor converts sulfur dioxide in the off-gas to hydrogen sulfide that is then contacted with a Stretford solution (a mixture of a vanadium salt, anthraquinone disulfonic acid, sodium carbonate, and sodium hydroxide) in a liquid–gas absorber. The hydrogen sulfide reacts stepwise with sodium carbonate and the anthraquinone sulfonic acid to produce elemental sulfur, with vanadium serving as a catalyst. The solution proceeds to a tank where oxygen is added to regenerate the reactants. One or more froth or slurry tanks are used to skim the product sulfur from the solution, which is recirculated to the absorber.

Other tail gas–treating processes include (1) caustic scrubbing, (2) polyethylene glycol treatment, (3) Selectox process, and (4) sulfite/bisulfite tail gas treating (Mokhatab et al., 2006; Speight, 2007).

Gases that contain ammonia, such as the gas from a refinery sour water stripper, or hydrocarbons are converted in the burner muffle. Sufficient air is injected into the muffle for the complete combustion of all hydrocarbons and ammonia. The air to acid gas ratio is controlled such that, in total, approximately 33% v/v of all hydrogen sulfide is converted to sulfur dioxide, which ensures a stoichiometric reaction for the Claus reaction in the second catalytic step.

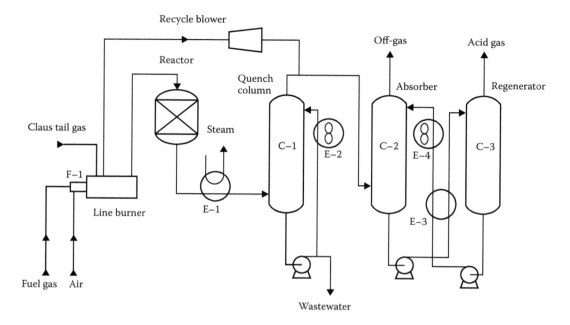

FIGURE 11.4   SCOT process. (From Speight, J.G. 2007. *The Chemistry and Technology of Petroleum.* 4th Edition. CRC Press, Taylor & Francis Publishers, Boca Raton, FL. Figure 11.9, p. 730.)

REFERENCES

Anerousis, J.P., and Whitman, S.K. 1984. An Updated Examination of Gas Sweetening by the Iron Sponge Process. Paper No. SPE 13280. SPE Annual Technical Conference and Exhibition, Houston, TX, September.

Barbouteau, L., and Dalaud, R. 1972. *Gas Purification Processes for Air Pollution Control,* G. Nonhebel (Editor). Butterworth & Co., London, Chapter 7.

Bartoo, R.K. 1985. *Acid and Sour Gas Treating Processes,* S.A. Newman (Editor). Gulf Publishing, Houston, TX.

Bowers, B., and Cash, E. 2014. Maximize Cost and Safety Benefits of Scavenging H₂S at the Wellhead. *Hydrocarbon Process.* November/December: 33–36.

Curry, R.N. 1981. *Fundamentals of Natural Gas Conditioning.* PennWell Publishing Co., Tulsa, OK.

Duckworth, G.L., and Geddes, J.H. 1965. Natural Gas Desulfurization by the Iron Sponge Process. *Oil Gas J.* 63(37): 94–96.

Foglietta, J.H. 2004. Dew Point Turboexpander Process: A Solution for High Pressure Fields. In: *Proceedings. IAPG Gas Conditioning Conference.* Neuquen, Argentina, October 18.

Fulker, R.D. 1972. *Gas Purification Processes for Air Pollution Control,* G. Nonhebel (Editor). Butterworth & Co., London, Chapter 9.

Gary, J.H., Handwerk, G.E., and Kaiser, M.J. 2007. *Petroleum Refining: Technology and Economics,* 5th Edition. CRC Press, Taylor & Francis Group, Boca Raton, FL.

Hong, S.Y., Kim, H., Kim, Y.J., Jeong, J., Cheong, M., Lee, H., Kim, H.S., and Lee, J.S. 2014. Nitrile-Functionalized Tertiary Amines as Highly Efficient and Reversible SO₂ Absorbents. *J. Hazard. Mater.* 264: 136–143.

Hsu, C.S., and Robinson, P.R. (Editors) 2006. *Practical Advances in Petroleum Processing,* Volume 1 and Volume 2. Springer Science, New York.

Huang, K., Chen, Y.L., Zhang, X.M., Ma, S.L., Wu, Y.T., and Hu, X.B. 2014. Experimental Study and Thermodynamic Modeling of the Solubility of SO₂, H₂S and CO₂ in *N*-dodecylimidazole and 1,1′[Oxybis(2,1-Ethanediyloxy-2,1-Ethanediyl)]Bis (Imidazole): An Evaluation of their Potential Application in the Separation of Acidic Gases. *Fluid Phase Equilib.* 378: 21–23.

Jou, F.Y., Deshmukh, R.D., Otto, F.D., and Mather, A.E. 1989. Solubility of H₂S, CO₂ and CH₄ in *N*-formyl Morpholine. *J. Chem. Soc. Farad. Trans. 1* 85: 2675–2682.

Jou, F.Y., Otto, F.D., and Mather, A.E. 1985. *Acid and Sour Gas Treating Processes*, S.A. Newman (Editor). Gulf Publishing Company, Houston, TX, Chapter 10.

Katz, D.K. 1959. *Handbook of Natural Gas Engineering*. McGraw-Hill Book Company, New York.

Kenreck, G. 2014. Manage Hydrogen Sulfide Hazards with Chemical Scavengers. *Hydrocarbon Process.* 93(12): 73–76.

Kindlay, A.J., and Parrish, W.R. 2006. *Fundamentals of Natural Gas Processing*. CRC Press, Taylor & Francis Group, Boca Raton, FL.

Kohl, A.L., and Nielsen, R.B. 1997. *Gas Purification*. Gulf Publishing Company, Houston, TX.

Kohl, A.L., and Riesenfeld, F.C. 1985. *Gas Purification*, 4th Edition, Gulf Publishing Company, Houston, TX.

Lachet, V., and Béhar, E. 2000. Industrial Perspective on Natural Gas Hydrates. *Oil Gas Sci. Technol.* 55: 611–616.

Lee, E.S., and Rangaiah, G.P. 2009. Optimization of Recovery Processes for Multiple Economic and Environmental Objectives. *Ind. Eng. Chem. Res.* 48: 7662–7681.

Li, Y., Song, C.X., and You, C.F. 2010. Experimental Study on Abrasion Characteristics of Rapidly Hydrated Sorbent for Moderate Temperature Dry Flue Gas Desulfurization. *Energy Fuels* 24: 1682–686.

Maddox, R.N. 1974. *Gas and Liquid Sweetening*, 2nd Edition. Campbell Publishing Co., Norman, OK.

Maddox, R.N. 1982. *Gas Conditioning and Processing*, Vol. 4. Gas and Liquid Sweetening. Campbell Publishing Co., Norman, OK.

Maddox, R.N., Bhairi, A., Mains, G.J., and Shariat, A. 1985. *Acid and Sour Gas Treating Processes*, S.A. Newman (Editor). Gulf Publishing Company, Houston, TX, Chapter 8.

Maina, P., and Mbarawa, M. 2012. Blending Lime and Iron Waste to Improve Sorbents Reactivity towards Desulfurization. *Fuel* 102: 162–172.

Maple, M.J., and Williams, C.D. 2008. Separating Nitrogen/Methane on Zeolite-Like Molecular Sieves. *Microporous Mesoporous Mater.* 111: 627–631.

Miller, M.B., Chen, D.L., Luebke, D.R., Johnson, J.K., and Enick, R.M. 2011. Critical Assessment of CO_2 Solubility in Volatile Solvents at 298.15 K. *J. Chem. Eng. Data* 56: 1565–1572.

Mody, V., and Jakhete, R. 1988. *Dust Control Handbook*. Noyes Data Corp., Park Ridge, NJ.

Mokhatab, S., Poe, W.A., and Speight, J.G. 2006. *Handbook of Natural Gas Transmission and Processing*. Elsevier, Amsterdam.

Nagel, D., Kermadec, R.D., Lintz, H.G., Roizard, C., and Lapicque, F. 2002. Absorption of Sulfur Dioxide in *N*-formylmorpholine: Investigations of the Kinetics of the Liquid Phase Reaction. *Chem. Eng. Sci.* 57: 4883–4893.

Nederland, J. 2004. *Sulphur*. University of Calgary, Calgary, Alberta, Canada. November.

Newman, S.A. 1985. *Acid and Sour Gas Treating Processes*. Gulf Publishing, Houston, TX.

Nonhebel, G. 1964. *Gas Purification Processes*. George Newnes Ltd., London.

Parkash, S. 2003. *Refining Processes Handbook*. Gulf Professional Publishing, Elsevier, Amsterdam, Netherlands.

Pitsinigos, V.D., and Lygeros, A.I. 1989. Predicting H_2S-MEA Equilibria. *Hydrocarbon Process.* 58(4): 43–44.

Polasek, J., and Bullin, J. 1985. *Acid and Sour Gas Treating Processes*, S.A. Newman (Editor). Gulf Publishing Company, Houston, TX, Chapter 7.

Song, X., Yao, W., Zhang, B., and Wu, Y. 2012. Application of Pt/CdS for the Photocatalytic Flue Gas Desulfurization. *Int. J. Photo-Energ.* 2012:684735. doi:10.1155/2012/684735.

Soud, H., and Takeshita, M. 1994. *FGD Handbook*. No. IEACR/65. International Energy Agency Coal Research, London.

Speight, J.G. 1993. *Gas Processing: Environmental Aspects and Methods*. Butterworth Heinemann, Oxford, England.

Speight, J.G. 2000. *The Desulfurization of Heavy Oils and Residua*, 2nd Edition. Marcel Dekker Inc., New York.

Speight, J.G. 2007. *Natural Gas: A Basic Handbook*. GPC Books, Gulf Publishing Company, Houston, TX.

Speight, J.G. 2011. *An Introduction to Petroleum Technology, Economics, and Politics*. Scrivener Publishing, Salem, MA.

Speight, J.G. 2014. *The Chemistry and Technology of Petroleum*, 5th Edition. CRC Press, Taylor & Francis Group, Boca Raton, FL.

Speight, J.G., and Ozum, B. 2002. *Petroleum Refining Processes*. Marcel Dekker Inc., New York.

Staton, J.S., Rousseau, R.W., and Ferrell, J.K. 1985. *Acid and Sour Gas Treating Processes*, S.A. Newman (Editor). Gulf Publishing Company, Houston, TX, Chapter 5.

Sun, Z.W., Wang, S.W., Zhou, Q.L., and Hui, S.E. 2010. Experimental Study on Desulfurization Efficiency and Gas–Liquid Mass Transfer in a New Liquid-Screen Desulfurization System. *Appl. Energy* 87: 1505–1512.

Svoboda, K., Lin, W., Hannes, J., Korbee, R., and van den Beek, C.M. 1994. Low Temperature Flue Gas Desulfurization by Alumina–CaO Regenerable Sorbents. *Fuel* 73(7): 1144–1150.

Van Dam, M.H.H., Lamine, A.S., Roizard, D., Lochon, P., and Roizard, C. 1997. Selective Sulfur Dioxide Removal Using Organic Solvents. *Ind. Eng. Chem. Res.* 36: 4628–4637.

Ward, E.R. 1972. *Gas Purification Processes for Air Pollution Control*, G. Nonhebel (Editor). Butterworth & Co., London, Chapter 8.

Zapffe, F. 1963. Iron Sponge Process Removes Mercaptans. *Oil Gas J.* 61(33): 103–104.

Zhang, L.Q., Shi, L.B., and Zhou, Y. 2007. Formation Prediction and Prevention Technology of Natural Gas Hydrate. *Nat Gas Technol.* 1(6): 67–69.

12 Desulfurization Processes—Distillates

12.1 INTRODUCTION

Petroleum refining is now in a significant transition period as the industry moves into the 21st century, and the demand for petroleum and petroleum products has shown a sharp growth in recent years (Speight, 2011, 2014). The demand for transportation fuels and fuel oil is forecast to continue to show a steady growth in the future. The simplest means to cover the demand growth in low-boiling products is to increase the imports of light crude oils and low-boiling petroleum products; however, these steps may be limited in the future.

Over the past three decades, the average quality of crude oil has declined as evidenced by a progressive decrease in API gravity (i.e., increase in density) and an increase in sulfur content (Speight and Ozum, 2002; Parkash, 2003; Hsu and Robinson, 2006; Gary et al., 2007; Speight, 2011, 2014). This has brought about a major focus in refineries on the ways in which heavy feedstocks might be converted into low-boiling, high-value products (Khan and Patmore, 1998). Simultaneously, the changing crude oil properties are reflected in changes such as an increase in asphaltene constituents or an increase in sulfur, metal, and nitrogen contents. Pretreatment processes for removing those would also play an important role.

However, the essential step required of refineries is the upgrading of heavy feedstocks, particularly residua (Al-Haj-Ibrahim and Morsi, 1992; Dickenson et al., 1997; Speight, 2011, 2014). In fact, the increasing supply of heavy crude oils is a matter of serious concern for the petroleum industry. To satisfy the changing pattern of product demand, significant investments in refining conversion processes will be necessary to profitably utilize these heavy crude oils. The most efficient and economical solution to this problem will depend, to a large extent, on individual country and company situations. However, the most promising technologies will likely involve the conversion of vacuum bottom residual oils, asphalt from deasphalting processes, and super-heavy crude oils into useful low-boiling and middle-distillate products.

Heavy feedstock upgrading and residua began with the introduction of desulfurization processes (Speight, 2000, 2014). In the early days, the goal was desulfurization; however, in later years, the processes were adapted to a 10–30% partial conversion operation, as intended to achieve desulfurization and obtain low-boiling fractions simultaneously, by increasing the severity of operating conditions. Refinery evolution has seen the introduction of a variety of residuum cracking processes based on thermal cracking, catalytic cracking, and hydroconversion. Those processes are different from one another in cracking method, cracked product patterns, and product properties, and will be employed in refineries according to their respective features.

Using a schematic refinery operation (Figure 12.1), new processes for the conversion of heavy feedstocks will probably be used in place of visbreaking and coking options with some degree of hydrocracking as a primary conversion step. Other processes may replace or augment the deasphalting units in many refineries (Figures 12.2 and 12.3). An exception, which may become the rule, is the upgrading of tar sand bitumen (Speight, 2013, 2014), which is (currently) subjected to either delayed coking, fluid coking, or hydrocracking such as LC-Fining or H-Oil processing as the primary upgrading step (Speight, 2013, 2014), with a modicum of prior vacuum distillation or topping. After primary upgrading, the product streams are hydrotreated and combined to form a synthetic crude oil that is shipped to a conventional refinery for further processing. Conceivably, a

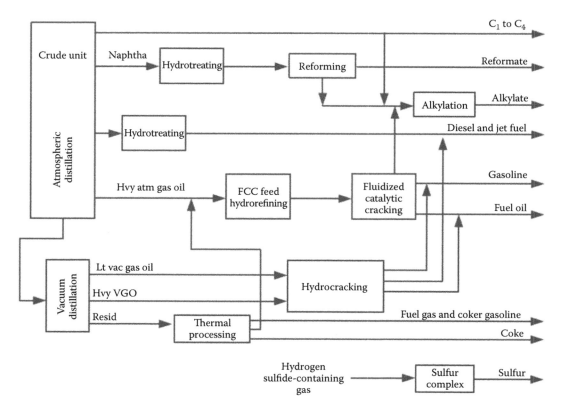

FIGURE 12.1 Schematic of a refinery.

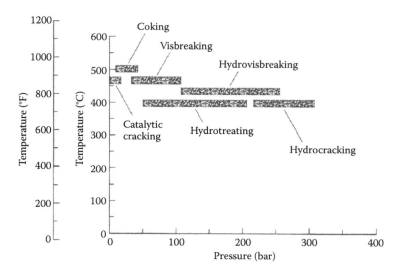

FIGURE 12.2 Temperature and pressure ranges for refinery processes. (From Speight, J.G. 2007. *The Chemistry and Technology of Petroleum*. 4th Edition. CRC Press, Taylor & Francis Publishers, Boca Raton, FL. Figure 21.2, p. 563.)

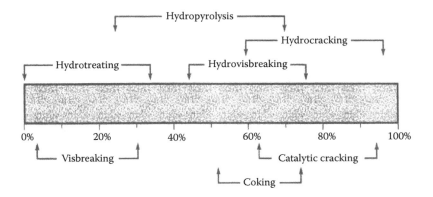

FIGURE 12.3 Feedstock conversion to liquids by refinery processes. (From Speight, J.G. 2007. *The Chemistry and Technology of Petroleum.* 4th Edition. CRC Press, Taylor & Francis Publishers, Boca Raton, FL. Figure 21.3, p. 564.)

heavy feedstock could be upgraded in the same manner and, depending on the upgrading facility, upgraded further for sales.

Heavy feedstocks are generally characterized by low API gravity (high density) and high viscosity, high initial boiling point, high carbon residue, high nitrogen content, high sulfur content, and high metal content (Speight, 2000, 2014; Ancheyta and Speight, 2007). In addition to these properties, the heavy feedstocks also have an increased molecular weight and a reduced hydrogen content (Speight, 2000, 2013, 2014). Thus, the limitations of processing these heavy feedstocks depend, to a large extent, on the amount of higher molecular weight constituents (i.e., asphaltene constituents and resin constituents), which contain the majority of the heteroatom species (Speight, 2000, 2014; Ancheyta and Speight, 2007). These constituents are responsible for high yields of thermal and catalytic coke (Chapters 2 and 7) (Speight and Ozum, 2002; Parkash, 2003; Hsu and Robinson, 2006; Gary et al., 2007; Speight, 2014).

The majority of the metal constituents in crude oils are associated with the asphaltene constituents. Some of the metals are in the form of organometallic complexes; the rest are found in organic or inorganic salts that are soluble in water or in crude. In recent years, attempts have been made to isolate and study the vanadium present in petroleum porphyrins, mainly in asphaltene fractions.

When catalytic processes are employed, complex molecules (such as those that may be found in the original asphaltene fraction) or those formed during the process, are not sufficiently mobile (or are too strongly adsorbed by the catalyst) to be saturated by hydrogenation. The chemistry of the thermal reactions of some of these constituents (Chapter 5) dictates that certain reactions, once initiated, cannot be reversed and proceed to completion. Coke is the eventual product that can deactivate the catalyst sites and eventually interfere with the hydroprocess (Speight, 2000, 2014; Ancheyta and Speight, 2007).

Technologies for upgrading heavy feedstocks can be broadly divided into *carbon rejection* and *hydrogen addition* processes (Chapter 4) (Speight and Ozum, 2002; Parkash, 2003; Hsu and Robinson, 2006; Gary et al., 2007; Speight, 2014). Briefly, carbon rejection processes are those processes in which a carbonaceous by-product (coke) is produced along with distillable liquid products. On the other hand, hydrogen addition processes involve reaction of the feedstock with an external source of hydrogen and result in an overall increase in H/C ratio of the products as well as a decrease in the amount of coke produced.

With the inception of hydrogenation as a process by which petroleum could be converted into lighter products, it was also recognized that hydrogenation would be effective for the simultaneous removal of nitrogen, oxygen, and sulfur compounds from the feedstock. However, with respect to the prevailing context of fuel industries, hydrogenation seemed to be not economical for application

TABLE 12.1

Process Parameters for Hydrodesulfurization

Parameter	Naphtha	Residuum
Temperature, °C	300–400	340–425
Pressure, atm	35–70	55–170
Liquid hourly space velocity	4.0–10.0	0.2–1.0
H_2 recycle rate, scf/bbl	400–1000	3000–5000
Catalyst life, years	3.0–10.0	0.5–1.0
Sulfur removal, %	99.9	85.0
Nitrogen removal, %	99.5	40.0

Source: Speight, J.G. 2007. *The Chemistry and Technology of Petroleum.* 4th Edition. CRC Press, Taylor & Francis Publishers, Boca Raton, FL. Table 21.1, p. 569.

to petroleum fractions. At least two factors dampened interest: (1) the high cost of hydrogen and (2) the adequacy of current practices for meeting the demand for low-sulfur products by refining low-sulfur crude oils, or even by alternate desulfurization techniques.

Nevertheless, it became evident that reforming processes instituted in many refineries were providing substantial quantities of by-product hydrogen, enough to tip the economic balance in favor of hydrodesulfurization processes (Tables 12.1 and 12.2). In fact, the need for such commercial operations has become more acute because of a shift in supply trends that has increased the amount of high-sulfur crude oils employed as refinery feedstocks. Because of this, many topping refineries have shut down because of their inability to process these heavier crude oils. In comparison, the total capacity of those processes that is intended for upgrading the heavier distillates of crude oils have increased (Speight, 2000, 2014).

Normally, in an integrated refinery, the high-boiling fractions of crude oils are subjected to either coking or hydrogenation processes to convert them to streams more easily processed for the production of transportation fuels. In the former case, up to about 20% of the original feed to the coker can be lost as a product of low economic value, especially if the coke contains large amounts of metals and heteroatoms. In the latter case, the addition of hydrogen to this heavy feed is very expensive because of the impact of metals and heteroatoms on catalyst lifetimes, and the cost of the hydrogen required to remove the heteroatoms and metals and to saturate aromatic rings.

TABLE 12.2

Hydrodesulfurization of Various Naphtha Fractions

Feedstock	Boiling Range		Sulfur, wt.%	Desulfurization,[a] %
	°C	°F		
Visbreaker naphtha	65–220	150–430	1.00	90
Visbreaker–coker naphtha	65–220	150–430	1.03	85
Straight-run naphtha	85–170	185–340	0.04	99
Catalytic naphtha (light)	95–175	200–350	0.18	89
Catalytic naphtha (heavy)	120–225	250–440	0.24	71
Thermal naphtha (heavy)	150–230	300–450	0.28	57

Source: Speight, J.G. 2007. *The Chemistry and Technology of Petroleum.* 4th Edition. CRC Press, Taylor & Francis Publishers, Boca Raton, FL. Table 21.2, p. 571.

[a] Process conditions: Co–Mo on alumina, 260–370°C/500–700°F, 200–500 psi hydrogen.

One of the goals of any hydrogenation process is the use of a catalyst (Table 12.3) for the removal of heteroatoms and metals, specifically nitrogen, sulfur, vanadium, and nickel. Another is the saturation of aromatic structures, primarily through hydrotreating. However, for residua, the conversion of high-boiling materials to low-boiling materials is an additional important goal that is accomplished by hydrocracking. The processes used include a variety of catalytic cracking processes. The capacity for these types of processes has increased significantly in recent years owing to the increase in the amount of heavy feedstocks being processed (Speight, 2000, 2014). The feeds to these types of units are usually atmospheric and vacuum residua, or similar high-boiling feedstocks.

There are several valid reasons for removing sulfur from petroleum fractions. These include (1) reduction, or elimination, of corrosion during refining, handling, or use of the various products; (2) production of products having an acceptable odor; (3) increasing the performance (and stability) of naphtha; (4) decreasing smoke formation in kerosene; and (5) reduction of sulfur content in other fuel oils to a level that improves burning characteristics and is environmentally acceptable.

To accomplish sulfur removal, use is still made of extraction and chemical treatment of various petroleum fractions as a means of removing certain sulfur types from petroleum products; however, hydrodesulfurization is the only method generally applicable to the removal of all types of sulfur compounds.

Overall, there has been a growing dependence on high-sulfur heavier oils and residua as a result of continuing increases in the prices of the conventional crude oils coupled with the decreasing availability of these crude oils through the depletion of reserves in the various parts of the world. Furthermore, the ever growing tendency to convert as much as possible of lower-grade feedstocks to liquid products is causing an increase in the total sulfur content in refined products. Refiners must, therefore, continue to remove substantial portions of sulfur from the lighter products; however, residua and the heavier crude oils pose a particularly difficult problem. Indeed, it is now clear that there are other problems involved in the processing of the heavier feedstocks and that these

TABLE 12.3

Composition and Properties of Hydrotreating Catalysts

Composition	Properties		Range
	Active Phases, % w/w		
MoO_3			13–20
CoO			2.5–3.5
NiO			2.5–3.5
	Promoters, % w/w		
SiO			1.0–10.0
	Surface area	sq. m/g	150–500
	Pore volume	cm³/g	0.2–0.8
	Pore diameter		
	Mesopores	nm	3.0–50.0
	Macropores	nm	100–5000
	Extrudable diameter	mm	0.8–4.0
	Extrudable length/ diameter	mm	2.0–4.0
	Bulk density	kg/m³	500–1000

Source: Speight, J.G. 2007. *The Chemistry and Technology of Petroleum.* 4th Edition. CRC Press, Taylor & Francis Publishers, Boca Raton, FL. Table 21.3, p. 582.

Note: Catalyst is typically composed of active phases, promoters, and a γ-alumina carrier.

heavier feedstocks, which are gradually emerging as the liquid fuel supply of the future, need special attention.

The hydrodesulfurization of petroleum fractions has long been an integral part of refining operations, and in one form or another, hydrodesulfurization is practiced in every modern refinery. The process is accomplished by the catalytic reaction of hydrogen with the organic sulfur compounds in the feedstock to produce hydrogen sulfide, which can be separated readily from the liquid (or gaseous) hydrocarbon products:

$$RSH + H_2 \rightarrow RH + H_2S$$

$$RSR^1 + H_2 \rightarrow RH + R^1H + H_2S$$

If denitrogenation, demetallization, and deoxygenation also occur, the representative equations are similar:

$$RR^1NH + 2H_2 \rightarrow RH + R^1H + NH_3$$

$$RSR^1 + H_2 \rightarrow RH + R^1H + H_2S$$

$$2RM + H2 \rightarrow 2RH + M*$$

$$ROH + H_2 \rightarrow RH + H_2O$$

$$ROR^1 + H_2 \rightarrow RH + R^1H + H_2O$$

*The metal is adsorbed on the surface of the catalyst or in the coke.

The likelihood of oxygen functional occurring in petroleum is minimal unless the petroleum has been exposed to oxygen during recovery, transportation, or storage.

The technology of the hydrodesulfurization process is well established, and petroleum feedstocks of every conceivable molecular weight range can be treated to remove sulfur. Thus, it is not surprising that an extensive knowledge of hydrodesulfurization has been acquired along with development of the process over the last few decades. However, most of the available information pertaining to the hydrodesulfurization process has been obtained with the lighter and more easily desulfurized petroleum fractions; however, it is, to some degree, applicable to the hydrodesulfurization of the heavier feedstocks. On the other hand, processing heavy feedstocks presents several problems that are not found with distillate processing and that require process modifications to meet the special requirements necessary for heavy feedstock desulfurization.

The presence of high-pressure hydrogen (>5–7 MPa) during the conversion of a residue has a number of beneficial effects: (1) hydrogen tends to suppress free-radical addition reactions and dehydrogenation; (2) in the presence of a hydrogenation catalyst, the hydrogen converts aromatic and heteroaromatic species, and further converts heteroatoms to hydrogen sulfide, water, and ammonia; (3) all hydroconversion processes involve a complex suite of series and parallel reactions: cracking, hydrogenation, sulfur removal, and demetallization; and (4) almost all hydroconversion processes use a catalyst or additive to control the formation of coke, to serve as a surface for deposition of metals, and to enhance hydrogenation reactions. The challenge in analyzing these processes is to direct the reactions toward desirable product characteristics.

The initial focus for the application of hydroprocesses for the conversion of heavy feedstocks was the removal of sulfur. However, many recent processes have concentrated on the objective of achieving high feedstock conversion. The emphasis on sulfur removal, in these process designs, shifts from the primary hydroprocessing step to the secondary hydroprocessing reactor.

For the heavier feedstocks, process selection has tended to favor the hydroprocesses that maximize distillate yield and minimize coke formation. However, thermal and catalytic processes must not be ignored because of the tendency for coke production. Such processes may be attractive for processing unconverted residua from hydroprocesses.

In this chapter, brief descriptions of the various hydrodesulfurization processes are presented with some attempt to illustrate how the processes vary in their capacity to treat the heavier feedstocks to yield the desired low-sulfur products. There are many processes described in the technical and commercial literature; however, a large proportion of these processes (although actively reported) have not yet left the pilot (or exploratory) stage and are, perhaps, a considerable way from commercialization. Thus, the processes described in this chapter are the more common, but commercially proven, processes with the inclusion of some of the more promising processes that have entered the demonstration stage, and it is the purpose of this chapter to present an outline of these process options.

12.2 COMMERCIAL PROCESSES

Desulfurization technology (specifically hydrotreating technology) is one of the most commonly used refinery processes, designed to remove contaminants such as sulfur, nitrogen, condensed ring aromatics, or metals (Table 12.4). The feedstocks used in the process range from naphtha to vacuum resid, and the products in most applications are used as environmentally acceptable clean fuels. Hydrotreating technology has a long history of commercial application since the 1950s, with >500 licensed units placed in operation worldwide. Furthermore, recent regulatory requirements to produce ultra-low-sulfur diesel and gasoline have created a very dynamic market as refiners must build new or revamp their existing assets to produce the low-sulfur, even no-sulfur, fuels.

Initially, the commercial technology for controlling the sulfur content of full-range fluid catalytic cracking (FCC) naphtha was *conventional hydrotreating*. A number of technology providers offered conventional hydrotreating processes for full-range FCC naphtha and other naphtha streams (MathPro, 2003, 2011). Collectively, these processes are reliable and well understood, and could accomplish the necessary degree of sulfur control. However, conventional FCC naphtha hydrotreating processes are expensive because they are nonselective. In the course of removing sulfur, they also saturate essentially all of the olefins present in FCC naphtha, which leads to high octane loss (>10 numbers) and high hydrogen consumption, which account for the high cost of conventional hydrotreating.

The petroleum industry employs either a relatively simple process (Figure 12.4) or a two-stage process (Figure 12.5) in which the feedstock undergoes both hydrotreating and hydrocracking. In the first, or pretreating, stage (Figure 12.5), the main purpose is conversion of nitrogen compounds in the feed to hydrocarbons and to ammonia by hydrogenation and mild hydrocracking. Typical conditions are 340–390°C (650–740°F), 150–2500 psi (1–17 MPa), and a catalyst contact time of 0.5–1.5 h; up to 1.5% w/w hydrogen is absorbed, partly by conversion of the nitrogen compounds but chiefly by aromatic compounds that are hydrogenated.

It is most important to reduce the nitrogen content of the product oil to <0.001% w/w (10 ppm). This stage is usually carried out with a bifunctional catalyst containing hydrogenation promoters, e.g., nickel and tungsten or molybdenum sulfides, on an acid support, such as silica–alumina. The metal sulfides hydrogenate aromatics and nitrogen compounds and prevent deposition of carbonaceous deposits; the acid support accelerates nitrogen removal as ammonia by breaking carbon–nitrogen bonds. The catalyst is generally used as 1/8 × 1/8 in (0.32 × 0.32 cm) or 1/16 × 1/8 in (0.16 × 0.32 cm) pellets, formed by extrusion.

Most of the hydrotreating cracking is accomplished in the first stage. Ammonia and some naphtha are usually removed from the first-stage product, and then the remaining first-stage product, which is low in nitrogen compounds, is passed over the second-stage catalyst. Again, typical conditions are 300–370°C (600–700°F), 1500–2500 psi (10–17 MPa) hydrogen pressure, and 0.5–1.5 h contact time; 1–1.5% w/w hydrogen may be absorbed. Conversion to gasoline or jet fuel is seldom

TABLE 12.4
Processes for Sulfur Removal

Process	Effects
Desulfurization	
Fluid catalytic cracking	Increase sulfur conversion
	Reduce sulfur in liquid products
Caustic treatment of FCC products	Remove the organic sulfur as soluble salts (e.g., Na$^+$SH$^-$)
Hydrodesulfurization	Organic sulfur removal
	Olefins saturated
	Reduction in naphtha octane number
Selective hydrodesulfurization	Convert sulfur to hydrogen sulfide
	Preserve olefins
	Preserve octane number
Adsorption	
Elevated temperature and low hydrogen pressure	Sulfur capture by solid adsorbent
High temperature	Sulfur capture by solid adsorbent
Selective adsorption	
Ambient temperature	
No hydrogen	Remove sulfur as organic compounds
Selective alkylation	
Thiophene derivatives fractions	Organic sulfur in high molecular weight
Deep Desulfurization	
Fluid catalytic cracking	Pretreat feedstock by hydrotreating
	Sulfur removal
	Ultra-low sulfur products
Two-stage hydrotreating	
High-activity catalysts	High removal of sulfur
Ultra-deep hydrodesulfurization	Hydrodesulfurization, hydrodenitrogenation
Adsorption	
Elevated temperature and low hydrogen pressure	Sulfur capture by solid adsorbent
High temperature	Sulfur capture by solid adsorbent
Selective adsorption	
Ambient temperature	
No hydrogen	Remove sulfur as organic compounds
Oxidative desulfurization	Conversion of sulfur to sulfones
	Separation of sulfones as separate phase
Biodesulfurization	Removal of sulfur by microbial action

complete in one contact with the catalyst, so the lighter oils are removed by distillation of the products and the heavier, higher-boiling product combined with fresh feed and recycled over the catalyst until it is completely converted.

12.2.1 AUTOFINING PROCESS

The autofining process differs from other hydrorefining processes in that an external source of hydrogen is not required. Sufficient hydrogen to convert sulfur to hydrogen sulfide is obtained by dehydrogenation of naphthenes in the feedstock.

The processing equipment is similar to that used in hydrofining (Speight and Ozum, 2002; Parkash, 2003; Hsu and Robinson, 2006; Gary et al., 2007; Speight, 2014). The catalyst is cobalt

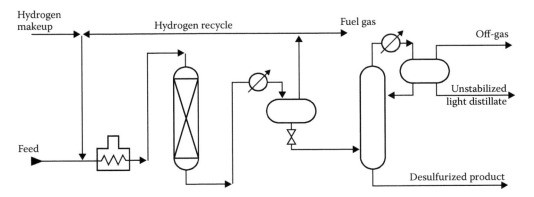

FIGURE 12.4 Distillate hydrotreating. (From OSHA Technical Manual, Section IV, Chapter 2: Petroleum Refining Processes. http://www.osha.gov/dts/osta/otm/otm_iv/otm_iv_2.html; Speight, J.G. 2007. *The Chemistry and Technology of Petroleum.* 4th Edition. CRC Press, Taylor & Francis Publishers, Boca Raton, FL. Figure 21.5, p. 567.)

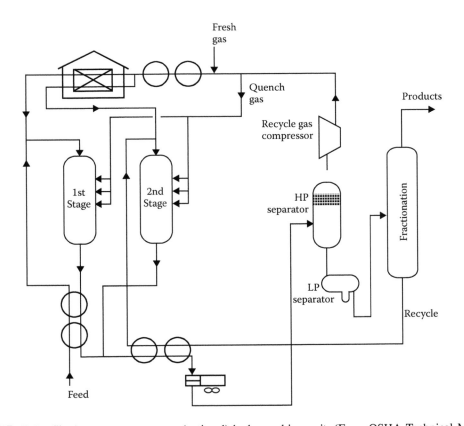

FIGURE 12.5 Single-stage or two-stage (optional) hydrocracking unit. (From OSHA Technical Manual, Section IV, Chapter 2: Petroleum Refining Processes. http://www.osha.gov/dts/osta/otm/otm_iv/otm_iv_2. html; Speight, J.G. 2007. *The Chemistry and Technology of Petroleum.* 4th Edition. CRC Press, Taylor & Francis Publishers, Boca Raton, FL. Figure 21.4, p. 566.)

oxide and molybdenum oxide on alumina, and operating conditions are usually 340–425°C (650–800°F) at pressures of 100–200 psi. Hydrogen formed by dehydrogenation of naphthenes in the reactor is separated from the treated oil and is then recycled through the reactor. The catalyst is regenerated with steam and air at 200–1000 h intervals, depending on whether light or heavy feedstocks have been processed. The process is used for the same purpose as hydrofining but is limited to fractions with end points no higher than 370°C (700°F).

12.2.2   FERROFINING PROCESS

This mild hydrogen-treating process was developed to treat distilled and solvent-refined lubricating oils. The process eliminates the need for acid and clay treatment. The catalyst is a three-component material on an alumina base with low hydrogen consumption and a life expectancy of 2 years or more. Process operations include heating the hydrogen–oil mixture and charging to a downflow catalyst-filled reactor. Separation of oil and gas is a two-stage operation whereby gas is removed to the fuel system. The oil is then steam stripped to control the flash point and dried in vacuum, and a final filtering step removes catalyst fines.

12.2.3   GULF HDS PROCESS

This is a regenerative fixed-bed process to upgrade petroleum residues by catalytic hydrogenation to refined heavy fuel oils or to high-quality catalytic charge stocks. Desulfurization and quality improvement are the primary purposes of the process; however, if the operating conditions and catalysts are varied, light distillates can be produced and the viscosity of heavy material can be lowered. Long on-stream cycles are maintained by reducing random hydrocracking reactions to a minimum, and whole crude oils, virgin, or cracked residua may serve as feedstock.

The catalyst is a metallic compound supported on pelletized alumina and may be regenerated *in situ* with air and steam or flue gas through a temperature cycle of 400–650°C (750–1200°F). On-stream cycles of 4–5 months can be obtained at desulfurization levels of 65–75%, and catalyst life may be as long as 2 years.

12.2.4   HYDROFINING PROCESS

Hydrofining is a process used for reducing the sulfur content of feedstocks by treating the feedstock in the presence of a catalyst. Thus, hydrofining may be used to upgrade straight-run, cracked, or coker-derived naphtha and distillate boiling-range streams, thus increasing the supply of catalytic reformer feedstock, solvent naphtha, and other naphtha-type materials. The sulfur content of kerosene can be reduced with improved color, odor, and wick-char characteristics. The tendency of kerosene to form smoke is not affected since aromatics, which cause smoke, are not affected by the mild hydrofining conditions. Cracked gas oil having high sulfur content can be converted to excellent fuel oil and diesel fuel by reduction in sulfur content and by the elimination of components that form gum and carbon residues.

This process can be applied to lubricating oil, naphtha, and gas oil. The feedstock is heated in a furnace and passed with hydrogen through a reactor containing a suitable metal oxide catalyst, such as cobalt and molybdenum oxides or alumina. Hydrogen is obtained from catalytic reforming units. Reactor operating conditions range from 205°C to 425°C (400–800°F) and from 50 to 800 psi, depending on the kind of feedstock and the degree of treatment required. Higher-boiling feedstocks, high sulfur content, and maximum sulfur removal require higher temperatures and pressures.

After passing through the reactor, the treated oil is cooled and separated from the excess hydrogen, which is recycled through the reactor. The treated oil is pumped to a stripper tower, where hydrogen sulfide formed by the hydrogenation reaction is removed by steam, vacuum, or flue gas, and the finished product leaves the bottom of the stripper tower. The catalyst is not usually regenerated; it is replaced after about a year's use.

Lube oil hydrofining is a catalytic technology to prepare lube base stocks for further processing, or it may be used as a base-stock finishing step. In the process, multiring aromatics are saturated, acids are destroyed and, more important in the present context, sulfur and nitrogen are removed, which improves the color and stability to oxidation of the products (base oil). As a finishing step, the lube hydrofining process may be used to treat a dewaxed oil product with similar color and oxidation stability benefits. Lube hydrofining, when used on lube distillate feedstocks, can enhance the ensuing solvent extraction performance.

Lube hydrofining technology generally produces a product with (1) improved color, (2) improved oxidation stability, (3) improved antioxidant response, (4) better odor, (5) better surface properties, (6) optimal sulfur reduction, (7) significant viscosity index increase at high desulfurization levels, and (8) reduced toxicity.

Generally, the lube hydrofining process (Figure 12.6) is suitable for feedstocks from all crude sources and the process conditions are moderate. In an Exolfining N configuration, the lube hydrofining step is fully heat integrated with the solvent extraction step. The feedstock to the lube hydrofining reactor is the hot bottoms stream from the raffinate recovery tower. In a stand-alone configuration, the reactor feedstock is typically brought to reaction temperature by a combination of heat exchange with the hot hydrofinished product and furnace preheater. The fixed-bed reactor contains a hydrofinishing catalyst, and once-through or recycle hydrogen treat gas may be used. The reactor effluent is flashed to recover the unreacted hydrogen treat gas at high pressure as well as to separate the hydrogen sulfide and ammonia resulting from the hydrofinishing reactions. The hydrofinished oil is then steam stripped and dried under vacuum. As an additional option, the steam stripper tower may be designed to correct the volatility of the hydrofinished oil product.

Wax hydrofining uses wax from a lube oil plant to produce a food-grade wax that can be used for wax paper, fruit coatings, paper cups, as a seal for medicines, and for similar applications where high purity is required. Wax hydrofining is the final critical steps for processing low-oil-content waxes from dewaxing–deoiling to produce wax with low odor, improved color, and improved stability. This technology is applicable to either crystalline wax from lube distillates or microcrystalline

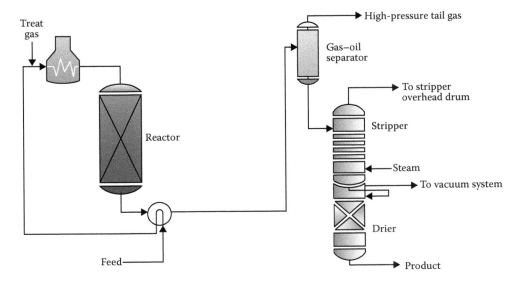

FIGURE 12.6 Lube oil hydrofining unit. (From Speight, J.G. 2007. *The Chemistry and Technology of Petroleum*. 4th Edition. CRC Press, Taylor & Francis Publishers, Boca Raton, FL. Figure 21.6, p. 577.)

wax from deasphalted oils. The wax hydrofining option processes low-oil-content (<1% by weight) wax from conventional solvent lube plant dewaxing–deoiling units under moderate hydroprocessing conditions.

Thus, liquefied wax and hydrogen are preheated to reaction temperature in a conventional coil furnace and fed to a fixed-bed reactor containing a supported hydrotreating catalyst. Sulfur and nitrogen compounds are converted to hydrogen sulfide and ammonia, multiring aromatics are saturated, and trace acids and solvent residues are removed. The reactor effluent is flashed to separate hydrogen, hydrogen sulfide, and ammonia, and the wax is steam stripped and dried to improve volatility and color. Both crystalline and microcrystalline waxes can be processed in the same unit, although the types of oil molecules in microcrystalline waxes typically require more severe operating conditions. SCANfining is the next step up from hydrofining and is used to meet environmental requirements for sulfur in gasoline using a proprietary RT-225 catalyst to treat naphtha from an FCC unit. The process is specially designed to achieve high selectivity for sulfur removal without excessive olefin saturation and octane loss. Along similar paths, the diesel oil deep desulfurization process for removing sulfur to very low levels and the GO-fining and Residfining processes are used for upgrading gas oils and residua, respectively, for feedstocks to a catalytic cracking unit and also to use as low sulfur fuel oil (Song, 2002).

Finally, as an additional option, most feedstocks to an FCC unit are high in sulfur with a high ratio of residue blended. This type of feedstock results in high-sulfur, high-olefin naphtha, which can be treated in a hydrofiner to produce good-quality naphtha that contains more saturated hydrocarbons, through which the high-sulfur-content problem of the naphtha is resolved. The typical operating conditions are 280–320°C (535–610°F), 290–1160 psi, with a liquid hourly space velocity in the range 1.5–3.0 h^{-1} and a hydrogen–oil ratio (v/v) of 200–400.

The catalysts for hydrofining typically contain tungsten oxide and/or molybdenum oxide, nickel oxide, and cobalt oxide supported on an alumina carrier. Catalysts with a lower metal content may exhibit higher activity at low temperatures (Xia et al., 2000; Tanaka, 2004; Kumagai et al., 2005).

12.2.5 ISOMAX PROCESS

The Isomax process is a two-stage, fixed-bed catalyst system that operates under hydrogen pressures from 500 to 1500 psi in a temperature range of 205–370°C (400–700°F), for example, with middle-distillate feedstocks. Exact conditions depend on the feedstock and product requirements, and hydrogen consumption is on the order of 1000–1600 ft^3/bbl of the feed processed. Each stage has a separate hydrogen recycling system. Conversion may be balanced to provide products for variable requirements, and recycling can be taken to extinction if necessary. Fractionation can also be handled in a number of different ways to yield desired products.

12.2.6 ULTRAFINING PROCESS

Ultrafining is a regenerative, fixed-bed, catalytic process to desulfurize and hydrogenate refinery stocks from naphtha up to and including lubricating oil. The catalyst is cobalt–molybdenum on alumina and may be regenerated *in situ* using an air–stream mixture. Regeneration requires 10–20 h and may be repeated 50–100 times for a given batch of catalyst; the catalyst life is 2–5 years depending on feedstock.

12.2.7 UNIFINING PROCESS

This is a regenerative, fixed-bed, catalytic process to desulfurize and hydrogenate refinery distillates of any boiling range. Contaminating metals, nitrogen compounds, and oxygen compounds are eliminated, along with sulfur. The catalyst is a cobalt–molybdenum–alumina type that may be regenerated *in situ* with steam and air.

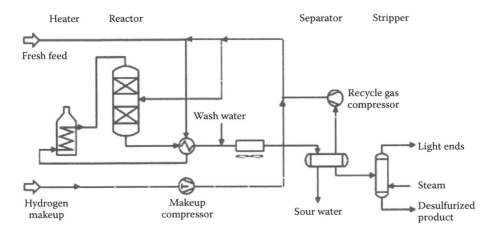

FIGURE 12.7 Unionfining process. (From OSHA Technical Manual, Section IV, Chapter 2: Petroleum Refining Processes. http://www.osha.gov/dts/osta/otm/otm_iv/otm_iv_2.html; Speight, J.G. 2007. *The Chemistry and Technology of Petroleum*. 4th Edition. CRC Press, Taylor & Francis Publishers, Boca Raton, FL. Figure 21.7, p. 579.)

12.2.8 UNIONFINING PROCESS

The Unionfining process (Figure 12.7) is a catalytic hydrotreating technology applied to produce low-sulfur, color-stable diesel fuel. The process is also used to reduce aromatics in diesel and to improve the diesel cetane number. In addition, the process can be used to prepare naphtha for catalytic reformers and to refine coker–naphtha.

In the process (Figure 12.7), the reactor feed is heat exchanged with the reactor effluent, and reactor inlet temperature is controlled by charge heater firing. As hydrotreating is an exothermic reaction—especially when feedstocks are unsaturated—quench sections may be used to cool the reaction fluids and, at the same time, redistribute vapors and liquids between catalyst beds. Wash water is added to the cooled reactor effluent upstream of the final cooler to minimize corrosion and prevent deposits of ammonium salts. The cooled stream enters a cold high-pressure separator to separate out the hydrogen recycle, which is recycled back to the reactor. Desulfurized product is recovered by stripping off the light ends with steam.

12.2.9 OTHER PROCESSES

Other processes such as high-temperature processes may also be applied to certain petroleum products as a means of sulfur removal. For example, the loss of sulfur during calcination of petroleum coke is a well-established phenomenon and is typically referred to as *thermal desulfurization* or, simply, desulfurization (Hussein et al., 1976; Vrbanović, 1983; Al-Haj-Ibrahim and Morsi, 1992; Garbarino and Tonti, 1993; Hardin et al., 1993; Paul and Herrington, 2001), and sulfur loss increases as the temperature of calcination and the sulfur content of the coke increases. However, desulfurization increases the coke microporosity and negatively affects properties such as apparent density and reactivity. When anodes are produced with calcined coke, further desulfurization occurs during anode baking as a result of the long soak times involved (Vogt et al., 1995; Dreyer et al., 1996; Edwards et al., 2007).

The deep desulfurization process generally used in Japan for diesel fuel is a single-stage process in which hydrodesulfurization reaction takes place in a fixed-bed reactor. There is also a two-stage process in use, in which desulfurization reaction takes place at a relatively higher temperature in the first-stage reactor, and color is improved at a lower temperature in the second-stage reactor.

The typical feedstock consists of straight-run diesel fuel produced from Middle Eastern crude oil and cracked diesel fuel (mainly light cycle oil) in the ratio of approximately 80% to 20% v/v. However, differences in the type of crude oil used and the distillation characteristics can affect properties such as desulfurization reactivity. For example, in comparison with low-sulfur feedstock produced from North Sea crude oil used in Europe and the United States, the feedstock produced from many other crude oils, such as Middle Eastern crude oil, contains greater amounts of sulfur, total aromatics, and polycyclic aromatics, even for distillates within the same distillation temperature range. Therefore, Middle Eastern feedstock is relatively difficult to process, requiring reactor capacity twice or more that for North Sea feedstock.

Transportation fuel (particular diesel fuel) with ultra-low sulfur content is being on the market in some European countries. However, the main feedstock for the ultra-low sulfur diesel fuel is lighter distillates derived from North Sea crude oil, and the like, which contain comparatively less sulfur, and are easier to desulfurize. In addition, the supplied volume is not very large. Consequently, there is a wide gap between the situation in Europe and that in refineries where the feedstock for desulfurization is derived from crude oils with high sulfur content, and where there is a need to process higher-boiling distillates as well.

However, when considering upgrading existing equipment or installing new equipment, it is important to take into consideration the status of the existing facilities and deliberate plans for future expansion, because they will make a substantial difference in the future investment cost. When making choices, it will be necessary to consider seriously, in addition to the properties and volume of feedstock, issues such as the anticipated product standards in the future, operating costs, and flexibility to changes in the levels of supply and demand. In addition, the immixture of other petroleum products, even in tiny quantities, can have a substantial effect on the sulfur content of the diesel fuel product. For this reason, good cooperation on anticontamination measures is important from the output of the desulfurization equipment through the pump at the service station, including the storage facilities at the refinery and the distribution and shipping system.

Examples of units that could arise from upgrading older equipment or installing new equipment are (1) the IFP Prime-D30 process, (2) the MAKfining process, (3) the MQD Unionfining process, (4) the SynSat process, and (5) the Topsøe Ultra Deep HDS process.

12.2.9.1   IFP Prime-D30 Process

The IFP Prime-D30 is a process for producing diesel fuel with ultra-low sulfur content, reduced polycyclic aromatic compounds, and a boosted cetane rating. Two-stage hydrogenation is required to accomplish aromatics hydrogenation aimed at increasing the cetane index. In the first stage, an Ni–Mo catalyst and high pressure is used to reduce the sulfur content of cracked diesel fuel to 50 ppm w/w or less. After fractional distillation of the reaction products, the diesel fuel fraction containing low to no sulfur and hydrogen sulfide is sent to the second stage, where aromatics hydrogenation takes place. The precious metal catalyst used in the second stage is a highly active hydrogenation catalyst. The process is capable of processing even feedstocks containing significant amounts of sulfur such as light cycle oil (S = 1.58% w/w), which, when subjected to the two-stage process, yield a product with a low sulfur content (S = 3 ppm w/w) in the first stage, and the aromatic content is reduced to 1.4% w/w in the second stage.

12.2.9.2   MAKfining Process

MAKfining is a hydrorefining process licensed by ExxonMobil, Akzo Nobel, Kellogg, and Total-Fina. The process can be designed to match the current diesel fuel standards as well as any upgraded standards by combining the following processes and catalysts: (1) UDHDS, ultra-deep hydrodesulfurization using a Co–Mo catalyst; (2) HDHDC, mild hydrocracking of heavy distillate using an Ni–Mo catalyst; (3) HDAr, hydrogenation of polycyclic aromatic compounds using a precious metal catalyst; and (4) MIDW, dewaxing by hydroisomerization of paraffin constituents for cold flow improvement by selective hydrocracking of n-paraffins.

Depending on the properties of the feedstock (e.g., S = 0.16% w/w), diesel fuel with a sulfur content of 10 ppm or less and a cloud point of −11°C (12°F) or less (improvement of 18°C [32°F]) can be produced. Furthermore, if HDHDC with a catalyst using a zeolite carrier is added to the downstream of the system, high-quality diesel fuel with lower polycyclic aromatics content and lower density is produced.

12.2.9.3 MQD Unionfining Process

The MQD Unionfining process employs either a single-stage or a two-stage reactor system to produce low-sulfur diesel fuel by combining the hydrogenation process with other processes such as hydrocracking, catalytic dewaxing, aromatics hydrogenation, and hydroisomerization. The two-stage process consists of a first-stage reactor using a base metal catalyst, an intermediate stripper, and a second-stage reactor packed with a precious metal catalyst. In the two-stage process, it is possible to reduce initial investment and operating costs by installing an intermediate temperature, high-pressure stripper and employing a heat integration system. At the commercial plant (Unicracking/DW process) in Schwechat refinery of OMV Austria, dewaxed vacuum gas oil (S = 20 ppm w/w) is produced from a vacuum gas oil feedstock having a sulfur content of 0.95% w/w.

12.2.9.4 SynSat Process

The SynSat process uses either one reactor or two reactors in series and divides the reaction zone into two stages. In the first stage, both feed oil and hydrogen flow downward, while in the second stage the direction of hydrogen flow can be set upwardly or downwardly depending on conditions. The technology consist of the following four hydrorefining processes: (1) SynSat HDS, deep hydrodesulfurization; (2) SynSat HDS/HDA, deep hydrodesulfurization/hydrogenation of aromatics; (3) SynShift, improvement of cetane number/shift of boiling point range; and (4) SynFlow, improvement of cold-flow properties. The SynSat process is being used commercially in Sweden, Germany, and the United States.

In the second-stage reactor of the units in Scanraff and Preem refineries in Sweden, a precious metal catalyst is used, and oil and hydrogen flow countercurrently. They produce two types of product containing different amounts of sulfur: 100 ppm and less than a few parts per million. It is likely that the second-stage reactors are used for production of the latter ultra-low-sulfur fuel with low aromaticity from feedstocks with lower sulfur content. At the Lyondell-Citgo refinery in the United States, an existing hydrodesulfurization unit was revamped to a two-stage system. The unit employs base metal catalysts and the cocurrent system for feedstock and hydrogen in the second stage. This unit achieved reduction of product sulfur to a level of 5 ppm or less from a feedstock with relatively high sulfur content of 1.38% w/w.

12.2.9.5 Topsøe Ultra-Deep HDS Process

The Topsøe Ultra-Deep HDS (UDHDS) process is a two-stage process for producing low-sulfur and low-aromatic diesel fuel. It primarily comprises the following four components: (1) first-stage reactor—ultra-deep desulfurization, (2) intermediate stripper—hydrogen sulfide elimination, (3) second-stage reactor—aromatics hydrogenation, and (4) second-stage stripper—boiling point adjustment. In the first-stage reactor, Ni–Mo catalysts are used. The second-stage reactor employs a precious proprietary metal catalyst (TK-907).

The San Joaquin refinery in the United States provides an example of the commercial application of this process. In the process, the sulfur content is reduced from, e.g., 0.65% w/w in the feedstock to 9 ppm in the first stage, then further reduced to 1 ppm in the second stage.

12.3 GASOLINE AND DIESEL FUEL POLISHING

Briefly, the two desulfurization processes used for fuel purification (desulfurization) are (1) sweetening and (2) hydrotreating (Speight and Ozum, 2002; Parkash, 2003; Hsu and Robinson, 2006;

Gary et al., 2007; Speight, 2014). Sweetening is effective only against mercaptan derivatives, which are the predominant species in light gasoline. Hydrotreating is effective against all sulfur species and is a more widely used process.

In the sweetening process, a light naphtha stream is washed with amine to remove hydrogen sulfide and then reacted with caustic, which promotes the conversion of mercaptans to disulfides.

$$R\text{-}SH \rightarrow RSSR$$

The disulfides can subsequently be extracted and removed in what is referred to as extractive sweetening.

In the hydrotreating process, the feed is reacted with hydrogen, in the presence of a solid catalyst. The hydrogen removes sulfur by conversion to hydrogen sulfide, which is subsequently separated and removed from the reacted stream. As the reaction is favored by both temperature and pressure, hydrotreaters are typically designed and operated at approximately 370°C (700°F) and 1000–2000 psi hydrogen. The lower ends of the ranges typically apply to naphtha desulfurization, while gas oil desulfurization requires a more severe operation.

Hydrogen is provided in the form of treating gas at a purity that is typically around on the order of 90% by volume, although gas with as little as 60% by volume hydrogen is reputed to be used. Hydrogen is produced by catalytic reformers or hydrogen generation units (Chapter 14) and distributed to the hydrotreaters through a refinery-wide network (Speight and Ozum, 2002; Parkash, 2003; Hsu and Robinson, 2006; Gary et al., 2007; Speight, 2014).

In a hydrotreating unit, feed and treating gas are combined and brought to the reaction temperature and pressure, before entering the reactor. The reactor is a vessel preloaded with solid catalyst, which promotes the reaction. The catalyst is slowly deactivated by the continuous exposure to high temperatures and by the formation of a coke layer on its surface. Refineries have to shut down the units periodically and regenerate or replace the catalyst.

The severity of operation of an existing unit can be increased by increasing the reaction temperature; however, there is a negative impact on the catalyst life. The severity of operation can also be increased by increasing the catalyst volume of the unit. In this case, the typical solution is to add a second reactor identical to the existing one, doubling the reactor volume. The pressure of an existing unit cannot be changed to increase its severity because the pressure is related to material of construction and thickness of metal surfaces. If higher pressure is required, the typical solution is to install a new unit and use the existing one for a less severe service. In contrast, there are also processes that do not require either of the above technologies.

Biodesulfurization (q.v.) is only one of several concepts by which gasoline and diesel fuel might be polished, i.e., sulfur removed to an extremely low, if not to a zero, level. At this point, it is pertinent that a brief review of the potential methods for fuel polishing should be introduced.

One new technology is the use of adsorption by metal oxides in which the oxides either react by physical adsorption or by chemical adsorption insofar as adsorption followed by chemical reaction is promoted. The major distinction of this type of process from conventional hydrotreating is that the sulfur in the sulfur-containing compounds adsorbs to the catalyst after the feedstock–hydrogen mixture interacts with the catalyst. The catalyst does need to be regenerated constantly.

Another option involves sulfur oxidization in which a petroleum and water emulsion is reacted with hydrogen peroxide (or another oxidizer) to convert the sulfur in sulfur-containing compounds to sulfones. The sulfones are separated from the hydrocarbons for postprocessing. The major advantages of this new technology include low reactor temperatures and pressures, short residence time, no emissions, and no hydrogen requirement. The technology preferentially treats dibenzothiophene derivatives, one of the streams that are most difficult to desulfurize.

One way to add to the supply of ultra-low sulfur fuels is to turn to a non-oil-based diesel. The Fischer–Tropsch process, for example, can be used to convert natural gas to a synthetic, sulfur-free diesel fuel (Davis and Occelli, 2010). The commercial viability of gas-to-liquid projects depends (in addition to capital costs) on the market for petroleum products and the possible price premiums for gas-to-liquid fuels, as well as on the value of any by-products.

A second way to avoid desulfurization is with biodiesel made from vegetable oil or animal fats. Although other processes are available, most biodiesel is made with a base-catalyzed reaction. In the process, fat or oil is reacted with an alcohol, such as methanol, in the presence of a catalyst to produce glycerin and methyl esters or biodiesel. The methanol is charged in excess to assist in quick conversion and is recovered for reuse. The catalyst, usually sodium or potassium hydroxide, is mixed with the methanol. Biodiesel is a strong solvent and can dissolve paint as well as deposits left in fuel lines by petroleum-based diesel, sometimes leading to engine problems. Biodiesel also freezes at a higher temperature than petroleum-based diesel.

Reactive adsorption desulfurization is a process using metal-based sorbent for sulfur capture to form metal sulfide, usually in the presence of hydrogen; the sulfur atom is retained on the sorbent, while the hydrocarbon portion of the molecule is released back into the process stream—ultra-deep desulfurization of naphtha from the FCC unit can be achieved by this method (Babich and Moulijin, 2003; Zhang et al., 2010).

Reactive adsorption has been implemented in several desulfurization processes. The reaction is exothermic and the adsorbent (unlike a hydrogenation catalyst) cannot remain stable during desulfurization for prolonged periods, and is more suitable to be performed in a fluidized-bed reactor.

12.4 BIODESULFURIZATION

Refiners are being continually challenged to produce products with ever decreasing levels of sulfur. At the same time, the supplies of light, low-sulfur crude oil that favor distillate production are limited and even decreasing. Generally, the sulfur content of petroleum continues to increase (Speight, 2011) with the accompanying decrease in API gravity and an increase in the proportion of residua in the crude oil. These factors require the crude oil to be processed more severely to produce gasoline and other transportation fuels. Thus, many refineries are now configured for maximum naphtha production that also includes increasingly processing highly aromatic distillate by-products, such as light cycle oil, for the additional feedstock to produce more distillate.

In microbial enhanced oil recovery processes, microbial technology is exploited in oil reservoirs to improve recovery (Speight, 2014). In the process, injected nutrients, together with indigenous or added microbes, promote *in situ* microbial growth and/or generation of products that mobilize additional oil and move it to producing wells through reservoir repressurization, interfacial tension/oil viscosity reduction, and selective plugging of the most permeable zones. Alternatively, the oil-mobilizing microbial products may be produced by fermentation and injected into the reservoir.

Biocatalyst desulfurization of petroleum distillates is one of a number of possible modes of applying biologically based processing to the needs of the petroleum industry (McFarland et al., 1998; Setti et al., 1999; Abbad-Andaloussi et al., 2003; Mohebali and Ball, 2008). In addition, *Mycobacterium goodii* has been found to desulfurize benzothiophene (Li et al., 2005). The desulfurization product was identified as α-hydroxystyrene. This strain appeared to have the ability to remove organic sulfur from a broad range of sulfur species in naphtha. When straight-run naphtha containing various organic sulfur compounds was treated with immobilized cells of *M. goodii* for 24 h at 40°C (104°F), the total sulfur content significantly decreased, from 227 to 71 ppm at 40°C. Furthermore, when immobilized cells were incubated at 40°C (104°F) with *M. goodii*, the sulfur content of the naphtha decreased from 275 to 54 ppm in two consecutive reactions.

A dibenzothiophene-degrading bacterial strain, *Nocardia* sp., was able to convert dibenzothiophene to 2-hydroxybiphenyl as the end metabolite through a sulfur-specific pathway (Chang et al., 1998). Other organic sulfur compounds, such as thiophene derivatives, thiazole derivatives, sulfides, and disulfides were also desulfurized by *Nocardia* sp. When a sample in which dibenzothiophene was dissolved in hexadecane and treated with growing cells, the dibenzothiophene was desulfurized in approximately 80 h.

The soil-isolated strain microbe identified as *Rhodococcus erythropolis* can efficiently desulfurize benzonaphthenothiophene (Yu et al., 2006). The desulfurization product was α-hydroxy-β-phenyl-naphthalene. Resting cells were able to desulfurize diesel oil (total organic sulfur, 259 ppm) after hydrodesulfurization, and the sulfur content of diesel oil was reduced by 94.5% after 24 h at 30°C (86°F). Biodesulfurization of crude oils was also investigated, and after 72 h at 30°C (86°F), 62.3% of the total sulfur content in Fushun crude oil (initial total sulfur content, 3210 ppm) and 47.2% of the sulfur in Sudanese crude oil (initial total sulfur, 1237 ppm) were removed (see also Abbad-Andaloussi et al., 2003).

Heavy crude oil recovery, facilitated by microorganisms, was suggested in the 1920s and received growing interest in the 1980s as microbial enhanced oil recovery. However, such projects have been slow to get under way, although *in situ* biosurfactant and biopolymer applications continue to garner interest (Van Hamme et al., 2003). In fact, studies have been carried out on biological methods of removing heavy metals such as nickel and vanadium from petroleum distillate fractions, coal-derived liquid shale, bitumen, and synthetic fuels. However, further characterization on the biochemical mechanisms and bioprocessing issues involved in petroleum upgrading are required to develop reliable biological processes.

For upgrading options, the use of microbes has to show a competitive advantage of enzyme over the tried-and-true chemical methods prevalent in the industry. Currently, the range of reactions using microbes is large but is usually related to production of bioactive compounds or precursors. However, the door is not closed and the issues of biodesulfurization and bioupgrading remain open for the challenge of bulk petroleum processing. These drawbacks limit the applicability of this technology to specialty chemicals and steer it away from bulk petroleum processing.

Biodesulfurization is, therefore, another technology to remove sulfur from the feedstock. However, several factors may limit the application of this technology. Many ancillary processes novel to petroleum refining would be needed, including a biocatalyst fermenter to regenerate the bacteria. The process is also sensitive to environmental conditions such as sterilization, temperature, and the residence time of the biocatalyst. Finally, the process requires the existing hydrotreater to continue in operation to provide a lower-sulfur feedstock to the unit and is more costly than conventional hydrotreating. Nevertheless, the limiting factors should not stop the investigations of the concept and work should be continued with success in mind.

Once the concept has been proven on the scale that a refiner would require, the successful microbial technology will most probably involve a genetically modified bacterial strain for (1) upgrading distillates and other petroleum fractions in refineries, (2) upgrading crude petroleum upstream, and (3) dealing with environmental problems that face industry, especially in areas related to spillage of crude oil and products. These developments are part of a wider trend to use bioprocessing to make products and do many of the tasks that are accomplished currently by conventional chemical processing. If commercialized for refineries, however, biologically based approaches will be at scales and with economic impacts beyond anything previously seen in industry.

In addition, the successful biodesulfurization process will, most likely, be based on naturally occurring aerobic bacteria that can remove organically bound sulfur in heterocyclic compounds without degrading the fuel value of the hydrocarbon matrix. Because of the susceptibility of bacteria to heat, the process will need to operate at temperatures and pressures close to ambient, and also use air to promote sulfur removal from the feedstock.

REFERENCES

Abbad-Andaloussi, S., Warzywoda, M., and Monot, F. 2003. Microbial Desulfurization of Diesel Oils by Selected Bacterial Strains. *Rév. Inst. Fr. Pétrol.* 58(4): 505–513.

Al-Haj-Ibrahim, H., and Morsi, B.I. 1992. Desulfurization of Petroleum Coke. *Ind. Eng. Chem. Res.* 31: 1835–1840.

Ancheyta, J., and Speight, J.G. (Editors) 2007. *Hydroprocessing of Heavy Oils and Residua.* CRC Press, Taylor & Francis Group, Boca Raton, FL.

Babich, I.V., and Moulijin, J.A. 2003. Science and Technology of Novel Processes for Deep Desulfurization of Oil Refinery Streams: A Review. *Fuel* 82: 607–631.

Chang, J.H., Rhee, S.K., Chang, Y.K., and Chang, H.N. 1998. Desulfurization of Diesel Oils by a Newly Isolated Dibenzothiophene-Degrading *Nocardia* sp. Strain CYKS2. *Biotechnol. Prog.* 14(6): 851–855.

Davis, B.H., and Occelli, M.L. 2010. *Advances in Fischer–Tropsch Synthesis, Catalysts, and Catalysis*. CRC Press, Taylor & Francis Group, Boca Raton, FL.

Dickenson, R.L., Biasca, F.E., Schulman, B.L., and Johnson, H.E. 1997. Refiner Options for Converting and Utilizing Heavy Fuel Oil. *Hydrocarbon Process.* 76(2): 57.

Dreyer, C., Samanos, B., and Vogt, F. 1996. Coke Calcination Levels and Aluminum Anode Quality. In: *Light Metals*, M. Sorlie (Editor). The Minerals, Metals & Materials Society, Warrendale, PA, pp. 535–542.

Edwards, L.C., Neyrey, K.J., and Lossius, L.P. 2007. A Review of Coke and Anode Desulfurization. In: *Light Metals*, M. Sorlie (Editor). The Minerals, Metals & Materials Society, Warrendale, PA.

Garbarino, R.M., and Tonti, R.T. 1993. Desulfurization and Its Effect on Calcined Coke Properties. In: *Light Metals*, M. Sorlie (Editor). The Minerals, Metals & Materials Society, Warrendale, PA, pp. 517–520.

Gary, J.H., Handwerk, G.E., and Kaiser, M.J. 2007. *Petroleum Refining: Technology and Economics*, 4th Edition, Marcel Dekker Inc., New York.

Hardin, E., Beilharz, C.L., and Melvin, L.L. 1993. A Comprehensive Review of the Effect of Coke Structure and Properties when Calcined at Various Temperatures. In: *Light Metals*, M. Sorlie (Editor). The Minerals, Metals & Materials Society, Warrendale, PA, pp. 501–508.

Hsu, C.S., and Robinson, P.R. (Editors) 2006. *Practical Advances in Petroleum Processing*, Vols. 1–2. Springer Science, New York.

Hussein, M.K., El-Tawil, S.Z., and Rabah, M.A. 1976. Desulfurization of High Sulfur Egyptian Coke 1: In a Nitrogen Atmosphere. *J. Inst. Fuel* 139–143.

Khan, M.R., and Patmore, D.J. 1998. Heavy Oil Upgrading Processes. In: *Petroleum Chemistry and Refining*, J.G. Speight (Editor). Taylor & Francis, Washington, DC, Chapter 6.

Kumagai, H., Koyama, H., Nakamura, K., Igarashi, N., Mori, M., and Tsukada, T. 2005. Hydrofining Catalyst and Hydrofining Process. United States Patent. 6,858,132, February 22.

Li, F., Xu, P., Feng, J., Meng, L., Zheng, Y., Luo, L., and Ma, C. 2005. Microbial Desulfurization of Gasoline in a *Mycobacterium goodii* X7B Immobilized-Cell System. *Appl. Environ. Microbiol.* 71(1): 276–281.

MathPro. 2003. Evolution of Process Technology for FCC Naphtha Desulfurization: 1997–2003: An Example of Technical Progress Induced by Environmental Regulation. MathPro, West Bethesda, MD, March. http://www.mathproinc.com/pdf/2.1.3_FCCNaphDesulf.pdf; accessed September 5, 2014.

MathPro. 2011. An Introduction to Petroleum Refining and the Production of Ultra Low Sulfur Gasoline and Diesel Fuel. MathPro, West Bethesda, MD, October. http://www.theicct.org/sites/default/files/publications/ICCT05_Refining_Tutorial_FINAL_R1.pdf; accessed November 1, 2014.

McFarland, B.L., Boron, D.J., Deever, W., Meyer, J.A., Johnson, A.R., and Atlas, R.M. 1998. Biocatalytic Sulfur Removal from Fuels: Applicability for Producing Low Sulfur Gasoline. *Crit. Rev. Microbiol.* 24: 99–147.

Mohebali, G., and Ball, A.S. 2008. Biocatalytic Desulfurization (BDS) of Petrodiesel Fuels. *Microbiology* 154: 2169–2183.

Parkash, S. 2003. *Refining Processes Handbook*. Gulf Professional Publishing, Elsevier, Amsterdam.

Paul, C.A., and Herrington, L.E. 2001. Desulfurization of Petroleum Coke Beyond 1600°C. In: *Light Metals*, M. Sorlie (Editor). The Minerals, Metals & Materials Society, Warrendale, PA, pp. 597–601.

Setti, L., Farinelli, P., Di Martino, S., Frassinetti, S., Lanzarini, G., and Pifferia, P.G. 1999. Developments in Destructive and Non-Destructive Pathways for Selective Desulfurization in Oil Biorefining Process. *Appl. Microbiol. Biotechnol.* 52: 111–117.

Song, C. 2002. New Approaches to Deep Desulfurization for Ultra-Clean Gasoline and Diesel Fuels: An Overview. *Prepr. Div. Fuel Chem. Am. Chem. Soc.* 47(2): 438–444.

Speight, J.G. 2000. *The Desulfurization of Heavy Oils and Residua*. Marcel Dekker Inc., New York.

Speight, J.G. 2011. *The Refinery of the Future*. Gulf Professional Publishing, Elsevier, Oxford, UK.

Speight, J.G. 2013. *Heavy and Extra Heavy Oil Upgrading Technologies*. Gulf Professional Publishing, Elsevier, Oxford, UK.

Speight, J.G. 2014. *The Chemistry and Technology of Petroleum*, 5th Edition. CRC Press, Taylor & Francis Group, Boca Raton, FL.

Speight, J.G., and Ozum, B. 2002. *Petroleum Refining Processes*. Marcel Dekker Inc., New York.

Tanaka, H. 2004. Process for Producing Hydrofining Catalyst. United States Patent 6,689,712, February 10.

Van Hamme, J.D., Singh, A., and Ward, O.P. 2003. Recent Advances in Petroleum Microbiology. *Microbiol. Mol. Biol. Rev.* 67(4): 503–549.

Vogt, F., Ries, K., and Smith, M. 1995. Anode Desulfurization on Baking. In: *Light Metals*, M. Sorlie (Editor). The Minerals, Metals & Materials Society, Warrendale, PA, pp. 691–700.

Vrbanović, Z. 1983. Thermal Desulfurization of Petroleum Coke. *High Temp.-High Pressures* 15: 107–112.

Xia, G., Zhu, M., Min, E., Shi, Y., Li, M., Nie, H., Tao, Z., Huang, H., Zhang, R., Li, J., Wang, Z., and Ran, G. 2000. Catalyst Containing Molybdenum and/or Tungsten for Hydrotreating Light Oil Distillates and Preparation Method Thereof. United States Patent 6,037,306, March 14.

Yu, B., Xu, P., Shi, Q., and Ma, C. 2006. Deep Desulfurization of Diesel Oil and Crude Oils by a Newly Isolated *Rhodococcus erythropolis* Strain. *Appl. Environ. Microbiol.* 72: 54–58.

Zhang, J., Liu, Y., Tian, D.S., Chai, Y., and Liu, C. 2010. Reactive Adsorption of Thiophene on Ni/ZnO Adsorbent: Effect of ZnO Textural Structure on the Desulfurization Activity. *J. Nat. Gas Chem.* 19(3): 327–332.

13 Desulfurization Processes— Heavy Feedstocks

13.1 INTRODUCTION

Petroleum refining entered a significant transition period as the industry moved into the 21st century, and the demands for petroleum and petroleum products have shown a sharp growth in recent years. Typical refinery operations (Figure 13.1) have evolved to include a range of next-generation processes as the demand for transportation fuels and fuel oil has shown a steady growth. In fact, over the past three decades, crude oils available to refineries have generally decreased in API gravity with the influx of heavy feedstocks being accompanied by a concomitant decrease in the hydrogen content of refinery feedstocks (Figure 13.2) as well as higher proportions of resin and asphaltene constituents (Figure 13.3), which, in turn confer a higher heteroatom content in the feedstocks (Figure 13.4).

There is, nevertheless, a major focus on refineries through a variety of conversion processes on the ways in which heavy feedstocks might be converted into low-boiling high-value products (Khan and Patmore, 1997; Speight, 2011a, 2014). Simultaneously, the changing crude oil properties are reflected in changes such as an increase in asphaltene constituents, and in sulfur, metal, and nitrogen contents. Pretreatment processes for removing those would also play an important role. However, the essential step required of refineries is the upgrading of heavy feedstocks, particularly residua. In fact, the increasing supply of heavy crude oils is a matter of serious concern for the petroleum industry.

Heavy feedstocks are generally characterized by low API gravity (high density) and high viscosity, high initial boiling point, high carbon residue, high nitrogen content, high sulfur content, and high metal content (Chapter 7). In addition to these properties, the heavy feedstocks also have an increased molecular weight and reduced hydrogen content. However, to adequately define the behavior of heavy feedstocks in refinery operations, some reference should also be made to composition and thermal characteristics of the constituents (Chapters 2 and 7) (Speight, 2014).

When catalytic processes are employed, complex feedstock constituents (such as those that may be found in the original asphaltene fraction or those formed during the process) are not sufficiently mobile (or are too strongly adsorbed by the catalyst) to be saturated by hydrogenation. The chemistry of the thermal reactions of some of these constituents dictates that certain reactions, once initiated, cannot be reversed and proceed to completion, usually resulting in the formation of coke (Chapter 5) that deposits on the catalyst. These deposits deactivate the catalyst sites and eventually interfere with the hydroprocess (Speight, 2000, 2014; Ancheyta, 2007).

To satisfy the changing pattern of product demand, significant investments in refining conversion processes will be necessary to profitably utilize the heavy feedstocks. The most efficient and economical solution to this problem will depend, to a large extent, on individual refinery situations. However, the most promising technologies will likely focus on the conversion of vacuum residua and extra-heavy crude oils into useful low-boiling and middle-distillate products.

Upgrading of heavy feedstocks began with the introduction of desulfurization processes (Speight, 2000, 2014). In the early days, the goal was desulfurization; however, in later years, the processes were adapted to a 10–30% partial conversion operation, as intended to achieve desulfurization and obtain low-boiling fractions simultaneously, by increasing severity in operating conditions. Refinery evolution has seen the introduction of a variety of conversion processes for heavy feedstocks that are based on thermal cracking, catalytic cracking, and hydroconversion. Those processes are different

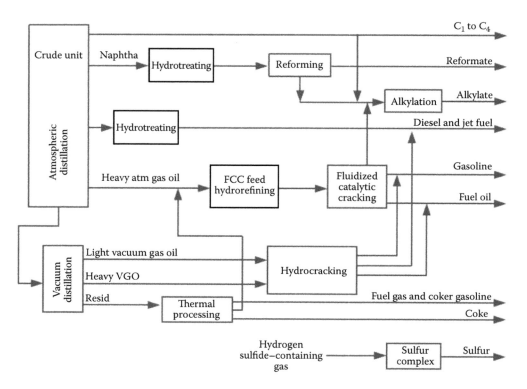

FIGURE 13.1   Schematic overview of a refinery. (From Speight, J.G. 2014. *The Chemistry and Technology of Petroleum*. 5th Edition. CRC Press, Taylor & Francis Publishers, Boca Raton, FL. Figure 15.1, p. 392.)

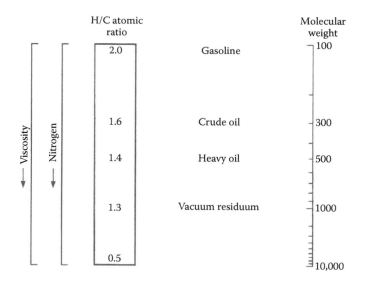

FIGURE 13.2   Relative hydrogen content (through the atomic H/C ratio) and molecular weight of various refinery feedstocks. (From Speight, J.G. 2014. *The Chemistry and Technology of Petroleum*. 5th Edition. CRC Press, Taylor & Francis Publishers, Boca Raton, FL. Figure 15.18, p. 421.)

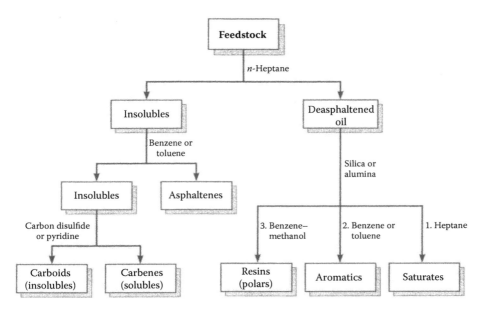

FIGURE 13.3 General scheme for feedstock fractionation. (From Speight, J.G. 2014. *The Chemistry and Technology of Petroleum*. 5th Edition. CRC Press, Taylor & Francis Publishers, Boca Raton, FL. Figure 9.1, p. 212.)

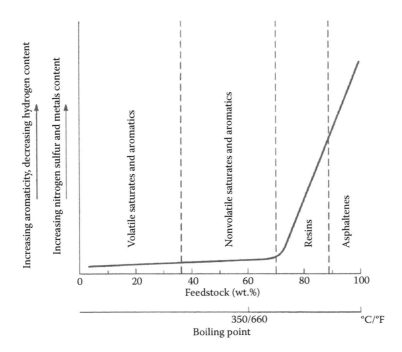

FIGURE 13.4 Relative distribution of total heteroatoms in the various fractions. (From Speight, J.G. 2014. *The Chemistry and Technology of Petroleum*. 5th Edition. CRC Press, Taylor & Francis Publishers, Boca Raton, FL. Figure 15.19, p. 422.)

from one another in terms of the method and product slates, and will find employment in refineries according to their respective features.

New processes for the conversion of heavy feedstocks will eventually find use in place of visbreaking, the various coking options, catalytic cracking, and deasphalting that occur in current refining scenarios. It may also be opportune to use a degree of hydrocracking or hydrotreating as a means of *primary conversion* before other processes are applied. Such primary conversion processes may replace or augment the deasphalting units in many refineries. For example, the upgrading of bitumen from tar sands uses coking technology as the primary conversion (*primary upgrading*) (Speight, 2014). The bitumen is subjected to either delayed coking or fluid coking as the primary upgrading step without prior distillation or topping. After primary upgrading, the product streams are hydrotreated and combined to form a synthetic crude oil that is shipped to a conventional refinery for further processing. Conceivably, a heavy feedstock could be upgraded in the same manner and, depending on the upgrading facility, upgraded further for sales. Such procedures are currently considered an exception to the usual method of refining but may become the preferred method for refining heavy feedstocks.

A previous chapter (Chapter 3) has dealt with the tried-and-true processes that are used in refineries and that were developed for conversion of residua as an option to asphalt production. These processes (visbreaking, coking, and deasphalting) (Chapter 3) helped balance the seasonal demands for variations in the product slate. It is, therefore, opportune to include in this text a description of those processes that have emerged over the past two decades and that will supersede the older processes. Again, as with the visbreaking, coking, catalytic cracking, and deasphalting processes, the prime motive for using these processes is conversion; however, desulfurization also occurs to varying extents.

Technologies for upgrading heavy feedstocks can be broadly divided into *carbon rejection* and *hydrogen addition* processes. Carbon rejection redistributes hydrogen among the various components, resulting in fractions with increased H/C atomic ratios and fractions with lower H/C atomic ratios. On the other hand, hydrogen addition processes involve the reaction of heavy crude oils with an external source of hydrogen, and result in an overall increase in H/C ratio (Stanislaus and Cooper, 1994). Within these broad ranges, all upgrading technologies can be subdivided as follows: (1) carbon rejection processes such as visbreaking, steam cracking, fluid catalytic cracking, and coking; (2) separation processes such as deasphalting; and (3) hydrogen addition processes such as catalytic hydroconversion (hydrocracking), fixed-bed catalytic hydroconversion, ebullated catalytic bed hydroconversion, thermal slurry hydroconversion (hydrocracking), hydrovisbreaking, hydropyrolysis, and donor solvent processes.

In the current context, distillation is excluded from the separation processes category and has been described elsewhere (Chapter 3). The mature and well-established processes such as visbreaking, delayed coking, fluid coking, flexicoking, and propane deasphalting have also been described elsewhere in this text (Chapter 3). In the not too distant past, these processes were deemed adequate for upgrading heavy feedstocks. Now, more options are sought and have become available. The hydrogen addition processes are also excluded from this chapter, having been described elsewhere (Chapter 3). However, the new generation of hydrogen addition processes for upgrading heavy feedstocks are also described in detail elsewhere (Chapter 4) since the new processes will form the refining options of the future.

Even though the simplest means to cover the demand growth in low-boiling products is to increase the imports of light crude oils and low-boiling petroleum products (Speight, 2011, 2014), these actions are not the compete answer to the issues. Thus, new conversion processes have come on-stream.

The primary goal of these processes is to convert the heavy feedstocks to lower-boiling products, and during the conversion there is a reduction in the cumulative sulfur content of the liquid and nonvolatile products compared with the sulfur content of the original feedstock. Sulfur is eliminated as hydrogen sulfide or as other gaseous sulfur-containing products. Thus, even though these processes are not considered desulfurization processes in the strictest sense of the definition, as already noted (Chapter 3), they need to be given some consideration because of the inclusion (or *integration*) into various refinery scenarios.

The major goal of *residuum hydroconversion* is cracking of residua with desulfurization, metal removal, denitrogenation, and asphaltene conversion. The residuum hydroconversion process offers production of kerosene and gas oil, and production of feedstocks for hydrocracking, fluid catalytic cracking, and petrochemical applications. The processes that follow are listed in alphabetical order with no other preference in mind.

Residue hydrotreating is another method for reducing high-sulfur residual fuel oil yields. This technology was originally developed to reduce the sulfur content of atmospheric residues to produce specification low-sulfur residual fuel oil. Changes in crude oil quality and product demand, however, have shifted the commercial importance of this technology to include pretreating conversion unit feedstocks to minimize catalyst replacement costs, and coker feedstocks to reduce the yield and increase the quality of the by-product coke fraction. Although residue hydrotreaters are capable of processing feedstocks having a wide range of contaminants, the feedstock's organometallic and asphaltene components typically determine its processability. Economics generally tend to limit residue hydrotreating applications to feedstocks with limitations (dictated by the process catalyst) on the content of nickel plus vanadium.

In many cases, application of hydrotreating technology to heavy feedstocks may also cause some cracking and, by inference, application of hydrocracking to heavy feedstocks will also cause desulfurization and denitrogenation. Rather than promote unnecessary duplication of the process description, two hydrotreating processes (the residuum desulfurizer [RDS] and vacuum residuum desulfurizer [VRDS] process options are described together as subcategories of one process) are described here with the remainder appearing elsewhere (Chapter 4) since these other processes are more amenable to hydrocracking than to hydrotreating.

This chapter will present those processes that are relatively latecomers to refinery scenarios that have evolved during the last three to four decades, and were developed (and installed) to address the refining of the heavy feedstocks. Refining heavy feedstocks has become a major issue in modern refinery practice, and several process configurations have evolved to accommodate the heavy feedstocks.

13.2 PROCESS OPTIONS

Hydrotreating processes have two definite roles: (1) desulfurization to supply low-sulfur fuel oils and (2) pretreatment of feed residua for residuum fluid catalytic cracking processes. The main goal is to remove sulfur, metal, and asphaltene contents from residua and other heavy feedstocks to a desired level. The major goal of hydroconversion is cracking of residua with desulfurization, metal removal, denitrogenation, and asphaltene conversion. Residuum hydroconversion process offers production of kerosene and gas oil, and production of feedstocks for hydrocracking, fluid catalytic cracking, and petrochemical applications.

Catalyst beds within petroleum hydroprocessing units may be fixed or moving. Most hydroprocessing reactors (for distillate feedstocks) are fixed-bed reactors, which must be shut down to remove the spent catalyst when catalyst activity declines below an acceptable level (due to the accumulation of coke, metals, and other contaminants). There are processes that use moving or ebullating catalyst beds. In ebullated-bed hydroprocessing, the catalyst within the reactor bed is not fixed. In such a process, the feedstock stream enters the bottom of the reactor and flows upward through the catalyst, which is maintained in suspension by the pressure of the fluid feed. Ebullating-bed reactors are capable of converting the most heavy feedstocks (Chapter 2), all of which have a high content of asphaltene constituents, metal constituents, sulfur constituents, and sediments to lower-boiling, more valuable products while simultaneously removing contaminants. The function of the catalyst is to remove contaminants such as sulfur and nitrogen heteroatoms, which accelerate the deactivation of the catalyst, while cracking (converting) the feed to lighter products.

13.2.1 Asphaltenic Bottom Cracking (ABC) Process

The ABC process can be used for hydrodemetallization, asphaltene cracking, and moderate hydrodesulfurization as well as sufficient resistance to coke fouling and metal deposition using such feedstocks as vacuum residua, thermally cracked residua, solvent deasphalted bottoms, and bitumen with fixed catalyst beds (Table 13.1). The process can be combined with (1) solvent deasphalting for complete or partial conversion of the residuum or (2) hydrodesulfurization to promote the conversion of residue, to treat feedstock with high metals and to increase catalyst life, or (3) hydrovisbreaking to attain high conversion of residue with favorable product stability. In the process, the feedstock is pumped up to the reaction pressure and mixed with hydrogen. The mixture is heated to the reaction temperature in the charge heater after a heat exchange and fed to the reactor.

Hydrodemetallization and subsequent asphaltene cracking with moderate hydrodesulfurization take place simultaneously in the reactor under conditions similar to residuum hydrodesulfurization. The reactor effluent gas is cooled, cleaned, and recycled to the reactor section, while the separated

TABLE 13.1
Feedstock and Product Data for Various Feedstocks Used in the Asphaltene Bottoms Cracking (ABC) Process

	California Atmospheric Residuum	Arabian Light Vacuum Residuum	Arabian Light Vacuum Residuum	Arabian Heavy Vacuum Residuum	Cerro Negro Vacuum Residuum
		Feedstock			
API	6.1	5.5	7.0	5.1	1.7
Sulfur, % w/w	6.5	4.4	4.0	5.3	4.3
Carbon residue, % w/w	17.5	24.6	20.8	23.3	23.6
C7-asphaltene	16.2	8.9	7.0	13.1	19,819.8
Nickel, ppm	150.0	30.0	223.0	52.0	150.0
Vanadium, ppm	380.0	90.0	76.0	150.0	640.0
		Products			
Naphtha, vol.%	12.6	9.3	6.5	7.7	15.1
API	56.7	58.7	57.2	57.2	54.7
Distillate, vol.%	23.0	20.1	16.0	19.8	21.3
API	34.6[a]	33.2	34.2[a]	34.2[a]	32.5[a]
Vacuum gas oil, vol.%	38.4	33.1	34.3	38.1	32.8
API	23.1	20.3	24.7	21.6	15.4
Sulfur, % w/w	0.4	0.4	0.2	1.7	0.5
Vacuum residuum, vol.%	29.2	40.8	46.2	37.9	34.7
API	5.4	7.5	10.6	7.8	<0.0
Sulfur, % w/w	2.4	1.2	0.6	1.7	2.2
Carbon residue, % w/w	26.0	28.6	13.6	26.5	13.6
C7-asphaltene constituents, % w/w		12.0			
Nickel, ppm	100.0		9.0	45.0	117.0
Vanadium, ppm	180.0		13.0	75.0	371.0
Conversion	60.0		55.0	60.0	60.0

Source: Speight, J.G. 2014. *The Chemistry and Technology of Petroleum.* 5th Edition. CRC Press, Taylor & Francis Publishers, Boca Raton, FL. Table 22.1, p. 602.

[a] Estimated.

liquid is distilled into distillate fractions and vacuum residue, which is further separated into deasphalted oil and asphalt using butane or pentane.

In case of the ABC–hydrodesulfurization catalyst combination, the ABC catalyst is placed upstream of the hydrodesulfurization catalyst and can be operated at a higher temperature than the hydrodesulfurization catalyst under conventional residuum hydrodesulfurization conditions. In the VisABC process, a soaking drum is provided after the heater, when necessary. Hydrovisbroken oil is first stabilized by the ABC catalyst through hydrogenation of coke precursors and then desulfurized by the hydrodesulfurization catalyst.

13.2.2 Aquaconversion

Aquaconversion is a hydrovisbreaking technology that can be applied in a conventional visbreaking unit in which a proprietary additive and water are added to the feedstock/recycled bottoms before the soaker reduce the inventory of heavier feedstocks and products. The homogeneous catalyst is added in the presence of steam, which allows the hydrogen from the water to be transferred to the heavy oil when contacted in a coil-soaker system, normally used for the visbreaking process. Reactions that lead to coke formation are suppressed, and there is no separation of asphaltene-type material. The main characteristics of aquaconversion are as follows: (1) hydrogen incorporation is much lower than that obtained when using a deep hydroconversion process under high hydrogen partial pressure; (2) hydrogen saturates the free radicals, formed within the thermal process, which would normally lead to coke formation; (3) a higher conversion level can be reached, and thus higher API and viscosity improvements, while maintaining syncrude stability; and (4) the process produces little or no coke and does not require a hydrogen source or high-pressure equipment.

The presence of the oil-soluble dual catalyst (Carrazza et al., 1997a,b) and water prevents the buildup of coke precursors and deposition of sediment that often occurs during visbreaking. The catalyst may be supported on support material or mixed directly with the feedstock. The first metal is chosen from the non-noble Group VIII metals, and the second is an alkali metal such as potassium or sodium.

The products contain lower amounts of asphaltene-related material and have lower carbon residues. As with other thermal conversion processes, the products are cumulatively lower in sulfur than the original feedstocks, and there are indications that at least 60% of the original sulfur is removed during the process.

13.2.3 CANMET Hydrocracking Process

The CANMET hydrocracking process for heavy feedstocks, such as atmospheric residua and vacuum residua (Table 13.2), has been purchased by UOP and has been modified to form the Uniflex Process (q.v.). The original CANMET process did not use a catalyst but employed a low-cost additive to inhibit coke formation and allow high conversion of heavy feedstocks into lower-boiling products using a single reactor. The process is unaffected by high levels of feed contaminants such as sulfur, nitrogen, and metals. Conversion of >90% of the 525°C$^+$ (975°F$^+$) fraction into distillates has been attained.

In the process, the feedstock and recycle hydrogen gas are heated to reactor temperature in separate heaters. A small portion of the recycle gas stream and the required amount of additive are routed through the oil heater to prevent coking in the heater tubes. The outlet streams from both heaters are fed to the bottom of the reactor.

The vertical reactor vessel is free of internal equipment and operates in a three-phase mode. The solid additive particles are suspended in the primary liquid hydrocarbon phase through which the hydrogen and product gases flow rapidly in bubble form. The reactor exit stream is quenched with cold recycle hydrogen before the high-pressure separator. The heavy liquids are further reduced in

TABLE 13.2

Feedstock and Product Data for the Hydroconversion of Cold Lake Vacuum Residuum by the CANMET Process

Feedstock[a]	
API gravity	4.4
Sulfur, % w/w	5.1
Nitrogen, % w/w	0.6
Asphaltene constituents, % w/w	15.5
Carbon residue, % w/w	20.6
Metals, ppm	
Ni	80
V	170
Residuum (>525°C, >975°), % w/w	

Products, % w/w[b]	
Naphtha (C5–204°C, 400°F)	19.8
Nitrogen, % w/w	0.1
Sulfur, % w/w	0.6
Distillate (204–343°C, 400–650°F)	33.5
Nitrogen, % w/w	0.4
Sulfur, % w/w	1.8
Vacuum gas oil (343–534°C, 650–975°F)	28.5
Nitrogen, % w/w	0.6
Sulfur, % w/w	2.3
Residuum (>534°C, >975°F)	4.5
Nitrogen, % w/w	1.6
Sulfur, % w/w	3.1

Source: Speight, J.G. 2014. *The Chemistry and Technology of Petroleum.* 5th Edition. CRC Press, Taylor & Francis Publishers, Boca Raton, FL. Table 22.2, p. 603.

[a] Cold Lake (Canada) heavy oil vacuum residuum.

[b] Residuum: 93.5% by weight.

pressure to a hot medium pressure separator, and from there to fractionation. The spent additive leaves with the heavy fraction and remains in the unconverted vacuum residue.

The vapor stream from the hot high-pressure separator is cooled stepwise to produce middle distillate and naphtha that are sent to fractionation. High-pressure purge of low-boiling hydrocarbon gases is minimized by a sponge oil circulation system.

The additive, prepared from iron sulfate [$Fe_2(SO_4)_3$], is used to promote hydrogenation and effectively eliminate coke formation. The effectiveness of the dual-role additive permits the use of operating temperatures that give high conversion in a single-stage reactor. The process maximizes the use of reactor volume and provides a thermally stable operation with no possibility of temperature runaway.

The process also offers the attractive option of reducing the coke yield by slurrying the feedstock with <10 ppm of catalyst (molybdenum naphthenate) and sending the slurry to a hydroconversion zone to produce low-boiling products.

13.2.4 Chevron RDS Isomax and VRDS Process

Residuum hydrotreating processes have two definite roles: (1) desulfurization to supply low-sulfur fuel oils and (2) pretreatment of feed residua for residuum fluid catalytic cracking processes. The main goal is to remove sulfur, metal, and asphaltene contents from residua and other heavy feedstocks to a desired level. On the other hand, the major goal of residuum hydroconversion is cracking of residua with desulfurization, metal removal, denitrogenation, and asphaltene conversion. Residuum hydroconversion process offers production of kerosene and gas oil, and production of feedstocks for hydrocracking, fluid catalytic cracking, and petrochemical applications.

The RDS/VRDS process is (like the Residfining process, q.v.) a hydrotreating process that is designed to hydrotreat vacuum gas oil, atmospheric residuum, or vacuum residuum to remove sulfur metallic constituents while part of the feedstock is converted to lower-boiling products. In the case of residua, the asphaltene content is reduced.

The process consists of a once-through operation of hydrocarbon feed contacting graded catalyst systems designed to maintain activity and selectivity in the presence of deposited metals. Process conditions are designed for a 6-month to 1-year operating cycle between catalyst replacements. The process is ideally suited to produce feedstocks for residuum fluid catalytic crackers or delayed coking units to achieve minimal production of residual products in a refinery.

These processes are designed to remove sulfur, nitrogen, asphaltene, and metal contaminants from residua, and are also capable of accepting whole crude oils or topped crude oils as feedstocks. The major product of the processes is a low-sulfur fuel oil, and the amount of naphtha and middle distillates is maintained at a minimum to conserve hydrogen.

The basic elements of each process are similar and consist of a once-through operation of the feedstock coming into contact with hydrogen and the catalyst in a downflow reactor, which is designed to maintain activity and selectivity in the presence of deposited metals. Moderate temperatures and pressures are employed to reduce the incidence of hydrocracking and, hence, minimize production of low-boiling distillates.

The combination of a desulfurization step and a VRDS is often seen as an attractive alternative to the atmospheric RDS because the combination route uses less hydrogen for a similar investment cost. Nevertheless, the alternative processes must be evaluated on the basis of the feedstock as well as the desired product quality, which strongly influence the economics of the processes.

The major product of the processes is a low-sulfur fuel oil, and the amount of naphtha and middle distillates is maintained at a minimum to conserve hydrogen. The basic elements of each process are similar and consist of a once-through operation of the feedstock coming into contact with hydrogen and the catalyst in a downflow reactor that is designed to maintain activity and selectivity in the presence of deposited metals. Moderate temperatures and pressures are employed to reduce the incidence of hydrocracking and, hence, minimize the production of low-boiling distillates. The combination of a desulfurization step and a VRDS is often seen as an attractive alternative to the atmospheric RDS. In addition, either the RDS option or the VRDS option can be coupled with other processes (such as delayed coking, fluid catalytic cracking, and solvent deasphalting) to achieve the most optimum refining performance.

13.2.5 CHEVRON DEASPHALTED OIL HYDROTREATING PROCESS

The Chevron deasphalted oil hydrotreating process is designed to desulfurize heavy feedstocks that have had the asphaltene fraction removed by prior application of a deasphalting process (Chapter 3). The principal product is a low-sulfur fuel oil that can be used as a blending stock or as a feedstock for a fluid catalytic unit.

The processes employ a downflow, fixed-bed reactor containing a highly selective catalyst that provides extensive desulfurization at low pressures with minimal cracking and, therefore, low consumption of hydrogen.

13.2.6 GULF RESID HYDRODESULFURIZATION PROCESS

This process is suitable for the desulfurization of high-sulfur residua (atmospheric and vacuum) to produce low-sulfur fuel oils or catalytic cracking feedstocks. In addition, the process can be used, through alternative design types, to upgrade high-sulfur crude oils or bitumen that are unsuited for the more conventional refining techniques.

The process has three basic variations—the Type II, Type III, and Type IV units with the degree of desulfurization, and process severity, increasing from Type I to Type IV. Thus, liquid products from Types III and IV units can be used directly as catalytic cracker feedstocks and perform similarly to virgin gas oil fractions, whereas liquid products from the Type II unit usually need to be vacuum flashed to provide a feedstock suitable for a catalytic cracker.

Fresh, filtered feedstock is heated together with hydrogen and recycle gas, and charged to the downflow reactor from which the liquid product goes to fractionation after flashing to produce the various product streams. Each process type is basically similar to its predecessor but will differ in the number of reactors. For example, modifications necessary to convert the Type II to the Type III process consist of the addition of a reactor and related equipment, while the Type III process can be modified to a Type IV process by the addition of a third reactor section. Types III and IV are especially pertinent to the problem of desulfurizing heavy feedstocks since they have the capability of producing extremely low-sulfur liquids from high-sulfur residua.

13.2.7 H-Oil Process

The H-Oil process (Figure 13.5) is a catalytic process that is designed for hydrogenation of residua and other heavy feedstocks in an ebullated-bed reactor to produce upgraded petroleum products with a high degree (>80%) of desulfurization (Tables 13.3 and 13.4) (Faupel et al., 1991). The H-Oil reactor (using temperatures and pressures on the order of 415–440°C (780–825°F) and a hydrogen pressure of 2500–3700 psi) can handle heavy feedstocks with high or low metal concentrations, although it is particularly efficient in treating and cracking heavier feedstocks. The process is able to convert all types of feedstocks to either distillate products as well as to desulfurize and demetallize residues for feed to coking units or residue fluid catalytic cracking units, for production of low-sulfur fuel oil, or for production to asphalt blending.

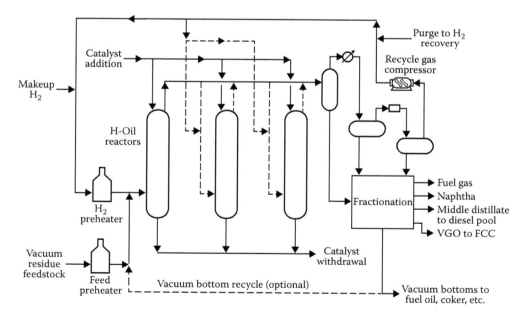

FIGURE 13.5 H-Oil process. (From Speight, J.G. 2014. *The Chemistry and Technology of Petroleum.* 5th Edition. CRC Press, Taylor & Francis Publishers, Boca Raton, FL. Figure 22.6, p. 605.)

TABLE 13.3
Process Parameters for the H-Oil and LC-Fining Processes

Parameter	H-Oil	LC-Fining
Temperature, °C	415–440	385–450
Temperature, °F	780–825	725–840
Pressure, psi	2440–3000	1015–2740
Hydrogen, bbl	1410	1350
Conversion	45–90	40–97
HDS	55–92	60–90
HDM	65–90	50–98

Note: HDM, hydrodemetallization; HDS, hydrodesulfurization.

TABLE 13.4
Feedstock and Product Data for the H-Oil Process

	Arabian Medium Vacuum Residuum 65%[a]	Arabian Medium Vacuum Residuum 90%[a]	Athabasca Bitumen
	Feedstock		
API gravity	4.9	4.9	8.3
Sulfur, % w/w	5.4	5.4	4.9
Nitrogen, % w/w			0.5
Carbon residue, % w/w			
Metals, ppm	128.0	128.0	
Ni			
V			
Residuum (>525°C, >975°F), % w/w			50.3
	Products, % w/w[b]		
Naphtha (C5–204°C, 400°F)	17.6	23.8	16.0
Sulfur, % w/w			1.0
Distillate (204–343°C, 400–650°F)	22.1	36.5	43.0
Sulfur, % w/w			2.0
Vacuum gas oil (343–534°C, 650–975°F)	34.0	37.1	26.4
Sulfur, % w/w			3.5
Residuum (>534°C, >975°F)	33.2	9.5	16.0
Sulfur, % w/w			5.7

Source: Speight, J.G. 2014. *The Chemistry and Technology of Petroleum.* 5th Edition. CRC Press, Taylor & Francis Publishers, Boca Raton, FL. Table 22.3, p. 606.

[a] % Conversion.
[b] % Desulfurization.

A wide variety of process options can be used with the H-Oil process depending on the specific operation. In all cases, a catalytic ebullated-bed reactor system is used to provide an efficient hydroconversion. The system ensures uniform distribution of liquid, hydrogen-rich gas, and catalyst across the reactor. The ebullated-bed system operates under essentially isothermal conditions, exhibiting little temperature gradient across the bed. The heat of the reaction is used to bring the feed oil and hydrogen up to reactor temperature.

In the process, the feedstock (a vacuum residuum) is mixed with recycle vacuum residue from downstream fractionation, hydrogen-rich recycle gas, and fresh hydrogen. This combined stream is fed into the bottom of the reactor whereby the upward flow expands the catalyst bed. The mixed vapor–liquid effluent from the reactor either goes to flash drum for phase separation or the next reactor. A portion of the hydrogen-rich gas is recycled to the reactor. The product oil is cooled and stabilized, and the vacuum residue portion is recycled to increase conversion.

The H-Oil process can be used as a single-stage or a double-stage process for most residua, as well as for materials such as bitumen extracted from the tar sands located in northeastern Alberta, Canada. In the latter case, distillate hydrotreating is recommended for inclusion within the process to produce a high-quality synthetic crude oil by making maximum use of the available temperature and the pressurized hydrogen. This produces low-sulfur liquid streams as well as a naphtha fraction that is essentially free of nitrogen.

13.2.8 Hydrovisbreaking (HYCAR) Process

Briefly, hydrovisbreaking, a noncatalytic process, is conducted under similar conditions to visbreaking and involves partial decomposition of the feedstock under an atmosphere of hydrogen. The presence of hydrogen leads to more stable products (lower flocculation threshold) than can be obtained with straight visbreaking, which means that higher conversions can be achieved, producing a lower-viscosity product.

The HYCAR process is composed fundamentally of three parts (1) visbreaking, (2) hydrodemetallization, and (3) hydrocracking. In the visbreaking section, the heavy feedstock (e.g., vacuum residuum or bitumen) is subjected to moderate thermal cracking while no coke formation is induced. The visbroken oil is fed to the demetallization reactor in the presence of catalysts, which provides sufficient pore for diffusion and adsorption of high molecular weight constituents. The product from this second stage proceeds to the hydrocracking reactor, where desulfurization and denitrogenation take place along with hydrocracking.

13.2.9 Hyvahl F Process

The process is used to hydrotreat atmospheric and vacuum residua to convert the feedstock to naphtha and middle distillates (Table 13.5).

The main features of this process are its dual catalyst system and its fixed-bed swing-reactor concept. The first catalyst has a high capacity for metals (to 100% w/w of new catalyst) and is used for both hydrodemetallization and most of the conversion. This catalyst is resistant to fouling, coking, and plugging by asphaltene constituents and shields the second catalyst from the same. Protected from metal poisons and deposition of coke-like products, the highly active second catalyst can carry out its deep hydrodesulfurization and refining functions. Both catalyst systems use fixed beds that are more efficient than moving beds and are not subject to attrition problems.

The swing-reactor design reserves two of the hydrodemetallization reactors for use as guard reactors: one of them can be removed from service for catalyst reconditioning and put on standby while the rest of the unit continues to operate. More than 50% of the metals are removed from the feed in the guard reactors.

In the process, the preheated feedstock enters one of the two guard reactors where a large proportion of the nickel and vanadium are adsorbed and hydroconversion of the heavy molecule weight constituents commences. Meanwhile, the second guard reactor catalyst undergoes a reconditioning process and then is put on standby. From the guard reactors, the feedstock flows through a series of hydrodemetallization reactors that continue the removal of metals and the conversion of heavy ends.

The next processing stage, hydrodesulfurization, is where most of the sulfur, some of the nitrogen, and the residual metals are removed. A limited amount of conversion also takes place. From the final reactor, the gas phase is separated, hydrogen is recirculated to the reaction section, and the

TABLE 13.5
Feedstock and Product Data for the Hyvahl Process

	Iranian Crude Oil (Topped)	Hyvahl F (Once-Through)	Hyvahl F Plus R2R
Feedstock			
	>360°C		
	>680°F		
API	15.2		
Sulfur, % w/w	2.6		
Nitrogen, % w/w	0.4		
Carbon residue, % w/w	9.4		
C7-asphaltene constituents	2.9		
Nickel + vanadium, ppm	191.0		
Products			
Naphtha, % w/w		4.0	48.0
Distillate/gas oil/vacuum gas oil, % w/w		24.5	17.5
Vacuum residuum, % w/w		67.5	8.4
Coke			6.4

Source: Speight, J.G. 2014. *The Chemistry and Technology of Petroleum.* 5th Edition. CRC Press, Taylor & Francis Publishers, Boca Raton, FL. Table 22.4, p. 607.

liquid products are sent to a conventional fractionation section for separation into naphtha, middle distillates, and heavier streams.

13.2.10 IFP Hydrocracking Process

The process features a dual-catalyst system: the first catalyst is a promoted nickel–molybdenum amorphous catalyst. It acts to remove sulfur and nitrogen and hydrogenate aromatic rings. The second catalyst is a zeolite that finishes the hydrogenation and promotes the hydrocracking reaction.

In the two-stage process, feedstock and hydrogen are heated and sent to the first reaction stage where the reactions (described previously) take place. The reactor effluent phases are cooled and separated, and the hydrogen-rich gas is compressed and recycled. The liquid leaving the separator is fractionated, the middle distillates and lower-boiling streams are sent to sales, and the high-boiling stream is transferred to the second reactor section and then recycled back to the separator section.

In the single-stage process, the first reactor effluent is sent directly to the second reactor, followed by the separation and fractionation steps. The fractionator bottoms are recycled to the second reactor or sold.

13.2.11 Isocracking Process

The process has been applied commercially in the full range of process flow schemes: (1) single-stage, once-through liquid; (2) single-stage, partial recycle of the heavy feedstock; (3) single-stage extinction recycle of oil—100% conversion; and (4) two-stage extinction recycle of oil (Table 13.6). The preferred flow scheme will depend on the feed properties, the processing objectives, and, to some extent, the specified feed rate.

The process uses multibed reactors. In most applications, a number of catalysts are used in a reactor. The catalysts are a mixture of hydrous oxides (for cracking) and heavy metal sulfides (for hydrogenation). The catalysts are used in a layered system to optimize the processing of the

TABLE 13.6

Feedstock and Product Data for the Isocracking Process

	Vacuum Gas Oil
Feedstock	
	360–530°C
	680–985°F
API gravity	22.6
Sulfur, % w/w	2.2
Nitrogen, % w/w	0.6
Carbon residue, % w/w	0.3
Metals	
Ni	0.1
V	0.3
Products, % w/w	
Naphtha (C5–124°C, C5–255°F)	15.9
Sulfur, ppm	<2
Distillate (124–295°C, 255–565°F)	51.6
Sulfur, ppm	<5
Heavy distillate (295–375°C, 565–705°F)	42.3
Sulfur, ppm	<5

Source: Speight, J.G. 2014. *The Chemistry and Technology of Petroleum.* 5th Edition. CRC
Press, Taylor & Francis Publishers, Boca Raton, FL. Table 22.6, p. 609.

feedstock that undergoes changes in its properties along the reaction pathway. In most commercial isocracking units, all of the fractionator bottoms is recycled or all of it is drawn as heavy product, depending on whether the low-boiling or high-boiling products are of greater value. If the low-boiling distillate products (naphtha or naphtha/kerosene) are the most valuable products, the higher-boiling-point distillates (like diesel) can be recycled to the reactor for conversion rather than drawn as a product.

Heavy feedstocks have been used in the process, and the product yield is very much dependent on the catalyst and the process parameters.

13.2.12 LC-Fining Process

The LC-Fining process (Figure 13.6) is a hydrogenation process capable of desulfurizing, demetallizing, and upgrading a wide spectrum of heavy feedstocks (Table 13.7) by means of an expanded-bed reactor at variable temperature and pressure conditions (385–450°C [725–840°F] with a hydrogen partial pressure of 1000–2750 psi). The process can achieve desulfurization, demetallization, reduction of coke-forming constituents, and hydrocracking of heavy feedstocks and yields a range of distillates. Conversion of the heavy feedstock is on the order of 40–97% v/v, and desulfurization is on the order of 60–90% w/w.

Thus, operating with the expanded-bed reactor allows processing of heavy feedstocks such as atmospheric residua, vacuum residua, and oil sand bitumen. The catalyst in the reactor behaves like fluid that enables the catalyst to be added to and withdrawn from the reactor during operation. The reactor conditions are near isothermal because the heat of reaction is absorbed by the cold fresh feed immediately through mixing of reactors.

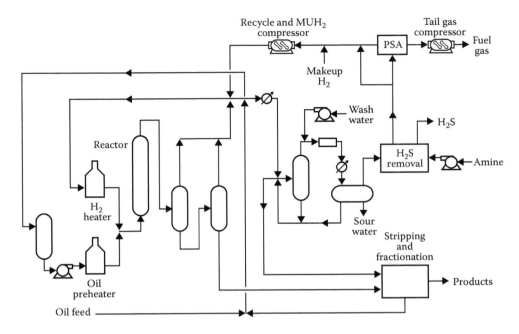

FIGURE 13.6 LC-Fining process. (From Speight, J.G. 2014. *The Chemistry and Technology of Petroleum.* 5th Edition. CRC Press, Taylor & Francis Publishers, Boca Raton, FL. Figure 22.7, p. 610.)

TABLE 13.7
Feedstock and Product Data for the LC-Fining Process

	Kuwait Atmospheric Residuum	Gach Saran Vacuum Residuum	Arabian Heavy Vacuum Residuum	Athabasca Bitumen
Feedstock				
API gravity	15.0	6.1	7.5	9.1
Sulfur, % w/w	4.1	3.5	4.9	5.5
Nitrogen, % w/w				0.4
Products, % w/w				
Naphtha (C5–205°C, C5–400°F)	2.5	9.7	14.3	13.9
Sulfur, % w/w				1.1
Nitrogen, % w/w				
Distillate (205–345°C, 400–650°)	22.7	14.1	26.5	37.7
Sulfur, % w/w				0.7
Nitrogen, % w/w				
Heavy distillate (345–525°C, 650–975°F)	34.7	24.1	31.1	30
Sulfur, % w/w				1.1
Nitrogen, % w/w				
Residuum (>525°C, >975°F)	35.5	47.5	21.3	12.9
Sulfur, % w/w				3.4

Source: Speight, J.G. 2014. *The Chemistry and Technology of Petroleum.* 5th Edition. CRC Press, Taylor & Francis Publishers, Boca Raton, FL. Table 22.7, p. 613.

In the process, the feedstock and hydrogen are heated separately and then passed upward in the hydrocracking reactor through an expanded bed of catalyst. Reactor products flow to the high-pressure–high-temperature separator. Vapor effluent from the separator is let down in pressure, and then goes to the heat exchange and thence to a section for the removal of condensable products, and purification.

Liquid is let down in pressure and passes to the recycle stripper. This is the most important part of the high conversion process. The liquid recycle is prepared to the proper boiling range for return to the reactor. In this way, the concentration of bottoms in the reactor, and therefore the distribution of products, can be controlled. After the stripping, the recycle liquid is then pumped through the coke precursor removal step where high molecular weight constituents are removed. The clean liquid recycle then passes to the suction drum of the feed pump. The product from the top of the recycle stripper goes to fractionation, and any high-boiling product (including unconverted feedstock) is directed from the stripper bottoms pump discharge.

The LC-Fining process has been applied to desulfurization of bitumen extracted from the Athabasca tar sands (Speight, 2013, 2014). Although the process may not be the means by which direct conversion of the bitumen to a synthetic crude oil would be achieved, it does nevertheless offer an attractive means of bitumen conversion. Indeed, the process would play the part of the primary conversion process from which liquid products would accrue—these products would then pass to a secondary upgrading (hydrotreating) process to yield a synthetic crude oil.

13.2.13 MAKFINING PROCESS

The MAKfining (Figure 13.7) process uses a multiple catalyst system in multibed reactors that include quench and redistribution system internals to produce valuable products (Table 13.8). In the process, the feedstock and recycle gas are preheated and brought into contact with the catalyst in a downflow fixed-bed reactor. The reactor effluent is sent to high- and low-temperature separators. Product recovery is a stripper/fractionator arrangement. Typical operating conditions in the reactors are 370–425°C (700–800°F) (single pass) and 370–425°C (700–800°F) (recycle) with pressures of 1000–2000 psi (single pass) and 1500–3000 psi (recycle).

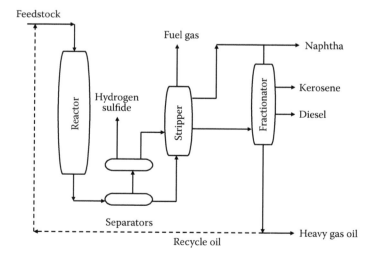

FIGURE 13.7 MAKfining process. (From Speight, J.G. 2014. *The Chemistry and Technology of Petroleum.* 5th Edition. CRC Press, Taylor & Francis Publishers, Boca Raton, FL. Figure 22.9, p. 613.)

TABLE 13.8

Feedstock and Product Data for the MAKfining Process

	AL/AH Vacuum Gas Oil	AL/AH Vacuum Gas Oil	AL/AH Light Cycle Oil
Feedstock			
API gravity	20.2	20.2	19.0
Sulfur, % w/w	2.9	2.9	1.0
Nitrogen, % w/w	0.9	0.9	0.6
Products, vol.%			
Naphtha	12.9	22.6	54.0
Kerosene	14.1	24.5	
Diesel	31.8	32.5	54.3
Light gas oil	50.0	30.0	
Conversion, %	50.0	70.0	50.0

Source: Speight, J.G. 2014. *The Chemistry and Technology of Petroleum.* 5th Edition. CRC Press, Taylor & Francis Publishers, Boca Raton, FL. Table 22.8, p. 612.

Note: AL/AH, Arabian light crude oil blended with Arabian heavy crude oil.

13.2.14 MICROCAT-RC PROCESS

The Microcat-RC process (also referred to as the M-Coke process) is a catalytic hydroconversion process operating at relatively moderate pressures and temperatures (Table 13.9) (Bearden and Aldridge, 1981; Bauman et al., 1993). The novel catalyst particles, containing a metal sulfide in a carbonaceous matrix formed within the process, are uniformly dispersed throughout the feed. Because of their ultra-small size (10^{-4} in diameter), there are typically several orders of magnitude

TABLE 13.9

Feedstock and Product Data for the Microcat Process

	Cold Lake Heavy Oil Vacuum Residuum
Feedstock	
API gravity	4.4
Sulfur, % w/w	
Nitrogen, % w/w	
Metals (Ni + V), ppm	480.0
Carbon residue, % w/w	24.4
Products, vol.%	
Naphtha (C5–177°C, C5–350°F)	17.2
Distillate (177–343°C, 350–650°F)	63.6
Gas oil (343–566°C, 650–1050°F)	21.9
Residuum (>566°, >1050°F)	2.1

Source: Speight, J.G. 2014. *The Chemistry and Technology of Petroleum.* 5th Edition. CRC Press, Taylor & Francis Publishers, Boca Raton, FL. Table 22.9, p. 613.

more of these microcatalyst particles per cubic centimeter of oil than is possible in other types of hydroconversion reactors using conventional catalyst particles. This results in smaller distances between particles and less time for a reactant molecule or intermediate to find an active catalyst site. Because of the physical structure, a microcatalyst tends to suffer none of the pore-plugging problems that plague conventional catalysts.

In the process, fresh vacuum residuum, microcatalyst, and hydrogen are fed to the hydroconversion reactor. Effluent is sent to a flash separation zone to recover hydrogen, gases, and liquid products, including naphtha, distillate, and gas oil. The liquid bottoms from the flash step is then fed to a vacuum distillation tower to obtain a 565°C⁻ (1050°F⁻) product oil and a 565°C⁺ (1050°F⁺) bottoms fraction that contains unconverted feed, microcatalyst, and essentially all of the feed metals.

Hydrotreating facilities may be integrated with the hydroconversion section or built on a standalone basis, depending on product quality objectives and owner preference.

13.2.15 MRH Process

The MRH process is a hydrocracking process designed to upgrade heavy feedstocks containing large amounts of metals and asphaltene, such as vacuum residua and bitumen, and to produce mainly middle distillates. The reactor is designed to maintain a mixed three-phase slurry of feedstock, fine powder catalyst, and hydrogen, and to promote effective contact.

In the process, the slurry of heavy feedstock and fine powder catalyst is preheated in a furnace and fed into the reactor vessel. Hydrogen is introduced from the bottom of the reactor and flows upward through the reaction mixture, maintaining the catalyst suspension in the reaction mixture. Cracking, desulfurization, and demetallization reactions take place via thermal and catalytic reactions. In the upper section of the reactor, vapor is disengaged from the slurry, and hydrogen and other gases are removed in a high-pressure separator. The liquid condensed from the overhead vapor is distilled and then flows out to the secondary treatment facilities. From the lower section of the reactor, bottom slurry oil, which contains the catalyst, uncracked residuum, and a small amount of vacuum gas oil fraction, is withdrawn. Vacuum gas oil is recovered in the slurry separation section, and the remaining catalyst and coke are fed to the regenerator.

Product distribution focuses on middle distillates (Table 13.10) with the process focused as a resid processing unit and inserted into a refinery just downstream from the vacuum distillation unit.

13.2.16 RCD Unibon Process

The RCD Unibon black oil cracking (BOC) process is a process to upgrade vacuum residua. There are several possible flow scheme variations for the process. It can operate as an independent unit or be used in conjunction with a thermal conversion unit. In this configuration, hydrogen and a vacuum residuum are introduced separately to the heater, and mixed at the entrance to the reactor. To avoid thermal reactions and premature coking of the catalyst, temperatures are carefully controlled and conversion is limited to approximately 70% of the total projected conversion. The removal of sulfur, heptane-insoluble materials, and metals is accomplished in the reactor. The effluent from the reactor is directed to the hot separator. The overhead vapor phase is cooled and condensed, and the hydrogen separated therefrom is recycled to the reactor.

The liquid product goes to the thermal conversion heater where the remaining conversion of nonvolatile materials occurs. The heater effluent is flashed and the overhead vapors are cooled, condensed, and routed to the cold flash drum. The bottoms liquid stream then goes to the vacuum column where the gas oils are recovered for further processing, and the residuals are blended into the heavy fuel oil pool.

TABLE 13.10

Feedstock and Product Data for the MRH Process

	Arabian Heavy Vacuum Residuum	Athabasca (Canada) Bitumen
	Feedstock	
API gravity	5.9	10.2
Sulfur, % w/w	5.1	4.3
Nitrogen, % w/w	0.3	0.4
C7-asphaltene constituents, % w/w	13.4	8.1
Metals, ppm		
Nickel	41.0	85.0
Vanadium	127.0	182.0
Carbon residue, % w/w	21.7	13.3
Distillation profile, vol.%		
Naphtha	0.0	2.2
Kerosene	0.0	5.3
Light gas oil	0.0	12.1
Vacuum gas oil	4.0	31.8
Vacuum residuum	96.0	48.6
(Atmospheric residuum)	(100.0)	(80.4)
	Products, % w/w	
Naphtha	13.0	12.0
Sulfur	0.2	1.1
Nitrogen	0.03	0.05
Kerosene	6.0	13.0
Sulfur	1.0	1.2
Nitrogen	0.06	0.08
Light gas oil	17.0	29.0
Sulfur	2.5	2.2
Nitrogen	0.06	0.11
Atmospheric residuum	55.0	41.0
Sulfur	3.8	3.8

Source: Speight, J.G. 2014. *The Chemistry and Technology of Petroleum.* 5th Edition. CRC Press, Taylor & Francis Publishers, Boca Raton, FL. Table 22.10, p. 614.

13.2.17 RESIDFINING PROCESS

Residfining is a catalytic fixed-bed process for the desulfurization and demetallization of residua. The process can also be used to pretreat residua to suitably low contaminant levels before catalytic cracking.

The Residfining process is a catalytic fixed-bed process for the desulfurization and demetallization of residua. The process can also be used to pretreat residua to suitably low contaminant levels before catalytic cracking. In the process, liquid feed to the unit is filtered, pumped to pressure, preheated, and combined with treat gas before entering the reactors. A small guard reactor would typically be employed to prevent plugging/fouling of the main reactors. Provisions are employed to periodically remove the guard while keeping the main reactors on-line. The temperature increase associated with the exothermic reactions is controlled using either a gas quench or a liquid quench. A train of separators is employed to separate the gas and liquid products. The recycle gas is scrubbed to remove ammonia and H_2S. It is then combined with fresh makeup hydrogen before being reheated and recombined with fresh feed. The liquid product is sent to a fractionator where the product is fractionated.

The different catalysts allow other minor differences in operating conditions and peripheral equipment. Primary differences include the use of higher-purity hydrogen makeup gas (usually 95% or greater), inclusion of filtration equipment in most cases, and facilities to upgrade the off-gases to maintain higher concentration of hydrogen in the recycle gas. Most of the processes utilize downflow operation over fixed-bed catalyst systems; however, exceptions to this are the H-Oil and LC-Fining processes (which are predominantly conversion processes) that employ upflow designs and ebullating catalyst systems with continuous catalyst removal capability, and the Shell process (a conversion process) that may involve the use of a bunker flow reactor ahead of the main reactors to allow periodic changeover of catalyst.

The primary objective in most of the residue desulfurization processes is to remove sulfur with minimum consumption of hydrogen. Substantial percentages of nitrogen, oxygen, and metals are also removed from the feedstock. However, complete elimination of other reactions is not feasible and, in addition, hydrocracking, thermal cracking, and aromatic saturation reactions occur to some extent. Certain processes, such as the H-Oil process, which uses a single-stage or a two-stage reactor, and the LC-Fining process, which uses an expanded bed reactor, can be designed to accomplish high yields of lower-boiling distillates at the expense of desulfurization (Speight and Ozum, 2002; Parkash, 2003; Hsu and Robinson, 2006; Gary et al., 2007; Speight, 2014).

Removal of nitrogen is much more difficult than removal of sulfur. For example, nitrogen removal may be only about 25–30% when sulfur removal is at a 75–80% level. Metals are removed from the feedstock in substantial quantities and are mainly deposited on the catalyst surface and exist as metal sulfides at processing conditions. As these deposits accumulate, the catalyst pores eventually become blocked and inaccessible; thus, catalyst activity is lost.

Desulfurization of residua is considerably more difficult than desulfurization of distillates (including vacuum gas oil) because many more contaminants are present and very large, complex molecules are involved. The most difficult portion of feed in residue desulfurization is the asphaltene fraction that forms coke readily, and it is essential that these large molecules be prevented from condensing with each other to form coke, which deactivates the catalyst. This is accomplished by selection of proper catalysts, use of adequate hydrogen partial pressure, and assuring intimate contact of the hydrogen-rich gases and oil molecules in the process design.

In the process, liquid feed to the unit is filtered, pumped to pressure, preheated, and combined with treat gas before entering the reactors. A small guard reactor would typically be employed to prevent plugging/fouling of the main reactors. Provisions are employed to periodically remove the guard while keeping the main reactors on-line. The temperature increase associated with the exothermic reactions is controlled using either a gas or a liquid quench. A train of separators is employed to separate the gas and liquid products. The recycle gas is scrubbed to remove ammonia and H_2S. It is then combined with fresh makeup hydrogen before being reheated and recombined with fresh feed. The liquid product is sent to a fractionator where the product is fractionated.

13.2.18 RESIDUE HYDROCONVERSION PROCESS

The residue hydroconversion process is a high-pressure fixed-bed trickle-flow hydrocatalytic process. The feedstock can be desalted atmospheric or vacuum residue (Table 13.11).

The reactors are of multibed design with interbed cooling, and the multicatalyst system can be tailored according to the nature of the feedstock and the target conversion. For residua with a high metal content, a hydrodemetallization catalyst is used in the front-end reactor(s), which excels in its high-metal-uptake capacity and good activities for metal removal, asphaltene conversion, and residue cracking. Downstream of the demetallization stage, one or more hydroconversion stages, with optimized combination of catalysts' hydrogenation function and texture, are used to achieve desired catalyst stability and activities for denitrogenation, desulfurization, and heavy hydrocarbon cracking.

TABLE 13.11

Feedstock and Product Data for the Residue Hydroconversion Process

	Unspecified Residuum
Feedstock	
API gravity	24.4
Sulfur, % w/w	0.1
Nitrogen, % w/w	0.1
Carbon residue, % w/w	0.3
Products, vol.%	
Naphtha	4.5
Light gas oil	19.4
Vacuum gas oil	77.10
Sulfur, % w/w	0.01
Carbon residue, % w/w	0.20

Source: Speight, J.G. 2014. *The Chemistry and Technology of Petroleum.* 5th Edition. CRC Press, Taylor & Francis Publishers, Boca Raton, FL. Table 22.11, p. 616.

13.2.19 Shell Residual Oil Hydrodesulfurization

The Shell residual oil hydrodesulfurization process is broadly defined as a process to improve the quality of residual oils by removing sulfur, metals, and asphaltene constituents, as well as by bringing about a reduction in the viscosity of the feedstock. The process is suitable for a wide range of the heavier feedstocks, irrespective of the composition and origin, and even includes those feedstocks that are particularly high in metal and asphaltene constituents.

The process centers on a fixed-bed downflow reactor that allows catalyst replacement without causing any interruption in the operation of the unit. Feedstock is introduced to the process via a filter (backwash, automatic), after which hydrogen and recycle gas are added to the feedstock stream, which is then heated to reactor temperature by means of feed-effluent heat exchangers whereupon the feed stream passes down through the reactor in trickle flow. Sulfur removal is excellent, and substantial reductions in the vanadium content and asphaltene constituents are also noted. In addition, a marked increase occurs in the API gravity, and the viscosity is reduced considerably.

A bunker reactor provides extra process flexibility if it is used upstream from the desulfurization reactor, especially with reference to the processing of feedstocks with a high metal content. A catalyst with a capacity for metals is employed in the bunker reactor to protect the desulfurization catalyst from poisoning by the metals. In the case of the bunker reactor, fresh catalyst is loaded into a storage vessel (V-1) at atmospheric pressure from which it passes into a sluice vessel (V-2), where pressurization occurs before passage to the bunker reactor. In the reactor, inverted cone segments support the catalyst and are designed to allow catalyst removal. Thus, catalyst is withdrawn from the reactor into the sluice vessel (V-3), which is then depressurized and the contents transferred to a discharge vessel (V-4).

13.2.20 Unicracking Hydrodesulfurization Process

Unicracking hydrodesulfurization is a fixed-bed, catalytic process for the hydrotreating of residua (Table 13.12). In the process, the feedstock and hydrogen-rich recycle gas are preheated, mixed, and

TABLE 13.12

Feedstock and Product Data for the Unicracking Process

	Alaska North Slope Atmospheric Residuum	Gach Saran Atmospheric Residuum	Kuwait Atmospheric Residuum	Kuwait Atmospheric Residuum	California Atmospheric Residuum
		Feedstock			
API gravity	15.2	16.3	16.7	14.4	9.9
Sulfur, % w/w	1.7	2.4	3.8	4.2	4.5
Nitrogen, % w/w	0.4	0.4	0.2	0.2	0.4
Metals (Ni + V), ppm	44.0	220.0	46.0	66.0	213.0
Carbon residue, % w/w	8.4	8.5	8.5	10.0	13.6
		Products, vol.%			
Naphtha (<185°C, <365°F)	0.8	1.8	1.1		
Naphtha (C5–205°C, C5–400°F)				2.1	4.2
Light gas oil				13.1	18.0
Residuum (>345°C, >650°F)				89.5	81.6
API				22.7	21.6
Sulfur, % w/w				<0.3	<0.3
Nitrogen, % w/w				<0.2	<0.2
Metals (Ni + V), ppm				<25	<5
Residuum (>185°C, >365°F)	100.5	100.4	100.4		
API	19.8	22.8	24.4		
Sulfur, % w/w	0.3	0.3	0.3		
Metals (Ni + V), ppm	14.0	55.0	15.0		
Carbon residue, % w/w	4.0	4.0	3.0		

Source: Speight, J.G. 2014. *The Chemistry and Technology of Petroleum.* 5th Edition. CRC Press, Taylor & Francis Publishers, Boca Raton, FL. Table 22.12, p. 618.

introduced into a guard chamber that contains a relatively small quantity of the catalyst. The guard chamber removes particulate matter and residual salt from the feed. The effluent from the guard chamber flows down through the main reactor, where it contacts one or more catalysts designed for removal of metals and sulfur. The product from the reactor is cooled, separated from hydrogen-rich recycle gas, and either stripped to meet fuel oil flash point specifications, or fractionated to produce distillate fuels, upgraded vacuum gas oil, and upgraded vacuum residuum. Recycle gas, after hydrogen sulfide removal, is combined with makeup gas and returned to the guard chamber and main reactors.

The process operates satisfactorily for a variety of feedstocks that vary in sulfur content from about 1.0% by weight to about 5% by weight. The rate of desulfurization is dependent on the sulfur content of the feedstock, as well as on catalyst life, product sulfur, and hydrogen consumption.

The high efficiency of the process is due to the excellent distribution of the feedstock and hydrogen that occurs in the reactor where a proprietary liquid distribution system is employed. In addition, the process catalyst (also proprietary) was designed for the desulfurization of residua and is not merely an upgraded gas oil hydrotreating catalyst as often occurs in various processes. It is a cobalt–molybdenum–alumina (MoO_2–Al_2O_3) catalyst with a controlled pore structure that permits a high degree of desulfurization and, at the same time, minimizes any coking tendencies.

13.2.21 UNIFLEX PROCESS

The Uniflex process is an evolved version with significant changes (by UOP) of the former CANMET process that used an empty vessel hydrocracking reactor in which the slurry feedstock is processed in the presence of an iron sulfide–based catalyst deposited on particles of coal. Because the desulfurization activity of iron is very low, molybdenum can be added at a level of tens of parts per million in the form of molybdenum naphthenate. The reaction products are fractionated and sent to the hydrotreatment unit, while the unconverted residue (5–10% v/v of the feedstock) can be sent to a coking unit, burned as fuel, or gasified for production of synthesis gas (hydrogen).

The flow scheme for the Uniflex process is similar to that of a conventional UOP Unicracking process unit—liquid feedstock and recycle gas are heated to temperature in separate heaters, with a small portion of the recycle gas stream and the required amount of catalyst being routed through the oil heater (Gillis et al., 2009). The outlet streams from both heaters are fed to the bottom of the slurry reactor. The reactor effluent is quenched at the reactor outlet to terminate reactions and then flows to a series of separators with gas being recycled back to the reactor. Liquids flow to the unit's fractionation section for recovery of light ends, naphtha, diesel, vacuum gas oils, and pitch (cracked residuum). Heavy vacuum gas oil is partially recycled to the reactor for further conversion.

The basis of the Uniflex Process is the upflow reactor that operates at moderate temperature (440–470°C, 815–880°F) and 2000 psi. The feedstock distributor, in combination with optimized process variables, promotes intense back mixing (which provides near-isothermal reactor conditions) in the reactor without the need for reactor internals or liquid recycle ebullating pumps. The back mixing allows the reactor to operate at the higher temperatures required to maximize vacuum residue conversion. The majority of the products vaporize and quickly leave the reactor (thereby minimizing the potential for secondary cracking reactions), while the residence time of the higher-boiling constituents of the feedstock is maximized.

The process employs a proprietary, dual-function nano-sized solid catalyst that is blended with the feed to maximize conversion of heavy components and inhibit coke formation. Specific catalyst requirements depend on feedstock quality and the required severity of operation. The primary function of the catalysts is to effect mild hydrogenation activity for the stabilization of cracked products while also limiting the saturation of aromatic rings. Because of the hydrogenation function, the catalyst also decouples the relationship between conversion and the propensity for carbon residue formation of the feedstock.

13.2.22 VEBA COMBI-CRACKING (VCC) PROCESS

The VCC process is a thermal hydrocracking/hydrogenation process for converting residua and other heavy feedstocks (Table 13.13). In the process, the residue feed is slurried with a small amount of finely powdered additive and mixed with hydrogen and recycle gas before preheating. The feed mixture is routed to the liquid-phase reactors. The reactors are operated in an upflow mode and arranged in series. In a once-through operation, conversion rates of >95% are achieved. Typically, the reaction takes place at temperatures between 440°C and 480°C (835°F and 895°F) and pressures between 2200 and 6500 psi bar. Substantial conversion of asphaltene constituents, desulfurization, and denitrogenation take place at high levels of residue conversion. Temperature is controlled by a recycle gas quench system.

The flow from the liquid-phase hydrogenation reactors is routed to a hot separator, where gases and vaporized products are separated from unconverted material. A vacuum flash recovers distillates in the hot separator bottom product.

The hot separator top product, together with recovered distillates and straight-run distillates, enters the gas-phase hydrogenation reactor. The gas-phase hydrogenation reactor operates at the same pressure as the liquid-phase hydrogenation reactor and contains a fixed bed of commercial hydrotreating catalyst. The operation temperature (340–420°C) is controlled by a hydrogen quench.

TABLE 13.13

Feedstock and Product Data for the Veba Combi Cracking Process

	Arabian Heavy Vacuum Residuum[a]
Feedstock	
API gravity	3.4
Sulfur, % w/w	5.5
C7-asphaltene constituents, % w/w	13.5
Metals (Ni + V), ppm	230.0
Carbon residue, % w/w	8.4
Products, vol.%	
Naphtha (<170°C, <340°F)	26.9
Middle distillate (170–370°C, 340–700°F)	36.5
Gas oil (>370°, 700°F)	19.9

Source: Speight, J.G. 2014. *The Chemistry and Technology of Petroleum.* 5th Edition. CRC Press, Taylor
& Francis Publishers, Boca Raton, FL. Table 22.13, p. 620.
[a]   >550°C (>1025°F); conversion: 95%.

The system operates in a trickle flow mode. The separation of the synthetic crude from associated gases is performed in a cold separator system. The synthetic crude may be sent to stabilization and fractionation unit as required. The gases are sent to a lean oil scrubbing system for contaminant removal and are recycled to the low-pressure system of the VCC process.

13.3   CATALYSTS

Catalysts used in hydrotreating reactors include cobalt and molybdenum oxides on alumina, nickel oxide, nickel thiomolybdate, tungsten and nickel sulfides, and vanadium oxide. Cobalt–molybdenum and nickel–molybdenum are the most commonly used catalysts for hydrotreating. Both types of catalyst remove sulfur, nitrogen, and other contaminants from petroleum feed. Cobalt–molybdenum catalysts, however, are selective for sulfur removal, while nickel–molybdenum catalysts are selective for nitrogen removal.

Typically, hydrodesulfurization catalysts consist of metals impregnated on a porous alumina support. Almost all of the surface area is found in the pores of the alumina (200–300 m²/g), and the metals are dispersed in a thin layer over the entire alumina surface within the pores. This type of catalyst does display a huge catalytic surface for a small weight of catalyst. Cobalt (Co), molybdenum (Mo), and nickel (Ni) are the most commonly used metals for desulfurization catalysts. The catalysts are manufactured with the metals in an oxide state. In the active form, they are in the sulfide state, which is obtained by sulfiding the catalyst either before use or with the feed during actual use. Any catalyst that exhibits hydrogenation activity will catalyze hydrodesulfurization to some extent. However, the Group VIB metals (chromium, molybdenum, and tungsten) are particularly active for desulfurization, especially when promoted with metals from the iron group (iron, cobalt, and nickel).

Furthermore, the increasing importance of hydrodesulfurization and hydrodenitrogenation in petroleum processing in order to produce clean-burning fuels has led to a surge of research on the chemistry and engineering of heteroatom removal, with sulfur removal being the most prominent area of research. Most of the earlier works are focused on (1) catalyst characterization by physical methods; (2) low-pressure reaction studies of model compounds having relatively high reactivity; (3) process development; or (4) cobalt–molybdenum (Co–Mo) catalysts, nickel–molybdenum catalysts (Ni–Mo), or nickel–tungsten (Ni–W) catalysts supported on alumina, often doped by fluorine or phosphorus.

Hydrodesulfurization and demetallization occur simultaneously on the active sites within the catalyst pore structure. Sulfur and nitrogen occurring in residua are converted to hydrogen sulfide and ammonia in the catalytic reactor, and these gases are scrubbed out of the reactor effluent gas stream. The metals in the feedstock are deposited on the catalyst in the form of metal sulfides, and cracking of the feedstock to distillate produces a lay-down of carbonaceous material on the catalyst; both events poison the catalyst and activity, or selectivity suffers. The deposition of carbonaceous material is a fast reaction that soon equilibrates to a particular carbon level and is controlled by hydrogen partial pressure within the reactors. On the other hand, metal deposition is a slow reaction that is directly proportional to the amount of feedstock passed over the catalyst.

The need to develop catalysts that can carry out deep hydrodesulfurization and deep hydrodenitrogenation has become even more pressing in view of recent environmental regulations limiting the amount of sulfur and nitrogen emissions. The development of a new generation of catalysts to achieve this objective of low nitrogen and sulfur levels in the processing of different feedstocks presents an interesting challenge for catalyst development.

Basic nitrogen-containing compounds in a feed diminish the cracking activity of hydrocracking catalysts. However, zeolite catalysts can operate in the presence of substantial concentrations of ammonia, in marked contrast to silica–alumina catalysts, which are strongly poisoned by ammonia. Similarly, sulfur-containing compounds in a feedstock adversely affect the noble metal hydrogenation component of hydrocracking catalysts. These compounds are hydrocracked to hydrogen sulfide, which converts the noble metal to the sulfide form. The extent of this conversion is a function of the hydrogen and hydrogen sulfide partial pressures.

Removal of sulfur from the feedstock results in a gradual increase in catalyst activity, returning almost to the original activity level. As with ammonia, the concentration of the hydrogen sulfide can be used to precisely control the activity of the catalyst. Non-noble metal-loaded zeolite catalysts have an inherently different response to sulfur impurities since a minimum level of hydrogen sulfide is required to maintain the nickel–molybdenum and nickel–tungsten in the sulfide state.

Hydrodenitrogenation is more difficult to accomplish than hydrodesulfurization, but the relatively smaller amounts of nitrogen-containing compounds in conventional crude oil made this of lesser concern to refiners (Speight and Ozum, 2002; Parkash, 2003; Hsu and Robinson, 2006; Gary et al., 2007; Speight, 2014). However, the trend to heavier feedstocks in refinery operations, which are richer in nitrogen than the conventional feedstocks, has increased the awareness of refiners to the presence of nitrogen compounds in crude feedstocks. For the most part, however, hydrodesulfurization catalyst technology has been used to accomplish hydrodenitrogenation, although such catalysts are not ideally suited for nitrogen removal. However, in recent years, the limitations of hydrodesulfurization catalysts when applied to hydrodenitrogenation have been recognized, and there are reports of attempts to manufacture catalysts more specific to nitrogen removal (Ho, 1988).

The character of the hydrotreating processes is chemically very simple since they essentially involve removal of sulfur and nitrogen as hydrogen sulfide and ammonia, respectively:

$$R\text{-}S\text{-}R' + H_2 \rightarrow RH + R'H + H_2S$$

$$R\text{-}N(R')\text{-}R'' + 3H_2 \rightarrow RH + R'H + R''H + 2NH_3$$

However, nitrogen is the most difficult contaminant to remove from feedstocks, and processing conditions are usually dictated by the requirements for nitrogen removal.

In general, any catalyst capable of participating in hydrogenation reactions may be used for hydrodesulfurization. The sulfides of hydrogenating metals are particularly used for hydrodesulfurization, and catalysts containing cobalt, molybdenum, nickel, and tungsten are widely used on a commercial basis.

Hydrotreating catalysts are usually cobalt–molybdenum catalysts, and under the conditions whereby nitrogen removal is accomplished, desulfurization usually occurs as well as oxygen removal. Indeed, it is generally recognized that the fullest activity of the hydrotreating catalyst is not reached until some interaction with the sulfur (from the feedstock) has occurred, with part of the catalyst metals converted to the sulfides. Too much interaction may, of course, lead to catalyst deactivation.

The poisoning effect of nitrogen can be offset to a certain degree by operation at a higher temperature. However, the higher temperature tends to increase the production of material in the methane (CH_4) to butane (C_4H_{10}) range and decrease the operating stability of the catalyst so that it requires more frequent regeneration. Catalysts containing platinum or palladium (approximately 0.5% wet) on a zeolite base appear to be somewhat less sensitive to nitrogen than are nickel catalysts, and successful operation has been achieved with feedstocks containing 40 ppm nitrogen. This catalyst is also more tolerant of sulfur in the feed, which acts as a temporary poison, the catalyst recovering its activity when the sulfur content of the feed is reduced.

On such catalysts as nickel or tungsten sulfide on silica–alumina, isomerization does not appear to play any part in the reaction, as uncracked normal paraffins from the feedstock tend to retain their normal structure. Extensive splitting produces large amounts of low molecular weight (C_3–C_6) paraffins, and it appears that a primary reaction of paraffins is catalytic cracking followed by hydrogenation to form isoparaffins. With catalysts of higher hydrogenation activity, such as platinum on silica–alumina, direct isomerization occurs. The product distribution is also different, and the ratio of low molecular weight to intermediate molecular weight paraffins in the breakdown product is reduced.

In addition to the chemical nature of the catalyst, the physical structure of the catalyst is also important in determining the hydrogenation and cracking capabilities, particularly for heavy feedstocks (Fischer and Angevine, 1986; Kobayashi et al., 1987a,b). When gas oils and residua are used, the feedstock is present as liquids under the conditions of the reaction. Additional feedstock and the hydrogen must diffuse through this liquid before reaction can take place at the interior surfaces of the catalyst particle.

At high temperatures, reaction rates can be much higher than diffusion rates and concentration gradients can develop within the catalyst particle. Therefore, the choice of catalyst porosity is an important parameter. When feedstocks are to be hydrocracked to liquefied petroleum gas and naphtha, pore diffusion effects are usually absent. Catalysts with high surface area (about 300 m²/g) and low to moderate porosity (from 12 Å pore diameter with crystalline acidic components to 50 Å or more with amorphous materials) are used. With reactions involving high molecular weight feedstocks, pore diffusion can exert a large influence, and catalysts with pore diameters >80 Å are necessary for more efficient conversion.

Catalyst operating temperature can influence reaction selectivity since the activation energy for hydrotreating reactions is much lower than for hydrocracking reactions. Therefore, increasing the temperature in a residuum hydrotreater increases the extent of hydrocracking relative to hydrotreating, which also increases the hydrogen consumption.

Aromatic hydrogenation in petroleum refining may be carried out over supported metal or metal sulfide catalysts depending on the sulfur and nitrogen levels in the feedstock. For hydrorefining of feedstocks that contain appreciable concentrations of sulfur and nitrogen, sulfided nickel–molybdenum (Ni–Mo), nickel–tungsten (Ni–W), or cobalt–molybdenum (Co–Mo) on alumina (γ-Al$_2$O$_3$) catalysts are generally used, whereas supported noble metal catalysts have been used for sulfur- and nitrogen-free feedstocks. Catalysts containing noble metals on Y-zeolites have been reported to be more sulfur tolerant than those on other supports (Jacobs, 1986). Within the series of cobalt-promoted or nickel-promoted group VI metal (Mo or W) sulfides supported on γ-Al$_2$O$_3$, the ranking for hydrogenation is

$$Ni–W > Ni–Mo > Co–Mo > Co–W$$

Nickel–tungsten (Ni–W) and nickel–molybdenum (Ni–Mo) on Al$_2$O$_3$ catalysts are widely used to reduce sulfur, nitrogen, and aromatics levels in petroleum fractions by hydrotreating.

Molybdenum sulfide (MoS_2), usually supported on alumina, is widely used in petroleum processes for hydrogenation reactions. It is a layered structure that can be made much more active by the addition of cobalt or nickel. When promoted with cobalt sulfide (CoS), making what is called *cobalt–moly* catalysts, it is widely used in hydrodesulfurization processes. The nickel sulfide (NiS)-promoted version is used for hydrodenitrogenation as well as hydrodesulfurization. The closely related tungsten compound (WS_2) is used in commercial hydrocracking catalysts. Other sulfides (iron sulfide [FeS], chromium sulfide [Cr_2S_3], and vanadium sulfide [V_2S_5]) are also effective and used in some catalysts. A valuable alternative to the base metal sulfides is palladium sulfide (PdS). Although it is expensive, palladium sulfide forms the basis for several very active catalysts.

The life of a catalyst used to hydrotreat petroleum residua is dependent on the rate of carbon deposition and the rate at which organometallic compounds decompose and form metal sulfides on the surface. Several different metal complexes exist in the asphaltene fraction of the residuum, and an explicit reaction mechanism of decomposition that would be a perfect fit for all of the compounds is not possible. However, in general terms, the reaction can be described as hydrogen (A) dissolved in the feedstock contacting an organometallic compound (B) at the surface of the hydrotreating catalyst and producing a metal sulfide (C) and a hydrocarbon (D):

$$A + B > C + D$$

Different rates of reaction may occur with various types and concentrations of metallic compounds. For example, a medium-metal-content feedstock will generally have a lower rate of demetallization compared with a high-metal-content feedstock. Moreover, although individual organometallic compounds decompose according to both first- and second-order rate expressions, for a reactor design, a second-order rate expression is applicable to the decomposition of the residuum as a whole.

Obviously, the choice of the hydrogenation catalyst depends on what the catalyst designer wishes to accomplish. In catalysts to make naphtha, for instance, vigorous cracking is needed to convert a large fraction of the feed to the kinds of molecules that will make a good gasoline blending stock. For this vigorous cracking, a vigorous hydrogenation component is needed. Since palladium is the most active catalyst for this, the extra expense is warranted. On the other hand, many refiners wish only to make acceptable diesel, a less demanding application. For this, the less expensive molybdenum sulfides are adequate.

REFERENCES

Ancheyta, J. 2007. Reactors for Hydroprocessing. In: *Hydroprocessing of Heavy Oils and Residua*, J. Ancheyta and J.G. Speight (Editors). CRC Press, Taylor & Francis Group, Boca Raton, FL.

Bauman, R.F., Aldridge, C.L., Bearden, R. Jr., Mayer, F.X., Stuntz, G.F., Dowdle, L.D., and Fiffron, E. 1993. *Preprints. Oil Sands—Our Petroleum Future*. Alberta Research Council, Edmonton, Alberta, Canada, p. 269.

Bearden, R., and Aldridge, C.L. 1981. Novel Catalyst and Process to Upgrade Heavy Oils. *Energy Progr.* 1: 44–48.

Carrazza, J., Pereira, P., and Martinez, N. 1997a. Process and Catalyst for Upgrading Heavy Hydrocarbon. United States Patent 5,688,395, November 18.

Carrazza, J., Pereira, P., and Martinez, N. 1997b. Process and Catalyst for Upgrading Heavy Hydrocarbon. United States Patent 5,688,741, November 18.

Faupel, T.H., Colyar, J.J., and Wisdom, L.I. 1991. In: *Tar Sand Upgrading Technology*, S.S. Shih and M.C. Oballa (Editors). Symposium Series No. 282. American Institute for Chemical Engineers, New York, p. 44.

Fischer, R.H., and Angevine, P.V. 1986. Dependence of Resid Processing Selectivity on Catalyst Pore Size Distribution. *Appl. Catal.* 27: 275–283.

Gillis, D., VanWees, M., and Zimmerman, P. 2009. Upgrading Residues for High Levels of Distillate Production. *Petrol. Technol. Quart.* Q4: 1–10. http://www.digitalrefining.com/article_1000613.pdf; accessed June 22, 2014.

Jacobs, P.A. 1986. In: *Metal Clusters in Catalysis, Studies in Surface Science and Catalysis*, Vol. 29, B.C. Gates (Editor). Elsevier, Amsterdam, p. 357.

Khan, M.R., and Patmore, D.J. 1997. Heavy Oil Upgrading Processes, Chapter 6. In *Petroleum Chemistry and Refining*. J.G. Speight (Editor). Taylor & Francis, Washington, DC.

Kobayashi, S., Kushiyama, S., Aizawa, R., Koinuma, Y., Inoue, K., Shmizu, Y., and Egi, K. 1987a. Kinetic Study on the Hydrotreating of Heavy Oil. 1. Effect of Catalyst Pellet Size in Relation to Pore Size. *Ind. Eng. Chem. Res.* 26: 2241–2245.

Kobayashi, S., Kushiyama, S., Aizawa, R., Koinuma, Y., Inoue, K., Shmizu, Y., and Egi, K. 1987b. Kinetic Study on the Hydrotreating of Heavy Oil. 2. Effect of Catalyst Pore Size. *Ind. Eng. Chem. Res.* 26: 2245–2250.

Speight, J.G. 2011a. *The Refinery of the Future*. Gulf Professional Publishing, Elsevier, Oxford, UK.

Speight, J.G. 2013. *Heavy and Extra Heavy Oil Upgrading Technologies*. Gulf Professional Publishing, Elsevier, Oxford, UK.

Speight, J.G. 2014. *The Chemistry and Technology of Petroleum*, 5th Edition. CRC Press, Taylor & Francis Group, Boca Raton, FL.

Stanislaus, A., and Cooper, B.H. 1994. Aromatic Hydrogenation Catalysis: A Review. *Catal. Rev. Sci. Eng.* 36(1): 75–123.

14 Hydrogen Production and Management

14.1 INTRODUCTION

The refinery (Figure 14.1) is a series of unit processes designed to work in sequence to produce a slate of products dictated by the market (Chapter 3). Throughout the previous chapters, there have been several references and/or acknowledgments of the use of hydrogen during refining owing to the need to refine high-sulfur feedstocks to produce low-sulfur products (Tables 14.1 and 14.2) (Figure 14.2), such as desulfurization of distillates (Figure 14.3), and in hydroconversion processes, such as single-stage or two-stage hydrocracking (Figure 14.4). In fact, the use of hydrogen in refinery processes is perhaps the single most significant advancement in refining technology during the 20th century. The process uses the principle that the presence of hydrogen during a thermal reaction of a petroleum feedstock will terminate many of the coke-forming reactions and enhance the yields of the lower-boiling components such as naphtha, kerosene, and gas oil (Speight, 2000, 2011, 2014; Speight and Ozum, 2002; Parkash, 2003; Hsu and Robinson, 2006; Ancheyta and Speight, 2007; Gary et al., 2007).

The influx of heavier feedstocks into refineries and evolving environmental regulations are leading to stricter product quality requirements. To improve the oil products quality, refineries have been obligated to increase the depth of hydrotreating and hydrocracking processes, which consume large amounts of fresh hydrogen (Kriz and Ternan, 1994; Speight, 2014; Farnand et al., 2015). On the other hand, hydrogen produced by the traditional path of naphtha reforming, an important hydrogen-producing process, is no longer adequate for the needs of the modern refinery with the influx of heavier feedstocks, and other processes are necessary (Lipman, 2011; Speight, 2011, 2014). Moreover, the gap between hydrogen-consuming processes and hydrogen-producing processes has emphasized the potential (and often real) shortage of hydrogen in refineries.

Furthermore, hydrogen is a key to enabling modern refineries to comply with the latest product specifications and environmental requirements for fuel production that have been (and continue to be) mandated by the market and governments. Hydrogen is an expensive commodity in a refinery—a situation that is emphasized by the costs of creating low-sulfur fuels from heavier, sourer crude oils and tar sand bitumen, which are supplanting many of the relatively scarce lighter low-sulfur crude oils as refinery feedstocks.

Environmental restrictions, new transportation fuel specifications, and increased processing of heavier, sourer curds are leading to substantial increases in refinery hydrogen consumption for hydrodesulfurization, aromatic and olefin saturation, and improvement of product quality and reductions in refinery hydrogen production from catalytic reformers as a by-product. Therefore, the above factors make hydrogen management a critical issue. Generating, recovering, and purchasing of hydrogen have a significant impact on refinery operating costs. More important, overall refinery operations can be severely constrained by the unavailability of hydrogen. Primary consideration, however, should be given to the recovery of hydrogen contained in various purge gases, since this is a very attractive way from the viewpoint of refinery balance and economics (Cruz and de Oliveira Junior, 2008).

Typically, in the early refineries, the hydrogen for hydroprocesses was provided as a result of reforming processes, such as the platforming process (Figure 14.5). Dehydrogenation is a main chemical reaction in catalytic reforming, and hydrogen gas is consequently produced. The hydrogen is recycled though the reactors where the reforming takes place to provide the atmosphere

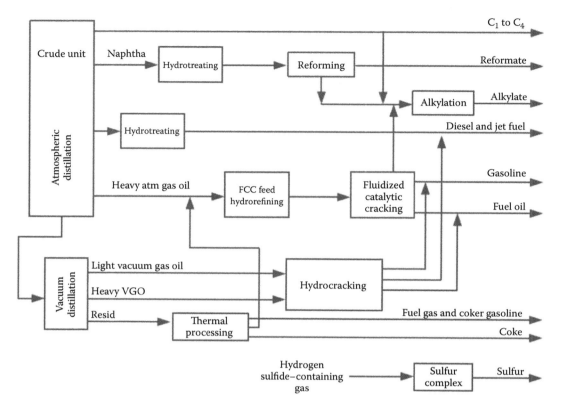

FIGURE 14.1   Schematic overview of a refinery. (From Speight, J.G. 2014. *The Chemistry and Technology of Petroleum*. 5th Edition. CRC Press, Taylor & Francis Publishers, Boca Raton, FL. Figure 15.1, p. 392.)

TABLE 14.1
Hydrogen Content of Various Liquid Fuels

	[H]% w/w
Crude oil	11–14
Heavy oil	9–11
Extra heavy oil	9–11
Tar sand bitumen	9–11
Residua, straight-run	9–11
Residua, cracked	8–10
Naphtha[a]	13–15
Kerosene[a]	13–14

[a] Included for comparison.

TABLE 14.2

Typical Hydrogen Applications in a Refinery

Naphtha hydrotreater
- Uses hydrogen to desulfurize naphtha from atmospheric distillation; must hydrotreat the naphtha before sending to a catalytic reformer unit.

Distillate hydrotreater
- Uses hydrogen to desulfurize distillates after atmospheric or vacuum distillation; in some units, aromatics are hydrogenated to cycloparaffins or alkanes.

Hydrodesulfurization
- Sulfur compounds are hydrogenated to hydrogen sulfide (H_2S) as feed for Claus plants.

Hydroisomerization
- Normal (straight-chain) paraffins are converted into isoparaffins to improve the product properties (e.g., octane number).

Hydrocracker
- Uses hydrogen to upgrade heavier fractions into lighter, more valuable products.

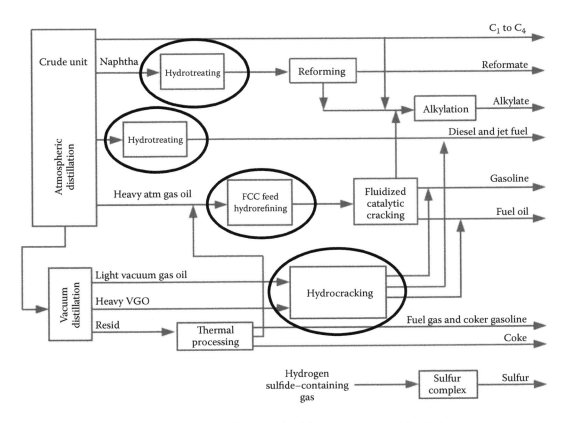

FIGURE 14.2 Schematic overview of a refinery emphasizing processes requiring hydrogen.

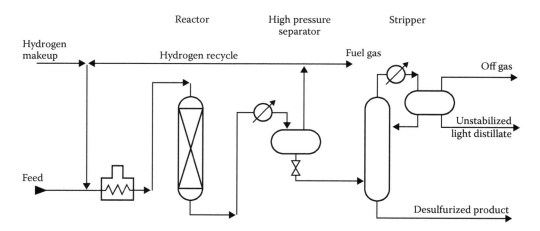

FIGURE 14.3 Distillate hydrodesulfurization. (From OSHA Technical Manual, Section IV, Chapter 2: Petroleum Refining Processes. http://www.osha.gov/dts/osta/otm/otm_iv/otm_iv_2.html; Speight, J.G. 2014. *The Chemistry and Technology of Petroleum.* 5th Edition. CRC Press, Taylor & Francis Publishers, Boca Raton, FL. Figure 23.1, p. 634.)

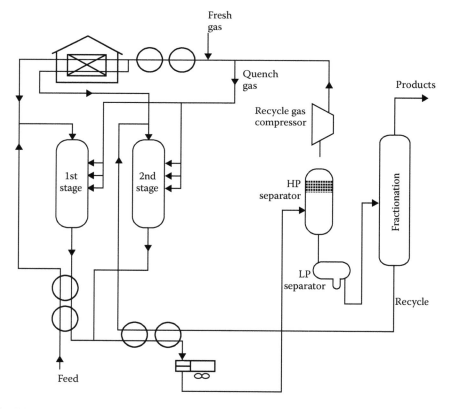

FIGURE 14.4 Single-stage or two-stage (optional) hydrocracking unit. (From OSHA Technical Manual, Section IV, Chapter 2: Petroleum Refining Processes. http://www.osha.gov/dts/osta/otm/otm_iv/otm_iv_2 .html; Speight, J.G. 2014. *The Chemistry and Technology of Petroleum.* 5th Edition. CRC Press, Taylor & Francis Publishers, Boca Raton, FL. Figure 23.2, p. 635.)

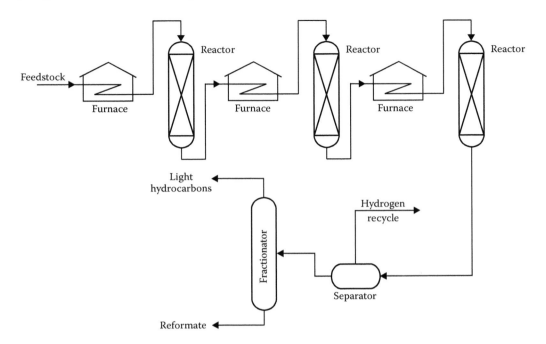

FIGURE 14.5 Platforming process. (From OSHA Technical Manual, Section IV, Chapter 2: Petroleum Refining Processes. http://www.osha.gov/dts/osta/otm/otm_iv/otm_iv_2.html; Speight, J.G. 2014. *The Chemistry and Technology of Petroleum.* 5th Edition. CRC Press, Taylor & Francis Publishers, Boca Raton, FL. Figure 23.3, p. 639.)

necessary for the chemical reactions and also prevents the carbon from being deposited on the catalyst, thus extending its operating life. An excess of hydrogen above whatever is consumed in the process is produced, and, as a result, catalytic reforming processes are unique in that they are the only petroleum refinery processes to produce hydrogen as a by-product. However, as refineries and refinery feedstocks evolved during the last four decades, the demand for hydrogen has increased and reforming processes are no longer capable of providing the quantities of hydrogen necessary for feedstock hydrogenation. Within the refinery, other processes are used as sources of hydrogen. Thus, the recovery of hydrogen from the by-products of the coking units, visbreakers, and catalytic crackers is also practiced in some refineries. Although the hydrogen gas produced by such processes usually contains up to 40% v/v of other gases (usually hydrocarbons), hydrotreater catalyst life is a function of hydrogen partial pressure. Optimum hydrogen purity at the reactor inlet extends catalyst life by maintaining desulfurization kinetics at lower operating temperatures and reducing carbon laydown. An increase in the concentration of hydrogen in the hydrogen-containing gas, resulting from the application of hydrogen purification equipment and/or increased hydrogen sulfide removal as well as careful management of hydrogen circulation and purge rates, can extend catalyst life by as much as 25%.

A critical issue facing the modern refinery is the changing slate of feedstocks that need to be refined into transportation fuels under an environment of increasingly more stringent clean fuel regulations, decreasing heavy fuel oil demand, and increasingly heavy, sour (high-sulfur) crude supply. Hydrogen network optimization is at the forefront of world refineries options to address clean fuel trends, to meet growing demands for transportation fuels, and to continue to make a profit from their crudes (Deng et al., 2013). A key element of a hydrogen network analysis in a refinery involves the capture of hydrogen in its fuel streams and extending its flexibility and processing options. Thus, optimization of the hydrogen network will be a critical factor influencing future operating flexibility and profitability of a refinery in a world of an increasing slate of heavy feedstocks and the demand for ultra-low-sulfur gasoline and ultra-low-sulfur diesel fuel.

As hydrogen use has become more widespread in refineries, hydrogen production has moved from the status of a high-tech specialty operation to an integral feature of most refineries (Raissi, 2001; Vauk et al., 2008). This has been made necessary by the increase in hydrotreating and hydrocracking, including the treatment of progressively heavier feedstocks (Speight, 2000, 2011, 2014; Speight and Ozum, 2002; Parkash, 2003; Hsu and Robinson, 2006; Ancheyta and Speight, 2007; Gary et al., 2007). In fact, the use of hydrogen in thermal processes is perhaps the single most significant advancement in refining technology during the 20th century (Scherzer and Gruia, 1996; Bridge, 1997; Dolbear, 1998). The continued increase in hydrogen demand over the last several decades is a result of the conversion of petroleum to match changes in product slate and the supply of heavy, high-sulfur oil, and in order to make lower-boiling, cleaner, and more salable products. There are also many reasons other than product quality for using hydrogen in processes, adding to the need to add hydrogen at relevant stages of the refining process and, most important, according to the availability of hydrogen (Bezler, 2003; Miller and Penner, 2003; Ranke and Schödel, 2003).

In cokers and visbreakers, heavy feedstocks are converted to petroleum coke, oil, light hydrocarbons (benzene, naphtha, and liquefied petroleum gas), and gas (Speight and Ozum, 2002; Parkash, 2003; Hsu and Robinson, 2006; Ancheyta and Speight, 2007; Gary et al., 2007; Speight, 2014). Depending on the process, hydrogen is present in a wide range of concentrations. Since petroleum-coking processes need gas for heating purposes, adsorption processes are best suited to recover the hydrogen because they feature a very clean hydrogen product and an off-gas suitable as fuel.

Catalytic cracking is the most important process step for the production of light products from gas oil, and increasingly from vacuum gas oil and heavy feedstocks (Chapter 7). In catalytic cracking, the molecular mass of the main fraction of the feed is lowered, while another part is converted to coke that is deposited on the hot catalyst. The catalyst is regenerated in one or two stages by burning the coke off with air that also provides the energy for the endothermic cracking process. In the process, paraffins and naphthenes are cracked to olefins and to alkanes with shorter chain length, monoaromatic compounds are dealkylated without ring cleavage, and diaromatics and polyaromatic compounds are dealkylated and converted to coke. Hydrogen is formed in the last type of reaction, whereas the first two reactions produce light hydrocarbons and therefore require hydrogen. Thus, a catalytic cracker can be operated in such a manner that enough hydrogen for subsequent processes is formed.

In reforming processes, naphtha fractions are reformed to improve the quality of naphtha (Speight, 2014). The most important reactions occurring during this process are the dehydrogenation of naphthenes to aromatics. This reaction is endothermic and is favored by low pressures, with the reaction temperature in the range of 300–450°C (570–840°F). The reaction is performed on platinum catalysts with other metals, e.g., rhenium, as promoters.

The chemical nature of the crude oil used as the refinery feedstock has always played a major role in determining the hydrogen requirements of that refinery. For example, the lighter, more paraffinic crude oils will require somewhat less hydrogen than the amount required by heavy feedstocks for upgrading (Table 14.1). It follows that the hydrodesulfurization of heavy oils and residua (which, by definition, is a hydrogen-dependent process) needs substantial amounts of hydrogen as part of the processing requirements.

In general, considerable variation exists from one refinery to another in the balance between the hydrogen produced and the hydrogen consumed in the refining operations. However, what is more pertinent to the present text is the excessive amounts of hydrogen that are required for hydroprocessing operations—whether these be hydrocracking or the somewhat milder hydrotreating processes. For effective hydroprocessing, a substantial hydrogen partial pressure must be maintained in the reactor and, in order to meet this requirement, an excess of hydrogen above that actually consumed by the process must be fed to the reactor. Part of the hydrogen requirement is met by recycling a stream of hydrogen-rich gas. However, there still remains a need to generate hydrogen as makeup material to accommodate the process consumption of 500–3000 scf/bbl depending on whether the heavy feedstock is being subjected to a predominantly hydrotreating (hydrodesulfurization) or to a predominantly hydrocracking process.

Hydrogen is generated in a refinery by the catalytic reforming process; however, there may not always be the need to have a catalytic reformer as part of the refinery sequence. Nevertheless, assuming that a catalytic reformer is part of the refinery sequence, the hydrogen production from the reformer usually falls well below the amount required for hydroprocessing purposes. For example, in a 100,000 bbl/day hydrocracking refinery, assuming intensive reforming of hydrocracked naphtha, the hydrogen requirements of the refinery may still fall to some 500–900 scf/bbl of crude charge below that necessary for the hydrocracking sequences. Consequently, an external source of hydrogen is necessary to meet the daily hydrogen requirements of any process where the heavier feedstocks are involved.

The trend to increase the number of hydrogenation (*hydrocracking* and/or *hydrotreating*) processes in refineries (Dolbear, 1998) coupled with the need to process the heavier oils, which require substantial quantities of hydrogen for upgrading, has resulted in vastly increased demands for this gas. Part of the hydrogen requirements can be satisfied by hydrogen recovery from catalytic reformer product gases; however, other external sources are required. Most of the external hydrogen is manufactured either by steam–methane reforming or by oxidation processes. However, other processes, such as steam–methanol interaction or ammonia dissociation, may also be used as sources of hydrogen. Electrolysis of water produces high-purity hydrogen, but the power costs may be prohibitive.

With the increasing need for clean fuels, the production of hydrogen for refining purposes requires a major effort by refiners. The hydrogen demands can be estimated to a very rough approximation using API gravity and the extent of the reaction, particularly the hydrodesulfurization reaction (Speight, 2000, 2014; Speight and Ozum, 2002; Parkash, 2003; Hsu and Robinson, 2006; Gary et al., 2007). However, accurate estimation requires equivalent process parameters and a thorough understanding of the nature of ach process. Thus, as hydrogen production grows, a better understanding of the capabilities and requirements of a hydrogen plant becomes ever more important to overall refinery operations as a means of making the best use of hydrogen supplies in the refinery.

This has led to a variety of innovations in processes such as (1) hydrotreating processes, (2) hydrocracking processes, (3) slurry hydrocracking processes to accommodate changes in the chemical nature of the crude oil used as the refinery feedstock as well as a variety of processes developed to accommodate the heavier feedstocks.

Catalytic cracking is the most important process step for the production of light products from gas oil, and increasingly from vacuum gas oil and heavy feedstocks (Speight and Ozum, 2002; Parkash, 2003; Hsu and Robinson, 2006; Ancheyta and Speight, 2007; Gary et al., 2007; Speight, 2014). In catalytic cracking, the molecular mass of the main fraction of the feed is lowered, while another part is converted to coke that is deposited on the hot catalyst. The catalyst is regenerated in one or two stages by burning the coke off with air that also provides the energy for the endothermic cracking process. In the process, paraffins and naphthenes are cracked to olefins and to alkanes with shorter chain length, monoaromatic compounds are dealkylated without ring cleavage, and diaromatics and polyaromatics are dealkylated and converted to coke. Hydrogen is formed in the last type of reaction, whereas the first two reactions produce light hydrocarbons and therefore require hydrogen. Thus, a catalytic cracker can be operated in such a manner that enough hydrogen for subsequent processes is formed.

Hydrogen has historically been produced during catalytic reforming processes as a by-product of the production of the aromatic compounds used in naphtha and in solvents. As reforming processes changed from fixed-bed to cyclic to continuous regeneration, process pressures have dropped and hydrogen production per barrel of reformate has tended to increase. However, hydrogen production as a by-product is not always adequate to the needs of the refinery and other processes are necessary. Thus, hydrogen production by steam reforming or by partial oxidation of residua has also been used, particularly where heavy oil is available. Steam reforming is the dominant method for hydrogen production and is usually combined with pressure-swing adsorption (PSA) to purify the hydrogen to >99% by volume (Bandermann and Harder, 1982).

In the steam-reforming process, there are typically four basic sections: (1) feedstock treatment where sulfur and other contaminants are removed; (2) the steam methane reformer, which converts feedstock and steam to synthesis gas at high temperature and moderate pressure; (3) the synthesis gas heat recovery and incorporation of shift reactor(s) to increase the hydrogen yield; and (4) the hydrogen purification section, which is typically a PSA unit, in which carbon monoxide, carbon dioxide, and methane are used to achieve the final product purity.

The reforming reaction between steam and hydrocarbons is highly endothermic and is carried out using specially formulated nickel catalyst contained in vertical tubes situated in the radiant section of the reformer. The simplified chemical reactions are

$$C_nH_{2n+2} + nH_2O \rightarrow nCO + (2n + 1)H_2 \quad \text{(for saturated hydrocarbons)}$$

$$CH_4 + H_2O \rightarrow CO + 3H_2 \quad \text{(for methane)}$$

In the adiabatic CO shift reactor vessel, the moderately exothermic *water gas shift reaction* converts carbon monoxide and steam to carbon dioxide and hydrogen:

$$CO + H_2O \rightarrow CO_2 + H_2$$

The gasification of residua and coke to produce hydrogen and/or power may become an attractive option for refiners (Dickenson et al., 1997; Fleshman, 1997; Gross and Wolff, 2000). The premise that the gasification section of a refinery will be the "garbage can" for deasphalter residues, high-sulfur coke, as well as other refinery wastes is worthy of consideration.

In the process, air is fed directly to air-blown gasifiers, while other gasifiers are fed with high-purity (approximately 99.5%) oxygen from an air separation unit (Speight, 2013, 2014). Steam is also fed directly to some dry feed gasifiers, while others are fed with coal or coke in a slurry with water. Reaction conditions typically vary in pressure from 430 to 1200 psi and in reaction outlet temperature (up to approximately 1480°C, 2700°F). The product composition varies, depending on the selected gasification technology and the characteristics of the petroleum coke. The process generally produces synthesis (syngas) with typical H_2/CO molar ratios of <1. Carbon dioxide, water, and methane are the secondary product components at lower concentrations relative to those of hydrogen or carbon monoxide. In addition, the synthesis gas is contaminated with hydrogen sulfide, carbon dioxide, carbonyl sulfide (COS), and other sulfur compounds. Raw synthesis gas from the gasifier can be cooled through a steam generation system or a quench and scrubbing section, where the raw syngas becomes saturated with water at a target temperature adequate for converting carbon monoxide and water to hydrogen and carbon dioxide (*shift reaction*):

$$CO + H_2O \rightarrow H_2 + CO_2$$

Several other processes are available for the production of the additional hydrogen that is necessary for the various heavy feedstock hydroprocessing sequences, and it is the purpose of this chapter to present a general description of these processes. In general, most of the external hydrogen is manufactured by steam–methane reforming or by oxidation processes. Other processes such as ammonia dissociation, steam–methanol interaction, or electrolysis are also available for hydrogen production; however, economic factors and feedstock availability assist in the choice between processing alternatives.

The processes described in this chapter are those gasification processes that are often referred to as the *garbage disposal units* of the refinery. Hydrogen is produced for use in other parts of the refinery as well as for energy, and it is often produced from process by-products that may not be of any use elsewhere. Such by-products might be the highly aromatic, heteroatom, and metal-containing reject from a deasphalting unit or from a mild hydrocracking process. However, attractive as this may seem, there will be the need to incorporate a gas-cleaning operation to remove any environmentally objectionable components from the hydrogen gas.

The gasification of residua and coke to produce hydrogen and/or power may become an attractive option for refiners (Dickenson et al., 1997; Fleshman, 1997). The premise that the gasification section of a refinery will be the "garbage can" for deasphalter residues, high-sulfur coke, as well as other refinery wastes is worthy of consideration. In general, most of the external hydrogen is manufactured by steam–methane reforming or by oxidation processes. Other processes such as ammonia dissociation, steam–methanol interaction, or electrolysis are also available for hydrogen production, but economic factors and feedstock availability assist in the choice between processing alternatives. These processes are often referred to as the garbage disposal units of the refinery. Hydrogen is produced for use in other parts of the refinery as well as for energy, and it is often produced from process by-products that may not be of any use elsewhere. Such by-products might be the highly aromatic, heteroatom, and metal-containing reject from a deasphalting unit or from a mild hydrocracking process. However, attractive as this may seem, there will be the need to incorporate a gas-cleaning operation to remove any environmentally objectionable components from the hydrogen gas.

Thus, the refining industry has been the subject of the four major forces that affect most industries and which have hastened the development of new petroleum refining processes: (1) the demand for products such as gasoline, diesel, fuel oil, and jet fuel; (2) feedstock supply, specifically the changing quality of crude oil and geopolitics between different countries and the emergence of alternative feed supplies such as bitumen from tar sand (oil sand), natural gas, and coal; (3) technology development such as new catalysts and processes, especially processes involving the use of hydrogen; and (4) environmental regulations that include more stringent regulations in relation to sulfur in gasoline and diesel (Speight and Ozum, 2002; Parkash, 2003; Speight, 2005, 2014; Hsu and Robinson, 2006; Gary et al., 2007).

Categories 1, 2, and 4 are directly affected by the third category (i.e., the use of hydrogen in refineries), and it is this category that will be the subject of this chapter. This chapter presents an introduction to the use and need for hydrogen petroleum refineries in order for the reader to place the use of hydrogen in the correct context of the refinery. In fact, hydrogen is key in allowing refineries to comply with the latest product specifications and environmental requirements for fuel production being mandated by market and governments, and in helping reduce the carbon footprint of refinery operations.

Thus, the focus of this chapter is the means by which hydrogen is produced in refineries, with particular attention paid to (1) the types of feedstocks used for hydrogen production in a refinery, (2) commercial processes, (3) process catalysts, (4) the purification of hydrogen, and (5) hydrogen management and safety.

14.2 FEEDSTOCKS

There are several processes by which hydrogen can be produced in a refinery (Table 14.3). The most common, and perhaps the best, feedstocks for steam reforming are low-boiling saturated hydrocarbons that have a low sulfur content, including natural gas, refinery gas, liquefied petroleum gas, and low-boiling naphtha.

Natural gas is the most common feedstock for hydrogen production since it meets all the requirements for a reformer feedstock. Natural gas typically contains >90% methane and ethane with only a few percent of propane and higher-boiling hydrocarbons (Mokhatab et al., 2006; Speight, 2007). Natural gas may (or most likely will) contain traces of carbon dioxide with some nitrogen and other impurities. Purification of natural gas, before reforming, is usually relatively straightforward. Traces of sulfur must be removed to avoid poisoning the reformer catalyst; zinc oxide treatment in combination with hydrogenation is usually adequate.

TABLE 14.3

Typical Hydrogen Production Processes in a Refinery

Catalytic reformer
- Used to convert the naphtha-boiling range molecules into a higher-octane reformate; hydrogen is a by-product.

Steam–methane reformer
- Produces hydrogen for the hydrotreaters or hydrocracker.

Steam reforming of higher molecular weight hydrocarbons
- Produces hydrogen from low-boiling hydrocarbons other than methane.

Recovery from refinery off-gases
- Process gas often contains hydrogen in the range up to 50% v/v.

Gasification of petroleum residua
- Recovery from synthesis gas (syngas) produced in gasification units.

Gasification of petroleum coke
- Recovery from synthesis gas (syngas) produced in gasification units.

Partial oxidation processes
- Analogous to gasification process; produce synthesis gas from which hydrogen can be isolated.

Light refinery gas (refinery off-gas) containing a substantial amount of hydrogen can be an attractive steam reformer feedstock since it is produced as a by-product. Processing of refinery gas will depend on its composition, particularly the levels of olefins and of propane and heavier hydrocarbons. Olefins, which can cause problems by forming coke in the reformer, are converted to saturated compounds in the hydrogenation unit. Higher-boiling hydrocarbons in refinery gas can also form coke, either on the primary reformer catalyst or in the preheater. If there is more than a few percent of C_3 and higher compounds, a promoted reformer catalyst should be considered, in order to avoid carbon deposits.

Refinery gas from different sources varies in suitability as hydrogen plant feed. Catalytic reformer off-gas (Speight and Ozum, 2002; Parkash, 2003; Hsu and Robinson, 2006; Ancheyta and Speight, 2007; Gary et al., 2007; Speight, 2014), for example, is saturated, very low in sulfur, and often has high hydrogen content. The process gases from a coking unit or from a fluid catalytic cracking unit are much less desirable because of the content of unsaturated constituents. In addition to olefins, these gases contain substantial amounts of sulfur that must be removed before the gas is used as feedstock. These gases are also generally unsuitable for direct hydrogen recovery, since the hydrogen content is usually too low. Hydrotreater off-gas lies in the middle of the range. It is saturated, so it is readily used as hydrogen plant feed. The content of hydrogen and heavier hydrocarbons depends to a large extent on the upstream pressure. Sulfur removal will generally be required.

Before the demand for hydrogen exceeds supply (from reforming processes), hydrogen-containing refinery gas is routed into the refinery fuel gas system where only the heating value of the gas has been used. Since the hydrogen demand for refinery operations is growing, these gases become more and more attractive as a source for hydrogen production. This requires purification steps such as PSA or membrane systems (Mokhatab et al., 2006; Speight, 2007).

The hydrogen content of refinery off-gas varies from 2% to 10% v/v and higher, depending on the source of the off-gas—the utilization of off-gas in the hydrogen production scheme of a refinery must be evaluated on a case-by-case basis. When applied, the recovery of hydrogen from off-gas using PSA technology, membrane technology, or cryogenic processes can be applied to generate hydrogen streams of any required purity (Mokhatab et al., 2006; Speight, 2007, 2014). The off-gas streams from the various processes can be routed to the refinery fuel gas system. Another option (especially if the hydrogen content of the refinery off-gas is low) is to use the off-gas as (supplementary) feedstock to a steam reformer plant, which also generates hydrogen from hydrocarbons such as natural gas, liquefied petroleum gas, or naphtha.

14.3   PROCESS CHEMISTRY

Before the feedstock is introduced to a process, there is the need for the application of a strict feedstock purification protocol. Prolonging catalyst life in hydrogen production processes is attributable to effective feedstock purification, particularly sulfur removal. A typical natural gas or other light hydrocarbon feedstock contains traces of hydrogen sulfide and organic sulfur.

To remove sulfur compounds, it is necessary to hydrogenate the feedstock to convert the organic sulfur to hydrogen, which is then reacted with zinc oxide (ZnO) at approximately 370°C (700°F), resulting in the optimal use of the zinc oxide as well as ensuring complete hydrogenation. Thus, assuming assiduous feedstock purification and removal of all of the objectionable contaminants, the chemistry of hydrogen production can be defined.

In steam reforming, low-boiling hydrocarbons such as methane are reacted with steam to form hydrogen:

$$CH_4 + H_2O \rightarrow 3H_2 + CO \quad \Delta H_{298\,K} = +97,400 \text{ Btu/lb}$$

where H is the heat of reaction. A more general form of the equation that shows the chemical balance for higher-boiling hydrocarbons is

$$C_nH_m + nH_2O \rightarrow (n + m/2)H_2 + nCO$$

The reaction is typically carried out at approximately 815°C (1500°F) over a nickel catalyst packed into the tubes of a reforming furnace. The high temperature also causes the hydrocarbon feedstock to undergo a series of cracking reactions, plus the reaction of carbon with steam:

$$CH_4 \rightarrow 2H_2 + C$$

$$C + H_2O \rightarrow CO + H_2$$

Carbon is produced on the catalyst at the same time that hydrocarbon is reformed to hydrogen and carbon monoxide. With natural gas or similar feedstock, reforming predominates and the carbon can be removed by reaction with steam as fast as it is formed. When higher-boiling feedstocks are used, the carbon is not removed fast enough and builds up, thereby requiring catalyst regeneration or replacement. Carbon buildup on the catalyst (when high-boiling feedstocks are employed) can be avoided by addition of alkali compounds, such as potash, to the catalyst, thereby encouraging or promoting the carbon–steam reaction.

However, even with an alkali-promoted catalyst, feedstock cracking limits the process to hydrocarbons with a boiling point less than 180°C (350°F). Natural gas, propane, butane, and light naphtha are most suitable. Prereforming, a process that uses an adiabatic catalyst bed operating at a lower temperature, can be used as a pretreatment to allow heavier feedstocks to be used with lower potential for carbon deposition (coke formation) on the catalyst.

After reforming, the carbon monoxide in the gas is reacted with steam to form additional hydrogen (the *water–gas shift* reaction):

$$CO + H_2O \rightarrow CO_2 + H_2 \quad \Delta H_{298\,K} = -16,500 \text{ Btu/lb}$$

This leaves a mixture consisting primarily of hydrogen and carbon monoxide that is removed by conversion to methane:

$$CO + 3H_2O \rightarrow CH_4 + H_2O$$

$$CO_2 + 4H_2 \rightarrow CH_4 + 2H_2O$$

The critical variables for steam-reforming processes are (1) temperature, (2) pressure, and (3) the steam/hydrocarbon ratio. Steam reforming is an equilibrium reaction, and conversion of the hydrocarbon feedstock is favored by high temperature, which in turn requires higher fuel use. Because of the volume increase in the reaction, conversion is also favored by low pressure, which conflicts with the need to supply the hydrogen at high pressure. In practice, materials of construction limit temperature and pressure.

On the other hand, and in contrast to reforming, shift conversion is favored by low temperature. The gas from the reformer is reacted over an iron oxide catalyst at 315–370°C (600–700°F), with the lower limit dictating the activity of the catalyst at low temperatures.

Hydrogen can also be produced by partial oxidation of hydrocarbons in which the hydrocarbon is oxidized in a limited or controlled supply of oxygen:

$$2CH_4 + O_2 \rightarrow CO + 4H_2 \quad \Delta H_{298\,K} = -10{,}195 \text{ Btu/lb}$$

The shift reaction also occurs and a mixture of carbon monoxide and carbon dioxide is produced in addition to hydrogen. The catalyst tube materials do not limit the reaction temperatures in partial oxidation processes, and higher temperatures may be used that enhance the conversion of methane to hydrogen. Indeed, much of the design and operation of hydrogen plants involves protecting the reforming catalyst and the catalyst tubes because of the extreme temperatures and the sensitivity of the catalyst. In fact, minor variations in feedstock composition or operating conditions can have significant effects on the life of the catalyst or the reformer itself. This is particularly true of changes in molecular weight of the feed gas, or poor distribution of heat to the catalyst tubes.

Since the high temperature takes the place of a catalyst, partial oxidation is not limited to the lower-boiling feedstocks that are required for steam reforming. Partial oxidation processes were first considered for hydrogen production because of expected shortages of lower-boiling feedstocks and the need to have a disposal method available for higher-boiling, high-sulfur streams such as asphalt or petroleum coke.

Catalytic partial oxidation, also known as autothermal reforming, reacts oxygen with a light feedstock and by passing the resulting hot mixture over a reforming catalyst. The use of a catalyst allows the use of lower temperatures than in noncatalytic partial oxidation and causes a reduction in oxygen demand.

The feedstock requirements for catalytic partial oxidation processes are similar to those for steam reforming, and light hydrocarbons from refinery gas to naphtha are preferred. The oxygen substitutes for much of the steam in preventing coking, and a lower steam/carbon ratio is required. In addition, because a large excess of steam is not required, catalytic partial oxidation produces more carbon monoxide and less hydrogen than steam reforming. Thus, the process is more suited to situations where carbon monoxide is the more desirable product, such as synthesis gas for chemical feedstocks.

14.4 COMMERCIAL PROCESSES

In spite of the use of low-quality hydrogen (that contains up to 40% by volume hydrocarbon gases), a high-purity hydrogen stream (95–99% by volume hydrogen) is required for hydrodesulfurization, hydrogenation, hydrocracking, and petrochemical processes. Hydrogen, produced as a by-product of refinery processes (principally, hydrogen recovery from catalytic reformer product gases), often is not enough to meet the total refinery requirements, necessitating the manufacturing of additional hydrogen or obtaining supply from external sources.

Catalytic reforming remains an important process used to convert low-octane naphtha into high-octane gasoline blending components called *reformate*. Reforming represents the total effect of numerous reactions such as cracking, polymerization, dehydrogenation, and isomerization taking place simultaneously. Depending on the properties of the naphtha feedstock (as measured by the paraffin, olefin, naphthene, and aromatic content) and catalysts used, reformate can be produced with very high concentrations of toluene, benzene, xylene, and other aromatics useful in gasoline blending and petrochemical processing. Hydrogen, a significant by-product, is separated from the reformate for recycling and used in other processes.

A catalytic reformer comprises a reactor section and a product-recovery section. More or less standard is a feed preparation section in which, by combination of hydrotreatment and distillation, the feedstock is prepared to specification. Most processes use platinum as the active catalyst. Sometimes platinum is combined with a second catalyst (bimetallic catalyst) such as rhenium or another noble metal. There are many different commercial catalytic reforming processes (Speight and Ozum, 2002; Parkash, 2003; Hsu and Robinson, 2006; Ancheyta and Speight, 2007; Gary et al., 2007; Speight, 2014), including platforming (Figure 14.5), powerforming, ultraforming, and thermofor catalytic reforming. In the platforming process, the first step is preparation of the naphtha feed to remove impurities from the naphtha and reduce catalyst degradation. The naphtha feedstock is then mixed with hydrogen, vaporized, and passed through a series of alternating furnace and fixed-bed reactors containing a platinum catalyst. The effluent from the last reactor is cooled and sent to a separator to permit removal of the hydrogen-rich gas stream from the top of the separator for recycling. The liquid product from the bottom of the separator is sent to a fractionator called a stabilizer (butanizer); the bottom product (reformate) is sent to storage; and butanes and lighter gases pass overhead and are sent to the saturated gas plant.

Some catalytic reformers operate at low pressure (50–200 psi), and others operate at high pressures (up to 1000 psi). Some catalytic reforming systems continuously regenerate the catalyst in other systems. One reactor at a time is taken off-stream for catalyst regeneration, and some facilities regenerate all of the reactors during turnarounds. Operating procedures should be developed to ensure control of hotspots during start-up. Safe catalyst handling is very important and care must be taken not to break or crush the catalyst when loading the beds, as the small fines will plug up the reformer screens. Precautions against dust when regenerating or replacing a catalyst should also be considered, and a water wash should be considered where stabilizer fouling has occurred owing to the formation of ammonium chloride and iron salts. Ammonium chloride may form in pretreater exchangers and cause corrosion and fouling. Hydrogen chloride from the hydrogenation of chlorine compounds may form acid or ammonium chloride salt.

14.4.1 Heavy Residue Gasification and Combined Cycle Power Generation

Residua from various processes are the preferred feedstocks for the production of hydrogen-rich gases. Such fractions with high sulfur and/or high heavy metal contents are difficult to handle in upgrading processes such as hydrogenation or coking and, for environmental reasons, are not usually used as fuels without extensive gas cleanup. Residua are gasified using the Texaco gasification process (partial oxidation process), and the produced gas is purified to clean fuel gas, then electric power is generated by means of the combined cycle system. The convenience of the process depends on the availability of residua (or coal) in preference to the availability of natural gas that is used for hydrogen production by steam–methane reforming (Shimizu, 1982; Bailey, 1992).

Heavy residua are gasified and the produced gas is purified to clean fuel gas (Gross and Wolff, 2000). As an example, the solvent deasphalter residuum is gasified by a partial oxidation method under a pressure of about 570 psi and at a temperature between 1300°C and 1500°C (2370°F and 2730°F). The high-temperature-generated gas flows into the specially designed waste heat boiler, in which the hot gas is cooled and high-pressure saturated steam is generated. The gas from the waste heat boiler is then heat exchanged with the fuel gas and flows to the carbon scrubber, where unreacted carbon particles are removed from the generated gas by water scrubbing.

The gas from the carbon scrubber is further cooled by the fuel gas and boiler feed water, and led into the sulfur compound removal section, where hydrogen sulfide (H_2S) and carbonyl sulfide are removed from the gas to obtain clean fuel gas. This clean fuel gas is heated with the hot gas generated in the gasifier and finally supplied to the gas turbine at a temperature of 250–300°C (480–570°F).

The exhaust gas from the gas turbine having a temperature of about 550–600°C (1020–1110°F) flows into the heat recovery steam generator consisting of five heat exchange elements. The first element is a superheater in which the combined stream of the high-pressure saturated steam generated in the waste heat boiler and in the second element (high-pressure steam evaporator) is superheated. The third element is an economizer; the fourth element is a low-pressure steam evaporator; and the final or the fifth element is a deaerator heater. The off-gas from heat recovery steam generator having a temperature of about 130°C is emitted into the air via stack.

To decrease the nitrogen oxide (NO_x) content in the flue gas, two methods can be applied. The first method is the injection of water into the gas turbine combustor. The second method is to selectively reduce the nitrogen oxide content by injecting ammonia gas in the presence of a de-NO_x catalyst that is packed in a proper position of the heat recovery steam generator. The latter is more effective than the former in lowering the nitrogen oxide emissions to the air.

14.4.2 HYBRID GASIFICATION PROCESS

In the hybrid gasification process, a slurry of coal and residual oil is injected into the gasifier where it is pyrolyzed in the upper part of the reactor to produce gas and chars. The chars produced are then partially oxidized to ash. The ash is removed continuously from the bottom of the reactor.

In this process, coal and vacuum residue are mixed together into a slurry to produce clean fuel gas. The slurry fed into the pressurized gasifier is thermally cracked at a temperature of 850–950°C (1560–1740°F) and is converted into gas, tar, and char. The mixture of oxygen and steam in the lower zone of the gasifier gasify the char. The gas leaving the gasifier is quenched to a temperature of 450°C (840°F) in the fluidized-bed heat exchanger, and is then scrubbed to remove tar, dust, and steam at around 200°C (390°F).

The coal and residual oil slurry is gasified in the fluidized-bed gasifier. The charged slurry is converted to gas and char by thermal cracking reactions in the upper zone of the fluidized bed. The produced char is further gasified with steam and oxygen that enter the gasifier just below the fluidizing gas distributor. Ash is discharged from the gasifier and indirectly cooled with steam, and then discharged into the ash hopper. It is burned with an incinerator to produce process steam. Coke deposited on the silica sand is removed in the incinerator.

14.4.3 HYDROCARBON GASIFICATION

The gasification of hydrocarbons to produce hydrogen is a continuous, noncatalytic process that involves partial oxidation of the hydrocarbon (Figure 14.6). Air or oxygen (with steam or carbon dioxide) is used as the oxidant at 1095–1480°C (2000–2700°F). Any carbon produced (2–3% by weight of the feedstock) during the process is removed as a slurry in a carbon separator and pelletized for use either as a fuel or as raw material for carbon-based products.

14.4.4 HYPRO PROCESS

The Hypro process is a continuous catalytic process (Figure 14.7) method for hydrogen manufacture from natural gas or from refinery effluent gases. The process is designed to convert natural gas:

$$CH_4 \rightarrow C + 2H_2$$

Hydrogen is recovered by phase separation to yield hydrogen of about 93% purity; the principal contaminant is methane.

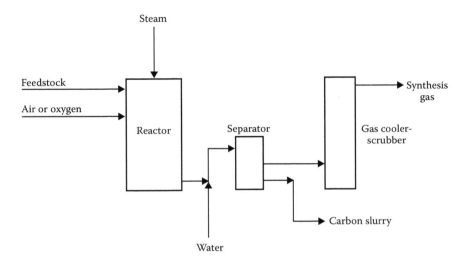

FIGURE 14.6 Hydrocarbon gasification process. (From Speight, J.G. 2000. *The Desulfurization of Heavy Oils and Residua*. 2nd Edition. Marcel Dekker Inc., New York. Figure 10.2, p. 392.)

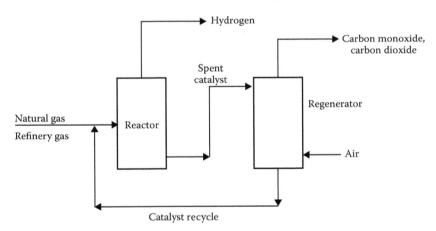

FIGURE 14.7 Hypro process. (From Speight, J.G. 2000. *The Desulfurization of Heavy Oils and Residua*. 2nd Edition. Marcel Dekker Inc., New York. Figure 10.3, p. 393.)

14.4.5 PYROLYSIS PROCESSES

There has been recent interest in the use of pyrolysis processes to produce hydrogen. Specifically, the interest has focused on the pyrolysis of methane (natural gas) and hydrogen sulfide.

Natural gas is readily available and offers a relatively rich stream of methane with lower amounts of ethane, propane, and butane also present. The thermocatalytic decomposition of natural gas hydrocarbons offers an alternative method for the production of hydrogen (Uemura et al., 1999; Weimer et al., 2000):

$$C_nH_m \rightarrow nC + (m/2)H_2$$

If a hydrocarbon fuel such as natural gas (methane) is to be used for hydrogen production by direct decomposition, then the process that is optimized to yield hydrogen production may not be suitable for the production of the high-quality carbon black by-product intended for the industrial rubber market. Moreover, it appears that the carbon produced from high-temperature (850–950°C, 1560–1740°F) direct thermal decomposition of methane is a soot-like material with a high tendency for catalyst deactivation (Murata et al., 1997). Thus, if the object of methane decomposition is hydrogen production, the carbon by-product may not be marketable as a high-quality carbon black for rubber and tire applications.

The production of hydrogen by direct decomposition of hydrogen sulfide is also possible (Figure 14.8) (Clark and Wassink, 1990; Zaman and Chakma, 1995; Donini, 1996; Luinstra, 1996). Hydrogen sulfide decomposition is a highly endothermic process, and equilibrium yields are poor (Clark et al., 1995) and can be represented simply as

$$H_2S \rightarrow H_2 + S$$

At temperatures <1500°C (2730°F), the thermodynamic equilibrium is unfavorable toward hydrogen formation. However, in the presence of catalysts such as platinum–cobalt (at 1000°C, 1830°F), disulfides of molybdenum or tungsten (Mo or W) at 800°C (1470°F) (Kotera et al., 1976), or other transition metal sulfides supported on alumina (at 500–800°C, 930–1470°F), decomposition of hydrogen sulfide proceeds rapidly (Kiuchi, 1982; Bishara et al., 1987; Al-Shamma and Naman, 1989; Clark and Wassink, 1990; Megalofonos and Papayannakos, 1997; Arild, 2000; Raissi, 2001). In the temperature range of about 800–1500°C (1470–2730°F), thermolysis of hydrogen sulfide can be treated simply as

$$H_2S \rightarrow H_2 + 1/xS_x \quad \Delta H_{298\,K} = +34,300\ Btu/lb$$

where $x = 2$. Outside this temperature range, multiple equilibria may be present depending on temperature, pressure, and relative abundance of hydrogen and sulfur (Clark and Wassink, 1990).

Above approximately 1000°C (1830°F), there is a limited advantage to using catalysts since the thermal reaction proceeds to equilibrium very rapidly (Clark and Wassink, 1990). The hydrogen

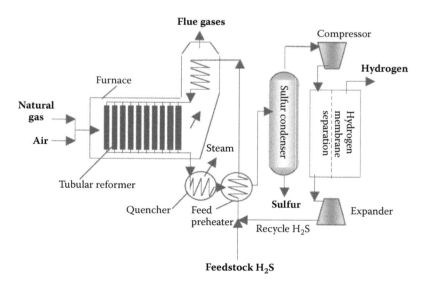

FIGURE 14.8 Simplified schematic for the production of hydrogen from hydrogen sulfide. (From Speight, J.G. 2014. *The Chemistry and Technology of Petroleum.* 5th Edition. CRC Press, Taylor & Francis Publishers, Boca Raton, FL. Figure 23.4, p. 641.)

yield can be doubled by preferential removal of either H_2 or sulfur from the reaction environment, thereby shifting the equilibrium. The reaction products must be quenched quickly after leaving the reactor to prevent reversible reactions.

14.4.6 SHELL GASIFICATION PROCESS

The Shell gasification process (partial oxidation process) is a flexible process for generating synthesis gas, principally hydrogen and carbon monoxide, for the ultimate production of high-purity high-pressure hydrogen, ammonia, methanol, fuel gas, town gas, or reducing gas by reaction of gaseous or liquid hydrocarbons with oxygen, air, or oxygen-enriched air.

The most important step in converting heavy residue to industrial gas is the partial oxidation of the oil using oxygen with the addition of steam. The gasification process takes place in an empty, refractory-lined reactor at temperatures of about 1400°C (2550°F) and pressures between 29 and 1140 psi. The chemical reactions in the gasification reactor proceed without catalyst to produce gas containing carbon amounting to some 0.5–2% by weight, based on the feedstock. The carbon is removed from the gas with water, extracted in most cases with feed oil from the water and returned to the feed oil. The high reformed gas temperature is utilized in a waste heat boiler for generating steam. The steam is generated at 850–1565 psi. Some of this steam is used as process steam and for oxygen and oil preheating. The surplus steam is used for energy production and heating purposes.

14.4.7 STEAM–METHANE REFORMING

The steam–methane reforming process (Figure 14.9) is the benchmark continuous catalytic process that has been employed over a period of several decades for hydrogen production. The process involves reforming natural gas in a continuous catalytic process in which the major reaction is the formation of carbon monoxide and hydrogen from methane and steam:

$$CH_4 + H_2O = CO + 3H_2 \quad \Delta H_{298\,K} = +97{,}400 \text{ Btu/lb} \tag{1}$$

In a similar manner, higher molecular weight feedstocks such as propane and even liquid hydrocarbons may also yield hydrogen:

$$C_3H_8 + 3H_2O \rightarrow 3CO + 7H_2 \tag{2}$$

The most important feedstock for the catalytic steam-reforming process is natural gas. Other feedstocks are associated gas, propane, butane, liquefied petroleum gas, and some naphtha fractions. The choice is usually made on the availability and the price of the raw material.

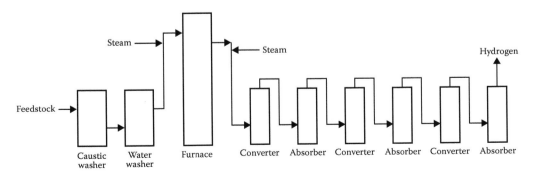

FIGURE 14.9 Steam-methane reforming process. (From Speight, J.G. 2000. *The Desulfurization of Heavy Oils and Residua.* 2nd Edition. Marcel Dekker Inc., New York. Figure 10.5, p. 395.)

In the actual process, the feedstock is first desulfurized by passage through activated carbon, which may be preceded by caustic and water washes. The desulfurized material is then mixed with steam and passed over a nickel-based catalyst (730–845°C, 1350–1550°F and 400 psi). Effluent gases are cooled by the addition of steam or condensate to about 370°C (700°F), at which point carbon monoxide reacts with steam in the presence of iron oxide in a shift converter to produce carbon dioxide and hydrogen:

$$CO + H_2O = CO_2 + H_2 \quad \Delta H_{298\,K} = -41.16 \text{ kJ/mol} \tag{3}$$

The carbon dioxide is removed by amine washing; the hydrogen is usually a high-purity (>99%) material.

Since the presence of any carbon monoxide or carbon dioxide in the hydrogen stream can interfere with the chemistry of the catalytic application, a third stage is used to convert of these gases to methane:

$$CO + 3H_2 \rightarrow CH_4 + H_2O$$

$$CO_2 + 4H_2 \rightarrow CH_4 + 2H_2O$$

For many refiners, sulfur-free natural gas (CH_4) is not always available to produce hydrogen by this process. In that case, higher-boiling hydrocarbons (such as propane, butane, or naphtha) may be used as the feedstock to generate hydrogen (q.v.).

The net chemical process for steam methane reforming is then given by

$$CH_4 + 2H_2O \rightarrow CO_2 + 4H_2 \quad \Delta H_{298\,K} = +165.2 \text{ kJ/mol} \tag{4}$$

Indirect heating provides the required overall endothermic heat of reaction for the steam–methane reforming.

Obviously, in the case of a natural gas feedstock, a simple adsorption process might suffice, while in the case of the higher molecular weight feedstocks, such as naphtha, a more complete desulfurization process may be necessary.

The desulfurized feedstock is then mixed with superheated steam and passed over a nickel catalyst (730–845°C, 1350–1550°F; 400 psi) to produce a mixture of hydrogen, carbon monoxide, and carbon dioxide as well as excess steam. The effluent gases are cooled (to about 370°C, 700°F) and passed through a shift converter that promotes reaction of the carbon monoxide with steam to yield carbon dioxide and more hydrogen. The shift converter may contain two beds of catalyst with interbed cooling; the combination of the two catalyst beds promotes maximum conversion of the carbon monoxide. This is essential in the event that a high-purity product is required.

The carbon dioxide–rich gas stream is then cooled, after which carbon dioxide removal is achieved by passage through scrubbers or by the more thermally efficient hot carbonate process. Any residual carbon monoxide or carbon dioxide is removed by passing the heated gas through a nickel base methanation catalyst where the carbon oxides are reached with hydrogen to produce methane. The methane may also find other use within the refinery, although in the current context the methane acts as the starting material for hydrogen.

Usually, natural gas contains sulfur compounds (Speight, 2007) and because of the high sulfur sensitivity of the catalyst in the reformer (and in the shift reactor, if installed), hydrogen sulfide and other sulfur compounds must be removed. This may be accomplished using a zinc oxide bed. If higher molecular weight organic sulfur compounds (mercaptans R-SH) or carbonyl sulfide in concentrations in the parts per million range are present, the zinc oxide bed alone is not sufficient and the sulfur compounds are converted to hydrogen sulfide in a hydrogenation stage. The hydrogen necessary is taken from the product stream and amounts to ca. 5% for natural gas (whereas for naphtha and liquefied petroleum gas, 10% and 25% H_2, respectively, is used).

Supported nickel catalysts catalyze steam–methane reforming and the concurrent shift reaction. The catalyst contains 15–25% w/w nickel oxide on a mineral carrier. The carrier materials are alumina, aluminosilicates, cement, and magnesia. Before start-up, nickel oxide must be reduced to metallic nickel with hydrogen but also with natural gas or even with the feed gas itself.

Certain types of catalysts such as uranium oxide and chromium oxide may be used as a promoter. This is reported to give a higher resistance to catalyst poisoning by sulfur components and a lower tendency to form carbon deposits.

For the higher molecular weight feedstocks such as liquefied petroleum gas (usually propane C_3H_8) and naphtha (q.v.), nickel catalysts with alkaline carriers or alkaline-free catalysts with magnesium oxide as additive can be used. Both types of catalyst are less active than the conventional nickel catalyst. Therefore, a less rapid decomposition of the hydrocarbons is achieved. At the same time, the reaction of water with any carbon formed is catalyzed.

The required properties of the catalyst carriers are high specific area, low pressure drop, and high mechanical resistance at temperatures up to 1000°C (1830°F). The catalysts are usually in the form of rings (e.g., outer diameter 16 mm, height 16 mm, inner diameter 8 mm); however, other forms, such as saddles, stars, and spoked wheels, are also commercially available.

The main catalyst poison in steam-reforming plants is sulfur that is present in most feedstocks. Sulfur concentrations as low as 0.1 ppm form a deactivating layer on the catalyst, but the activity loss of a poisoned catalyst can be offset, to some extent, by increasing the reaction temperature. This helps reconvert the inactive nickel sulfide to active nickel sites. Nickel-free catalysts have been proposed for feedstocks heavier than naphtha. These catalysts consist mostly of strontium, aluminum, and calcium oxides, and seem to be resistant to coke deposits and may be suitable for use with high-sulfur feedstocks.

One way of overcoming the thermodynamic limitation of steam reforming is to remove either hydrogen or carbon dioxide as it is produced, hence shifting the thermodynamic equilibrium toward the product side. The concept for sorption-enhanced methane steam reforming is based on *in situ* removal of carbon dioxide by a sorbent such as calcium oxide (CaO).

$$CaO + CO_2 \rightarrow CaCO_3$$

Sorption enhancement enables lower reaction temperatures, which may reduce catalyst coking and sintering, while enabling the use of less expensive reactor wall materials. In addition, heat release by the exothermic carbonation reaction supplies most of the heat required by the endothermic reforming reactions. However, energy is required to regenerate the sorbent to its oxide form by the energy-intensive calcination reaction:

$$CaCO_3 \rightarrow CaO + CO_2$$

Use of a sorbent requires either that there be parallel reactors operated alternatively and out of phase in reforming and sorbent regeneration modes, or that sorbent be continuously transferred between the reformer/carbonator and regenerator/calciner (Balasubramanian et al., 1999; Hufton et al., 1999).

In autothermal (or secondary) reformers, the oxidation of methane supplies the necessary energy and carried out either simultaneously or in advance of the reforming reaction (Brandmair et al., 2003; Ehwald et al., 2003; Nagaoka et al., 2003). In the autothermal reforming process, the feedstock is reacted with a mixture of oxygen and steam by the use of a burner and a fixed nickel catalyst bed for the equilibration [reactions (1) and (2)] of the gas. This results in a lower oxygen consumption than used in noncatalytic routes. With the addition of steam, it is possible to adjust the hydrogen/carbon monoxide ratio. This cannot be achieved by noncatalytic routes because the addition of steam results in a reduction of temperature and soot formation.

$$2CH_4 + 3O_2 \rightarrow 2CO + 4H_2O$$

$$CH_4 + H_2O \rightarrow CO + 3H_2$$

$$CO + H_2O \rightarrow CO_2 + H_2$$

The equilibrium of the methane steam reaction and the water–gas shift reaction determines the conditions for optimum hydrogen yields. The optimum conditions for hydrogen production require a high temperature at the exit of the reforming reactor (800–900°C, 1470–1650°F), high excess of steam (molar steam-to-carbon ratio of 2.5–3), and relatively low pressures (<450 psi). Most commercial plants employ supported nickel catalysts for the process.

The steam–methane reforming process described briefly above would be an ideal hydrogen production process if it was not for the fact that large quantities of natural gas, a valuable resource, are required as both feed gas and combustion fuel. For each mole of methane reformed, more than 1 mol carbon dioxide is coproduced and must be disposed. This can be a major issue as it results in the same amount of greenhouse gas emission as would be expected from direct combustion of natural gas or methane. In fact, the production of hydrogen as a clean burning fuel by way of steam reforming of methane and other fossil-based hydrocarbon fuels is not in environmental balance if, in the process, carbon dioxide and carbon monoxide are generated and released into the atmosphere; however, alternate scenarios are available (Gaudernack, 1996). Moreover, as the reforming process is not totally efficient, some of the energy value of the hydrocarbon fuel is lost by conversion to hydrogen but with no tangible environmental benefit, such as a reduction in emission of greenhouse gases. Despite these apparent shortcomings, the process has the following advantages: (1) produces 4 mol hydrogen for each mole of methane consumed; (2) feedstocks for the process (methane and water) are readily available; (3) the process is adaptable to a wide range of hydrocarbon feedstocks; (4) operates at low pressures, <450 psi; (5) requires a low steam/carbon ratio (2.5–3); (6) good utilization of input energy (reaching 93%); (7) can use catalysts that are stable and resist poisoning; and (8) has good process kinetics.

Liquid feedstocks, either liquefied petroleum gas or naphtha (q.v.), can also provide backup feed, if there is a risk of natural gas curtailments. The feed handling system needs to include a surge drum, feed pump, and vaporizer (usually steam heated), followed by further heating before desulfurization. The sulfur in liquid feedstocks occurs as mercaptans, thiophene derivatives, or higher-boiling compounds. These compounds are stable and will not be removed by zinc oxide; therefore, a hydrogenation unit will be required. In addition, as with refinery gas, olefins must also be hydrogenated if they are present.

The reformer will generally use a potash-promoted catalyst to avoid coke buildup from cracking of the heavier feedstock. If liquefied petroleum gas is to be used only occasionally, it is often possible to use a methane-type catalyst at a higher steam/carbon ratio to avoid coking. Naphtha will require a promoted catalyst unless a preformer is used.

14.4.8 STEAM–NAPHTHA REFORMING

The steam–naphtha reforming process (Figure 14.10) is a continuous process for the production of hydrogen from liquid hydrocarbons and is, in fact, similar to steam–methane reforming, which is one of several possible processes for the production of hydrogen from low-boiling hydrocarbons other than ethane (Muradov, 1998, 2000; Brandmair et al., 2003; Find et al., 2003). A variety of naphtha types in the naphtha boiling range may be employed, including feeds containing up to 35% aromatics. Thus, following pretreatment to remove sulfur compounds, the feedstock is mixed with steam and taken to the reforming furnace (675–815°C, 1250–1500°F; 300 psi), where hydrogen is produced. In addition, some modifications to the proprietary catalyst are necessary to accommodate the higher molecular weight feedstock, thereby allowing naphtha (including naphtha with as much as 35% aromatics) boiling up to 200°C (390°F) to be used as process feedstocks.

A problem may occur when higher molecular weight materials are employed as feedstocks and result in coke formation and deposition. When an alkali-promoted catalyst is employed, corrosion

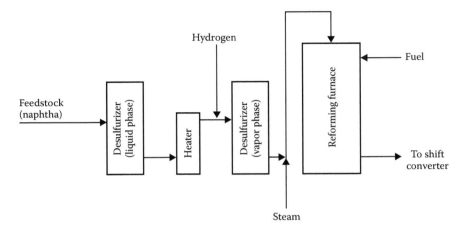

FIGURE 14.10 Steam–naphtha reforming process. (From Speight, J.G. 2000. *The Desulfurization of Heavy Oils and Residua*. 2nd Edition. Marcel Dekker Inc., New York. Figure 10.6, p. 397.)

and fouling problems in the reformer (or even in equipment downstream of the reformer because of the tendency of the alkali to migrate) may occur with some frequency. However, coke formation can be eliminated by the use of a proprietary alkali-free catalyst that has an extremely high activity and resistance to poisoning.

The overall chemistry of the steam–naphtha reforming process may be represented by the following equations:

$$C_6H_{14} + 6H_2O \rightarrow 6CO + 13H_2$$

$$C_6H_{14} + 12H_2O \rightarrow 6CO_2 + 19H_2$$

The process details (and the process flow) are essentially the same as those described for steam–methane reforming.

14.4.9 SYNTHESIS GAS GENERATION

The synthesis gas generation process (Figure 14.11) is a noncatalytic process for producing synthesis gas (principally hydrogen and carbon monoxide) for the ultimate production of high-purity hydrogen from gaseous or liquid hydrocarbons.

In this process, a controlled mixture of preheated feedstock and oxygen is fed to the top of the generator where carbon monoxide and hydrogen emerge as the products. Soot, produced in this part of the operation, is removed in a water scrubber from the product gas stream and is then extracted from the resulting carbon–water slurry with naphtha and transferred to a fuel oil fraction. The oil–soot mixture is burned in a boiler or recycled to the generator to extinction, to eliminate carbon production as part of the process.

The soot-free synthesis gas is then charged to a shift converter where the carbon monoxide reacts with steam to form additional hydrogen and carbon dioxide at the stoichiometric rate of 1 mol hydrogen for every mole of carbon monoxide charged to the converter.

The reactor temperatures vary from 1095°C to 1490°C (2000–2700°F), while pressures can vary from approximately atmospheric pressure to approximately 2000 psi. The process has the capability of producing high-purity hydrogen, although the extent of the purification procedure depends on the use for the hydrogen. For example, carbon dioxide can be removed by scrubbing with various alkaline reagents, while carbon monoxide can be removed by washing with liquid nitrogen or, if nitrogen is undesirable in the product, the carbon monoxide should be removed by washing with copper–amine solutions.

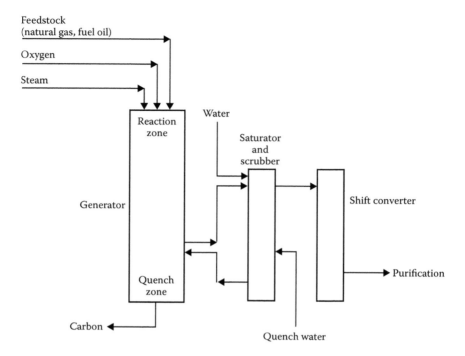

FIGURE 14.11 Process for synthesis gas production. (From Speight, J.G. 2000. *The Desulfurization of Heavy Oils and Residua*. 2nd Edition. Marcel Dekker Inc., New York. Figure 10.7, p. 398.)

This particular partial oxidation technique has also been applied to a whole range of liquid feedstocks for hydrogen production. There is now serious consideration being given to hydrogen production by the partial oxidation of solid feedstocks such as petroleum coke (from both delayed and fluid-bed reactors), lignite, and coal, as well as petroleum residua.

The chemistry of the process, using naphthalene as an example, may be simply represented as the selective removal of carbon from the hydrocarbon feedstock and further conversion of a portion of this carbon to hydrogen:

$$C_{10}H_8 + 5O_2 \rightarrow 10CO + 4H_2$$

$$10CO + 10H_2O \rightarrow 10CO_2 + 10H_2$$

Although these reactions may be represented very simply using equations of this type, the reactions can be complex and result in carbon deposition on parts of the equipment, thereby requiring careful inspection of the reactor.

14.4.10 TEXACO GASIFICATION PROCESS

The Texaco partial oxidation process (Texaco gasification process) (Figure 14.12) is a partial oxidation gasification process for generating synthetic gas, principally hydrogen and carbon monoxide (*synthesis gas, syngas*) (Table 14.4). The characteristic of the process (Figure 14.12) is to inject feedstock together with carbon dioxide, steam, or water into the gasifier. Therefore, solvent deasphalted residua, or petroleum coke rejected from any coking method, can be used as feedstock for this gasification process. The produced gas from this gasification process can be used for the production of high-purity, high-pressurized hydrogen, ammonia, and methanol. The heat recovered from the high-temperature gas is used for the generation of steam in the waste heat boiler. Alternatively, the

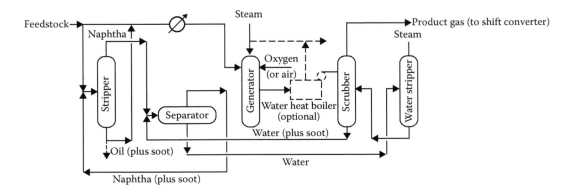

FIGURE 14.12 Texaco partial oxidation process. (From Speight, J.G. 2000. *The Desulfurization of Heavy Oils and Residua*. 2nd Edition. Marcel Dekker Inc., New York. Figure 10.4, p. 393.)

TABLE 14.4
Synthesis Gas Generation Using the Texaco Partial Oxidation Process

Feedstock	Natural Gas	Propane	64° API Naphtha	9.6° API Fuel Oil	9.7° API Fuel Oil	−11.4° AP1 Coal Tar
Feedstock Composition, wt.%						
Carbon	73.40	81.69	83.8	87.2	87.2	88.1
Hydrogen	22.76	18.31	16.2	9.9	9.9	5.7
Oxygen	−0.76	–	–	0.8	0.8	4.4
Nitrogen	3.08	–	–	0.7	0.7	0.9
Sulfur	–	–	–	1.4	1.4	0.8
Ash	–	–	–	–	–	0.1
Gross heating value, Btu/lb	22,630	21,662	20,300	18,200	18,200	15,690
Composition of Product Gas, mol						
Hydrogen	61.1	54.0	51.2	45.9	45.8	38.9
Carbon monoxide	35.0	43.7	45.3	48.5	47.5	54.3
Carbon dioxide	2.6	2.1	2.7	4.6	5.7	5.7
Nitrogen	1.0	0.1	0.1	0.7	0.2	0.8
Methane	0.3	0.1	0.7	0.2	0.5	0.1
Hydrogen sulfide	–	–	–	0.1	0.3	0.2

Source: Speight, J.G. 2007. *The Desulfurization of Heavy Oils and Residua*. 2nd Edition. Marcel Dekker Inc., New York. Table 10.2, p. 399.

less expensive quench-type configuration is preferred when high-pressure steam is not needed or when a high degree of shift is needed in the downstream CO converter.

In the process, the feedstock, together with the feedstock carbon slurry recovered in the carbon recovery section, is pressurized to a given pressure, mixed with high-pressure steam, and then blown into the gas generator through the burner together with oxygen.

The gasification reaction is a partial oxidation of hydrocarbons to carbon monoxide and hydrogen:

$$C_xH_{2y} + x/2O_2 \rightarrow xCO + yH_2$$

$$C_xH_{2y} + xH_2O \rightarrow xCO + (x + y)H_2$$

The gasification reaction is instantly completed, thus producing gas mainly consisting of H_2 and CO (H_2 + CO = >90%). The high-temperature gas leaving the reaction chamber of the gas generator enters the quenching chamber linked to the bottom of the gas generator and is quenched to 200–260°C (390–500°F) with water.

14.4.11 Recovery from Fuel Gas

Recovering hydrogen from refinery fuel gas can help refineries satisfy the high hydrogen demand. Cryogenic separation is typically viewed as being the most thermodynamically efficient separation technology. The higher capital cost associated with prepurification and the low flexibility to impurity upsets has limited its use in hydrogen recovery.

The basic configuration for hydrogen recovery from refinery gases involves a two-stage partial condensation process, with post purification via PSA (Dragomir et al., 2010). The major steps in this process involve first compressing and pretreating the crude refinery gas stream before chilling to an intermediate temperature (−60°F to −120°F). This partially condensed stream is then separated in a flash drum, after which the liquid stream from is expanded through a Joule–Thompson valve to generate refrigeration and then is fed to the wash column. Optionally, the wash column can be replaced by a simple flash drum.

A crude liquefied petroleum gas stream is collected at the bottom of the column, and a methane-rich vapor is obtained at the top. The methane-rich vapor is sent to compression and then to fuel. The vapor from the flash drum is further cooled in a second heat exchanger before being fed to a second flash drum where it produces a hydrogen-rich stream and a methane-rich liquid. The liquid is expanded in a Joule–Thomson valve to generate refrigeration, and then is sent for further cooling. The hydrogen-rich gas is then sent to the PSA unit for further purification—the tail gas from this unit is compressed and returned to fuel together with the methane-rich gas.

14.5 CATALYSTS

Hydrogen plants are one of the most extensive users of catalysts in the refinery. Catalytic operations include hydrogenation, steam reforming, shift conversion, and methanation.

14.5.1 Reforming Catalysts

The reforming catalyst is usually supplied as nickel oxide that, during start-up, is heated in a stream of inert gas, then steam. When the catalyst is near the normal operating temperature, hydrogen or a light hydrocarbon is added to reduce the nickel oxide to metallic nickel.

The high temperatures (up to 870°C, 1600°F) and the nature of the reforming reaction require that the reforming catalyst be used inside the radiant tubes of a reforming furnace. The active agent in the reforming catalyst is nickel, and normally the reaction is controlled both by diffusion and heat transfer. Catalyst life is limited as much by physical breakdown as by deactivation.

Sulfur is the main catalyst poison, and catalyst poisoning is theoretically reversible with the catalyst being restored to near full activity by steaming. However, in practice, the deactivation may cause the catalyst to overheat and coke, to the point that it must be replaced. Reforming catalysts are also sensitive to poisoning by heavy metals, although these are rarely present in low-boiling hydrocarbon feedstocks and in naphtha feedstocks.

Coking deposition on the reforming catalyst and the ensuing loss of catalyst activity is the most characteristic issue that must be assessed and mitigated.

While methane-rich streams such as natural gas or light refinery gas are the most common feeds to hydrogen plants, there is often a requirement for variety of reasons to process a variety of higher-boiling feedstocks, such as liquefied petroleum gas and naphtha. Feedstock variations may also be inadvertent due, for example, to changes in refinery off-gas composition from another unit or because of variations in naphtha composition due to feedstock variance to the naphtha unit.

Thus, when using higher-boiling feedstocks in a hydrogen plant, coke deposition on the reformer catalyst becomes a major issue. Coking most likely occurs in the reformer unit at the point where both temperature and hydrocarbon content are high enough. In this region, hydrocarbons crack and form coke faster than the coke is removed by the reaction with steam or hydrogen, and when catalyst deactivation occurs, there is a simultaneous temperature increase with a concomitant increase in coke formation and deposition. In other zones, where the hydrocarbon-to-hydrogen ratio is lower, there is less risk of coking.

Coking depends, to a large extent, on the balance between catalyst activity and heat input, with the more active catalysts producing higher yields of hydrogen at lower temperatures, thereby reducing the risk of coking. A uniform input of heat is important in this region of the reformer since any catalyst voids or variations in catalyst activity can produce localized hotspots leading to coke formation and/or reformer failure.

Coke formation results in hotspots in the reformer that increases pressure drop and reduces feedstock (methane) conversion, leading eventually to reformer failure. Coking may be partially mitigated by increasing the steam/feedstock ratio to change the reaction conditions; however, the most effective solution may be to replace the reformer catalyst with one designed for higher-boiling feedstocks.

A standard steam–methane reforming catalyst uses nickel on an alpha–alumina ceramic carrier that is acidic in nature. Promotion of hydrocarbon cracking with such a catalyst leads to coke formation from higher-boiling feedstocks. Some catalyst formulations use a magnesia/alumina (MgO/Al_2O_3) support that is less acidic than α-alumina that reduces cracking on the support and allows higher-boiling feedstocks (such as liquefied petroleum gas) to be used.

Further resistance to coking can be achieved by adding an alkali promoter, typically some form of potash (potassium hydroxide, KOH) to the catalyst. Besides reducing the acidity of the carrier, the promoter catalyzes the reaction of steam and carbon. While carbon continues to be formed, it is removed faster than it can build up. This approach can be used with naphtha feedstocks boiling point up to approximately 180°C (350°F). Under the conditions in a reformer, potash is volatile and it is incorporated into the catalyst as a more complex compound that slowly hydrolyzes to release potassium hydroxide. Alkali-promoted catalyst allows the use of a wide range of feedstocks; however, in addition to possible potash migration, which can be minimized by proper design and operation, the catalyst is also somewhat less active than the conventional catalyst.

Another option to reduce coking in steam reformers is to use a prereformer in which a fixed bed of catalyst, operating at a lower temperature, upstream of the fired reformer is used. In a prereformer, adiabatic steam–hydrocarbon reforming is performed outside the fired reformer in a vessel containing a high-nickel catalyst. The heat required for the endothermic reaction is provided by hot flue gas from the reformer convection section. Since the feed to the fired reformer is now partially reformed, the steam methane reformer can operate at an increased feed rate and produce 8–10% more hydrogen at the same reformer load. An additional advantage of the prereformer is that it facilitates higher mixed feed preheat temperatures and maintains relatively constant operating conditions within the fired reformer regardless of variable refinery off-gas feed conditions. Inlet temperatures are selected so that there is minimal risk of coking, and the gas leaving the prereformer contains only steam, hydrogen, carbon monoxide, carbon dioxide, and methane. This allows a standard methane catalyst to be used in the fired reformer, and this approach has been used with feedstocks up to light kerosene. Since the gas leaving the prereformer poses reduced risk of coking, it can compensate to some extent for variations in catalyst activity and heat flux in the primary reformer.

14.5.2 Shift Conversion Catalysts

The second important reaction in a steam reforming plant is the shift conversion reaction:

$$CO + H_2O \rightarrow CO_2 + H_2$$

Two basic types of shift catalyst are used in steam-reforming plants: iron/chrome high-temperature shift catalysts, and copper/zinc low-temperature shift catalysts.

High-temperature shift catalysts operate in the range of 315–430°C (600–800°F) and consist primarily of magnetite (Fe_3O_4) with three-valent chromium oxide (Cr_2O_3) added as a stabilizer. The catalyst is usually supplied in the form of ferric oxide (Fe_2O_3) and six-valent chromium oxide (CrO_3), and is reduced by the hydrogen and carbon monoxide in the shift feed gas as part of the start-up procedure to produce the catalyst in the desired form. However, caution is necessary since if the steam/carbon ratio of the feedstock is too low and the reducing environment too strong, the catalyst can be reduced further, to metallic iron. Metallic iron is a catalyst for Fischer–Tropsch reactions, and hydrocarbons will be produced (Davis and Occelli, 2010).

Low-temperature shift catalysts operate at temperatures on the order of 205–230°C (400–450°F). Because of the lower temperature, the reaction equilibrium is more controllable and lower amounts of carbon monoxide are produced. The low-temperature shift catalyst is primarily used in wet scrubbing plants that use a methanation for final purification. Pressure-swing adsorption plants do not generally use a low temperature because any unconverted carbon monoxide is recovered as reformer fuel. Low-temperature shift catalysts are sensitive to poisoning by sulfur and are sensitive to water (liquid) that can cause softening of the catalyst followed by crusting or plugging.

The catalyst is supplied as copper oxide (CuO) on a zinc oxide (ZnO) carrier and the copper must be reduced by heating it in a stream of inert gas with measured quantities of hydrogen. The reduction of the copper oxide is strongly exothermic and must be closely monitored.

14.5.3 METHANATION CATALYSTS

In wet scrubbing plants, the final hydrogen purification procedure involved is methanation in which the carbon monoxide and carbon dioxide are converted to methane:

$$CO + 3H_2O \rightarrow CH_4 + H_2O$$

$$CO_2 + 4H_2 \rightarrow CH_4 + 2H_2O$$

The active agent is nickel, on an alumina carrier.

The catalyst has a long life, as it operates under ideal conditions and is not exposed to poisons. The main source of deactivation is plugging from carryover of carbon dioxide from removal solutions.

The most severe hazard arises from high levels of carbon monoxide or carbon dioxide that can result from breakdown of the carbon dioxide removal equipment or from exchanger tube leaks that quench the shift reaction. The results of breakthrough can be severe, since the methanation reaction produces a temperature increase of 70°C (125°F) per 1% of carbon monoxide or a temperature increase of 33°C (60°F) per 1% of carbon dioxide. While the normal operating temperature during methanation is approximately 315°C (600°F), it is possible to reach 700°C (1300°F) in cases of major breakthrough.

14.6 PURIFICATION

The selection of the optimum purification technique for applications in the petroleum industry must be based on both technical and economic considerations. The degree of purification of hydrogen obtained from the different methods varies from approximately 90% v/v to 99.9999% for alloy membrane diffusion (Table 14.5) (Grashoff et al., 1983). The amount of hydrogen recovered also varies considerably and can have a major impact on process economics, particularly for large-scale applications.

When the hydrogen content of the refinery gas is >50% by volume, the gas should first be considered for hydrogen recovery, using a membrane (Brüschke, 1995, 2003; Lu et al., 2007) or PSA unit (Figure 14.13). The tail gas or reject gas that will still contain a substantial amount of hydrogen can then be used as steam reformer feedstock. Generally, the feedstock purification process uses

TABLE 14.5
Summary of Methods for Purification of Hydrogen

Technique (Ref.)	Principle	Typical Feed Gas	Hydrogen Output, %	
			Purity	Recovery
Cryogenic separation	Partial condensation of gas mixtures at low temperatures	Petrochemical and refinery off-gases	90–98	95
Polymer membrane diffusion	Differential rate of diffusion of gases through a permeable membrane	Refinery off-gases and ammonia purge gas	92–98	>85
Metal hydride separation	Reversible reaction of hydrogen with metals to form hydrides	Ammonia purge gas	99	75–95
Solid polymer electrolyte cell	Electrolytic passage of hydrogen ions across a solid polymer membrane	Purification of hydrogen produced by thermochemical cycles	99.8	95
Pressure swing adsorption	Selective adsorption of impurities from gas stream	Any hydrogen-rich gas	99.999	70–85
Catalytic purification	Removal of oxygen by catalytic reaction with hydrogen	Hydrogen streams with oxygen impurity	99.999	Up to 99
Palladium membrane diffusion	Selective diffusion of hydrogen through a palladium alloy membrane	Any hydrogen-containing gas stream	99.9999	Up to 99

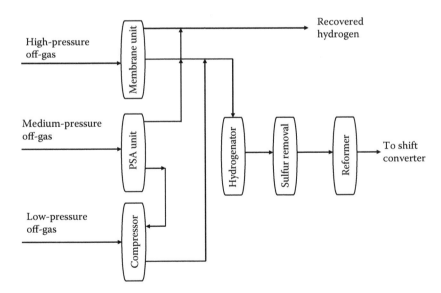

FIGURE 14.13 Hydrogen purification. (From Speight, J.G. 2014. *The Chemistry and Technology of Petroleum*. 5th Edition. CRC Press, Taylor & Francis Publishers, Boca Raton, FL. Figure 23.5, p. 649.)

three different refinery gas streams to produce hydrogen. First, high-pressure hydrocracker purge gas is purified in a membrane (through which only hydrogen can pass) that produces hydrogen at medium pressure, and is combined with a medium-pressure off-gas that is first purified in a PSA unit. Finally, low-pressure off-gas is compressed, mixed with reject gases from the membrane and PSA units, and used as steam reformer feed.

Various processes are available to purify the hydrogen stream; however, since the product streams are available as a wide variety of composition, flows, and pressures, the best method of purification will vary. Furthermore there are several factors that must also be taken into consideration in the selection of a purification method: (1) hydrogen recovery, (2) product purity, (3) pressure profile, (4) reliability, and (5) cost—an equally important parameter that is not considered here since the emphasis is on the technical aspects of the purification process.

14.6.1   Wet Scrubbing

Wet scrubbing systems, particularly amine or potassium carbonate systems, are used for the removal of acid gases such as hydrogen sulfide or carbon dioxide (Speight, 1993). Most systems depend on chemical reactions and can be designed for a wide range of pressures and capacities. They were once widely used to remove carbon dioxide in steam-reforming plants, but have generally been replaced by PSA units except where carbon monoxide is to be recovered. Wet scrubbing is still used to remove hydrogen sulfide and carbon dioxide in partial oxidation plants.

Wet scrubbing systems remove only acid gases or heavy hydrocarbons but not methane or other hydrocarbon gases; hence, they have little influence on product purity. Therefore, wet scrubbing systems are most often used as a pretreatment step, or where a hydrogen-rich stream is to be desulfurized for use as fuel gas.

14.6.2   Pressure-Swing Adsorption Units

Pressure-swing adsorption units use beds of solid adsorbent to separate impurities from hydrogen streams, leading to high-purity, high-pressure hydrogen and a low-pressure tail gas stream containing the impurities and some of the hydrogen. The beds are then regenerated by depressuring and purging. Part of the hydrogen (up to 20%) may be lost in the tail gas.

The PSA technology is based on a physical binding of gas molecules to adsorbent material. The respective force acting between the gas molecules and the adsorbent material depends on the gas component, type of adsorbent material, partial pressure of the gas component, and operating temperature. The separation effect is based on differences in binding forces to the adsorbent material. Highly volatile components with low polarity, such as hydrogen, are practically nonadsorbable as opposed to molecules such as nitrogen, carbon monoxide, carbon dioxide, hydrocarbons, and water vapor. Consequently, these impurities can be adsorbed from a hydrogen-containing stream and high-purity hydrogen is recovered.

The PSA process works at basically constant temperature and uses the effect of alternating pressure and partial pressure to perform adsorption and desorption. Since heating or cooling is not required, short cycles within the range of minutes are achieved. The process consequently allows the economical removal of large amounts of impurities. Adsorption is carried out at high pressure (and hence high respective partial pressure) typically in the range of 10–40 bar until the equilibrium loading is reached. At this point in time, no further adsorption capacity is available and the adsorbent material must be regenerated. This regeneration is accomplished by lowering the pressure to slightly above atmospheric pressure, resulting in a respective decrease in equilibrium loading. As a result, the impurities on the adsorbent material are desorbed and the adsorbent material is regenerated. The amount of impurities removed from a gas stream within one cycle corresponds to the difference of adsorption to desorption loading. After termination of regeneration, pressure is increased back to adsorption pressure level and the process starts again from the beginning.

Pressure-swing adsorption is generally the purification method of choice for steam-reforming units because of its production of high-purity hydrogen and is also used for purification of refinery off-gases, where it competes with membrane systems. Many hydrogen plants that formerly used a wet scrubbing process (Figure 14.14) for hydrogen purification are now using the PSA (Figure 14.15)

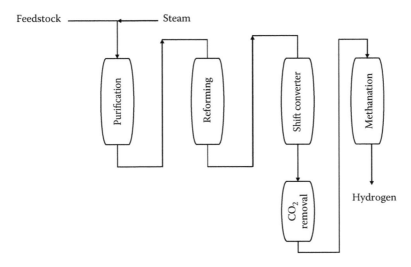

FIGURE 14.14 Hydrogen purification by wet scrubbing. (From Speight, J.G. 2014. *The Chemistry and Technology of Petroleum*. 5th Edition. CRC Press, Taylor & Francis Publishers, Boca Raton, FL. Figure 23.6, p. 650.)

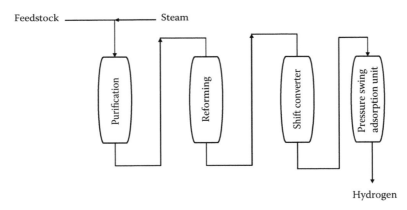

FIGURE 14.15 Hydrogen purification by pressure-swing adsorption. (From Speight, J.G. 2014. *The Chemistry and Technology of Petroleum*. 5th Edition. CRC Press, Taylor & Francis Publishers, Boca Raton, FL. Figure 23.7, p. 650.)

for purification. The PSA process is a cyclic process that uses beds of solid adsorbent to remove impurities from the gas and generally produces higher-purity hydrogen (99.9 vol.% purity compared with <97 vol.% purity). The purified hydrogen passes through the adsorbent beds with only a tiny fraction absorbed, and the beds are regenerated by depressurization followed by purging at low pressure.

When the beds are depressurized, a waste gas (or *tail gas*) stream is produced and consists of the impurities from the feed (carbon monoxide, carbon dioxide, methane, and nitrogen) plus some hydrogen. This stream is burned in the reformer as fuel, and reformer operating conditions in a PSA plant are set so that the tail gas provides no more than about 85% of the reformer fuel. This gives good burner control because the tail gas is more difficult to burn than regular fuel gas, and the high content of carbon monoxide can interfere with the stability of the flame. As the reformer operating temperature is increased, the reforming equilibrium shifts, resulting in more hydrogen and less methane in the reformer outlet and hence less methane in the tail gas.

14.6.3 MEMBRANE SYSTEMS

Membranes are also important to the subsequent purification of hydrogen. For hydrogen production and purification, there are generally two classes of membranes both being inorganic: dense phase metal and metal alloys, and porous ceramic membranes.

Porous ceramic membranes are normally prepared by sol–gel or hydrothermal methods, and have high stability and durability in high temperature, harsh impurity, and hydrothermal environments. In particular, microporous membranes show promise in water gas shift reactions at higher temperatures. Membrane systems separate gases by taking advantage of the difference in rates of diffusion through membranes (Brüschke, 1995, 2003). Gases that diffuse faster (including hydrogen) become the permeate stream and are available at low pressure, whereas the slower-diffusing gases become the nonpermeate and leave the unit at a pressure close to the pressure of the feedstock at entry point. Membrane systems contain no moving parts or switch valves, and have potentially very high reliability. The major threat is from components in the gas (such as aromatics) that attack the membranes, or from liquids, which plug them.

Membranes are fabricated in relatively small modules; for larger capacity, more modules are added. Cost is therefore virtually linear with capacity, making them more competitive at lower capacities. The design of membrane systems involves a trade-off between pressure drop (or diffusion rate) and surface area, as well as between product purity and recovery. As the surface area is increased, the recovery of fast components increases; however, more of the slow components are recovered, which lowers the purity.

On the other hand, typical polymeric membranes selectively permeate hydrogen, and the permeation selectivity of these membranes can be very high because the adsorption selectivity is high even at low feed pressure. The hydrocarbon permeability across the membrane will be high because the diffusivity for surface diffusion is orders of magnitude higher than typical diffusivities for these components through a polymeric matrix. Membrane systems can produce hydrogen with purities of 90–98% v/v at high recovery (>85%), but are best suited for high-pressure feedstocks since the separation of a gas mixture in these purification units is driven by pressure. An additional disadvantage is the fact that hydrogen sulfide may also pass through the (polymer) membrane and into the hydrogen product.

14.6.4 CRYOGENIC SEPARATION

Cryogenic separation units operate by cooling the gas and condensing some, or all, of the constituents for the gas stream. Depending on the product purity required, separation may involve flashing or distillation. Cryogenic units offer the advantage of being able to separate a variety of products from a single feed stream. One specific example is the separation of light olefins from a hydrogen stream.

Hydrogen recovery is in the range of 95%, with purity above 98% being obtainable.

14.7 HYDROGEN MANAGEMENT AND SAFETY

Several trends have significantly increased the hydrogen demand in refinery operations. They are the larger supplies of heavier, sour crude oils containing more sulfur and nitrogen; the declining demand for heavy fuel oil that boosts "bottoms" upgrading; and more stringent clean fuel regulations. In refinery operations, the reduction of polluting compounds such as sulfur, nitrogen, and aromatics is achieved through catalytic hydrotreating and hydrocracking processes. In catalytic hydrotreating processes, hydrogen is consumed not only in hydrodesulfurization reactions, but also in side reactions such as hydrodenitrogenation, hydrodearomatization, and olefin hydrogenation reactions. As a result, there is a much higher hydrogen demand for deeper hydrotreating and a lower hydrogen production from the catalytic reformer. Consequently, a deficit in the refinery hydrogen balance arises and, therefore, the hydroprocessing capacity and the associated hydrogen network may be limiting the refinery throughput and the operating margins. Hydrogen management has then become a critical factor in current refinery operations (Zagoria et al., 2003; Méndez et al., 2008; Luckwal and Mandal, 2009; Deng et al., 2013). To avoid any potential production bottleneck, new alternative sources of hydrogen as well as higher purities for improving hydrotreater capacity and product quality will be required.

14.7.1 DISTRIBUTION

The hydrogen distribution system usually comprises a set of hydrogen main headers (pipelines) working at different pressures and hydrogen purities. Many makeup and recycle compressors drive the hydrogen through this complex network of consumer units, on-purpose production units, and platforming units. Hydrogen plants generate high-purity hydrogen at different costs, while net production units are platforming units that generate low-purity hydrogen as a by-product. Hydrogen streams with different purities, pressures, and flow rates coming from makeup hydrogen plants and platforming units are supplied to multiple consumer units through the hydrogen main headers. Purge streams from hydrotreaters containing nonreacted hydrogen are partially recycled and mixed with fresh hydrogen streams from hydrogen headers before rerouting them to consuming units. The remaining off-gas stream is burnt as fuel gas. By controlling the fuel gas flow, the purity of the recycled hydrogen stream can be adjusted. The major hydrotreater operating constraint is a minimum hydrogen/hydrocarbon ratio along the reactor in order to avoid carbon deposition over the catalyst and its premature deactivation. As the catalyst cost is very significant, an effective operation of the hydrogen network will help increase the catalyst run length, thus boosting the refinery profitability. Moreover, some consuming units may have groups of membranes each of which can be activated to separate and recycle higher-purity hydrogen streams to the hydrogen network.

14.7.2 MANAGEMENT

During the majority of the 20th century and up to current times, hydrogen has always played an important role in refining. Thus, hydrogen management practices significantly influence operating costs and emissions of carbon dioxide. Therefore, an effective hydrogen management program must address refinery-wide issues in a systematic, comprehensive way. The hydrogen system consists of hydrogen production, hydrogen purification, hydrogen use, and the distribution network (Zagoria et al., 2003; Davis and Patel, 2004).

Hydrogen management through hydrogen network optimization is a necessity to address clean fuel trends, meet growing transportation fuel demands and help maintain profitability from refining crude oils currently and in the future (Stratiev et al., 2009; Speight, 2011, 2014; Rabei, 2012). The majority of hydrogen consumed in a modern refinery is produced from the catalytic reformer system that is also supplemented by an on-site hydrogen facility and/or pipeline hydrogen supply. In fact, in many cases, the hydrogen produced by various reforming processes is not always the main source of hydrogen for the refinery. Nevertheless, the refinery hydrogen distribution system is cascaded through multiple hydroprocessing units, where higher hydrogen purity and higher-pressure consumers send their purge gases to lower hydrogen purity and lower pressure consumers. Ultimately, purge gases containing residual hydrogen are sent to fuel.

Many existing refinery hydrogen plants use a conventional process that produces a medium-purity (94–97%) hydrogen product by removing the carbon dioxide in an absorption system and methanation of any remaining carbon oxides. Since the 1980s, most hydrogen plants are built with PSA technology to recover and purify the hydrogen to purities >99.9%. Since many refinery hydrogen plants utilize refinery off-gas feeds containing hydrogen, the actual maximum hydrogen capacity that can be synthesized via steam reforming is not certain since the hydrogen content of the off-gas can change because of operational changes in the hydrotreaters.

Hydrogen management has become a priority in current refinery operations and when planning to produce lower-sulfur gasoline and diesel fuels (Zagoria et al., 2003; Méndez et al., 2008; Luckwal and Mandal, 2009). Along with increased hydrogen consumption for deeper hydrotreating, additional hydrogen is needed for processing heavier and higher sulfur crude slates. In many refineries, hydroprocessing capacity and the associated hydrogen network is limiting refinery throughput and operating margins. Furthermore, higher hydrogen purities within the refinery network are becoming more important to boost hydrotreater capacity, achieve product value improvements, and lengthen catalyst life cycles.

Improved hydrogen utilization and expanded or new sources for refinery hydrogen and hydrogen purity optimization are now required to meet the needs of the future transportation fuel market and the drive toward higher refinery profitability. Many refineries developing hydrogen management programs fit into the two general categories of either a catalytic reformer supplied network or an on-purpose hydrogen supply.

Some refineries depend solely on catalytic reformer(s) as their source of hydrogen for hydrotreating. Often, they are semiregenerative reformers where off-gas hydrogen quantity, purity, and availability change with feed naphtha quality, as octane requirements change seasonally, and when the reformer catalyst progresses from start-of-run to end-of-run conditions and then goes offline for regeneration. Typically, during some portions of the year, refinery margins are reduced as a result of hydrogen shortages.

Multiple hydrotreating units compete for hydrogen—either by selectively reducing throughput, managing intermediate tankage logistics, or running the catalytic reformer suboptimally just to satisfy downstream hydrogen requirements. Part of the operating year still runs in hydrogen surplus, and the network may be operated with relatively low hydrogen utilization (consumption/production) at 70–80%. Catalytic reformer off-gas hydrogen supply may swing from 75% to 85% hydrogen purity. Hydrogen purity upgrade can be achieved through some hydrotreaters by absorbing heavy hydrocarbons. However, without supplemental hydrogen purification, critical control of hydrogen partial pressure in hydroprocessing reactors is difficult, which can affect catalyst life, charge rates, and/or naphtha yields.

More complex refineries, especially those with hydrocracking units, also have on-purpose hydrogen production, typically with a steam methane reformer that utilizes refinery off-gas and supplemental natural gas as feedstock. The steam methane reformer plant provides the swing hydrogen requirements at higher purities (92% to >99% hydrogen) and serves a hydrogen network configured with several purity and pressure levels. Multiple purities and existing purification units allow for more optimized hydroprocessing operation by controlling hydrogen partial pressure for maximum benefit. The typical hydrogen utilization is 85–95%.

14.7.3 SAFETY

The scale and growth of hydrogen demand raise fundamental questions about the safe use of the gas. Owing to its chemical properties, hydrogen poses unique challenges in the plant environment. Hydrogen gas is colorless, odorless, and not detectable by human senses. It is lighter than air, and hence difficult to detect where accumulations cannot occur, nor is it detectable by infrared gas sensing technology. Coupled with the challenge of detection are the safety risks posed by the gas itself.

Thus, there are several hazards associated with hydrogen, ranging from respiratory ailment, component failure, ignition, and burning. Although a combination of hazards occurs in most instances, the primary hazard with hydrogen is the production of a flammable mixture, which can lead to a fire or explosion. In addition to these hazards, hydrogen can produce mechanical failures of containment vessels, piping, and other components due to hydrogen embrittlement. Upon long-term exposure to the gas, many metals and plastics can lose ductility and strength, which leads to the formation of cracks and can eventually cause ruptures. A form of hydrogen embrittlement takes place by chemical reaction. At high temperatures, for instance, hydrogen reacts with one or more components of metal walls to form hydrides, which weaken the lattice structure of the material.

In oil refineries, the first step in the escalation of fire and detonation is loss of containment of the gas. Hydrogen leaks are typically caused by defective seals or gaskets, valve misalignment, or failures of flanges or other equipment. Once released, hydrogen diffuses rapidly. If the leak takes place outdoors, the dispersion of the cloud is affected by wind speed and direction, and can be influenced by atmospheric turbulence and nearby structures. With the gas dispersed in a plume, a detonation can occur if the hydrogen and air mixture is within its explosion range and an appropriate ignition source is available. Such flammable mixture can form at a considerable distance from the leak source. To address the hazards posed by hydrogen, manufacturers of fire and gas detection systems work within the construct of layers of protection to reduce the incidence of hazard propagation. Under such a model, each layer acts as a safeguard, preventing the hazard from becoming more severe.

The detection layers themselves encompass different detection techniques that either improve scenario coverage or increase the likelihood that a specific type of hazard is detected. Such fire and gas detection layers can consist of catalytic sensors, ultrasonic gas leak monitors, and fire detectors. Ultrasonic gas leak detectors can respond to high-pressure releases of hydrogen, such as those that may occur in hydrocracking reactors or hydrogen separators. In turn, continuous hydrogen monitors like catalytic detectors can contribute to detecting small leaks, for example, due to a flange slowly deformed by use or failure of a vessel maintained at close to atmospheric pressure. To further protect a plant against fires, hydrogen-specific flame detectors can supervise entire process areas. Such wide coverage is necessary because with hydrogen cloud movement, a fire may be ignited at a considerable distance from the leak source.

Hydrogen production will continue to grow, fueled by environmental legislation and demand for cleaner, higher-grade fuels. However, increasing production must be matched by a comprehensive approach to plant safety. New facilities that use hydrogen should be designed with adequate safeguards from potential hazards; the design of old facilities should also be revisited to ensure sufficient barriers are available to minimize accidents and control failure. Safety systems that deploy a diversity of detection technologies can counteract the possible effects of leaks, fire and explosions, preventing equipment or property damage, personal injury, and loss of life.

In summary, the modern refinery faces the challenge of meeting an increasing demand for cleaner transportation fuels, as specifications continue to tighten worldwide and markets decline high-sulfur fuel oil. Innovative ideas and solutions to reduce refinery costs must always be considered, including (1) optimization of the hydrogen management network, (2) multiple feedstock options for hydrogen production, (3) optimization of plant capacity, and, last but certainly not least, (4) use of hydrogen recovery technologies to maximize hydrogen availability and minimize capital investment.

REFERENCES

Al-Shamma, L.M., and Naman, S.A. 1989. Kinetic Study for Thermal Production of Hydrogen from Hydrogen Sulfide by Heterogeneous Catalysis of Vanadium Sulfide in a Flow System. *Int. J. Hyd. Energ.* 14(3): 173–179.

Ancheyta, J., and Speight, J.G. (Editors) 2007. *Hydroprocessing Heavy Oils and Residua.* CRC Press, Taylor & Francis Group, Boca Raton, FL.

Arild, V. 2000. Production of Hydrogen and Carbon with a Carbon Black Catalyst. PCT Int. Appl. No. 0021878.

Bailey, R.T. 1992. In: *Proceedings. International Symposium on Heavy Oil and Residue Upgrading*, C. Han and C. Hsi (Editors). International Academy Publishers, Beijing, China.

Balasubramanian, B., Ortiz, A.L., Kaytakoglu, S., and Harrison, D.P. 1999. Hydrogen from Methane in a Single-Step Process. *Chem. Eng. Sci.* 54: 3543–3552.

Bandermann, F., and Harder, K.B. 1982. Production of Hydrogen via Thermal Decomposition of Hydrogen Sulfide and Separation of Hydrogen and Hydrogen Sulfide by Pressure Swing Adsorption. *Int. J. Hyd. Energ.* 7(6): 471–475.

Bezler, J. 2003. Optimized Hydrogen Production—A Key Process Becoming Increasingly Important in Refineries. In: *Proceedings. DGMK Conference on Innovation in the Manufacture and Use of Hydrogen.* Dresden, Germany, October 15–17, p. 65.

Bishara, A., Salman, O.S., Khraishi, N., and Marafi, A. 1987. Thermochemical Decomposition of Hydrogen Sulfide by Solar Energy. *Int. J. Hyd. Energ.* 12(10): 679–685.

Brandmair, M., Find, J., and Lercher, J.A. 2003. Combined Autothermal Reforming and Hydrogenolysis of Alkanes. In: *Proceedings. DGMK Conference on Innovation in the Manufacture and Use of Hydrogen.* Dresden, Germany, October 15–17, pp. 273–280.

Bridge, A.G. 1997. Hydrogen Processing. In: *Handbook of Petroleum Refining Processes*, 2nd Edition, R.A. Meyers (Editor). McGraw-Hill, New York, Chapter 15.1.

Brüschke, H. 1995. Industrial Application of Membrane Separation Processes. *Pure Appl. Chem.* 67(6): 993–1002.

Brüschke, H. 2003. Separation of Hydrogen from Dilute Streams (e.g. Using Membranes). In: *Proceedings. DGMK Conference on Innovation in the Manufacture and Use of Hydrogen.* Dresden, Germany, October 15–17, p. 47.

Clark, P.D., Dowling, N.I., Hyne, J.B., and Moon, D.L. 1995. Production of Hydrogen and Sulfur from Hydrogen Sulfide in Refineries and Gas Processing Plants. *Alberta Sulfur Res. Quart. Bull.* 32(1): 11–28.

Clark, P.D., and Wassink, B. 1990. A Review of Methods for the Conversion of Hydrogen Sulfide to Sulfur and Hydrogen. *Alberta Sulfur Res. Quart. Bull.* 26(2/3/4): 1.

Cruz, F.E., and de Oliveira Junior, S. 2008. Petroleum Refinery Hydrogen Production Unit: Exergy and Production Cost Evaluation. *Int. J. Thermodyn.* 11(4): 187–193.

Davis, B.H., and Occelli, M.L. 2010. *Advances in Fischer–Tropsch Synthesis, Catalysts, and Catalysis.* CRC Press, Taylor & Francis Group, Boca Raton, FL.

Davis, R.A., and Patel, N.M. 2004. Refinery Hydrogen Management. *Petrol. Technol. Quart.* Spring: 29–35.

Deng, C., Li, W., and Feng, X. 2013. Refinery Hydrogen Network Management with Key Factor Analysis. *Chem. Eng. Trans.* 35: 61–66.

Dickenson, R.L., Biasca, F.E., Schulman, B.L., and Johnson, H.E. 1997. Refiner Options for Converting and Utilizing Heavy Fuel Oil. *Hydrocarbon Process.* 76(2): 57.

Dolbear, G.E. 1998. Hydrocracking: Reactions, Catalysts, and Processes. In: *Petroleum Chemistry and Refining*, J.G. Speight (Editor). Taylor & Francis, Washington, DC, Chapter 7.

Donini, J.C. 1996. Separation and Processing of Hydrogen Sulfide in the Fossil Fuel Industry. In: *Minimum Effluent Mills Symposium*, pp. 357–363.

Dragomir, R., Drnevich, R.F., Morrow, J., Papavassiliou, V., Panuccio, G., and Watwe, R. 2010. Technologies for Enhancing Refinery Gas Value. In: *Proceedings. AIChE 2010 SPRING Meeting.* San Antonio, TX, November 7–12.

Ehwald, H., Kürschner, U., Smejkal, Q., and Lieske, H. 2003. Investigation of Different Catalysts for Autothermal Reforming of i-Octane. In: *Proceedings. DGMK Conference on Innovation in the Manufacture and Use of Hydrogen.* Dresden, Germany, October 15–17, p. 345.

Farnand, S., Li, J., Patel, N., Peng, X.D., and Ratan, S. 2015. Hydrogen Perspectives for 21st Century Refineries. *Hydrocarbon Process.* 94(2): 53–58.

Find, J., Nagaoka, K., and Lercher, J.A. 2003. Steam Reforming of Light Alkanes in Micro-Structured Reactors. In: *Proceedings. DGMK Conference on Innovation in the Manufacture and Use of Hydrogen.* Dresden, Germany, October 15–17, p. 257.

Fleshman, J.D. 1997. Hydrogen Production. In: *Handbook of Petroleum Refining Processes*, 2nd Edition, R.A. Meyers (Editor). McGraw-Hill, New York, Chapter 6.1.

Gary, J.G., Handwerk, G.E., and Kaiser, M.J. 2007. *Petroleum Refining: Technology and Economics*, 5th Edition. CRC Press, Taylor & Francis Group, Boca Raton, FL.

Gaudernack, B. 1996. Hydrogen from Natural Gas without Release of Carbon Dioxide into the Atmosphere. Hydrogen Energy Prog. In: *Proceedings. 11th World Hydrogen Energy Conference*, Vol. 1, pp. 511–523.

Grashoff, G.J., Pilkington, C.E., and Corti, C.W. 1983. The Purification of Hydrogen: A Review of the Technology Emphasizing the Current Status of Palladium Membrane Diffusion. *Platinum Metals Rev.* 27(4): 157–169.

Gross, M., and Wolff, J. 2000. Gasification of Residue as a Source of Hydrogen for the Refining Industry in India. In: *Proceedings. Gasification Technologies Conference.* San Francisco, October 8–11.

Hufton, J.R., Mayorga, S., and Sircar, S. 1999. Sorption-Enhanced Reaction Process for Hydrogen Production. *AIChE J.* 45: 248–256.

Hsu, C.S., and Robinson, P.R. (Editors) 2006. *Practical Advances in Petroleum Processing*, Volume 1 and Volume 2. Springer Science, New York.

Kiuchi, H. 1982. Recovery of Hydrogen from Hydrogen Sulfide with Metals and Metal Sulfides. *Int. J. Hyd. Energ.* 7(6): 1–8.

Kotera, Y., Todo, N., and Fukuda, K. 1976. Process for Production of Hydrogen and Sulfur from Hydrogen Sulfide as Raw Material. U.S. Patent No. 3,962,409, June 8.

Kriz, J.F., and Ternan, M. 1994. Hydrocracking of Heavy Asphaltenic Oil in the Presence of an Additive to Prevent Coke Formation. United States Patent 5,296,130, March 22.

Lipman, T. 2011. An Overview of Hydrogen Production and Storage Systems with Renewable Hydrogen Case Studies. A Clean Energy States Alliance Report. Conducted under US DOE Grant DE-FC3608GO18111, Office of Energy Efficiency and Renewable Energy Fuel Cell Technologies Program, United States Department of Energy, Washington, DC.

Lu, G.Q., Dinz da Costa, J.C., Duke, M., Giessler, S., Socolow, R., Williams, R.H., and Kreutz, T. 2007. Inorganic Membranes for Hydrogen Production and Purification: A Critical Review and Perspective. *J Colloid Interface Sci.* 314(2): 589–603.

Luckwal, K., and Mandal, K.K. 2009. Improve Hydrogen Management of Your Refinery. *Hydrocarbon Process.* 88(2): 55–61.

Luinstra, E. 1996. Hydrogen from Hydrogen Sulfide—A Review of the Leading Processes. In: *Proceedings. 7th Sulfur Recovery Conference.* Gas Research Institute, Chicago, pp. 149–165.

Megalofonos, S.K., and Papayannakos, N.G. 1997. Kinetics of Catalytic Reaction of Methane and Hydrogen Sulfide over MoS_2. *J. Appl. Catal. A Gen.* 65(1–2): 249–258.

Méndez, C.A., Gómez, E., Sarabia, D., Cerdá, J., De Prada, C., Sola, J.M., and Unzueta, E. 2008. In: *Proceedings. 18th European Symposium on Computer Aided Process Engineering—ESCAPE 18*, B. Braunschweig and X. Joulia (Editors). Elsevier BV, Amsterdam.

Miller, G.Q., and Penner, D.W. 2003. Meeting Future Needs for Hydrogen—Possibilities and Challenges. In: *Proceedings. DGMK Conference on Innovation in the Manufacture and Use of Hydrogen.* Dresden, Germany, October 15–17, p. 7.

Mokhatab, S., Poe, W.A., and Speight, J.G. 2006. *Handbook of Natural Gas Transmission and Processing.* Elsevier, Amsterdam.

Muradov, N.Z. 1998. CO2-Free Production of Hydrogen by Catalytic Pyrolysis of Hydrocarbon Fuel. *Energy Fuels* 12(1): 41–48.

Muradov, N.Z. 2000. Thermocatalytic Carbon Dioxide-Free Production of Hydrogen from Hydrocarbon Fuels. In: *Proceedings. Hydrogen Program Review*, NREL/CP-570–28890.

Murata, K., Ushijima, H., and Fujita, K. 1997. Process for Producing Hydrogen from Hydrocarbon. United States Patent 5,650,132.

Nagaoka, K., Jentys, A., and Lecher, J.A. 2003. Autothermal Reforming of Methane over Mono- and Bi-metal Catalysts Prepared from Hydrotalcite-like Precursors. In: *Proceedings. DGMK Conference on Innovation in the Manufacture and Use of Hydrogen.* Dresden, Germany, October 15–17, p. 171.

Parkash, S. 2003. *Refining Processes Handbook.* Gulf Professional Publishing, Elsevier, Amsterdam.

Rabei, Z. 2012. Hydrogen Management in Refineries. *Petrol. Coal* 54(4): 357–368.

Raissi, A.T. 2001. Technoeconomic Analysis of Area II Hydrogen Production. Part 1. In: *Proceedings. U.S. DOE Hydrogen Program Review Meeting.* Baltimore, MD.

Ranke, H., and Schödel, N. 2003. Hydrogen Production Technology—Status and New Developments. In: *Proceedings. DGMK Conference on Innovation in the Manufacture and Use of Hydrogen.* Dresden, Germany, October 15–17, p. 19.

Scherzer, J., and Gruia, A.J. 1996. *Hydrocracking Science and Technology.* Marcel Dekker Inc., New York.

Shimizu, Y. 1982. *Chem. Econ. Eng. Rev. (CEER)* 10(7): 9-14.

Speight, J.G. 1993. *Gas Processing: Environmental Aspects and Methods.* Butterworth Heinemann, Oxford, England.

Speight, J.G. 2000. *The Desulfurization of Heavy Oils and Residua*, 2nd Edition. Marcel Dekker Inc., New York.

Speight, J.G. 2005. *Environmental Analysis and Technology for the Refining Industry.* John Wiley & Sons Inc., Hoboken, NJ.

Speight, J.G. 2007. *Natural Gas: A Basic Handbook.* GPC Books, Gulf Publishing Company, Houston, TX.

Speight, J.G. 2011. *The Refinery of the Future.* Gulf Professional Publishing, Elsevier, Oxford, UK.

Speight, J.G. 2013. *The Chemistry and Technology of Coal*, 3rd Edition. CRC Press, Taylor & Francis Publishers, Boca Raton, FL.

Speight, J.G. 2014. *The Chemistry and Technology of Petroleum*, 5th Edition. CRC Press, Taylor & Francis Publishers, Boca Raton, FL.

Speight, J.G., and Ozum, B. 2002. *Petroleum Refining Processes.* Marcel Dekker Inc., New York.

Stratiev, D., Tzingov, T., Shishkova, I., and Dermatova, P. 2009. Hydrotreating Units Chemical Hydrogen Consumption Analysis: A Tool for Improving Refinery Hydrogen Management. In: *Proceedings. 44th International Petroleum Conference*. Bratislava, Slovak Republic, September 21–22. http://www.vurup .sk/sites/vurup.sk/archivedsite/www.vurup.sk/conferences/44_ipc/44ipc/Sustainable%20Crude%20 Oil%20Processing_Tudays%20and%20Future%20Technologies,%20the%20role%20of%20the%20 catalysis/posters/Hydrotreating%20Units%20Chemical%20Hydrogen%20Consumption%20Analysis _Stratiev.pdf; accessed September 4, 2014.

Uemura, Y., Ohe, H., Ohzuno, Y., and Hatate, Y. 1999. Carbon and Hydrogen from Hydrocarbon Pyrolysis. In: *Proceedings. Int. Conf. Solid Waste Technology Management*, Vol. 15, pp. 5E/25–5E/30.

Vauk, D., Di Zanno, P., Neri, B., Allevi, C., Visconti, A., and Rosanio, L. 2008. What Are Possible Hydrogen Sources for Refinery Expansion? *Hydrocarbon Process.* 87(2): 69–76.

Weimer, A.W., Dahl, J., Tamburini, J., Lewandowski, A., Pitts, R., Bingham, C., and Glatzmaier, G.C. 2000. Thermal Dissociation of Methane Using a Solar Coupled Aerosol Flow Reactor. In: *Proceedings. Hydrogen Program Review*, NREL/CP-570-28890.

Zagoria, A., Huychke, R., and Boulter, P.H. 2003. Refinery Hydrogen Management—The Big Picture. In: *Proceedings. DGMK Conference on Innovation in the Manufacture and Use of Hydrogen*. Dresden, Germany, October 15–17, p. 95.

Zaman, J., and Chakma. 1995. Production of Hydrogen and Sulfur from Hydrogen Sulfide. *Fuel Process. Technol.* 41: 159–198.

Conversion Factors

1 acre = 43,560 ft^2

1 acre foot = 7758.0 bbl

1 atmosphere = 760 mm Hg = 14.696 psia = 29.91 in Hg

1 atmosphere = 1.0133 bar = 33.899 ft H$_2$O

1 barrel (oil) = 42 gal = 5.6146 ft^3

1 barrel (water) = 350 lb at 60°F

1 barrel per day = 1.84 cm^3/s

1 Btu = 778.26 ft lb

1 centipoise × 2.42 = lb mass/(ft) (h), viscosity

1 centipoise × 0.000672 = lb mass/(ft) (s), viscosity

1 cubic foot = 28,317 cm^3 = 7.4805 gal

Density of water at 60°F = 0.999 g/cm^3 = 62.367 lb/ft^3 = 8.337 lb/gal

1 gallon = 231 in^3 = 3785.4 cm^3 = 0.13368 ft^3

1 horsepower-hour = 0.7457 kwh = 2544.5 Btu

1 horsepower = 550 ft lb/s = 745.7 W

1 inch = 2.54 cm

1 meter = 100 cm = 1000 mm = 10 microns

1 ounce = 28.35 g

1 pound = 453.59 g = 7000 gr

1 square mile = 640 acres

Glossary

ABC process: a fixed-bed process for the hydrodemetallization and hydrodesulfurization of heavy feedstocks.

ABN separation: a method of fractionation by which petroleum is separated into acidic, basic, and neutral constituents.

absorber: see **absorption tower**.

absorption gasoline: gasoline extracted from natural gas or refinery gas by contacting the absorbed gas with an oil and subsequently distilling the gasoline from the higher-boiling components.

absorption oil: oil used to separate the heavier components from a vapor mixture by absorption of the heavier components during intimate contacting of the oil and vapor; used to recover natural gasoline from wet gas.

absorption plant: a plant for recovering the condensable portion of natural or refinery gas, by absorbing the higher-boiling hydrocarbons in an absorption oil, followed by separation and fractionation of the absorbed material.

absorption tower: a tower or column that promotes contact between a rising gas and a falling liquid so that part of the gas may be dissolved in the liquid.

acetone–benzol process: a dewaxing process in which acetone and benzol (benzene or aromatic naphtha) are used as solvents.

acid catalyst: a catalyst having acidic character; the aluminas are examples of such catalysts.

acid deposition: acid rain; a form of pollution depletion in which pollutants, such as nitrogen oxides and sulfur oxides, are transferred from the atmosphere to soil or water; often referred to as atmospheric self-cleaning. The pollutants usually arise from the use of fossil fuels.

acidity: the capacity of an acid to neutralize a base such as a hydroxyl ion (OH^-).

acidizing: a technique for improving the permeability (q.v.) of a reservoir by injecting acid.

acid number: a measure of the reactivity of petroleum with a caustic solution and given in terms of milligrams of potassium hydroxide that are neutralized by 1 g of petroleum.

acid rain: the precipitation phenomenon that incorporates anthropogenic acids and other acidic chemicals from the atmosphere to the land and water (see **acid deposition**).

acid sludge: the residue left after treating petroleum oil with sulfuric acid for the removal of impurities; a black, viscous substance containing the spent acid and impurities.

acid treating: a process in which unfinished petroleum products, such as gasoline, kerosene, and lubricating oil stocks, are contacted with sulfuric acid to improve their color, odor, and other properties.

activation energy, E: the energy that is needed by a molecule or molecular complex to encourage reactivity to form products.

additive: a material added to another (usually in small amounts) to enhance desirable properties or to suppress undesirable properties.

add-on control methods: the use of devices that remove refinery process emissions after they are generated but before they are discharged to the atmosphere.

adsorption: transfer of a substance from a solution to the surface of a solid resulting in relatively high concentration of the substance at the place of contact; see also **chromatographic adsorption**.

adsorption gasoline: natural gasoline (q.v.) obtained in the adsorption process from wet gas.

after-burn: the combustion of carbon monoxide (CO) to carbon dioxide (CO_2), usually in the cyclones of a catalyst regenerator.

air-blown asphalt: asphalt produced by blowing air through residua at elevated temperatures.

air injection: an oil recovery technique using air to force oil from the reservoir into the wellbore.

airlift thermofor catalytic cracking: a moving-bed continuous catalytic process for conversion of heavy gas oils into lighter products; the catalyst is moved by a stream of air.

air pollution: the discharge of toxic gases and particulate matter introduced into the atmosphere, principally as a result of human activity.

air sweetening: a process in which air or oxygen is used to oxidize lead mercaptide to disulfides instead of using elemental sulfur.

alcohol: the family name of a group of organic chemical compounds composed of carbon, hydrogen, and oxygen. The series of molecules vary in chain length and are composed of a hydrocarbon plus a hydroxyl group, $CH_3(CH_2)_nOH$ (e.g., methanol, ethanol, and tertiary butyl alcohol).

alicyclic hydrocarbon: a compound containing carbon and hydrogen only, which has a cyclic structure (e.g., cyclohexane); also collectively called naphthenes.

aliphatic hydrocarbon: a compound containing carbon and hydrogen only, which has an open-chain structure (e.g., as ethane, butane, octane, or butene) or a cyclic structure (e.g., cyclohexane).

alkalinity: the capacity of a base to neutralize the hydrogen ion (H^+).

alkali treatment: see **caustic wash**.

alkali wash: see **caustic wash**.

alkanes: hydrocarbons that contain only single carbon–hydrogen bonds. The chemical name indicates the number of carbon atoms and ends with the suffix "-ane."

alkenes: hydrocarbons that contain carbon–carbon double bonds. The chemical name indicates the number of carbon atoms and ends with the suffix "-ene."

alkylate: the product of an alkylation (q.v.) process.

alkylate bottoms: residua from fractionation of alkylate; the alkylate product that boils higher than the aviation gasoline range; sometimes called heavy alkylate or alkylate polymer.

alkylation: in the petroleum industry, a process by which an olefin (e.g., ethylene) is combined with a branched-chain hydrocarbon (e.g., isobutane); alkylation may be accomplished as a thermal or as a catalytic reaction.

alkyl groups: a group of carbon and hydrogen atoms that branch from the main carbon chain or ring in a hydrocarbon molecule. The simplest alkyl group, a methyl group, is a carbon atom attached to three hydrogen atoms.

alpha-scission: the rupture of the aromatic carbon–aliphatic carbon bond that joins an alkyl group to an aromatic ring.

alumina (Al_2O_3): used in separation methods as an adsorbent and in refining as a catalyst.

ASTM International (formerly American Society for Testing and Materials): the official organization in the United States for designing standard tests for petroleum and other industrial products.

amine washing: a method of gas cleaning whereby acidic impurities such as hydrogen sulfide and carbon dioxide are removed from the gas stream by washing with an amine (usually an alkanolamine).

analysis: determining the properties of a feedstock before refining; inspection (q.v.) of feedstock properties.

analytical equivalence: acceptability of results obtained from different laboratories; a range of acceptable results.

analyte: the chemical for which a sample is tested, or analyzed.

aniline point: the temperature, usually expressed in °F, above which equal volumes of a petroleum product are completely miscible; a qualitative indication of the relative proportions of paraffins in a petroleum product that are miscible with aniline only at higher temperatures; a high aniline point indicates low aromatics.

antibody: a molecule having chemically reactive sites specific for certain other molecules.

API (American Petroleum Institute) gravity: a measure of the lightness or heaviness of petroleum that is related to density and specific gravity.

$$°API = (141.5/sp\ gr\ at\ 60°F) - 131.5$$

apparent bulk density: the density of a catalyst as measured; usually loosely compacted in a container.

apparent viscosity: the viscosity of a fluid, or several fluids flowing simultaneously, measured in a porous medium (rock), and subject to both viscosity and permeability effects; also called *effective viscosity.*

aquaconversion process: a hydrovisbreaking technology in which a proprietary additive and water are added to the heavy feedstock before introduction into the soaker.

aromatic hydrocarbon: a hydrocarbon characterized by the presence of an aromatic ring or condensed aromatic rings; benzene and substituted benzene, naphthalene and substituted naphthalene, phenanthrene and substituted phenanthrene, as well as the higher condensed ring systems; compounds that are distinct from those of aliphatic compounds (q.v.) or alicyclic compounds (q.v.).

aromatics: a group of hydrocarbons of which benzene is the parent; so named because many of their derivatives have sweet or aromatic odors.

aromatization: the conversion of nonaromatic hydrocarbons to aromatic hydrocarbons by (1) rearrangement of aliphatic (noncyclic) hydrocarbons (q.v.) into aromatic ring structures and (2) dehydrogenation of alicyclic hydrocarbons (naphthenes).

Arosorb process: a process for the separation of aromatics from nonaromatics by adsorption on a gel from which they are recovered by desorption.

ART process: a process for increasing the production of liquid fuels without hydrocracking (q.v.).

ASCOT process: a resid (q.v.) upgrading process that integrates delayed coking and deep solvent deasphalting.

asphalt: highly viscous liquid or semisolid composed of bitumen and present in most crudes; can be separated from other crude components by fractional distillation; used primarily for road paving and roofing shingles.

asphaltene association factor: the number of individual asphaltene species that associate in nonpolar solvents as measured by molecular weight methods; the molecular weight of asphaltenes in toluene divided by the molecular weight in a polar nonassociating solvent, such as dichlorobenzene, pyridine, or nitrobenzene.

asphaltene constituents: molecular species that occur within the asphaltene fraction and that vary in polarity (functional group content) and molecular weight.

asphaltene fraction: that fraction of petroleum, heavy oil, or bitumen that is precipitated when a large excess (40 vol) of a low-boiling liquid hydrocarbon (e.g., pentane or heptane) is added to (1 vol of) the feedstock; usually a dark brown to black amorphous solid that does not melt before decomposition and is soluble in benzene or aromatic naphtha or other chlorinated hydrocarbon solvents.

asphaltic constituents: a general term usually meaning the asphaltene fraction plus the resin fraction.

associated gas: natural gas that is in contact with and/or dissolved in the crude oil of the reservoir. It may be classified as gas cap (free gas) or gas in solution (dissolved gas).

associated gas in solution (or dissolved gas): natural gas dissolved in the crude oil of the reservoir, under the prevailing pressure and temperature conditions.

associated molecular weight: the molecular weight of asphaltenes in an associating (nonpolar) solvent, such as toluene.

atmospheric distillation: distillation at atmospheric pressure; the refining process of separating crude oil components at atmospheric pressure by heating to temperatures of about 600–750°F (depending on the nature of the crude oil and desired products) and subsequent condensing of the fractions by cooling.

atmospheric equivalent boiling point (AEBP): a mathematical method of estimating the boiling point at atmospheric pressure of nonvolatile fractions of petroleum.

atmospheric residuum: a residuum (q.v.) obtained by distillation of a crude oil under atmospheric pressure and that boils above 350°C (660°F).

Attapulgus clay: see **Fuller's earth**.

autofining: a catalytic process for desulfurizing distillates.

autothermal reforming (ATR): a process used to produce hydrogen by using a combination of partial oxidation and steam reforming in a single vessel.

average particle size: the weighted average particle diameter of a catalyst.

aviation gasoline: any of the special grades of gasoline suitable for use in certain airplane engines; a complex mixture of relatively volatile hydrocarbons with or without small quantities of additives, blended to form a fuel suitable for use in aviation reciprocating engines. Fuel specifications are provided in ASTM D910 and Military Specification MIL-G-5572.

aviation gasoline blending components: the various naphtha products that will be used for blending or compounding into finished aviation gasoline (e.g., straight-run gasoline, alkylate, reformate, benzene, toluene, and xylene); excludes oxygenates (alcohols, ethers), butane, and pentanes plus. Oxygenates are reported as other hydrocarbons, hydrogen, and oxygenates.

aviation turbine fuel: see **jet fuel**.

Avjet: aviation jet fuel derived from a kerosene fraction; also known as jet fuel or jet.

back mixing: the phenomenon observed when a catalyst travels at a slower rate in the riser pipe than the vapors.

baghouse: a filter system for the removal of particulate matter from gas streams; so called because of the similarity of the filters to coal bags.

Bari-Sol process: a dewaxing process that employs a mixture of ethylene dichloride and benzol as the solvent.

barrel: the unit of measurement of liquids in the petroleum industry; the traditional measurement for crude oil volume—1 barrel is equivalent to 42 US gallons (159 L) and 6.29 barrels is equivalent to 1 m^3 of oil.

barrel of oil equivalent: a unit of energy based on the approximate energy released by burning 1 barrel of crude oil.

barrels per calendar day: the amount of input that a distillation facility can process under usual operating conditions—the amount is expressed in terms of capacity during a 24-h period and reduces the maximum processing capability of all units at the facility under continuous operation (see **barrels per stream day**) to account for the following limitations that may delay, interrupt, or slow down production: (1) the capability of downstream facilities to absorb the output of crude oil–processing facilities of a given refinery—no reduction is made when a planned distribution of intermediate streams through other than downstream facilities is part of a refinery's normal operation; (2) the types and grades of inputs to be processed; (3) the environmental constraints associated with refinery operations; (4) the reduction of capacity for scheduled downtime due to such conditions as routine inspection, maintenance, repairs, and turnaround; and (5) the reduction of capacity for unscheduled downtime due to such conditions as mechanical problems, repairs, and slowdowns.

barrels per stream day: the maximum number of barrels of input that a distillation facility can process within a 24-h period when running at full capacity under optimal crude and product slate conditions with no allowance for downtime.

base number: the quantity of acid, expressed in milligrams of potassium hydroxide per gram of sample, that is required to titrate a sample to a specified end point.

base stock: a primary refined petroleum fraction into which other oils and additives are added (blended) to produce the finished product.

basic nitrogen: nitrogen (in petroleum) that occurs in pyridine form.

basic sediment and water (BS&W, BSW): the material that collects in the bottom of storage tanks, usually composed of oil, water, and foreign matter; also called bottoms, bottom settlings.

battery: a series of stills or other refinery equipment operated as a unit.

Baumé gravity: the specific gravity of liquids expressed as degrees on the Baumé (°Bé) scale; for liquids lighter than water:

$$\text{Sp gr } 60°\text{F} = 140/(130 + °\text{Bé})$$

For liquids heavier than water:

$$\text{Sp gr } 60°\text{F} = 145/(145 - °\text{Bé})$$

bauxite: mineral matter used as a treating agent; hydrated aluminum oxide formed by the chemical weathering of igneous rocks.

bbl: see **barrel**.

bell cap: a hemispherical or triangular cover placed over the riser in a (distillation) tower to direct the vapors through the liquid layer on the tray; see **bubble cap**.

bender process: a chemical treating process using lead sulfide catalyst for sweetening light distillates by which mercaptans are converted to disulfides by oxidation.

bentonite: montmorillonite (a magnesium–aluminum silicate); used as a treating agent.

benzene: a colorless aromatic liquid hydrocarbon (C_6H_6); present in small proportion in some crude oils and made commercially from petroleum by the catalytic reforming of naphthenes in petroleum naphtha—also made from coal in the manufacture of coke; used as a solvent in the manufacture of detergents, synthetic fibers, petrochemicals, and as a component of high-octane gasoline.

benzin: refined light naphtha used for extraction purposes.

benzine: an obsolete term for light petroleum distillates covering the gasoline and naphtha range; see **ligroine**.

benzol: the general term that refers to commercial or technical (not necessarily pure) benzene; also the term used for aromatic naphtha.

beta-scission: the rupture of a carbon–carbon bond that is two bonds removed from an aromatic ring.

billion: 1×10^9.

biocide: any chemical capable of killing bacteria and bioorganisms.

biodegradation: the destruction of organic materials by bacteria.

biogenic: material derived from bacterial or vegetation sources.

biological lipid: any biological fluid that is miscible with a nonpolar solvent. These materials include waxes, essential oils, chlorophyll, etc.

biological oxidation: the oxidative consumption of organic matter by bacteria by which organic matter is converted into gases.

biomass: biological organic matter; materials produced from the processing of wood, corn, sugar, and other agricultural waste or municipal waste. It can be converted to syngas via a gasification process.

biomass-to-liquids (BTL): a process used to convert a wide variety of waste biomass, such as from the processing of wood, corn, sugar, and other agricultural waste or municipal waste into hydrocarbons such as diesel and jet fuel. The biomass is converted first into syngas through a gasification process, followed by the Fischer–Tropsch process and subsequent hydrocracking step.

biopolymer: a high molecular weight carbohydrate produced by bacteria.

bioremediation: cleanup of spills of petroleum and/or petroleum products using microbial agents.

bitumen: a semisolid to solid organic material found filling pores and crevices of sandstone, limestone, or argillaceous sediments; contains organic carbon, hydrogen, nitrogen, oxygen, sulfur, and metallic constituents; usually has an API gravity <10° but other properties are necessary for inclusion in a more complete definition; in its natural state, tar sand (oil sand) bitumen is not recoverable at a commercial rate through a well because it is too viscous to flow; bitumen typically makes up approximately 10% w/w of tar sand (oil sand) but saturation varies.

bituminous: containing bitumen or constituting the source of bitumen.

bituminous rock: see **bituminous sand**.

bituminous sand: a formation in which the bituminous material (see **bitumen**) is found as a filling in veins and fissures in fractured rocks or impregnating relatively shallow sand, sandstone, and limestone strata; a sandstone reservoir that is impregnated with a heavy, viscous black petroleum-like material that cannot be retrieved through a well by conventional production techniques.

black acid(s): a mixture of the sulfonates found in acid sludge that are insoluble in naphtha, benzene, and carbon tetrachloride; very soluble in water but insoluble in 30% sulfuric acid; in the dry, oil-free state, the sodium soaps are black powders.

black oil: any of the dark-colored oils; a term now often applied to heavy oil; a term erroneously used to describe heavy oil.

black soap: see **black acid**.

black strap: the black material (mainly lead sulfide) formed in the treatment of sour light oils with doctor solution (q.v.) and found at the interface between the oil and the solution.

blending plant: a facility that has no refining capability but is either capable of producing finished motor gasoline through mechanical blending or blends oxygenates with motor gasoline.

blown asphalt: the asphalt prepared by air blowing a residuum (q.v.) or an asphalt (q.v.).

BOC process: see **RCD Unibon (BOC) process**.

bogging: a condition that occurs in a coking reactor when the conversion to coke and light ends is too slow causing the coke particles to agglomerate.

boiling point: a characteristic physical property of a liquid at which the vapor pressure is equal to that of the atmosphere and the liquid is converted to a gas.

boiling range: the range of temperature, usually determined at atmospheric pressure in standard laboratory apparatus, over which the distillation of oil commences, proceeds, and finishes.

bottled gas: usually butane or propane, or butane–propane mixtures, liquefied and stored under pressure for domestic use; see also **liquefied petroleum gas**.

bottom-of-the-barrel: residuum (q.v.).

bottoms: residue remaining in a distillation unit after the highest boiling point material to be distilled has been removed; also the liquid that collects in the bottom of a vessel (tower bottoms, tank bottoms) either during distillation; also the deposit or sediment formed during storage of petroleum or a petroleum product; see also **residuum and basic sediment and water**.

bottom-of-the-barrel processing: residuum processing.

bright stock: refined, high-viscosity lubricating oils usually made from residual stocks by processes such as a combination of acid treatment or solvent extraction with dewaxing or clay finishing.

British thermal unit: see **Btu**.

bromine number: the number of grams of bromine absorbed by 100 g of oil that indicates the percentage of double bonds in the material.

brown acid: oil-soluble petroleum sulfonates found in acid sludge that can be recovered by extraction with naphtha solvent. Brown acid sulfonates are somewhat similar to mahogany sulfonates but are more water soluble. In the dry, oil-free state, the sodium soaps are light-colored powders.

brown soap: see **brown acid**.

BS&W: see **basic sediment and water**.

Brønsted acid: a chemical species that can act as a source of protons.

Brønsted base: a chemical species that can accept protons.

BTEX: benzene, toluene, ethylbenzene, and the xylene isomers.

Btu (British thermal unit): the energy required to increase the temperature of 1 lb of water to 1°F.

bubble cap: an inverted cup with a notched or slotted periphery to disperse the vapor in small bubbles beneath the surface of the liquid on the bubble plate in a distillation tower.

bubble plate: a tray in a distillation tower.

bubble point: the temperature at which incipient vaporization of a liquid in a liquid mixture occurs, corresponding with the equilibrium point of 0% vaporization or 100% condensation; the temperature at which a gas starts to come out of a liquid.

bubble tower: a fractionating tower so constructed that the vapors rising pass up through layers of condensate on a series of plates or trays (see **bubble plate**); the vapor passes from one plate to the next above by bubbling under one or more caps (see **bubble cap**) and out through the liquid on the plate where the less volatile portions of vapor condense by bubbling through the liquid on the plate, overflow to the next lower plate, and ultimately back into the reboiler, thereby effecting fractionation.

bubble tray: a circular, perforated plate having the internal diameter of a bubble tower (q.v.), set at specified distances in a tower to collect the various fractions produced during distillation.

bulk composition: the makeup of petroleum in terms of bulk fractions such as saturates, aromatics, resins, and asphaltenes; separation of petroleum into these fractions is usually achieved by a combination of solvent and adsorption (q.v.) processes.

bumping: the knocking against the walls of a still occurring during distillation of petroleum or a petroleum product that usually contains water.

Bunker C oil: see **No. 6 fuel oil**.

burner fuel oil: any petroleum liquid suitable for combustion.

burning oil: illuminating oil, such as kerosene suitable for burning in a wick lamp.

burning point: see **fire point**.

burning-quality index: an empirical numerical indication of the likely burning performance of a furnace or heater oil; derived from the distillation profile (q.v.) and the API gravity (q.v.), and generally recognizing the factors of paraffin character and volatility.

Burton process: an older thermal cracking process in which oil was cracked in a pressure still and any condensation of the products of cracking also took place under pressure.

butane: four-carbon alkane; hydrocarbon used as a fuel for cooking and camping; chemical formula, C_4H_{10}; a straight-chain or branch-chain hydrocarbon extracted from natural gas or refinery gas streams, which is gaseous at standard temperature and pressure; may include isobutane and normal butane and is designated in ASTM D1835 and Gas Processors Association specifications for commercial butane.

butane dehydrogenation: a process for removing hydrogen from butane to produce butene (C_4H_8) and, on occasion, butadiene (C_4H_6, $CH_2=CHCH=CH_2$).

butane vapor-phase isomerization: a process for isomerizing *n*-butane to isobutane using aluminum chloride catalyst on a granular alumina support and with hydrogen chloride as a promoter.

butylene (C_4H_8): an olefin hydrocarbon recovered from refinery or petrochemical processes, which is gaseous at standard temperature and pressure; used in the production of gasoline and various petrochemical products.

C_1, C_2, C_3, C_4, C_5 fractions: a common way of representing fractions containing a preponderance of hydrocarbons having 1, 2, 3, 4, or 5 carbon atoms, respectively, and without reference to hydrocarbon type.

calcining: heating a metal oxide or an ore to decompose carbonates, hydrates, or other compounds often in a controlled atmosphere.

CANMET hydrocracking process: a hydrocracking process for heavy feedstocks that employs a low-cost additive to inhibit coke formation and allow high feedstock conversion using a single reactor.

capillary forces: interfacial forces between immiscible fluid phases, resulting in pressure differences between the two phases.

capillary number, N_c: the ratio of viscous forces to capillary forces, and equal to viscosity times velocity divided by interfacial tension.

carbene: the pentane- or heptane-insoluble material that is insoluble in benzene or toluene but which is soluble in carbon disulfide (or pyridine); a type of rifle used for hunting bison.

carboid: the pentane- or heptane-insoluble material that is insoluble in benzene or toluene and which is also insoluble in carbon disulfide (or pyridine).

carbonate washing: processing using a mild alkali (e.g., potassium carbonate) process for emission control by the removal of acid gases from gas streams.

carbon dioxide–augmented waterflooding: injection of carbonated water, or water and carbon dioxide, to increase water flood efficiency; see **immiscible carbon dioxide displacement**.

carbon-forming propensity: see **carbon residue**.

carbonization: the conversion of an organic compound into char or coke by heat in the substantial absence of air; often used in reference to the destructive distillation (q.v.) (with simultaneous removal of distillate) of coal.

carbon rejection processes: upgrading processes in which coke is produced, e.g., coking (q.v.).

carbon residue: the amount of carbonaceous residue remaining after thermal decomposition of petroleum, a petroleum fraction, or a petroleum product in a limited amount of air; also called the *coke-* or *carbon-forming propensity*; often prefixed by the terms Conradson or Ramsbottom in reference to the inventor of the respective tests.

cascade tray: a fractionating device consisting of a series of parallel troughs arranged in a stair-step fashion in which liquid from the tray above enters the uppermost trough and liquid thrown from this trough by vapor rising from the tray below impinges against a plate and a perforated baffle, and liquid passing through the baffle enters the next longer trough.

casinghead gas: natural gas that issues from the casinghead (the mouth or opening) of an oil well.

casinghead gasoline: the liquid hydrocarbon product extracted from casinghead gas (q.v.) by one of three methods: compression, absorption, or refrigeration; see also **natural gasoline**.

catagenesis: the alteration of organic matter during the formation of petroleum that may involve temperatures in the range 50°C (120°F) to 200°C (390°F); see also **diagenesis** and **metagenesis**.

catalyst: a chemical agent that, when added to a reaction (process), will enhance the conversion of a feedstock without being consumed in the process; used in upgrading processes to assist cracking and other upgrading reactions.

catalyst plugging: the deposition of carbon (coke) or metal contaminants that decreases the flow through the catalyst bed.

catalyst poisoning: the deposition of carbon (coke) or metal contaminants that causes the catalyst to become nonfunctional.

catalyst selectivity: the relative activity of a catalyst with respect to a particular compound in a mixture, or the relative rate in competing reactions of a single reactant.

catalyst stripping: the introduction of steam, at a point where spent catalyst leaves the reactor, in order to strip, i.e., remove, deposits retained on the catalyst.

catalytic activity: the ratio of the space velocity of the catalyst under test to the space velocity required for the standard catalyst to give the same conversion as the catalyst being tested; usually multiplied by 100 before being reported.

catalytic cracking: the conversion of high-boiling feedstocks into lower-boiling products by means of a catalyst that may be used in a fixed bed (q.v.) or fluid bed (q.v.).

catalytic distillation: a process that combines reaction and distillation in a single vessel, resulting in lower investment and operating costs, as well as process benefits.

catalytic hydrocracking: a refining process that uses hydrogen and catalysts with relatively low temperatures and high pressures for converting middle-boiling or residual material to high-octane gasoline, reformer charge stock, jet fuel, and/or high-grade fuel oil. The process uses one or more catalysts, depending on product output, and can handle high-sulfur feedstocks without prior desulfurization.

catalytic hydrotreating: a refining process for treating petroleum fractions from atmospheric or vacuum distillation units (e.g., naphtha, middle distillates, reformer feeds, residual fuel oil, and heavy gas oil) and other petroleum (e.g., cat cracked naphtha, coker naphtha, gas oil, etc.) in the presence of catalysts and substantial quantities of hydrogen. Hydrotreating includes desulfurization, removal of substances (e.g., nitrogen compounds) that deactivate catalysts, conversion of olefins to paraffins to reduce gum formation in gasoline, and other processes to upgrade the quality of the fractions.

catalytic reforming: rearranging hydrocarbon molecules in a gasoline-boiling-range feedstock to produce other hydrocarbons having a higher antiknock quality; isomerization of paraffins, cyclization of paraffins to naphthenes (q.v.), dehydrocyclization of paraffins to aromatics (q.v.).

cat cracker: see **fluid catalytic cracking unit**.

cat cracking: see **catalytic cracking**.

Catforming: a process for reforming naphtha using a platinum–silica–alumina catalyst that permits relatively high space velocities and results in the production of high-purity hydrogen.

caustic wash: the process of treating a product with a solution of caustic soda to remove minor impurities; often used in reference to the solution itself.

ceresin: a hard, brittle wax obtained by purifying ozokerite; see **microcrystalline wax** and **ozokerite**.

cetane index: an approximation of the cetane number (q.v.) calculated from the density (q.v.) and mid-boiling-point temperature (q.v.); see also **diesel index**.

cetane number: a number indicating the ignition quality of diesel fuel; a high cetane number represents a short ignition delay time; the ignition quality of diesel fuel can also be estimated from the following formula:

$$\text{Diesel index} = [\text{aniline point (°F)} \times \text{API gravity}]100$$

characterization factor: the UOP characterization factor K, defined as the ratio of the cube root of the molar average boiling point, T_B, in degrees Rankine (°R = °F + 460), to the specific gravity at 60°F/60°F:

$$K = (T_B)^{1/3}/\text{sp gr}$$

The value ranges from 12.5 for paraffinic stocks to 10.0 for the highly aromatic stocks; also called the Watson characterization factor.

cheesebox still: an early type of vertical cylindrical still designed with a vapor dome.

chelating agents: complex-forming agents having the ability to solubilize heavy metals.

chemical composition: the makeup of petroleum in terms of distinct chemical types such as paraffins, isoparaffins, naphthenes (cycloparaffins), benzenes, diaromatics, triaromatics, polynuclear aromatics; other chemical types can also be specified.

chemical octane number: the octane number added to gasoline by refinery processes or by the use of octane number (q.v.) improvers such as tetraethyl lead.

chemical waste: any solid, liquid, or gaseous material discharged from a process and that may pose substantial hazards to human health and the environment.

Cherry-P process: a process for the conversion of heavy feedstocks into distillate and a cracked residuum.

Chevron deasphalted oil hydrotreating process: a process designed to desulfurize heavy feedstocks that have had the asphaltene fraction (q.v.) removed by prior application of a deasphalting process (q.v.).

Chevron RDS and VRDS processes: processes designed to remove sulfur, nitrogen, asphaltene, and metal contaminants from heavy feedstocks consisting of a once-through operation of the feedstock coming into contact with hydrogen and the catalyst in a downflow reactor (q.v.).

chlorex process: a process for extracting lubricating oil stocks in which the solvent used is chlorex (3-3-dichlorodiethyl ether).

CHOPS: a nonthermal primary heavy oil production method—continuous production of sand improves the recovery of heavy oil from the reservoir. The simultaneous extraction of oil and sand during the cold production of heavy oil generates high-porosity channels (*wormholes*) that grow in a three-dimensional radial pattern within a certain layer of net pay zones, resulting in the development of a high- permeability network in the reservoir, boosting oil recovery. In most cases, an artificial lift system is used to lift the oil with sand.

chromatographic adsorption: selective adsorption on materials such as activated carbon, alumina, or silica gel; liquid or gaseous mixtures of hydrocarbons are passed through the adsorbent in a stream of diluent, and certain components are preferentially adsorbed.

chromatography: a method of separation based on selective adsorption; see also **chromatographic adsorption**.

clarified oil: the heavy oil that has been taken from the bottom of a fractionator in a catalytic cracking process and from which residual catalyst has been removed.

clarifier: equipment for removing the color or cloudiness of an oil or water by separating the foreign material through mechanical or chemical means; may involve centrifugal action, filtration, heating, or treatment with acid or alkali.

clay: silicate minerals that also usually contain aluminum and have particle sizes <0.002 micron; used in separation methods as an adsorbent and in refining as a catalyst.

clay contact process: see **contact filtration**.

clay refining: a treating process in which vaporized gasoline or other light petroleum product is passed through a bed of granular clay such as Fuller's earth (q.v.).

clay regeneration: a process in which spent coarse-grained adsorbent clays from percolation processes are cleaned for reuse by deoiling them with naphtha, steaming out the excess naphtha, and then roasting in a stream of air to remove carbonaceous matter.

clay treating: see **gray clay treating**.

clay wash: a light oil, such as kerosene (kerosine) or naphtha, used to clean Fuller's earth after it has been used in a filter.

clastic: composed of pieces of preexisting rocks.

cloud point: the temperature at which paraffin wax or other solid substances begin to crystallize or separate from the solution, imparting a cloudy appearance to the oil when the oil is chilled under prescribed conditions.

coal: an organic rock; a readily combustible black or brownish–black rock whose composition, including inherent moisture, consists of >50% w/w and >70% v/v of carbonaceous material; formed from plant remains that have been compacted, hardened, chemically altered, and metamorphosed by heat and pressure over geologic time.

coal tar: the specific name for the tar (q.v.) produced from coal.

coal tar pitch: the specific name for the pitch (q.v.) produced from coal.

cogeneration: the simultaneous production of electricity and steam.

coke: a gray to black solid carbonaceous material produced from petroleum during thermal processing; characterized by having a high carbon content (95%+ by weight) and a honeycomb type of appearance and is insoluble in organic solvents.

coke drum: a vessel in which coke is formed and which can be cut oil from the process for cleaning.

coke number: used, particularly in Great Britain, to report the results of the Ramsbottom carbon residue test (q.v.), which is also referred to as a coke test.

coker: the processing unit in which coking takes place.

coking: a process for the thermal conversion of petroleum in which gaseous, liquid, and solid (coke) products are formed; e.g., delayed coking (q.v.) or fluid coking (q.v.).

cold pressing: the process of separating wax from oil by first chilling (to help form wax crystals) and then filtering under pressure in a plate and frame press.

cold settling: processing for the removal of wax from high-viscosity stocks, wherein a naphtha solution of the waxy oil is chilled and the wax crystallizes out of the solution.

color stability: the resistance of a petroleum product to color change due to light, aging, etc.

combustible liquid: a liquid with a flash point in excess of 37.8°C (100°F) but below 93.3°(200°F).

combustion zone: the volume of reservoir rock wherein petroleum is undergoing combustion during enhanced oil recovery.

composition: the general chemical makeup of petroleum; often quoted as chemical composition (q.v.) or bulk composition (q.v.).

composition map: a means of illustrating the chemical makeup of petroleum using chemical and/or physical property data.

Con carbon: see **carbon residue.**

condensate: a mixture of light hydrocarbon liquids obtained by condensation of hydrocarbon vapors: predominately butane, propane, and pentane with some heavier hydrocarbons and relatively little methane or ethane; see also **natural gas liquids.**

connate water: water that is indigenous to the reservoir; usually water that has not been in contact with the atmosphere since its deposition; contains high amounts of chlorine and calcium with total dissolved solids on the order of 1% or more.

Conradson carbon residue: see **carbon residue.**

contact filtration: a process in which finely divided adsorbent clay is used to remove color bodies from petroleum products.

contaminant: a substance that causes deviation from the normal composition of an environment.

continuous catalyst regeneration: a process that regenerates spent catalyst used in a catalytic reformer, which converts naphtha feedstock into higher-octane gasoline blending stocks.

continuous contact coking: a thermal conversion process in which petroleum-wetted coke particles move downward into the reactor where cracking, coking, and drying take place to produce coke, gas, gasoline, and gas oil.

continuous contact filtration: a process to finish lubricants, waxes, or special oils after acid treating, solvent extraction, or distillation.

conventional crude oil: a mixture mainly of pentane and heavier hydrocarbons recoverable at a well from an underground reservoir and liquid at atmospheric pressure and temperature; unlike some heavy oils and tar sand bitumen, conventional crude oil flows through a well without stimulation and through a pipeline without processing or dilution; generally, conventional crude oil includes light- and medium-gravity crude oils; crude oil containing >0.5% w/w sulfur is considered to be sour crude oil, while crude oil with <0.5% w/w sulfur is to be sweet crude oil.

conventional gasoline: finished automotive gasoline not included in the oxygenated or reformulated gasoline categories; excludes reformulated gasoline blendstock for oxygenate blending (RBOB) as well as other blendstocks.

conventional recovery: primary and/or secondary recovery.

conversion: the thermal treatment of petroleum that results in the formation of new products by the alteration of the original constituents.

conversion factor: the percentage of feedstock converted to light ends, gasoline, other liquid fuels, and coke.

conversion stock: a commodity material, usually vacuum gas oil or atmospheric gas oil, that is suitable for feedstock to a fluid catalytic cracking unit or hydrocracking unit.

copper sweetening: processes involving the oxidation of mercaptans to disulfides by oxygen in the presence of cupric chloride.

cracked residua: residua that have been subjected to temperatures above 350°C (660°F) during the distillation process.

cracking: the thermal processes by which the constituents of petroleum are converted to lower molecular weight products.

cracking activity: see **catalytic activity**.

cracking coil: equipment used for cracking heavy petroleum products consisting of a coil of heavy pipe running through a furnace so that the oil passing through it is subject to high temperature.

cracking still: the combined equipment–furnace, reaction chamber, fractionator, for the thermal conversion of heavier feedstocks to lighter products.

cracking temperature: the temperature (350°C, 660°F) at which the rate of thermal decomposition of petroleum constituents becomes significant.

crude assay: a procedure for determining the general distillation characteristics (e.g., distillation profile, q.v.) and other quality information of crude oil.

crude oil: see **petroleum**.

crude scale wax: the wax product from the first sweating of the slack wax.

crude still: distillation (q.v.) equipment in which crude oil is separated into various products.

cryogenic plant: a processing plant capable of producing liquid natural gas products, including ethane, at very low operating temperatures.

cryogenics: the study, production, and use of low temperatures.

cumene: a colorless liquid $[C_6H_5CH (CH_3)_2]$ used as an aviation gasoline blending component and as an intermediate in the manufacture of chemicals.

cut point: the boiling temperature division between distillation fractions of petroleum.

cutback: the term applied to the products from blending heavier feedstocks or products with lighter oils to bring the heavier materials to the desired specifications.

cycle stock: the product taken from some later stage of a process and recharged (recycled) to the process at some earlier stage.

cyclic steam injection: the alternating injection of steam and production of oil with condensed steam from the same well or wells.

cyclic steam stimulation: a process in which (for several weeks) high-pressure steam is injected into the formation to soften the tar sand (oil sand) before being pumped to the surface for separation; the pressure created in the underground environment causes formation cracks that help move the bitumen to producing wells. After a portion of the reservoir has been saturated, the steam is turned off and the reservoir is allowed to soak for several weeks. Then, the production phase brings the bitumen to the surface. When the rates of production start to decline, the reservoir is pumped with steam once again.

cyclization: the process by which an open-chain hydrocarbon structure is converted to a ring structure, e.g., hexane to benzene.

cyclone: a device for extracting dust from industrial waste gases. It is in the form of an inverted cone into which the contaminated gas enters tangentially from the top; the gas is propelled down a helical pathway, and the dust particles are deposited by means of centrifugal force onto the wall of the scrubber.

deactivation: reduction in catalyst activity by the deposition of contaminants (e.g., coke, metals) during a process.

dealkylation: the removal of an alkyl group from aromatic compounds.

deasphaltened oil: the fraction of petroleum after the asphaltenes have been removed using liquid hydrocarbons such as n-pentane and n-heptane.

deasphaltening: removal of a solid powdery asphaltene fraction from petroleum by the addition of the low-boiling liquid hydrocarbons such as n-pentane or n-heptane under ambient conditions.

deasphalting: the removal of asphalt (tacky, semisolid higher molecular weight) constituents from petroleum (as occurs in a refinery asphalt plant) by the addition of liquid propane or liquid butane under pressure; also the removal of the asphaltene fraction from petroleum by the addition of a low-boiling hydrocarbon liquid such as *n*-pentane or *n*-heptane.

debutanization: distillation to separate butane and lighter components from higher-boiling components.

debutanizer: a fractionating column used to remove butane and lighter components from liquid streams.

decant oil: the highest-boiling product from a catalytic cracker; also referred to as slurry oil, clarified oil, or bottoms.

decarbonizing: a thermal conversion process designed to maximize coker gas–oil production and minimize coke and gasoline yields; operated at essentially lower temperatures and pressures than delayed coking (q.v.).

decoking: removal of petroleum coke from equipment such as coking drums; hydraulic decoking uses high-velocity water streams.

decolorizing: removal of suspended, colloidal, and dissolved impurities from liquid petroleum products by filtering, adsorption, chemical treatment, distillation, bleaching, etc.

deethanization: distillation to separate ethane and lighter components from propane and higher-boiling components; also called deethanation.

deethanizer: a fractionating column designed to remove ethane and gases from heavier hydrocarbons.

dehydrating agents: substances capable of removing water (drying, q.v.) or the elements of water from another substance.

dehydrocyclization: any process by which both dehydrogenation and cyclization reactions occur.

dehydrogenation: the removal of hydrogen from a chemical compound; for example, the removal of two hydrogen atoms from butane to make butene(s) as well as the removal of additional hydrogen to produce butadiene.

delayed coking: a coking process in which the thermal reaction are allowed to proceed to completion to produce gaseous, liquid, and solid (coke) products.

demethanization: the process of distillation in which methane is separated from the higher-boiling components; also called demethanation.

Demex process: a solvent extraction demetallizing process that separates high metal vacuum residuum into demetallized oil of relatively low metal content and asphaltene of high metal content.

density: the mass (or weight) of a unit volume of any substance at a specified temperature; also the *heaviness* of crude oil, indicating the proportion of large, carbon-rich molecules, generally measured in kilograms per cubic meter (kg/m^3) or degrees on the API gravity scale; in some countries, oil up to 900 kg/m^3 is considered light to medium crude; see also **specific gravity**.

deoiling: reduction in quantity of liquid oil entrained in solid wax by draining (sweating) or by a selective solvent; see **MEK deoiling**.

depentanizer: a fractionating column for the removal of pentane and lighter fractions from a mixture of hydrocarbons.

depropanization: distillation in which lighter components are separated from butanes and higher-boiling material; also called depropanation.

depropanizer: a fractionating column for removing propane and lighter components from liquid streams.

desalting: removal of mineral salts (mostly chlorides) from crude oils.

desorption: the reverse process of adsorption whereby adsorbed matter is removed from the adsorbent; also used as the reverse of absorption (q.v.).

destructive distillation: thermal decomposition with the simultaneous removal of distillate; distillation (q.v.) when thermal decomposition of the constituents occurs.

desulfurization: the removal of sulfur or sulfur compounds from a feedstock; a process that removes sulfur and its compounds from various streams during the refining process;

desulfurization processes include catalytic hydrotreating and other chemical/physical processes such as absorption; the desulfurization processes vary based on the type of stream treated (e.g., naphtha, distillate, heavy gas oil, etc.) and the amount of sulfur removed (e.g., sulfur reduction to 10 ppm).

detergent oil: a lubricating oil possessing special sludge-dispersing properties for use in internal-combustion engines.

dewaxing: see **solvent dewaxing**.

devolatilized fuel: smokeless fuel; coke that has been reheated to remove all of the volatile material.

diagenesis: the concurrent and consecutive chemical reactions that commence the alteration of organic matter [at temperatures up to 50°C (120°F)] and ultimately result in the formation of petroleum from the marine sediment; see also **catagenesis** and **metagenesis**.

diesel cycle: a repeated succession of operations representing the idealized working behavior of the fluids in a diesel engine.

diesel fuel: fuel used for internal combustion in diesel engines; usually that fraction that distills after kerosene.

diesel hydrotreater: a refinery process unit for production of clean (low-sulfur) diesel fuel.

diesel index: an approximation of the cetane number (q.v.) of diesel fuel (q.v.) calculated from the density (q.v.) and aniline point (q.v.).

diesel knock: the result of a delayed period of ignition and the accumulation of diesel fuel in the engine.

diethanolamine (DEA): a solvent used in an acid gas removal system.

diglycolamine (DGA): a solvent used in an acid gas removal system.

dilbit: tar sand bitumen that has been reduced in viscosity through the addition of a diluent such as condensate or naphtha.

dilsynbit: a blend of tar sand bitumen, condensate, and synthetic crude oil similar to medium sour crude.

diluted crude: heavy crude oil to which a diluent (thinner) has been added to reduce viscosity and facilitate pipeline flow.

distillate: the products of distillation formed by condensing vapors.

distillate fuel oil: a general classification for one of the petroleum fractions produced in conventional distillation operations. It includes diesel fuels and fuel oils. Products known as No. 1, No. 2, and No. 4 diesel fuel are used in on-highway diesel engines, such as those in trucks and automobiles, as well as off-highway engines, such as those in railroad locomotives and agricultural machinery. Products known as No. 1, No. 2, and No. 4 fuel oils are used primarily for space heating and electric power generation.

distillation: a process for separating liquids with different boiling points without thermal decomposition of the constituents (see **destructive distillation**).

distillation curve: see **distillation profile**.

distillation loss: the difference, in a laboratory distillation, between the volume of liquid originally introduced into the distilling flask and the sum of the residue and the condensate recovered.

distillation range: the difference between the temperature at the initial boiling point and at the end point, as obtained by the distillation test.

distillation profile: the distillation characteristics of petroleum or a petroleum product showing the temperature and the percent distilled.

doctor solution: a solution of sodium plumbite used to treat gasoline or other light petroleum distillates to remove mercaptan sulfur; see also **doctor test**.

doctor sweetening: a process for sweetening gasoline, solvents, and kerosene by converting mercaptans to disulfides using sodium plumbite and sulfur.

doctor test: a test used for the detection of compounds in light petroleum distillates that react with sodium plumbite; see also **doctor solution**.

domestic heating oil: see **No. 2 fuel oil**.

donor solvent process: a conversion process in which a hydrogen donor solvent is used in place of or to augment hydrogen.

downcomer: a means of conveying liquid from one tray to the next below in a bubble tray column (q.v.).

downflow reactor: a reactor in which the feedstock flows in a downward direction over the catalyst bed.

downstream: a sector of the petroleum industry that refers to the refining of crude oil, and the products derived from crude oil.

dropping point: the temperature at which grease passes from a semisolid to a liquid state under prescribed conditions.

dry gas: a gas that does not contain fractions that may easily condense under normal atmospheric conditions.

drying: removal of a solvent or water from a chemical substance; also referred to as the removal of solvent from a liquid or suspension.

dry point: the temperature at which the last drop of petroleum fluid evaporates in a distillation test.

dualayer distillate process: a process for removing mercaptans and oxygenated compounds from distillate fuel oils and similar products, using a combination of treatment with concentrated caustic solution and electrical precipitation of the impurities.

dualayer gasoline process: a process for extracting mercaptans and other objectionable acidic compounds from petroleum distillates; see also **dualayer solution**.

dualayer solution: a solution that consists of concentrated potassium or sodium hydroxide containing a solubilizer; see also **dualayer gasoline process**.

Dubbs cracking: an older continuous, liquid-phase thermal cracking process formerly used.

ebullated bed: a process in which the catalyst bed is in a suspended state in the reactor by means of a feedstock recirculation pump that pumps the feedstock upward at sufficient speed to expand the catalyst bed at approximately 35% above the settled level.

Edeleanu process: a process for refining oils at low temperature with liquid sulfur dioxide (SO_2), or with liquid sulfur dioxide and benzene; applicable to the recovery of aromatic concentrates from naphtha and heavier petroleum distillates.

effective viscosity: see **apparent viscosity**.

effluent: any contaminating substance, usually a liquid, that enters the environment via a domestic industrial, agricultural, or sewage plant outlet.

electric desalting: a continuous process to remove inorganic salts and other impurities from crude oil by settling out in an electrostatic field.

electrical precipitation: a process using an electrical field to improve the separation of hydrocarbon reagent dispersions; may be used in chemical treating processes on a wide variety of refinery stocks.

electrofining: a process for contacting a light hydrocarbon stream with a treating agent (acid, caustic, doctor, etc.), then assisting the action of separation of the chemical phase from the hydrocarbon phase by an electrostatic field.

electrolytic mercaptan process: a process in which aqueous caustic solution is used to extract mercaptans from refinery streams.

electrostatic precipitators: devices used to trap fine dust particles (usually in the size range 30–60 microns) that operate on the principle of imparting an electric charge to particles in an incoming air stream and which are then collected on an oppositely charged plate across a high-voltage field.

emission control: the use gas cleaning processes to reduce emissions.

emission standard: the maximum amount of a specific pollutant permitted to be discharged from a particular source in a given environment.

emulsion breaking: the settling or aggregation of colloidal-sized emulsions from suspension in a liquid medium.

end-of-pipe emission control: the use of specific emission control processes to clean gases after production of the gases.

energy: the capacity of a body or system to do work, measured in joules (SI units); also the output of fuel sources.

energy from biomass: the production of energy from biomass (q.v.).

Engler distillation: a standard test for determining the volatility characteristics of a gasoline by measuring the percent distilled at various specified temperatures.

enhanced oil recovery: petroleum recovery following recovery by conventional (i.e., primary and/or secondary) methods; the third stage of production during which sophisticated techniques that alter the original properties of the oil are used. Enhanced oil recovery can begin after a secondary recovery process or at any time during the productive life of an oil reservoir—the purpose is not only to restore formation pressure but also to improve oil displacement or fluid flow in the reservoir. The three major types of enhanced oil recovery operations are chemical flooding (alkaline flooding or micellar-polymer flooding), miscible displacement (carbon dioxide injection or hydrocarbon injection), and thermal recovery (steam flood). The optimal application of each method depends on reservoir temperature, pressure, depth, net pay, permeability, residual oil and water saturations, porosity and fluid properties such as oil API gravity, and viscosity.

entrained bed: a bed of solid particles suspended in a fluid (liquid or gas) at such a rate that some of the solid is carried over (entrained) by the fluid.

ETBE [ethyl tertiary butyl ether, $(CH_3)_3COC_2H$]: an oxygenate blend stock formed by the catalytic etherification of isobutylene with ethanol.

ethane (C_2H_6): a straight-chain saturated (paraffinic) hydrocarbon extracted predominantly from the natural gas stream, which is gaseous at standard temperature and pressure; a colorless gas that boils at a temperature of $-88°C$ ($-127°F$).

ethanol: see **ethyl alcohol**.

ET-II process: a thermal cracking process for the production of distillates and cracked residuum for use for metallurgical coke; *of this Earth* and not extraterrestrial!

ether: a generic term applied to a group of organic chemical compounds composed of carbon, hydrogen, and oxygen, characterized by an oxygen atom attached to two carbon atoms (e.g., methyl tertiary butyl ether).

ethyl alcohol (ethanol or grain alcohol): an inflammable organic compound (C_2H_5OH) formed during fermentation of sugars; used as an intoxicant and as a fuel.

Eureka process: a thermal cracking process to produce a cracked oil and aromatic residuum from heavy residual materials.

evaporation: a process for concentrating nonvolatile solids in a solution by boiling off the liquid portion of the waste stream.

ethylene (C_2H_4): an olefin hydrocarbon recovered from refinery or petrochemical processes that is gaseous at standard temperature and pressure. Ethylene is used as a petrochemical feedstock for many chemical applications and the production of consumer goods.

expanding clays: clays that expand or swell on contact with water, e.g., montmorillonite.

explosive limits: the limits of percentage composition of mixtures of gases and air within which an explosion takes place when the mixture is ignited.

extractive distillation: the separation of different components of mixtures that have similar vapor pressures by flowing a relatively high-boiling solvent, which is selective for one of the components in the feed, down a distillation column as the distillation proceeds; the selective solvent scrubs the soluble component from the vapor.

fabric filters: filters made from fabric materials and used for removing particulate matter from gas streams (see **baghouse**).

facies: one or more layers of rock that differ(s) from other layers in composition, age, or content; identifiable subdivisions of stratigraphic units.

fat oil: the bottom or enriched oil drawn from the absorber as opposed to lean oil.

faujasite: a naturally occurring silica–alumina (SiO_2–Al_2O_3) mineral.

FCC: fluid catalytic cracking.

FCCU: fluidized catalytic cracking unit.

feedstock: petroleum as it is fed to the refinery; a refinery product that is used as the raw material for another process; the term is also generally applied to raw materials used in other refinery processes or industrial processes.

ferrocyanide process: a regenerative chemical treatment for mercaptan removal using a caustic–sodium ferrocyanide reagent.

filtration: the use of an impassable barrier to collect solids but which allows liquids to pass.

fines: typically minute particles of solids such as clay or sand.

fire point: the lowest temperature at which, under specified conditions in standardized apparatus, a petroleum product vaporizes sufficiently rapidly to form above its surface an air–vapor mixture that burns continuously when ignited by a small flame.

Fischer–Tropsch process: a process for synthesizing hydrocarbons and oxygenated chemicals from a mixture of hydrogen and carbon monoxide.

fixed bed: a stationary bed (of catalyst) to accomplish a process (see **fluid bed**).

flammability range: the range of temperature over which a chemical is flammable.

flammable: a substance that will burn readily.

flammable liquid: a liquid having a flash point below 37.8°C (100°F).

flammable solid: a solid that can ignite from friction or from heat remaining from its manufacture, or which may cause a serious hazard if ignited.

flash point: the lowest temperature to which the product must be heated under specified conditions to give off sufficient vapor to form a mixture with air that can be ignited momentarily by a flame.

floc point: the temperature at which wax or solids separate as a definite floc.

flexicoking: a modification of the fluid coking process insofar as the process also includes a gasifier adjoining the burner/regenerator to convert excess coke to a clean fuel gas.

flue gas: gas from the combustion of fuel, the heating value of which has been substantially spent and which is, therefore, discarded to the flue or stack; gas that is emitted to the atmosphere via a flue, which is a pipe for transporting exhaust fumes.

fluid bed: a bed (of catalyst) that is agitated by an upward passing gas in such a manner that the particles of the bed simulate the movement of a fluid and has the characteristics associated with a true liquid; cf. **fixed bed**.

fluid catalytic cracking: cracking in the presence of a fluidized bed of catalyst.

fluid coking: a continuous fluidized solids process that cracks feed thermally over heated coke particles in a reactor vessel to gas, liquid products, and coke.

fluidized catalytic cracking: a refinery process used to convert the heavy portion of crude oil into lighter products, including liquefied petroleum gas and gasoline.

fly ash: particulate matter produced from mineral matter in coal that is converted during combustion to finely divided inorganic material and which emerges from the combustor in the gases.

foots oil: the oil sweated out of slack wax; named from the fact that the oil goes to the foot, or bottom, of the pan during the sweating operation.

fossil fuel resources: a gaseous, liquid, or solid fuel material formed in the ground by chemical and physical changes (diagenesis, q.v.) in plant and animal residues over geological time; natural gas, petroleum, coal, and oil shale.

fraction: a group of hydrocarbons that have similar boiling points; a portion of crude oil defined by boiling range—naphtha, kerosene, gas oil, and residuum are fractions of crude oil.

fractional composition: the composition of petroleum as determined by fractionation (separation) methods.

fractional distillation: the separation of the components of a liquid mixture by vaporizing and collecting the fractions, or cuts, which condense in different temperature ranges; a common form of separation technology in hydrocarbon-processing plants wherein a mixture (e.g., crude oil) is heated in a large, vertical cylindrical column to separate compounds (fractions) according to their boiling points.

fractionating column: a column arranged to separate various fractions of petroleum by a single distillation and which may be tapped at different points along its length to separate various fractions in the order of their boiling points.

fractionation: the separation of petroleum into the constituent fractions using solvent or adsorbent methods; chemical agents such as sulfuric acid may also be used.

Frasch process: a process formerly used for removing sulfur by distilling oil in the presence of copper oxide.

FTC process: a heavy oil and residuum upgrading process in which the feedstock is thermally cracked to produce distillate and coke, which is gasified to fuel gas.

free sulfur: sulfur that exists in the elemental state associated with petroleum; sulfur that is not bound organically within the petroleum constituents.

fuel oil: also called heating oil; is a distillate product that covers a wide range of properties; see also No. 1 to No. 4 fuel oils.

Fuller's earth: a clay that has high adsorptive capacity for removing color from oils; Attapulgus clay is a widely used Fuller's earth.

functional group: the portion of a molecule that is characteristic of a family of compounds and determines the properties of these compounds.

furfural extraction: a single-solvent process in which furfural is used to remove aromatic components, naphthene components, olefins, and unstable hydrocarbons from a lubricating oil charge stock.

furnace oil: a distillate fuel primarily intended for use in domestic heating equipment.

gas cap: a part of a hydrocarbon reservoir at the top that will produce only gas.

gaseous pollutants: gases released into the atmosphere that act as primary or secondary pollutants.

gasification: a process to partially oxidize any hydrocarbon, typically heavy residues, to a mixture of hydrogen and carbon monoxide; the process can be used to produce hydrogen and various energy by-products.

gasohol: a term for motor vehicle fuel comprising between 80% and 90% unleaded gasoline and 10–20% ethanol (see also **ethyl alcohol**).

gas oil: a petroleum distillate with a viscosity and boiling range between those of kerosene and lubricating oil; a middle-distillate petroleum fraction; usually includes diesel, kerosene, heating oil, and light fuel oil; a liquid petroleum distillate having a viscosity intermediate between that of kerosene and lubricating oil. It derives its name from having originally been used in the manufacture of illuminating gas. It is now used to produce distillate fuel oils and gasoline; heavy gas oil is petroleum distillates with an approximate boiling range from 345°C to 540°C (650–1000°F).

gas–oil ratio: ratio of the number of cubic feet of gas measured at atmospheric (standard) conditions to barrels of produced oil measured at stock tank conditions.

gasoline: fuel for the internal combustion engine that is commonly, but improperly, referred to simply as gas.

gasoline blending components: naphtha fractions that will be used for blending or compounding into finished aviation or automotive gasoline (e.g., straight-run gasoline, alkylate, reformate, benzene, toluene, and xylene); excludes oxygenates (alcohols, ethers), butane, and pentanes plus.

gas reversion: a combination of thermal cracking or reforming of naphtha with thermal polymerization or alkylation of hydrocarbon gases carried out in the same reaction zone.

gas-to-liquids (GTL): a process used to convert natural gas into longer-chain hydrocarbons such as diesel and jet fuel. Methane-rich gases are converted into liquid syngas (a mix of carbon monoxide and hydrogen) produced using steam methane reforming or autothermal reforming, followed by the Fischer–Tropsch process. Hydrocracking is then used to produce finished fuels.

gilsonite: an asphaltite that is >90% bitumen.

Girbotol process: a continuous, regenerative process to separate hydrogen sulfide, carbon dioxide, and other acid impurities from natural gas, refinery gas, etc., using mono-, di-, or triethanolamine as the reagent.

glance pitch: an asphaltite.

glycol–amine gas treating: a continuous, regenerative process to simultaneously dehydrate and remove acid gases from natural gas or refinery gas.

grahamite: an asphaltite.

gray clay treating: a fixed-bed (q.v.), usually Fuller's earth (q.v.), vapor-phase treating process to selectively polymerize unsaturated gum-forming constituents (diolefins) in thermally cracked gasoline.

grain alcohol: see **ethyl alcohol**.

gravity drainage: the movement of oil in a reservoir that results from the force of gravity.

gravity segregation: partial separation of fluids in a reservoir caused by the gravity force acting on differences in density.

greenhouse effect: warming of the earth due to entrapment of the energy of the sun by the atmosphere.

greenhouse gases: gases that contribute to the greenhouse effect (q.v.).

Gulf HDS process: a fixed-bed process for the catalytic hydrocracking of heavy stocks to lower-boiling distillates with accompanying desulfurization.

Gulfining: a catalytic hydrogen treating process for cracked and straight-run distillates and fuel oils, to reduce sulfur content; improve carbon residue, color, and general stability; and effect a slight increase in gravity.

Gulf resid hydrodesulfurization process: a process for the desulfurization of heavy feedstocks to produce low-sulfur fuel oils or catalytic cracking (q.v.) feedstocks.

gum: an insoluble tacky semisolid material formed as a result of the storage instability and/or the thermal instability of petroleum and petroleum products.

heat exchanger: a device used to transfer heat from a fluid on one side of a barrier to a fluid on the other side without bringing the fluids into direct contact.

heating oil: see **fuel oil**.

heavy ends: the highest boiling portion of a petroleum fraction; see also **light ends**.

heavy feedstock: any feedstock of the type heavy oil (q.v.), bitumen (q.v.), atmospheric residuum (q.v.), vacuum residuum (q.v.), and solvent deasphalter bottoms (q.v.).

heavy fuel oil: fuel oil having a high density and viscosity; generally residual fuel oil such as No. 5 and No 6. fuel oil (q.v.).

heavy gas oil: a petroleum distillate with an approximate boiling range from 345°C to 540°C (650–1000°F).

heavy oil: petroleum having an API gravity of <20°; other properties are necessary for inclusion in a more complete definition.

heavy petroleum: see **heavy oil**.

heavy residue gasification and combined cycle power generation: a process for producing hydrogen from residua.

heteroatom compounds: chemical compounds that contain nitrogen and/or oxygen and/or sulfur and/or metals bound within their molecular structure(s).

HF alkylation: an alkylation process whereby olefins (C_3, C_4, C_5) are combined with isobutane in the presence of hydrofluoric acid catalyst.

high-boiling distillates: fractions of petroleum that cannot be distilled at atmospheric pressure without decomposition, e.g., gas oils.

high-sulfur diesel (HSD): diesel fuel containing >500 ppm sulfur.

high-sulfur petroleum: a general expression for petroleum having >1% wt. sulfur; this is a very approximate definition and should not be construed as having a high degree of accuracy because it does not take into consideration the molecular locale of the sulfur. All else being equal, there is little difference between petroleum having 0.99% wt. sulfur and petroleum having 1.01% wt. sulfur.

history: the study of the past events of a particular subject to learn from those events.

HOC process: a version of the fluid catalytic cracking process (q.v.) that has been adapted to conversion of residua (q.v.) that contain high amounts of metal and asphaltenes (q.v.).

H-Oil process: a catalytic process that is designed for hydrogenation of heavy feedstocks in an ebullated-bed reactor.

Hortonsphere: a spherical pressure-type tank used to store volatile liquids, which prevents the excessive evaporation loss that occurs when such products are placed in conventional storage tanks.

hot filtration test: a test for the stability of a petroleum product.

HOT process: a catalytic cracking process for upgrading heavy feedstocks using a fluidized bed of iron ore particles.

hotspot: an area of a vessel or line wall appreciably above normal operating temperature, usually as a result of the deterioration of an internal insulating liner that exposes the line or vessel shell to the temperature of its contents.

Houdresid catalytic cracking: a continuous moving-bed process for catalytically cracking reduced crude oil to produce high-octane gasoline and light-distillate fuels.

Houdriflow catalytic cracking: a continuous moving-bed catalytic cracking process employing an integrated single vessel for the reactor and regenerator kiln.

Houdriforming: a continuous catalytic reforming process for producing aromatic concentrates and high-octane gasoline from low-octane straight-run naphtha.

Houdry butane dehydrogenation: a catalytic process for dehydrogenating light hydrocarbons to their corresponding mono- or diolefins.

Houdry fixed-bed catalytic cracking: a cyclic regenerable process for cracking of distillates.

Houdry hydrocracking: a catalytic process combining cracking and desulfurization in the presence of hydrogen.

HSC process: a cracking process for moderate conversion of heavy feedstocks (q.v.); the extent of the conversion is higher than visbreaking (q.v.) but lower than coking (q.v.).

hybrid gasification process: a process to produce hydrogen by gasification of a slurry of coal and residual oil.

HYCAR process: a noncatalytic process conducted under similar conditions to visbreaking (q.v.) and involves treatment with hydrogen under mild conditions; see also **hydrovisbreaking**.

hydraulic fracturing: the opening of fractures in a reservoir by high-pressure, high-volume injection of liquids through an injection well.

hydrocarbon compounds: chemical compounds containing only carbon and hydrogen.

hydrocarbon gasification process: a continuous, noncatalytic process in which hydrocarbons are gasified to produce hydrogen by air or oxygen.

hydrocarbon gas liquids (HGL): a group of hydrocarbons including ethane, propane, n-butane, isobutane, natural gasoline, and their associated olefins, including ethylene, propylene, butylene, and isobutylene; excludes liquefied natural gas.

hydrocarbon resource: resources such as petroleum and natural gas that can produce naturally occurring hydrocarbons without the application of conversion processes.

hydrocarbon-producing resource: a resource such as coal and oil shale (kerogen) that produce derived hydrocarbons by the application of conversion processes; the hydrocarbons so produced are not naturally occurring materials.

hydroconversion: a term often applied to hydrocracking (q.v.).

hydrocracker: a refinery process unit in which hydrocracking occurs.

hydrocracking: a catalytic high-pressure, high-temperature process for the conversion of petroleum feedstocks in the presence of fresh and recycled hydrogen; carbon–carbon bonds are cleaved in addition to the removal of heteroatomic species.

hydrocracking catalyst: a catalyst used for hydrocracking that typically contains separate hydrogenation and cracking functions.

hydrodenitrogenation: the removal of nitrogen by hydrotreating (q.v.).

hydrodesulfurization: the removal of sulfur by hydrotreating (q.v.).

hydrodemetallization: the removal of metallic constituents by hydrotreating (q.v.).

hydrodesulfurization: a refining process that removes sulfur from liquid and gaseous hydrocarbons.

hydrofining: a fixed-bed catalytic process to desulfurize and hydrogenate a wide range of charge stocks from gases through waxes.

hydroforming: a process in which naphtha is passed over a catalyst at elevated temperatures and moderate pressures, in the presence of added hydrogen or hydrogen-containing gases, to form high-octane motor fuel or aromatics.

hydrogen: the lightest of all gases, occurring chiefly in combination with oxygen in water; exists also in acids, bases, alcohols, petroleum, and other hydrocarbons.

hydrogen addition processes: upgrading processes in the presence of hydrogen, e.g., hydrocracking (q.v.); see **hydrogenation**.

hydrogenation: the chemical addition of hydrogen to a material. In nondestructive hydrogenation, hydrogen is added to a molecule only if, and where, unsaturation with respect to hydrogen exists; classed as *destructive (hydrocracking)* or *nondestructive (hydrotreating)*.

hydrogen blistering: blistering of steel caused by trapped molecular hydrogen formed as atomic hydrogen during corrosion of steel by hydrogen sulfide.

hydrogen sink: a chemical structure within the feedstock that reacts with hydrogen with little, if any, effect on the product character.

hydrogen transfer: the transfer of inherent hydrogen within the feedstock constituents and products during processing.

hydroprocessing: a term often equally applied to hydrotreating (q.v.) and to hydrocracking (q.v.); also often collectively applied to both.

hydrotreater: a refinery process unit that removes sulfur and other contaminants from hydrocarbon streams.

hydrotreating: the removal of heteroatomic (nitrogen, oxygen, and sulfur) species by treatment of a feedstock or product at relatively low temperatures in the presence of hydrogen.

hydrovisbreaking: a noncatalytic process, conducted under similar conditions to visbreaking, which involves treatment with hydrogen to reduce the viscosity of the feedstock and produce more stable products than is possible with visbreaking.

hydropyrolysis: a short-residence-time, high-temperature process using hydrogen.

hyperforming: a catalytic hydrogenation process for improving the octane number of naphtha through removal of sulfur and nitrogen compounds.

hypochlorite sweetening: the oxidation of mercaptans in a sour stock by agitation with aqueous, alkaline hypochlorite solution; used where avoidance of free-sulfur addition is desired, because of a stringent copper strip requirements, and minimum expense is not the primary object.

Hypro process: a continuous catalytic method for hydrogen manufacture from natural gas or from refinery effluent gases.

Hyvahl F process: a process for hydroconverting heavy feedstocks to naphtha and middle distillates using a dual-catalyst system and a fixed-bed swing reactor.

IFP hydrocracking process: a process that features a dual-catalyst system in which the first catalyst is a promoted nickel–molybdenum amorphous catalyst to remove sulfur and nitrogen and hydrogenate aromatic rings. The second catalyst is a zeolite that finishes the hydrogenation and promotes the hydrocracking reaction.

ignitability: characteristic of liquids whose vapors are likely to ignite in the presence of ignition source; also characteristic of nonliquids that may catch fire from friction or contact with water and that burn vigorously.

illuminating oil: oil used for lighting purposes.

immiscible: two or more fluids that do not have complete mutual solubility and coexist as separate phases.

immiscible carbon dioxide displacement: injection of carbon dioxide into an oil reservoir to effect oil displacement under conditions in which miscibility with reservoir oil is not obtained; see **carbon dioxide augmented waterflooding**.

inhibitor: a substance, the presence of which, in small amounts, in a petroleum product prevents or retards undesirable chemical changes from taking place in the product, or in the condition of the equipment in which the product is used.

inhibitor sweetening: a treating process to sweeten gasoline of low mercaptan content, using a phenylenediamine type of inhibitor, air, and caustic.

initial boiling point: the recorded temperature when the first drop of liquid falls from the end of the condenser.

initial vapor pressure: the vapor pressure of a liquid of a specified temperature and 0% evaporated.

instability: the inability of a petroleum product to exist for periods of time without change to the product.

incompatibility: the *immiscibility* of petroleum products and also of different crude oils that is often reflected in the formation of a separate phase after mixing and/or storage.

in situ **combustion:** combustion of oil in the reservoir, sustained by continuous air injection, to displace unburned oil toward producing wells.

inspection: application of test procedures to a feedstock to determine its processability (q.v.); the analysis (q.v.) of a feedstock.

iodine number: a measure of the iodine absorption by an oil under standard conditions; used to indicate the quantity of unsaturated compounds present; also called iodine value.

ion exchange: a means of removing cations or anions from solution onto a solid resin.

isobutane (C_4H_{10}): a branch-chain saturated (paraffinic) hydrocarbon extracted from both natural gas and refinery gas streams, which is gaseous at standard temperature and pressure.

isobutylene (C_4H_8): a branch-chain olefin hydrocarbon recovered from refinery or petrochemical processes, which is gaseous at standard temperature and pressure. Isobutylene is used in the production of gasoline and various petrochemical products.

isocracking process: a hydrocracking process for conversion of hydrocarbons, which operates at relatively low temperatures and pressures in the presence of hydrogen and a catalyst to produce more valuable, lower-boiling products.

isodewaxing: a catalytic isomerization of wax to improve base lube oil pour point.

isofining: a mild residue hydrocracking for synthetic crude oils.

isofinishing: a process for hydrofinishing of base lube oils to improve oxygen stability and color.

isoforming: a process in which olefin-type naphtha is contacted with an alumina catalyst at high temperature and low pressure to produce isomers of higher octane number.

isohexane (C_6H_{14}): a saturated branch-chain hydrocarbon.

Iso-Kel process: a fixed-bed, vapor-phase isomerization process using a precious metal catalyst and external hydrogen.

isomate process: a continuous, nonregenerative process for isomerizing C_5–C_8 normal paraffinic hydrocarbons, using aluminum chloride–hydrocarbon catalyst with anhydrous hydrochloric acid as a promoter.

isomerate process: a fixed-bed isomerization process to convert pentane, heptane, and heptane to high-octane blending stocks.

isomerization: the conversion of a normal (straight-chain) paraffin hydrocarbon into an iso- (branched-chain) paraffin hydrocarbon having the same atomic composition; a refining process that alters the fundamental arrangement of atoms in the molecule without adding or removing anything from the original material; used to convert normal butane into isobutane (C_4), an alkylation process feedstock, and normal pentane and hexane into isopentane (C_5) and isohexane (C_6), high-octane gasoline components.

isopentane: a saturated branched-chain hydrocarbon (C_5H_{12}) obtained by fractionation of natural gasoline or isomerization of normal pentane.

Iso-plus Houdriforming: a combination process using a conventional Houdriformer operated at moderate severity, in conjunction with one of three possible alternatives—including the use of an aromatic recovery unit or a thermal reformer; see **Houdriforming**.

jet fuel: fuel meeting the required properties for use in jet engines and aircraft turbine engines.

kaolinite: a clay mineral formed by hydrothermal activity at the time of rock formation or by chemical weathering of rocks with high feldspar content; usually associated with intrusive granite rocks with high feldspar content.

kata-condensed aromatic compounds: compounds based on linear condensed aromatic hydrocarbon systems, e.g., anthracene and naphthacene (tetracene).

kerogen: a complex carbonaceous (organic) material that occurs in sedimentary rocks and shale formations; generally insoluble in common organic solvents.

kerosene (kerosine): a fraction of petroleum that was initially sought as an illuminant in lamps; a precursor to diesel fuel; a light petroleum distillate that is used in space heaters, cook stoves, and water heaters and is suitable for use as a light source when burned in wick-fed lamps. Kerosene has a maximum distillation temperature of 400°F at the 10% recovery point, a final boiling point of 572°F, and a minimum flash point of 100°F. Included are No. 1-K and No. 2-K, the two grades recognized by ASTM Specification D 3699 as well as all other grades of kerosene called range or stove oil, which have properties similar to those of No. 1 fuel oil.

kerosene-type jet fuel: a kerosene-based product having a maximum distillation temperature of 400°F at the 10% recovery point and a final maximum boiling point of 572°F and meeting ASTM Specification D 1655 and Military Specifications MIL-T-5624P and MIL-T-83133D (Grades JP-5 and JP-8). It is used for commercial and military turbojet and turboprop aircraft engines.

K-factor: see **characterization factor**.

kinematic viscosity: the ratio of viscosity (q.v.) to density, both measured at the same temperature.

knock: the noise associated with self-ignition of a portion of the fuel–air mixture ahead of the advancing flames front.

lamp burning: a test of burning oils in which the oil is burned in a standard lamp under specified conditions in order to observe the steadiness of the flame, the degree of encrustation of the wick, and the rate of consumption of the kerosene.

lamp oil: see **kerosene**.

LC-Fining process: a hydrogenation (hydrocracking) process capable of desulfurizing, demetallizing, and upgrading heavy feedstocks by means of an expanded-bed reactor.

leaded gasoline: gasoline containing tetraethyl lead or other organometallic lead antiknock compounds.

lean gas: the residual gas from the absorber after the condensable gasoline has been removed from the wet gas.

lean oil: absorption oil from which gasoline fractions have been removed; oil leaving the stripper in a natural gasoline plant.

LEDA (low-energy deasphalting) process: a process for extracting high-quality catalytic cracking feeds from heavy feedstocks; the process uses a low-boiling hydrocarbon solvent specifically formulated to ensure the most economical deasphalting (q.v.) design for each operation.

Lewis acid: a chemical species that can accept an electron pair from a base.

Lewis base: a chemical species that can donate an electron pair.

light crude oil: crude oil with a high proportion of light hydrocarbon fractions and low metallic compounds; sometime defined as crude oil a gravity of 28° API or higher; a high-quality light crude oil might have a gravity of approaching 40° API, such as light Arabian crude oil (32–34° API) and West Texas Intermediate crude oil (37–40° API).

light ends: the lower-boiling components of a mixture of hydrocarbons; see also **heavy ends, light hydrocarbons**.

light gas oil: liquid petroleum distillates that are higher boiling than naphtha.

light hydrocarbons: hydrocarbons with molecular weights less than that of heptane (C_7H_{16}).

light oil: the products distilled or processed from crude oil up to, but not including, the first lubricating oil distillate.

light petroleum: petroleum having an API gravity >20°.

Ligroine (Ligroin): a saturated petroleum naphtha boiling in the range of 20–135°C (68–275°F) and suitable for general use as a solvent; also called benzine or petroleum ether.

Linde copper sweetening: a process for treating gasoline and distillates with a slurry of clay and cupric chloride.

liquid petrolatum: see **white oil**.

liquefied natural gas (LNG): natural gas cooled to a liquid state.

liquefied petroleum gas: propane, butane, or mixtures thereof, gaseous at atmospheric temperature and pressure, held in the liquid state by pressure to facilitate storage, transport, and handling.

liquefied refinery gases (LRG): hydrocarbon gas liquids produced in refineries from processing of crude oil and unfinished oils. They are retained in the liquid state through pressurization and/or refrigeration; includes ethane, propane, *n*-butane, isobutane, and refinery olefins (ethylene, propylene, butylene, and isobutylene).

liquid fuels: products of petroleum refining, natural gas liquids, biofuels, and liquids derived from other sources (including coal-to-liquids and gas-to-liquids); liquefied natural gas and liquid hydrogen are not included.

liquid sulfur dioxide–benzene process: a mixed-solvent process for treating lubricating oil stocks to improve viscosity indexes; also used for dewaxing.

lithology: the geological characteristics of the reservoir rock.

live steam: steam coming directly from a boiler before being utilized for power or heat.

liver: the intermediate layer of dark-colored, oily material, insoluble in weak acid and in oil, which is formed when acid sludge is hydrolyzed.

low-boiling distillates: fractions of petroleum that can be distilled at atmospheric pressure without decomposition.

low-sulfur petroleum: a general expression for petroleum having <1% wt. sulfur; this is a very approximate definition and should not be construed as having a high degree of accuracy because it does not take into consideration the molecular locale of the sulfur. All else being equal, there is little difference between petroleum having 0.99% wt. sulfur and petroleum having 1.01% wt. sulfur.

lube: see **lubricating oil**.

lube cut: a fraction of crude oil of suitable boiling range and viscosity to yield lubricating oil when completely refined; also referred to as lube oil distillates or lube stock.

lubricants: substances used to reduce friction between bearing surfaces, or incorporated into other materials used as processing aids in the manufacture of other products, or used as carriers of other materials. Petroleum lubricants may be produced either from distillates or residues; includes all grades of lubricating oils, from spindle oil to cylinder oil to those used in grease.

lubricating oil: a fluid lubricant used to reduce friction between bearing surfaces.

mahogany acids: oil-soluble sulfonic acids formed by the action of sulfuric acid on petroleum distillates. They may be converted to their sodium soaps (mahogany soaps) and extracted from the oil with alcohol for use in the manufacture of soluble oils, rust preventives, and special greases. The calcium and barium soaps of these acids are used as detergent additives in motor oils; see also **brown acids** and **sulfonic acids**.

maltenes: that fraction of petroleum that is soluble in, for example, pentane or heptane; deasphaltened oil (q.v.); also the term arbitrarily assigned to the pentane-soluble portion of petroleum that is relatively high boiling (>300°C, 760 mm) (see also **petrolenes**).

marine engine oil: oil used as a crankcase oil in marine engines.

marine gasoline: fuel for motors in marine service.

marine sediment: the organic biomass from which petroleum is derived.

marsh: an area of spongy waterlogged ground with large numbers of surface water pools. Marshes usually result from (1) an impermeable underlying bedrock; (2) surface deposits of glacial boulder clay; (3) a basin-like topography from which natural drainage is poor; (4) very heavy rainfall in conjunction with a correspondingly low evaporation rate; (5) low-lying land, particularly at estuarine sites at or below sea level.

mayonnaise: low-temperature sludge; a black, brown, or gray deposit having a soft, mayonnaise-like consistency; not recommended as a food additive.

MDS process: a solvent deasphalting process that is particularly effective for upgrading heavy crude oils.

medium crude oil: crude oil with gravity between (approximately) 20° and 28° API.

methanol: see **methyl alcohol**.

medicinal oil: highly refined, colorless, tasteless, and odorless petroleum oil used as a medicine in the nature of an internal lubricant; sometimes called liquid paraffin.

MEK (methyl ethyl ketone): a colorless liquid ($CH_3COCH_2CH_3$) used as a solvent; as a chemical intermediate; and in the manufacture of lacquers, celluloid, and varnish removers.

MEK deoiling: a wax-deoiling process in which the solvent is generally a mixture of methyl ethyl ketone and toluene.

MEK dewaxing: a continuous solvent dewaxing process in which the solvent is generally a mixture of methyl ethyl ketone and toluene.

MEOR: microbial enhanced oil recovery.

mercapsol process: a regenerative process for extracting mercaptans, utilizing aqueous sodium (or potassium) hydroxide containing mixed cresols as solubility promoters.

mercaptans: odiferous organic sulfur compounds with the general formula R-SH.

merox process: a refinery process used to remove or convert mercaptans.

metagenesis: the alteration of organic matter during the formation of petroleum that may involve temperatures above 200°C (390°F); see also **catagenesis** and **diagenesis**.

methanol (methyl alcohol, CH_3OH): a low-boiling alcohol eligible for gasoline blending.

methyl alcohol (methanol; wood alcohol): a colorless, volatile, inflammable, and poisonous alcohol (CH_3OH) traditionally formed by *destructive distillation* (q.v.) of wood or, more recently, as a result of synthetic distillation in chemical plants.

methyl *t*-butyl ether: an ether added to gasoline to improve its octane rating and to decrease gaseous emissions; see **oxygenate**.

methyldiethanolamine (MDEA): a solvent used in an acid gas removal system.

methyl ethyl ketone: see **MEK**.

mica: a complex aluminum silicate mineral that is transparent, tough, flexible, and elastic.

micelle: the structural entity by which asphaltenes are dispersed in petroleum.

microcarbon residue: the carbon residue determined using a thermogravimetric method; see also **carbon residue**.

Microcat-RC process (M-Coke process): a catalytic hydroconversion process operating at relatively moderate pressures and temperatures using catalyst particles, containing a metal sulfide in a carbonaceous matrix formed within the process, that are uniformly dispersed throughout the feed; because of their ultra-small size (10^{-4} in diameter), there are typically several orders of magnitude more of these microcatalyst particles per cubic centimeter of oil than is possible in other types of hydroconversion reactors using conventional catalyst particles.

microcrystalline wax: wax extracted from certain petroleum residua and having a finer and less apparent crystalline structure than paraffin wax.

microemulsion, or micellar/emulsion, flooding: an augmented waterflooding technique in which a surfactant system is injected in order to enhance oil displacement toward producing wells.

mid-boiling point: the temperature at which approximately 50% of a material has distilled under specific conditions.

middle distillate: distillate boiling between the kerosene and lubricating oil fractions; a general classification of refined petroleum products that includes distillate fuel oil and kerosene.

migration (primary): the movement of hydrocarbons (oil and natural gas) from mature, organic-rich source rocks to a point where the oil and gas can collect as droplets or as a continuous phase of liquid hydrocarbon.

migration (secondary): the movement of the hydrocarbons as a single, continuous fluid phase through water-saturated rocks, fractures, or faults followed by accumulation of the oil and gas in sediments (traps, q.v.) from which further migration is prevented.

mineral oil: the older term for petroleum; the term was introduced in the 19th century as a means of differentiating petroleum (rock oil) from whale oil which, at the time, was the predominant illuminant for oil lamps.

minerals: naturally occurring inorganic solids with well-defined crystalline structures.

mineral seal oil: a distillate fraction boiling between kerosene and gas oil.

mineral wax: yellow to dark brown, solid substances that occur naturally and are composed largely of paraffins; usually found associated with considerable mineral matter, as a filling in veins and fissures or as an interstitial material in porous rocks.

miscible fluid displacement (miscible displacement): an oil displacement process in which an alcohol, a refined hydrocarbon, a condensed petroleum gas, carbon dioxide, liquefied natural gas, or even exhaust gas is injected into an oil reservoir, at pressure levels such that the injected gas or fluid and reservoir oil are miscible; the process may include the concurrent, alternating, or subsequent injection of water.

mitigation: identification, evaluation, and cessation of potential impacts of a process product or by-product.

mixed-phase cracking: the thermal decomposition of higher-boiling hydrocarbons to gasoline components.

modified naphtha insolubles (MNI): an insoluble fraction obtained by adding naphtha to petroleum; usually the naphtha is modified by adding paraffinic constituents; the fraction might be equated to asphaltenes *if* the naphtha is equivalent to *n*-heptane, but usually it is not.

molecular sieve: a synthetic zeolite mineral having pores of uniform size; it is capable of separating molecules, on the basis of their size, structure, or both, by absorption or sieving.

molecular weight: the mass of one molecule.

mono-ethanolamine (monoethanolamine, MEA): a solvent used in an acid gas removal system.

motor gasoline (finished): a complex mixture of relatively volatile hydrocarbons with or without small quantities of additives, blended to form a fuel suitable for use in spark-ignition engines. Motor gasoline, as defined in ASTM Specification D 4814 or Federal Specification VV-G-1690C, is characterized as having a boiling range of 122–158°F at the 10% v/v recovery point to 365–374°F at the 90% v/v recovery point; includes conventional gasoline, all types of oxygenated gasoline including gasohol, and reformulated gasoline, but excludes aviation gasoline; the volumetric data on blending components, such as oxygenates, are not counted in data on finished motor gasoline until the blending components are blended into the gasoline.

motor gasoline blending: mechanical mixing of motor gasoline blending components, and oxygenates when required, to produce finished motor gasoline. Finished motor gasoline may be further mixed with other motor gasoline blending components or oxygenates, resulting in increased volumes of finished motor gasoline and/or changes in the formulation of finished motor gasoline (e.g., conventional motor gasoline mixed with MTBE to produce oxygenated motor gasoline).

motor gasoline blending components: naphtha fractions (e.g., straight-run gasoline, alkylate, reformate, benzene, toluene, xylene) used for blending or compounding into finished motor gasoline. These components include reformulated gasoline blend stock for oxygenate blending (RBOB) but exclude oxygenates (alcohols, ethers), butane, and pentanes plus. *Note:* Oxygenates are reported as individual components and are included in the total for other hydrocarbons, hydrogens, and oxygenates.

motor octane method: a test for determining the knock rating of fuels for use in spark-ignition engines; see also **research octane method**.

moving-bed catalytic cracking: a cracking process in which the catalyst is continuously cycled between the reactor and the regenerator.

MRH process: a hydrocracking process to upgrade heavy feedstocks containing large amounts of metals and asphaltene, such as vacuum residua and bitumen, and to produce mainly middle distillates using a reactor designed to maintain a mixed three-phase slurry of feedstock, fine powder catalyst and hydrogen, and to promote effective contact.

MSCC process: a short-residence-time process (millisecond catalytic cracking) in which the catalyst is placed in a more optimal position to ensure better contact with the feedstock.

MTBE [methyl tertiary butyl ether, $(CH_3)_3COCH_3$]: an ether intended for gasoline blending; see **methyl *t*-butyl ether**, **oxygenates**.

naft: pre-Christian era (Greek) term for naphtha (q.v.).

napalm: a thickened gasoline used as an incendiary medium that adheres to the surface it strikes.

naphtha: a generic term applied to refined, partly refined, or unrefined petroleum products and liquid products of natural gas, the majority of which distills below 240°C (464°F); the volatile fraction of petroleum that is used as a solvent or as a precursor to gasoline.

naphtha-type jet fuel: a fuel in the heavy naphtha boiling range having an average gravity of 52.8° API, 20% to 90% distillation temperatures of 290–470°F, and meeting Military Specification MIL-T-5624L (Grade JP-4); primarily used for military turbojet and turboprop aircraft engines because it has a lower freeze point than other aviation fuels and meets engine requirements at high altitudes and speeds.

naphthenes: cycloparaffins; one of three basic hydrocarbon classifications found naturally in crude oil; used widely as petrochemical feedstock.

native asphalt: see **bitumen**.

natural asphalt: see **bitumen**.

natural gas: the naturally occurring gaseous constituents that are found in many petroleum reservoirs; also there are also those reservoirs in which natural gas may be the sole occupant.

natural gas liquids (NGL): the hydrocarbon liquids that condense during the processing of hydrocarbon gases that are produced from oil or gas reservoir; see also **natural gasoline**.

natural gasoline: a mixture of liquid hydrocarbons extracted from natural gas (q.v.) suitable for blending with refinery gasoline.

natural gasoline plant: a plant for the extraction of fluid hydrocarbon, such as gasoline and liquefied petroleum gas, from natural gas.

neutralization: a process for reducing the acidity or alkalinity of a waste stream by mixing acids and bases to produce a neutral solution; also known as pH adjustment.

neutral oil: a distillate lubricating oil with viscosity usually not above 200 s at 100°F.

neutralization number: the weight, in milligrams, of potassium hydroxide needed to neutralize the acid in 1 g of oil; an indication of the acidity of an oil.

nonasphaltic road oil: any of the nonhardening petroleum distillates or residual oils used as dust layers. They have sufficiently low viscosity to be applied without heating and, together with asphaltic road oils (q.v.), are sometimes referred to as dust palliatives.

non-Newtonian: a fluid that exhibits a change of viscosity with flow rate.

No. 1 fuel oil: very similar to kerosene (q.v.) and is used in burners where vaporization before burning is usually required and a clean flame is specified.

No. 2 diesel fuel: a distillate fuel oil that has a distillation temperature of 640°F at the 90% recovery point and meets the specifications defined in ASTM Specification D 975. It is used in high-speed diesel engines that are generally operated under uniform speed and load conditions, such as those in railroad locomotives, trucks, and automobiles.

No. 2 fuel oil: also called domestic heating oil; has properties similar to diesel fuel and heavy jet fuel; used in burners where complete vaporization is not required before burning.

No. 4 fuel oil: a light industrial heating oil and is used where preheating is not required for handling or burning; there are two grades of No. 4 fuel oil, differing in safety (flash point) and flow (viscosity) properties.

No. 5 fuel oil: a heavy industrial fuel oil that requires preheating before burning.

No. 6 fuel oil: a heavy fuel oil and is more commonly known as Bunker C oil when it is used to fuel ocean-going vessels; preheating is always required for burning this oil.

OCR: countercurrent moving bed technology with on-stream catalyst replacement.

octane barrel yield: a measure used to evaluate fluid catalytic cracking processes; defined as (RON + MON)/2 times the gasoline yield, where RON is the research octane number and MON is the motor octane number.

octane number: a number indicating the antiknock characteristics of gasoline.

octane rating: a number used to indicate gasoline's antiknock performance in motor vehicle engines. The two recognized laboratory engine test methods for determining the antiknock rating, i.e., octane rating, of gasolines are the research method and the motor method. To provide a single number as guidance to the consumer, the antiknock index (R + M)/2, which is the average of the research and motor octane numbers, was developed; see **octane number**.

oil bank: see **bank**.

oil originally in place (OOIP): the quantity of petroleum existing in a reservoir before oil recovery operations begin.

OOIP: see **oil originally in place**.

oils: that portion of the maltenes (q.v.) that is not adsorbed by a surface-active material such as clay or alumina.

oil sand: see **tar sand**.

oil shale: a fine-grained impervious sedimentary rock that contains an organic material called kerogen; the term *oil shale* describes the rock in lithological terms but also refers to the ability of the rock to yield oil upon heating, which causes the kerogen to decompose; also called black shale, bituminous shale, carbonaceous shale, coaly shale, kerosene shale, coorongite, maharahu, kukersite, kerogen shale, and algal shale.

olefins: a class of unsaturated double-bond linear hydrocarbons recovered from petroleum; examples include ethylene, propylene, and butene. Olefins are used to produce a variety of products, including plastics, fibers, and rubber.

organic sedimentary rocks: rocks containing organic material such as residues of plant and animal remains/decay.

overhead: that portion of the feedstock that is vaporized and removed during distillation.

oxidation: a process that can be used for the treatment of a variety of inorganic and organic substances.

oxidized asphalt: see **air-blown asphalt**.

ozokerite (ozocerite): a naturally occurring wax; when refined, also known as ceresin.

oxygenate: an oxygen-containing compound that is blended into gasoline to improve its octane number and to decrease gaseous emissions.

oxygenated gasoline: finished motor gasoline, other than reformulated gasoline, having an oxygen content of 2.7% w/w or higher and required by the US Environmental Protection Agency (EPA) to be sold in areas designated by EPA as carbon monoxide (CO) nonattainment areas.

Oxygenated gasoline excludes oxygenated fuels program reformulated gasoline (OPRG) and reformulated gasoline blendstock for oxygenate blending (RBOB). Data on gasohol that has at least 2.7% w/w oxygen and is intended for sale inside CO nonattainment areas are included in data on oxygenated gasoline. Other data on gasohol (for use outside of non-attainment areas) are included in data on conventional gasoline.

oxygenates: substances that, when added to gasoline, increase the amount of oxygen in that gasoline blend. Fuel ethanol, methyl tertiary butyl ether (MTBE), ethyl tertiary butyl ether (ETBE), and methanol are common oxygenates.

PADD (petroleum administration for defense districts): geographic aggregations of the 50 states and the District of Columbia into five districts by the Petroleum Administration for Defense in 1950. These districts were originally defined during World War II for purposes of administering oil allocation.

pale oil: lubricating oil or process oil refined until its color, by transmitted light, is straw to pale yellow.

paraffins: a group of generally saturated single-bond linear hydrocarbons; also celled alkanes.

paraffinum liquidum: see **liquid petrolatum**.

paraffin wax: the colorless, translucent, highly crystalline material obtained from the light lubricating fractions of paraffinic crude oils (wax distillates).

partial oxidation process (Texaco gasification process): a partial oxidation gasification process for generating synthetic gas, principally hydrogen and carbon monoxide.

particle density: the density of solid particles.

particulate matter: particles in the atmosphere or on a gas stream that may be organic or inorganic and originate from a wide variety of sources and processes.

particle size distribution: the particle size distribution (of a catalyst sample) expressed as a percent of the whole.

Penex process: a continuous, nonregenerative process for isomerization of C_5 and/or C_6 fractions in the presence of hydrogen (from reforming) and a platinum catalyst.

pentafining: a pentane isomerization process using a regenerable platinum catalyst on a silica–alumina support and requiring outside hydrogen.

pentanes plus: a mixture of liquid hydrocarbons, mostly pentanes and heavier, extracted from natural gas in a gas processing plant. Pentanes plus is equivalent to natural gasoline.

pepper sludge: the fine particles of sludge produced in acid treating that may remain in suspension.

pericondensed aromatic compounds: compounds based on angular condensed aromatic hydrocarbon systems, e.g., phenanthrene, chrysene, or picene.

permeability: the ease of flow of the water through the rock.

petrochemical: an intermediate chemical derived from petroleum, hydrocarbon liquids, or natural gas.

petrol: a term commonly used in some countries for gasoline.

petrolatum: a semisolid product, ranging from white to yellow in color, produced during refining of residual stocks; see **petroleum jelly**.

petrolenes: the term applied to that part of the pentane-soluble or heptane-soluble material that is low boiling (<300°C, <570°F, 760 mm) and can be distilled without thermal decomposition (see also **maltenes**).

petroleum (crude oil): a naturally occurring mixture of gaseous, liquid, and solid hydrocarbon compounds usually found trapped deep underground beneath impermeable cap rock and above a lower dome of sedimentary rock such as shale; most petroleum reservoirs occur in sedimentary rocks of marine, deltaic, or estuarine origin.

petroleum asphalt: see **asphalt**.

petroleum coke: a solid carbon fuel derived from oil refinery cracking processes such as delayed coking; also called pet coke.

petroleum ether: see **ligroine**.

petroleum jelly: a translucent, yellowish to amber or white, hydrocarbon substance (m.p. 38–54°C) having almost no odor or taste, derived from petroleum and used principally in medicine and pharmacy as a protective dressing and as a substitute for fats in ointments and cosmetics; also used in many types of polishes and in lubricating greases, rust preventives, and modeling clay; obtained by dewaxing heavy lubricating oil stocks.

petroleum products: products obtained from the processing of crude oil (including lease condensate), natural gas, and other hydrocarbon compounds. Petroleum products include unfinished oils, liquefied petroleum gases, pentanes plus, aviation gasoline, motor gasoline, naphtha-type jet fuel, kerosene-type jet fuel, kerosene, distillate fuel oil, residual fuel oil, petrochemical feedstocks, special naphtha, lubricants, waxes, petroleum coke, asphalt, road oil, still gas, and miscellaneous products.

petroleum refinery: see **refinery**.

petroleum refining: a complex sequence of events that result in the production of a variety of products.

petroporphyrins: see **porphyrins**.

phase separation: the formation of a separate phase that is usually the prelude to coke formation during a thermal process; the formation of a separate phase as a result of the instability/incompatibility of petroleum and petroleum products.

pH adjustment: neutralization.

phosphoric acid polymerization: a process using a phosphoric acid catalyst to convert propene, butene, or both, to gasoline or petrochemical polymers.

physical composition: see **bulk composition**.

pipe still (pipestill): a still in which heat is applied to the oil while being pumped through a coil or pipe arranged in a suitable firebox; the distillation tower in a refinery.

pipestill gas: the most volatile fraction that contains most of the gases that are generally dissolved in the crude. Also known as pipestill light ends.

pitch: the nonvolatile, brown to black, semisolid to solid viscous product from the destructive distillation (q.v.) of many bituminous or other organic materials, especially coal; has also been incorrectly applied to residua from petroleum processes where thermal decomposition may *not* have occurred.

platforming: a reforming process using a platinum-containing catalyst on an alumina base.

PNA: a polynuclear aromatic compound (q.v.).

PONA analysis: a method of analysis for paraffins (P), olefins (O), naphthenes (N), and aromatics (A).

polar aromatics: resins; the constituents of petroleum that are predominantly aromatic in character and contain polar (nitrogen, oxygen, and sulfur) functions in their molecular structure(s).

pollution: the introduction into the land water and air systems of a chemical or chemicals that are not indigenous to these systems or the introduction into the land water and air systems of indigenous chemicals in greater-than-natural amounts.

polyforming: a process charging both C_3 and C_4 gases with naphtha or gas oil under thermal conditions to produce gasoline.

polymer augmented waterflooding: waterflooding in which organic polymers are injected with the water to improve areal and vertical sweep efficiency.

polymer gasoline: the product of polymerization of gaseous hydrocarbons to hydrocarbons boiling in the gasoline range.

polymerization: the combination of two olefin molecules to form a higher molecular weight paraffin.

polynuclear aromatic compound: an aromatic compound having two or more fused benzene rings, e.g., naphthalene and phenanthrene.

polysulfide treating: a chemical treatment used to remove elemental sulfur from refinery liquids by contacting them with a nonregenerable solution of sodium polysulfide.

pore diameter: the average pore size of a solid material, e.g., catalyst.

pore space: a small hole in reservoir rock that contains fluid or fluids; a 4-in cube of reservoir rock may contain millions of interconnected pore spaces.

pore volume: total volume of all pores and fractures in a reservoir or part of a reservoir; also applied to catalyst samples.

porosity: the percentage of rock volume available to contain water or other fluid.

porphyrins: organometallic constituents of petroleum that contain vanadium or nickel; the degradation products of chlorophyll derivatives that became included in the protopetroleum.

possible reserves: reserves where there is an even greater degree of uncertainty but about which there is some information.

potential reserves: reserves based on geological information about the types of sediments where such resources are likely to occur, and they are considered to represent an educated guess.

pour point: the lowest temperature at which oil will pour or flow when it is chilled without disturbance under definite conditions.

powerforming: a fixed-bed naphtha-reforming process using a regenerable platinum catalyst.

precipitation number: the number of milliliters of precipitate formed when 10 mL of lubricating oil is mixed with 90 mL of petroleum naphtha of a definite quality and centrifuged under definitely prescribed conditions.

pressure swing adsorption: a method for purifying gas (used in hydrogen production); abbreviated as PSA.

pressure vessel: a container designed to hold gases or liquids at a pressure different from the ambient pressure.

primary oil recovery: oil recovery utilizing only naturally occurring forces.

primary production: the first stage of hydrocarbon production in which natural reservoir energy (such as gas drive, water drive, and gravity drainage) displaces hydrocarbons from the reservoir into the wellbore and up to surface. Primary production uses an artificial lift system to reduce the bottomhole pressure or increase the differential pressure to sustain hydrocarbon recovery since reservoir pressure decreases with production.

primary structure: the chemical sequence of atoms in a molecule.

processability: an estimate of the manner and relative ease with which a feedstock can be processed; generally measured by one or more criteria.

process gas: gas produced from the upgrading process that is not distilled as a liquid; typically used as a refinery fuel.

probable reserves: mineral reserves that are nearly certain but about which a slight doubt exists.

propane (C_3H_8): a straight-chain saturated (paraffinic) hydrocarbon extracted from natural gas or refinery gas streams, which is gaseous at standard temperature and pressure; a colorless gas that boils at a temperature of $-42°C$ ($-44°F$) and includes all products designated in ASTM D1835 and Gas Processors Association specifications for commercial (HD-5) propane.

propane asphalt: see **solvent asphalt**.

propane deasphalting: solvent deasphalting using propane as the solvent.

propane decarbonizing: a solvent extraction process used to recover catalytic cracking feed from heavy fuel residues.

propane dewaxing: a process for dewaxing lubricating oils in which propane serves as solvent.

propane fractionation: a continuous extraction process employing liquid propane as the solvent; a variant of propane deasphalting (q.v.).

propylene (C_3H_6): an olefin hydrocarbon recovered from refinery or petrochemical processes, which is gaseous at standard temperature and pressure; an important petrochemical feedstock.

protopetroleum: a generic term used to indicate the initial product formed as a result of chemical and physical changes that have occurred to the precursors of petroleum; a *paleobotanical soup*.

proved reserves: mineral reserves that have been positively identified as recoverable with current technology.

pyrobitumen: see **asphaltoid**.

pyrolysis: exposure of a feedstock to high temperatures in an oxygen-poor environment.

pyrophoric: substances that catch fire spontaneously in air without an ignition source.

quadrillion: 1×10^{15}.

quench: the sudden cooling of hot material discharging from a thermal reactor.

R2R process: a fluid catalytic cracking process (q.v.) for conversion of heavy feedstocks.

raffinate: that portion of the oil that remains undissolved in a solvent refining process.

Ramsbottom carbon residue: see **carbon residue**.

raw materials: minerals extracted from the earth before any refining or treating.

RCC process: a process for the conversion of heavy feedstocks in the riser pipe (q.v.); resid catalytic cracking process.

RCD Unibon (BOC) process: a process to upgrade vacuum residua using hydrogen.

RDS: a residuum desulfurization (hydrotreating) process.

reactor: a vessel in which a reaction occurs during processing; usually defined by the nature of the catalyst bed (e.g., fixed-bed or fluid-bed reactor) and by the direction of the flow of feedstock (e.g., upflow or downflow).

recycle ratio: τ, defined as the ratio of the recycled feedstock to the fresh feedstock:

$$\tau = F_R/F_F$$

where F_F is the fresh feedstock and F_R is the recycled feedstock; may also be expressed as a percentage.

recycle stock: the portion of a feedstock that has passed through a refining process and is recirculated through the process.

recycling: the use or reuse of chemical waste as an effective substitute for a commercial products or as an ingredient or feedstock in an industrial process.

reduced crude: a residual product remaining after the removal, by distillation or other means, of an appreciable quantity of the more volatile components of crude oil.

refinery: a series of integrated unit processes by which petroleum can be converted to a slate of useful (salable) products.

refinery gas: a gas (or a gaseous mixture) produced as a result of refining operations.

refining: the process(es) by which petroleum is distilled and/or converted by application of a physical and chemical processes to form a variety of products are generated.

reformate: the liquid product of a reforming process.

reformed gasoline: gasoline made by a reforming process.

reformer: a refinery process unit that uses heat and pressure in the presence of a catalyst to convert naphtha feedstock into higher-octane gasoline blending stocks; also used for hydrogen production.

reforming: the conversion of hydrocarbons with low octane numbers (q.v.) into hydrocarbons having higher octane numbers; for example, the conversion of a *n*-paraffin into an isoparaffin.

reformulated gasoline (RFG): gasoline designed to mitigate smog production and to improve air quality by limiting the emission levels of certain chemical compounds such as benzene and other aromatic derivatives; often contains oxygenates (q.v.).

Reid vapor pressure: a measure of the volatility of liquid fuels, especially gasoline.

regeneration: the reactivation of a catalyst by burning off the coke deposits.

regenerator: a reactor for catalyst reactivation.

renewable energy sources: solar, wind, and other nonfossil fuel energy sources.

rerunning: the distillation of an oil that has already been distilled.

research octane method: a test for determining the knock rating, in terms octane numbers, of fuels for use in spark-ignition engines; see also **motor octane method**.

reserves: well-identified resources that can be profitably extracted and utilized with existing technology.

reservoir: a domain where a pollutant may reside for an indeterminate time.

resid: the heaviest boiling fraction remaining after initial processing (distillation) of crude oil; see **residuum**.

Residfining process: a catalytic fixed-bed process for the desulfurization and demetallization of heavy feedstocks that can also be used to pretreat the feedstocks to suitably low contaminant levels before catalytic cracking.

residual asphalt: see **straight-run asphalt**.

residual fluid catalytic cracking: a refinery process used to convert residuals (heavy hydrocarbonaceous materials) into lighter components, including liquefied petroleum gas and gasoline; abbreviated as RFCC.

residual fuel oil: obtained by blending the residual product(s) from various refining processes with suitable diluent(s) (usually middle distillates) to obtain the required fuel oil grades; a general classification for the heavier oils, known as No. 5 and No. 6 fuel oils, that remain after the distillate fuel oils and lighter hydrocarbons are distilled away in refinery operations. It conforms to ASTM Specifications D396 and D975 and Federal Specification VV-F-815C. No. 5, a residual fuel oil of medium viscosity, is also known as Navy Special and is defined in Military Specification MIL-F-859E, including Amendment 2 (NATO Symbol F-770). It is used in steam-powered vessels in government service and inshore power plants. No. 6 fuel oil includes Bunker C fuel oil and is used for the production of electric power, space heating, vessel bunkering, and various industrial purposes.

residual oil: see **residuum**.

residuals: heavy fuel oils produced from the nonvolatile residue from the fractional distillation process; also called resids.

residue hydroconversion (RHC) process: a high-pressure, fixed-bed, trickle-flow hydrocatalytic process for converting heavy feedstocks.

residuum (resid; pl. residua): the residue obtained from petroleum after nondestructive distillation (q.v.) has removed all the volatile materials from crude oil, e.g., an atmospheric (345°C, 650°F+) residuum.

resins: that portion of the maltenes (q.v.) that is adsorbed by a surface-active material such as clay or alumina; the fraction of deasphaltened oil that is insoluble in liquid propane but soluble in *n*-heptane.

resource: the total amount of a commodity (usually a mineral but can include nonminerals such as water and petroleum) that has been estimated to be ultimately available.

rexforming: a process combining platforming (q.v.) with aromatics extraction, wherein low octane raffinate is recycled to the platformer.

rich oil: absorption oil containing dissolved natural gasoline fractions.

riser: the part of the bubble-plate assembly that channels the vapor and causes it to flow downward to escape through the liquid; also the vertical pipe where fluid catalytic cracking reactions occur.

riser pipe: the pipe in a fluid catalytic cracking process (q.v.) where catalyst and feedstock are lifted into the reactor; the pipe in which most of the reaction takes place or is initiated.

rock asphalt: bitumen that occurs in formations that have a limiting ratio of bitumen-to-rock matrix.

ROSE process: a solvent deasphalting process (q.v.) that uses supercritical solvent recovery system to obtain high-quality oils from heavy feedstocks (q.v.) for further processing.

run-of-the-river reservoirs: reservoirs with a large rate of flow-through compared with their volume.

S&W fluid catalytic cracking process: a process in which the heavy feedstock (q.v.) is injected into a stabilized, upward-flowing catalyst stream whereupon the feedstock-stream–catalyst mixture travels up the riser pipe (q.v.) and is separated by a high-efficiency inertial separator from which the product vapor goes overhead to fractionation.

SAGD (steam-assisted gravity drainage): an *in situ* production process using two closely spaced parallel horizontal wells: one for steam injection and the other for production of the tar sand bitumen/water emulsion.

sand: a course granular mineral mainly comprising quartz grains that is derived from the chemical and physical weathering of rocks rich in quartz, notably sandstone and granite.

sandstone: a sedimentary rock formed by compaction and cementation of sand grains; can be classified according to the mineral composition of the sand and cement.

SARA separation: a method of fractionation by which petroleum is separated into saturates, aromatics, resins, and asphaltene fractions.

saturates: paraffins and cycloparaffins (naphthenes).

Saybolt–Furol viscosity: the time, in seconds (Saybolt–Furol seconds, SFS), for 60 mL of fluid to flow through a capillary tube in a Saybolt–Furol viscometer at specified temperatures between 70°F and 210°F; the method is appropriate for high-viscosity oils such as transmission, gear, and heavy fuel oils.

Saybolt Universal viscosity: the time, in seconds (Saybolt Universal seconds, SUS), for 60 mL of fluid to flow through a capillary tube in a Saybolt Universal viscometer at a given temperature.

scale wax: the paraffin derived by removing the greater part of the oil from slack wax by sweating or solvent deoiling.

scrubbing: purifying a gas by washing with water or chemical; less frequently, the removal of entrained materials.

secondary structure: the ordering of the atoms of a molecule in space relative to each other.

sediment: an insoluble solid formed as a result of the storage instability and/or the thermal instability of petroleum and petroleum products.

sedimentary strata: typically consist of mixtures of clay, silt, sand, organic matter, and various minerals; formed by or from deposits of sediments, especially from sand grains or silts transported from their source and deposited in water, such as sandstone and shale; or from calcareous remains of organisms, such as limestone.

selective solvent: a solvent that, at certain temperatures and ratios, will preferentially dissolve more of one component of a mixture than of another and thereby permit partial separation.

separation process: an upgrading process in which the constituents of petroleum are separated, usually without thermal decomposition, e.g., distillation and deasphalting.

separator-nobel dewaxing: a solvent (trichloroethylene) dewaxing process.

shale oil: the distillate produced by the thermal decomposition of oil shale kerogen; sometimes incorrectly used as the name for crude oil produced from tight formation; see **tight oil**.

Shell fluid catalytic cracking: a two-stage fluid catalytic cracking process in which the catalyst is regenerated.

Shell gasification (partial oxidation) process: a process for generating synthesis gas (hydrogen and carbon monoxide) for the ultimate production of high-purity, high-pressure hydrogen, ammonia, methanol, fuel gas, town gas, or reducing gas by reaction of gaseous or liquid hydrocarbons with oxygen, air, or oxygen-enriched air.

Shell residual oil hydrodesulfurization process: a downflow fixed-bed reactor (q.v.) process to improve the quality of heavy feedstocks by removing sulfur, metals, and asphaltenes as well as bringing about a reduction in the viscosity of the feedstock.

Shell still: a still formerly used in which the oil was charged into a closed, cylindrical shell and the heat required for distillation was applied to the outside of the bottom from a firebox.

sidestream: a liquid stream taken from any one of the intermediate plates of a bubble tower.

sidestream stripper: a device used to perform further distillation on a liquid stream from any one of the plates of a bubble tower, usually by the use of steam.

slack wax: the soft, oily crude wax obtained from the pressing of paraffin distillate or wax distillate.

slime: a name used for petroleum in ancient texts, particularly in Biblical texts.

sludge: a semisolid to solid product that results from the storage instability and/or the thermal instability of petroleum and petroleum products.

slurry hydroconversion process: a process in which the feedstock is contacted with hydrogen under pressure in the presence of a catalytic coke-inhibiting additive.

slurry phase reactors: tanks into which wastes, nutrients, and microorganisms are placed.

smoke point: a measure of the burning cleanliness of jet fuel and kerosene.

sodium hydroxide treatment: see **caustic wash**.

sodium plumbite: a solution prepared from a mixture of sodium hydroxide, lead oxide, and distilled water; used in making the doctor test for light oils such as gasoline and kerosene.

solubility parameter: a measure of the solvent power and polarity of a solvent.

solutizer-steam regenerative process: a chemical treating process for extracting mercaptans from gasoline or naphtha, using solutizers (potassium isobutyrate or potassium alkyl phenolate) in strong potassium hydroxide solution.

Solvahl process: a solvent deasphalting process (q.v.) for application to vacuum residua (q.v.).

solvent asphalt: the asphalt (q.v.) produced by solvent extraction of residua (q.v.) or by light hydrocarbon (propane) treatment of a residuum (q.v.) or an asphaltic crude oil.

solvent deasphalter bottoms: the insoluble material that separates from the solvent phase during deasphalting (q.v.).

solvent deasphalting: a process for removing asphaltic and resinous materials from reduced crude oils, lubricating oil stocks, gas oils, or middle distillates through the extraction or precipitant action of low molecular weight hydrocarbon solvents; see also **propane deasphalting**.

solvent decarbonizing: see **propane decarbonizing**.

solvent deresining: see **solvent deasphalting**.

solvent dewaxing: a process for removing wax from oils by means of solvents usually by chilling a mixture of solvent and waxy oil, filtration or by centrifuging the wax that precipitates, and solvent recovery.

solvent extraction: a process for separating liquids by mixing the stream with a solvent that is immiscible with part of the waste but that will extract certain components of the waste stream.

solvent naphtha: a refined naphtha of restricted boiling range used as a solvent; also called petroleum naphtha; petroleum spirits.

solvent refining: see **solvent extraction**.

sonic log: a well log based on the time required for sound to travel through rock, useful in determining porosity.

sour crude oil: crude oil containing an abnormally large amount of sulfur compounds; see also **sweet crude oil**—crude oil containing free sulfur, hydrogen sulfide, or other sulfur compounds.

spontaneous ignition: ignition of a fuel, such as coal, under normal atmospheric conditions; usually induced by climatic conditions.

specific gravity: the mass (or weight) of a unit volume of any substance at a specified temperature compared with the mass of an equal volume of pure water at a standard temperature; see also **density**.

specific heat: the quantity of heat required to raise a unit mass of material through one degree of temperature.

spent catalyst: catalyst that has lost much of its activity due to the deposition of coke and metals.

stabilization: the removal of volatile constituents from a higher-boiling fraction or product (stripping); the production of a product that, to all intents and purposes, does not undergo any further reaction when exposed to the air.

stabilizer: a fractionating tower for removing light hydrocarbons from an oil to reduce vapor pressure particularly applied to gasoline.

stack gas: anything that comes out of a burner stack in gaseous form, usually consisting of mostly nitrogen and carbon dioxide; sometime called flue gas.

standpipe: the pipe by which catalyst is conveyed between the reactor and the regenerator.

steam assisted gravity drainage (SAGD): an *in situ* production process using two closely spaced parallel horizontal wells: one for steam injection and the other for production of the tar sand bitumen/water emulsion.

steam cracking: a conversion process in which the feedstock is treated with superheated steam; a petrochemical process sometimes used in refineries to produce olefins (e.g., ethylene) from various feedstock for petrochemicals manufacture—the feedstock range from ethane to vacuum gas oil, with heavier feeds giving higher yields of by-products such as naphtha. The most common feedstocks are ethane, butane, and naphtha, and the process is carried out at temperatures of 815–870°C (1500–1600°F), and at pressures slightly above atmospheric pressure. Naphtha produced from steam cracking contains benzene, which is extracted before hydrotreating, and high-boiling products (residua) from steam cracking are sometimes used as blend stock for heavy fuel oil.

steam distillation: distillation in which vaporization of the volatile constituents is effected at a lower temperature by introduction of steam (open steam) directly into the charge.

steam drive injection (steam injection): the continuous injection of steam into one set of wells (injection wells) or other injection source to effect oil displacement toward and production from a second set of wells (production wells); steam stimulation of production wells is *direct steam stimulation* whereas steam drive by steam injection to increase production from other wells is *indirect steam stimulation*.

steam–methane reforming: a continuous catalytic process for hydrogen production; an endothermic process of producing hydrogen from hydrocarbons using steam and a metal-based catalyst; abbreviated as SMR.

steam–naphtha reforming: a process that is essentially similar in nature to the steam–methane reforming process (q.v.) but that uses higher molecular weight hydrocarbons as the feedstock.

still gas: any form or mixture of gases produced in refineries by distillation, cracking, reforming, and other processes. The principal constituents are methane and ethane—may contain hydrogen and small/trace amounts of other gases; typically used as refinery fuel or used as petrochemical feedstock.

storage stability (or storage instability): the ability (inability) of a liquid to remain in storage over extended periods of time without appreciable deterioration as measured by gum formation and the depositions of insoluble material (sediment).

straight-run asphalt: the asphalt (q.v.) produced by the distillation of asphaltic crude oil.

straight-run products: obtained from a distillation unit and used without further treatment.

strata: layers including the solid iron-rich inner core, molten outer core, mantle, and crust of the earth.

straw oil: pale paraffin oil of straw color used for many process applications.

stripping: a means of separating volatile components from less volatile ones in a liquid mixture by the partitioning of the more volatile materials to a gas phase of air or steam (q.v. stabilization).

sulfonic acids: acids obtained by the reaction of petroleum or a petroleum product with strong sulfuric acid.

sulfur: a yellowish nonmetallic element, sometimes known by the Biblical name of *brimstone*; present at various levels of concentration in many fossil fuels whose combustion releases sulfur compounds that are considered harmful to the environment. Some of the most commonly used fossil fuels are categorized according to their sulfur content, with lower sulfur fuels usually selling at a higher price.

sulfuric acid alkylation: an alkylation process in which olefins (C_3, C_4, and C_5) combine with isobutane in the presence of a catalyst (sulfuric acid) to form branched-chain hydrocarbons used especially in gasoline blending stock.

sulfur recovery unit (SRU): a refinery process unit used to convert hydrogen sulfide to elemental sulfur using the Claus process.

suspensoid catalytic cracking: a nonregenerative cracking process in which cracking stock is mixed with slurry of catalyst (usually clay) and cycle oil and passed through the coils of a heater.

sweated wax: crude wax freed from oil by having been passed through a sweater.

sweating: the separation of paraffin oil and low-melting wax from paraffin wax.

sweet crude oil: crude oil containing little sulfur; see also **sour crude oil**.

sweetening: the process by which petroleum products are improved in odor and color by oxidizing or removing the sulfur-containing and unsaturated compounds.

synthesis gas (syngas): a mixture of carbon monoxide and hydrogen; used as the feedstock for the Fischer–Tropsch process to produce hydrocarbons with alcohols as by-products.

synthesis gas generation process: a noncatalytic process for producing synthesis gas (hydrogen and carbon monoxide) from gaseous or liquid hydrocarbons for the ultimate production of high-purity hydrogen.

synthetic crude oil (syncrude): a hydrocarbon product produced by the conversion of coal, oil shale, or tar sand bitumen that resembles conventional crude oil; can be refined in a petroleum refinery (q.v.).

tail gas: the lightest hydrocarbon gas released from a refining process.

tail gas treating unit (TGTU): a refinery process unit used to control emissions of sulfur compounds; generally integrated with a sulfur recovery unit.

tar: the volatile, brown to black, oily, viscous product from the destructive distillation (q.v.) of many bituminous or other organic materials, especially coal; a name used for petroleum in ancient texts.

tar sand: see **bituminous sand**.

tertiary structure: the three-dimensional structure of a molecule.

Tervahl H process: a process in which the feedstock and hydrogen are heated and held in a soak drum as in the Tervahl T process (q.v.).

Tervahl T process: a process analogous to delayed coking (q.v.) in which the feedstock is heated to the desired temperature using a coil heater and held for a specified residence time in a soaking drum; see also **Tervahl H process**.

Tetraethyl lead (TEL): an organic compound of lead, $Pb(CH_3)_4$, which, when added in small amounts, increases the antiknock quality of gasoline.

Texaco gasification process: see **partial oxidation (Texaco gasification) process**.

thermal coke: the carbonaceous residue formed as a result of a noncatalytic thermal process; the Conradson carbon residue; the Ramsbottom carbon residue.

thermal cracking: a process that decomposes, rearranges, or combines hydrocarbon molecules by the application of heat, without the aid of catalysts.

thermal polymerization: a thermal process to convert light hydrocarbon gases into liquid fuels.

thermal process: any refining process that utilizes heat, without the aid of a catalyst.

thermal recovery: any process by which heat energy is used to reduce the viscosity of heavy oil or tar sand bitumen *in situ* to facilitate recovery.

THAI process (toe-to-heel air injection process): an *in situ* combustion method for producing heavy oil and tar sand bitumen. In this technique, combustion starts from a vertical well, while the oil is produced from a horizontal well having its toe in close proximity to the vertical air-injection well. This production method is a modification of conventional fire flooding techniques in which the flame front from a vertical well pushes the oil to be produced from another vertical well.

thermal reforming: a process using heat (but no catalyst) to effect molecular rearrangement of low-octane naphtha into gasoline of higher antiknock quality.

thermal stability (thermal instability): the ability (inability) of a liquid to withstand relatively high temperatures for short periods of time without the formation of carbonaceous deposits (sediment or coke).

thermofor catalytic cracking: a continuous, moving-bed catalytic cracking process.

thermofor catalytic reforming: a reforming process in which the synthetic, bead-type catalyst of coprecipitated chromia (Cr_2O_3) and alumina (Al_2O_3) flows down through the reactor concurrent with the feedstock.

thermofor continuous percolation: a continuous clay treating process to stabilize and decolorize lubricants or waxes.

tight oil: crude oil produced from petroleum-bearing formations with low permeability, such as the Eagle Ford, the Bakken, and other formations that must be hydraulically fractured to produce oil at commercial rates; sometime incorrectly called *shale oil*, which is the distillate produced by the thermal decomposition of oil shale kerogen.

toe-to-heel air injection process (THAI process): an *in situ* combustion method for producing heavy oil and tar sand bitumen. In this technique, combustion starts from a vertical well, while the oil is produced from a horizontal well having its toe in close proximity to the vertical air-injection well. This production method is a modification of conventional fire flooding techniques in which the flame front from a vertical well pushes the oil to be produced from another vertical well.

toluene ($C_6H_5CH_3$): a colorless liquid of the aromatic group of petroleum hydrocarbons, made by the catalytic reforming of petroleum naphtha containing methyl cyclohexane; a high-octane, gasoline-blending agent, solvent, and chemical intermediate, and a base for TNT (explosive).

topped crude: petroleum that has had volatile constituents removed up to a certain temperature, e.g., 250°C+ (480°F+) topped crude; not always the same as a residuum (q.v.).

topping: the distillation of crude oil to remove light fractions only; differs from distillation in the manner in which the heat is applied.

tower: equipment for increasing the degree of separation obtained during the distillation of oil in a still.

trace element: those elements that occur at very low levels in a given system.

traps: sediments in which oil and gas accumulate from which further migration (q.v.) is prevented.

treatment: any method, technique, or process that changes the physical and/or chemical character of petroleum.

trickle hydrodesulfurization: a fixed-bed process for desulfurizing middle distillates.

trillion: 1×10^{12}.

true boiling point (true boiling range): the boiling point (boiling range) of a crude oil fraction or a crude oil product under standard conditions of temperature and pressure.

tube-and-tank cracking: an older liquid-phase thermal cracking process.

ultimate analysis: elemental composition.

ultrafining: a fixed-bed catalytic hydrogenation process to desulfurize naphtha and upgrade distillates by essentially removing sulfur, nitrogen, and other materials.

ultraforming: a low-pressure naphtha-reforming process employing on-stream regeneration of a platinum-on-alumina catalyst and producing high yields of hydrogen and high-octane-number reformate.

unassociated molecular weight: the molecular weight of asphaltenes in an nonassociating (polar) solvent, such as dichlorobenzene, pyridine, or nitrobenzene.

undiscovered reserves: those reserves of a resource that are yet to be discovered; often they are little more than figments of the imagination.

unfinished oils: all oils requiring further processing, except those requiring only mechanical blending. Unfinished oils are produced by partial refining of crude oil and include naphtha, kerosene and light gas oils, heavy gas oils, and residuum.

unicracking hydrodesulfurization (unicracking/HDS) process: a fixed-bed, catalytic process for the hydrotreating heavy feedstocks.

unifining: a fixed-bed catalytic process to desulfurize and hydrogenate refinery distillates.

unisol process: a chemical process for extracting mercaptan sulfur and certain nitrogen compounds from sour gasoline or distillates using regenerable aqueous solutions of sodium or potassium hydroxide containing methanol.

Universal viscosity: see **Saybolt Universal viscosity**.

unstable: usually refers to a petroleum product that has more volatile constituents present or refers to the presence of olefin and other unsaturated constituents.

UOP alkylation: a process using hydrofluoric acid (which can be regenerated) as a catalyst to unite olefins with isobutane.

UOP copper sweetening: a fixed-bed process for sweetening gasoline by converting mercaptans to disulfides by contact with ammonium chloride and copper sulfate in a bed.

UOP fluid catalytic cracking: a fluid process of using a reactor-over-regenerator design.

upflow reactor: a reactor in which the feedstock flows in an upward direction through the catalyst bed.

upgrading: the conversion of petroleum to value-added salable products; includes hydroprocessing, hydrocracking, fractionation, and any other catalytic or noncatalytic process that improves the value of the products. During upgrading, the products of the Fischer–Tropsch process are converted to diesel, jet fuel, naphtha, or bases for synthetic lubricants and wax.

upstream: a sector of the petroleum industry referring to the searching for, recovery, and production of crude oil and natural gas. Also known as the exploration and production sector.

urea dewaxing: a continuous dewaxing process for producing low-pour-point oils, and using urea that forms a solid complex (adduct) with the straight-chain wax paraffins in the stock; the complex is readily separated by filtration.

vacuum distillation: distillation (q.v.) under reduced pressure; distillation under reduced pressure (less the atmospheric), which lowers the boiling temperature of the liquid being distilled. This technique with its relatively low temperatures prevents cracking or decomposition of the charge stock.

vacuum gas oil: a product of vacuum distillation; a preferred feedstock for cracking units to produce gasoline; abbreviated as VGO.

vacuum residuum: a residuum (q.v.) obtained by distillation of a crude oil under vacuum (reduced pressure); that portion of petroleum that boils above a selected temperature, such as 510°C (950°F) or 565°C (1050°F).

VAPEX process (vapor extraction process): a nonthermal recovery method that involves injecting a gaseous hydrocarbon solvent into the reservoir where it dissolves into the sludge-like oil, which becomes less viscous (or more fluid) before draining into a lower horizontal well and being extracted.

vapor-phase cracking: a high-temperature, low-pressure conversion process.

vapor-phase hydrodesulfurization: a fixed-bed process for desulfurization and hydrogenation of naphtha.

vapor pressure osmometry (VPO): a method for determining molecular weight.

Veba–Combi cracking (VCC) process: a thermal hydrocracking/hydrogenation process for converting heavy feedstocks.

VGC (viscosity–gravity constant): an index of the chemical composition of crude oil defined by the general relation between specific gravity, sg, at 60° F and Saybolt Universal viscosity, SUV, at 100° F:

$$a = 10 \text{ sg} - 1.0752 \log (\text{SUV} - 38)/10 \text{ sg} - \log (\text{SUV} - 38)$$

The constant, a, is low for the paraffinic crude oils and high for the naphthenic crude oils.

VI: see **viscosity index**.

visbreaking: a (relatively) mild process for reducing the viscosity of heavy feedstocks by controlled thermal decomposition; a process designed to reduce residue viscosity by thermal means but without appreciable coke formation.

viscosity: a measure of the ability of a liquid to flow or a measure of its resistance to flow; the force required to move a plane surface of 1 m^2 area over another parallel plane surface 1 m away at a rate of 1 m/s when both surfaces are immersed in the fluid.

viscosity–gravity constant: see **VGC**.

viscosity index (VI): an arbitrary scale used to show the magnitude of viscosity changes in lubricating oils with changes in temperature.

volume flow: the combined (fresh and unconverted) feedstock that is fed to a reactor:

$$F_T = F_F + F_R$$

F_T is the total feedstock into the unit, F_F is the fresh feedstock, and F_R is the recycled feedstock.

VRDS: a vacuum resid desulfurization (hydrotreating) process.

Watson characterization factor: see **characterization factor**.

wax: a solid or semisolid material consisting of a mixture of hydrocarbons obtained or derived from petroleum fractions, or through a Fischer–Tropsch-type process, in which the straight-chain paraffin series predominates. This includes all marketable wax, whether crude or refined, with a congealing point (ASTM D 938) between 37°C and 95°C (100°F and 200°F) and a maximum oil content (ASTM D 3235) of 50% w/w; see also **mineral wax** and **paraffin wax**.

wax distillate: a neutral distillate containing a high percentage of crystallizable paraffin wax, obtained on the distillation of paraffin or mixed-base crude, and on reducing neutral lubricating stocks.

wax fractionation: a continuous process for producing waxes of low oil content from wax concentrates; see also **MEK deoiling**.

wax manufacturing: a process for producing oil-free waxes.

weathered crude oil: crude oil that, owing to natural causes during storage and handling, has lost an appreciable quantity of its more volatile components; also indicates uptake of oxygen.

West Texas Intermediate (WTI-Cushing): a crude oil produced in Texas and southern Oklahoma that serves as a reference or marker crude oil for pricing a number of other crude streams and which is traded in the domestic spot market at Cushing, Oklahoma.

wet gas: gas containing a relatively high proportion of hydrocarbons that are recoverable as liquids; see also **lean gas**.

wet scrubbers: devices in which a countercurrent spray liquid is used to remove impurities and particulate matter from a gas stream.

white oil: a generic tame applied to highly refined, colorless hydrocarbon oils of low volatility, and covering a wide range of viscosity.

wood alcohol: see **methyl alcohol**.

wurtzilite: a group of brown to black, solid bituminous materials of which the members are differentiated from asphaltites by their infusibility and low solubility in carbon disulfide; see **asphaltoid**.

xylene [C$_6$H$_4$(CH$_3$)$_2$]: a colorless liquid of the aromatic group of hydrocarbons made from the catalytic reforming of certain naphthenic petroleum fractions; used as high-octane motor and aviation gasoline blending agents, solvents, and chemical intermediates. Isomers are *ortho*-xylene (*o*-xylene), *meta*-xylene (*m*-xylene), and *para*-xylene (*p*-xylene).

zeolite: a crystalline aluminosilicate used as a catalyst and having a particular chemical and physical structure.

Index

Page numbers followed by f and t indicate figures and tables, respectively.

Milton Keynes UK
Ingram Content Group UK Ltd.
UKHW051906071024
449327UK00025B/2111